AutoCAD and Its Applications
BASICS

2012

by

Terence M. Shumaker
Faculty Emeritus
Former Chairperson
Drafting Technology
Autodesk Premier Training Center
Clackamas Community College, Oregon City, Oregon

David A. Madsen
President, Madsen Designs Inc.
Faculty Emeritus, Former Department Chairperson Drafting Technology
Autodesk Premier Training Center
Clackamas Community College, Oregon City, Oregon
Director Emeritus, American Design Drafting Association

David P. Madsen
President, Engineering Drafting & Design, Inc.
Vice President, Madsen Designs Inc.
Computer-Aided Design and Drafting Consultant and Educator
Autodesk Developer Network Member
American Design Drafting Association Member

19th Edition

Publisher
Goodheart-Willcox Company, Inc.
Tinley Park, IL
www.g-w.com

Manufactured in the United States of America.

Library of Congress Catalog Card Number 2011007997

ISBN 978-1-60525-561-3

1 2 3 4 5 6 7 8 9 – 12 – 16 15 14 13 12 11

The Goodheart-Willcox Company, Inc. Brand Disclaimer: Brand names, company names, and illustrations for products and services included in this text are provided for educational purposes only and do not represent or imply endorsement or recommendation by the author or the publisher.

The Goodheart-Willcox Company, Inc. Safety Notice: The reader is expressly advised to carefully read, understand, and apply all safety precautions and warnings described in this book or that might also be indicated in undertaking the activities and exercises described herein to minimize risk of personal injury or injury to others. Common sense and good judgment should also be exercised and applied to help avoid all potential hazards. The reader should always refer to the appropriate manufacturer's technical information, directions, and recommendations; then proceed with care to follow specific equipment operating instructions. The reader should understand these notices and cautions are not exhaustive.

The publisher makes no warranty or representation whatsoever, either expressed or implied, including but not limited to equipment, procedures, and applications described or referred to herein, their quality, performance, merchantability, or fitness for a particular purpose. The publisher assumes no responsibility for any changes, errors, or omissions in this book. The publisher specifically disclaims any liability whatsoever, including any direct, indirect, incidental, consequential, special, or exemplary damages resulting, in whole or in part, from the reader's use or reliance upon the information, instructions, procedures, warnings, cautions, applications, or other matter contained in this book. The publisher assumes no responsibility for the activities of the reader.

The Goodheart-Willcox Company, Inc. Internet Disclaimer: The Internet resources and listings in this Goodheart-Willcox Publisher product are provided solely as a convenience to you. These resources and listings were reviewed at the time of publication to provide you with accurate, safe, and appropriate information. Goodheart-Willcox Publisher has no control over the referenced websites and, due to the dynamic nature of the Internet, is not responsible or liable for the content, products, or performance of links to other websites or resources. Goodheart-Willcox Publisher makes no representation, either expressed or implied, regarding the content of these websites, and such references do not constitute an endorsement or recommendation of the information or content presented. It is your responsibility to take all protective measures to guard against inappropriate content, viruses, or other destructive elements.

Library of Congress Cataloging-in-Publication Data

Shumaker, Terence M.
AutoCAD and its applications. Basics 2012 / by Terence M. Shumaker, David A. Madsen, David P. Madsen. -- 19th ed.
p. cm.
Includes bibliographical references and index.
ISBN 978-1-60525-561-3
1. Computer graphics. 2. AutoCAD. I. Madsen, David A. II. Madsen, David P. III. Title.
T385.S461466 2012
620'.00420285536--dc22

2011007997

Introduction

AutoCAD and Its Applications—Basics is a textbook providing complete instruction in mastering fundamental AutoCAD® 2011 tools and drawing techniques. Typical applications of AutoCAD are presented with basic drafting and design concepts. The topics are covered in an easy-to-understand sequence and progress in a way that allows you to become comfortable with the tools as your knowledge builds from one chapter to the next. *AutoCAD and Its Applications—Basics* offers the following features:
- Step-by-step use of AutoCAD tools
- In-depth explanations of how and why tools function as they do
- Extensive use of font changes to specify certain meanings
- Examples and descriptions of industry practices and standards
- Screen captures of AutoCAD features and functions
- Professional tips explaining how to use AutoCAD effectively and efficiently
- More than 280 exercises to reinforce the chapter topics and build on previously learned material
- Chapter tests for review of tools and key AutoCAD concepts
- Practice questions and problems for Autodesk's AutoCAD Certified Associate and AutoCAD Certified Professional certification exams
- A large selection of drafting problems supplementing each chapter

With *AutoCAD and Its Applications—Basics*, you learn AutoCAD tools and become acquainted with information in other areas:
- Preliminary planning and sketches
- Drawing geometric shapes and constructions
- Parametric drawing techniques
- Special editing operations that increase productivity
- Placing text and tables according to accepted industry practices
- Making multiview drawings (orthographic projection)
- Dimensioning techniques and practices, based on accepted standards
- Drawing section views and designing graphic patterns
- Creating shapes and symbols
- Creating and managing symbol libraries
- Plotting and printing drawings

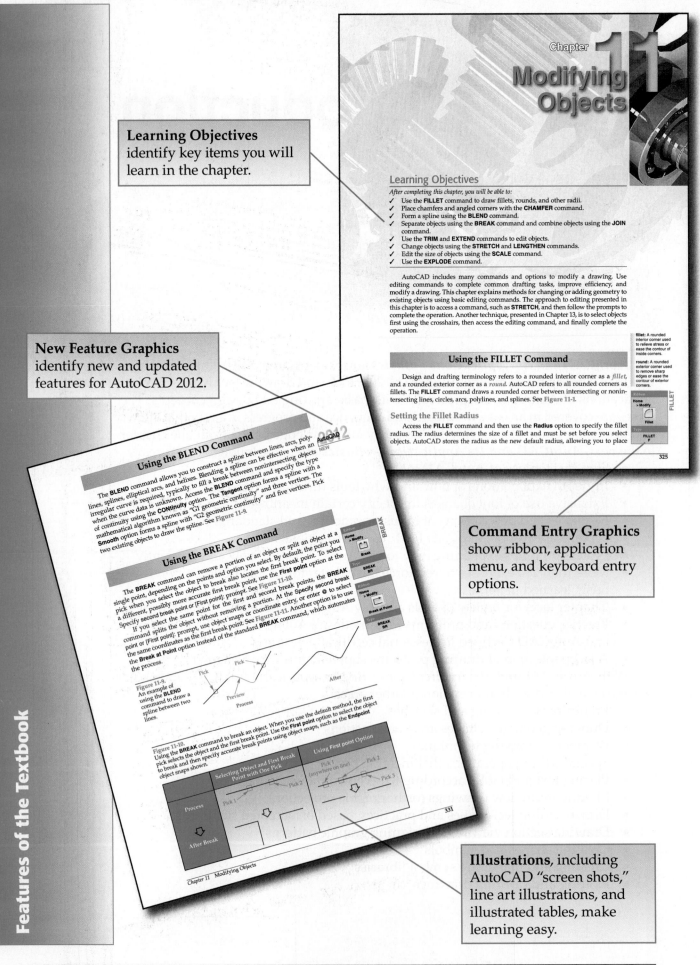

Learning Objectives
identify key items you will
learn in the chapter.

New Feature Graphics
identify new and updated
features for AutoCAD 2012.

Command Entry Graphics
show ribbon, application
menu, and keyboard entry
options.

Illustrations, including
AutoCAD "screen shots,"
line art illustrations, and
illustrated tables, make
learning easy.

Features of the Textbook

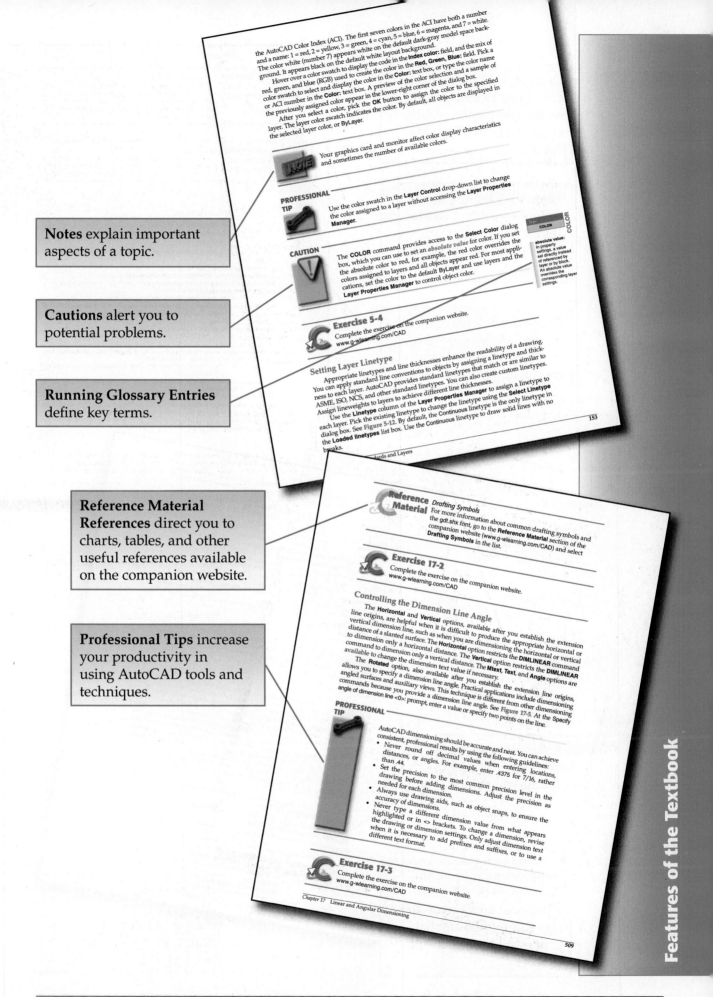

Notes explain important aspects of a topic.

Cautions alert you to potential problems.

Running Glossary Entries define key terms.

Reference Material References direct you to charts, tables, and other useful references available on the companion website.

Professional Tips increase your productivity in using AutoCAD tools and techniques.

Features of the Textbook

Express Tool References direct you to information on the companion website about AutoCAD Express Tools.

Template Development References direct you to Template Development material on the companion website.

Chapter Reviews reinforce the knowledge gained by reading the chapter and completing the exercises.

Exercise References direct you to step-by-step tutorial exercises on the companion website.

Supplemental Material References direct you to additional material on the companion website that is relevant to the current chapter.

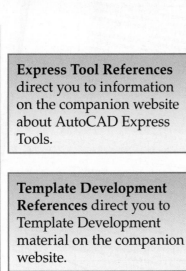

Express Tools
Chapter 29

Layout Express Tools
The **Layout** panel of the **Express Tools** ribbon tab includes additional layout commands. For information about the most useful layout express tools, go to the companion website (www.g-wlearning.com/CAD), select this chapter, and select **Layout Express Tools**.

Template Development
Chapter 29

Adding Layouts
For detailed instructions on adding layouts to each drawing template, go to the companion website (www.g-wlearning.com/CAD), select this chapter, and select **Template Development**.

Chapter Review

Answer the following questions. Write your answers on a separate sheet of paper or complete the electronic chapter review on the companion website.
www.g-wlearning.com/CAD

1. Name the two types of content that are brought together to create a complete drawing.
2. What commands can you use to modify the boundary of a floating viewport?
3. Briefly explain how to create a polygonal viewport.
4. How can you convert an object created in paper space into a floating viewport?
5. How do you activate a floating viewport?
6. How can you tell that a viewport is active in paper space?
7. How do you reactivate paper space after activating a floating viewport for editing?
8. How does the scale you assign to a floating viewport compare with the drawing scale?
9. To what value should the **CELTSCALE**, **PSLTSCALE**, and **MSLTSCALE** system variables be set so that the **LTSCALE** value will be applied correctly in model space and paper space?
10. Viewport edges may cut off the drawing when the viewport is correctly scaled. List three options to display the entire view.
11. Why should you lock a viewport after you adjust the drawing in the viewport to reflect the proper scale and view?
12. Give an example of why you would hide objects in a floating viewport without removing the viewport.
13. What is a plot stamp?
14. If you make changes to the page setup using the **Plot** dialog box, how can you save these changes to the page setup so that the changes apply to future plots?
15. Give at least two reasons why you should always preview a plot before sending the information to the plot device.

927

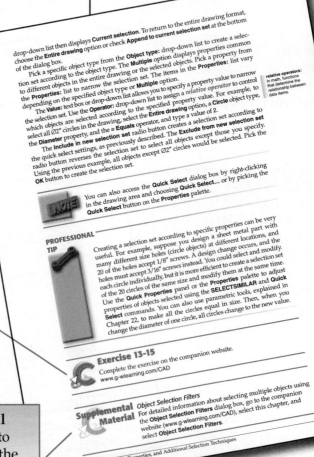

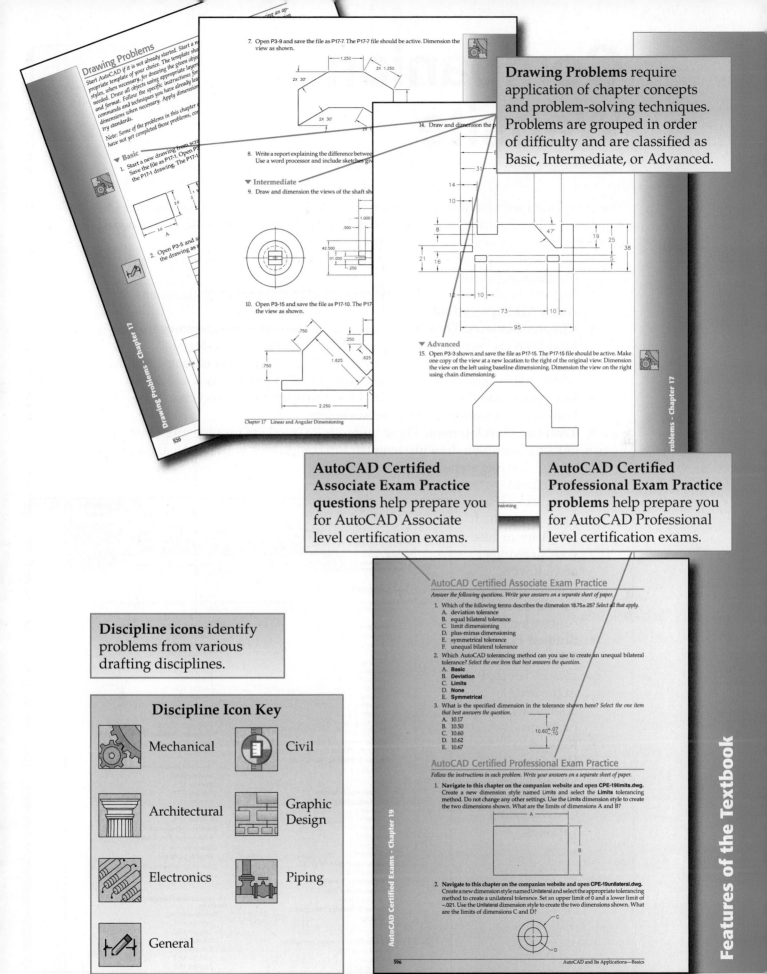

Drawing Problems require application of chapter concepts and problem-solving techniques. Problems are grouped in order of difficulty and are classified as Basic, Intermediate, or Advanced.

AutoCAD Certified Associate Exam Practice questions help prepare you for AutoCAD Associate level certification exams.

AutoCAD Certified Professional Exam Practice problems help prepare you for AutoCAD Professional level certification exams.

Discipline icons identify problems from various drafting disciplines.

Discipline Icon Key

Mechanical

Civil

Architectural

Graphic Design

Electronics

Piping

General

Companion Website

The companion website (www.g-wlearning.com/CAD) provides additional resources to help you get the most from the *AutoCAD and Its Applications* textbook. The following components are available on the companion website:

- **Exercises.** Over 280 step-by-step tutorial exercises are provided on the companion website.
- **Express Tools Material.** This reference material provides explanations of AutoCAD Express Tools.
- **Chapter Reviews.** The Chapter Reviews at the end of the chapters are also located on the companion website in Microsoft Word DOC format.
- **Supplemental Materials.** Organized by chapter, these documents provide additional information about topics discussed in the textbook.
- **Reference Materials.** You will find these articles, tables, and charts useful both in the classroom and in the workplace.
- **Template Development.** These in-depth instructions provide guidelines for creating your own drawing templates in compliance with ASME and other related drafting standards.
- **Predefined Templates.** Use these predefined templates to base your drawings on industry-related drawing standards and conventions.
- **Student Practice Files.** Use these files as directed in the text-book drawing problems.
- **Related Websites.** Use this to access a wide variety of CAD/drafting websites.

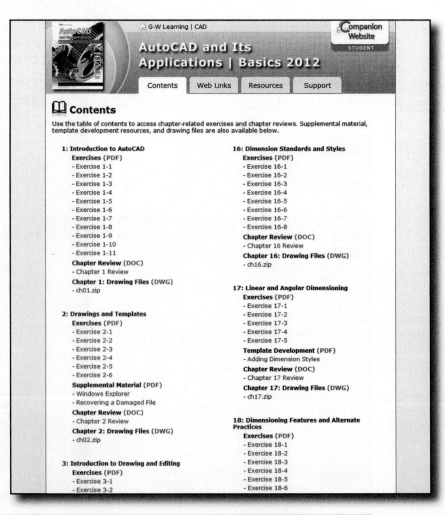

Fonts Used in This Textbook

Different typefaces are used throughout this textbook to define terms and identify AutoCAD commands. The following typeface conventions are used in this textbook:

Text Element	Example
AutoCAD commands	**LINE** command
AutoCAD menu browser menus	**Draw > Arc > 3 Points**
AutoCAD system viarables	**LTSCALE** system variable
AutoCAD toolbars and buttons	**Quick Access** toolbar, **Undo** button
AutoCAD dialog boxes	**Insert Table** dialog box
Keyboard entry (in text)	Type LINE
Keyboard keys	[Ctrl]+[1] key combination
File names, folders, and paths	C:\Program Files\AutoCAD 2011\mydrawing.dwg
Microsoft Windows features	Start menu, Programs folder
Prompt sequence	Command:
Keyboad input at prompt sequence	Command: **L** *or* **LINE** ↵
Comment at a prompt sequence	Specify first point: (*pick a point or press* [Enter])

Other Text References

For additional information, standards from organizations such as ANSI (American National Standards Institute) and ASME (American Society of Mechanical Engineers) are referenced throughout the textbook. Use these standards to create drawings that follow industry, national, and international practices.

Also for your convenience, other Goodheart-Willcox textbooks are referenced. Referenced textbooks include *AutoCAD and Its Applications—Advanced* and *Geometric Dimensioning and Tolerancing*. These textbooks can be ordered directly from Goodheart-Willcox.

AutoCAD and Its Applications—Basics covers basic AutoCAD applications. For a textbook covering the advanced AutoCAD applications, please refer to *AutoCAD and Its Applications—Advanced*.

Contents in Brief

About the Authors

Terence M. Shumaker is Faculty Emeritus, the former Chairperson of the Drafting Technology Department, and former Director of the Autodesk Premier Training Center at Clackamas Community College in Oregon City, Oregon. Terence taught at the community college level for over 28 years. He has professional experience in surveying, civil drafting, industrial piping, and technical illustration. He is the author of Goodheart-Willcox's *Process Pipe Drafting* and coauthor of the *AutoCAD and Its Applications* series and *AutoCAD Essentials*.

David A. Madsen is the president of Madsen Designs Inc. (www.madsendesigns.com) and an Authorized Autodesk Author. David is Faculty Emeritus of Drafting Technology and the Autodesk Premier Training Center at Clackamas Community College in Oregon City, Oregon. David was an instructor and department Chairperson at Clackamas Community College for nearly 30 years. In addition to community college experience, David was a Drafting Technology instructor at Centennial High School in Gresham, Oregon. David is a former member of the American Design Drafting Association (ADDA) Board of Directors, and was honored by the ADDA with Director Emeritus status at the annual conference in 2005. David has extensive experience in mechanical drafting, architectural design and drafting, and building construction. David holds a Master of Education degree in Vocational Administration and a Bachelor of Science degree in Industrial Education. David is the author of *Geometric Dimensioning and Tolerancing* and coauthor of *Architectural AutoCAD, Architectural Desktop and its Applications, Architectural Drafting Using AutoCAD, AutoCAD and Its Applications: Basics, Advanced, and Comprehensive, AutoCAD Essentials*, and other textbooks in the areas of architectural drafting, mechanical drafting, engineering drafting, civil drafting, architectural print reading, and mechanical printing reading.

David P. Madsen is the president of Engineering Drafting & Design, Inc., the vice president of Madsen Designs Inc. (www.madsendesigns.com), an Authorized Autodesk Author, and a SolidWorks Research Associate. Dave provides drafting and design consultation and training for all disciplines. Dave has been a professional design drafter since 1996, and has extensive experience in a variety of drafting, design, and engineering disciplines. Dave has provided drafting and computer-aided design and drafting instruction to secondary and postsecondary learners since 1999, and has considerable curriculum, program coordination, and development experience. Dave holds a Master of Science degree in Educational Policy, Foundations, and Administrative Studies with a specialization in Postsecondary, Adult, and Continuing Education; a Bachelor of Science degree in Technology Education; and an Associate of Science degree in General Studies and Drafting Technology. Dave is the author of *Inventor and its Applications*, and coauthor of *Architectural Drafting Using AutoCAD, AutoCAD and Its Applications: Basics and Comprehensive, Geometric Dimensioning and Tolerancing*, and other textbooks in the areas of architectural drafting, mechanical drafting, engineering drafting, civil drafting, architectural print reading, and mechanical printing reading.

Acknowledgments

Technical Assistance and Contribution of Materials

Margo Bilson of Willamette Industries, Inc.
Fitzgerald, Hagan, & Hackathorn
Bruce L. Wilcox, Johnson and Wales University School of Technology

Contribution of Technical Information

Arthur Baker
Autodesk
CADalyst magazine
CADENCE magazine
Chris Lindner
EPCM Services, Ltd.
Harris Group, Inc.

International Source for Ergonomics
Jim Webster
Kunz Associates
Myonetics, Inc.
Norwest Engineering
Schuchart & Associates, Inc.
Willamette Industries, Inc.

Trademarks

Contents

Companion Website Content

Exercises
Supplemental Materials
Reference Materials
Template Development
Express Tools
Activity and Practice Drawing Files
Predefined Templates
Chapter Reviews
Related Websites

Introduction to AutoCAD

Learning Objectives

After completing this chapter, you will be able to:

✓ Define computer-aided design and drafting.
✓ Describe typical AutoCAD applications.
✓ Explain the value of planning your work and system management.
✓ Describe the purpose and importance of drawing standards.
✓ Demonstrate how to start and exit AutoCAD.
✓ Recognize the AutoCAD interface and access AutoCAD commands.
✓ Use help resources.

Computer-aided design and drafting (CADD) is the process of using a computer with CADD software to design and produce drawings and models according to specific industry and company standards. The terms *computer-aided design (CAD)* and *computer-aided drafting (CAD)* refer to specific aspects of the CADD process. This chapter introduces the AutoCAD CADD system. You will begin working with AutoCAD and learn to control the AutoCAD environment.

> **computer-aided design and drafting (CADD):** The process of using a computer with CADD software to design and produce drawings and models.

AutoCAD Applications

AutoCAD *commands* and *options* allow you to draw objects of any size or shape. Use AutoCAD to prepare two-dimensional (2D) drawings, three-dimensional (3D) models, and animations. AutoCAD is a universal CADD program that applies to any drafting, design, or engineering discipline. For example, use AutoCAD to design and document mechanical parts and assemblies, architectural buildings, civil and structural engineering projects, and electronics.

> **command:** An instruction issued to the computer to complete a specific task. For example, use the **LINE** command to draw lines.

> **option:** A choice associated with a command, or an alternative function of a command.

2D Drawings

2D drawings display an object's length and width, width and height, or height and length in a flat (2D) form. 2D drawings are the established design and drafting format and are common in all engineering and architectural industries and related disciplines. A complete 2D drawing typically includes dimensions, notes, and text that describe view features and details. This practice results in a document used to

Figure 1-1.
AutoCAD provides commands and options to create 2D drawings accurately, such as this architectural floor plan of a home.

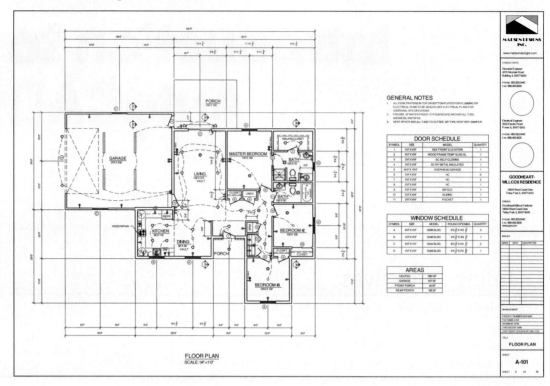

FLOOR PLAN
SCALE: 1/4" = 1'-0"

wireframe model: The most basic 3D model. It contains only information about object edges and the points where edges intersect, known as *vertices*, and describes the appearance of the model as if it were constructed from wires.

surface model: A 3D model that contains information about object edges, vertices, and the outer boundaries of the object, known as *surfaces*; surface models have zero thickness, lack mass, and may not enclose a volume.

solid model: The most complex 3D model; contains information about object edges, vertices, surfaces, and mass; encloses a volume.

walkthrough: A computer simulation that replicates walking through or around a 3D model.

flythrough: A computer simulation that replicates flying through or around a 3D model.

manufacture or construct a product. 2D drawings are the conventional and often required method of communicating a project. **Figure 1-1** shows an example of a 2D architectural floor plan created using AutoCAD. Use this textbook to learn how to construct, design, dimension, and annotate 2D AutoCAD drawings.

3D Models

3D models allow for advanced visualization, simulation, and analysis typically not possible with 2D drawings. AutoCAD provides commands and options for developing *wireframe*, *surface*, and *solid models*. An accurate solid model is an exact digital representation of a product. Add color, lighting, and texture to display a realistic view of the model. See **Figure 1-2A**. Use view tools to rotate and adjust a model to view it from any direction. See **Figure 1-2B**. Apply animation to a model to show product design or function. For example, you can perform a *walkthrough* of a model home or a *flythrough* of a civil engineering project. *AutoCAD and Its Applications—Advanced* provides detailed instruction on 3D modeling and rendering.

Reference Material

Glossaries

For detailed glossaries of CADD, AutoCAD, and computer terms, go to the **Reference Material** section of the companion website (www.g-wlearning.com/CAD) and select **Glossary of Computer Terms** or **Glossary of CADD Terms**.

Figure 1-2.
A 3D AutoCAD model of a mechanical assembly. A—A wireframe model (left) with realistic colors and textures added (right). B—You can rotate, zoom in and out, and view a model from any location in 3D space.

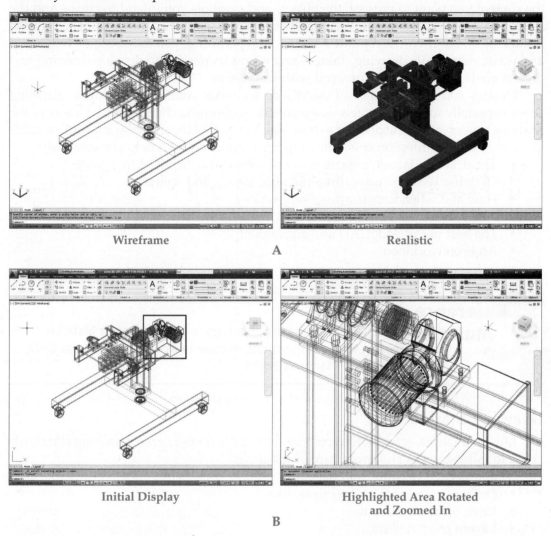

Wireframe   Realistic

A

Initial Display   Highlighted Area Rotated
                         and Zoomed In

B

Before You Begin

Designing and drafting effectively with a computer requires a skilled CADD operator. To be a proficient AutoCAD user, you must have detailed knowledge of AutoCAD commands and processes, and know which command and process is best suited for a specific task. You must also understand and be able to apply design and drafting systems and conventions when using AutoCAD.

As you begin your CADD training, develop effective methods for managing your work. First, plan your *drawing sessions* thoroughly to organize your thoughts. Second, learn and use industry, classroom, or office standards. Third, save your work often. If you follow these procedures, you will find it easier to use AutoCAD commands and methods, and your drawing experience will be more productive and enjoyable.

drawing sessions: Time spent working on a drawing project, including analyzing design parameters and using AutoCAD.

Planning Your Work

A drawing plan involves thinking about the entire process or project in which you are involved and determining how to approach it. Your drawing plan focuses on the content you want to present, the objects and symbols you intend to create, and the appropriate use of standards. You may want processes to be automatic or to happen immediately, but if you hurry and do little or no planning, you may become frustrated and waste time while drawing. Take as much time as needed to develop drawing and project goals so that you can proceed with confidence.

During your early stages of AutoCAD training, consider creating a planning sheet, especially for your first few assignments. A planning sheet should document the drawing session and all aspects of a drawing. A sketch of the drawing is also a valuable element of the planning process. The drawing plan and sketch help you establish:
- The drawing layout: area, number of views, and required free space
- Drawing settings: units, drawing aids, layers, and styles
- How and when to perform specific tasks
- What objects and symbols to draw
- The best use of AutoCAD and equipment
- An even workload

Reference Material   *Planning Sheet*
For a sample planning sheet, go to the **Reference Material** section of the companion website (www.g-wlearning.com/CAD) and select **Planning Sheet**.

Drawing Standards

standards: Guidelines that specify drawing requirements, appearance, techniques, operating procedures, and record-keeping methods.

drawing template (template): A file that contains standard drawing settings and objects for use in new drawings.

Most industries, schools, and companies establish *standards*. Drawing standards apply to most settings and procedures, including:
- File storage, naming, and backup
- *Drawing template*, or *template*, files
- Units of measurement
- Layout characteristics
- Borders and title blocks
- Symbols
- Layers
- Text, dimension, multileader, and table styles
- Plot styles and plotting

Company or school drawing standards should follow appropriate national industry standards whenever possible. Although standards vary in content, the most important aspect is that standards exist and are understood and used by all CADD personnel. When you follow drawing standards, your drawings are consistent, you become more productive, and the classroom or office functions more efficiently.

This textbook presents mechanical drafting standards developed by the American Society of Mechanical Engineers (ASME) and accredited by the American National Standards Institute (ANSI). This textbook also references International Standards Organization (ISO) mechanical drafting standards and discipline-specific standards when appropriate, including the United States National CAD Standard® (NCS) and American Welding Society (AWS) standards.

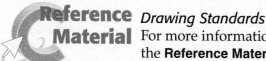

Reference Material *Drawing Standards*
For more information about drawing standards, go to the **Reference Material** section of the companion website (www.g-wlearning.com/CAD) and select **Drawing Standards**.

You may consider other drafting standards when preparing drawings, such as the *BSI*, *DIN*, *GB*, *GOST*, and *JIS* standards.

BSI: The British Standards Institution.

DIN: Deutsches Institut Für Normung, established by the German Institute for Standardization.

GB: Chinese Guóbiāo standard.

GOST: Gosudarstvennyy standart, maintained by the Euro-Asian Council for Standardization.

JIS: Japanese Industrial Standards.

Saving Your Work

Drawings are lost due to software error, hardware malfunction, power failure, or accident. Prepare for such an event by saving your work frequently. Develop a habit of saving your work at least every 10 to 15 minutes. You can set the automatic save option, described in Chapter 2, to save drawings automatically at set intervals. However, you should also frequently save your work manually.

Working Procedures Checklist

Proficient use of AutoCAD requires several skills. Use the following checklist to become comfortable with AutoCAD, and to help you work quickly and efficiently:

✓ Carefully plan your work
✓ Frequently check object and drawing settings, such as layers, styles, and properties, to see which object characteristics and drawing options are in effect
✓ Follow the prompts, tooltips, notifications, and *alerts* that appear as you work
✓ Constantly check for the correct options, instructions, or keyboard entry
✓ *Right-click* to access shortcut menus and review available options
✓ Think ahead to prepare for each stage of the drawing session
✓ Learn commands, tools, and options that increase your speed and efficiency
✓ Save your work at least every 10 to 15 minutes
✓ Learn to use available resources, such as this textbook, to help solve problems and answer questions

alert: A pop-up that indicates a required action or potential problem.

right-click: Press the right mouse button.

Exercise 1-1

Complete the exercise on the companion website.
www.g-wlearning.com/CAD

double-click: Quickly press the left mouse button twice.

icon: Small graphic representing an application, file, or command.

pick (click): Press the left mouse button.

Starting AutoCAD

One of the quickest methods to start AutoCAD is to *double-click* on the AutoCAD 2012 Windows desktop *icon*. A second option is to *pick* the Start *button* in the lower-left corner of the Windows desktop, then *hover* over or pick Programs. Then select Autodesk, followed by AutoCAD 2012, and finally AutoCAD 2012.

button: A "hot spot" on the screen that you pick to access an application, command, or option.

hover: Pause the cursor over an item to display information or options.

AutoCAD 2012 operates with Windows 7 and specific versions of Windows Vista and Windows XP. Do not be concerned if you see illustrations in this textbook that appear slightly different from those on your screen.

Exiting AutoCAD

Access the **EXIT** command to end an AutoCAD session. Pick the program **Close** button, located in the upper-right corner of the AutoCAD window. Other ways to close AutoCAD are to double-click the **Application Menu** button in the upper-left corner of the AutoCAD window, select the **Exit AutoCAD** button in the **Application Menu**, or, with a file open, type EXIT or QUIT and press [Enter]. See **Figure 1-3**.

If you attempt to exit before saving your work, AutoCAD prompts you to save or discard changes.

Exercise 1-2

Complete the exercise on the companion website.
www.g-wlearning.com/CAD

Figure 1-3.
Use any of several techniques to exit AutoCAD when you finish a drawing session.

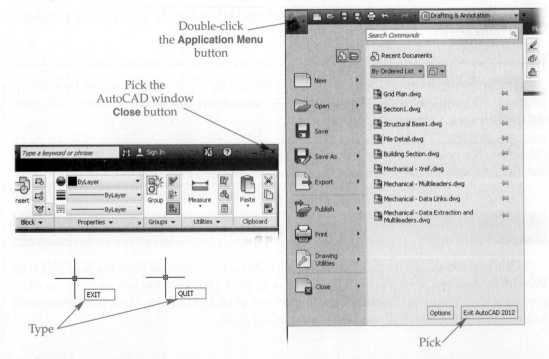

The AutoCAD Interface

Interface items include devices to input data, such as the keyboard and mouse, and devices to receive computer outputs, such as the monitor. AutoCAD uses a Windows-style *graphical user interface (GUI)* with an **Application Menu**, ribbon, dialog boxes, and AutoCAD-specific items. See **Figure 1-4**. You will explore specific elements of the unique AutoCAD interface in this chapter and throughout this textbook. Learn the format, appearance, and proper use of interface items to help quickly master AutoCAD.

interface: Items that allow you to input data to and receive outputs from a computer system.

graphical user interface (GUI): On-screen features that allow you to interact with a software program.

As you learn AutoCAD, you may want to customize the graphical user interface according to common tasks and specific applications. *AutoCAD and Its Applications—Advanced* explains customizing the user interface.

Autodesk Exchange

The **Autodesk Exchange** window appears by *default* when you first launch AutoCAD. See **Figure 1-4**. The **Autodesk Exchange** window provides access to AutoCAD resources online, including the AutoCAD help system. The default **Home** page provides links to a variety of resources, including announcements, featured videos and topics, and information about product updates, subscriptions, and product support. Choose the **Help** button to access the online AutoCAD help system, described later in this chapter. Deselect the **Show this window at startup** *check box* to prevent the **Autodesk Exchange** window from appearing the next time you launch AutoCAD. Then pick the **Close** button of the **Autodesk Exchange** window to exit the screen and begin working.

NEW

default: A value maintained by the computer until changed.

check box: A selectable box that turns an item on (when checked) or off (when unchecked).

The **Autodesk Exchange** window appears when you access the AutoCAD help system, or when you pick the **Exchange** button in the **InfoCenter**, described later in this chapter.

Workspaces

The **Drafting & Annotation** *workspace*, shown in **Figure 1-4**, is active by default when you launch AutoCAD. The **Drafting & Annotation** workspace displays interface features above and below a large *drawing window*, also called the *graphics window*, and contains the commands and options most often used for 2D drawing. To activate a different workspace, pick the **Workspace** *flyout* on the **Quick Access** toolbar or the **Workspace Switching** button on the status bar and select a different workspace. See **Figure 1-5**.

The **3D Basics** and **3D Modeling** workspaces provide commands and options appropriate for 3D modeling. The **AutoCAD Classic** workspace displays the traditional AutoCAD menu bar, toolbars, and tool palettes. Custom workspaces saved by users also appear in the list.

workspace: A preset work environment containing specific interface items.

drawing window (graphics window): The largest area in the AutoCAD window, where drawing and modeling occurs.

flyout: A set of related buttons that appears when you pick the arrow next to certain tool buttons.

Figure 1-4.
The default AutoCAD window with the **Drafting & Annotation** workspace active.

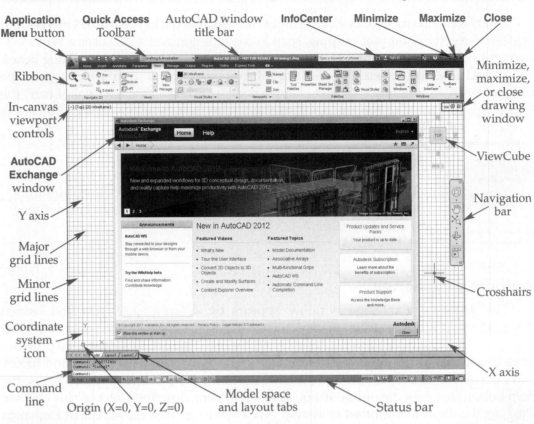

Application Menu button — Quick Access Toolbar — AutoCAD window title bar — InfoCenter — Minimize — Maximize — Close

Ribbon

In-canvas viewport controls

AutoCAD Exchange window

Y axis

Major grid lines

Minor grid lines

Coordinate system icon

Command line

Origin (X=0, Y=0, Z=0)

Minimize, maximize, or close drawing window

ViewCube

Navigation bar

Crosshairs

X axis

Model space and layout tabs

Status bar

Figure 1-5.
Use the **Workspace Switching** flyout on the **Quick Access** toolbar or the **Workspace Switching** button on the status bar to change to a different workspace, create a new workspace, or customize the user interface.

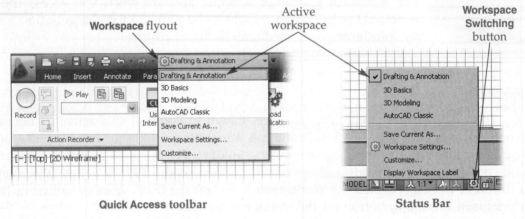

Workspace flyout

Active workspace

Workspace Switching button

Quick Access toolbar

Status Bar

This textbook focuses on the default **Drafting & Annotation** workspace, except in specific situations that require additional interface items. The default model space drawing window background color is dark gray, but this textbook shows a white background for clarity. Add items and AutoCAD tools to the interface as needed. *AutoCAD and Its Applications—Advanced* details the **3D Modeling** workspace and explains how to customize a workspace.

Reload the **Drafting & Annotation** workspace to return interface items to their default locations. You can reload the workspace by picking the **Workspace Switching** button on the status bar and selecting the **Drafting & Annotation** option.

Exercise 1-3

Complete the exercise on the companion website.
www.g-wlearning.com/CAD

Crosshairs and Cursor

The AutoCAD crosshairs is the primary means of pointing to and selecting objects or locations within a drawing window. The crosshairs changes to the familiar Windows cursor when you move it outside of the drawing area or over an interface item, such as the status bar.

Control crosshairs length using the *text box* or *slider* found in the **Crosshair size** area on the **Display** tab of the **Options** dialog box. Longer crosshairs can help to reference alignment between objects. To display the **Options** dialog box, pick the **Options** button at the bottom of the **Application Menu**, or right-click in the drawing area and select **Options**.

text box: A box in which you type a name, number, or single line of information.

slider: A movable bar that increases or decreases a value when you slide the bar.

Tooltips

A *tooltip* displays when you hover over most interface items. See **Figure 1-6**. Tooltip content varies depending on the item. Many tooltips expand as you continue to hover. The initial tooltip might display the command name and a brief description of the command. As you continue to hover, an explanation, illustration, or short video on how to use the command may appear.

tooltip: A pop-up that provides information about the item over which you hover.

Shortcut Menus

AutoCAD uses *shortcut menus*, also known as *cursor menus*, *right-click menus*, or *pop-up menus*, to simplify and accelerate command and option access. When you right-click in the drawing area while a command is not active, the first item in the shortcut menu is typically an option to repeat the previous command or operation. If you right-click while a command is active, the shortcut menu contains *context-sensitive* menu options. See **Figure 1-7**. Some menu options have a small arrow to the right of the option name. Hover over the option to display a *cascading menu*, also known as a *cascading submenu*. The **Recent Input** cascading menu shows a list of recently used commands, options, or values, depending on the shortcut menu. Pick from the list to reuse a function or value.

shortcut menu (cursor menu, right-click menu, pop-up menu): A context-sensitive menu available by right-clicking on interface items or objects.

context-sensitive: Specific to the active command or option.

cascading menu (cascading submenu): A menu of options related to the chosen menu item.

Figure 1-6.
Examples of tooltips that appear when you hover over an item.

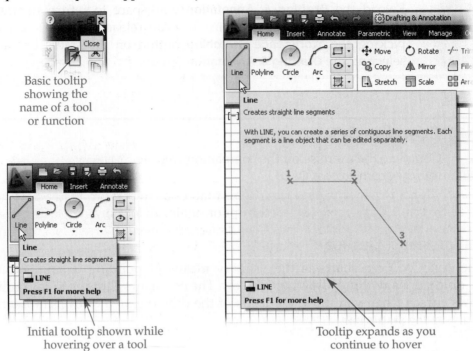

Basic tooltip showing the name of a tool or function

Initial tooltip shown while hovering over a tool

Tooltip expands as you continue to hover

Figure 1-7.
Shortcut menus provide instant access to general or context-sensitive commands and options.

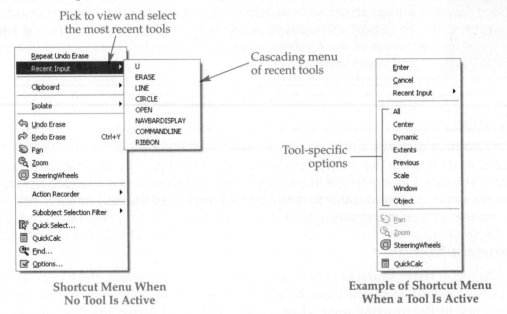

Pick to view and select the most recent tools

Cascading menu of recent tools

Tool-specific options

Shortcut Menu When No Tool Is Active

Example of Shortcut Menu When a Tool Is Active

Controlling Windows

Control the AutoCAD and drawing windows using the same methods you use to control other windows within the Windows operating system. To minimize, maximize, restore, or close the AutoCAD window or individual drawing windows, pick the appropriate button in the upper-right corner of the window. You can also adjust the AutoCAD window by right-clicking on the title bar and choosing from the standard window control menu. Window sizing operations are also the same as those for other windows within the Windows operating system.

Floating and Docking

Several interface items, including the AutoCAD and drawing windows, can *float* or *dock*. Some items, such as the drawing window, have a title bar at the top or side. You can move and resize floating windows in the same manner as other windows. However, drawing windows can only be moved and resized within the AutoCAD window. Different options are available depending on the particular interface item and the float or dock status of the item. Typically, the close and minimize or maximize options are available. Some floating items, such as sticky panels, include *grab bars*.

Locking

You can lock certain interface items to prevent them from moving accidentally in either a floating or a docked state. To access locking options, pick the **Toolbar/Window Positions** button on the status bar to access the menu shown in **Figure 1-8**.

Select an option to lock the interface items that reside in that group in a floating or docked state. To unlock a group, select the option again. To lock or unlock all interface items, select **Locked** or **Unlocked** from the **All** cascading menu. Move a locked feature without unlocking it by holding down [Ctrl] while moving the feature.

> **float:** Describes interface items that appear within a border and can be resized or moved.
>
> **dock:** Describes interface items locked into position on an edge of the AutoCAD window (top, bottom, left, or right).
>
> **grab bars:** Two thin bars at the top or left edge of a docked or floating feature; used to move the feature.

Exercise 1-4

Complete the exercise on the companion website.
www.g-wlearning.com/CAD

Application Menu

The **Application Menu** is a menu system that provides access to application- and file-related commands and settings. The **Application Menu** displays when you pick the **Application Menu** button, located in the upper-left corner of the AutoCAD window. See **Figure 1-9**.

Using the Buttons and Menus

Items on the left side of the **Application Menu** function as buttons to activate common application commands, and except for the **Save** button, they also display menus. For example, press the **New** button to begin a new file using the **QNEW** command. To display a menu, hover over the menu name, or pick the arrow on the right side of the button. Long menus include small arrows at the top and bottom for scrolling through selections. Some options have a small arrow to the right of the item name that, when selected or hovered over, expands to provide a submenu. Pick an option from the list to activate the command.

A command or option accessible from the **Application Menu** appears as a graphic in the margin of this textbook. The graphic represents the process of picking the **Application Menu** button, then selecting a menu button or hovering over a menu and picking a menu or submenu option. The example shown in this margin illustrates accessing the **PAGESETUP** command from the **Application Menu**, as shown in **Figure 1-9**.

> Application Menu
> **Print**
> **> Page Setup**

Figure 1-8.
You can lock some or all interface items in position.

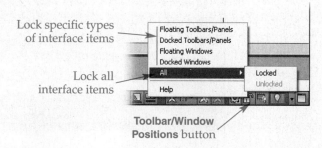

Lock specific types of interface items

Lock all interface items

Floating Toolbars/Panels
Docked Toolbars/Panels
Floating Windows
Docked Windows
All — Locked / Unlocked
Help

Toolbar/Window Positions button

Figure 1-9.
Use the **Application Menu** to access common application and file management commands and settings, search for commands, and view open and recently used documents.

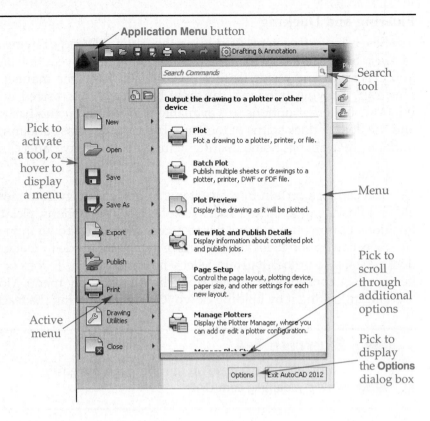

Application Menu button

Search tool

Pick to activate a tool, or hover to display a menu

Menu

Pick to scroll through additional options

Active menu

Pick to display the **Options** dialog box

Searching for Commands

Use the **Application Menu** search tool to locate and access any AutoCAD command listed in the Customize User Interface (CUI) file. Type a command name in the **Search** text box. Commands that match the letters you enter appear as you type. Type additional letters to narrow the search, with the best-matched command listed first. **Figure 1-10** shows using the **Search** text box to locate the **SAVE** command for saving a file. Pick a command from the list to activate the command.

Figure 1-10.
Use the **Application Menu** to search for a command. Pick the command from the list to activate the command.

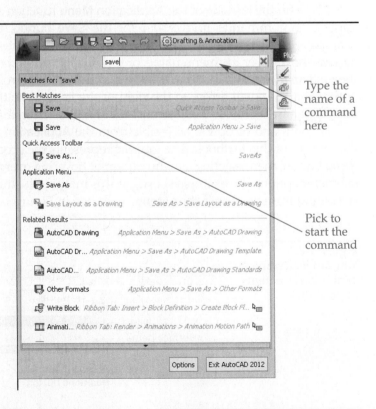

Type the name of a command here

Pick to start the command

The **Recent Documents** and **Open Documents** features of the **Application Menu** provide access to recent and active files, as described in Chapter 2.

Exercise 1-5

Complete the exercise on the companion website.
www.g-wlearning.com/CAD

Quick Access Toolbar

Toolbars contain *tool buttons*. Each tool button includes an icon that represents an AutoCAD command or option. As you move the cursor over a tool button, the button highlights and may display a border and tooltip. Use the tooltip to become familiar with the command. Select a tool button to activate the associated command. Some tool buttons include flyouts. Select a flyout and then pick from the list to activate the command.

The default **Quick Access** toolbar appears on the title bar in the upper-left corner of the AutoCAD window, to the right of the **Application Menu** button. See **Figure 1-11**. The **Quick Access** toolbar provides fast, convenient access to several common commands. One or two picks activate a command from the **Quick Access** toolbar. Most other interface items require two or more picks to activate a command.

toolbars: Interface items that contain tool buttons or drop-down lists.

tool buttons: Interface items used to start commands.

Figure 1-11.
Use the **Quick Access** toolbar to access commonly used commands. Pick a command button to activate the corresponding command or pick a flyout to access related or alternative commands.

Pick to display a flyout

Pick to display options for customizing the **Quick Access** toolbar

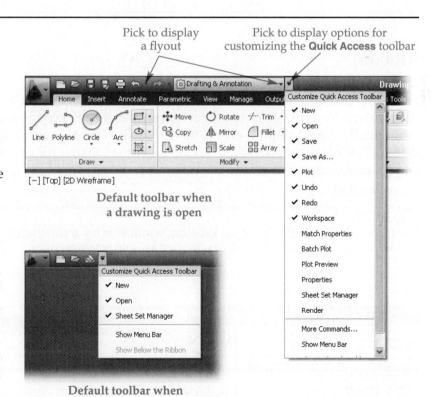

Default toolbar when a drawing is open

Default toolbar when no drawing is open

REDO

Quick Access

Redo

When a drawing is open and the default **Drafting & Annotation** workspace is active, the **Quick Access** toolbar contains **New, Open, Save, Save As, Plot, Undo, Redo,** and **Workspace** buttons. When no drawings are open, the **New, Open,** and **Sheet Set Manager** buttons display. The **Quick Access** toolbar is customizable by adding, removing, and relocating tool buttons. To make basic adjustments, pick the **Customize Quick Access Toolbar** flyout on the right side of the toolbar. *AutoCAD and Its Applications—Advanced* further explains customizing the user interface.

A command or option accessible from the **Quick Access** toolbar appears as a graphic in the margin of this textbook. The graphic represents the process of picking a **Quick Access** toolbar button from the toolbar or flyout. The example shown in this margin illustrates accessing the **REDO** command from the **Quick Access** toolbar to redo a previously undone operation.

> **NOTE**
> Several toolbars appear in the **AutoCAD Classic** workspace. These toolbars are usually application- or task-specific. The **Application Menu, Quick Access** toolbar, and ribbon replace classic toolbars in all other workspaces. Refer to *AutoCAD and Its Applications—Advanced* for information about displaying toolbars and interface customization.

Exercise 1-6

Complete the exercise on the companion website.
www.g-wlearning.com/CAD

Ribbon

The ribbon docks horizontally below the AutoCAD window title bar by default and is the primary means of accessing commands and options. See **Figure 1-12**. The ribbon provides a convenient location from which to select commands and options that traditionally would require access by extensive typing, multiple toolbars, or several menus. The ribbon allows you to spend less time looking for commands and options and reduces clutter in the AutoCAD window.

Figure 1-12.
The ribbon docked at the top of the drawing window is the most often used palette. Palettes provide access to commands, options, properties, and settings.

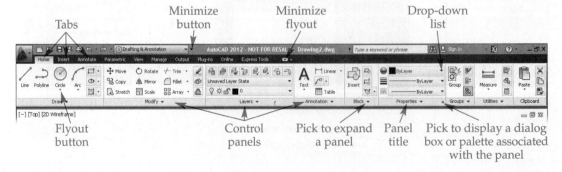

The ribbon appears by default in all workspaces except the **AutoCAD Classic** workspace. Use the *tabs* along the top of the ribbon to access collections of related *ribbon panels*, or *panels*. Each panel houses groups of similar commands. For example, the **Annotate** tab includes several panels, each with specific commands for creating, modifying, and formatting annotations, such as text. The tabs and panels shown when the **Drafting & Annotation** workspace is active provide access to 2D drawing commands. Highlighted, context-sensitive tabs appear when some commands, such as the **HATCH** command, are active or you are when working in a unique environment, such as the **Block Editor**.

A command or option accessible from the ribbon appears in a graphic located in the margin of this textbook. The graphic identifies the tab and panel where the command is located. You may need to expand the panel or pick a flyout to locate the command. The example shown in this margin illustrates using the ribbon to access the **LINE** command.

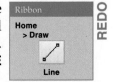

Ribbon Panels

The large tool button in a panel typically signifies the most often used panel command. In addition to tool buttons, panels can contain flyouts, *drop-down lists*, and other items. Some panels have a triangle, or arrow, next to the panel name. If you see this arrow, pick the title at the bottom of the panel to display additional related commands and functions. See **Figure 1-13**. To show the expanded list on-screen at all times, select the pushpin button in the lower-left corner of the expanded panel.

When you pick an option from a ribbon flyout, the option becomes the new default and appears in the ribbon. This makes it easier to select the same option the next time you use the command.

Some panels include a small arrow in the lower-right corner of the panel. Pick this arrow to access a dialog box or palette closely associated with the panel. For example, pick the arrow in the lower-right corner of the **Annotate** tab, **Dimensions** panel, as shown in **Figure 1-13**, to display the **Dimension Style Manager** dialog box used to control the format of dimensions.

Basic Adjustment

The ribbon appears maximized by default. You can minimize the display to show only tabs, panel titles, or panel buttons by repeatedly pressing the **Minimize** button to the right of the tabs, or by selecting the appropriate option from the **Minimize** flyout. Picking the **Minimize** button corresponds to the **Cycle through All** flyout selection. When **Minimize to Tabs** is active, pick a tab to show all panels in the tab. When **Minimize to Panel Titles** is active, pick a panel title to display the panel. When **Minimized to Panel Buttons** is active, pick a panel button to display the panel.

Figure 1-13.
An expanded panel provides additional, related commands and functions. This example shows the expanded list of dimensioning commands found in **Dimensions** panel.

Pick to pin the expanded list to the screen

Pick to display the **Dimension Style Manager**

Right-click on a portion of the ribbon unoccupied by a panel to access the options briefly described in **Figure 1-14**. *AutoCAD and Its Applications—Advanced* provides additional information on customizing the ribbon, including repositioning tabs and panels, options available when floating the ribbon, and creating and using a *sticky panel*.

sticky panel: A ribbon panel moved out of a tab and made to float in the drawing window.

The **Application Menu**, **Quick Access** toolbar, and ribbon replace the traditional menu bar in workspaces other than the **AutoCAD Classic** workspace. To display the menu bar, pick the **Customize Quick Access Toolbar** flyout on the right side of the **Quick Access** toolbar and choose **Show Menu Bar**.

Palettes

palette (modeless dialog box): Special type of window containing tool buttons and features common to dialog boxes. Palettes can remain open while other commands are active.

list box: A boxed area that contains a list of items or options from which to select.

scroll bar: A bar tipped with arrow buttons used to scroll through a list of options or information.

Palettes, also known as *modeless dialog boxes*, control many AutoCAD functions. Palettes may look like extensive toolbars or more like dialog boxes, depending on the function and floating or docked state. You can consider the ribbon a palette used to access commands and options. Palettes contain tool buttons, flyouts, dropdown lists, and many other features, such as *list boxes* and *scroll bars*. Unlike a dialog box, palettes need not be closed to use other commands and work on the drawing. Like the ribbon, panels divide some palettes into groups of commands. Large palettes are divided into separate pages or windows, which you commonly access using tabs.

To display a palette, pick a palette button from the **Palettes** panel in the **View** ribbon tab. You can also display most palettes using palette-specific access techniques. For example, to access the **Properties** palette, pick the arrow in the lower-left corner of the **Properties** panel in the **Home** ribbon tab; double-click on most objects in the drawing window; select an object, right-click and select **Properties**; or type **PROPERTIES**.

When you display a palette for the first time, it is often in a floating state, although you can dock some palettes. Right-click on the palette title bar or pick the **Properties** button to select from a list of undocked palette control options. The **Auto-hide** option allows the palette to minimize when the cursor is away from the palette, conserving drawing space.

Figure 1-14.
Right-click options for displaying and organizing ribbon elements.

Selection	Result
Show Related Tool Palette Group	Displays tool palette groups customized to associate with a ribbon tab.
Tool Palette Group	Allows you to select which related tool palette groups to show.
Show Tabs	Allows you to choose which tabs to display; also available by right-clicking on a panel.
Show Panels	Allows you to select which panels to display; also available by right-clicking on a panel.
Show Panel Titles	Uncheck to hide panel titles.
Undock	Changes the ribbon to a floating state. Double-click the ribbon title bar or drag and drop to dock the floating ribbon.
Close	Closes the ribbon. Use the **RIBBON** tool to redisplay the ribbon.

Deselect the **Allow Docking** palette property or menu option to disable the ability to dock palettes. The **Properties** button or shortcut menu on some palettes includes other functions, such as the **Transparency...** option. This option makes the palette transparent, allowing you to view drawing geometry behind the palette. See **Figure 1-15**.

PROFESSIONAL TIP

Resize a floating palette using the resizing arrows that appear when you move the cursor over the edge. Then pick the **Auto-hide** button to have quick access to the palette while displaying the largest possible drawing area.

Exercise 1-7

Complete the exercise on the companion website.
www.g-wlearning.com/CAD

Status Bars

AutoCAD provides an application status bar and a drawing status bar. The application status bar applies to all open files. The drawing status bar, when activated, appears above the *command line* and is specific to each file. Status bars are the quickest and most effective way to manage certain drawing settings.

command line:
Area where you can type commands (command names) and options.

Application Status Bar

The application status bar appears along the bottom of the AutoCAD window. See **Figure 1-16**. The application status bar includes areas that display and control a variety of drawing aids and commands. The coordinate display field, located on the left side of the application status bar, shows the location, or coordinates, of the crosshairs in drawing

Figure 1-15.
A—Pick the **Properties** button or right-click in the title bar and select **Transparency...** to access the **Transparency** dialog box. B—The transparent **Layer Properties Manager** palette positioned over a commercial building floor plan.

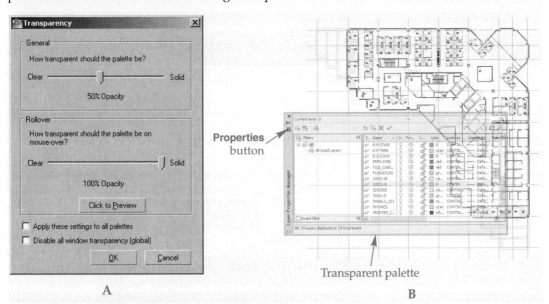

Transparent palette

A   B

space. *Status toggle buttons* appear next to the coordinate display field. Status toggle buttons appear as icons by default. To change the display from icons to names, right-click on any status toggle button and deselect **Use Icons**.

The items on the right side of the application status bar control windows and the drawing environment, activate commands, and adjust annotation scaling. This area also includes a notification tray that identifies the status of and provides notifications for some AutoCAD commands and processes.

Right-click on the application status bar, away from the coordinate display field or a button, or pick the **Application Status Bar Menu** flyout to access options for modifying the display of the application status bar. Uncheck an item on the list to hide the item from the status bar. *AutoCAD and Its Applications—Advanced* further describes how to customize the status bar.

Drawing Status Bar

Select the **Drawing Status Bar** option from the application status bar shortcut menu or the **Display** tab of the **Options** dialog box to display a separate drawing status bar under the drawing window. The **Annotation Scale**, **Annotation Visibility**, and **AutoScale** commands and the notification tray move from the application status bar to the drawing status bar. A **Drawing Status Bar Menu** flyout also appears. See **Figure 1-17**. Drawing status bar settings are unique to each open file.

Figure 1-16.
Picking buttons on the application status bar is the quickest and most effective way to manage certain drawing settings.

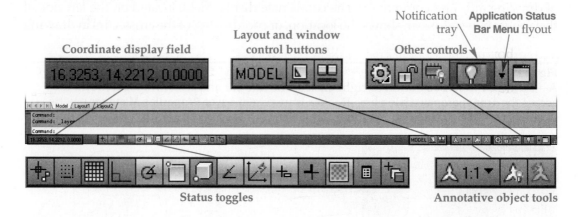

Figure 1-17.
The drawing status bar, when displayed, is specific to the current file. Each open file has its own drawing status bar.

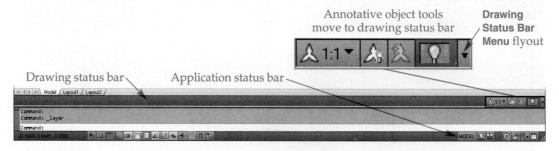

PROFESSIONAL TIP

Right-click on the coordinate display field or a button in the application or drawing status bar to view a shortcut menu specific to the item. Picking options from a status bar shortcut menu is often the most efficient method of controlling drawing settings.

Exercise 1-8

Complete the exercise on the companion website.
www.g-wlearning.com/CAD

Dialog Boxes

You will see many *dialog boxes* during a drawing session, including those used to create, save, and open files. Dialog boxes contain many of the same features found in other interface items, such as icons, text, buttons, and flyouts. **Figure 1-18** shows the dialog box that appears when you use the **INSERT** command. The **Insert** dialog box includes many common dialog box elements.

dialog box: A window-like item that contains various settings and information.

A dialog box appears when you pick any menu selection or button displaying an ellipsis (…).

Use the cursor to set variables and select items in a dialog box. Many dialog boxes include icons, images, *preview boxes*, or other cues to help you to select appropriate options. When you pick a button in a dialog box that includes an ellipsis (…), another dialog box appears. You must make a selection from the second dialog box before returning to the original dialog box. A button with an arrow icon requires you to select in the drawing area.

preview box: An area in a dialog box that shows the results of the options and settings you select.

Figure 1-18.
A dialog box appears when you pick a menu item with a name followed by an ellipsis (...) or a button displaying an ellipsis. The dialog box shown here displays when you issue the **INSERT** command.

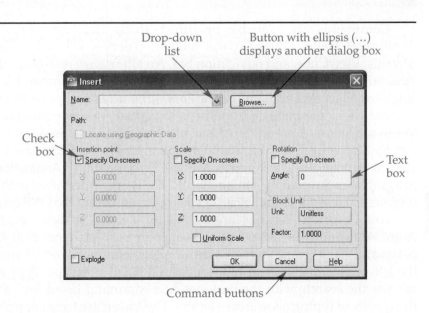

The AutoCAD interface includes several other unique items, such as the in-canvas viewport controls, ViewCube, navigation bar, and **Model** and **Layout** tabs. Refer to **Figure 1-4** to recognize these features. You will explore these features and their specific control operations throughout this textbook. Do not use or adjust these tools until you learn about their function, because doing so can unexpectedly change the interface display and operation. Ensure that the **Model** tab is active.

Exercise 1-9

Complete the exercise on the companion website.
www.g-wlearning.com/CAD

System Options

The **Options** dialog box contains AutoCAD system options. System options apply to the entire program and are not specific to a file. Many system options help configure the work environment, such as the background color of the drawing window. This textbook focuses on the default system options and references the **Options** dialog box when applicable.

You can also access the **Options** dialog box by right-clicking when no command is active and selecting **Options....**

Accessing Commands

dynamic input: Area near the crosshairs where you can type commands and options and view context-oriented information.

command alias: Abbreviated command name entered at the keyboard.

Commands are available by direct access from the ribbon, shortcut menus, **Application Menu**, Quick Access toolbar, palettes, status bar, in-canvas viewport controls, ViewCube, and navigation bar. An alternative is to enter the command using *dynamic input* or the command line. To activate a command by typing, type the single-word command name or the *command alias* and press [Enter] or the space bar, or right-click. You can use uppercase, lowercase, or a combination of uppercase and lowercase letters. You can only issue one command at a time.

You can activate any command or option by typing. Each command name and alias, along with other access techniques available in the **Drafting & Annotation** workspace, appear in a graphic in the margin of this textbook. The example displayed in this margin shows the command name (**LINE**) and alias (**L**) you can use to access the **LINE** command.

AutoComplete settings are active by default to help locate and access any AutoCAD command listed in the Customize User Interface (CUI) file. Begin typing a command name using dynamic input or the command line. Commands that match the letters you enter appear in a suggestion list as you type. Type additional letters to narrow the search, with the best-matched command listed first. **Figure 1-19A** shows the results of typing a lowercase letter l. The lowercase l auto-appends to an uppercase

Figure 1-19.
A—Use AutoComplete settings to help access commands when typing. Notice that the suggestion list displays the command alias and name. B—Options for controlling AutoComplete.

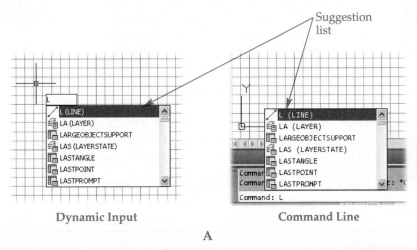

Dynamic Input   Command Line

A

Selection	Result	Default
Auto-Append	Changes the entry to the preferred format.	Selected
Suggestion List	Displays a list of recommended commands.	Selected
Display Icons	Icons appear with command names in the suggestion list.	Selected
Display System Variables	Includes system variables in the suggestion list.	Selected
Delay Time	Specifies the seconds before the suggestion list appears.	0.30 seconds

B

L by default. Hover over a command in the suggestion list to display a tooltip, and select the command you want to activate.

Use the **AUTOCOMPLETE** command to adjust AutoComplete preferences. An easy way to access **AUTOCOMPLETE** command settings is to right-click on the dynamic input suggestion list or on the command line and choose from the **AutoComplete** cascading menu. **Figure 1-19B** briefly describes AutoComplete settings.

Deactivate the AutoComplete **Auto-Append** and **Suggestion List** options to turn off AutoComplete. You can also use the **ON** and **OFF** options of the **AUTOCOMPLETE** command to toggle the tool on and off.

A major benefit of accessing a command using a method other than typing is that you do not need to memorize command names or aliases. Another advantage is that commands, options, and your drawing activities appear on-screen as you work, using visual icons, tooltips, and prompts. As you work with AutoCAD, you will become familiar with the display and location of commands. Decide which command selection techniques work best for you. A combination of command selection methods often proves most effective.

AUTOCOMPLETE

Even though you may not choose to access commands by typing command aliases or names, you must still type certain values, as explained in Chapter 3. For example, you may have to type the diameter of a circle or radius of an arc.

Reference Material

Common Command Aliases
For a detailed list of command aliases, go to the **Reference Material** section of the companion website (www.g-wlearning.com/CAD) and select **Common Command Aliases**.

Dynamic Input

Dynamic input allows you to keep your focus at the crosshairs while you draw. When dynamic input is on, a temporary input area appears in the drawing window, below and to the right of the crosshairs by default. See **Figure 1-20**.

Depending on the command in progress, different information and options appear in the dynamic input area. For example, **Figure 1-21** shows the display after starting the **RECTANGLE** command. The first portion of the dynamic input area is the prompt, which reads Specify first corner point or. In this case, to draw a rectangle, you need to pick in the drawing window or enter *coordinates* to specify the first corner of the rectangle, or access other options as suggested by the "or" portion of the prompt.

Press the down arrow key to display available command options. See **Figure 1-22**. Select an option using the cursor, or press the down arrow again to cycle through the options. Press [Enter] to select the highlighted option. You can also choose an option by right-clicking and picking an option from the shortcut menu. The information displayed in the dynamic input area changes while you work with a command, depending on the actions you choose. **Figure 1-23** shows the dynamic input display when the **LINE** command is active.

coordinates:
Numerical values used to locate a point in the drawing area.

Toggle dynamic input on and off by picking the **Dynamic Input** button on the status bar or pressing [F12]. You can issue commands without dynamic input on.

Figure 1-20.
Use dynamic input to type or select commands and values from a temporary input area next to the crosshairs.

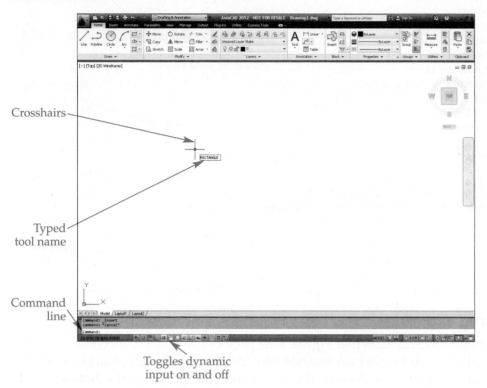

Crosshairs

Typed tool name

Command line

Toggles dynamic input on and off

Figure 1-21.
The dynamic input fields that appear after you enter the **RECTANGLE** command.

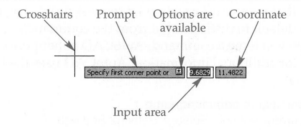

Crosshairs   Prompt   Options are available   Coordinate

Input area

Figure 1-22.
Press the down arrow key to display command options. Pick an option with the cursor, or use the up and down arrow keys to highlight the desired option and press [Enter] to select.

Pick this arrow or press the down arrow on the keyboard to display options

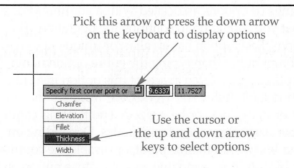

Use the cursor or the up and down arrow keys to select options

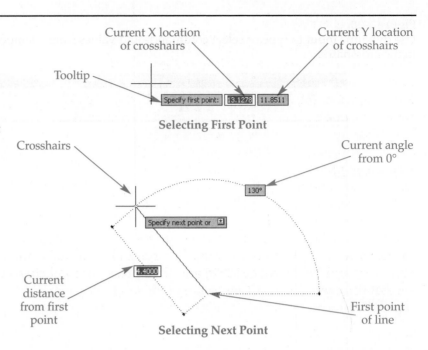

Figure 1-23.
Dynamic input fields change while a command is in use. In this example, the crosshairs coordinates appear first. Once you select the first endpoint, the distance and angle of the crosshairs relative to the first endpoint are displayed.

Current X location of crosshairs

Current Y location of crosshairs

Tooltip

Specify first point: 13.1278 11.8511

Selecting First Point

Crosshairs

Current angle from 0°

130°

Specify next point or

Current distance from first point

4.4000

First point of line

Selecting Next Point

Command Line

The command line, shown in **Figure 1-20**, provides the same function as dynamic input, but allows you to enter commands and context-specific information in a traditional window format. By default, the command line docks at the bottom of the AutoCAD window, above the status bar. It acts like a palette, displays the Command: prompt, and reflects any commands you issue. The command line also displays prompts that supply information or request input. The active prompt background is white by default to distinguish it from the command history, which is gray by default.

When you issue a command, AutoCAD either performs the specified operation or prompts for additional information. AutoCAD uses the following standard command line format:

Command: **commandname**↵
Current settings: Setting1 Setting2 Setting3
Prompt [Option1/oPtion2/opTion3/…] <default option or value>:

Settings or options associated with a command display as shown. The prompt indicates what you should do to continue the operation. The square brackets ([]) contain available options. Each option has an alias, or unique uppercase character(s), that you can enter at the prompt rather than typing the entire option name. Angle brackets (<>) surround the default option; press [Enter] to accept this option instead of typing the value again.

Each default AutoCAD workspace includes the command line. The command line can float or dock, lock, or be resized. The floating command line contains the **Auto-hide** and **Properties** buttons found on palettes. Depending on your working preference, use the command line with dynamic input, or disable the command line if you use only dynamic input. To hide the command line, pick the **Close** button on the command line title bar, right-click on the command line and pick **Close**, type COMMANDLINEHIDE, or press [Ctrl]+[9].

PROFESSIONAL
TIP

While learning AutoCAD, pay close attention to prompts in the dynamic input area and at the command line. Prompts guide you through the operation.

Keyboard Shortcuts

Many keys on the keyboard, known as *shortcut keys* or *keyboard shortcuts*, allow you to perform AutoCAD functions quickly. Become familiar with these keys to improve your AutoCAD performance. To cancel a command or exit a dialog box, press the *escape key* [Esc]. Some command sequences require that you press [Esc] twice to cancel the operation.

When no command is active, press the up arrow key as many times as necessary to cycle through the sequence of previously used command names. Use the down arrow key to return to a later command in the list. If dynamic input is active, previously used commands appear near the crosshairs by default. To display previously used commands at the command line, pick the command line before pressing the up arrow, or turn off dynamic input. Press [Enter] to activate the displayed command.

Function keys provide instant access to commands and are programmable to perform a series of commands. Control and shift key combinations require that you press and hold [Ctrl] or [Shift] and then press a second character. You can activate several commands using [Ctrl] combinations. A tooltip or a display in a shortcut menu typically indicates if a key combination is available for a command.

> **shortcut key (keyboard shortcut):** Single key or key combination used to issue a command or select an option.
>
> **escape key:** Keyboard key used to cancel a command or exit a dialog box.
>
> **function keys:** The keys labeled [F1] through [F12] along the top of the keyboard.

Reference Material

Shortcut Keys

For a complete list of keyboard shortcuts, go to the **Reference Material** section of the companion website (www.g-wlearning.com/CAD) and select **Shortcut Keys**.

Exercise 1-10

Complete the exercise on the companion website.
www.g-wlearning.com/CAD

Getting Help

If you need help with a specific command, option, or AutoCAD feature, use this textbook as a guide. AutoCAD also includes a help system. The **Autodesk Exchange** window you see when you launch AutoCAD for the first time provides access to the online AutoCAD help system, as shown in **Figure 1-24**. The graphic shown in this margin identifies other ways to access help and the **Autodesk Exchange** window. You can also open the **Autodesk Exchange** window from the **InfoCenter**, described later in this chapter, or by selecting **Help** from a shortcut menu.

Help system files are installed with AutoCAD, allowing you to view the **Autodesk Exchange** window offline. If you cannot view the Web version, force AutoCAD to display the installed help system by deselecting the **Access online content (including help) when available** check box in the **Autodesk Exchange** area on the **System** tab of the **Options** dialog box.

The **Autodesk Exchange** window uses a format similar to a typical website with menus of links, navigation options, and a search function. Select a topic from the main menu on the left side of the window to access links to AutoCAD help data. Smaller menus are also available and provide direct access to content such as getting started videos for new AutoCAD users and Autodesk support. To search the help system index

Figure 1-24.
The **Autodesk Exchange** window and the **InfoCenter** found in the AutoCAD window.

Autodesk
Exchange window

InfoCenter

Pick to display
the **Autodesk
Exchange** window

Expanded
Help flyout

Search for
a help topic

Select
an option to
display
related help
content

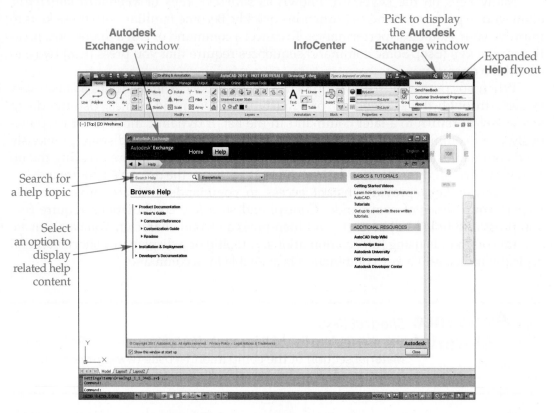

for a specific topic, such as a command or option, type in the **Search Help** box, located near the upper-left corner of the **Help** page. Use the drop-down list to filter the search.

PROFESSIONAL TIP

context-oriented
help: Help
information for the
active command.

Press [F1] while using a command to display *context-oriented help*. This saves time when you are looking for help with the current command and drawing task.

InfoCenter

The **InfoCenter**, located on the right side of the title bar in Figure 1-24, allows you to search for help topics without first displaying the **Autodesk Exchange** window. Enter a topic in the text box to search for related information. The **Autodesk Exchange** window appears with the search results. If you have an Autodesk account or want to create one, pick the **Sign In** button to access Autodesk products, services, support, and community resources. Pick the **Autodesk Exchange** button to display the **Autodesk Exchange** window. Select the **Help** button to access the **Help** page of the **Autodesk Exchange** window, or choose a help system option from the **Help** flyout.

Exercise 1-11

Complete the exercise on the companion website.
www.g-wlearning.com/CAD

Chapter Review

1. Describe at least one application for AutoCAD software.
2. Briefly explain what is involved in planning a drawing.
3. What are drawing standards?
4. Why should you save your work every 10 to 15 minutes?
5. What is the quickest method of starting AutoCAD?
6. Name one method of exiting AutoCAD.
7. What is the name for an interface that includes on-screen features?
8. Define or explain the following terms:
   A. Default
   B. Pick (or click)
   C. Hover
   D. Button
   E. Function key
   F. Option
   G. Command
9. What is a workspace?
10. What is a flyout?
11. How do you change from one workspace to another?
12. How do you access a shortcut menu?
13. What does it mean when a shortcut menu is described as context-sensitive?
14. What is the difference between a docked interface item and a floating interface item?
15. How do you select the locking options to lock the interface items in either their floating or docked state?
16. Explain the basic function of the **Application Menu**.
17. Describe the **Application Menu** search tool and briefly explain how to use it.
18. Briefly describe an advantage of using the ribbon.
19. What is the function of tabs in the ribbon?
20. What is another name for a palette?
21. Describe the function of the application status bar.
22. What is the meaning of the … (ellipsis) in a menu option or button?
23. Describe two methods for accessing AutoCAD commands and list interface items associated with each.
24. Briefly describe the function of dynamic input.
25. Briefly explain the function of the [Esc] key.
26. How do you access previously used commands when dynamic input is on?
27. Name the function keys that execute the following tasks. (Refer to the Shortcut Keys document in the **Reference Material** section on the companion website.)
    A. Snap mode (toggle)
    B. Grid mode (toggle)
    C. Ortho mode (toggle)
28. Describe two ways to access the **Autodesk Exchange** window.
29. What is context-oriented help, and how is it accessed?
30. Describe the purpose of the **InfoCenter**, and explain how to use the **InfoCenter** text box.

Problems

Start AutoCAD if it is not already started. Follow the specific instructions for each problem.

▼ Basic

1. Perform the following tasks:
   A. Open the **Autodesk Exchange** window.
   B. Access the **Help** page.
   C. Pick the **User's Guide** link.
   D. Pick the **Get Information** topic.
   E. Pick the **Find the Information You Need** topic.
   F. Pick the **Access Help and Other Sources of Information** topic.
   G. Read the information provided.
   H. Close the **Autodesk Exchange** window, and then close AutoCAD.

2. Perform the following tasks:
   A. Move the cursor over the buttons in the status bar and read the tooltip for each.
   B. Slowly move the cursor over each of the ribbon panels and read the tooltips.
   C. Pick the **Application Menu** to display it. Hover over the **File** menu, and then use the right arrow key to move to the **File** options. Then use the down arrow key to move through all the menu options.
   D. Press [Esc] to dismiss the menu.
   E. Close AutoCAD.

▼ Intermediate

3. Interview your drafting instructor or supervisor and try to determine what type of drawing standards exist at your school or company. Write them down and keep them with you as you learn AutoCAD. Make notes as you progress through this textbook on how you use these standards. Also, note how you could change the standards to match the capabilities of AutoCAD.

4. Research your drawing department standards. If you do not have a copy of the standards, acquire one. If AutoCAD standards exist, make notes as to how you can use these in your projects. If no standards exist in your department or company, make notes about how you can help develop standards. Write a report on why your school or company should create CAD standards and how to use the standards. Describe who should be responsible for specific tasks. Recommend procedures, techniques, and forms, if necessary. Develop this report as you progress through your AutoCAD instruction and as you read this textbook.

5. Develop a drawing planning sheet for use in your school or company. List items you think are important for planning a CAD drawing. Make changes to this sheet as you learn more about AutoCAD.

6. Create a freehand sketch of the default AutoCAD window with the **Drafting & Annotation** workspace active. Label each of the screen areas. To the side of the sketch, write a short description of the function of each screen area.

7. Create a freehand sketch showing three examples of tooltips displayed as you hover over an item. To the side of the sketch, write a short description of each example's function.

8. Using the **Application Menu** search tool, type the letter C and review the information provided in the **Application Menu**. Then add the letter L. How does the information change? Continue typing O, S, and E to complete the **CLOSE** command. Write a short paragraph explaining how you might use this search tool to find a command if you are unsure how the command is spelled or where it is located.

Drawing Problems - Chapter 1

▼ Advanced

9. Research and write a report of approximately 250 words covering the American Society of Mechanical Engineers (ASME) standards accredited by the American National Standards Institute (ANSI).

10. Research and write a report of approximately 250 words covering the International Standards Organization (ISO) drafting standards.

11. Research and write a report of approximately 250 words covering the United States National CAD Standard (NCS).

12. Research and write a report of approximately 250 words covering workplace ethics, especially as related to CADD applications and CADD-related software.

13. Research and write a report of approximately 150 words covering an ergonomically designed CADD workstation. Include a sketch of what you consider a high-quality design for a workstation and label its characteristics.

14. Go to the Autodesk Education Community website at students.autodesk.com/, and register to join the Autodesk Education Community. After you register, download a student version of AutoCAD to your home or laptop computer. Use your copy of AutoCAD to complete assignments and study AutoCAD when you are unable to access a CADD lab. The Autodesk Education Community website provides complete information on the registration and download process. If you do not have a school-supplied e-mail account, your instructor can register and invite you to participate.

AutoCAD Certified Associate Exam Practice

Answer the following questions. Write your answers on a separate sheet of paper.

1. Which workspaces are available by default in the AutoCAD 2012 software? *Select all that apply.*
   A. **3D Basics**
   B. **3D Animation**
   C. **3D Modeling**
   D. **AutoCAD Classic**
   E. **2D Modeling**
   F. **Drafting & Annotation**

2. Which of the following is a method of starting the AutoCAD software? *Select the one item that best answers the question.*
   A. Access the Windows Start menu, select Run…, and enter autocad.exe
   B. Right-click on the Windows desktop and select AutoCAD 2012 from the shortcut menu
   C. Double-click the **AutoCAD 2012** icon on the Windows desktop
   D. Navigate to autodesk.com and double-click **AutoCAD 2012**

3. If you cannot view the Web version of the **Autodesk Exchange** window, you can force AutoCAD to display the installed help system by deselecting the **Access online content (including help) when available** check box in which of the following locations? *Select the one item that best answers the question.*
   A. **InfoCenter**
   B. **Options** dialog box, **System** tab
   C. **Application Menu**, **Drawing Utilities** menu
   D. Application status bar

AutoCAD Certified Professional Exam Practice

Follow the instructions in each problem. Write your answers on a separate sheet of paper.

1. **Start AutoCAD using one of the methods described in the chapter.**
   Turn off dynamic input. At the Command: prompt, type the letters LA and press [Enter]. What command does AutoCAD execute? (The command appears at the command line in capital letters after the LA you typed.)

2. **Start AutoCAD using one of the methods described in the chapter.**
   Access the **Application Menu** search tool and type the letters C, L, and O. Review the results of the search. What entry appears in the Ribbon Tab: Home category?

Drawings and Templates

Learning Objectives

After completing this chapter, you will be able to:

✓ Begin a new drawing.
✓ Save your work.
✓ Close files.
✓ Open saved files.
✓ Work with multiple open documents.
✓ View and adjust drawing properties.
✓ Determine and specify drawing units and limits.
✓ Create drawing template files.

In this chapter, you will learn how to start new drawings, save drawings, open existing drawings, and begin the process of preparing drawing templates. This chapter also explains commands and options that assist in organizing and setting up a drawing session, including basic drawing settings. You will find the drawing settings described in this chapter very useful as you begin working with drawing and drawing template files.

Beginning a New Drawing

AutoCAD uses several different types of files for specific functions. The primary file types are *drawing files*, which have a .dwg extension, and *drawing template files*, also known as *templates*, which have a .dwt extension. You typically begin a new drawing by referencing a template that includes standard drawing settings and objects. A new drawing based on a template includes all of the template settings and content. To help avoid confusion as you learn AutoCAD, remember that a new drawing file references a drawing template file, but the drawing file is where you draw.

A new drawing file appears by default when you launch AutoCAD. This drawing references the acad.dwt template installed with AutoCAD by default. The drawing is appropriate for initial drawing applications and uses decimal unit settings with a U.S. Customary (inch) unit preference. You are ready to draw, then save and close the file. To start another new drawing, use a template or start from scratch.

drawing files: Files you use to create and store drawings.

drawing template files (templates): Files you reference to develop new drawings; contain standard drawing settings and objects.

Starting from a Template

The **QNEW** command is the primary means of starting a new drawing. By default, the **Select template** dialog box appears when you access the **QNEW** command. See **Figure 2-1**. The **Select template** dialog box lists the templates found in the specified drawing template folder. The default template folder shown in **Figure 2-1** includes a variety of templates supplied with AutoCAD.

> **NOTE**
> All file navigation dialog boxes, including the **Select template** dialog box and those used to save, close, and open files, support auto-complete. Type in the **File Name** text box to view a list of files matching the characters you enter.

The tutorial templates for manufacturing and architecture include a border and title block. All of the other supplied templates are blank, but include drawing settings specific to the requirements of a certain industry or drawing. For general 2D drawing applications, use the default acad.dwt template, or select the acadiso.dwt template, which uses basic decimal unit settings according to metric ISO standards. To use a template to begin a new drawing, double-click on the file name, right-click on the file and pick **Select**, or select the file and pick the **Open** button.

> **NOTE**
> You can also access the **QNEW** command from the **Quick View Drawings** tool described later in this chapter.

Figure 2-1.
Use the **Select template** dialog box to begin a new drawing. Choose a template to reference, or pick the arrow next to the **Open** button and select an **Open with no Template** option to start a drawing "from scratch".

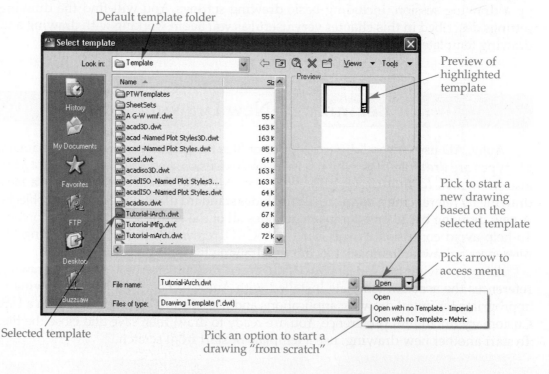

Use the **Options** dialog box to change the default drawing template folder displayed in the **Select template** dialog box. In the **Files** tab, expand **Template Settings**, and then expand **Drawing Template File Location**. Pick the **Browse...** button to select a folder.

Starting from Scratch

For a new AutoCAD user, an effective way to begin a new drawing is to start "from scratch" using a blank drawing file without a border, title block, modified layouts, or customized drawing settings. Start from scratch when you just begin to learn AutoCAD, when you do not yet know the units and drawing settings to be used, and when you create your own templates. To start from scratch, pick the flyout next to the **Open** button in the **Select template** dialog box. See **Figure 2-1**. Then, to begin a drawing using basic inch unit settings, pick **Open with no Template-Imperial**, or to begin a drawing using basic metric unit settings, pick **Open with no Template-Metric**.

Setting the Quick Start Template

The **Options** dialog box provides a quick start feature that allows you to begin a drawing using the **QNEW** command and a specific template, skipping the **Select template** dialog box. Pick the **Files** tab, expand the **Template Settings** option, and then expand the **Default Template File Name for QNEW** function. See **Figure 2-2**. **None** displays by default and causes the **Select template** dialog box to appear. Pick the **Browse...** button to select a specific template to launch each time you use the **QNEW** command.

Figure 2-2.
Specifying a template for the **QNEW** command.

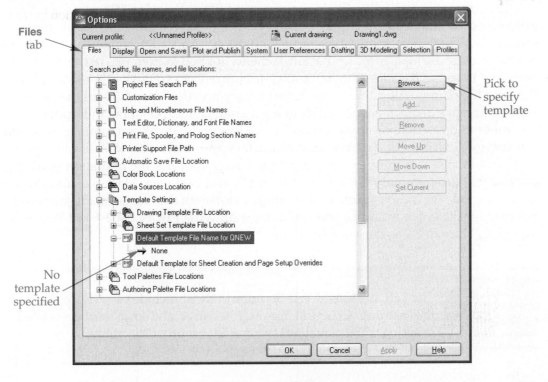

Use the **NEW** command to override the quick start template and display the **Select template** dialog box. This allows you to use the **QNEW** command to begin a drawing by referencing the template you use most frequently, but access other templates as needed.

Saving Your Work

You should save your drawing or template immediately after you begin work. Then save at least every 10 to 15 minutes while working to avoid the possibility of losing your work. Several AutoCAD commands allow you to save your work. In addition, when you close a file or exit AutoCAD, an alert appears asking if you want to save changes. This gives you a final opportunity to save or discard changes.

File Storage and Naming

Store drawings on your computer or a network drive according to school or company practice. Back up files to a removable storage device or other system to help ensure that information is not permanently lost during a system failure.

Develop an organized structure of file folders, and use subfolders as needed to help organize each project. Using a standard naming system allows you to find drawings quickly and easily by content, category, or other criteria. For example, you might save all of your textbook exercises to an Exercises folder, or create chapter-specific folders to store exercises, tests, and problems. An example of a mechanical drafting project is a main folder titled ACME.4001 based on the company name ACME, Inc. and project and assembly number 4001. An example of an architectural drafting project is a main folder titled JAMES0111 to identify the first James Residence project of 2011.

file properties: Values used to define a variety of file and design characteristics.

The name you assign to a file is one of several *file properties*. File naming is typically based on a specific system associated with the product and approved drawing standards. A file name should be concise and allow you to determine the content of the file. A basic example is the file naming scheme applied to exercises in this textbook. You save Exercise 2-1, for example, as EX2-1. **Figure 2-3** provides other examples of drawing file names. File name characteristics vary greatly depending on the product, specific drawing requirements, and drawing standard interpretation and options.

Rules and restrictions apply to naming folders and files. When naming files, you can use most alphabetic and numeric characters and spaces, as well as most punctuation symbols. You can use 255 characters. You cannot use the quotation mark ("), asterisk (*), question mark (?), forward slash (/), and backward slash (\). You do not have to include the file extension, such as .dwg or .dwt, with the file name. File names are not case sensitive. For example, you can name a drawing PROBLEM 2-1, but Windows interprets Problem 2-1 as the same file name.

The U.S. National CAD Standard (NCS) includes a comprehensive file naming structure for architectural and construction-related drawings. You can adapt the system for other disciplines.

Figure 2-3.
Examples of mechanical, architectural, and structural drawing file naming schemes. The architectural and structural examples follow the U.S. National CAD Standard (NCS) file naming format.

Mechanical
Drawing of a compressor housing
ACME.4001.15A.C
— Revision level C
— Compressor housing part number 15A
— Compressor project/assembly number 4001
— Manufactured by ACME, Inc.

Mechanical
Drawing of a seat bracket
MDI-101065-023
— Seat bracket part number 023
— Seat project/assembly number 101065
— Manufactured by Madsen Designs, Inc.

Architectural
Drawing of a first floor plan
A-101
— First floor of the building
— Sheet type designation, plan
— Architectural designation

Structural
Drawing of framing details
SF501
— First sheet
— Sheet type designation, details
— Structural framing designation

Using the QSAVE Command

If you have not yet saved a file, the **QSAVE** command displays the **Save Drawing As** dialog box. See **Figure 2-4.** To save a file, first choose the type of file to save from the **Files of type:** drop-down list. Select **AutoCAD 2012 Drawing (*.dwg)** for most applications.

Quick Access
Save

Application Menu
Save

Type
QSAVE
[Ctrl]+[S]

QSAVE

Figure 2-4.
The **Save Drawing As** dialog box is a standard file selection dialog box.

Select folder where drawing will be saved

Move up one level from current folder

Create a new folder

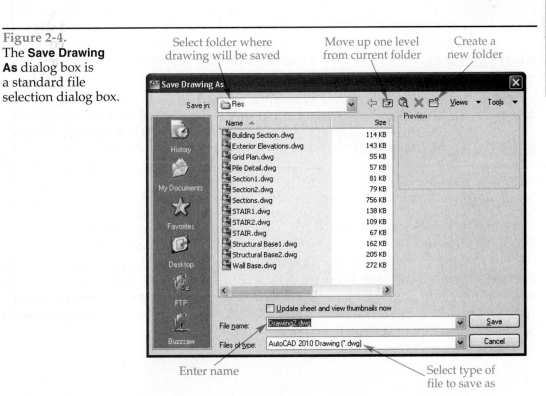

Enter name

Select type of file to save as

Use **AutoCAD Drawing Template (*.dwt)** to save a template file. To share a file with someone using an older version of AutoCAD, choose the appropriate older AutoCAD version from the list. You also have the option of saving a *drawing exchange file (DXF)* or *drawing standards file (DWS)*.

Next, select the folder in which to store the file from the **Save in:** drop-down list. To move upward from the current folder, pick the **Up one level** button. To create a new folder in the current location, pick the **Create New Folder** button and type a new folder name.

The name Drawing1 appears in the **File name:** text box if the file is the first file since you launched AutoCAD. Change the name to the desired file name. After you specify the correct location and file name, pick the **Save** button to save the file. You can also press [Enter] to activate the **Save** button.

If you saved the file previously, the **QSAVE** command updates, or resaves, the file based on the current file state. In this situation, **QSAVE** issues no prompts and displays no dialog box.

Using the SAVEAS Command

SAVEAS

Quick Access
Save As

Application Menu
Save As

Type
SAVEAS

Use the **SAVEAS** command to save a copy of a file using a different name or file type. You can also use the **SAVEAS** command when you open a drawing template file to use as a basis for another drawing. This leaves the template unchanged and ready to use for starting other drawings.

The **SAVEAS** command always displays the **Save Drawing As** dialog box. The location and name of an existing file appear. Confirm that the **Files of type:** drop-down list displays the desired file type and that the **Save in:** drop-down list displays the correct drive and folder. Type the new file name in the **File name:** text box and pick the **Save** button.

NOTE

Pick an option from the **Save As** menu in the **Application Menu** to preset the file type in the **Save Drawing As** dialog box.

PROFESSIONAL TIP

When you save a version of a drawing in an earlier format, give the file a name that is different from the AutoCAD 2012 version to prevent accidentally overwriting your working drawing with the older format.

Automatic Saves

AutoCAD provides an *automatic save* function that automatically creates a temporary backup file while you work. Settings in the **File Safety Precautions** area of the **Open and Save** tab in the **Options** dialog box control automatic saves. See **Figure 2-5.** Automatic save is on by default and saves every 10 minutes. Type the number of minutes between saves in the **Minutes between saves** text box. By default, AutoCAD names automatically saved files *FileName_n_n_nnnn.sv$* in the C:\Documents and Settings\user\Local Settings\Temp folder.

Figure 2-5.
Use the **Open and Save** tab in the **Options** dialog box to set up the automatic save feature and backup files.

Activates autosave

Autosave timer setting

Creates backup copies

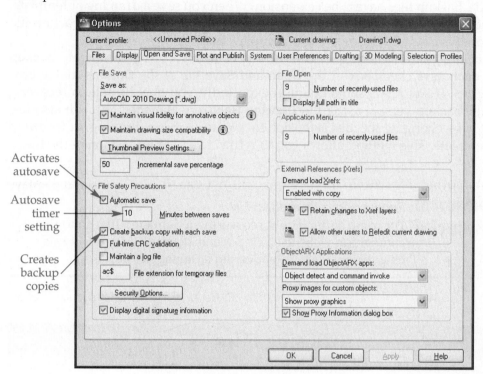

The automatic save timer starts as soon as you make a change to the file and resets when you save the file. The file saves automatically when you start the first command after reaching the automatic save time. Keep this in mind if you let the computer remain idle; an automatic save does not execute until you return and issue a command. Be sure to save your file manually if you plan to be away from your computer for an extended period.

The automatic save file is available for use if AutoCAD shuts down unexpectedly. Therefore, when you close a file normally, AutoCAD deletes the automatic save file. However, after a system failure, the **Drawing Recovery Manager** displays the next time you open AutoCAD. The **Drawing Recovery Manager** contains a node for every file that was open at the time of the system failure. It displays all of the available versions of each file: the original file, the recovered file saved at the time of the system failure, the automatic save file, and the .bak file. Pick a version in the **Drawing Recovery Manager** to view, determine which version you want to save, and then save that file. You can save the recovered file over the original file name.

Application Menu
Drawing
> Open the
Drawing
Recovery
Manager

DRAWING RECOVERY

The **Automatic Save File Location** item in the **Files** tab of the **Options** dialog box determines the folder in which automatic save files are stored.

Backup Files

By default, AutoCAD saves a backup file in the same folder as the drawing or template file. Backup files have a .bak extension. When you save a drawing or template, the file updates, and the old file overwrites the backup file. Therefore, the backup file is always one save behind the drawing or template file.

The backup feature is on by default and is controlled using the **Create backup copy with each save** check box in the **Open and Save** tab of the **Options** dialog box. See **Figure 2-5**. If AutoCAD shuts down unexpectedly, you may be able to recover a file from the backup version using the **Drawing Recovery Manager**. You can also use a backup file by changing the .bak extension to .dwg to restore a drawing, or to .dwt to restore a template. This method allows you to return to an earlier version of the file.

Supplemental Material

Recovering a Damaged File
For information about recovering a damaged file, go to the companion website (www.g-wlearning.com/CAD), select this chapter, and select **Recovering a Damaged File**.

Closing Files

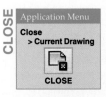

CLOSE

Application Menu
Close
> Current Drawing
CLOSE

Use the **CLOSE** command to close the current file without exiting AutoCAD. One of the quickest methods of closing a file is to pick the **Close** button from the title bar of a drawing window. If you close a file before saving, AutoCAD prompts you to save or discard changes. Pick the **Yes** button to save the file, or pick the **No** button to discard any changes made to the file since the previous save. Pick the **Cancel** button if you decide not to close the drawing.

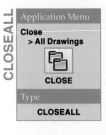

CLOSEALL

Application Menu
Close
> All Drawings
CLOSE

Type
CLOSEALL

AutoCAD allows you to have multiple files open at the same time. Use the **CLOSEALL** command to close all open files. AutoCAD prompts you to save each file to which you made changes.

You can also close files using the **Quick View Drawings** tool, described later in this chapter.

Exercise 2-1

Complete the exercise on the companion website.
www.g-wlearning.com/CAD

Opening a Saved File

Use the **OPEN** command to open a saved file. You can also open a recent document from the **Application Menu** or open a file from Windows Explorer.

Using the OPEN Command

Access the **OPEN** command to display the **Select File** dialog box shown in Figure 2-6. The buttons on the left side of the dialog box provide instant access to certain folders. See Figure 2-7. Double-click on a folder to display the contents. Pick a file to view an image of the file in the **Preview** area. The preview provides an easy way for you to identify the content of a file without loading it into AutoCAD. You can quickly preview other files using the up and down arrow keys to move vertically, and the left and right arrow keys to move horizontally. To open the highlighted file, double-click on a file, pick the **Open** button, or press [Enter].

You can also access the **Open** dialog box by picking the **Open...** button in the **Quick View Drawings** toolbar, described later in this chapter.

Figure 2-6.
The **Select File** dialog box is a standard file selection dialog box that allows you to open an AutoCAD file. The AutoCAD 2012\Sample\Mechanical Sample folder is open in this illustration. The Mechanical – Multileaders drawing is selected and appears in the **File name:** text box and in the preview area.

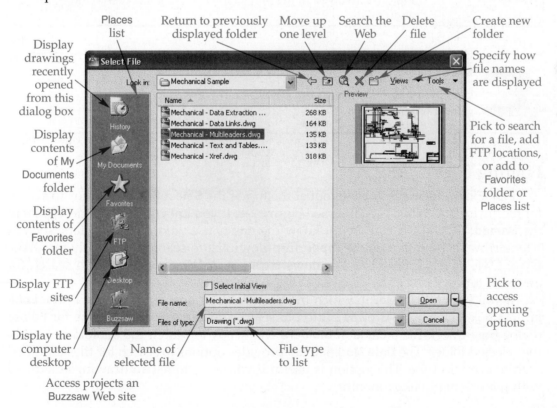

Figure 2-7.
Additional features in the **Select File** dialog box. Right-click on the list to access options for adding, removing, and adjusting folders, and for restoring the original folders.

Button	Description
History	Lists drawing files recently opened from the **Select File** dialog box.
My Documents	Displays the files and folders contained in the My Documents folder for the current user.
Favorites	Displays files and folders located in the Favorites folder for the current user. To add the folder displayed in the **Look in:** box to the favorites list, select **Add to Favorites** from the **Tools** flyout.
FTP	Displays available FTP (file transfer protocol) sites. To add or modify the listed FTP sites, select **Add/Modify FTP Locations** from the **Tools** flyout.
Desktop	Lists the files, folders, and drives located on the computer desktop.
Buzzsaw	Displays projects on the Buzzsaw.com Web site. After setting up a project hosting account, users can access drawings from a given construction project on the Web site. This allows the various companies involved in the project to have instant access to the drawing files.

Exercise 2-2

Complete the exercise on the companion website.
www.g-wlearning.com/CAD

Finding Files

Pick **Find...** from the **Tools** flyout at the top of the **Select File** dialog box to search for files using the **Find** dialog box. See **Figure 2-8**. If you know the file name, type it in the **Named:** text box. If you do not know the name, use wildcard characters, such as *, to narrow the search. Use the **Type:** drop-down list to search specifically for DWG, DWS, DXF, or DWT files. Use Windows Explorer or Windows Search to search for other file types.

If you know the folder in which the file is located, specify the folder in the **Look in:** text box, or pick the **Browse** button to select a folder from the **Browse for Folder** dialog box. Check the **Include subfolders** check box to search the subfolders within the selected folder. The **Date Modified** tab provides options to search for files modified within a certain time. This option is useful if you want to list all drawings modified within a specific week or month.

Figure 2-8.
Use the **Find** dialog
box to locate
AutoCAD files.

Enter file name
or wildcards

Select type of file
to search for

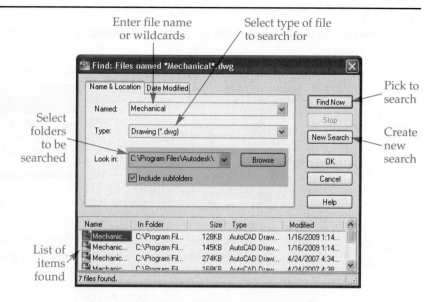

Pick to
search

Create
new
search

Select
folders
to be
searched

List of
items
found

Right-click on a folder or file to display a shortcut menu of options.
Pick somewhere off the menu to close the menu.

**PROFESSIONAL
TIP**

Certain file management capabilities are available in file dialog boxes,
similar to those in Windows Explorer. For example, to rename an
existing file or folder, slowly double-click on the name. This places the
name in a text box for editing. Type the new name and press [Enter].

CAUTION

Use caution when deleting or renaming files and folders. Never
delete or rename a file if you are not certain you should. If you are
unsure, ask your instructor or system administrator for assistance.

Exercise 2-3
Complete the exercise on the companion website.
www.g-wlearning.com/CAD

Using the Recent Documents Menu

Pick the **Recent Documents** button in the **Application Menu** to display a list of
recently opened documents. See **Figure 2-9**. The **Recent Documents** tool provides
convenient access to files that may be related to the current project. Nine of the most
recent AutoCAD files appear by default.

Use the **Ordered List** drop-down list to organize the files. Select **By Ordered List,
By Access Date, By Size**, or **By Type** according to how you want files arranged. Select

Figure 2-9.
Use the **Recent Documents** menu on the **Application Menu** to open a recent file.

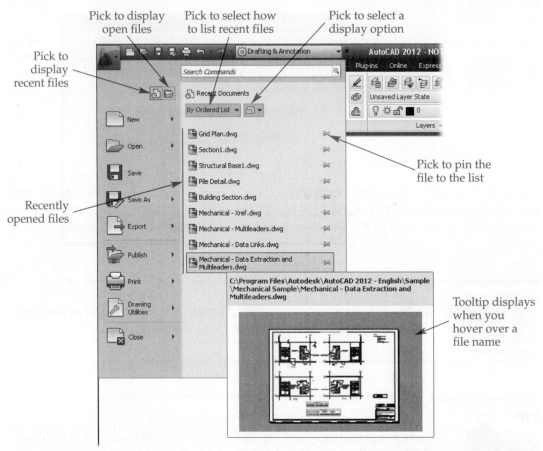

Pick to display open files

Pick to select how to list recent files

Pick to select a display option

Pick to display recent files

Recently opened files

Pick to pin the file to the list

C:\Program Files\Autodesk\AutoCAD 2012 - English\Sample\Mechanical Sample\Mechanical - Data Extraction and Multileaders.dwg

Tooltip displays when you hover over a file name

from the display options flyout to specify how recent files appear. Choose the appropriate option to display files as icons, small images, medium images, or large images.

Hover over a file in the list to display a tooltip with file information and a preview of the file. Pick a file from the list to open. Pick the pushpin icon to the right of the file to keep the file in the recent documents list. AutoCAD eventually removes unpinned files from the recent documents list as you open other files.

PROFESSIONAL TIP

Specify the number of files displayed in the **Recent Documents** list by accessing the **Open and Save** tab of the **Options** dialog box. The **Number of recently-used files** text box in the **Application Menu** area controls this function.

If you try to open a deleted or moved file, AutoCAD displays the message Cannot find the specified drawing file. Please verify that the file exists. AutoCAD then opens the **Select File** dialog box.

Using Windows Explorer

Double-click on an AutoCAD file from Windows Explorer to open the file. If AutoCAD is not already running, it starts and the file opens. You can also launch

AutoCAD, if it is not running, and open a file by dragging and dropping the file from Windows Explorer onto the AutoCAD 2012 desktop icon. If AutoCAD is running, you can drag and drop a file onto the command line to open the file.

CAUTION

Dragging and dropping an AutoCAD file into the drawing area inserts the file as a block into the existing file. You must drop the file onto the command line to open the file.

Opening Old Files

When using AutoCAD 2012, you can open AutoCAD files created in AutoCAD Release 12 or later. When you save a file from a previous release in AutoCAD 2012, AutoCAD automatically updates the file to the current file format. A file saved in AutoCAD 2012 displays in the **Preview** image tile in the **Select File** dialog box.

In order to view the original format of an older release file opened in AutoCAD 2012, you must use the **SAVEAS** command to save it back to the appropriate file format.

Opening as Read-Only or Partial Open

You can open files in various modes by selecting the appropriate option from the **Open** flyout in the **Select File** dialog box. When you open a file as *read-only*, you cannot save changes to the original file. However, you can make changes to the file and then use the **SAVEAS** command to save changes using a different file name. This technique ensures that the original file remains unchanged.

When opening a large drawing, you may choose to issue a *partial open*. This allows you to open a portion of a drawing by selecting specific views and layers to open. Views and layers are described in later chapters. You can also partially open a drawing in the read-only mode.

read-only: Describes a drawing file intended for viewing only. To keep any changes made to the drawing, use the **SAVEAS** command to save the file using a different name.

partial open: Describes opening a portion of a file by specifying only the views and layers you need to see.

Managing Multiple Documents

Most drafting projects include several closely related files. Each file presents or organizes a different aspect of the project. Examples of drawings for a mechanical drafting project include an assembly drawing, subassembly drawings, and detail drawings of each part. Examples of drawings for a basic architectural drafting project include a site plan, a floor plan, electrical and plumbing plans, and details. You can open multiple AutoCAD files at the same time to work with different portions of a project. Drag-and-drop and similar operations allow you to easily share content between documents.

Controlling Windows

Each file you start or open in AutoCAD appears in its own drawing window. The name of the active file appears on the AutoCAD window title bar and on a title bar of a floating drawing window. Control drawing windows using the minimize, maximize, restore, and close buttons located in the upper-right corner of each window. See **Figure 2-10**. Resize windows using standard window sizing operations.

Figure 2-10.
A—Float drawing windows when more than one file is open to work efficiently between drawings. B—Minimize drawing windows to display title bars only. Pick the title bar to display a window control menu.

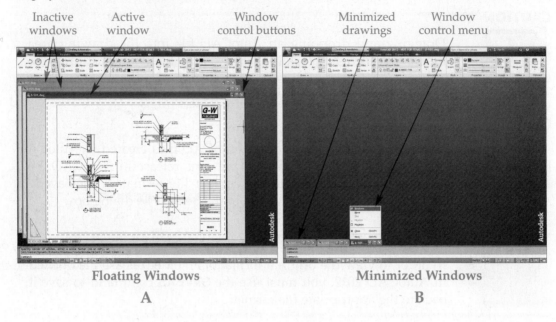

Floating Windows
A

Minimized Windows
B

The **Windows** panel of the **View** ribbon tab provides one method to access additional window management commands. Pick the **Switch Windows** flyout to display a list of all open files. Select a file from the flyout to activate. The **Tile Horizontally**, **Tile Vertically**, and **Cascade** buttons allow you to control the arrangement of floating drawing windows. Select the **Tile Vertically** button to tile drawings in a vertical arrangement, with the active window placed on the left. Pick the **Tile Horizontally** option to tile drawings in a horizontal arrangement, with the active window placed on top. Pick the **Cascade** option to arrange drawing windows in a cascading style. See Figure 2-11.

You can also switch drawing windows from the **Application Menu** by picking the **Open Documents** button shown in Figure 2-9. Files are listed in the order you opened them. You can display open files as icons, small images, medium images, or large images by picking the appropriate option from the display options flyout. Pick a file from the list to activate the corresponding drawing window.

Using the Quick View Drawings Tool

The **Quick View Drawings** tool allows you to see and control open AutoCAD files without changing drawing windows. The quickest way to access the **Quick View Drawings** tool is to pick the **Quick View Drawings** button on the status bar. The **Quick View Drawings** tool appears in the lower center of the AutoCAD window. See Figure 2-12.

A thumbnail image and file name identify each open file. Files are arranged in the order opened, with the file opened first on the left side of the row. The current file is highlighted when you initially access the **Quick View Drawings** tool. Move the cursor over a thumbnail to show additional options for controlling the drawing window. You can also roll the mouse wheel to move between thumbnails. Pick the thumbnail of a file to make the drawing window of the file current. To adjust the size of the thumbnails, press and hold [Ctrl] then roll the mouse wheel forward to enlarge, or back to reduce the size.

The **Quick View Drawings** tool includes a small toolbar below the file thumbnail images. See Figure 2-13. By default, the **Quick View Drawings** tool disappears when

Figure 2-11.
Examples of tiled and cascading floating drawing windows. The effect of tiling drawing windows varies depending on the number of windows and the tile option you choose.

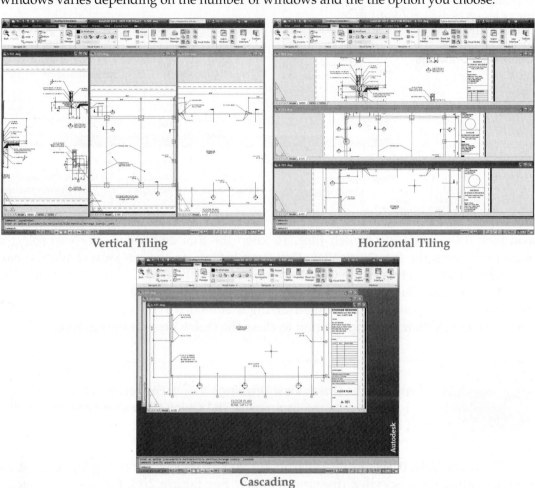

Vertical Tiling

Horizontal Tiling

Cascading

Figure 2-12.
The **Quick View Drawings** tool offers an effective visual method for switching and controlling open files.

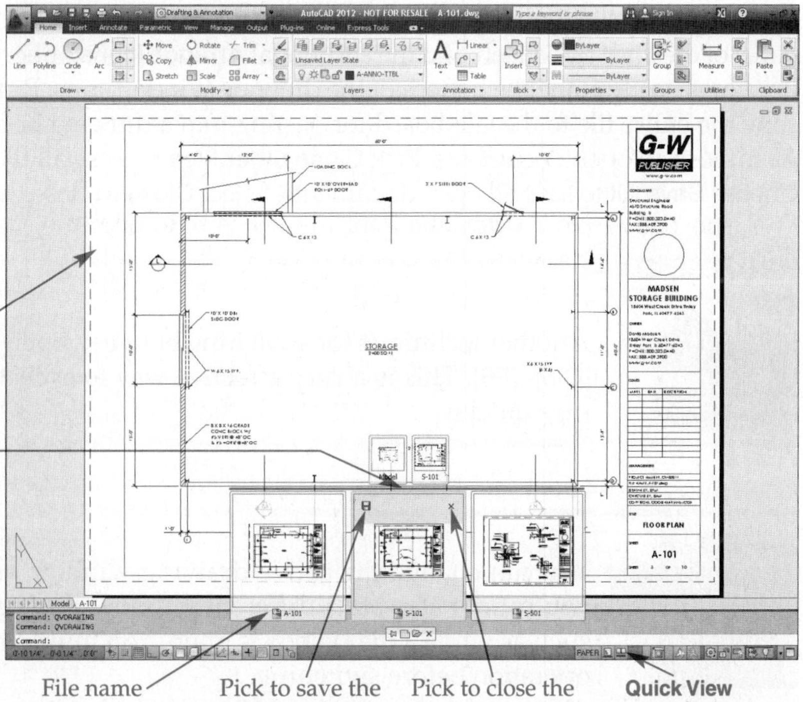

Active file

All files currently open

File name

Pick to save the highlighted file

Pick to close the highlighted file

Quick View Drawings button

Figure 2-13.
The **Quick View Drawings** toolbar provides access to basic functions directly from the **Quick View Drawings** tool.

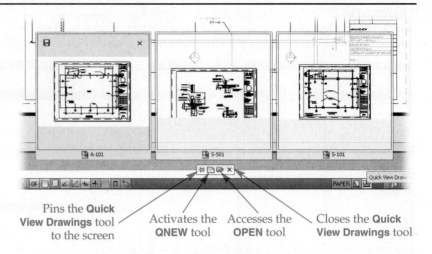

Pins the **Quick View Drawings** tool to the screen

Activates the **QNEW** tool

Accesses the **OPEN** tool

Closes the **Quick View Drawings** tool

you pick a thumbnail to switch files. To keep the tool on-screen after you select a thumbnail, pick the **Pin Quick View Drawings** button on the left side of the toolbar. The **New...** button activates the **QNEW** command, and the **Open...** button activates the **OPEN** command. The **New...** and **Open...** buttons are especially useful for starting a new drawing or opening an existing drawing that relates to the current project. Select the **Close Quick View Drawings** button to close the **Quick View Drawings** tool.

If you pin the **Quick View Drawings** tool to the screen, close the tool, and then access the tool again, the **Quick View Drawings** tool will still be in the pinned state.

The **Quick View Drawings** tool also allows you to display a drawing layout without first activating the associated drawing window. See **Figure 2-14.** Layouts are used to prepare a drawing for plotting, as fully described later in this textbook.

Right-click on a thumbnail image to access a shortcut menu of options for controlling open files. The **Windows** cascading menu provides the same features found in the **Windows** panel of the **View** ribbon tab, and the additional **Arrange Icons** option to arrange minimized drawings neatly along the bottom of the drawing window area. Pick **Copy File as a Link** to copy the entire file to the Clipboard for pasting into a drawing or document. Select **Close All** to close all open documents. Pick **Close other files** to close all files except the active file. Choose **Save All** to save all open documents. Select **Close** to close the active file.

PROFESSIONAL TIP

Another technique for switching between open drawings is to press [Ctrl]+[F6]. This is a very effective way to cycle through open drawings quickly.

You can change the active drawing window as desired. However, you cannot activate a different window during certain operations, such as while a dialog box is open. You must complete or cancel the operation before switching.

Figure 2-14.
A—The initial display when you hover over a file thumbnail.
B—Hover over the model space and layout thumbnails to enlarge.

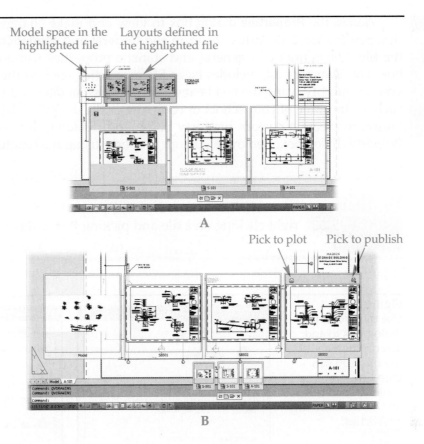

Model space in the highlighted file

Layouts defined in the highlighted file

A

Pick to plot   Pick to publish

B

Exercise 2-4

Complete the exercise on the companion website.
www.g-wlearning.com/CAD

Drawing Properties

An AutoCAD file stores drawing properties that identify the file. The file type and name are examples of file properties. Add values to drawing properties to record common drawing information, such as the title of the product represented by the drawing. You can then reference drawing properties for a variety of purposes, such as adding text to a title block, or creating a bill of materials or report. Additionally, content in the drawing that is linked to drawing properties updates when you make changes to values in the **Properties** dialog box.

The **Properties** dialog box is different from the **Properties** palette, which will be explained in later chapters. The **Properties** dialog box contains properties of the drawing file. The **Properties** palette provides access to the properties of objects created in the drawing file.

DWGPROPS

Application Menu

Drawing Utilities
> Drawing
Properties

Line

Type

DWGPROPS

Access the **Properties** dialog box to view and make changes to drawing properties. See **Figure 2-15**. Values form in the **General** and **Statistics** tabs when you save the file. You cannot edit general and statistic properties using the **Properties** dialog box. The **Summary** tab includes basic file properties, such as the title and author, that you can modify using the text boxes. Use the **Custom** tab to add properties to the file, such as the name of your school or company or the design revision level. Pick the **Add** button to create a custom property name and default value using the **Add Custom Property** dialog box. Remove a custom property using the **Delete** button.

> **NOTE**
> You can view some drawing properties in Windows Explorer by right-clicking on a file and picking **Properties**.

Figure 2-15.
Use the **Properties** dialog box to add information about the file. You can then link drawing content, such as text, to the drawing properties.

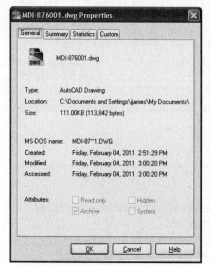

General Tab

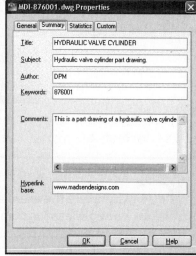

Summary Tab

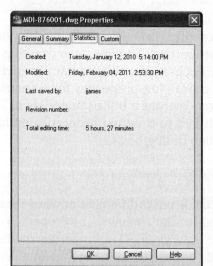

Statistics Tab

Pick to remove a custom property   Pick to create a custom property

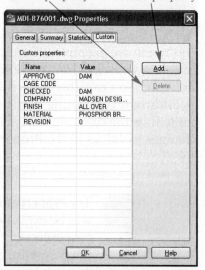

Custom Tab

Exercise 2-5

Complete the exercise on the companion website.
www.g-wlearning.com/CAD

Basic Drawing Settings

Drawing settings determine the general characteristics of a drawing, such as the units of measurement. You can change drawing settings as needed throughout the drawing process. However, you can specify appropriate drawing settings in template files to preset common drawing properties every time you begin a new file. You will create templates later in this chapter.

AutoCAD includes many drawing settings. The most basic drawing settings control the units of measurement and limits of the drawing area.

Introduction to Model Space

Before you adjust drawing settings and begin drawing, you should understand the two environments in which you can work. *Model space* is where you design and draft the *model* of a product. In mechanical drafting, for example, use model space to draw part and assembly views. In architectural drafting, use model space to draw building plans, elevations, sections, and details.

Once you complete the drawing or model in model space, you switch to *paper (layout) space*, where you prepare a *layout*. A layout represents the sheet of paper used to organize and scale, or lay out, and plot or export a drawing or model. Layouts typically include items such as a border, title block, and general notes. A single drawing can have multiple layouts.

You know you are in model space when you see the model space coordinate system icon, the active **Model** tab, the **MODEL** button on the status bar, and no representation of a sheet in the drawing area. See **Figure 2-16**. You know you are working with a layout when you see the paper space coordinate system icon, an active layout tab, the **PAPER** button on the status bar, and a representation of a sheet in the drawing area. See **Figure 2-17**. If you find that you are not in model space, pick the **Model** tab or the **PAPER** button on the status bar, or use **Quick View Drawings** or **Quick View Layouts**. You will explore **Quick View Layouts** later in this textbook.

model space: The environment in AutoCAD in which the majority of drawing usually occurs, including the design and drafting of drawing views.

model: A term that usually describes a 3D model, but in AutoCAD also refers to 2D drawing geometry, typically created at full size.

paper (layout) space: The environment in AutoCAD in which you create layouts for plotting and display purposes.

layout: An arrangement in paper space of sheet elements, typically including a border, title block, general notes, and a display of items drawn in model space.

CAUTION

This textbook explains paper space when appropriate. Until you are ready to lay out a drawing, all of your drawing and drawing setup should occur in model space. Model space is active by default when you start a drawing from scratch and when you begin a file using one of several available templates. Activate model space in all current drawings and custom templates.

Figure 2-16.
Model space is the environment in which you design and draft product geometry, such as the views of this hydraulic valve cylinder.

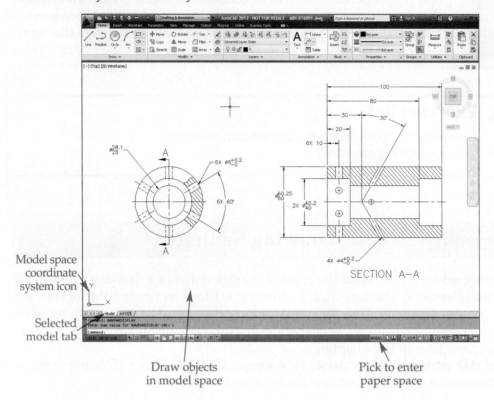

Model space coordinate system icon

Selected model tab

Draw objects in model space

Pick to enter paper space

Figure 2-17.
Paper space is the environment in which you lay out model space geometry to complete the drawing and prepare for plotting or export. This example shows common items added to a layout sheet, including an ASME border, title block, revision block, and general notes.

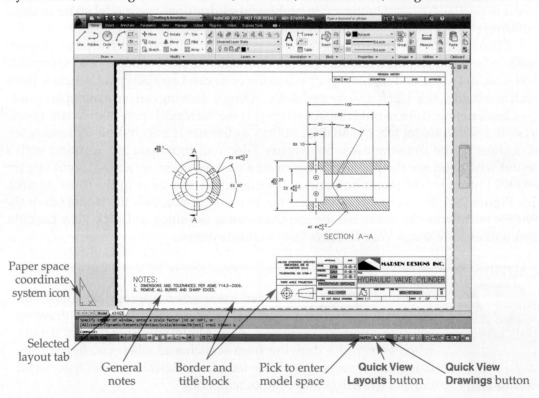

Paper space coordinate system icon

Selected layout tab

General notes

Border and title block

Pick to enter model space

Quick View Layouts button

Quick View Drawings button

Drawing Units

Drawing units define the linear and angular units of measurement used while drawing and the precision to which these measurements display. *You* determine what "1 unit" means in AutoCAD. For example, 1 unit can mean 1 inch, 1 millimeter, 1 meter, or 1 mile. Most AutoCAD users generally think of 1 unit to be 1 inch or 1 millimeter. Access the **Drawing Units** dialog box to set linear and angular units. See Figure 2-18. Specify linear unit characteristics in the **Length** area. Use the **Type:** drop-down list to set the linear units format, and use the **Precision:** drop-down list to specify the precision of linear units. Figure 2-19 describes linear unit formats.

drawing units:
The standard linear and angular units and precision of measurement.

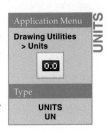

Figure 2-18.
Use the **Drawing Units** dialog box to set linear and angular unit values.

Select linear units

Select linear precision

Specifies the units used to insert objects, but does not preset the drawing to use a specific unit of measurement

Select angular units

Select angular precision

Pick to change direction of angular measurement

Access the **Direction Control** dialog box

Figure 2-19.
Linear unit formats available in the **Drawing Units** dialog box. Select the appropriate type and precision for the specific drawing application.

Type	Typical Applications	Characteristics	Example
Decimal	Mechanical, architectural, structural, civil (inch or metric)	• Decimal inches or millimeters. • Conforms to the ASME Y14.5 dimensioning and tolerancing standard. • Four decimal place default precision.	14.1655
Engineering	Civil—feet and inch	• Feet and decimal inches. • Four decimal place default precision.	1'-2.1655"
Architectural	Architectural, structural—feet and inch	• Feet, inches, and fractional inches. • 1/16" default precision.	1'-2 3/16"
Fractional	Mechanical—fractional	• Fractional parts of any common unit of measure. • 1/16" default precision.	14 3/16
Scientific	Chemical engineering, astronomy	• E+01 means the base number is multiplied by 10 to the first power. • Used when very large or small values are required. • Four decimal place default precision.	1.4166E+01

Set the angular unit format and precision in the **Type:** and **Precision:** drop-down lists in the **Angle** area of the **Drawing Units** dialog box. To change the direction for angular measurements to clockwise from the default setting of counterclockwise, select the **Clockwise** check box.

Pick the **Direction...** button to access the **Direction Control** dialog box. See **Figure 2-20.** Pick the **East**, **North**, **West**, or **South** *radio button* to set the compass orientation. The **Other** radio button activates the **Angle:** text box and the **Pick an angle** button. Enter an angle for zero direction in the **Angle:** text box. The **Pick an angle** button allows you to pick two points on-screen to establish the angle zero direction.

radio button: A selection that activates a single item in a group of options.

CAUTION

Use the default direction of **0° East** at all times, unless you have a specific need to change the compass direction angle, such as when measuring direction using azimuths (**0° North**). This textbook uses the **0° East** direction, and this default should be used in order to complete most exercises and problems correctly.

Figure 2-21 describes angular unit formats. After selecting the linear and angular units and precision, pick the **OK** button to exit the **Drawing Units** dialog box.

Exercise 2-6

Complete the exercise on the companion website.
www.g-wlearning.com/CAD

Drawing Limits

You prepare an AutoCAD drawing at actual size, or full scale, regardless of the type of drawing, the units used, or the size of the final layout on paper. Use model space to draw full-scale objects. AutoCAD allows you to specify the size of a virtual model space drawing area, known as the model space drawing limits, or *limits*. You typically set limits in a template, but you can change limits as necessary.

limits: The size of the virtual drawing area in model space.

The concept of limits is somewhat misleading, because the AutoCAD drawing area is infinite in size. For example, if you set limits to 17″ × 11″, you can still create objects that extend past the 17″ × 11″ area, such as a line that is 1200′ long. Therefore, you can choose not to consider limits while developing a template or creating a drawing. Conversely, as you learn AutoCAD, you may decide that setting appropriate drawing limits is helpful, especially when you are drawing large objects. Regardless of whether you choose to acknowledge limits, you should be familiar with the concept and recognize that some AutoCAD commands, such as **ZOOM** and **PLOT**, provide options associated with limits. The **ZOOM** and **PLOT** commands are explained in later chapters.

Figure 2-20.
The **Direction Control** dialog box.

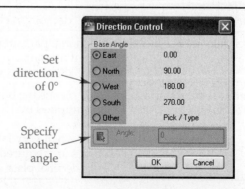

Set direction of 0°

Specify another angle

AutoCAD and Its Applications—Basics

Figure 2-21.
Angular unit formats available in the **Drawing Units** dialog box. Select the appropriate type and precision for the specific drawing application.

Type	Applications and Characteristics	Example
Decimal Degrees	• Mechanical, architectural, and structural drafting applications. • Degrees and decimal parts of a degree. • Initial default setting.	45°
Deg/Min/Sec	• Civil and sometimes mechanical, architectural, and structural drafting applications. • Degrees, minutes, and seconds. • 1 degree = 60 minutes; 1 minute = 60 seconds.	45°0'0"
Grads	• Grad is the abbreviation for *gradient*. • One-quarter of a circle has 100 grads; a full circle has 400 grads.	50.000g
Radians	• A radian is an angular unit of measurement in which 2π radians = 360° and π radians = 180°. Pi (π) is approximately equal to 3.1416. • A 90° angle has $\pi/2$ radians and an arc length of $\pi/2$. • Changing the precision displays the radian value rounded to the specified decimal place.	0.785r
Surveyor	• Civil drafting applications. • Degrees, minutes, and seconds. • Uses bearings. A bearing is the direction of a line with respect to one of the quadrants of a compass. Bearings are measured clockwise or counterclockwise (depending on the quadrant), beginning from either north or south. • An angle measuring 55°45'22" from north toward west is expressed as N55°45'22"W. • Set precision to degrees, degrees/minutes, degrees/minutes/seconds, or decimal display accuracy of the seconds part of the measurement.	N45°E

Access the **LIMITS** command to set model space drawing limits. The first prompt asks you to specify the coordinates for the lower-left corner of the drawing limits. Press [Enter] to accept the default 0,0 value or enter a new value and press [Enter]. The next prompt asks you to specify the coordinates for the upper-right corner of the virtual drawing area. For example, type 17,11 and then press [Enter]. The first value is the horizontal measurement of the limits, and the second value is the vertical measurement. A comma separates the values. This example creates a limits of 17 units × 11 units if you specify 0,0 as the lower-left corner.

In general, you should set limits larger than the objects you plan to draw. You can determine limits accurately by identifying the drawing scale, converting the scale to a scale factor, and then multiplying the scale factor by the size of sheet on which you plan to plot the drawing. For now, calculate the approximate total length and width of all objects you plan to draw, and add extra space for dimensions and notes. For example, when drawing a 48' × 24' building floor plan, allow 10' on each side for dimensions and notes to make a total virtual drawing area of 68' × 44'.

The **LIMITS** command provides a limits-checking feature that, when turned on, restricts your ability to draw outside of the drawing limits. Use the **ON** option of the **LIMITS** command to turn on limits checking and the **OFF** option to turn off limits checking.

Introduction to Templates

Drawing templates allow you to use an existing drawing as a starting point for a new drawing. A template includes drawing settings for specific applications that are preset each time you begin a new drawing. Templates are incredible productivity boosters, and can help ensure that everyone at your school or company uses the same drawing standards.

Templates usually include the following, set according to the drawing requirements:
- Units and limits
- Drawing aids, such as dynamic input, grid display, and snaps
- Layers with specific line standards
- Text, dimension, multileader, table, and plot styles
- Common symbols and blocks
- Layouts with a border, title block, and general notes

Reference Material

Drawing Sheets

sheet: The paper used to lay out and plot drawings.

sheet size: Size of the paper used to lay out and plot drawings.

For tables describing *sheet* characteristics, including *sheet size*, drawing scale, and drawing limits, go to the **Reference Material** section of the companion website (www.g-wlearning.com/CAD) and select **Drawing Sheets**. Additional information regarding sheet parameters and selection is described later in this textbook.

Template Development

Template Development

Chapter 2

Starting from scratch or using a template supplied with AutoCAD is an appropriate way to begin drawing as you first learn AutoCAD. However, you will soon find it necessary to create your own templates. Creating your own templates helps you reduce setup time significantly, increase productivity, and consistently adhere to drafting standards. Begin the basic process of template development now, and continue to add content to your templates while you learn AutoCAD.

You can save any drawing or template file as a template using the **SAVEAS** command and **Save Drawing As** dialog box. Pick AutoCAD Drawing Template (*.dwt) from the **Files of type:** drop-down list, and then specify a template name and storage location. Template names typically relate to the drawing application, such as Mechanical Template for mechanical part and assembly drawings, or Architectural floor plans for architectural floor plan drawings.

By default, the file list box in the **Save Drawing As** dialog box shows the drawing templates currently found in the **Template** folder. See **Figure 2-22A**. You can store templates in any appropriate location, but if you place them in the **Template** folder, they automatically appear in the **Select template** dialog box. Pick the **Save** button to save the template and display the **Template Options** dialog box. See **Figure 2-22B**. Type a brief description of the template file in the **Description** area. Specify **English** or **Metric** units in the **Measurement** drop-down list and pick the **OK** button.

Figure 2-22.
A—Save a template using the .dwt file extension. If desired, save it in the **AutoCAD Template** folder. B—Type a description of the new template in the **Template Options** dialog box.

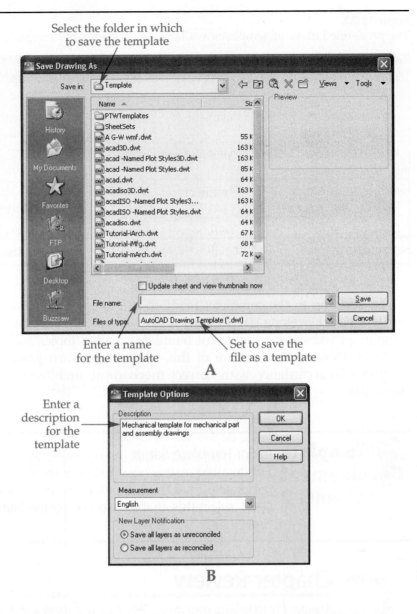

Select the folder in which to save the template

Enter a name for the template

Set to save the file as a template

A

Enter a description for the template

B

Pick **AutoCAD Drawing Template** from the **Save As** menu of the **Application Menu** to preset the template file type in the **Save Drawing As** dialog box.

PROFESSIONAL TIP

As you refine your setup procedure, open, revise, and then resave your template files. Use the **SAVEAS** command to save a new template from an existing template.

The companion website for this textbook contains several predefined templates that you can use to create drawings in accordance with correct mechanical, architectural, and civil drafting standards. **Figure 2-23** describes each available template. The mechanical drafting templates follow ASME and ISO drafting standards. The architectural and civil drafting templates follow appropriate architectural and civil drafting standards, including standards specified in the NCS.

Figure 2-23.
The predefined drawing templates available on the companion website.

Template File	Discipline	Units	Layout Sheet Sizes
MECHANICAL-INCH.dwt	Mechanical	Decimal inches	A-size, B-size, C-size, D-size
MECHANICAL-METRIC.dwt	Mechanical	Metric	A4-size, A3-size, A2-size, A1-size
ARCHITECTURAL-US.dwt	Architectural	Feet and inches	Architectural C-size, Architectural D-size
ARCHITECTURAL-METRIC.dwt	Architectural	Metric	Architectural A2-size, Architectural A1-size
CIVIL-US.dwt	Civil	Decimal inches	C-size, D-size
CIVIL-METRIC.dwt	Civil	Metric	A2-size, A1-size

In addition to the complete, ready-to-use templates on the companion website, the Template Development sections at the end of several chapters refer you to the companion website for important template creation topics and procedures. Use the Template Development feature of this textbook to learn gradually how to prepare templates in accordance with correct mechanical, architectural, and civil drafting standards.

Template Development Chapter 2

Initial Template Setup
For detailed instructions to begin the development of drawing templates, go to the companion website (www.g-wlearning.com/CAD), select this chapter, and select the **Template Development** link.

Chapter Review

Answer the following questions. Write your answers on a separate sheet of paper or complete the electronic chapter review on the companion website.
www.g-wlearning.com/CAD

1. What is a drawing template?
2. What is the name of the dialog box that opens by default when you pick the **New** button on the **Quick Access** toolbar?
3. Briefly explain how to start a drawing from scratch.
4. How often should you save your work?
5. Explain the benefits of using a standard system for naming drawing files.
6. Name the command that allows you to save your work quickly without displaying a dialog box.
7. What command allows you to save a drawing file in an older AutoCAD format?
8. How do you set AutoCAD to save your work automatically at designated intervals?
9. Identify the command you would use to exit a drawing file, but remain in the AutoCAD session.
10. How can you close all open drawing windows at the same time?
11. How can you set the number of recently opened files listed in the **Application Menu**?

12. What does the term *read-only* mean?
13. Describe the advantages of using the **Quick View Drawings** tool to work with multiple open drawings.
14. How do you keep the **Quick View Drawings** tool on-screen after you pick a thumbnail image?
15. What is the function of drawing properties?
16. What is the advantage of adding values to drawing properties?
17. Name three settings you can specify in the **Drawing Units** dialog box.
18. What is sheet size?
19. Explain the benefit of using a drawing template file to create a new drawing.
20. How can you convert a drawing file into a drawing template?

Drawing Problems

Start AutoCAD if it is not already started. Follow the specific instructions for each problem.

▼ Basic

1. Start a new drawing using the acad-Named Plot Styles template supplied by AutoCAD. Save the new drawing as a file named P2-1.dwg.

2. Start a new drawing using the Tutorial-iArch template supplied by AutoCAD. Save the new drawing as a file named P2-3.dwg.

▼ Intermediate

Problems 3 and 4 refer to drawings available at www.autodesk.com/autocad-samples.

3. Locate and click on the Line Weights drawing to open it. Type Z and press [Enter], then type A and press [Enter] to view the entire drawing. Describe the drawing.

4. Locate and click on the TrueType drawing to open it. Type Z and press [Enter], then type A and press [Enter] to view the entire drawing. Describe the drawing.

▼ Advanced

For Problems 5–7, create the specified template for possible future use.

5. Create a template with the following settings: decimal units with 0.0 precision, decimal degrees with 0.0 precision, default angle measure and orientation, and limits of 0,0 × 17,11. Save the template as P2-5.dwt. Enter an appropriate description for the template.

6. Create a template with the following settings: metric units with 0.0 precision, decimal degrees with 0.0 precision, default angle measure and orientation, and limits of 0,0 × 22,17. Save the template as P2-6.dwt. Enter an appropriate description for the template.

7. Create a template with architectural units with 0 -0 precision, decimal degrees with 0 precision, and limits of 0,0 × 1632,1056. Save the template as P2-7.dwt. Enter an appropriate description for the template.

8. Research the requirements for a set of working drawings of a mechanical assembly. Identify a mechanical assembly that consists of several parts and possibly subassemblies. Describe the product and provide a detailed list of each drawing required in the set of working drawings. Create, in writing only, a complete directory for storing and organizing the drawing files. Use an appropriate system to name each folder and file.

9. Research the requirements for a set of working drawings for a single-family home according to national and local codes. Provide a detailed list of each drawing required in the set of working drawings. Create, in writing only, a complete directory for storing and organizing the drawing files. Use an appropriate system to name each folder and file. Refer to the U.S. National CAD Standard if available.

AutoCAD Certified Associate Exam Practice

Answer the following questions. Write your answers on a separate sheet of paper.

1. Which of the following methods can you use to close an AutoCAD drawing file? *Select all that apply.*
   A. Enter **CLOSEFILE** at the keyboard
   B. Pick the **Close** button on the drawing window title bar
   C. Pick **Close** in the **Application Menu**
   D. Right-click in the drawing area and pick **Close...**
   E. Select **Close** and then select **Current Drawing** in the **Application Menu**

2. Which of the following tasks can you accomplish using the **Application Menu**? *Select all that apply.*
   A. Access the **Options** dialog box
   B. Open a new drawing
   C. Specify the limits for a drawing
   D. Specify the units and precision for a drawing
   E. Access **Quick View Drawings**

AutoCAD Certified Professional Exam Practice

Follow the instructions in each problem. Write your answers on a separate sheet of paper.

1. Open the db_samp.dwg file in the Database Connectivity subfolder of the AutoCAD Sample folder.
   Access the **LIMITS** command and press [Enter] repeatedly to cycle through the prompts. Do not change any of the values. What are the coordinates for the upper-right corner of the drawing limits for this drawing?

2. Open the Mechanical - Multileaders.dwg file in the **Mechanical Sample** subfolder of the **AutoCAD Sample** folder.
   Access the **Drawing Units** dialog box. Do not change any of the values. What type of linear drawing units are specified for this drawing, and at what precision?

Chapter

Introduction to Drawing and Editing

Learning Objectives

After completing this chapter, you will be able to:

✓ Use appropriate values when responding to prompts.
✓ Apply basic viewing methods.
✓ Draw given objects using the **LINE** command.
✓ Describe and use several point entry tools and methods.
✓ Describe and use basic drawing aids.
✓ Use the **ERASE**, **UNDO**, **U**, **REDO**, and **OOPS** commands appropriately.
✓ Create selection sets using various selection options.

This chapter introduces a variety of fundamental drawing and editing concepts and processes. You will learn to pick points, draw lines, select objects, and erase objects. You will also use basic drawing aids and explore several other primary AutoCAD commands and operations.

Responding to Prompts

AutoCAD commands prompt, or ask, you to perform a specific task. For example, when you draw a line, prompts ask you to specify line endpoints. When you erase an object, a prompt asks you to select objects to erase. You must understand how to respond to prompts before you begin drawing. Many prompts provide options that you can select instead of responding to the immediate request. For example, after you pick the first point of a line, a prompt asks you to select the next point, or you can choose the **Undo** option to remove the previous selection.

Responding with Numbers

Many commands require you to enter specific numerical data, such as the location of a point, or the radius of a circle. Acceptable values vary depending on the command and prompt. AutoCAD understands that a number is positive without the plus sign (+) in front of the value. However, you must add a hyphen (-) in front of a negative number, such as -6.375. Do not include a space when responding to a prompt, because the space bar functions like [Enter] to end the input.

The drawing units have some effect on values that you can enter when responding to prompts. All drawing length units accept decimal or fractional values. For example, you can type 2.5 or 2-1/2 to specify two and one half units. Remember, you are responsible for applying the appropriate units. Architectural units recognize this example value as 2 1/2", decimal units as 2.5 units, engineering units as 2.5", fractional units as 2 1/2 units, and scientific units as 2.5E+0 units, depending on the precision.

When a drawing is set up with architectural or engineering units, AutoCAD accepts only the inch (") and foot (') symbols and does not recognize other suffixes, such as mm. The numerator and denominator of fractions must be whole numbers greater than zero, such as 1/2 or 2/3. You must include a hyphen between a whole number and a fraction for values greater than one, such as 2-3/4, because of the function of the space bar. The numerator can be larger than the denominator, as in 3/2, but only if you do not include a whole number with the fraction. For example, 1-3/2 is not a valid input.

AutoCAD assumes inches when a drawing is set up with architectural and engineering length units. Do not add inch marks (") when specifying an inch value; they are unnecessary and time-consuming. Values can be whole numbers, decimals, or fractions. For measurements in feet, place the foot symbol (') after the number, as in 24'. Do not include a space or hyphen when specifying a value in feet and whole inches. For example, 24'6 is the proper input for the value 24'-6". Separate fractional inches with a hyphen, such as 24'6-1/2. Never mix feet with inch values greater than one foot. For example, 24'18 is invalid; type 25'6 instead.

Ending and Canceling Commands

Some AutoCAD commands remain active until stopped. For example, you can continue to pick points to create new line segments until you end the **LINE** command. You can usually end a command by pressing [Enter] or the space bar, or by right-clicking and selecting **Enter**. Press [Esc] to cancel an active command or abort data entry. It may be necessary to press [Esc] twice to cancel certain commands completely.

If you press the wrong key or misspell a word when answering a prompt, use [Backspace] to correct the error. This works only if you notice your mistake *before* you accept the value. If you enter an invalid value or option, AutoCAD usually responds with an error message. Access the **AutoCAD Text Window** to view lengthy error messages, or to review entries. Return to the graphics screen using the same method you used to access the text window, or pick any visible portion of the graphics screen.

TEXTSCR	
Ribbon	
View	
> Windows	
> User Interface	
> Text	
Window	
Type	
TEXTSCR	
[F2]	

PROFESSIONAL TIP

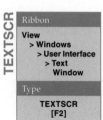

You can cancel the active command and access a new command at the same time by picking a ribbon button or an **Application Menu** option.

Introduction to Drawing

point entry:
Locating a point, such as the endpoint of a line, on the AutoCAD coordinate system.

Most object drawing commands use *point entry* to locate and size geometry. Examples of point entry include specifying the two endpoints of a line, locating the center and a point on the edge of a circle, and specifying the points at opposite corners of a rectangle. The most basic point entry technique is to pick a random location in space using the left mouse button. Specific coordinate entry methods and drawing aids allow for accurate point entry, as explained later in this textbook.

AutoCAD uses the *Cartesian (rectangular) coordinate system* of X, Y, and Z coordinate values. These values, called *rectangular coordinates*, locate any point in 3D space. In 2D drafting, the *origin* divides the coordinate system into four quadrants on the XY plane. See **Figure 3-1**. The origin is usually positioned at the lower-left corner of the drawing. This setup places all points in the upper-right quadrant of the XY plane, where both X and Y coordinate values are positive. See **Figure 3-2**. To locate a point in 3D space, a third dimension rises up from the surface of the XY plane along the Z axis.

Describe a coordinate location using the X value first, followed by the Y value, and finally the Z value. A comma separates each value. For example, the coordinate location of 3,1,6 represents a point that is three units from the origin in the X direction, one unit from the origin in the Y direction, and six units from the origin in the Z direction. *AutoCAD and Its Applications—Advanced* describes how to use the Z axis to construct 3D models.

Cartesian (rectangular) coordinate system: A system that locates points in space according to distances from three intersecting axes.

rectangular coordinates: A set of numerical values that identify the location of a point on the X, Y, and Z axes of the Cartesian coordinate system.

origin: The intersection point of the X, Y, and Z axes. The position of the default 2D origin is 0,0, where X = 0 and Y = 0.

Figure 3-1.
The 2D Cartesian coordinate system consists of X and Y axes. The origin is located at the intersection of the axes.

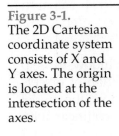
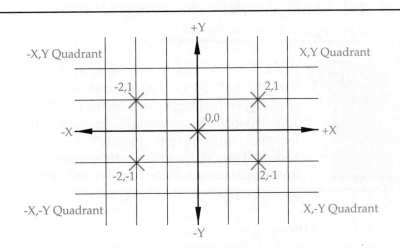

Figure 3-2.
By default, the upper-right quadrant of the Cartesian coordinate system fills the screen. The model space coordinate system, or UCS, icon identifies the coordinate system orientation.

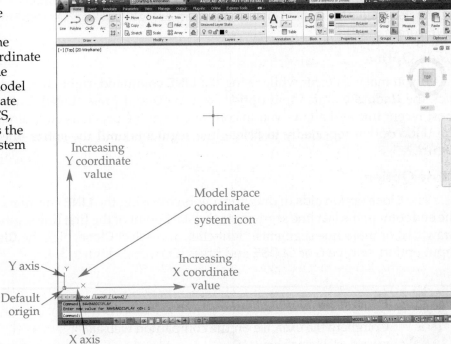

Exercise 3-1

Complete the exercise on the companion website.
www.g-wlearning.com/CAD

Basic Viewing Methods

zoom: Make objects appear bigger (zoom in) or smaller (zoom out) on the screen without affecting their actual size.

pan: Change the drawing display so that different portions of the drawing are visible on-screen.

zoom out: Change the display area to show a larger part of the drawing at a lower magnification.

zoom in: Change the display area to show a smaller part of the drawing at a higher magnification.

rubberband line: A reference line that extends from the crosshairs in certain drawing commands after you make the first selection.

View tools allow you to navigate the infinite AutoCAD drawing area and observe and work more efficiently with a specific portion of a drawing. Chapter 6 explains the many view tools available to adjust the display of a 2D drawing. However, as you begin drawing, you should be able to apply basic *zoom* and *pan* operations. To *zoom out*, roll the mouse wheel forward. To *zoom in*, roll the mouse wheel back. Double-click the wheel to zoom to the furthest extents of objects in the drawing. To pan, press and hold the mouse wheel while you move the mouse. For now, these basic zoom and pan functions will allow you to view drawings effectively.

Refer to the X and Y axis lines in the drawing window to help identify the location you are viewing, and the location of objects in space. If the origin is visible in the drawing window, the model space coordinate system icon appears at the origin, collinear with the axes, and displays a small box. If the current view does not show the origin, the coordinate system icon floats in space near the lower-right corner of the drawing window.

Drawing Lines

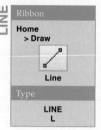

LINE

Ribbon

Home
> Draw

Line

Type

LINE
L

To draw a line, access the **LINE** command and specify a start point, which is the first line endpoint. As you move the crosshairs, a *rubberband line* appears connecting the first point and the crosshairs. Continue locating points to connect a series of line segments. Then press [Enter] or the space bar, or right-click and select **Enter** to end the **LINE** command.

Undo Option

If you make an error while using the **LINE** command, right-click and select **Undo**, pick the **Undo** dynamic input option, or type U and press [Enter]. This removes the most recent line and allows you to continue from the previous endpoint. You can use the **Undo** option repeatedly to delete line segments until the entire line is gone. See **Figure 3-3**.

Close Option

polygon: A closed plane figure with at least three sides, such as a triangle or rectangle.

The **Close** option aids in drawing a *polygon* using the **LINE** command. To connect the endpoint of the last line segment to the start point of the first line segment after you draw two or more line segments, right-click and select **Close**, pick the **Close** dynamic input option, or type C or CLOSE and press [Enter]. See **Figure 3-4**.

Exercise 3-2

Complete the exercise on the companion website.
www.g-wlearning.com/CAD

Figure 3-3.
Using the **Undo** option of the active **LINE** command. The dashed lines represent the undone lines, and do not appear when you use the **Undo** option.

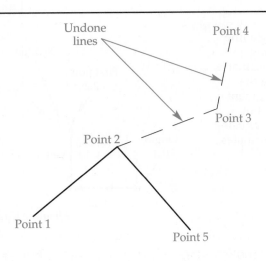

Figure 3-4.
Constructing the final segment of a rectangle using the **Close** option of the **LINE** command.

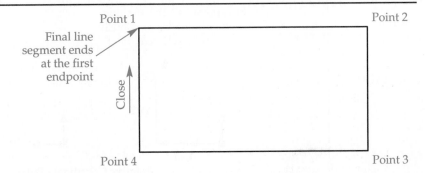

Coordinate Entry Methods

Creating an object by picking random points in space with the left mouse button is not accurate. Specific coordinate input is one way to locate points precisely to create accurate geometry. Another option is to use drawing aids, as explained later in this chapter.

Absolute Coordinates

Points located using *absolute coordinates* are measured from the origin (0,0). For example, a point located two units horizontally (X = 2) and two units vertically (Y = 2) from the origin is at the absolute coordinate 2,2. **Figure 3-5A** shows using absolute coordinates to draw a line starting at 2,2 and ending at 4,4. The first point of a line is often positioned using absolute coordinates. Remember, when using the absolute coordinate system, you locate each point from 0,0. If you enter negative X and Y values, the selection occurs outside of the upper-right XY plane quadrant.

absolute coordinates: Coordinate distances measured from the origin.

Relative Coordinates

When using *relative coordinates*, think of the previous point as the "temporary origin." The @ symbol specifies coordinates as relative. For example, input @2,2 to locate the second point shown in **Figure 3-5B** at the absolute coordinate 4,4.

relative coordinates: Coordinates specified from, or relative to, the previous position, rather than from the origin.

Polar Coordinates

To apply *polar coordinates*, specify the length of the line, followed by the less than (<) symbol, and then the angle at which the line is drawn. **Figure 3-6** shows the default angular values used for polar coordinate entry. Precede a polar coordinate with the @ symbol to locate the point relative to the previous point. Otherwise the coordinate is located relative to the origin. For example, to draw a line 2 units long at a 45° angle, starting 2 units from 0,0 at a 45° angle, type 2<45 for the first point and @2<45 for the second point. See **Figure 3-7**.

polar coordinates: Coordinates based on the distance from a fixed point at a given angle.

Figure 3-5.
A—Drawing a line segment using absolute coordinates. B—Drawing the same line segment shown in A, but locating points using relative coordinates.

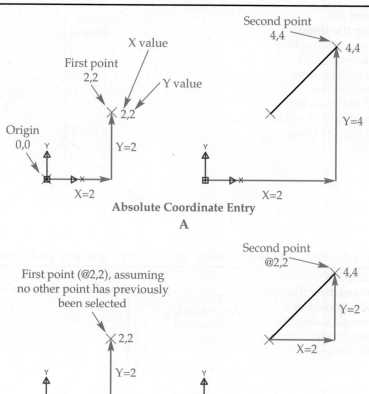

Absolute Coordinate Entry
A

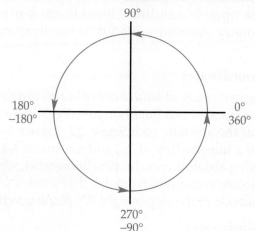

Relative Coordinate Entry
B

Figure 3-6.
The default angles used when entering polar coordinates. 0° is to the right, or east, and angles are measured counterclockwise. For specific applications, such as when measuring azimuths, you can adjust the direction and rotation using options in the **Drawing Units** dialog box.

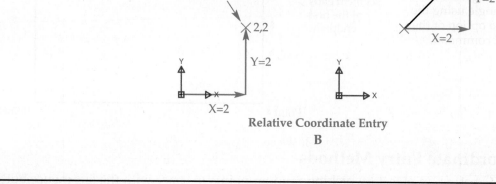

Crosshairs Coordinates

The area on the left side of the status bar shows the coordinate display field. The drawing units determine the format and precision shown. To toggle the display on or off, pick the display field or right-click and select **On** or **Off**. When coordinate display is on, the coordinates constantly change as the crosshairs moves. When coordinate display is off, the coordinates are "grayed out," but update to identify the location of the last point you select.

You can choose the coordinate display mode when a command such as **LINE** is active and you have selected a point. Right-click on the display field and choose **Relative** to view the coordinates of the crosshairs as polar coordinates relative to the previously

Figure 3-7.
Locating points
using polar
coordinates.

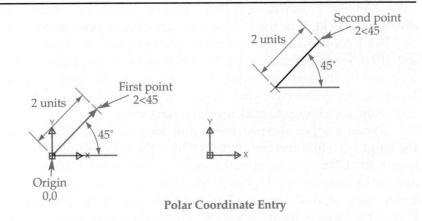

Polar Coordinate Entry

picked point. The coordinates update each time you pick a new point. Select **Absolute** to view the coordinates of the crosshairs relative to the origin. The **Geographic** mode is available if you specify the geographic drawing location, as explained in *AutoCAD and Its Applications—Advanced*.

Dynamic Input

Dynamic input is on by default and is a very effective way to enter coordinates. A quick way to toggle dynamic input on and off is to pick the **Dynamic Input** button on the status bar. Dynamic input provides the same function as the command line, but allows you to keep your focus at the crosshairs while you draw. Dynamic input also offers additional coordinate entry techniques. When dynamic input is active, point entry functions differently, according to dynamic input settings.

When you start the **LINE** command, dynamic input prompts you to specify the first point. The X coordinate input field is active, and the Y coordinate input field appears. See **Figure 3-8.** Use *pointer input* to specify the absolute, relative, or polar coordinates of the first point. Absolute coordinates are the default when you select the first point. Type the X

pointer input: The process of entering points using dynamic input.

Figure 3-8.
Dynamic input
displays these fields
after you start the
LINE command.
When you type
@ to use relative
coordinates, the
symbol appears in
a field to the right
of the prompt.
When you use polar
coordinates, the
less than symbol (<)
appears before the
angle input field.

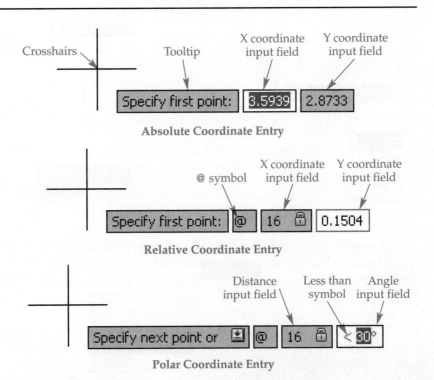

Polar Coordinate Entry

value, and then press [Tab] or type a comma to lock in the X value and move to the Y coordinate input field. Type the Y value and right-click or press [Enter] to select the point.

Polar coordinates are the other likely option for specifying the first point. To specify polar coordinates, type the distance from the origin in the distance input field, followed by the less than symbol (<) to move to the angle input field. Type the angle from the origin and right-click or press [Enter] to select the point. Dynamic input fields automatically change to anticipate the next entry.

Dynamic input also provides a *dimensional input* feature that allows you to enter the length of a line and the angle at which the line is drawn. To use dimensional input, access the **LINE** command and specify a start point. Distance and angle input fields appear by default. See **Figure 3-9A**. Move the crosshairs toward where you want to locate the endpoint, type the length of the line in the distance input field, and press [Tab]. This locks in the distance and moves the cursor to the angle input field. Type the angle of the line and right-click or press [Enter] to select the point. Dimensional input does not follow the same rules as polar coordinates, and drawing units do not control direction and rotation. All angles relate to the location of the crosshairs from 0° east. **Figure 3-9B** shows additional examples of specifying dimensional input angles.

dimensional input:
An instinctive dynamic input point entry technique, similar to polar coordinate entry.

Figure 3-9.
A—Steps for using dimensional input to define the length and angle of a line. B—Pay close attention to the location of the crosshairs and the reading in the angle input field. All angles relate to the location of the crosshairs from 0° east.

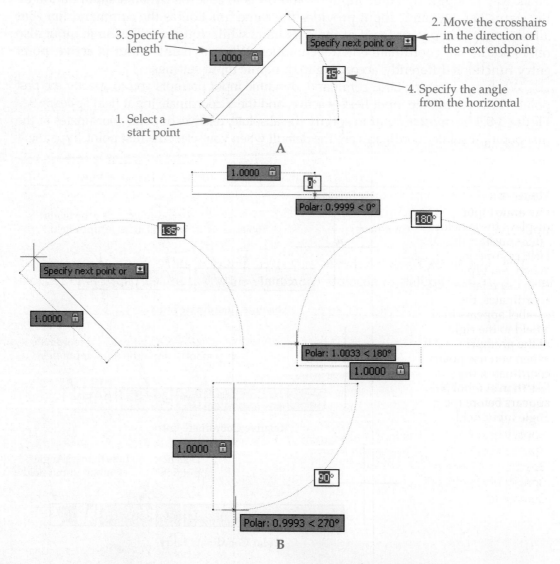

AutoCAD and Its Applications—Basics

You can also use pointer input to pick additional points once you select the start point of the line. Relative coordinates are active by default, which means you do not need to type @ before entering the X and Y values. To select the second point using a relative coordinate entry, type the X value in the distance input field, then type a comma to lock in the X value and move to the Y coordinate input field. Type the Y value and right-click or press [Enter] to select the point.

Dimensional input is on by default, but you can turn it off temporarily by typing the pound symbol (#) before entering values. When dimensional input is off, dynamic input defaults to polar format. Type the length of the line in the distance input field, and press [Tab] to lock in the length and move to the angle input field. Enter the angle of the line and right-click or press [Enter] to select the point. In order to use an absolute coordinate entry with the default settings, type #, enter the X coordinate in the active field, type a comma, type the Y value, and right-click or press [Enter] to select the point.

PROFESSIONAL TIP

Use [Tab] to cycle through dynamic input fields. You can make changes to values before accepting the coordinates.

Use the **Dynamic Input** tab of the **Drafting Settings** dialog box to configure dynamic input to meet your personal, school, or company standards. A quick way to access the **Drafting Settings** dialog box is to right-click on a status bar toggle button and select **Settings....**

Command Line

You can use the command line at the same time as dynamic input, or you can disable dynamic input to use only the command line. Another option is to hide the command line to free additional drawing space and focus on using dynamic input. Absolute, relative, and polar point entry methods accomplish the same tasks whether you enter them using dynamic input or the command line. However, dimensional input and quick input settings are unavailable with the command line, and coordinate entry is slightly different when using the command line.

The next three figures provide examples of point entry using the command line. **Figure 3-10** applies to absolute coordinate entry, **Figure 3-11** applies to relative

Figure 3-10.
Drawing a shape using the **LINE** command and absolute coordinates.

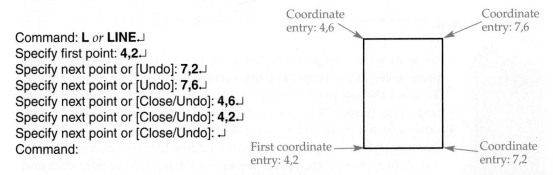

```
Command: L or LINE↵
Specify first point: 4,2↵
Specify next point or [Undo]: 7,2↵
Specify next point or [Undo]: 7,6↵
Specify next point or [Close/Undo]: 4,6↵
Specify next point or [Close/Undo]: 4,2↵
Specify next point or [Close/Undo]: ↵
Command:
```

Coordinate entry: 4,6
Coordinate entry: 7,6
First coordinate entry: 4,2
Coordinate entry: 7,2

Absolute Coordinate Entry

Figure 3-11.
Drawing a shape using the **LINE** command and a combination of absolute and relative coordinates. Notice that negative (–) values are used and the coordinates are entered counterclockwise from the first point, in this case 2,2.

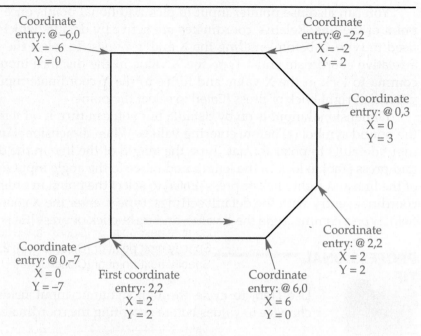

Coordinate entry: @ –6,0
X = –6
Y = 0

Coordinate entry:@ –2,2
X = –2
Y = 2

Coordinate entry: @ 0,3
X = 0
Y = 3

Coordinate entry: @ 2,2
X = 2
Y = 2

Coordinate entry: @ 0,–7
X = 0
Y = –7

First coordinate entry: 2,2
X = 2
Y = 2

Coordinate entry: @ 6,0
X = 6
Y = 0

Command: **L** *or* **LINE**↵
Specify first point: **2,2**↵
Specify next point or [Undo]: **@6,0**↵
Specify next point or [Undo]: **@2,2**↵
Specify next point or [Close/Undo]: **@0,3**↵
Specify next point or [Close/Undo]: **@-2,2**↵
Specify next point or [Close/Undo]: **@-6,0**↵
Specify next point or [Close/Undo]: **@0,-7**↵
Specify next point or [Close/Undo]: ↵
Command:

Relative Coordinate Entry

coordinate entry, and Figure 3-12 applies to polar coordinate entry. You can apply the same examples to dynamic input. Even if you choose not to use the command line, review these examples to help better understand point entry techniques. You must disable dynamic input in order for these exact command sequences to work properly.

Exercises 3-3, 3-4, and 3-5

Complete the exercises on the companion website.
www.g-wlearning.com/CAD

PROFESSIONAL TIP

To position the start point of a line at the last point entered, right-click or press [Enter] or the space bar when you see the Specify first point: prompt. To select the last point entered at any point selection prompt, type @ and press [Enter]. You can reference the coordinates of other previously selected points by pressing the up arrow key at a point selection prompt. Dynamic input or the command line lists the coordinates, and a symbol appears at each point on-screen. Press [Enter] or right-click and choose **Enter** to select the desired coordinates.

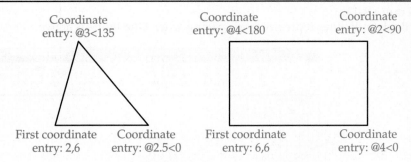

Figure 3-12.
Drawing shapes using the **LINE** command and a combination of absolute and polar coordinates.

Coordinate entry: @3<135

Coordinate entry: @4<180

Coordinate entry: @2<90

First coordinate entry: 2,6

Coordinate entry: @2.5<0

First coordinate entry: 6,6

Coordinate entry: @4<0

```
Command: L or LINE↵
Specify first point: 2,6↵
Specify next point or [Undo]: @2.5<0↵
Specify next point or [Undo]: @3<135↵
Specify next point or [Close/Undo]: 2,6↵
Specify next point or [Close/Undo]: ↵
Command: ↵
LINE Specify first point: 6,6↵
Specify next point or [Undo]: @4<0↵
Specify next point or [Undo]: @2<90↵
Specify next point or [Close/Undo]: @4<180↵
Specify next point or [Close/Undo]: @2<270↵
Specify next point or [Close/Undo]: ↵
Command:
```

Polar Coordinate Entry

Introduction to Drawing Aids

AutoCAD includes many drawing aids that increase accuracy and productivity. The crosshairs coordinates on the status bar and dynamic input are examples of drawing aids. This chapter describes other basic drawing aids and provides an overview of several important drawing aids that are explained in detail later in this textbook.

Grid Mode

Turn on **Grid** mode to display a *grid* on-screen. See **Figure 3-13**. The grid functions like virtual graph paper, but appears for reference only, as a visual aid to drawing layout. A quick way to toggle **Grid** mode on and off is to pick the **Grid Display** button on the status bar.

Use the options on the **Snap and Grid** tab of the **Drafting Settings** dialog box to adjust grid settings. See **Figure 3-14**. A quick way to access the **Snap and Grid** tab is to right-click on the **Grid Display** or **Snap Mode** button on the status bar and select **Settings...**. Turn the grid on and off using the **Grid On** check box. The grid appears as lines by default. Use the check boxes in the **Grid style** area to replace the grid lines with dots in selected environments.

Set the grid spacing in the **Grid spacing** area by typing values in the **Grid X spacing:** and **Grid Y spacing:** text boxes. For decimal units, set the grid spacing to standard decimal increments such as .125, .25, .5, or 1 for an inch drawing or 1, 10, 20, or 50 for a metric drawing, depending on the size of objects. For architectural units, use standard increments such as 1, 6, and 12 for inches, or 1, 2, 4, 5, and 10 for feet. A very large drawing might have a grid spacing of 12 (one foot), or 120 (ten feet), while a small drawing may use a spacing of .125 or less. Type the number of grid rows to display between the bold, major grid lines in the **Major line every** text box.

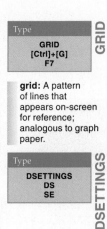

GRID

Type
GRID
[Ctrl]+[G]
F7

grid: A pattern of lines that appears on-screen for reference; analogous to graph paper.

DSETTINGS

Type
DSETTINGS
DS
SE

Figure 3-13.
By default, lines represent grid spacing when **Grid** mode is active.

Major grid lines

Minor grid lines

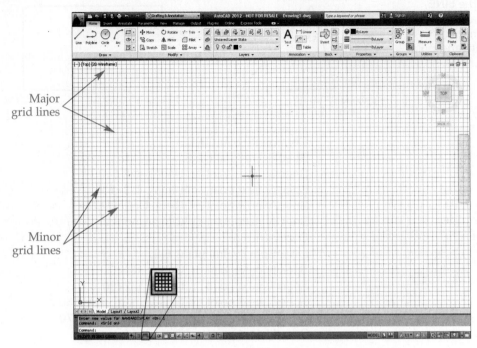

Figure 3-14.
Use the **Snap and Grid** tab of the **Drafting Settings** dialog box to specify grid and snap grid settings.

Turn snap on and off

Turn grid on and off

Select environments in which to display the grids as dots

Set snap spacing

Set grid spacing

Determines whether X and Y spacing can be different

Set occurrence of major gridlines

Select type of snap

Controls density of grid when zoomed out

Controls grid display beyond drawing limits

The options in the **Grid behavior** area determine how the grid appears on-screen. When **Adaptive Grid** is checked and you zoom out to a display where the grid spacing becomes too dense, AutoCAD adjusts the grid display to show the grid at a larger scale. The **Display grid beyond Limits** option determines whether the grid appears only within the drawing limits. The **Allow subdivision below grid spacing** and **Follow Dynamic UCS** options apply to 3D applications.

Snap Mode

Turn on Snap mode to activate the *snap grid*, also known as *snap resolution* or *snap*. A quick way to toggle snap on and off is to pick the **Snap Mode** button on the status bar. Snap is only noticeable when you are drawing or editing objects, such as constructing a line or moving a circle. By default, snap is off, allowing the crosshairs to move freely on-screen. Turn snap on to move the crosshairs in specific increments. Snap is different from the grid, because snap controls the movement of the cross-hairs, while the grid is only a visual guide. However, grid and snap settings typically complement each other.

The **Snap and Grid** tab of the **Drafting Settings** dialog box includes options for setting snaps. Refer again to **Figure 3-14**. Turn snap on or off using the **Snap On** check box. Set the snap increment in the **Snap spacing** area by typing values in the **Snap X spacing:** and **Snap Y spacing:** text boxes. For example, if you set the X and Y grid spacing to .5, an appropriate X and Y snap spacing is .125 or .25. With these settings, each mode plays a separate role in assisting drawing layout. Often the most effective use of grid and snap is to set equal X and Y spacing. However, if many horizontal features conform to one increment and most vertical features correspond to another, you may choose to set different X and Y values.

Use the **Snap type** area to control how snaps function. The default, previously described snap type is **Grid snap** with the **Rectangular snap** style. Select the **Grid snap** type and **Isometric snap** style to aid in creating isometric drawings. The **PolarSnap** type allows you to snap to precise distances along alignment paths when you use polar tracking, as explained in Chapter 7.

Adjust the grid and snap settings as needed, such as when larger or smaller values would assist you with a certain drawing task. Changing grid and snap settings does not affect the location of existing points or objects.

Exercise 3-6

Complete the exercise on the companion website.
www.g-wlearning.com/CAD

Supplemental Material

Introduction to Isometric Drawings
For an introduction to pictorial drawings and information about isometric snaps, go to the companion website (www.g-wlearning.com/CAD), select this chapter, and select **Introduction to Isometric Drawings**.

Polar Tracking

Polar tracking causes the drawing crosshairs to "snap" to predefined angle increments. Chapter 7 fully explains polar tracking, but because polar tracking is on by default, you should have a basic understanding of the tool. Turn polar tracking on or off by picking the **Polar Tracking** button on the status bar or by pressing [F10]. You can use polar tracking to draw lines at accurate lengths and angles using *direct distance entry*. As you move the crosshairs toward a polar tracking angle, AutoCAD displays an alignment path and tooltip. The default polar angle increments are 0°, 90°, 180°, and 270°.

To apply direct distance entry using polar tracking, access the **LINE** command and specify a start point. Then move the crosshairs in alignment with a polar tracking angle. Type the length of the line and press [Enter] or right-click and select **Enter**. See **Figure 3-15**.

Exercise 3-7

Complete the exercise on the companion website.
www.g-wlearning.com/CAD

Ortho Mode

ORTHO

ortho: From *orthogonal*, which means "at right angles."

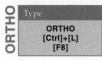

Type
ORTHO
[Ctrl]+[L]
[F8]

Ortho mode forces a horizontal or vertical line. Ortho mode is off by default. Pick the **Ortho Mode** button on the status bar to toggle **Ortho** mode on and off. If **Ortho** mode is off, you can temporarily turn it on while drawing by holding down [Shift]. You can use **Ortho** mode to draw accurate lengths of horizontal and vertical lines using direct distance entry.

To apply direct distance entry using **Ortho** mode, access the **LINE** command and specify a start point. Move the crosshairs to display a horizontal or vertical rubberband line in the direction you want to draw. Then type the length of the line and press [Enter] or right-click and select **Enter**. See **Figure 3-16**.

> **NOTE**
> You can use direct distance entry to specify the length of a line at any angle. However, direct distance entry itself is not very useful unless you incorporate drawing aids such as polar tracking or **Ortho** mode.

Figure 3-15.
Using polar tracking and direct distance entry to draw connected and perpendicular lines at specific lengths. 0° and 90° tracking are the defaults. This example shows dynamic input active, but the same technique applies when you use the command line.

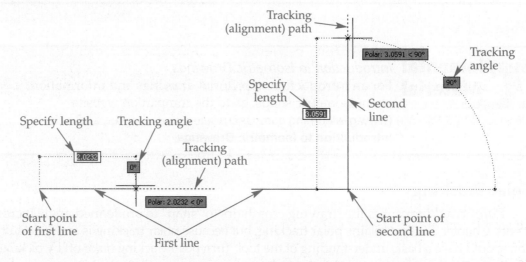

Figure 3-16.
Using **Ortho** mode and direct distance entry to construct a rectangle. Notice that the crosshairs specifies the general direction of the line endpoint and does not attach to the rubberband line. This example shows dynamic input active. The same technique applies when you use the command line.

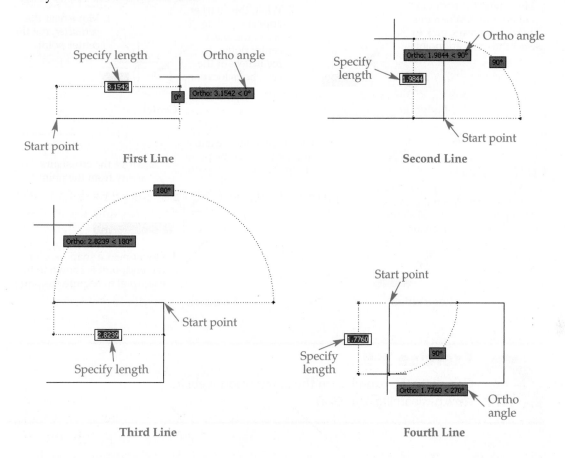

First Line

Second Line

Third Line

Fourth Line

Exercise 3-8
Complete the exercise on the companion website.
www.g-wlearning.com/CAD

Object Snap

Object snap increases drafting performance and accuracy through the concept of *snapping*. Chapter 7 fully explains object snap, but because *running object snaps* are on by default, you should have a basic understanding of how to use them. Turn running object snaps on or off by picking the **Object Snap** button on the status bar or pressing [F3]. Object snap modes identify the points on objects to which the crosshairs snap. The **Endpoint**, **Center**, **Intersection**, and **Extension** running object snap modes are active by default. The AutoSnap feature is also on by default, and displays *markers* at each active snap point. After a brief pause, a tooltip appears to indicate the object snap mode. See **Figure 3-17**.

You can snap to a point at any point selection prompt; for example, when picking the start point or endpoint of a line segment. **Figure 3-17** shows a basic example of locating the start point of a line using each default running object snap mode. Follow the instructions shown to snap to the corresponding point.

Figure 3-17.
Using the default **Endpoint**, **Center**, **Intersection**, and **Extension** running object snap modes to locate the start point of a line. When you see the correct AutoSnap marker and tooltip, pick to locate the point at the exact snap location.

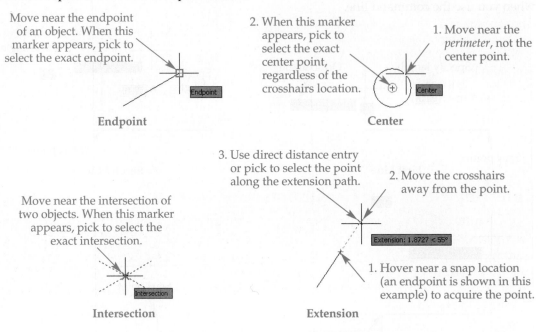

Move near the endpoint of an object. When this marker appears, pick to select the exact endpoint.

Endpoint

2. When this marker appears, pick to select the exact center point, regardless of the crosshairs location.

1. Move near the *perimeter*, not the center point.

Center

3. Use direct distance entry or pick to select the point along the extension path.

Move near the intersection of two objects. When this marker appears, pick to select the exact intersection.

2. Move the crosshairs away from the point.

Extension: 1.8727 < 55°

1. Hover near a snap location (an endpoint is shown in this example) to acquire the point.

Intersection

Extension

Exercise 3-9

Complete the exercise on the companion website.
www.g-wlearning.com/CAD

Object Snap Tracking

Object snap tracking has two requirements: running object snaps must be active, and the crosshairs must hover over the intended selection long enough to acquire the point. Chapter 7 fully explains object snap tracking, but because object snap tracking and running object snaps are on by default, you should have a basic understanding of these tools. Turn object snap on and off by picking the **Object Snap Tracking** button on the status bar or pressing [F11]. **Figure 3-18** shows an example of using the **Endpoint** running object snap with object snap tracking to locate the endpoint of a line exactly vertical to another endpoint.

PROFESSIONAL TIP

Practice using different point entry techniques and drawing aids, and decide which method works best for specific situations.

Exercise 3-10

Complete the exercise on the companion website.
www.g-wlearning.com/CAD

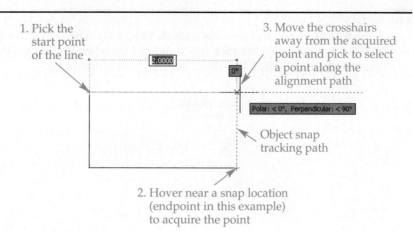

Figure 3-18.
Using the **Endpoint** running object snap and object snap tracking to construct the top side of a rectangle using the **LINE** command. Polar tracking is active to create a horizontal line.

1. Pick the start point of the line

2.0000

0°

3. Move the crosshairs away from the acquired point and pick to select a point along the alignment path

Polar: < 0°, Perpendicular: < 90°

Object snap tracking path

2. Hover near a snap location (endpoint in this example) to acquire the point

Inferring Geometric Constraints

Chapter 22 describes the process of constraining, or applying relationships between, objects. However, before continuing, you should be aware of a command that automatically forms, or *infers*, constraints. A constrained drawing functions differently than an unconstrained drawing. Until you have read Chapter 22 and are ready to constrain objects, confirm that the **Infer Constraints** command is off before drawing. The **Infer Constraints** button on the status bar toggles the **Infer Constraints** command on and off.

Introduction to Editing

AutoCAD includes many *editing* commands for making changes to a drawing and increasing productivity. One of the most basic ways to edit a drawing is to remove objects using the **ERASE** command. You will also learn to use the **OOPS**, **UNDO**, **U**, and **REDO** commands, which are commonly needed in the drawing and editing processes.

A common approach to editing is to access a command, such as **ERASE**, select the objects to modify, and then complete the operation by right-clicking, or pressing [Enter] or the space bar. Another approach is to select objects first using the crosshairs, then access the editing command, and finally complete the operation. The process of selecting objects is generally the same for both methods. Choose the technique you prefer, but selecting objects first is most appropriate when you are editing using *grips*. Chapter 14 explains grip editing.

ERASE Command

Access the **ERASE** command to remove objects from the drawing. The Select objects: prompt appears and an object selection target, or *pick box*, replaces the screen crosshairs. Move the pick box over the item to erase and pick. The object becomes highlighted and the Select objects: prompt remains active, allowing you to select additional objects to erase. When you finish selecting objects, erase the selection set by right-clicking, or pressing [Enter] or the space bar. See Figure 3-19. If you choose to select objects before accessing the **ERASE** command, you can erase the selected objects by pressing [Delete].

editing: A procedure used to modify an existing object.

grips: Small boxes that appear at strategic points on a selected object, allowing you to edit the object directly.

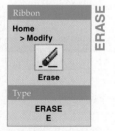

Ribbon
Home
> Modify

Erase

Type
ERASE
E

ERASE

pick box: A small box that replaces the crosshairs when the Select objects: prompt is active.

Figure 3-19.
Using the **ERASE** command to erase a single object. A—The initial display before you access the **ERASE** command. B—The pick box that appears when you access the **ERASE** command. C—Selecting a single object. D—The completed erase operation.

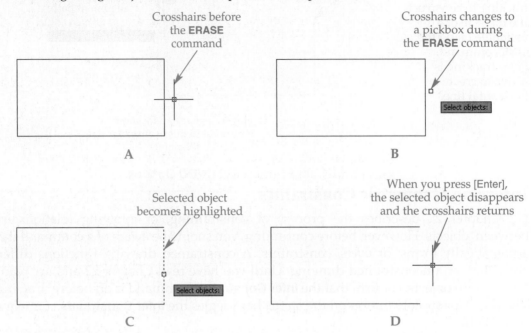

By default, when you hover over an object, the object changes to a thicker lineweight and appears dashed. Basic object properties also appear. When you move the crosshairs or pick box off the object, the object display returns to normal. This allows you to preview the object before you select. When a small area contains many objects, this feature helps you select the correct object the first time and often eliminates the need to cycle through stacked objects.

Exercise 3-11

Complete the exercise on the companion website.
www.g-wlearning.com/CAD

U Command

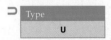

The **U** command undoes the effect of the last command you entered. You can reissue the **U** command to continue undoing command actions, but you can only undo one command at a time. Commands are undone in the order you used them, starting with the most recent.

You can also activate the **U** command by right-clicking in the drawing window and selecting **Undo current**.

UNDO Command

The **UNDO** command allows you to undo a single operation or a number of operations at once. The **UNDO** command is different from the **Undo** option of certain commands, such as the **LINE** command. The quickest way to use the **UNDO** command is to pick the **Undo** button on the **Quick Access** toolbar. Select the button as many times as needed to undo multiple operations. An alternative is to pick the flyout and select all of the commands to undo from the list.

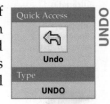

Supplemental Material *UNDO Options*

For detailed information about the options available when you access the **UNDO** command from a source other than the **Quick Access** toolbar, go to the companion website (www.g-wlearning.com/CAD), select this chapter, and select **UNDO Options**.

REDO Command

Use the **REDO** command to reverse the action of the **UNDO** and **U** commands. The **REDO** command works only *immediately* after you have undone something. The quickest way to use the **REDO** command is to pick the **Redo** button on the **Quick Access** toolbar. Select the button as many times as needed to redo multiple undone operations. An alternative is to pick the flyout and select one or more undone operations from the list to redo. The **REDO** command does not bring back line segments removed using the **Undo** option of the **LINE** command.

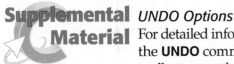

Complete the exercise on the companion website.
www.g-wlearning.com/CAD

OOPS Command

The **OOPS** command brings back the last object you *erased*. Unlike the **UNDO** and **U** commands, **OOPS** only returns objects erased in the most recent procedure. It has no effect on other modifications. If you erase several objects in the same command sequence, all of the objects return to the screen.

Object Selection

Editing commands and similar operations prompt you to select the objects to modify. So far, you have selected objects individually. However, when you need to select more than one object, a more efficient method is to create a *selection set*. This chapter describes several basic options for creating selection sets. You will explore additional selection techniques later in this textbook.

selection set:
A group of one or more selected objects, typically created to perform an editing operation on the selected objects.

Window and Crossing Selection

Window selection allows you to select objects by creating a box, or "window," around the objects. Only objects entirely within the window are selected. Crossing selection also requires you to create a box, but all objects within *and crossing* the box are selected. The default window selection box has a solid outline and light blue background to distinguish it from the crossing selection box, which has a dotted outline and light green background.

The quickest and most effective way to use window or crossing selection is through a feature known as *automatic windowing*, or *implied windowing*, which is on by default. You can apply automatic windowing at the Select objects: prompt, or when no command is active, such as when grip editing. To apply automatic window selection, pick a point clearly above or below and to the *left* of the objects to be selected. Then move the cursor to the right and up or down to position the opposite corner of the selection box to cover the objects you want to select. Pick to locate the second corner and select the objects. Right-click or press [Enter] or the space bar to complete the operation. See **Figure 3-20**.

To apply automatic crossing selection, pick a point clearly above or below and to the *right* of the objects to be selected. Then move the corner of the selection box to the left and up or down, across the objects you want to select. Right-click or press [Enter] or the space bar to complete the operation. See **Figure 3-21**.

Dragging a window or crossing selection box is another method of implied windowing that is available by default. To apply this technique, pick and hold down the left mouse button at the first corner of the window or crossing selection box. Then drag the opposite corner into position and release the left mouse button to make the selection. By default, if no command is active, such as when grip editing, you can press and hold the left mouse button at the first corner of the window or crossing selection box directly on an object without selecting the object. This can sometimes help when you are selecting specific objects that are near or overlapping other objects.

> **NOTE**
> You can type **W** or **WINDOW** at the Select objects: prompt to use manual window selection, or type **C** or **CROSSING** to use manual crossing selection. When you use manual window or crossing selection, the selection box uses the window or crossing format regardless of where you pick to create the box.

Figure 3-20.
Using window selection on a mechanical part drawing view to select and erase all objects that lie completely inside the window selection box. Notice that the selection set does not include the circles that are only partially inside the window.

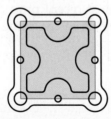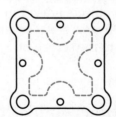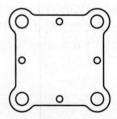

Figure 3-21.
Using crossing selection on an architectural floor plan to select all objects inside the selection box as well as all those that touch the sides of the box.

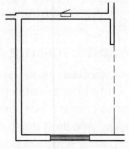

Exercise 3-13

Complete the exercise on the companion website.
www.g-wlearning.com/CAD

Window and Crossing Polygon Selection

The window and crossing polygon selection methods are useful for creating a selection set when it is difficult to use a standard rectangular window or crossing selection box. To use window polygon selection, type WPOLYGON or WP at the Select objects: prompt. Then pick points to draw a polygon enclosing the objects you want to select. See Figure 3-22A. To use crossing polygon selection, type CPOLYGON or CP at the Select objects: prompt. Then pick points to draw a polygon around and through the objects to select. See Figure 3-22B.

PROFESSIONAL TIP

In window or crossing polygon selection, AutoCAD does not allow you to select a point that causes the lines of the selection polygon to intersect each other. Pick locations that do not result in an intersection. Use the **Undo** option if you need to go back and relocate a previous pick point.

Fence Selection

Fence selection allows you to select all objects that contact a "fence" of connected segments you draw while using an edit command. To use fence selection, type FENCE or F at the Select objects: prompt. Then pick points to draw a fence through objects to select them. See Figure 3-23. Often you only need to draw a single segment to create a useful selection set.

Figure 3-22.
A—Using window polygon selection to erase only the wire reinforcement symbol from a structural slab detail. B—Using crossing polygon selection to erase multiple objects from the structural slab detail. Everything within and contacting the crossing polygon is selected.

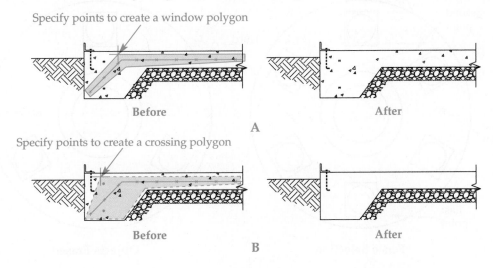

Fence, WPolygon, and **CPolygon** options are also available after you specify the first corner when you use implied windowing before accessing a command, such as when grip editing.

Exercise 3-14

Access the companion website (www.g-wlearning.com/CAD) and complete Exercise 3-14.

Last Selection

Type LAST or L at the Select objects: prompt to select the last object drawn. You must access a single command, such as **ERASE**, repeatedly and use **LAST** selection each time to select individual items in reverse order. Other selection options are usually much faster than **LAST** selection.

Previous Selection

Type Previous or P at the Select objects: prompt to reselect all the objects selected in the previous selection set. **Previous** selection is especially useful when you need to carry out more than one editing operation on a specific group of objects. Use **Previous** selection to reselect the objects you just edited.

Previous selection does not reselect erased objects.

Figure 3-23.
Using fence selection to erase specific objects from a mechanical part drawing view. The fence can be staggered, as shown, or a single straight segment.

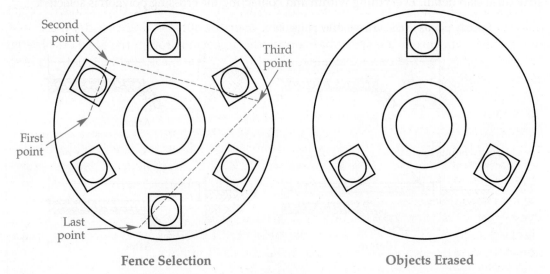

Second point

Third point

First point

Last point

Fence Selection

Objects Erased

Selecting All Objects

Use the **Select All** command to select every object in the drawing that is not on a frozen layer, including objects that are outside of the current drawing window display. Chapter 5 describes layers.

Ribbon

Home
> Utilities

Select All

Type

ALL
[Ctrl]+[A]

Changing the Selection Set

The quickest way to remove one or more objects from a selection set is to hold down [Shift] and reselect the objects. This is possible only for individual picks and automatic windowing. To change the selection set using automatic windowing, hold down [Shift], pick the first corner, then release [Shift] and pick the second corner. You can also drag the window or crossing selection box as previously described. Select objects as usual to add them back to the selection set.

Another option for removing objects from a selection set is to type REMOVE or R at the Select objects: prompt. This enters the **Remove** option and changes the Select objects: prompt to Remove objects:, allowing you to pick objects to remove from the selection set. To switch back to selection mode, type ADD or A at the Remove objects: prompt. This enters the **Add** option and restores the Select objects: prompt, allowing you to select additional objects.

PROFESSIONAL TIP

Removing items from a selection set is especially effective if you first use the **Select All** selection option. This allows you to keep a few specific objects while erasing everything else.

Exercise 3-15

Complete the exercise on the companion website.
www.g-wlearning.com/CAD

Cycling through Stacked Objects

While drawing, you will sometimes create *stacked objects*, intersecting objects, or objects that become very close together. To *cycle* through overlapping objects to find the object to select, first access an editing command, such as **ERASE**. When the Select objects: prompt appears, move the pick box over the intersecting objects, then hold down [Shift] and press the space bar repeatedly to cycle through the stacked objects. When the object you want to select is highlighted, release [Shift] and pick (left-click) to select. See **Figure 3-24**.

AutoCAD also includes a **Selection Cycling** tool for cycling through stacked objects before you access a command, which is common for grip editing. Pick the **Selection Cycling** button on the status bar to toggle selection cycling on and off. Move the crosshairs over stacked objects. When you see the **Selection Cycling** icon, pick using the left mouse button to display a list of stacked objects. Move the cursor over an object in the list to highlight the corresponding object in the drawing. Select an object from the list box or choose **None** to exit. See **Figure 3-25A**. Use the options on the **Selection Cycling** tab of the **Drafting Settings** dialog box to adjust selection cycling settings. See **Figure 3-25B**. A quick way to access the **Selection Cycling** tab is to right-click on the **Selection Cycling** button on the status bar and select **Settings...**.

stacked objects: Objects that overlap in a drawing. When you pick with the mouse, the topmost object is selected by default.

cycle: Repeatedly select a series of stacked objects until the desired object is highlighted.

Figure 3-24.
Access the **ERASE** command and then cycle through a series of stacked bushes on a site plan to locate a specific bush to erase.

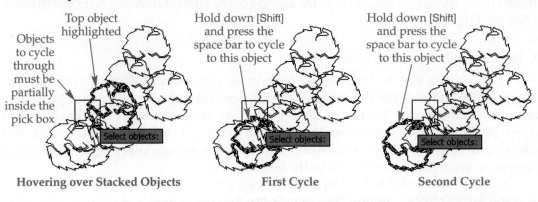

Objects to cycle through must be partially inside the pick box

Top object highlighted

Hovering over Stacked Objects

Hold down [Shift] and press the space bar to cycle to this object

First Cycle

Hold down [Shift] and press the space bar to cycle to this object

Second Cycle

Figure 3-25.
A—Turn on selection cycling to cycle though stacked objects before you access a command. B—Use the **Selection Cycling** tab of the **Drafting Settings** dialog box to specify selection cycling preferences.

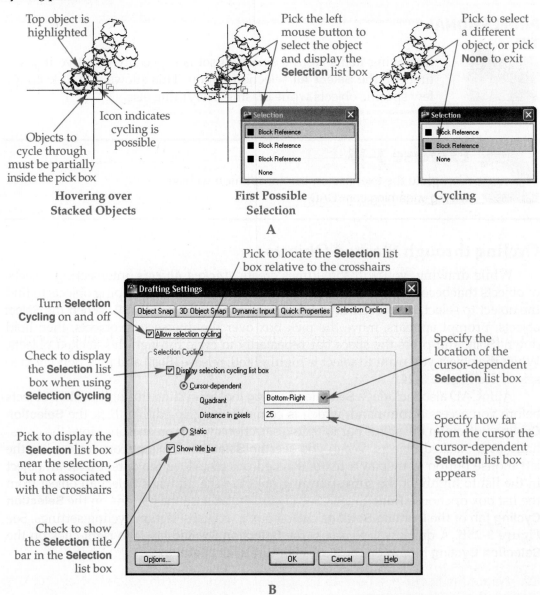

Top object is highlighted

Objects to cycle through must be partially inside the pick box

Icon indicates cycling is possible

Hovering over Stacked Objects

Pick the left mouse button to select the object and display the **Selection** list box

First Possible Selection

Pick to select a different object, or pick **None** to exit

Cycling

A

Pick to locate the **Selection** list box relative to the crosshairs

Turn **Selection Cycling** on and off

Check to display the **Selection** list box when using **Selection Cycling**

Pick to display the **Selection** list box near the selection, but not associated with the crosshairs

Check to show the **Selection** title bar in the **Selection** list box

Specify the location of the cursor-dependent **Selection** list box

Specify how far from the cursor the cursor-dependent **Selection** list box appears

B

You can only cycle through objects if the objects are close enough together that a portion of each object fits inside the pick box.

The **Selection** tab of the **Options** dialog box includes a **Selection Preview** area with settings for adjusting various selection display and preview options.

Template Development Chapter 3

Setting Grid and Snap
For detailed instructions on setting grid and snap values for specific drawing templates, go to the companion website (www.g-wlearning.com/CAD), select this chapter, and select **Template Development**.

Chapter Review

Answer the following questions. Write your answers on a separate sheet of paper or complete the electronic chapter review on the companion website.
www.g-wlearning.com/CAD

1. When you enter a fractional number in AutoCAD, why is a hyphen required between a whole number and its associated fraction?
2. Briefly describe the Cartesian coordinate system.
3. Explain how to use the wheel on a mouse to zoom in, zoom out, and pan.
4. List two ways to discontinue drawing a line.
5. What does the absolute coordinate display 5.250,7.875 mean?
6. What does the polar coordinate display @2.750<90 mean?
7. Name three types of coordinates used for point entry.
8. How can you turn on the coordinate display field if it is off?
9. What two general methods of point entry are available when dynamic input is active?
10. Explain how you can continue drawing another line segment from a previously drawn line.
11. Name two ways to access the **Drafting Settings** dialog box.
12. How do you activate **Snap** mode?
13. How do you set a grid spacing of .25?
14. Explain, in general terms, how direct distance entry works.
15. What are the default angle increments for polar tracking?
16. How can you turn on **Ortho** mode?
17. Which running object snap modes are active by default?
18. Name the drawing aids that must be active for object snap tracking to function.
19. When you access the **ERASE** command, what replaces the screen crosshairs?
20. How many command sequences can you undo at one time with the **U** command?
21. Name the command used to bring back an object that was previously removed using the **UNDO** command.
22. Name the command used to bring back the last object(s) erased before starting another command.

23. How does the appearance of window and crossing selection boxes differ?
24. List five ways to select an object to erase.
25. Define *stacked objects*.

Drawing Problems

Start AutoCAD if it is not already started. Follow the specific instructions for each problem. Use only drawing commands and techniques you have already learned. Do not draw dimensions or text. Use your own judgment and approximate dimensions when necessary.

▼ Basic

1. Start a new drawing from scratch or use a template of your choice. Use the status bar to turn off all drawing aids, including grid, snap, polar tracking, object snap tracking, ortho, and inferred constraints. Use the **LINE** command to draw the following objects as accurately as possible.
   • Right triangle
   • Isosceles triangle
   • Rectangle
   • Square
   Save the drawing as P3-1.

2. Start a new drawing from scratch or use a template of your choice. Draw the same objects specified in Problem 1, but this time, turn the snap grid on. Observe the difference between having snap mode on for this problem and off for the previous problem. Save the drawing as P3-2.

3. Start a new drawing from scratch or use a decimal-unit template of your choice. Draw an object by connecting the following point coordinates. Use dynamic input to enter the coordinates. Save your drawing as P3-3.

Point	Coordinates	Point	Coordinates
1	2,2	8	@-1.5,0
2	@1.5,0	9	@0,1.25
3	@.75<90	10	@-1.25,1.25
4	@1.5<0	11	@2<180
5	@0,-.75	12	@-1.25,-1.25
6	@3,0	13	@2.25<270
7	@1<90		

4. Start a new drawing from scratch or use an architectural-unit template of your choice. Draw the front and side views of a wide flange, similar to the wide flange shown. Use grid and snap modes, default running object snaps, and object snap tracking when possible. Save the drawing as P3-4.

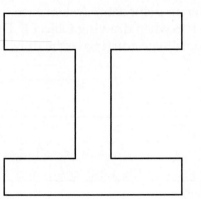
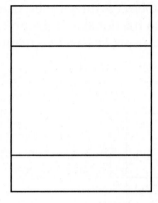

5. Start a new drawing from scratch or use a template of your choice. Draw the bar graph shown using direct distance entry and polar tracking. Each grid square represents one unit. Do not draw the grid lines. Save the drawing as P3-5.

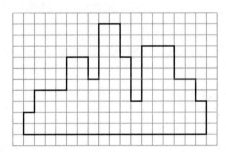

6. Start a new drawing from scratch or use a template of your choice. Draw the hexagon shown using the dimensional input feature of dynamic input. Each side of the hexagon is 2 units. Begin at the start point, and draw the lines in the direction indicated by the arrows. Save the drawing as P3-6.

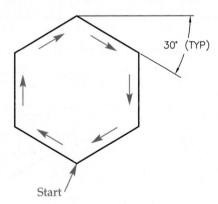

30° (TYP)

Start

Drawing Problems – Chapter 3

7. Start a new drawing from scratch or use a fractional-unit template of your choice. Draw the part views shown using absolute, relative, and polar coordinate entry methods. Set the units to decimal and the precision to 0.0 when drawing Object A. Draw Object A three times, using a different point entry system each time. Set the units to fractional and the precision to 1/16 when drawing Object B. Draw Object B once, using at least two methods of coordinate entry. Save the drawing as P3-7.

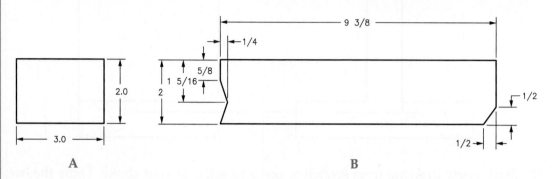

A   B

For Problems 8–9, start a new drawing from scratch or use a decimal template of your choice. Draw the part view shown. Save the drawings as **P3-8** *and* **P3-9**.

8.

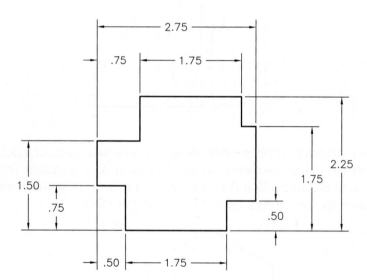

9.

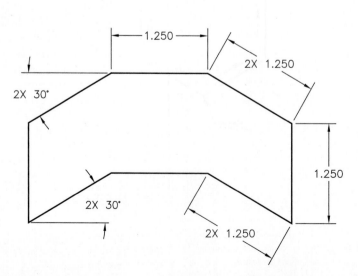

10. Start a new drawing from scratch or use a decimal template of your choice. Draw the part views shown in A and B. Begin at the start point and then discontinue the **LINE** command at the point shown. Complete each view by continuing from the previous endpoint. Save the drawing as P3-10.

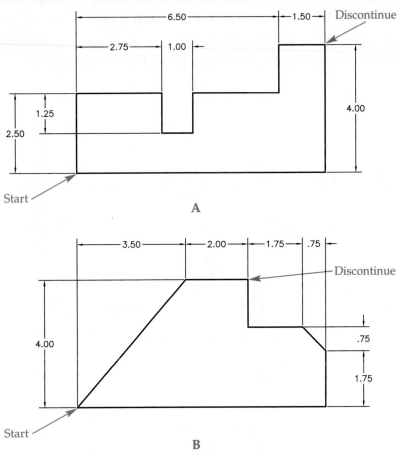

A

B

11. Sketch the X and Y axes on a sheet of paper. Label the origin, the positive values for X = 1 through X = 10, and the positive values for Y = 1 through Y = 10. Then sketch the object described by the following coordinate points:

2,2
8,2
8,7
7,7
7,3
6,3
6,6
4,6
4,3
3,3
3,7
2,7
2,2

▼ Advanced

12. Sketch the X and Y axes of the Cartesian coordinate system as you did for the previous problem. Then sketch an object outline of your choice within the axes. List, in order, the rectangular coordinates of the points a drafter would need to specify in AutoCAD to recreate the object in your sketch.

For Problems 13–15, start a new drawing from scratch or use a decimal template of your choice. Draw the part view shown. Save the drawings as P3-13, P3-14, *and* P3-15.

13.

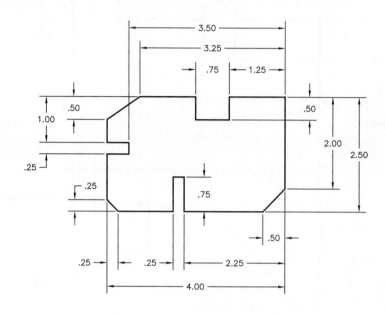

14.

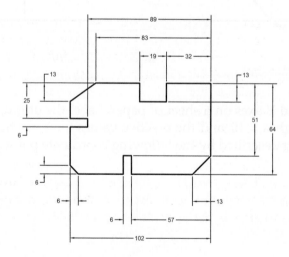

15.

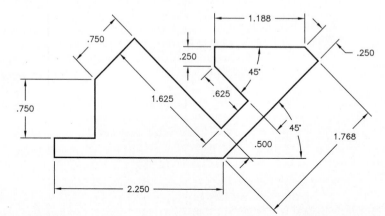

16. Start a new drawing from scratch or use an architectural template of your choice. Draw the window elevation symbol shown. Save the drawing as P3-16.

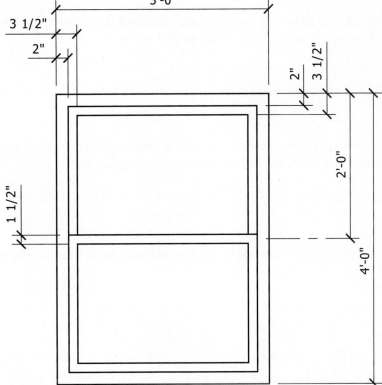

17. Create a 2D hand-drawn sketch of the view of your computer monitor you see when looking at the screen. Use available measuring devices, such as a tape measure and caliper, to dimension the size and location of each feature accurately. Convert any round objects to rectangular shapes that you can draw using the **LINE** command. Start a new drawing from scratch or use a decimal template of your choice. Draw the monitor from your sketch. Save the drawing as P3-17.

18. Create a dimensioned 2D sketch of the floor plan of a room in your school or company, complete with furniture. Use a tape measure to dimension the size and location of walls, doors, windows, and furniture accurately. Convert any round objects to rectangular shapes that you can draw using the **LINE** command. Start a new drawing from scratch or use an architectural template of your choice. Draw the room from your sketch. Save the drawing as P3-18.

Drawing Problems – Chapter 3

AutoCAD Certified Associate Exam Practice

Answer the following questions. Write your answers on a separate sheet of paper.

1. If a drawing is set to architectural units, which of the following can you enter to specify a line length of 11 3/4"? *Select all that apply.*
   A. 11.75"
   B. 11.75
   C. 11 3/4
   D. 11-3/4
   E. 11 3/4"

2. Which of the following can you enter at the Select objects: prompt to select objects that lie partially within the selection boundary? *Select all that apply.*
   A. C
   B. W
   C. CP
   D. WP

3. Which command or commands allow you to undo more than one operation at one time? *Select all that apply.*
   A. **OOPS**
   B. **REDO**
   C. **U**
   D. **UNDO**

AutoCAD Certified Professional Exam Practice

Follow the instructions in each problem. Write your answers on a separate sheet of paper.

1. **Navigate to this chapter on the companion website and open CPE-03line.dwg.**
   With dynamic input off, use the coordinates below to create the object shown. Do not change any settings in the drawing file. When you finish, use the **Endpoint** object snap mode to select point A. According to the coordinate display field in the lower-left corner of the screen, what are the coordinates of this point?

   Coordinate Values:
   2,3
   @6.25<0
   @4<50
   @3.5<90
   @4<130
   @6.25<180
   CLOSE

   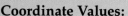

2. **Navigate to this chapter on the companion website and open CPE-03intersect.dwg.**
   With dynamic input off, use the coordinates below to create the object shown. Do not change any settings in the drawing file. Use the **Intersection** object snap mode to select point B. According to the coordinate display field in the lower-left corner of the screen, what are the coordinates of this point?

   Coordinate Values:
   5,2
   @3.45<60
   @4.75<180
   @8<−15

   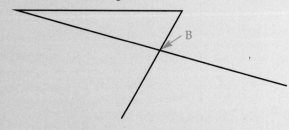

Basic Object Commands

Learning Objectives

After completing this chapter, you will be able to:

✓ Draw circles using **CIRCLE** command options.
✓ Draw arcs using **ARC** command options.
✓ Use the **ELLIPSE** command to draw ellipses and elliptical arcs.
✓ Use the **PLINE** command to draw polylines.
✓ Draw regular polygons using the **POLYGON** command.
✓ Draw rectangles using **RECTANGLE** command options.
✓ Draw donuts and filled circles using the **DONUT** command.
✓ Draw true spline curves using the **SPLINE** command.

This chapter describes several object commands and their options. It presents the ribbon as the primary way to access object commands, because the ribbon provides a direct link to many specific command options. Prompts associated with the options appear when you draw to automate the process. In contrast, when you issue a command using dynamic input or the command line, you must choose specific options while you draw to receive appropriate prompts for constructing the object. You can draw objects using point entry or drawing aids, similar to locating endpoints while using the **LINE** command. Several object commands also offer the option to input a direct value, such as the radius of a circle.

Drawing Circles

The **CIRCLE** command provides several options for drawing *circles*. Choose the appropriate option based on the information you know about locating and constructing the circle. The ribbon is an effective way to access **CIRCLE** command options. See **Figure 4-1**.

circle: A closed curve with a constant radius around a center point; usually dimensioned according to the diameter.

CIRCLE

Ribbon
Home
> Draw

Circle

Type
CIRCLE
C

Center, Radius Option

Access the **Center, Radius** option to specify the center of the circle, followed by the radius. Use point entry or drawing aids to locate the center point. If you know the radius, type a value and press [Enter] or the space bar, or right-click and pick **Enter**. You can also define the radius using point entry or drawing aids. See **Figure 4-2**.

Figure 4-1.
Select a **Circle** option from the **Draw** panel of the **Home** ribbon tab to preset the **CIRCLE** command to display appropriate prompts.

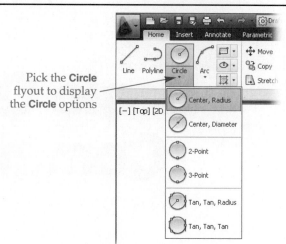

Pick the **Circle** flyout to display the **Circle** options

Figure 4-2.
Drawing a circle by specifying the center point and radius. Notice the rubberband line that appears when you move the crosshairs away from the center point.

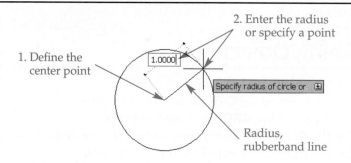

2. Enter the radius or specify a point

1. Define the center point

Radius, rubberband line

AutoCAD stores the radius of the circle you draw as the new default radius setting, allowing you to draw another circle quickly with the same radius.

Center, Diameter Option

Access the **Center, Diameter** option to specify the center of the circle, followed by the diameter. The **Center, Diameter** option is convenient for designs such as circular holes, shafts, and features sized according to diameter. Use point entry or drawing aids to locate the center point. If you know the diameter, type a value and press [Enter] or the space bar, or right-click and pick **Enter**. You can also define the diameter using point entry or drawing aids. See **Figure 4-3**.

If you use the **Center, Radius** option to draw a circle after using the **Diameter** option, AutoCAD changes the default to a radius measurement based on the previous diameter.

2-Point Option

quadrant: A point on the circumference at the horizontal or vertical quarter of a circle, arc, donut, or ellipse.

Access the **2-Point** option to specify diameter using two points at opposite *quadrants* of the circle. The **2-Point** option is useful when you know the diameter of the circle, but the center is difficult to locate. A common example is drawing a circle between two existing objects. Use point entry or drawing aids to locate the first and second points. See **Figure 4-4**.

Figure 4-3.
Drawing a circle by specifying the center point and diameter. Notice that the crosshairs measures the diameter, but the rubberband line passes midway between the center and the crosshairs.

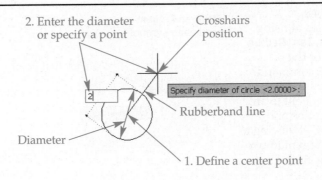

2. Enter the diameter or specify a point

Crosshairs position

Specify diameter of circle <2.0000>:

Rubberband line

Diameter

1. Define a center point

Figure 4-4.
Using the **2-Point** option of the **CIRCLE** command. A common application is drawing a circle between two existing objects, such as these lines.

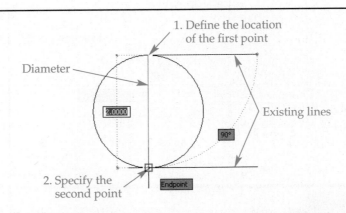

1. Define the location of the first point

Diameter

2.0000

Existing lines

90°

2. Specify the second point

Endpoint

3-Point Option

Select the **3-Point** option to draw a circle according to three known points on the circumference of the circle. The **3-Point** option is most commonly used when the location of the center point, the radius, and the diameter are unknown. Specify the three points in any order using point entry or drawing aids. See **Figure 4-5.**

Tan, Tan, Radius Option

Access the **Tan, Tan, Radius** option to pick two objects *tangent* to a circle and the circle radius. Hover the crosshairs over the first line, arc, or circle to which the new circle will be tangent. When you see the **Deferred Tangent** object snap marker, pick to select the first *point of tangency*. Repeat the process to select the second object to which the new circle will be tangent. The order in which you pick is not critical. If you know the radius, type a value and press [Enter] or the space bar, or right-click and pick **Enter**. You can also define the radius using point entry or drawing aids. See **Figure 4-6.**

tangent: A line, circle, or arc that meets another circle or arc at only one point.

point of tangency: The point shared by tangent objects.

Figure 4-5.
Using the **3-Point** option of the **CIRCLE** command. A common application is drawing a circle by referencing three known points, such as the endpoints of these lines.

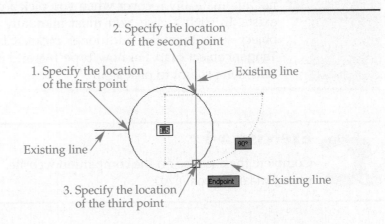

2. Specify the location of the second point

1. Specify the location of the first point

Existing line

1.5

90°

Existing line

3. Specify the location of the third point

Endpoint

Existing line

Figure 4-6.
Examples of using the **Tan, Tan, Radius** option of the **CIRCLE** command. A—Drawing a circle tangent to an existing line and circle. B—Drawing a circle tangent to two existing circles.

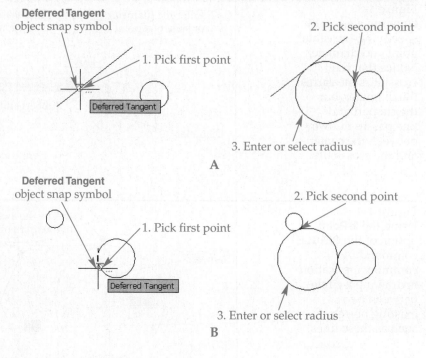

A

B

If the radius you enter while using the **Tan, Tan, Radius** option is too small, AutoCAD displays the message Circle does not exist.

Tan, Tan, Tan Option

Select the **Tan, Tan, Tan** option to draw a circle tangent to three existing objects. Hover the crosshairs over the first line, arc, or circle to which the new circle will be tangent. When you see the **Deferred Tangent** object snap marker, pick to select the first point of tangency. Repeat the process to select the second and third objects to which the new circle will be tangent. You must make selections when you see the **Deferred Tangent** object snap marker, but the order in which you pick is not critical. See **Figure 4-7.**

Unlike the **Tan, Tan, Radius** option, the **Tan, Tan, Tan** option does not automatically recover when you pick a point where no tangent exists. In such a case, you must manually reactivate the **Tangent** object snap to make additional picks. Chapter 7 describes the **Tangent** object snap. For now, type TAN and press [Enter] at the point selection prompt to pick again.

Exercise 4-1

Complete the exercise on the companion website.
www.g-wlearning.com/CAD

Figure 4-7.
Examples of using the **Tan, Tan, Tan** option of the **CIRCLE** command.
A—Drawing a circle tangent to three existing lines.
B—Drawing a circle tangent to two existing lines and a circle.

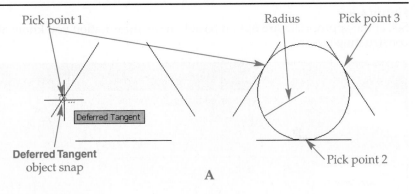

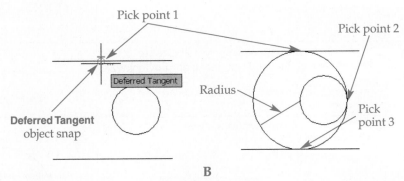

Drawing Arcs

ARC

The **ARC** command offers multiple options for drawing *arcs*. Select the appropriate option based on the information you know about locating and constructing the arc. The ribbon is an effective way to access arc command options. See **Figure 4-8**.

Figure 4-9 provides a step-by-step example of using each **ARC** command option. The selections and values you enter determine arc placement. Some options prompt for the *included angle*, and others prompt for the *chord length*. Locating points in a clockwise or counterclockwise pattern affects the result in most arc options. The values you specify, including the use of positive or negative numbers, also affects the result.

arc: Any portion of a circle; usually dimensioned according to the radius.

included angle: The angle formed between the center, start point, and endpoint of an arc.

chord length: The linear distance between two points on a circle or arc.

Figure 4-8.
Selecting an **Arc** option from the **Draw** panel of the **Home** ribbon tab to preset the **ARC** command to display appropriate prompts.

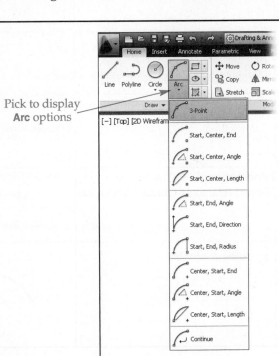

Figure 4-9.
Select the appropriate **Arc** option based on the information you know about locating and constructing the arc.

Option	Direction	Steps
3-Point	Clockwise or counterclockwise	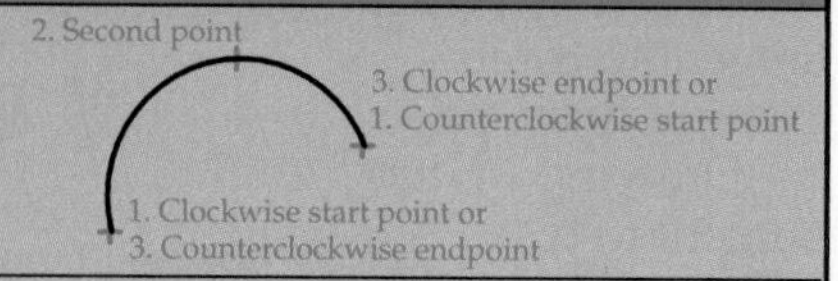 2. Second point; 3. Clockwise endpoint or 1. Counterclockwise start point; 1. Clockwise start point or 3. Counterclockwise endpoint
Start, Center, End	Counterclockwise	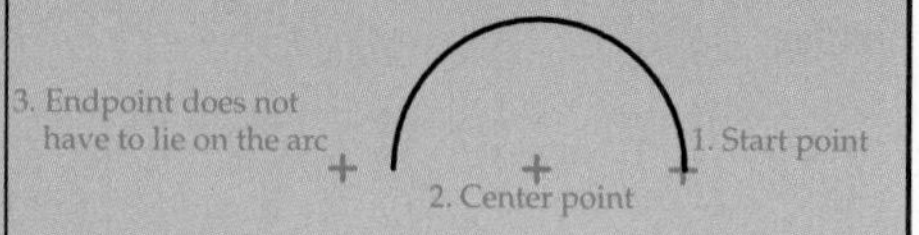 3. Endpoint does not have to lie on the arc; 1. Start point; 2. Center point
Start, Center, Angle	Positive angle = Clockwise Negative angle = Counterclockwise	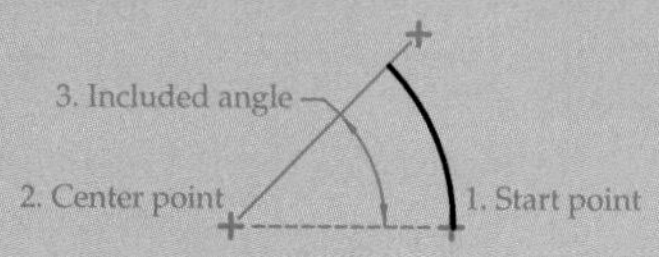 3. Included angle; 2. Center point; 1. Start point
Start, Center, Length	Counterclockwise	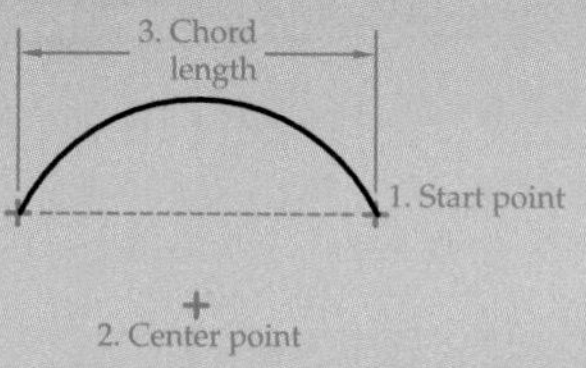 3. Chord length; 1. Start point; 2. Center point
Start, End, Angle	Positive angle = Clockwise Negative angle = Counterclockwise	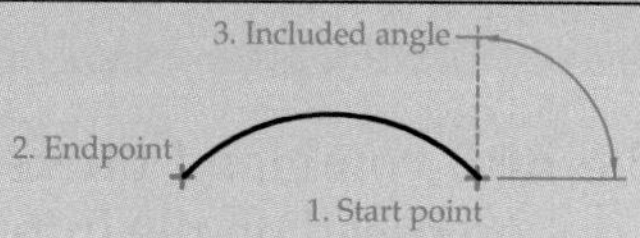 3. Included angle; 2. Endpoint; 1. Start point
Start, End, Direction	Tangent to specified direction	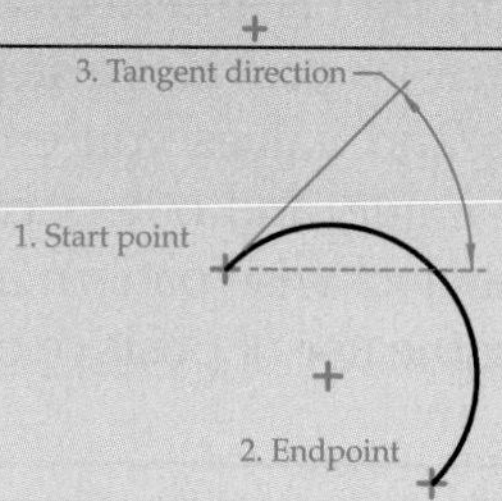 3. Tangent direction; 1. Start point; 2. Endpoint
Start, End, Radius	Counterclockwise	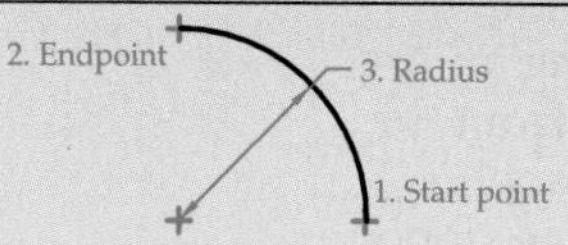 2. Endpoint; 3. Radius; 1. Start point
Center, Start, End	Counterclockwise	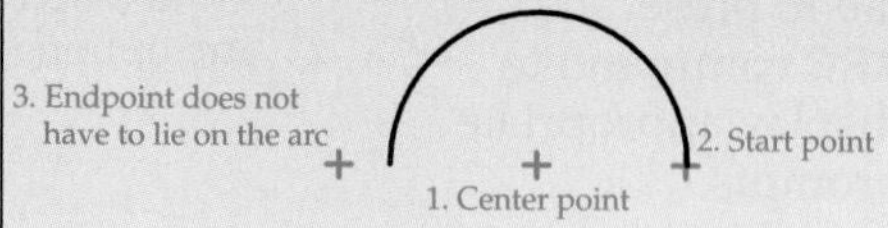 3. Endpoint does not have to lie on the arc; 2. Start point; 1. Center point
Center, Start, Angle	Positive angle = Clockwise Negative angle = Counterclockwise	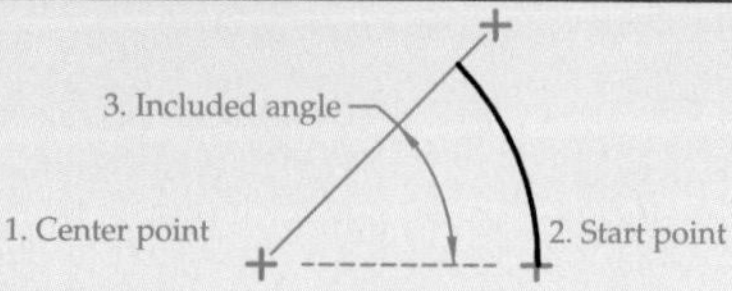 3. Included angle; 1. Center point; 2. Start point
Center, Start, Length	Counterclockwise	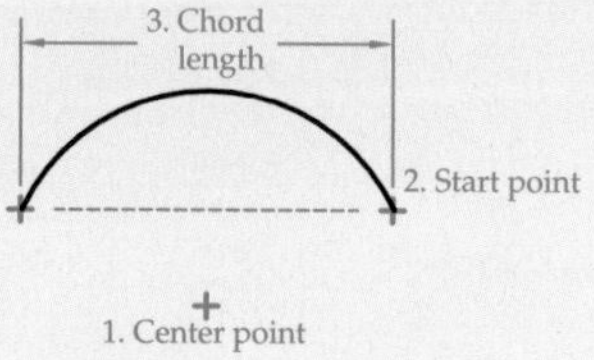 3. Chord length; 2. Start point; 1. Center point

 The **3-Point** option is default when you enter the **ARC** command at the keyboard.

 *Chord Length Table*
For a chord length table and other reference tables, go to the **Reference Material** section of the companion website (www.g-wlearning.com/CAD) and select **Standard Tables**.

 ## Exercises 4-2 and 4-3

Complete the exercises on the companion website.
www.g-wlearning.com/CAD

Continue Option

Use the **Continue** option to continue an arc from the endpoint of a previously drawn line or arc. The arc automatically attaches to the endpoint of the previously drawn line or arc, and the Specify endpoint of arc: prompt appears. Pick the second endpoint of the new arc to create the arc.

The **Continue** option is a quick way to draw an arc beginning at the endpoint of a previously drawn line, tangent to the line. See **Figure 4-10**. Use this technique for applications such as drawing slots. When you draw a series of arcs using the **Continue** option, each arc is tangent to the previous arc. The start point and direction are based on the endpoint and direction of the previous arc. See **Figure 4-11**.

Figure 4-10.
Continuing an arc from the previous line. Point 2 is the start of the arc, and Point 3 is the end of the arc. The arc and line are tangent at Point 2.

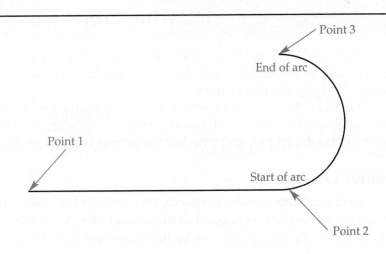

Figure 4-11.
Using the **Continue** option to draw three tangent arcs.

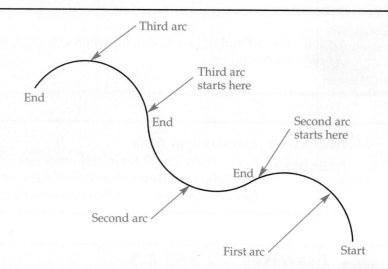

 You can also access the **Continue** option by beginning the **ARC** command and pressing [Enter] or the space bar, or by right-clicking and selecting **Enter** when prompted to specify the start point of the arc.

 ## Exercise 4-4

Complete the exercise on the companion website.
www.g-wlearning.com/CAD

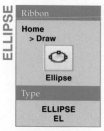

ELLIPSE

Ribbon

Home > Draw

Ellipse

Type

ELLIPSE EL

ellipse: An oval shape that contains two centers of equal radius.

major axis: The longer of the two axes in an ellipse.

minor axis: The shorter of the two axes in an ellipse.

Drawing Ellipses

An *ellipse* has a *major axis* and a *minor axis*. See Figure 4-12. A circle appears as an ellipse when you view the circle at an angle. For example, a 30° ellipse is a circle rotated 30° from the line of sight.

The **ELLIPSE** command offers several options for drawing elliptical shapes. Choose the appropriate option based on the information you know about locating and constructing the ellipse, and whether the ellipse is whole or an elliptical arc.

Center Option

Select the **Center** option to specify the center of the ellipse, then an endpoint of the first axis, and finally an endpoint of the second axis. Axis endpoints originate from the center of the ellipse, forming half of the major and minor axes. See Figure 4-13.

Figure 4-12.
The parts of an ellipse.

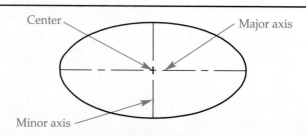

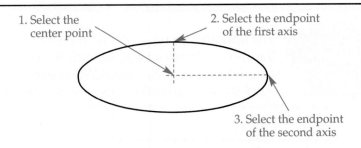

Figure 4-13.
Drawing an ellipse by picking the center and an endpoint for each axis. The order in which you specify axis endpoints is not critical. The distance from each endpoint to the center point determines the major and minor axes.

1. Select the center point
2. Select the endpoint of the first axis
3. Select the endpoint of the second axis

Axis, End Option

Choose the **Axis, End** option to specify the first endpoint of an axis, then the second endpoint of the same axis, and finally one endpoint of the second axis. The first axis can be the major or minor axis. See **Figure 4-14**.

Rotation Option

Use the **Rotation** option to create an ellipse by specifying the angle at which a circle rotates from the line of sight. For example, a 30° ellipse is a circle rotated 30° from the line of sight. Begin by constructing an ellipse as usual, but be sure to create the major axis when you specify the first axis endpoint. Then, when the Specify distance to other axis or: prompt appears, select the **Rotation** option instead of picking the second axis endpoint. Finally, enter the angle at which the circle rotates from the line of sight, such as 30 for a 30° rotation. **Figure 4-15** shows examples of rotation angles.

Figure 4-14.
Constructing the same ellipse by choosing different axis endpoints. Select points based on known information or the location of existing objects.

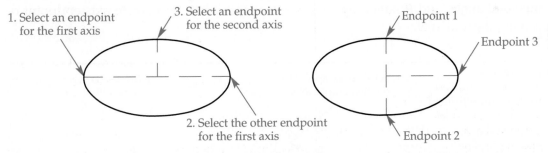

1. Select an endpoint for the first axis
3. Select an endpoint for the second axis
2. Select the other endpoint for the first axis

Endpoint 1
Endpoint 3
Endpoint 2

Figure 4-15.
The relationship among several ellipses having the same major axis length, but different rotation angles.

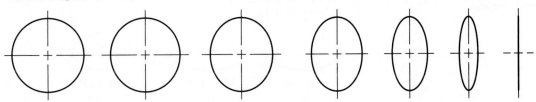

 The **Rotation** option works with the **Center** and **Axis, End** options. A 0 response draws an ellipse with the minor axis equal to the major axis, which is a circle. AutoCAD rejects any rotation angle between 89.99994° and 90.00006° and between 269.99994° and 270.00006°.

 Exercise 4-5

Complete the exercise on the companion website.
www.g-wlearning.com/CAD

Drawing Elliptical Arcs

Ribbon

Home
> Draw

Elliptical Arc

Use the **Arc** option of the **ELLIPSE** command to draw elliptical arcs. Drawing an elliptical arc is just like drawing an ellipse, but with two additional steps that define the beginning and end of the elliptical arc. Several options are available for defining the size and shape of an elliptical arc.

The default elliptical arc option is similar to the **Axis, End** ellipse option. Specify the first endpoint of an axis, then the second endpoint of the same axis, and then one endpoint of the second axis. Finally, select the start and end angles for the elliptical arc. See **Figure 4-16**. The start and end angles are the angular relationships between the center of the ellipse and the arc endpoints. The angle of the first axis establishes the angle of the elliptical arc. For example, a 0° start angle begins the arc at the first endpoint of the first axis. A 45° start angle begins the arc 45° counterclockwise from the first endpoint of the first axis. End angles are also calculated counterclockwise from the start point.

Figure 4-17 briefly describes additional elliptical arc options. Use the **Center** option when appropriate instead of the default axis endpoint method. The **Parameter**, **Included angle**, and **Rotation** options are available when you create axis endpoint or center elliptical arcs.

Figure 4-16.
The steps required to draw an elliptical arc using a 0° start angle and a 90° end angle. The three examples at the bottom were created using the same steps, but with different start and end angles.

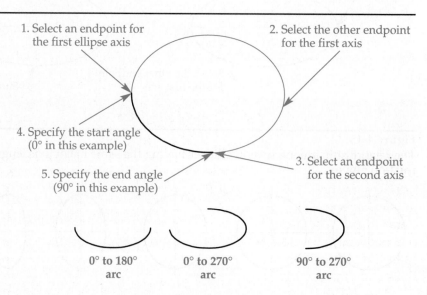

1. Select an endpoint for the first ellipse axis
2. Select the other endpoint for the first axis
4. Specify the start angle (0° in this example)
3. Select an endpoint for the second axis
5. Specify the end angle (90° in this example)

0° to 180° arc   0° to 270° arc   90° to 270° arc

Figure 4-17.
Additional options for drawing elliptical arcs.

Option	Application	Process
Center	Lets you establish the center of the elliptical arc. **Rotation**, **Parameter**, and **Included angle** options are available.	1. Select the ellipse center point. 2. Select the endpoint of one of the ellipse axes. 3. Pick the endpoint of the other axis to form the ellipse. 4. Enter the start angle for the elliptical arc. 5. Select the end angle.
Parameter	Use instead of picking the start angle of the elliptical arc. AutoCAD uses a different means of vector calculation to create the elliptical arc.	1. Specify the start parameter point. 2. Specify the end parameter point.
Included angle	Establishes an included angle beginning at the start angle.	1. Specify the included angle.
Rotation	Allows you to rotate the elliptical arc about the first axis by specifying a rotation angle. **Parameter** and **Included angle** options are available.	1. Specify the rotation around the major axis. 2. Specify the start angle for the elliptical arc. 3. Specify the end angle.

Exercise 4-6

Complete the exercise on the companion website.
www.g-wlearning.com/CAD

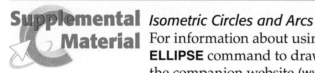

Supplemental Material *Isometric Circles and Arcs*
For information about using the **Isocircle** option of the **ELLIPSE** command to draw isometric circles and arcs, go to the companion website (www.g-wlearning.com/CAD), select this chapter, and select **Isometric Circles and Arcs**.

Drawing Polylines

Use the **PLINE** command to draw *polylines*. When you use the default polyline settings, drawing polyline segments is identical to drawing line segments using the **LINE** command. Access the **PLINE** command and use point entry or drawing aids to locate polyline endpoints. Press [Enter], the space bar, or [Esc], or right-click and select **Enter** to exit. The difference between a polyline and a line is that all of the segments of a polyline act as a single object. The **PLINE** command also provides more flexibility than the **LINE** command, allowing you to draw a single object composed of straight lines and arcs of varying thickness.

The **PLINE** command includes the same **Undo** and **Close** options available with the **LINE** command. Use the **Undo** option to remove the last segment of a polyline and continue from the previous endpoint without leaving the **PLINE** command. You can use the **Undo** option repeatedly to delete polyline segments until the entire object is

polyline: A series of lines and arcs that constitute a single object.

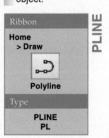

PLINE

Ribbon
Home > Draw
Polyline

Type
PLINE
PL

gone. Use the **Close** option to connect the endpoint of the last polyline segment to the start point of the first polyline segment.

Setting Polyline Width

The default polyline settings create a polyline with a constant width of 0. A polyline with a constant width of 0 is similar to a standard line and accepts the lineweight applied to the layer on which the polyline is drawn. Chapter 5 explains layers. Adjust the polyline width to create thick or tapered polyline objects.

To change the width of a polyline segment, access the **PLINE** command, select the first point, and choose the **Width** option. AutoCAD prompts you to specify the starting width of the line, followed by the ending width of the line. Enter the same starting and ending width value to draw a polyline with constant width. See **Figure 4-18A**. The rubberband line from the first point reflects the width settings. The location of the start point and endpoint is at the center of the segment width.

To create a tapered line segment for applications such as an arrowhead, enter different values for the starting and ending widths. See **Figure 4-18B**. To draw an arrowhead with a sharp point, use the **Width** option and specify 0 as the starting or ending width, and then use an appropriate value greater than 0 for the opposite width.

A starting or ending width value other than 0 overrides the lineweight applied to the layer on which you draw the polyline.

Setting Polyline Halfwidth

Choose the **Halfwidth** option to specify the width of the polyline from the center to one side, as opposed to the total width of the polyline defined using the **Width** option. Access the **PLINE** command, pick the first polyline endpoint, and then choose the **Halfwidth** option. Specify starting and ending values at the appropriate prompts. **Figure 4-19** shows a polyline drawn using the **Halfwidth** option and the same width values applied in **Figure 4-18B**, resulting in a polyline that is twice as wide.

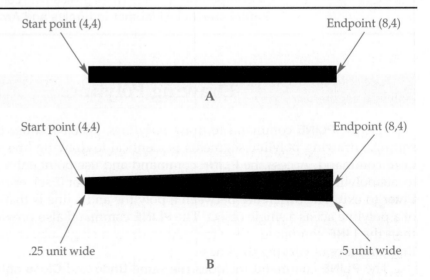

Figure 4-18.
A—A thick polyline drawn using the **Width** option of the **PLINE** command.
B—Using the **Width** option of the **PLINE** command to draw a tapered polyline.

Start point (4,4) Endpoint (8,4)

A

Start point (4,4) Endpoint (8,4)

.25 unit wide .5 unit wide

B

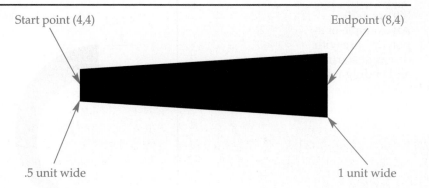

Figure 4-19.
Specifying the width of a polyline using the **Halfwidth** option. A starting value of .25 produces a polyline width of .5 unit, and an ending value of .5 produces a polyline width of 1 unit.

Start point (4,4)   Endpoint (8,4)

.5 unit wide   1 unit wide

All polyline objects with width—including polylines, polygons drawn using the **POLYGON** command, rectangles drawn using the **RECTANGLE** command, and donuts—can appear filled or empty. The **Apply solid fill** setting in the **Display performance** area of the **Display** tab in the **Options** dialog box controls the appearance. You can also type **FILL** or **FILLMODE** and use the **On** or **Off** option. Polyline objects are filled by default. The fill display for previously drawn polyline objects updates when the drawing regenerates. Type **REGEN** to regenerate the drawing manually.

Length Option

The **Length** option allows you to draw a polyline parallel to a previously drawn line or polyline. After you draw a line or polyline, access the **PLINE** command and pick a start point. Choose the **Length** option and specify the length. The resulting polyline is automatically drawn parallel to the previous line or polyline using the specified length.

Exercises 4-7 and 4-8

Complete the exercise on the companion website.
www.g-wlearning.com/CAD

Drawing Polyline Arcs

Use the **Arc** option to draw polyline arcs. Polyline arcs can continue from or to polyline segments drawn during the same operation to form a single object. You can use the **Width** or **Halfwidth** option to add width to a polyline arc, ranging from 0 to the radius of the arc. You can also set different starting and ending arc widths. See **Figure 4-20**. Enter the **Width** or **Halfwidth** and **Arc** options in either order. Use the **Line** option to return to the straight-line segment mode of the **PLINE** command.

In addition to the **Close**, **Undo**, **Width**, and **Halfwidth** options, the polyline **Arc** option includes functions for controlling the size and location of polyline arcs. Many of the polyline **Arc** options allow you to create polyline arcs using the same methods available for drawing arcs using the **ARC** command. Select the appropriate option and follow the prompts to create the polyline arc. Review the **ARC** command options described in **Figure 4-21** to help recognize the function of similar polyline **Arc** options.

Figure 4-20.
An example of a polyline arc with different starting and ending widths.

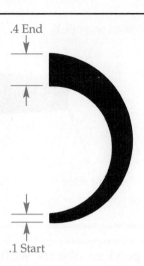

.4 End

.1 Start

Figure 4-21.
Additional options available for drawing polyline arcs.

Option	Application	Options for Completion
Angle	Specify the polyline arc size according to an included angle.	1. Specify an endpoint. 2. Use the **Center** option to select the center point. 3. Use the **Radius** option to enter the radius.
Center	Specify the location of the polyline arc center point, instead of allowing AutoCAD to calculate the location automatically.	1. Specify an endpoint. 2. Use the **Angle** option to specify the included angle. 3. Use the **Length** option to specify the chord length.
Direction	Alter the polyline arc bearing, or tangent direction, instead of allowing the polyline arc to form tangent to the last object drawn.	1. Specify an endpoint.
Radius	Specify the polyline arc radius.	1. Specify an endpoint. 2. Use the **Angle** option to specify the included angle.
Second point	Draw a three-point polyline arc.	1. Pick the second point, followed by the endpoint.

Exercise 4-9

Complete the exercise on the companion website.
www.g-wlearning.com/CAD

Multilines
AutoCAD also includes commands for working with objects made up of multiple lines. For information about drawing and editing multiline objects, go to the companion website (www.g-wlearning.com/CAD), select this chapter, and select **Multilines**.

Drawing Regular Polygons

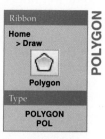

Ribbon

Home
> Draw

Polygon

Type

POLYGON
POL

POLYGON

Access the **POLYGON** command to draw any *regular polygon* with up to 1024 sides. Polygons drawn using the **POLYGON** command are single polyline objects. The first prompt asks for the number of sides. For example, to draw an octagon, which is a regular polygon with eight sides, enter 8. Next, decide how to describe the size and location of the polygon. The default setting involves choosing the center and radius of an imaginary circle. To use this method, specify a location for the polygon center point. A prompt then asks if you want to form an *inscribed polygon* or a *circumscribed polygon*. Select the appropriate option and specify the radius to create the polygon. See Figure 4-22.

regular polygon: A closed geometric figure with three or more equal sides and equal angles.

inscribed polygon: A polygon drawn inside an imaginary circle so that its corners touch the circle.

circumscribed polygon: A polygon drawn outside an imaginary circle so that the sides of the polygon are tangent to the circle.

The number of polygon sides you enter, the **Inscribed in circle** or **Circumscribed about circle** option you select, and the radius you specify are stored as the new default settings, allowing you to draw another polygon quickly with the same characteristics.

PROFESSIONAL TIP

Regular polygons, such as the *hexagons* commonly drawn to represent bolt heads and nuts on mechanical drawings, are normally dimensioned across the flats. Use the **Circumscribed about circle** option to draw a polygon dimensioned across the flats. The radius you enter is equal to one-half the distance across the flats. Use the **Inscribed in circle** option to dimension a polygon across the corners or to confine the polygon within a circular area.

hexagon: A six-sided regular polygon.

Edge Option

Use the **Edge** option to construct a polygon if you do not know the center point location or radius of the imaginary circle, but you do know the size and location of a polygon edge. After you access the **POLYGON** command and enter the number of sides, choose the **Edge** option at the Specify center of polygon or [Edge]: prompt. Specify a point for the first endpoint of one side, followed by the second endpoint of the side. See Figure 4-23.

Figure 4-22.
Regular polygons can be inscribed in a circle or circumscribed around a circle.

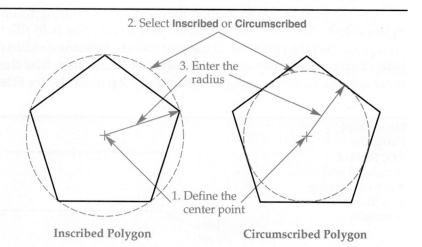

2. Select **Inscribed** or **Circumscribed**

3. Enter the radius

1. Define the center point

Inscribed Polygon   Circumscribed Polygon

Figure 4-23.
Use the **Edge** option of the **POLYGON** command to construct a regular polygon according to the location and size of an edge.

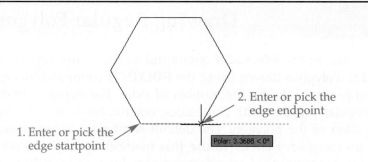

2. Enter or pick the edge endpoint

1. Enter or pick the edge startpoint

Polar: 3.3688 < 0°

Exercise 4-10

Complete the exercise on the companion website.
www.g-wlearning.com/CAD

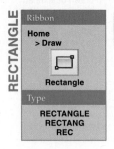

RECTANGLE

Ribbon
Home
> Draw

Rectangle

Type
RECTANGLE
RECTANG
REC

Drawing Rectangles

Use the **RECTANGLE** command to draw rectangles easily. Rectangles drawn using the **RECTANGLE** command are single polyline objects. To draw a rectangle using default settings, specify the point of one corner, followed by the point of the diagonally opposite corner. See **Figure 4-24**. By default, the **RECTANGLE** command draws a rectangle at a 0° angle with sharp corners.

Adding Chamfered Corners

chamfer: In mechanical drafting, a small angled surface used to relieve a sharp corner.

Use the **Chamfer** option to include *chamfered* corners during rectangle construction. See **Figure 4-25A**. When prompted, enter the first chamfer distance, followed by the second chamfer distance. Entering 0 at the first or second chamfer distance prompt creates a rectangle with sharp corners. After setting the distances, you can either draw the rectangle or set additional options. However, using the **Fillet** option overrides the **Chamfer** option.

The rectangle you draw must be large enough to accommodate the specified chamfer distances. Otherwise, the rectangle will have sharp corners. New rectangles are drawn with the specified chamfer until you reset the chamfer distances to 0 or use the **Fillet** option to create rounded corners.

Adding Rounded Corners

fillet: A rounded interior corner.

round: A rounded exterior corner.

Use the **Fillet** option to include rounded corners during rectangle construction. See **Figure 4-25B**. AutoCAD uses the term *fillet* to describe both *fillets* and *rounds*. When prompted, enter the radius for all fillets or rounds. Entering a radius of 0 creates a rectangle with sharp corners. After setting the radius, you can either draw the rectangle or set additional options. However, using the **Chamfer** option overrides the **Fillet** option.

Figure 4-24.
Using the **RECTANGLE** command and point entry or drawing aids to construct a rectangle.

Other corner

First corner

Figure 4-25.
A—Use the **Chamfer** option of the **RECTANGLE** command to add chamfers to rectangle corners during construction. B—Use the **Fillet** option of the **RECTANGLE** command to add rounds to rectangle corners during construction.

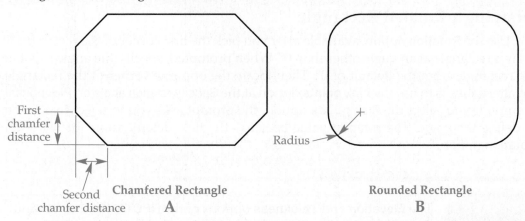

First chamfer distance

Second chamfer distance

Chamfered Rectangle
A

Radius

Rounded Rectangle
B

The rectangle you draw must be large enough to accommodate the specified fillet radii. Otherwise, the rectangle will have sharp corners. New rectangles are drawn with the specified fillets until you reset the fillet radius to 0 or use the **Chamfer** option to create chamfered corners.

This chapter introduces adding chamfers and rounds while creating rectangles. Chapter 12 covers adding chamfers using the **CHAMFER** command and rounds using the **FILLET** command.

Setting the Width

Select the **Width** option to adjust rectangle line width, or "boldness." Do not confuse width with lineweight, described in Chapter 5. A prompt asks you to enter the line width. For example, to create a rectangle with lines that are .5 wide, enter .5. After setting the rectangle width, you can either draw the rectangle or set additional options. All new rectangles are drawn using the specified width. Reset the **Width** option to 0 to create new rectangles using a standard "0-width" line.

Specifying the Area

The **Area** option is available after you pick the first corner point. This option is useful for drawing a rectangle when you know the area of the rectangle and the length of one side. Choose the **Area** option, and then specify the total area for the rectangle using a value that corresponds to the current units. For example, enter 45 to draw a rectangle with an area of 45 units. Next, choose the **Length** option if you know the length of a side (the X value), or choose the **Width** option if you know the width of a side (the Y value). When prompted, enter the length or width to complete the rectangle. AutoCAD calculates the unspecified dimension and draws the rectangle.

Specifying Rectangle Dimensions

The **Dimensions** option is available after you pick the first corner of the rectangle and allows you to specify the length and width of the rectangle. Choose the **Dimensions** option and specify the length of a side to indicate the X value. Next, enter the width of a side to indicate the Y value. AutoCAD then prompts for the other corner point.

To change the dimensions, select the **Dimensions** option again. If the dimensions are correct, specify another point to complete the rectangle. The second point determines which of four possible rectangles you draw. See Figure 4-26.

Drawing a Rotated Rectangle

Use the **Rotation** option, available after you pick the first corner of the rectangle, to draw a rectangle at an angle other than 0°. When prompted, specify the angle to rotate the rectangle from the default of 0°. Then locate the opposite corner of the rectangle. An alternative is to use the **Pick points** option at the Specify rotation angle or [Pick points]: prompt. If you select the **Pick points** option, the prompt asks you to select two points to define the angle. The rotation value becomes the new default angle for using the **Rotation** option.

NOTE

The **Elevation** and **Thickness** options of the **RECTANGLE** command are appropriate for 3D applications, as explained in *AutoCAD and Its Applications—Advanced*.

PROFESSIONAL TIP

You can use a combination of rectangle settings to draw a single rectangle. For example, you can enter a width value, chamfer distances, and length and width dimensions to create a rectangle.

Exercise 4-11

Complete the exercise on the companion website.
www.g-wlearning.com/CAD

Figure 4-26.
The second corner point, or quadrant, determines the orientation of the rectangle relative to the first corner point.

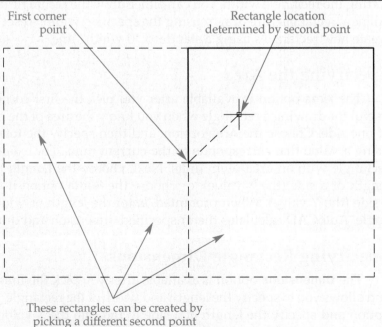

First corner point

Rectangle location determined by second point

These rectangles can be created by picking a different second point

Drawing Donuts and Filled Circles

The **DONUT** command allows you to draw a thick or filled circle. See **Figure 4-27**. A donut is a single polyline object. After activating the **DONUT** command, enter the inside diameter and then the outside diameter of the donut. Enter a value of 0 for the inside diameter to create a completely filled donut, or solid circle.

The center point of the donut attaches to the crosshairs, and the Specify center of donut or <exit>: prompt appears. Pick a location to place the donut. The **DONUT** command remains active until you right-click or press [Enter], the space bar, or [Esc]. This allows you to place multiple donuts of the same size using a single instance of the **DONUT** command.

Ribbon
Home
> Draw
Donut

Type
DONUT
DO

DONUT

Exercise 4-12

Complete the exercise on the companion website.
www.g-wlearning.com/CAD

Drawing True Splines

Access the **SPLINE** command to create a special type of curve using *non-uniform rational basis spline (NURBS)*, or *B-spline*, mathematics. A NURBS curve is a complex mathematical *spline* representation that includes control points. Examples of splines on a 2D drawing include curved edges on the drawing of an ergonomic consumer product and contour lines on a site plan.

To draw a default spline, specify fit points using point entry or drawing aids. By default, the spline forms, or fits, though the points. When you finish locating points, press [Enter] or the space bar, or right-click and select **Enter** to create the spline and exit the command. **Figure 4-28** shows a default spline drawn using absolute coordinates. Use the **Undo** option to remove the last segment of a spline without leaving the **SPLINE** command.

SPLINE

Ribbon
Home
> Draw
Spline

Type
SPLINE
SPL

non-uniform rational basis spline (NURBS, B-spline): The mathematics used by most surface modeling CADD systems to produce accurate curves and surfaces.

spline: A curve that uses a series of control points and other mathematical principles to define the location and form of the curve.

Figure 4-27.
The appearance of a donut depends on its inside and outside diameters and the current **FILL** mode.

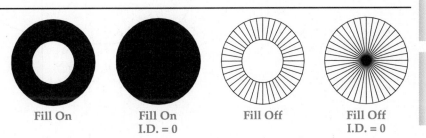

Fill On   Fill On   Fill Off   Fill Off
                 I.D. = 0   I.D. = 0

Figure 4-28.
A spline drawn using the default settings of the **SPLINE** command and three fit points.

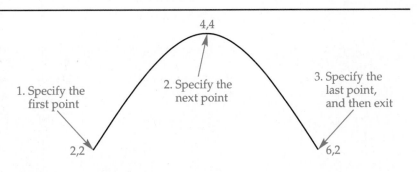

NOTE If you specify only two points for a spline curve and use default settings, an object that looks like a line forms, but the object is a spline.

Drawing Closed Splines

Use the **Close** option after locating at least two control points to connect the last point to the first point. See **Figure 4-29**. The **Close** option forms a smooth curve by creating what is known as a "periodic spline with C2 geometric continuity."

Exercise 4-13

Complete the exercise on the companion website.
www.g-wlearning.com/CAD

Supplemental Material *Spline Options*

For information about options available for drawing splines, go to the companion website (www.g-wlearning.com/CAD), select this chapter, and select **Spline Options**.

Figure 4-29.
Using the **Close** option of the **SPLINE** command with AutoCAD default tangents to draw a closed spline. Compare this spline to the spline shown in Figure 4-28.

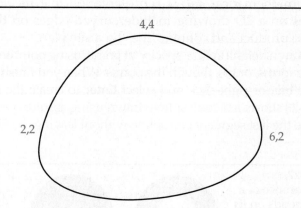

Chapter Review

Answer the following questions. Write your answers on a separate sheet of paper or complete the electronic chapter review on the companion website.
www.g-wlearning.com/CAD

1. When you use the **CIRCLE** command, what are the options for responding to the prompt Specify radius of circle?
2. Explain how to create a circle with a diameter of 2.5 units.
3. What option of the **CIRCLE** command creates a circle of a specific radius that is tangent to two existing objects?
4. Define the term *point of tangency*.
5. Explain how to draw a circle tangent to three objects.
6. Explain the procedure to draw an arc beginning with the center point and having a 60° included angle.
7. Define the term *included angle* as it applies to an arc.
8. What is the default option if you enter the **ARC** command at the keyboard?
9. List three input options that you can use to draw an arc tangent to the endpoint of a previously drawn arc.
10. Name the two axes found on an ellipse.
11. Briefly describe the procedure to draw an ellipse using the **Axis, End** option.
12. What **ELLIPSE** rotation angle results in a circle?
13. How do you draw a filled arrow using the **PLINE** command?
14. Which **PLINE** option allows you to specify the width from the center to one side?
15. Explain how to turn off **FILL** mode.
16. Briefly describe how to create a polyline parallel to a previously drawn line or polyline.
17. Explain how to draw a hexagon measuring 4″ (102 mm) across the flats.
18. Name at least three commands you could use to create a rectangle.
19. Name the command option used to draw rectangles with rounded corners.
20. Name the command option designed for drawing rectangles with a specific line thickness.
21. Explain how to draw a rectangle at an angle other than 0°.
22. Describe a method for drawing a solid circle.
23. Explain how to draw two donuts with an inside diameter of 6.25 and an outside diameter of 9.50.
24. Name the command you can use to create a true spline.
25. How do you accept the AutoCAD defaults for the start and end tangents of a spline?

Drawing Problems

Start AutoCAD if it is not already started. Start a new drawing from scratch or use an appropriate template of your choice. Follow the specific instructions for each problem. Use only drawing commands and techniques you have already learned. Do not draw dimensions or text. Use your own judgment and approximate dimensions when necessary.

▼ Basic

1. Use **LINE**, **CIRCLE**, and **RECTANGLE** commands to draw the objects shown. Save the drawing as P4-1.

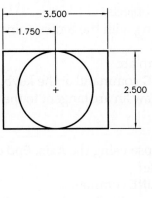

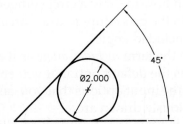

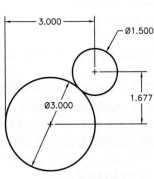

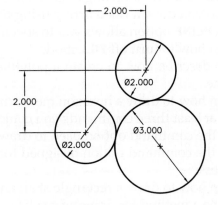

2. Use **CIRCLE** and **ARC** commands to draw the object shown. Save the drawing as P4-2.

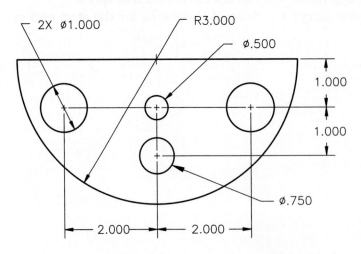

3. Draw the spacer shown. Save the drawing as P4-3.

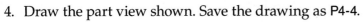
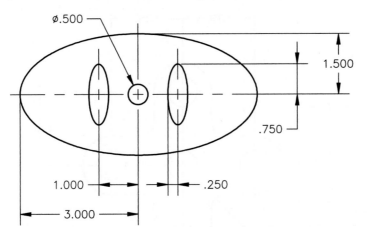

4. Draw the part view shown. Save the drawing as P4-4.

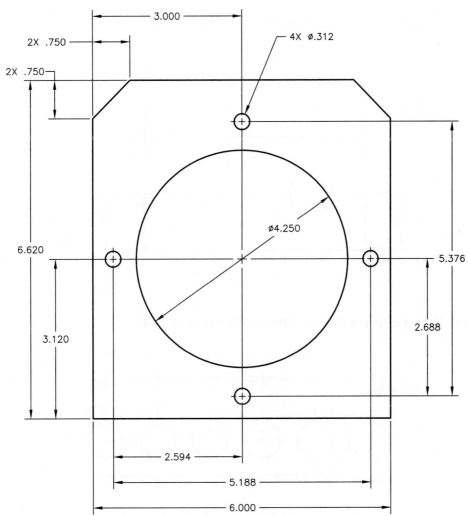

5. Draw the part view shown. Save the drawing as P4-5.

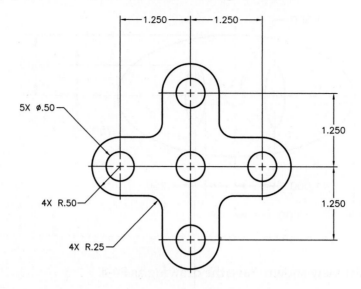

6. Draw the part view shown. Save the drawing as P4-6.

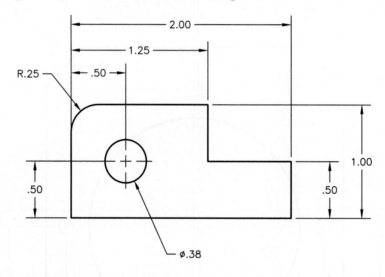

7. Draw the pipe spacer shown. Save the drawing as P4-7.

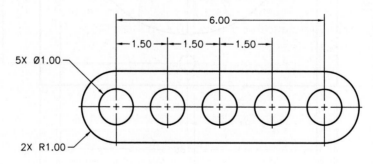

8. Use the **PLINE** command and a .032 width to draw the object shown. Save the drawing as P4-8.

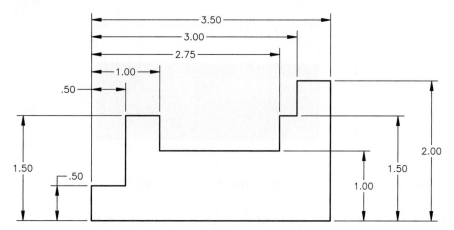

9. Use the **PLINE** command and a .032 width to draw the object shown. Save the drawing as P4-9.

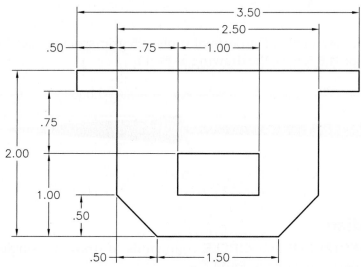

10. Use the **PLINE** command and a .032 width to draw the object shown.
    A. Deactivate solid fills and use the **REGEN** command.
    B. Reactivate solid fills and reissue the **REGEN** command.
    C. Observe the difference with solid fills enabled.
    D. Save the drawing as P4-10.

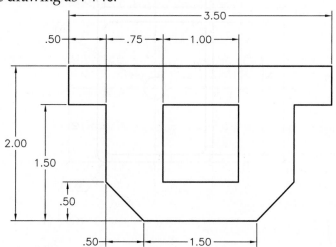

11. Use the **PLINE** command to draw the filled rectangle shown. Save the drawing as P4-11.

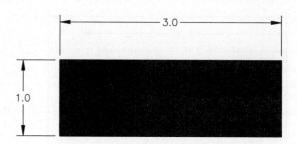

12. Draw the arrowheads shown. Save the drawing as P4-12.

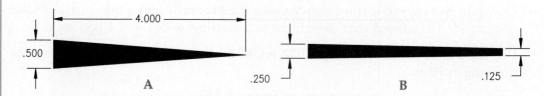

13. Draw the arrow shown. Set decimal units, .25 grid spacing, .0625 snap spacing, and limits of 11,8.5. Save the drawing as P4-13.

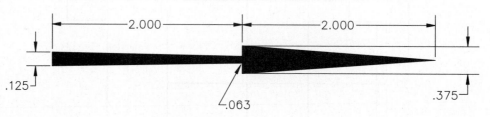

▼ Intermediate

14. Use the **RECTANGLE** and **CIRCLE** commands to draw the single kitchen sink shown. Save the drawing as P4-14.

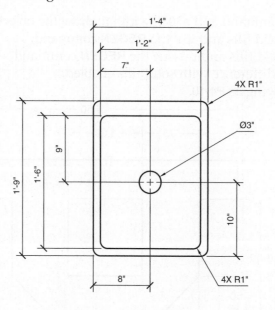

15. Draw the single polyline shown. Use the **Arc**, **Width**, and **Close** options of the **PLINE** command to complete the shape. Set the polyline width to 0, except at the points indicated. Save the drawing as P4-15.

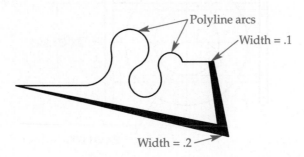

Polyline arcs

Width = .1

Width = .2

16. Draw the two curved arrows shown using the **Arc** and **Width** options of the **PLINE** command. The arrowheads should have a starting width of 1.4 and an ending width of 0. The body of each arrow should have a beginning width of .8 and an ending width of .4. Save the drawing as P4-16.

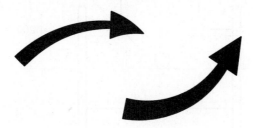

17. Draw the wrench shown. Save the drawing as P4-17.

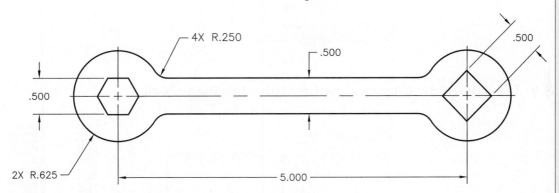

4X R.250

.500

.500

.500

2X R.625

5.000

18. Draw the pipe fitting shown. Save the drawing as P4-18.

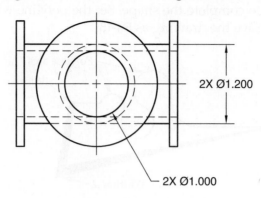

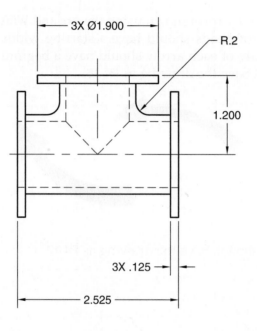

19. Draw the ellipse template shown. Save the drawing as P4-19.

ELLIPSE TABLE		
KEY	MAJOR DIA	MINOR DIA
A	.9951	.5745
B	1.0717	.6187
C	1.1482	.6629
D	1.2247	.7071
E	1.3013	.7513
F	1.3778	.7955
G	1.4544	.8397
H	1.5309	.8839
I	1.6075	.9281
J	1.6840	.9723
K	1.7606	1.0165
L	1.8371	1.0607
M	1.9902	1.1490
N	2.1433	1.2374
O	2.2964	1.3258
P	2.4495	1.4142

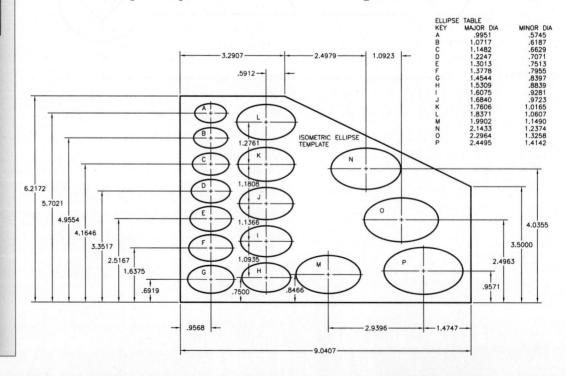

Drawing Problems - Chapter 4

AutoCAD and Its Applications—Basics

20. Draw the gasket shown. Save the drawing as P4-20.

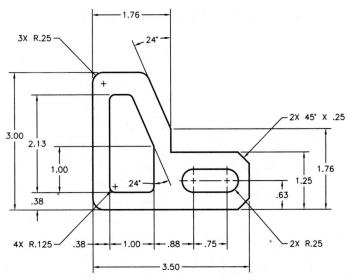

21. Use the **SPLINE** command to draw the curve for the cam displacement diagram. Use the following guidelines and the drawing shown to complete this problem:
   A. The total rise equals 2.000.
   B. The total displacement can be any length.
   C. Divide the total displacement into 30° increments.
   D. Draw a half circle divided into 6 equal parts on one end.
   E. Draw a horizontal line from each division of the half circle to the other end of the diagram.
   F. Draw the displacement curve with the **SPLINE** command by picking points where the horizontal and vertical lines cross.
   G. Label the displacement increments along the horizontal scale as shown. Save the drawing as P4-21.

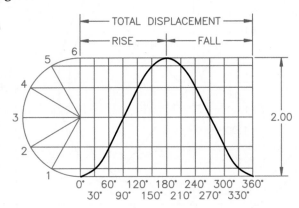

22. Draw the part view shown. Save the drawing as P4-22.

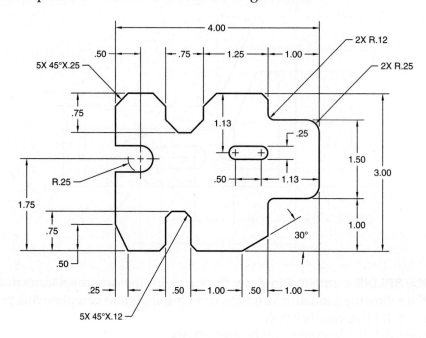

23. Draw the part view shown. Save the drawing as P4-23.

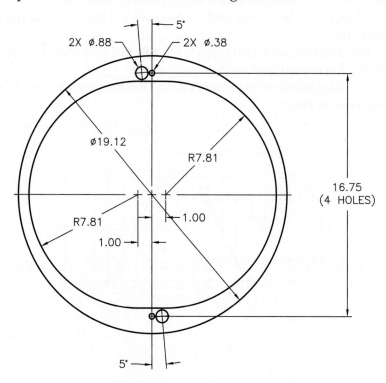

24. Create a drawing from the sketch of a car design shown. Use the **LINE** command and selected shape commands to draw the car using appropriate size and scale features. Use a tape measure to measure an actual car for reference if necessary. Consider the commands and techniques used to draw the car, and try to minimize the number of objects. Save your drawing as P4-24.

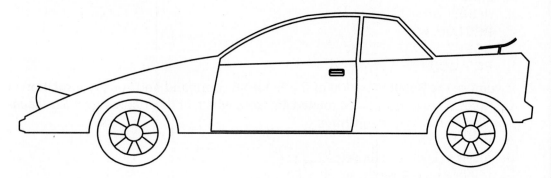

25. Draw the elevation shown using the **ARC**, **ELLIPSE**, **RECTANGLE**, and **DONUT** commands. Draw objects proportionate to the drawing shown. Use dimensions based on your experience, research, and measurements. Save the drawing as P4-25.

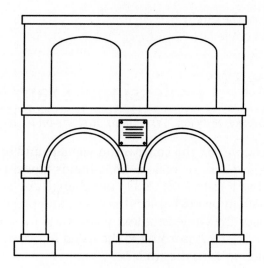

26. Research the design of an existing squeeze bottle with the following specifications: 8-ounce capacity; contaminant-resistant style; integral spout, nozzle, and draw tube are externally molded to the bottle; clear polyethylene material. Hand-draw a dimensioned 2D sketch of the existing design from the manufacturer's specifications, or take measurements from an actual squeeze bottle. Start a new drawing from scratch or use a decimal template of your choice. Draw the squeeze bottle from your sketch. Save the drawing as P4-26.

27. Find a door at your school, company, or home that includes features with several different shapes. Hand-draw a 2D sketch of the door elevation complete with casework and hardware. Use measuring devices such as a tape measure and caliper to dimension the size and location of door features accurately. Start a new drawing from scratch or use an architectural template of your choice. Draw the door from your sketch. Save the drawing as P4-27.

AutoCAD Certified Associate Exam Practice

Answer the following questions. Write your answers on a separate sheet of paper.

1. Which command allows you to draw a rectangle? *Select all that apply.*
   A. **LINE**
   B. **PLINE**
   C. **RECTANGLE**
   D. **DONUT**
   E. **POLYGON**

2. If you use the **Rotation** option of the **ELLIPSE** command and specify a rotation of 45, AutoCAD creates a circle rotated 45° from which of the following? *Select the one item that best answers the question.*
   A. line of sight
   B. major axis of the ellipse
   C. minor axis of the ellipse
   D. X axis
   E. Z axis

3. When you use the **2-Point** option of the **CIRCLE** command to create a circle, which of the following is defined by the points you specify? Select the one item that best answers the question.
   A. radius of the circle
   B. circumference of the circle
   C. diameter of the circle
   D. area of the circle

AutoCAD Certified Professional Exam Practice

Follow the instructions in each problem. Write your answers on a separate sheet of paper.

1. **Navigate to this chapter on the companion website and open CPE-04circle.dwg.** Use the **CIRCLE** command to create a circle that is tangent to both of the lines in the drawing and has a radius of 7.3. Do not change any settings in the drawing file. When you finish, type **LIST**, select the circle, and press [Enter] to display a list of information about the circle in a text window. (Chapter 16 explains the **LIST** command in more detail.) According to the listed information, what are the coordinates of the center point of the circle?

2. **Navigate to this chapter on the companion website and open CPE-04arc.dwg.** Use the **ARC** command to create the arc shown below. Specify the center point and start point of the arc using the **Endpoint** object snap, and specify an angle of 37°. Do not change any settings in the drawing file. Then use the **Endpoint** object snap mode to select the upper endpoint of the arc (Point A). According to the coordinate display field in the lower-left corner of the screen, what are the coordinates of this point?

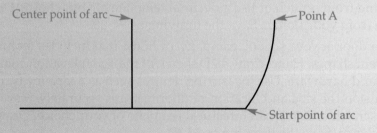

Line Standards and Layers

Learning Objectives

After completing this chapter, you will be able to:

✓ Describe basic line conventions.
✓ Explain the concept of layers in a CAD drawing.
✓ Describe how drawing layers are used in various drafting fields.
✓ Create and manage layers.
✓ Use **DesignCenter** to copy layers and linetypes between drawings.

AutoCAD has a layer system that allows you to organize and assign several properties to objects. *Layers* help you conform to drawing standards and conventions and create various displays, views, and sheets. This chapter introduces line conventions and the AutoCAD layer system. It also introduces **DesignCenter** as a tool for reusing drawing content.

layers: Components of the AutoCAD overlay system that allow you to separate objects into logical groups for formatting and display purposes.

Line Standards

Drafting is a graphic language that uses lines, symbols, and text to describe how to manufacture or construct a product. *Line conventions* provide a way to classify the content of a drawing to enhance readability. Layers allow you to apply line conventions while drawing with AutoCAD. Use line conventions as a guide to develop layers.

The ASME Y14.2 standard, *Line Conventions and Lettering*, recommends two line thicknesses to establish contrasting lines. See **Figure 5-1**. Lines are thick or thin. Thick lines are twice as thick as thin lines. The recommended thicknesses are 0.6 mm for thick lines and 0.3 mm for thin lines. **Figure 5-2** describes the most common linetypes. **Figure 5-3** shows an example of a drawing with several common linetypes.

The U.S. National CAD Standard (NCS) recommends a specific line thickness and characteristics for architectural and similar drawings. Thicknesses range from *extra fine* at 0.13 mm to *4X* at 2 mm. Use the range of NCS-recommended line thicknesses to provide accents to drawings as needed. A common practice is to select a few of the line thicknesses that correlate best to specific applications. See **Figure 5-4**.

line conventions: Standards related to line thickness, type, and purpose,

Figure 5-1.
Line conventions adapted from ASME Y14.2. Thick lines have a 0.6 mm width. Thin lines have a 0.3 mm line width.

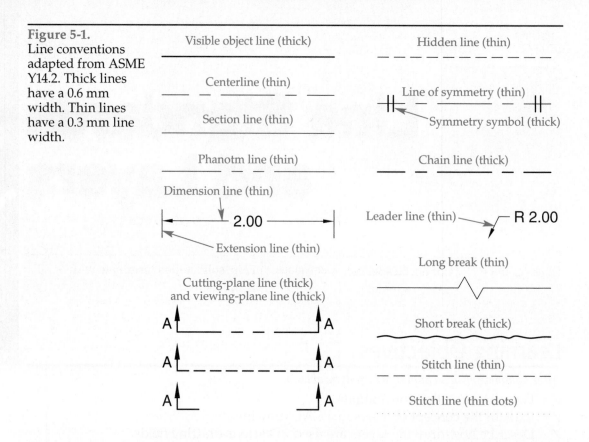

Figure 5-2.
Descriptions of common lines and line standards. Line characteristics and spacing are measured at full scale. Specifications vary according to drawing size.

Type	Purpose	Linetype	Standards
Object lines (visible, outline)	Show the contour or outline of objects.	Continuous	Thick, solid.
Hidden lines	Represent features that are hidden in the current view.	HIDDEN or DASHED	Thin, dashed. Dashes are .125″ (3 mm) long and are spaced .06″ (1.5 mm) apart.
Centerlines	Locate the centers of circles and arcs, and show the axis of cylindrical or symmetrical shapes.	CENTER	Thin. Extend .125″ to .25″ (3 mm to 6 mm) past objects. Centerlines consist of one .125″ dash alternating with one .75″ to 1.5″ (19 mm to 38 mm) dash. A .06″ (1.5 mm) space separates the dashes. Small centerline dashes should cross only at the center of a circle.
Extension lines	Show the extent of a dimension.	Continuous	Thin, solid. Begin .06″ (1.5 mm) from an object and extend .125″ (3 mm) beyond the last dimension line. Can cross object lines, hidden lines, and centerlines, but should not cross dimension lines. Centerlines become extension lines when used to show the extent of a dimension.

(Continued)

Figure 5-2.
(Continued)

Type	Purpose	Linetype	Standards
Dimension lines	Show the distance being measured.	Continuous	Thin, solid. Broken near the center for placement of the dimension numeral in mechanical drafting. Unbroken in architectural and structural drawings, with dimension placed on top of the dimension line. Arrows terminate the ends of dimension lines, except in architectural drafting, where slashes (ticks) or dots are often used.
Leader lines	Connect a specific note to a feature on a drawing.	Continuous	Thin, solid. Often terminate with an arrowhead at the feature. May be curved on architectural drawings. Straight leader lines often have a small shoulder at the note.
Cutting-plane lines	Identify the location and viewing direction of a section view.	PHANTOM or DASHED	Thick. Can be drawn in one of two ways.
Viewing-plane lines	Identify the location of a view	PHANTOM or DASHED	Thick. Can be drawn in one of two ways.
Section lines	In a section view, show where material has been cut away.	(Varies)	Thin, usually drawn in a pattern. Different linetypes can be used to indicate specific or different material.
Break lines	Show where a portion of an object has been removed for clarity or convenience.	Continuous	Thin or thick depending on the symbol, solid. Break representation is based on the object or material being broken.
Phantom lines	Identify repetitive details, show alternate positions of moving parts, and locate adjacent positions of related parts.	PHANTOM	Thin. Two .125″ (3 mm) dashes, alternating with one .75″ to 1.5″ (19 mm to 38 mm) dash. Spaces between dashes are .06″ (1.5 mm).
Chain lines	Indicate special features or unique treatment for a surface.	CENTER	Thick.

Figure 5-3.
An example of a mechanical assembly drawing with several common types of lines.

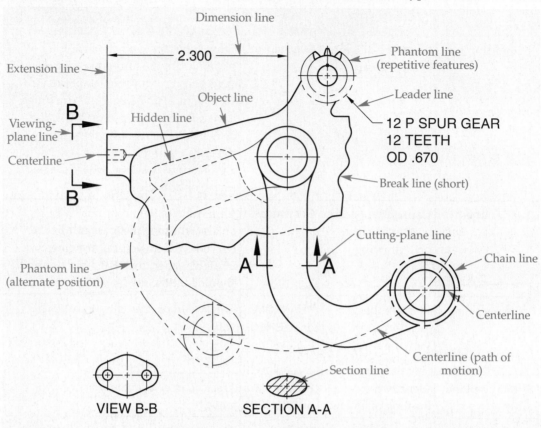

Dimension line

Extension line

2.300

Phantom line (repetitive features)

Leader line

12 P SPUR GEAR
12 TEETH
OD .670

Object line

Viewing-plane line

Hidden line

Centerline

Break line (short)

Cutting-plane line

Chain line

Phantom line (alternate position)

Centerline

VIEW B-B

SECTION A-A

Centerline (path of motion)

Section line

Figure 5-4.
A few of the most common U.S. National CAD Standard line thicknesses.

Thickness	Application
Thin, 0.25 mm	Dimension elements, phantom lines, hidden lines, centerlines, long break lines, schedule grid lines, and background objects.
Medium, 0.35 mm	Object lines, text for dimension values, notes and schedules, terminator marks, door and window elevations, and schedule grid accent lines.
Wide, 0.5 mm	Major object lines at elevation edges, cutting-plane lines, short break lines, title text, minor title underlines, and border lines.
Extra wide, 0.7 mm	Major title underlines, schedule outlines, large titles, special emphasis object lines, elevation and section grade lines, property lines, sheet borders, and schedule borders.

Many AutoCAD tools simplify and automate the process of applying correct line standards. You will learn applications and techniques for drawing specific types of lines in this textbook.

Introduction to Layers

In AutoCAD, you can use an *overlay system* of layers to separate different objects and elements of a drawing. For example, you might choose to draw all object lines on an Object layer and all dimensions on a Dimension layer. You can display both layers to show the complete drawing with dimensions, or hide the Dimension layer to show only the objects. The following is a list of ways you can use layers to increase productivity and add value to a drawing:

- Assign each layer a different color, linetype, and lineweight to correspond to line conventions and to help improve clarity.
- Make changes to layer properties to immediately update all objects drawn on the layer.
- Turn off or freeze selected layers to decrease the amount of information displayed on-screen or to speed screen regeneration.
- Plot each layer in a different color, linetype, or lineweight, or set a layer not to plot at all.
- Use separate layers to group specific information. For example, draw a floor plan using floor plan layers, an electrical plan using electrical layers, and a plumbing plan using plumbing layers.
- Create several sheets from the same drawing file by controlling layer visibility to separate or combine drawing information. For example, use layers to display a floor plan and electrical plan together to send to an electrical contractor, or display a floor plan and plumbing plan together to send to a plumbing contractor.

overlay system: A system of separating drawing components by layer.

Layers Used in Drafting Fields

The type of drawing typically determines the function of each layer. In mechanical drafting, you usually assign a specific layer to each different type of line or object. For example, draw object lines on an Object layer that is black in color, has a solid (Continuous) linetype, and is 0.6 mm wide. Draw hidden lines on a green Hidden layer that uses a 0.3 mm hidden (HIDDEN or DASHED) linetype.

Architectural and civil drawings may require hundreds of layers, each used to produce a specific item. For example, draw full-height floor plan walls on a black A-WALL-FULL layer that has a 0.5 mm solid (Continuous) linetype. Add plumbing fixtures to a floor plan on a blue P-FLOR-FIXT layer that has a 0.35 mm solid (Continuous) linetype.

You can create layers for any type of drawing: detail parts, assemblies, floor plans, foundation plans, partition layouts, plumbing systems, electrical systems, structural systems, roof drainage systems, reflected ceiling systems, HVAC systems, site plans, profiles, topographic maps, and details. Interior designers may use floor plan, interior partition, and furniture layers. Electronics drafters may draw each level of a circuit on its own layer.

Creating and Using Layers

Use the **LAYER** command to open the **Layer Properties Manager**, where you can create and control layers. See **Figure 5-5**. The columns in the list view pane on the right side of the **Layer Properties Manager** list layers and provide layer property controls. Properties in each column appear as an icon or as an icon and a name. See **Figure 5-6**. Pick a property to change the corresponding layer settings. The tree view pane on the left side of the **Layer Properties Manager** displays filters for limiting the number of layers displayed in the list view pane.

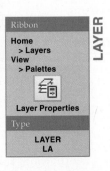

Ribbon

Home
> Layers
View
> Palettes

Layer Properties

Type

LAYER
LA

LAYER

Figure 5-5.
The **Layer Properties Manager**. Layer 0 is the default layer.

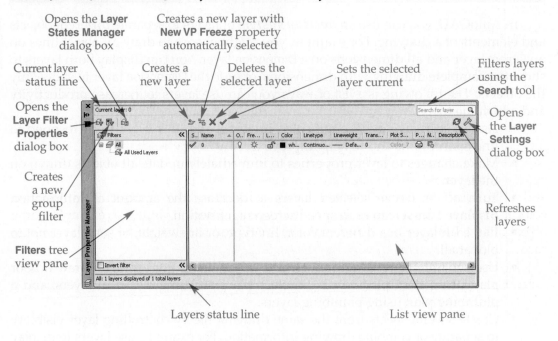

Opens the **Layer States Manager** dialog box

Creates a new layer with **New VP Freeze** property automatically selected

Current layer status line

Creates a new layer

Deletes the selected layer

Sets the selected layer current

Filters layers using the **Search** tool

Opens the **Layer Filter Properties** dialog box

Opens the **Layer Settings** dialog box

Creates a new group filter

Refreshes layers

Filters tree view pane

Layers status line

List view pane

Figure 5-6.
Pick an icon in the **Layer Properties Manager** to change layer settings.

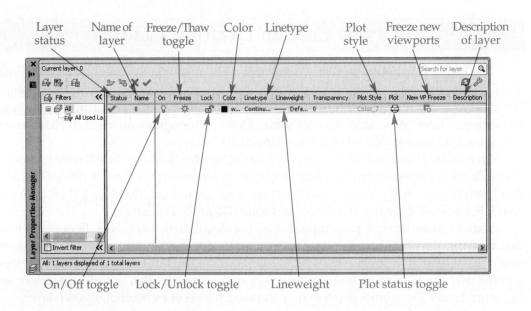

Layer status

Name of layer

Freeze/Thaw toggle

Color

Linetype

Plot style

Freeze new viewports

Description of layer

On/Off toggle

Lock/Unlock toggle

Lineweight

Plot status toggle

The default 0 layer is the only required layer in an AutoCAD drawing. You cannot delete, rename, or purge the 0 layer. However, the 0 layer is primarily reserved for drawing blocks, as described later in this textbook. Draw each object on a layer specific to the object. For example, draw object lines on an Object layer, draw floor plan walls on an A-WALL layer, and draw construction lines on a Construction or A-ANNO-NPLT layer. Make a conscious effort to assign appropriate layers to objects.

Adding Layers

Add layers to a drawing to meet the needs of the current drawing project. To add a new layer, select an existing layer with properties similar to those you want to assign to the new layer. Reference the 0 layer to create the first new layer using a default template. Then pick the **New Layer** button, right-click in the list view and select **New Layer**, or press [Alt]+[N]. A new layer appears, using a default name. The layer name is highlighted, allowing you to type a new name. See **Figure 5-7**. Pick away from the layer in the list or press [Enter] to accept the layer.

Layer Names

Name layers to reflect drawing content. Layer names can include letters, numbers, and certain other characters, including spaces. Layer names are usually set according to specific industry or company standards. **Figure 5-8** provides examples of typical mechanical, architectural, civil, and electronic drafting layer names.

However, simple or generic drawings may use a more basic naming system. For example, the name Continuous-White indicates a layer assigned a continuous linetype and white color. The name Object-7 identifies a layer for drawing object lines, assigned color 7. Another option is to assign the linetype a numerical value. For example, name object lines 1, hidden lines 2, and centerlines 3. If you use this method, keep a written record of the numbering system for reference.

More complex layer names are appropriate for some applications, and may include items such as drawing number, color code, and layer content. For example, the name Dwg100-2-Dimen refers to drawing DWG100, color 2, for use when adding dimensions. The American Institute of Architects (AIA) *CAD Layer Guidelines*, associated with the NCS, specifies a layer naming system for architectural and related drawings. The system uses a highly detailed layer naming process that assigns each layer a discipline designator and major group, and if necessary, one or two minor groups and a status field. The AIA system allows complete identification of drawing content.

Layer names are listed alphanumerically as you create new layers. See **Figure 5-9**. Pick any column heading in the list view to sort layer names in ascending or descending order according to that column. The **Layer Properties Manager** is a palette, so new

---

Figure 5-7.
AutoCAD names a new layer Layer*n* by default and provides the opportunity for you to change the name immediately.

Edit layer name

Figure 5-8.
Examples of typical layer names in common drafting fields.

Mechanical	Architectural	Civil	Electronic
Object	A-WALL-FULL	G-BLDG	Capacitor
Hidden	A-GLAZ	C-WATR	Coil
Center	A-DOOR	C-TOPO	Resistor
Dimension	E-LITE	C-PROP	Diode
Construction	P-FLOR-FIXT	C-NGAS	Transistor
Section	S-FNDN	C-SSWR	Notation
Border	M-HVAC	C-ELEV	Coupling

Figure 5-9.
Layer names are automatically listed in alphanumeric order when you create new layers or
change layer names.

Layer names sort
automatically

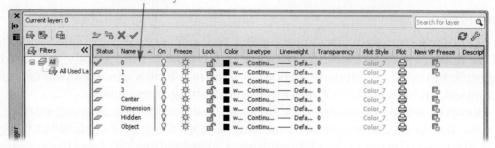

layers and changes made to existing layers are immediately applied to the drawing.
There is no need to "apply" changes or close the palette to see the effects of the layer
changes in the drawing.

PROFESSIONAL
TIP

If you need to create multiple layers, accelerate the process by
pressing the comma key [,] after typing each layer name to create
another new layer.

Renaming Layers

To change a layer name using the **Layer Properties Manager**, slowly double-click
on the existing name in the **Name** column to highlight it. Type the new name and press
[Enter] or pick outside of the text box. You can also rename a layer by picking the name
once to highlight it and then pressing [F2], or by right-clicking and selecting **Rename
Layer**. You cannot rename layer 0 or layers associated with an external reference.

Exercise 5-1

Complete the exercise on the companion website.
www.g-wlearning.com/CAD

Selecting Multiple Layers

Select multiple layers to speed the process of deleting or applying the same properties to several layers. Use standard selection practices or the shortcut menu to select
multiple layers. Hold [Shift] to select several consecutive layers, or hold [Ctrl] to select
several nonconsecutive layers. You can also use a window to select all the layers that
contact the window. The following selection options are available when you right-click
in the list view:
- **Select All**. Selects all layers.
- **Clear All**. Deselects all layers.
- **Select All but Current**. Selects all layers except the current layer.
- **Invert Selection**. Deselects all selected layers and selects all deselected layers.

Exercise 5-2

Complete the exercise on the companion website.
www.g-wlearning.com/CAD

Layer Status

The icon in the **Status** column describes the status, or use of a layer. A green check mark indicates the *current layer*. The status line at the top of the **Layer Properties Manager** also identifies the current layer.

A white sheet of paper, or **Not In Use** icon, in the **Status** column indicates a non-current layer that is not used by the drawing. A blue sheet of paper, or **In Use** icon, in the **Status** column means the layer is assigned to objects, but the layer is not current. The **In Use** icon also means that you cannot delete or purge the layer, even if no objects are assigned to the layer.

current layer: The active layer. Whatever you draw is placed on the current layer.

Current

Not in Use

In Use

> If the **Layer Properties Manager** does not indicate layers in use, pick the **Settings** button in the upper-right corner to display the **Layer Settings** dialog box and select the **Indicate layers in use** check box.

Setting the Current Layer

To set a layer current using the **Layer Properties Manager**, double-click the layer name, pick the layer name and select the **Set Current** button, or right-click on the layer and choose **Set Current**. You can also make a layer current without using the **Layer Properties Manager** by picking the layer name from the **Layer Control** drop-down list of the **Home** ribbon tab. See **Figure 5-10**. Use the vertical scroll bar to move up and down through a long list. The **Layer Control** drop-down list is the most effective way to activate and manage layers while drawing.

Figure 5-10.
The **Layer Control** drop-down list allows you to change the current layer and adjust specific layer properties.

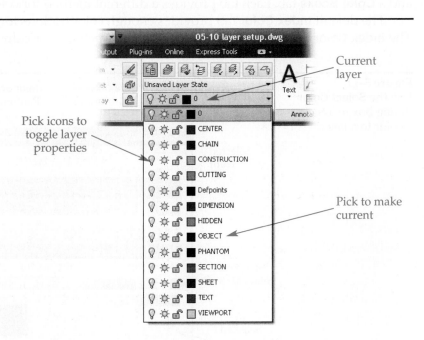

Current layer

Pick icons to toggle layer properties

Pick to make current

You can use the **Layer Properties Manager** or the **Layer Control** drop-down list to change the current layer or layer properties while a command is active. For example, draw a line segment using the current layer, and then, without exiting the **LINE** command, make a different layer current to draw the next line segment on a different layer.

Exercise 5-3

Access the companion website (www.g-wlearning.com/CAD) and complete Exercise 5-3.

Setting Layer Color

You can assign a unique color to each layer to help distinguish objects on-screen. You can also use layer colors to plot a drawing in color or to control object properties. Although plotting in color and controlling object properties using color are not common, assigning colors to layers is very important for on-screen drawing clarity, organization, workability, and format. Layer colors should highlight important features and symbols and not cause eyestrain.

Use the **Color** column of the **Layer Properties Manager** to assign a color to each layer. Pick the existing color swatch to change the color using the **Select Color** dialog box. See **Figure 5-11**. The **Select Color** dialog box includes an **Index Color** tab, a **True Color** tab, and a **Color Books** tab. Each tab provides a different method for color selection.

The default **Index Color** tab provides enough color options for most 2D drawings. The **Index Color** tab includes 255 color swatches, each numerically coded according to

Figure 5-11.
Use the **Select Color** dialog box to assign a color to a layer.

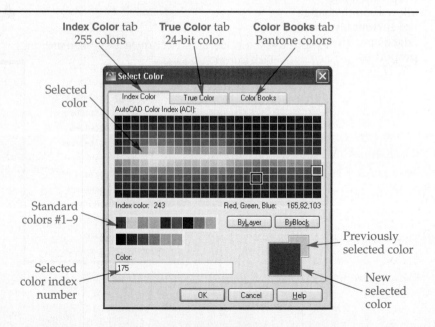

the AutoCAD Color Index (ACI). The first seven colors in the ACI have both a number and a name: 1 = red, 2 = yellow, 3 = green, 4 = cyan, 5 = blue, 6 = magenta, and 7 = white. The color white (number 7) appears white on the default dark-gray model space background. It appears black on the default white layout background.

Hover over a color swatch to display the code in the **Index color:** field, and the mix of red, green, and blue (RGB) used to create the color in the **Red, Green, Blue:** field. Pick a color swatch to select and display the color in the **Color:** text box, or type the color name or ACI number in the **Color:** text box. A preview of the color selection and a sample of the previously assigned color appear in the lower-right corner of the dialog box.

After you select a color, pick the **OK** button to assign the color to the specified layer. The layer color swatch indicates the color. By default, all objects are displayed in the selected layer color, or ByLayer.

 Your graphics card and monitor affect color display characteristics and sometimes the number of available colors.

PROFESSIONAL TIP

 Use the color swatch in the **Layer Control** drop-down list to change the color assigned to a layer without accessing the **Layer Properties Manager**.

CAUTION

 The **COLOR** command provides access to the **Select Color** dialog box, which you can use to set an *absolute value* for color. If you set the absolute color to red, for example, the red color overrides the colors assigned to layers and all objects appear red. For most applications, set the color to the default ByLayer and use layers and the **Layer Properties Manager** to control object color.

absolute value: In property settings, a value set directly instead of referenced by layer or by block. An absolute value overrides the corresponding layer settings.

 Exercise 5-4

Complete the exercise on the companion website.
www.g-wlearning.com/CAD

Setting Layer Linetype

Appropriate linetypes and line thicknesses enhance the readability of a drawing. You can apply standard line conventions to objects by assigning a linetype and thickness to each layer. AutoCAD provides standard linetypes that match or are similar to ASME, ISO, NCS, and other standard linetypes. You can also create custom linetypes. Assign lineweights to layers to achieve different line thicknesses.

Use the **Linetype** column of the **Layer Properties Manager** to assign a linetype to each layer. Pick the existing linetype to change the linetype using the **Select Linetype** dialog box. See **Figure 5-12**. By default, the Continuous linetype is the only linetype in the **Loaded linetypes** list box. Use the Continuous linetype to draw solid lines with no breaks.

Figure 5-12.

The **Select Linetype** dialog box allows you to load linetypes for use in the current drawing.

List of loaded linetypes

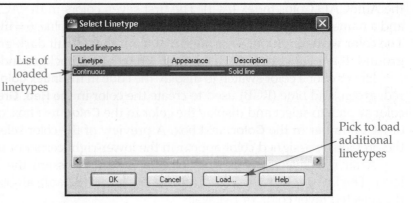

Pick to load additional linetypes

Loading Linetypes

AutoCAD maintains linetypes in external linetype definition files. Before you can apply a linetype other than Continuous to a layer, you must load the linetype into the **Select Linetype** dialog box. Pick the **Load...** button to display the **Load or Reload Linetypes** dialog box. See **Figure 5-13**. The acad.lin or acadiso.lin file is active, depending on the template you use to begin the drawing. The acad.lin and acadiso.lin files are identical except that the acadiso.lin file applies a 25.4 scale factor to non-ISO linetypes to convert inches to millimeters for metric drawings. Pick the **File...** button and use the **Select Linetype File** dialog box to select a different linetype definition file.

The **Available Linetypes** list displays the name and a description and image of each linetype available from the active linetype definition file. Use the scroll bar to view all available linetypes, and use the image in the **Description** column to aid in selecting the appropriate linetypes to load. Choose a single linetype, or select multiple linetypes using standard selection practices or the shortcut menu. Pick the **OK** button to return to the **Select Linetype** dialog box, where the linetypes you selected now appear. See **Figure 5-14**. In the **Select Linetype** dialog box, pick the linetype to assign to the layer, and then pick the **OK** button. **Figure 5-15** shows the HIDDEN linetype selected in **Figure 5-14** assigned to the layer named Hidden.

CAUTION

The **LINETYPE** command provides access to the **Linetype Manager**, which you can use to set an absolute value for linetype. If you set the absolute linetype to HIDDEN, for example, the HIDDEN linetype overrides the linetype assigned to layers and all objects appear in the hidden linetype. For most applications, set linetype to **ByLayer** and use layers and the **Layer Properties Manager** to control object linetype.

Figure 5-13.
The **Load or Reload Linetypes** dialog box displays linetypes available for loading.

Select file where linetype definitions are stored

Select linetypes to load into drawing

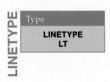

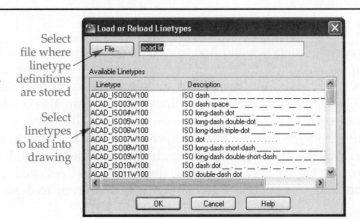

Figure 5-14.
Linetypes loaded
into the drawing
using the **Load or
Reload Linetypes**
dialog box appear in
the **Loaded** linetypes
list box.

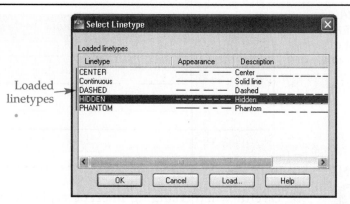

Loaded
linetypes

Figure 5-15.
Objects drawn on the Hidden layer will have the HIDDEN linetype.

Linetype changed
to HIDDEN

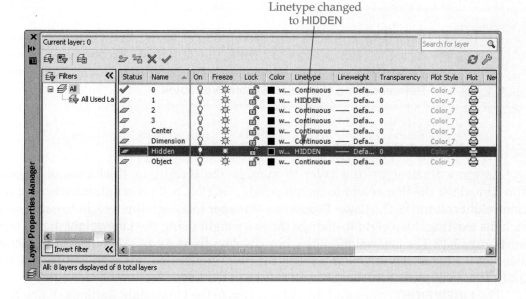

Exercise 5-5

Access the companion website (www.g-wlearning.com/CAD) and complete
Exercise 5-5.

Setting Linetype Scale

You can change the *linetype scale* to adjust the lengths of dashes and spaces in
linetypes to make a drawing more closely match standard drafting practices. Changing
the *global linetype scale* is the preferred method for adjusting linetype scale, but it is
possible to change the linetype scale of specific objects.

Use the **LTSCALE** *system variable* to make a global change to the linetype scale.
Enter **LTSCALE** at the keyboard and then enter a new value. The default global linetype scale factor is 1. A value less than 1 makes dashes and spaces smaller, and a value
greater than 1 makes them larger. See **Figure 5-16.** When you exit the **LTSCALE** system
variable, the drawing regenerates and the global linetype scale changes for all lines on
the drawing. Experiment with different linetype scales until you achieve the desired
results.

linetype scale:
The lengths of
dashes and spaces
in linetypes.

**global linetype
scale:** A linetype
scale applied to
every linetype in the
current drawing.

system variable:
A command that
configures AutoCAD
to accomplish a
specific task or
exhibit a certain
behavior. The value
of each variable
is saved with the
drawing, so the next
time the drawing is
opened, the value
remains the same.

Figure 5-16.
The CENTER
linetype at different
linetype scales.

Scale Factor	Line
0.5	
1.0	
1.5	

CAUTION

Be careful when changing linetype scales to avoid making your drawing look odd and not in accordance with drafting standards.

For a detailed listing and description of AutoCAD system variables, pick the **Help** button to access the **Autodesk Exchange** window. Under the **Product Documentation** heading, pick **Command Reference** and then pick **System Variables**.

Setting Layer Lineweight

lineweight: The assigned width of lines for display and plotting.

Assign a *lineweight* to a layer to manage the weight, or thickness, of objects. You can adjust the lineweight to match ASME, ISO, NCS, or other standards. Use the **Lineweight** column of the **Layer Properties Manager** to assign lineweight to each layer. Pick the existing lineweight to change the lineweight using the **Lineweight** dialog box. See **Figure 5-17**. The **Lineweight** dialog box displays fixed AutoCAD lineweights. Scroll through the **Lineweights:** list and select the lineweight to assign to the layer. Pick the **OK** button to apply the lineweight and return to the **Layer Properties Manager**.

Type
LINEWEIGHT
LWEIGHT
LW

The **LINEWEIGHT** command provides access to the **Lineweight Settings** dialog box, shown in **Figure 5-18**. Use the **Units for Listing** area to set the lineweight thickness to **Millimeters (mm)** or **Inches (in)**. The units apply only to values in the **Lineweight** and **Lineweight Settings** dialog boxes, helping you to select lineweights based on a known unit of measurement.

Figure 5-17.
Use the **Lineweight**
dialog box to assign
a lineweight to a
layer.

Select
lineweight
from list

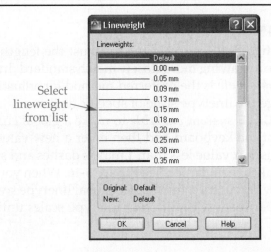

Figure 5-18.
The **Lineweight
Settings** dialog box.

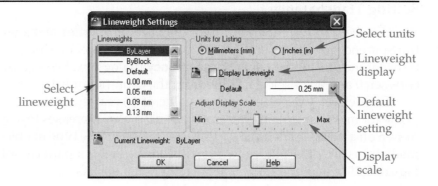

You can also access the **Lineweight Settings** dialog box by right-clicking on the **Show/Hide Lineweight** button on the status bar and selecting **Settings...**.

Check the **Display Lineweight** box to display object lineweight on-screen. Use the **Adjust Display Scale** slider to adjust the lineweight display scale to improve the appearance of lineweights when lineweight display is on. When lineweight display is off, all objects display a 0, or one-pixel, thickness regardless of the lineweight assigned to the layer. You can also toggle screen lineweight on and off using the **Show/Hide Lineweight** button on the status bar.

The Default drop-down list sets the value used when you assign the Default lineweight to a layer. The Default lineweight is an application setting and applies to any drawing you open. It is not template-specific and remains set until you change the value. Do not assign the Default lineweight to layers if you anticipate using a different default lineweight for different drawing applications. Assign a specific lineweight, other than Default, to each layer to maintain flexibility and consistency between drawings.

CAUTION

You can use the **Lineweights** area of the **Lineweight Settings** dialog box to set an absolute value for lineweight. If you set the absolute lineweight to 0.30 mm, for example, the 0.30 mm lineweight overrides the lineweight assigned to layers and all objects appear 0.30 mm thick. For most applications, set lineweight to **ByLayer** and use layers and the **Layer Properties Manager** to control object lineweight.

Exercise 5-6

Complete the exercise on the companion website.
www.g-wlearning.com/CAD

Layer Transparency

ASME standards recommend that all objects be opaque and dark for most applications. However, you can choose to draw transparent, or see-through, objects for specific drawing requirements, usually for architectural, civil, technical illustration, or related applications. For example, draw an existing building using transparent objects to highlight a proposed structure drawn using nontransparent objects.

Setting Transparency

Use the **Transparency** column of the **Layer Properties Manager** to assign a level of transparency to each layer. Pick the existing transparency value to change the level of transparency using the **Layer Transparency** dialog box. See **Figure 5-19**. Type a value between 0 and 90 or select a value from the drop-down list.

The default layer transparency value of 0 creates nontransparent objects, appropriate for most layers. A higher transparency value increases transparency. Any object drawn on a transparent layer appears transparent. One type of object that is commonly made transparent is a hatch, which fills an area with a pattern, solid, or gradient. See **Figure 5-20**. Chapter 23 explains creating hatch objects.

Showing and Hiding Transparency

Show or hide transparency using the **Show/Hide Transparency** button on the status bar. Transparency is on by default, and all objects drawn using a transparent layer appear at their transparent level. Disabling transparency using the **Show/Hide Transparency** button makes all transparent objects appear nontransparent, but does not change the layer transparency property.

Figure 5-19.
Use the **Layer Transparency** dialog box to assign a level of transparency to a layer.

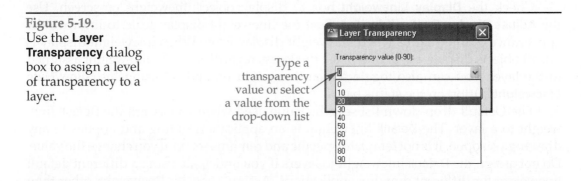

Type a transparency value or select a value from the drop-down list

Figure 5-20.
An example of a portion of a storm water pollution control plan with transparent solid hatch objects that represent different impervious and non-impervious surfaces.

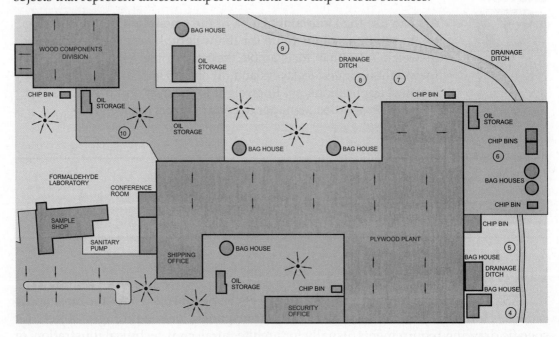

CAUTION

You can override layer transparency for specific objects, but for most applications, set transparency to **ByLayer** and use layers and the **Layer Properties Manager** to control object transparency.

PROFESSIONAL TIP

Layers should simplify and support drafting. Set color, linetype, lineweight, and transparency using layers, and do not override these properties for individual objects. Also, once you establish layers, avoid resetting and mixing color, linetype, lineweight, and transparency, which can lead to confusion and disorder.

Layer Plotting Properties

Use the **Plot Style** column of the **Layer Properties Manager** to assign a named plot style to each layer. AutoCAD uses color-dependent plot styles by default, so the plot style property is disabled. This textbook explains plot styles when appropriate.

Plot   No Plot

Use the **Plot** column of the **Layer Properties Manager** to disable a layer from printing or plotting. Pick the printer, or **Plot**, icon to change it to the **No Plot** icon if the layer should not plot. The layer displays on-screen and is selectable, but does not plot.

Adding a Layer Description

Use the **Description** column of the **Layer Properties Manager** to describe each layer. To add or change a description, slowly double-click on the blank area or existing description, type a description, and press [Enter] or pick outside of the **Description** text box. Another method is to right-click and select **Change Description**.

Turning Layers On and Off

Use the **On** column of the **Layer Properties Manager** to turn a layer on or off. The lit lightbulb, or **On** icon, indicates that the layer is turned on. Objects assigned to an "on" layer display on-screen and can be selected, regenerated, and plotted. Pick the lit lightbulb to turn the layer off, indicated by the gray lightbulb, or **Off** icon. Objects assigned to an "off" layer do not display on-screen and do not plot. You can select and edit objects that are off using advanced selection techniques. Objects that are off can also be regenerated.

On   Off

NOTE

Turn layers on or off using the **Layer Control** drop-down list in the **Layers** panel on the **Home** ribbon tab.

Freezing and Thawing Layers

Use the **Freeze** column of the **Layer Properties Manager** to freeze or thaw a layer. The sun, or **Thaw** icon, indicates a thawed layer. Objects assigned to a thawed layer display on-screen and can be selected, regenerated, and plotted. Pick the sun to freeze the layer, indicated by the snowflake, or **Freeze** icon. Objects assigned to a frozen layer do not display on-screen, and cannot be plotted or regenerated. You cannot select or edit objects that are frozen. Freeze layers to hide objects and ensure that you do not accidentally modify the objects.

Freeze   Thaw

VP Freeze

Use the **New VP Freeze** column to control thawing or freezing of layers when you create a new viewport. Additional layer functions also apply to layouts and viewports. This textbook explains layouts and viewports when appropriate.

VP Thaw

You can also freeze or thaw layers using the **Layer Control** drop-down list in the **Layers** panel of the **Home** ribbon tab. You cannot freeze the current layer or make a frozen layer current.

CAUTION

You cannot modify frozen objects, but you can modify objects that are off. For example, if you turn off layers and use the **All** selection option with the **Erase** command, even the objects assigned to the off layers are erased. However, if you freeze the layers, the objects are not selected or erased.

Locking and Unlocking Layers

Lock   Unlock

Use the **Lock** column of the **Layer Properties Manager** to lock or unlock a layer. The unlocked padlock, or **Unlock** icon, indicates an unlocked layer. Pick the unlocked padlock to lock the layer, indicated by the locked padlock, or **Lock** icon. A **Lock** icon also appears next to the cursor when you hover over an object on a locked layer. Objects assigned to a locked layer display on-screen, and you can use a locked layer to draw new objects. However, you cannot select or edit locked objects. Lock layers to display objects, but eliminate the possibility of selecting the objects.

You can also lock or unlock layers using the **Layer Control** drop-down list in the **Layers** panel on the **Home** ribbon tab.

Locked Layer Fading

By default, all locked layers fade, allowing unlocked layers to stand out on-screen. The quickest way to control locked layer fading is to use the options available in the expanded **Layers** panel of the **Home** ribbon tab. See **Figure 5-21**. Pick the **Locked layer fading** button to allow or disable locked layer fading. Use the **Locked Layer Fading** slider to increase or decrease fading, or type a fading percentage between 0 and 90. The default fade value is 50%. A higher fade value increases fading. **Figure 5-22** shows an example of using locked layer fading on an architectural drawing.

Transparency is a layer property that makes objects transparent. Locked layer fading is a function of the lock state, intended for on-screen drawing purposes only. For example, you can plot transparent objects, but you cannot plot the display created by locked faded layers.

Figure 5-21.
Locked layers fade by default. Increase or decrease fading and enable or disable locked layer fading as needed.

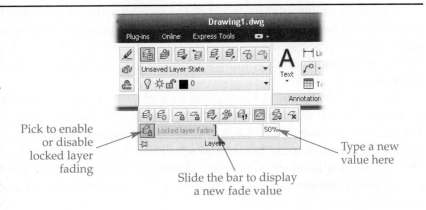

Pick to enable or disable locked layer fading

Slide the bar to display a new fade value

Type a new value here

Figure 5-22.
An example of a floor plan with all layers locked except the A-WALL-FULL layer, which contains the walls. A—Locked layer fading disabled. B—Locked layer fading enabled and set to a fade value of **75**.

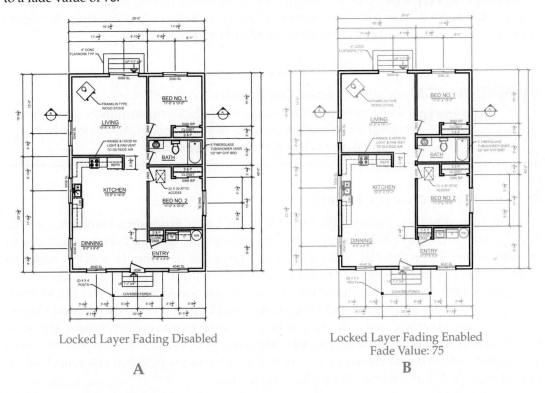

Locked Layer Fading Disabled
A

Locked Layer Fading Enabled
Fade Value: 75
B

Deleting Layers

To delete a layer using the **Layer Properties Manager**, select the layer and pick the **Delete Layer** button, or right-click on the layer and choose **Delete Layer**. You cannot delete or purge the 0 layer, the current layer, layers containing objects, or layers associated with external references.

Adjusting Property Columns

To resize a column in the **Layer Properties Manager**, move the cursor over the column edge to display the resize icon and drag the column width. Maximize column width to show the full heading or longest value in the column. To maximize the width of a specific column, right-click on the heading and select **Maximize column**. To maximize the width of all columns, right-click on any heading and select **Maximize all columns**.

Optimize column width to show the longest value in columns that list properties as text and reduce the width of columns that list properties as icons. To optimize the width of a specific column, right-click on the heading and select **Optimize column**. To optimize the width of all columns, right-click on any heading and select **Optimize all columns**.

By default, a vertical bar appears on the right side of the **Name** column. Any column to the left of the bar is "frozen" to remain in position when you move the scroll bar near the bottom of the **Layer Properties Manager**. Scroll columns to the right of the vertical bar using the horizontal scroll bar. To disable the column freeze function, right-click on a heading and select **Unfreeze column**. Right-click on a heading and select **Freeze column** to turn on the freeze function for every column to the left of the selected column.

To hide a column in the **Layer Properties Manager**, right-click on a heading and deselect the column name. Another option is to right-click on a heading and select **Customize...** to display the **Customize Layer Columns** dialog box. Deselect the check boxes corresponding to the columns to hide. To move a column left or right in the **Layer Properties Manager**, pick a column name and select the **Move Up** or **Move Down** button. Reset the display of all property columns to their default settings by right-clicking on a heading and selecting **Restore all columns to defaults**.

Supplemental Material

Additional Layer Commands

For information about additional layer commands, go to the companion website (www.g-wlearning.com/CAD), select this chapter, and select **Additional Layer Tools**.

Introduction to Layer Filters

The filter tree view pane on the left side of the **Layer Properties Manager** controls *layer filters*. See **Figure 5-23**. Layer filters are appropriate when it becomes difficult to manage a very large number of layers. Filter a large list of layers to make it easier to work with only those layers needed for a specific drawing task.

layer filters:
Settings that screen out, or filter, layers you do not want to display in the list view pane of the **Layer Properties Manager**.

Layer filters are listed in alphabetical order inside the All node. Select the All node to display all layers in the drawing. Pick the **All Used Layers** filter to display only layers used to create objects in the drawing. When you insert external references and save the drawing, an Xref filter node appears, allowing you to filter the display of layers associated with external references. This textbook explains external references when appropriate. You can create custom filters as needed.

Figure 5-23.
Create and restore layer filters using the filter tree view of the **Layer Properties Manager**.

Pick to create a new property filter

Pick to collapse the filter tree view pane

Pick to create a new group filter

Filter tree view pane

Pick to invert filter

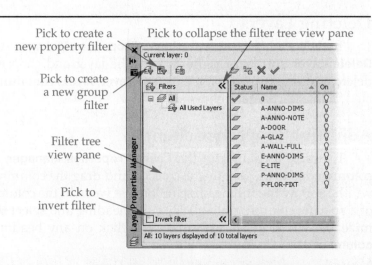

NOTE

Collapse the filter tree view of the **Layer Properties Manager** by picking the **Collapse Layer** filter tree button. To display all filters and layers in the list view, right-click in the layer list area and select **Show Filters in Layer List**.

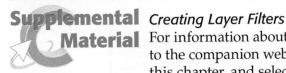
Supplemental Material

Creating Layer Filters

For information about creating and managing layer filters, go to the companion website (www.g-wlearning.com/CAD), select this chapter, and select **Creating Layer Filters**.

Layer States

Once you save a *layer state*, you can readjust layer settings to meet drawing tasks, with the option to restore a saved layer state when needed. For example, a basic architectural drawing might use the layers shown in **Figure 5-23**. You can use the drawing file to prepare a floor plan, a plumbing plan, and an electrical plan. **Figure 5-24** shows the layer settings for each of the three drawings.

Save each of the three groups of settings as an individual layer state. Then restore a layer state to return the layer settings for a specific drawing. This method is easier than changing the settings for each layer individually.

Use the **Layer States Manager**, shown in **Figure 5-25**, to create a new layer state. Pick the **New...** button to display the **New Layer State to Save** dialog box. See **Figure 5-26**. Type a name in the **New layer state name:** text box and a description in the **Description:** text box. Pick the **OK** button to save the new layer state. Once you create a layer state, you can adjust layer properties as needed. **Figure 5-27** describes the areas, options, and buttons available in the **Layer States Manager**.

> **layer state:** A saved setting, or state, of layer properties for all layers in the drawing.

Type
LAYERSTATE

LAYERSTATE

Figure 5-24.
Use layer states to manage the layers needed for several different plans.

Layer	Description	Floor Plan	Plumbing Plan	Electrical Plan
0		Off	Off	Off
A-NNO-DIM	**Floor Plan Dimensions**	On	Frozen	Frozen
A-ANNO-NOTE	**Floor Plan Notes**	On	Frozen	Frozen
A-DOOR	**Doors**	On	Frozen	Locked
A-GLAZ	**Windows**	On	Frozen	Locked
A-WALL-FULL	**Full Height Walls**	On	Locked	Locked
E-ANNO-DIMS	**Electrical Panel Dimensions**	Frozen	Frozen	On
E-LITE	**Electrical Plan Lights**	Frozen	Frozen	On
P-ANNO-DIMS	**Plumbing Plan Dimensions**	Frozen	On	Frozen
P-FLOR-FIXT	**Plumbing Plan Fixtures**	Locked	On	Locked

Figure 5-25.
The **Layer States Manager** allows you to save, restore, and manage layer settings.

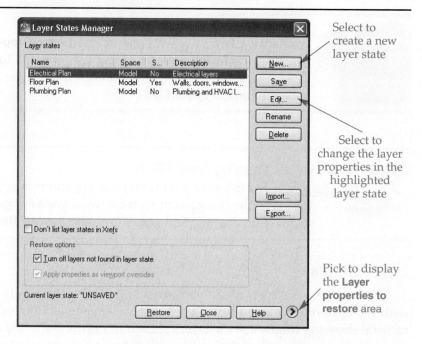

Select to create a new layer state

Select to change the layer properties in the highlighted layer state

Pick to display the **Layer properties to restore** area

Figure 5-26.
Creating a new layer state.

Enter the layer state name

Enter a description for the layer state

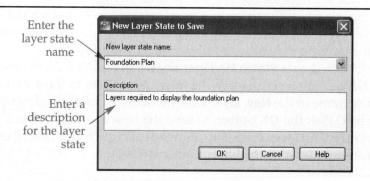

You can also access the **Layer States Manager** by picking the **Layer States Manager** button from the **Layer Properties Manager** or by right-clicking in the layer list view of the **Layer Properties Manager** and selecting **Restore Layer State**. To save a layer state outside the **Layer States Manager**, pick the **New Layer State...** option from the **Layer States** drop-down list in the **Layers** panel on the **Home** ribbon tab.

PROFESSIONAL TIP

Importing a layer state file (.las) to a drawing that does not contain any layers other than 0 adds the layers from the layer state file to the drawing.

After you create a layer state, you can restore layer properties to the settings saved in the layer state at any time. To activate a layer state using the ribbon, select it from the **Layer States** drop-down list in the **Layers** panel on the **Home** ribbon tab. You can also restore a layer state using the **Layer States Manager** by selecting the layer state from the list and picking the **Restore** button.

Figure 5-27.
Layer state options available in the **Layer States Manager**.

Item	Feature
Layer states	Displays saved layer states. The **Name** column provides the name of the layer state. The **Space** column indicates whether the layer state was saved in model space or paper space. The **Same as DWG** column indicates whether the layer state is the same as the current layer properties. The **Description** column lists the layer state description added when the layer state was saved.
Save	Pick to resave and override the selected layer state with the current layer properties.
Edit	Opens the **Edit Layer State** dialog box, where you can adjust the properties of each layer state without exiting the **Layer States Manager**.
Rename	Activates a text box that allows you to rename the current layer state.
Delete	Deletes the selected layer state.
Import	Opens the **Import layer** state dialog box, used to import an LAS file containing an existing layer state into the **Layer States Manager**.
Export	Opens the **Export layer** state dialog box, used to save a layer state as an LAS file. The file can be imported into other drawings, allowing you to share layer states between drawings containing identical layers.
Don't list layer states in Xrefs	Hides layer states associated with external reference drawings. External references are described in Chapter 30.
Restore options	Check the **Turn off layers not found in layer state** check box to turn off new layers or layers removed from a layer state when the layer state is restored. Check **Apply properties as viewport overrides** to apply layer viewport overrides when you are adjusting layer states within a layout.
Layer properties to restore	Check the layer properties that you want to restore when the layer state is restored. Pick the **Select All** button to pick all properties. Pick the **Clear All** button to deselect all properties.

Layer Settings

For information about options available in the **Layer Settings** dialog box, go to the companion website (www.g-wlearning.com/CAD), select this chapter, and select **Layer Settings**.

Reusing Drawing Content

In nearly every drafting discipline, many of the drawings created for a given project may share a number of common elements. All the drawings within a specific drafting project generally have the same set of standards. You often duplicate *drawing content*, such as layers, text and dimension characteristics, symbols, layouts, and details, in many different drawings. One of the most fundamental advantages of CADD is the ease with which you can share content between files. Once you create a common drawing element, you can reuse the item as needed in any number of drawings.

Drawing templates provide one way to reuse drawing content. Customized templates provide an effective way to start each new drawing using standard settings. Another way to reuse drawing content is to seek out data from existing files. This is a common requirement when you develop related drawings for a specific project,

drawing content: All of the objects, settings, and other components that make up a drawing.

or work on similar projects. Sharing drawing content is also common when revising drawings and when duplicating standards used by a consultant, vendor, or client. AutoCAD provides several ways to share drawing content, as described throughout this textbook. One of the most useful tools is **DesignCenter.**

Introduction to DesignCenter

DesignCenter is a palette for managing drawing content between files. See **Figure 5-28.** You can use **DesignCenter** to locate and reuse layers, linetypes, blocks, dimension styles, multileader styles, table styles, text styles, external references, and raster image files. **DesignCenter** allows you to load content from a file without actually opening the file.

DesignCenter floats by default. You can resize, move, auto-hide, or dock the floating **DesignCenter.**

Copying Layers and Linetypes

To copy content using **DesignCenter**, first pick a tab below the **DesignCenter** toolbar to locate a file with drawing content. Pick the **Folders** tab to explore the folders and files found on the hard drive and network, similar to using Windows Explorer. Pick the **Open Drawings** tab to list only drawings that are currently open. The **History** tab lists recently opened drawings and templates. Double-click on a recent file or right-click on the file and choose **Explore** to navigate to the file in the **Folders** tab.

The **Tree View** pane is displayed directly below the tabs by default. If the **Tree View** pane is not visible, toggle it on by picking the **Tree View Toggle** button from the **DesignCenter** toolbar. Use the **Tree View** pane with the **Folders** or **Open Drawings** tab to select a file with the content to reuse. Double-click the file or pick the plus sign (+) to view content categories. Pick **Layers** to load the **Content** pane with layers found in the selected drawing. See **Figure 5-29.**

Use a drag-and-drop operation to import content into the current drawing. Use standard selection practices to select multiple layers. Press and hold down the pick button on the layers to import, and then drag the cursor to the drawing window. See

---

Figure 5-28.
Use the **DesignCenter** to copy content from one drawing to another.

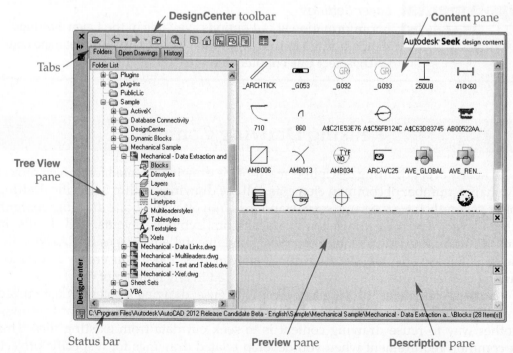

AutoCAD and Its Applications—Basics

Figure 5-29.
Using **DesignCenter** to display the layers found in a drawing.

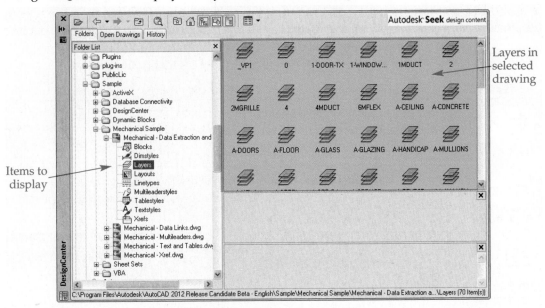

Items to display

Layers in selected drawing

Figure 5-30. Release the pick button to add the layers to the current file. An alternative is to right-click on the layers in the **Content** pane and pick **Add Layer(s)**.

Use **DesignCenter** to copy linetypes from one file to another using the same procedure as copying layers. In the **Tree View** pane, select and expand the file containing the linetypes to copy. Pick the **Linetypes** category to display the linetypes in the **Content** pane. Use drag-and-drop or the shortcut menu to add the linetypes to the current drawing. **Figure 5-31** briefly describes several additional **DesignCenter** features.

Figure 5-30.
To copy layers shown in **DesignCenter** into the current drawing, drag and drop the layers into the drawing area.

Select layer(s) to copy

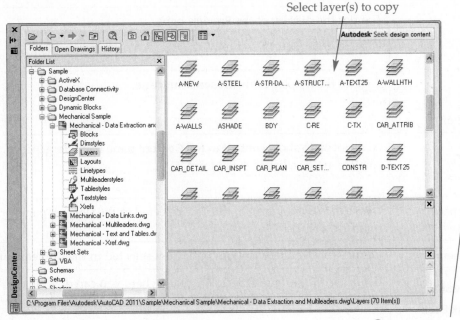

Cursor appearance during drag-and-drop operation

Figure 5-31.
Additional items available in **DesignCenter**.

Item	Function
Load	Displays the **Load** dialog box, which you can use to select a file with drawing content or download online content to **DesignCenter**.
Back	Select to show the content of the last selected file, or choose the flyout to access a list of previously viewed content.
Forward	Pick after using the **Back** button to show the content of the last selected file, or choose the flyout to access a list of previously viewed content.
Up	Moves up one level in the **Folder List** tree to a file, folder, drive, My Computer, or the desktop.
Search	Displays the **Search** dialog box, which you can use to locate drawing content on any drive that meets search criteria and load the content to **DesignCenter**.
Favorites	Navigates to the Autodesk folder in your Favorites folder. To add commonly used content to the Favorites folder, right-click on an item in the **Folder List** area and select **Add to Favorites**.
Home	Navigates to the **DesignCenter** folder. To change the "home" location, right-click on an item in the **Folder List** area and select **Set as Home**.
Tree View Toggle	Toggles the display of the **Tree View** pane.
Preview	Toggles the display of the **Preview** pane.
Description	Toggles the display of the **Description** pane.
Views	Allows you to show content in the **Content** pane using large icons, small icons, or a list or detail format.
Preview pane	Displays a saved image of the item selected in the **Content** pane, typically a block.
Description pane	Displays a saved description of the item selected in the **Content** pane.
Autodesk Seek — **Autodesk Seek**	Launches the Autodesk Seek Web site, where you can download content from contributing manufacturers and community members.

NOTE
You cannot import drawing content if the content uses the same name as existing content. For example, AutoCAD ignores a layer you try to import if the layer name already exists in the destination drawing. The existing settings for the layer are preserved, and a message at the command line indicates that duplicate settings were ignored.

Exercise 5-7
Complete the exercise on the companion website.
www.g-wlearning.com/CAD

Template Development Chapter 5

Adding Layers
For detailed instructions on adding layers to each of your drawing templates, go to the companion website (www.g-wlearning.com/CAD), select this chapter, and select **Template Development**.

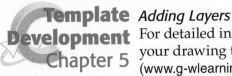

Chapter Review

Answer the following questions. Write your answers on a separate sheet of paper or complete the electronic chapter review on the companion website.
www.g-wlearning.com/CAD

1. Identify the following linetypes:

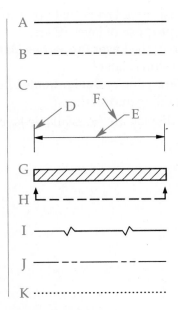

2. How can you tell if a layer is off, thawed, or unlocked by looking at the **Layer Properties Manager**?
3. Should you draw on layer 0? Explain.
4. Identify two ways to access the **Layer Properties Manager**.
5. How can you enter several new layer names consecutively in the **Layer Properties Manager** without using the **New Layer** button?
6. How do you make another layer current using the **Layer Properties Manager**?
7. How do you make another layer current using the ribbon?
8. How can you display the **Select Color** dialog box from the **Layer Properties Manager**?
9. List the seven standard color names and numbers.
10. How do you use the **Layer Properties Manager** to change the linetype assigned to a layer?
11. What is the default linetype in AutoCAD?
12. What condition must exist before you can assign a linetype to a layer?
13. Describe the basic procedure to change layer's linetype to HIDDEN.
14. What is the function of the linetype scale?
15. Explain the effects of using a global linetype scale.
16. Why do you have to be careful when changing linetype scales?
17. What is the state of a layer not displayed on-screen and not calculated by the computer when you regenerate the drawing?
18. Explain the purpose of locking a layer.
19. Explain the difference between a locked layer state and layer transparency.
20. Identify the following layer status icons:

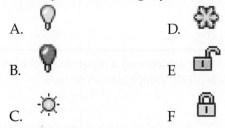

A.　　　　　D.

B.　　　　　E.

C.　　　　　F.

21. Identify at least three layers that you cannot delete from a drawing.
22. Describe the purpose of layer filters.
23. Which button in the **Layer Properties Manager** allows you to save layer settings so they can be restored later?
24. In the **Tree View** pane of **DesignCenter**, how do you view the content categories of one of the listed open drawings?
25. How do you display all the available layers in a drawing in **DesignCenter**?

Drawing Problems

Start AutoCAD if it is not already started. For each problem, start a new drawing from scratch or use an appropriate template of your choice. The template should include layers for drawing the given objects. Add layers as needed. Draw all objects using appropriate layers. Follow the specific instructions for each problem. Use only drawing commands and techniques you have already learned. Do not draw dimensions or text. Use your own judgment and approximate dimensions when necessary.

▼ Basic

1. Draw the hex head bolt pattern shown. Save the drawing as P5-1.

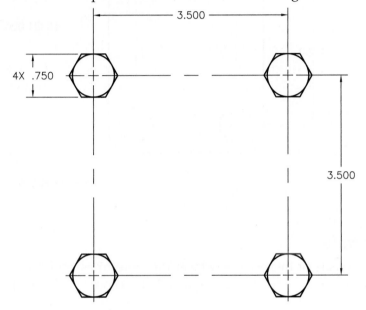

2. Draw the 1/2" hex nut with 3/4" across the flats and a .422" minor diameter. Save the drawing as P5-2.

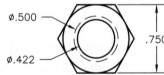

3. Draw the part view shown. Save the drawing as P5-3.

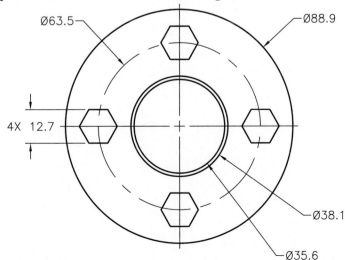

4. Draw the part view shown. Save the drawing as P5-4.

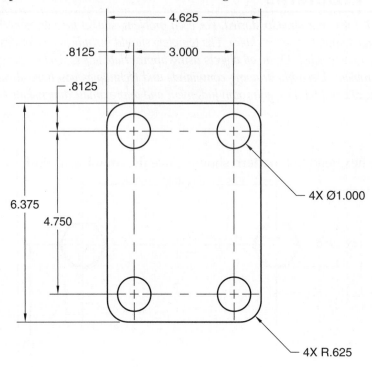

▼ Intermediate

5. Open P4-3, create or import a new layer for centerlines, and draw the centerlines. If you have not yet completed problem 4-3, do so now. Save the drawing as P5-5.

6. Open P4-7, create or import a new layer for centerlines, and draw the centerlines. If you have not yet completed problem 4-7, do so now. Save the drawing as P5-6.

▼ Advanced

7. Open P4-19, create or import a new layer for centerlines, and draw the centerlines. If you have not yet completed problem 4-19, do so now. Change the global linetype scale to achieve an effect similar to the centerlines shown in Chapter 4. Save the drawing as P5-7.

8. Draw the plot plan shown. Use the linetypes shown, which include Continuous, HIDDEN, PHANTOM, CENTER, FENCELINE2, and GAS_LINE. Draw objects proportionate to the drawing shown. Use dimensions based on your experience, research, or measurements. Save the drawing as P5-8.

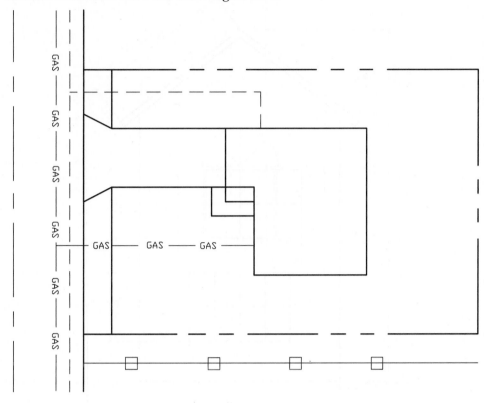

9. Draw the line chart shown. Use the linetypes shown, which include Continuous, HIDDEN, PHANTOM, CENTER, FENCELINE1, and FENCELINE2. Draw objects proportionate to the drawing. Save the drawing as P5-9.

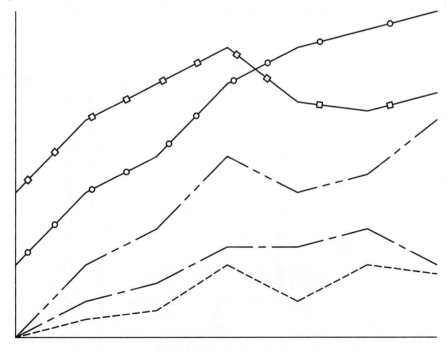

10. Draw the front elevation shown. Draw objects proportionate to the drawing shown. Use dimensions based on your experience, research, and measurements. Save the drawing as P5-10.

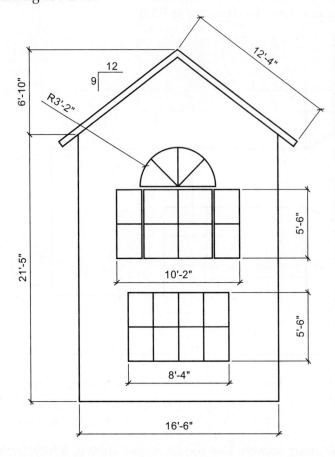

11. Create the controller integrated circuit diagram shown. Use a ruler or scale to keep the proportion as close as possible. Save the drawing as P5-11.

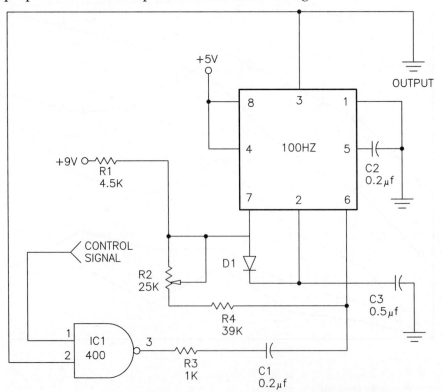

12. Draw a drift boat similar to the drift boat shown. Use dimensions based on your experience, research, or measurements. Use layers to group specific objects. Save the drawing as P5-12.

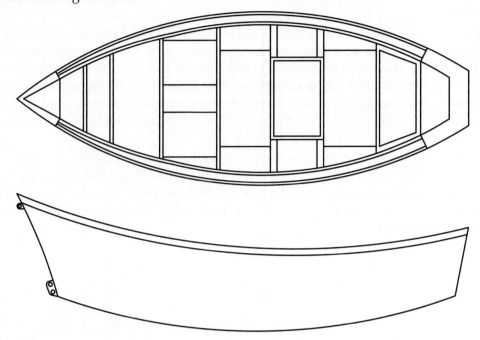

13. Draw a fishing boat similar to the fishing boat shown. Use the overall dimensions shown, and add dimensions based on your experience, research, or measurements. Use layers to group specific objects. Save the drawing as P5-13.

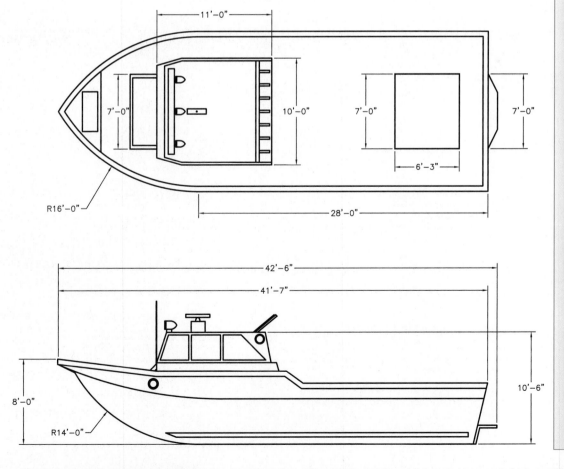

14. Research and write a report of approximately 500 words covering the American Institute of Architects (AIA) *CAD Layer Guidelines*. Include at least three lists of layers applied to specific disciplines, and explain what you would draw on each layer.

15. Research the design of an existing nut driver with the following specifications: fastens 1/2″ hex head screws and bolts, minimum 6″ overall length, solid zinc-plated steel shaft, plastic handle. Hand-draw a dimensioned 2D sketch of the existing design from the manufacturer's specifications, or from measurements taken from an actual nut driver. Start a new drawing from scratch or use a decimal template of your choice. Draw the nut driver from your sketch. Do not dimension the drawing. Save the drawing as P5-15.

16. Create a dimensioned 2D sketch of the floor plan of a kitchen, complete with fixtures and appliances. Use a tape measure to dimension the size and location of kitchen features accurately. Start a new drawing from scratch or use an architectural template of your choice. Draw the kitchen from your sketch. Save the drawing as P5-16.

Drawing Problems - Chapter 5

AutoCAD Certified Associate Exam Practice

Answer the following questions. Write your answers on a separate sheet of paper.

1. Which line widths are recommended by ASME Y14.2? *Select all that apply.*
   A. extra fine
   B. thin
   C. thick
   D. 3X
   E. 4X

2. Which of the following line characteristics would you typically assign to a layer for drawing object lines on a mechanical part view drawing? *Select the one item that best answers the question.*
   A. thick, HIDDEN linetype
   B. 4X, Continuous linetype
   C. extra fine, Continuous linetype
   D. 0.3 mm, Continuous linetype
   E. 0.6 mm, Continuous linetype

3. Which of the following operations are possible on a locked layer? *Select all that apply.*
   A. drawing new objects
   B. editing objects
   C. erasing objects
   D. selecting objects
   E. displaying objects on-screen

AutoCAD Certified Professional Exam Practice

Follow the instructions in each problem. Write your answers on a separate sheet of paper.

1. **Navigate to this chapter on the companion website and open CPE-05linetype.dwg.** Create a new layer called Centerline and assign it the color blue. Load the CENTER linetype and apply it to the Centerline layer. Then select the existing line and change it to the Centerline layer. Enter **LTSCALE** and specify a linetype scale of 2.0. The centerline linetype consists of long dashes alternating with short dashes. How many short dashes appear on the line at the current linetype scale?

2. **Navigate to this chapter on the companion website and open CPE-05layers.dwg.** Use the **Layer Control** drop-down list and the **Layer Properties Manager** to find the following information: On which layer was Line A drawn, and what line-weight is assigned to that layer?

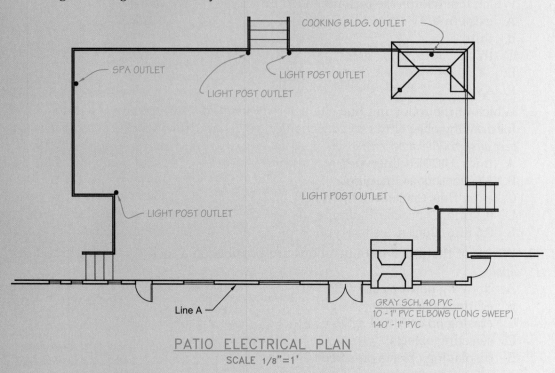

COOKING BLDG. OUTLET

SPA OUTLET

LIGHT POST OUTLET

LIGHT POST OUTLET

LIGHT POST OUTLET

LIGHT POST OUTLET

Line A

GRAY SCH. 40 PVC
10 - 1" PVC ELBOWS (LONG SWEEP)
140' - 1" PVC

PATIO ELECTRICAL PLAN
SCALE 1/8"=1'

Chapter 6

View Tools and Basic Plotting

Learning Objectives

After completing this chapter, you will be able to:

✓ Adjust the display window to view specific portions of a drawing.
✓ Use display commands transparently.
✓ Control object display order.
✓ Create named views for instant recall.
✓ Create multiple viewports in the drawing window.
✓ Print and plot drawings.

As you create drawings that are more complex and begin drawing both large and small objects, you will realize the importance of using view tools to adjust the drawing display. This chapter describes several view commands that you will use frequently during the drawing process. This chapter also introduces printing and plotting so that you can begin printing your drawings. Later chapters describe printing and plotting in more detail.

View Tools

View tools are available to help you navigate in drawings. They adjust the display without changing the objects in a drawing. View tools are available from the ribbon, keyboard entry, shortcut menus, in-canvas viewport controls, the navigation bar, SteeringWheels, and the ViewCube.

view tools: The various commands, options, and displays available in AutoCAD.

Navigation Bar

The navigation bar is a view toolbar positioned near the upper-right corner of the drawing window by default. See **Figure 6-1**. The navigation bar includes tool buttons for accessing common view commands associated with the current work environment. Right-click on a tool button or pick the **Settings** flyout to access options for changing the navigation bar.

A command or option accessible from the navigation bar appears as a graphic in the margin of this textbook. The graphic indicates picking a navigation bar button. The example shown in this margin illustrates accessing the **PAN** command from the navigation bar.

Ribbon
View
> Windows
> User Interface

Navigation Bar

Type
NAVBAR

Navigation Bar

Pan

179

Figure 6-1.
The default navigation bar in model space. The commands available from the navigation bar change when you enter a different work environment.

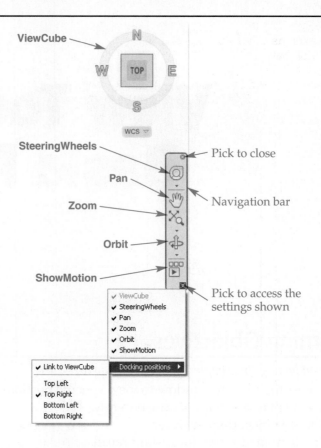

Some of the view commands available in the default **Drafting & Annotation** workspace are appropriate for 3D applications only. *AutoCAD and Its Applications—Advanced* explains 3D and animation view commands.

Zooming

zoom: Make objects appear bigger (zoom in) or smaller (zoom out) on the screen without affecting their actual size.

AutoCAD provides several methods for *zooming*. As described in Chapter 3, the easiest way to zoom is to use a mouse with a scroll wheel. To *zoom out*, roll the mouse wheel forward. To *zoom in*, roll the mouse wheel back. This technique also pans to the location of the crosshairs while zooming. Double-click the wheel to zoom to the furthest extents of objects in the drawing.

zoom out: Change the display area to show a larger part of the drawing at a lower magnification.

Additional zooming options are available from the **ZOOM** command. The ribbon and navigation bar provide effective ways to access a **ZOOM** command option, because they provide a direct link to the individual options. See **Figure 6-2**. In contrast, when you issue the **ZOOM** command using dynamic input or the command line, you must choose a specific option to receive appropriate prompts.

zoom in: Change the display area to show a smaller part of the drawing at a higher magnification.

Figure 6-2.
ZOOM options in the **Zoom** flyout on the **Navigate** panel of the **View** ribbon tab and on the navigation bar.

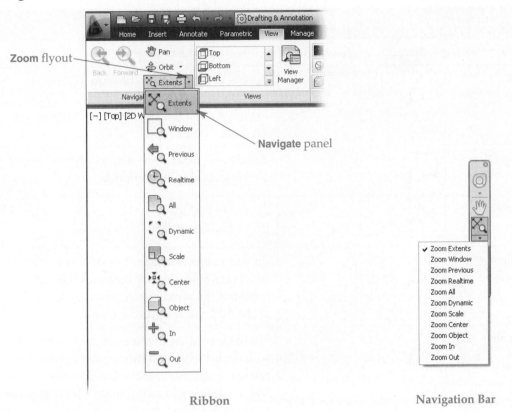

Ribbon   Navigation Bar

Realtime Zooming

Access the **Realtime** option to apply *realtime zooming*. The zoom cursor appears as a magnifying glass with plus and minus symbols. Press and hold the left mouse button and move the cursor up to zoom in or down to zoom out. Release the mouse button when you achieve the appropriate display. The **Realtime** option includes a shortcut menu that you can access while the zoom cursor is active. The menu provides a quick way to select an alternative view option, as described later in this chapter. When you are finished zooming, press [Esc], [Enter], or the space bar, or right-click and pick **Exit**.

realtime zoom: A zoom that you view as it occurs.

Additional Zoom Options

The mouse wheel and the **Realtime** option of the **ZOOM** command are effective for most zooming requirements. Apply one of the other zoom options to achieve a specific display. **Figure 6-3** provides a brief description of the most useful additional zoom options.

The **Previous, In, Out, Center**, and **Dynamic** options of the **ZOOM** command provide functions that you can achieve more easily using other viewing methods.

Exercise 6-1

Complete the exercise on the companion website.
www.g-wlearning.com/CAD

Figure 6-3.
Zooming options available from the ribbon, the navigation bar, and the shortcut menu that appears when you right-click while using the **Realtime** option of the **ZOOM** command.

Option	Button	Cursor	Function
The following options are available from the ribbon and the navigation bar.			
All			Zooms to the edges of the drawing limits, or to the edge of the geometry drawn past the limits. Use this option after changing the drawing limits.
Object			Zooms and centers the display on objects you select in the drawing window.
Extents*			Zooms to include all objects in the drawing (the "drawing extents").
Window*			Zoom to objects inside a window you create. Press and hold the pick button at the first corner of the window, and release the button at the diagonally opposite corner.
Scale			Zooms according to a specified magnification scale factor. The **nX** option scales the display relative to the current display. The **nXP** option scales a drawing in model space relative to paper space, as described later in this textbook.
The following options are available from the **Realtime** shortcut menu.			
Pan			Adjust the placement of the drawing on-screen using the **Realtime** option of the **PAN** tool.
Zoom			Return from the **Realtime** option of the **PAN** tool to the **Realtime** option of the **ZOOM** tool.
3D Orbit			View a 3D object as described in *AutoCAD and Its Applications—Advanced*.
Zoom Original		(No cursor)	Restores the previous display before any realtime zooming or panning occurred; useful if the modified display is not appropriate.

*Also available from the **RealTime shortcut menu.**

Panning

The easiest way to *pan* is to use a mouse with a scroll wheel. Press and hold the mouse wheel and then move the mouse. An alternative is to access the **PAN** command to apply *realtime panning*. The pan cursor appears as a hand. Press and hold the left mouse button and move the cursor in the direction to pan. Right-click to display the same shortcut menu available for realtime zooming. To exit realtime panning, press [Esc], [Enter], or the space bar, or right-click and pick **Exit**.

Another, but less effective, way to pan involves using drawing window scroll bars. To display the scroll bars, check **Display scroll bars** in drawing window in the **Window Elements** area of the **Display** tab in the **Options** dialog box.

pan: Change the drawing display so that different portions of the drawing are visible on-screen.

realtime panning: A panning operation in which you can see the drawing move on-screen as you pan.

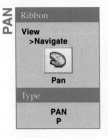

PAN

Ribbon
View
>Navigate

Pan

Type
PAN
P

AutoCAD supports some high-performance mice that include additional buttons and functions.

By default, when you use the **U** command after multiple zooming and panning operations, zooms and pans are grouped together, allowing you to return to the original view. To make each zoom and pan operation count individually, deselect the **Combine zoom and pan commands** check box in the **Undo/Redo** area of the **User Preferences** tab in the **Options** dialog box.

View Back and Forward

Issue the **VIEWBACK** command as needed to cycle back to prior views created by an operation such as zooming or panning. The **VIEWFORWARD** command is available after you issue the **VIEWBACK** command, and can be used to return to the previous view. The **VIEWBACK** and **VIEWFORWARD** commands are most effective in basic 2D design for working between specific displays quickly, such as viewing fine detail while zoomed in and then returning to the previous view of the drawing extents to view the overall design.

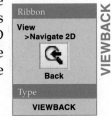

Exercise 6-2

Complete the exercise on the companion website.
www.g-wlearning.com/CAD

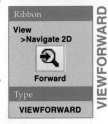

Introduction to SteeringWheels

AutoCAD SteeringWheels provide an alternate way to access and use certain view commands. A *navigation wheel* is a specific SteeringWheel. Some navigation wheels and many of the commands available from navigation wheels are most appropriate for 3D modeling. This textbook focuses on the default navigation wheels and the **ZOOM**, **CENTER**, **PAN**, and **REWIND** commands for 2D applications. *AutoCAD and Its Applications— Advanced* explains additional navigation wheel commands and settings.

The **Full Navigation Wheel** appears by default in model space, and the UCS icon changes to a 3D display. The **2D Navigation Wheel** is the only SteeringWheel available in layout space. See **Figure 6-4**. To access a different navigation wheel, select from the **SteeringWheels** flyout on the navigation bar, the shortcut menu that appears when you right-click while using a navigation wheel, or the flyout in the lower-right corner of a navigation wheel. The default format is big, but mini navigation wheels are available.

Navigation wheels display next to the cursor. *Wedges* contain the individual navigation commands and function similar to tool buttons. Hover over a wedge to highlight it. Some commands activate when you pick a wedge, and other commands require that you hold down the left mouse button on the wedge.

A navigation wheel remains on-screen until you close it. This allows you to use multiple navigation commands. To close a navigation wheel, pick the **Close** button in the upper-right corner of the wheel, press [Esc], [Enter], or the space bar, or right-click and pick **Close Wheel**.

navigation wheel: A SteeringWheel designed for use in a specific drawing setting or with a particular type of drawing.

wedges: The parts of a navigation wheel that contain navigation commands.

Figure 6-4.
The **Full Navigation Wheel** appears by default in model space. The **2D Wheel** is the only navigation wheel available in a layout. You can also display the **2D Wheel** in model space.

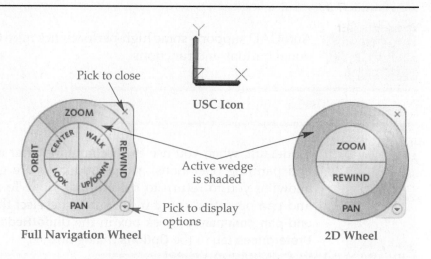

Pick to close

USC Icon

Active wedge is shaded

Pick to display options

Full Navigation Wheel

2D Wheel

The flyout in the lower-right corner of a navigation wheel provides access to other navigation wheels and SteeringWheel settings. The flyout also includes a **Fit to Window** option that zooms and pans to show all objects centered in the drawing window, and a **Close Wheel** option.

Zooming with the Navigation Wheel

The **ZOOM** navigation command offers realtime zooming. Press and hold the left mouse button on the **ZOOM** wedge on the **Full Navigation Wheel** to display the pivot point icon and zoom navigation cursor. See **Figure 6-5.** The pivot point is the location at which you press the **ZOOM** wedge. Move the zoom navigation cursor up or to the right to zoom in. Move the cursor down or to the left to zoom out. The pivot point icon also zooms in or out as a visual aid to zooming. Release the left mouse button when you achieve the appropriate display.

Using the Center Navigation Command

The **CENTER** navigation command centers the display screen at a picked point, without zooming. Press and hold the left mouse button on the **CENTER** wedge. The pivot point icon appears when you move the cursor over an object. Release the mouse button to pan so the object at the pivot point is displayed in the center of the drawing window when you release the mouse button. See **Figure 6-6.**

Figure 6-5.
To use the **ZOOM** navigation command, hold down the left mouse button and move the cursor up or to the right to zoom in, or down or to the left to zoom out.

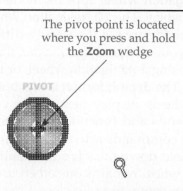

The pivot point is located where you press and hold the **Zoom** wedge

PIVOT

Zoom Tool

Figure 6-6.
To use the **CENTER** navigation command, hold down the left mouse button, move the cursor over an object at the point you want to center in the drawing area, and release the mouse button.

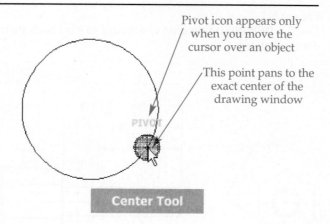

Pivot icon appears only when you move the cursor over an object

This point pans to the exact center of the drawing window

PIVOT

Center Tool

Panning with the Navigation Wheel

The **PAN** navigation command offers realtime panning. Press and hold the left mouse button on the **PAN** wedge to display the pan navigation cursor. Move the pan navigation cursor in the direction to pan. Release the mouse button when you achieve the desired display.

Rewinding

The **REWIND** navigation command allows you to observe display changes and return to a previous display. For example, if you use the navigation wheel to zoom in, then pan, then zoom out, you can rewind through each action and return to the original zoomed-in display, the panned display, and then back to the current zoomed-out display. You can rewind through view actions created using most view commands, not just navigation wheel commands.

Pick the **REWIND** command once to return to the previous display. Thumbnail images appear in frames as the previous view is restored. The thumbnail with an orange frame surrounded by brackets indicates the restored display and its location in the sequence of events. See **Figure 6-7**. Pick the **REWIND** button repeatedly to cycle back through prior views. Another option is to press and hold the left mouse button on the **REWIND** wedge to display the framed view thumbnails. Then, while still holding the left mouse button, move the brackets left over the thumbnails to cycle through earlier views, and right to return to later views. Release the mouse button when you achieve the desired display.

If you access and use a view command from a source other than the navigation wheel, such as the ribbon, a rewind icon appears in place of the thumbnail. A thumbnail displays as you move the brackets over the rewind icon.

 Exercise 6-3

Complete the exercise on the companion website.
www.g-wlearning.com/CAD

Figure 6-7.
Use the **REWIND** navigation command to step back through and restore previous display configurations.

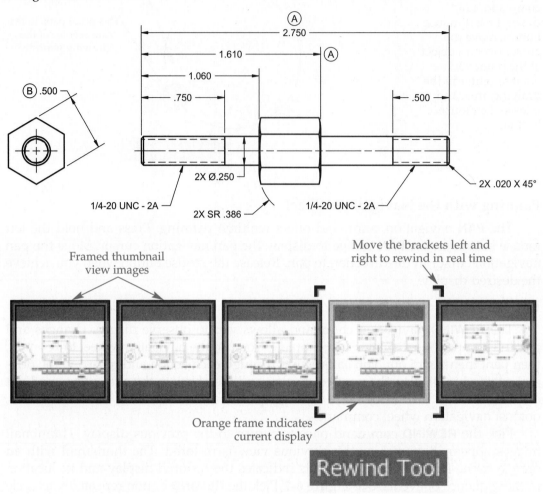

Framed thumbnail view images

Move the brackets left and right to rewind in real time

Orange frame indicates current display

Rewind Tool

Introduction to the ViewCube

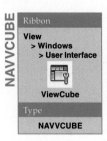
The ViewCube appears near the upper-right corner of the drawing window by default. See **Figure 6-8**. The ViewCube includes a WCS menu that is useful for some 2D drawing applications. However, the primary purpose of the ViewCube is to view a 3D model precisely using a labeled cube. Until you are ready to create 3D models, use the **NAVVCUBE** command to turn off the ViewCube. If you change the view orientation, pick the **Home** button and then the **TOP** face to return to the default 2D view. *AutoCAD and Its Applications—Advanced* provides complete information on the ViewCube.

CAUTION

Rotating the display of the 2D drawing plane using a tool such as the ViewCube is inappropriate for most 2D applications. The coordinate system may not behave as expected. For example, text, which should always be horizontal, may appear at an unacceptable angle. Use the **ROTATE** command to rotate objects as needed in 2D drawings, not just the display.

Figure 6-8.
This is how the
ViewCube should
look when you
create a 2D drawing.

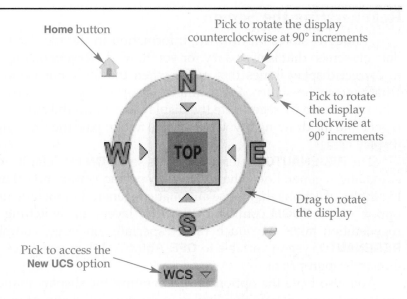

Home button

Pick to rotate the display
counterclockwise at 90° increments

Pick to rotate
the display
clockwise at
90° increments

Drag to rotate
the display

Pick to access the
New UCS option

Introduction to In-Canvas Viewport Controls

In-canvas viewport controls appear near the upper-left corner of the model space drawing window by default. See **Figure 6-9**. Pick a selection to display a menu of related options. The Viewport Controls menu provides options to toggle the display of the navigation bar, SteeringWheels, and ViewCube on and off.

The **Viewport Controls** menu also offers convenient access to tiled viewport controls, as explained later in this chapter. The **View Controls** menu contains options that are only appropriate for viewing a 3D model, as explained in *AutoCAD and Its Applications—Advanced*. The **View Controls** menu also includes an option for accessing the **View Manager** described later in this chapter. The **Visual Style Controls** menu lists options typically associated with 3D model display, as explained in *AutoCAD and Its Applications—Advanced*.

Use the check boxes in the **Display Tools in Viewport** area of the **3D Modeling** tab in the **Options** dialog box to control the display of in-canvas viewport controls.

Figure 6-9.
The in-canvas
viewport controls
in the model space
drawing window,
Viewport Controls
option selected.

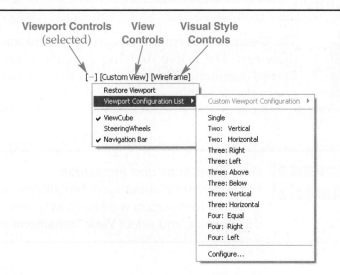

Viewport Controls
(selected)

View
Controls

Visual Style
Controls

[–] [Custom View] [Wireframe]

Restore Viewport
Viewport Configuration List ▶   Custom Viewport Configuration ▶

✔ ViewCube   Single
  SteeringWheels   Two: Vertical
✔ Navigation Bar   Two: Horizontal
                                    Three: Right
                                    Three: Left
                                    Three: Above
                                    Three: Below
                                    Three: Vertical
                                    Three: Horizontal
                                    Four: Equal
                                    Four: Right
                                    Four: Left

                                    Configure...

Regenerating the Screen

AutoCAD stores all drawing information in the file database, but only displays data on-screen that is necessary for you to work. *Regenerating* evaluates the drawing to correct display issues that result when the file database does not match exactly what you see on-screen. For example, if curved objects appear as straight segments when you zoom in, regenerate the display to smooth the curves. You may also need to regenerate the drawing if you are unable to pan past the drawing limits or the current display area.

The **REGENAUTO** system variable is set to **ON** by default. AutoCAD performs an automatic regeneration when you use specific commands that change the display. Examples of actions that cause automatic regeneration include using the **All** or **Extents** option of the **ZOOM** command, thawing layers, or switching layouts. If automatic regeneration takes too much time, especially on large, complex drawings, set the **REGENAUTO** system variable to **OFF**. AutoCAD then generates prompts when regeneration is appropriate.

You also have the option of regenerating the display manually. Use the **REGEN** command to regenerate the display in the current viewport only. Use the **REGENALL** command to regenerate the display in all viewports.

Type
REGEN

Type
REGENALL

Redrawing is different from regenerating. Redrawing is an old function that was important when computers and graphics were slower and less advanced. The **REDRAW** command redraws the display of the current viewport only. The **REDRAWALL** command redraws the display in all viewports. Neither of these commands recalculate the display based on the file database.

Cleaning the Screen

Type
[Ctrl]+[0] (zero)

The AutoCAD window can become crowded with multiple interface items, such as palettes, in the course of a drawing session. As the drawing area gets smaller, less of the drawing is visible, and drafting becomes more difficult. Use the **Clean Screen** command to clear the AutoCAD window of all palettes, toolbars, and title bars. See Figure 6-10. An easy way to toggle the **Clean Screen** command is to pick the **Clean Screen** button on the far right side of the status bar.

PROFESSIONAL TIP

The **Clean Screen** command can be helpful for displaying multiple drawings. The active drawing appears when you use the **Clean Screen** command. This allows you to work more efficiently within one of the drawings.

Supplemental Material

View Transitions and Resolution

For information about view transitions and view resolution, go to the companion website (www.g-wlearning.com/CAD), select this chapter, and select **View Transitions and Resolution**.

Figure 6-10.
Using the **Clean Screen** command.
A—Initial display with the **Properties** palette and **Layer Properties Manager** displayed.
B—Display after using the **Clean Screen** command.

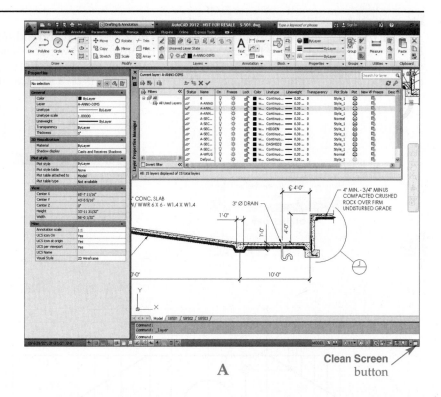

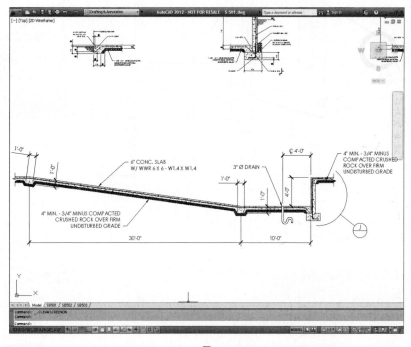

Clean Screen button

A

B

Isolating or Hiding Objects

Layer states allow you to use layers to control the display of objects. AutoCAD includes other commands that allow you to display (isolate) or hide only selected objects. These functions are view commands for temporarily isolating or hiding objects to clarify the drawing and focus on specific items.

Access the **ISOLATEOBJECTS** command to hide all objects except the objects you pick. An easy way to access the command is to pick the **Isolate Objects** button on the right end of the status bar and pick the **Isolate Objects** option, or right-click and select

Type
ISOLATEOBJECTS
ISOLATE

Figure 6-11.
A—Use the status bar or shortcut menu to access object isolation and hiding commands.
B—An example of using the **ISOLATEOBJECTS** command and window selection to show only objects within a portion of a mechanical part view. C—An example of using the **HIDEOBJECTS** command to hide several dimension and multileader objects on a structural detail.

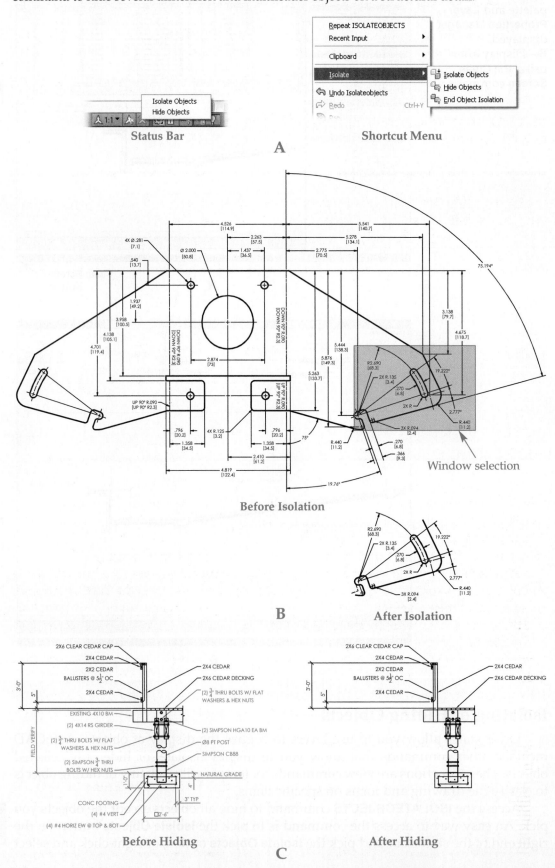

Status Bar

Shortcut Menu

A

Before Isolation

B

After Isolation

Before Hiding

After Hiding

C

Isolate Objects from the **Isolate** cascading menu. See **Figure 6-11A**. Pick the objects to isolate as shown in **Figure 6-11B**.

Access the **HIDEOBJECTS** command to hide the objects you pick. An easy way to access the command is to pick the **Isolate Objects** status bar button followed by the **Hide Objects** option, or right-click and select **Hide Objects** from the **Isolate** cascading menu. Pick the objects to hide as shown in **Figure 6-11C**.

You can use the **ISOLATEOBJECTS** and **HIDEOBJECTS** commands at the same time to isolate and hide objects. You can also add objects to the set of isolated or hidden objects as needed.

Use the **UNISOLATEOBJECTS** command to redisplay all hidden objects. An easy way to access the command is to pick the **Isolate Objects** status bar button followed by the **End Object Isolation** option, or right-click and select **End Object Isolation** from the **Isolate** cascading menu.

Type
HIDEOBJECTS

Type
UNISOLATEOBJECTS
UNISOLATE

The status bar tray icon indicates whether the drawing includes isolated or hidden objects. Use the **OBJECTISOLTATIONMODE** system variable to control whether isolation and hiding remains active when you close and reopen a drawing.

Exercise 6-4

Complete the exercise on the companion website.
www.g-wlearning.com/CAD

Using Commands Transparently

Activating a command usually cancels the command in progress and starts the new command. However, you can use some commands *transparently*. After completing the transparent operation, the interrupted command resumes. Therefore, it is not necessary to cancel the initial command. You can use many display commands transparently, including **ZOOM**, **PAN**, SteeringWheels, and the ViewCube.

An example of when transparent commands are useful is drawing a line when one end of the line is somewhere off the screen. One option is to cancel the **LINE** command, zoom out, and then reactivate the **LINE** command. A more efficient method is to use **PAN** or **ZOOM** transparently with the **LINE** command. To do so, begin the **LINE** command and pick the first point. At the Specify next point: prompt, pan or zoom to display the next line endpoint. You can use any access method, but using the mouse wheel is often quickest. When the correct view appears, pick the second line endpoint. You can also activate commands transparently by typing an apostrophe (') before the command name. For example, to enter the **ZOOM** command transparently, type 'Z or 'ZOOM.

transparently: When referring to command access, describes temporarily interrupting the active command to use a different command.

PROFESSIONAL TIP

You can activate and adjust drawing aids such as **Grid**, **Snap**, and **Ortho** transparently. However, the quickest activation and adjustment methods are using the appropriate button on the status bar or pressing a function key.

Controlling Draw Order

Drawings often include overlapping objects. The overlap is difficult to see when all objects have a thin lineweight and when lineweight display is off. To help understand display order, view an object that has width, such as the donuts shown in **Figure 6-12**. The donuts in this example were drawn after the other objects. You can change the drawing order of the donuts, and all other objects, to place the items above or below selected objects or to the front or back of all objects.

Use the **DRAWORDER** command to change the order of objects in a drawing. You can also set certain draw order options by selecting an object, right-clicking, and choosing an option from the **Draw Order** cascading menu. **Figure 6-13** describes draw order options.

DRAWORDER

Ribbon
Home
> Modify

Bring to Front

Type
DRAWORDER
DR

You may need to use the **DRAWORDER** command on several objects until the objects display correctly. Objects move to the front of the drawing when you modify them.

Figure 6-12.
Change the order of objects to place selected objects under or above other objects.

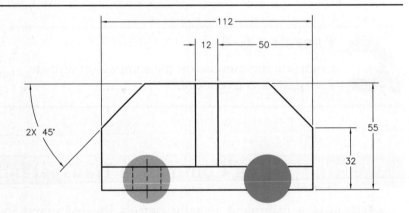

Figure 6-13.
The options available for rearranging the order of objects in a drawing. The text, dimension, and hatch options are available from the ribbon and apply to all text, dimension, or hatch objects in the drawing.

Option	Function
Bring to Front	Places the selected objects at the front of the drawing.
Send to Back	Places the selected objects at the back of the drawing.
Bring Above Objects	Moves the selected objects above the reference object.
Send Under Objects	Moves the selected objects below the reference object.
Bring Text to Front	Places all text objects above other objects.
Bring Dimensions to Front	Places all dimension objects above other objects.
Send Hatches to Back	Moves all hatch objects below other objects.

Exercise 6-5

Complete the exercise on the companion website.
www.g-wlearning.com/CAD

Named Views

Use the **View Manager** or the **View Controls** in-canvas viewport option to save a *named view*. A named view can be a portion of the drawing, such as a mechanical part drawing view or detail or a room or area on an architectural floor plan. A named view can also show an entire design or enlarged area. The advantage of naming views is that you can quickly recall a specific display, without searching for objects or using multiple view commands. Named views are also important to sheet sets, as explained later in this textbook.

Figure 6-14 shows the **View Manager** and an architectural example that includes named views. The tree on the left side of the **View Manager** lists each type of view. Pick a view type to display information about the type. The **Current** node lists the properties of the current display in the drawing window. Pick the plus sign (+) to list related views. The **Model Views** node lists named model views, and the **Layout Views** node lists named layout views. The **Preset Views** node lists all preset orthogonal and isometric views. This chapter focuses on the use of named model views. You will learn about layout views later in this textbook. *AutoCAD and Its Applications—Advanced* explains preset views.

Pick a model view to list its properties in the middle portion of the **View Manager**. You can modify some properties, such as **Name**. Other properties are read-only and are set during view configuration. The right side of the **View Manager** contains buttons to create and control views. The options are also available from the shortcut menu that appears when you right-click on a view.

> **named view:** A specific drawing display saved for easy recall and future use, analogous to taking a picture.

Ribbon
View
> Views
Named Views

Type
VIEW
V

The lower-right corner of the **View Manager** shows a preview of the selected model or layout view.

Figure 6-14.
Use the **View Manager** to assign new named views and to control existing named views. This example shows a named view of an overall floor plan titled FLOOR PLAN. The other named views display specific remodel areas for easy recall and future use.

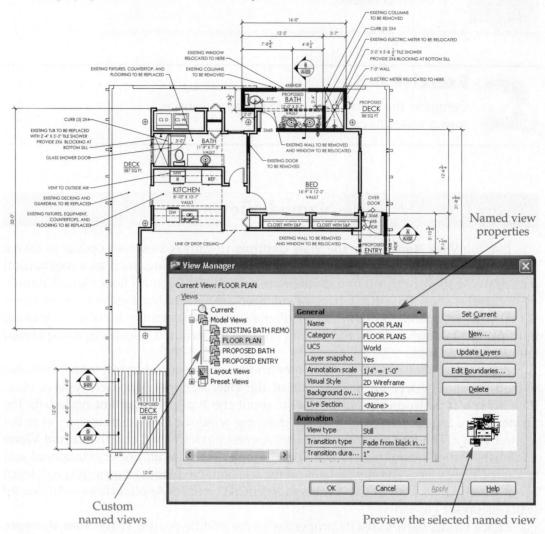

Named view properties

Custom named views

Preview the selected named view

New Views

To save a display as a named view, pick the **New...** button or right-click on a view node or view and select **New...** to access the **New View/Shot Properties** dialog box. See **Figure 6-15.** Type the view name in the **View name:** text box.

To create a basic 2D view, select **Still** from the **View type** drop-down list, and focus on the settings in the **View Properties** tab. The **Current display** radio button, which is the default, prepares the view boundary according to the current display in the drawing window. To construct a view boundary, pick the **Define window** radio button. The **New View/Shot Properties** dialog box disappears temporarily so you can pick two opposite corners to define a window around the area to display. See **Figure 6-16.** Press [Enter] or right-click to return to the **New View/Shot Properties** dialog box.

Select the **Save layer snapshot with view** check box to save the current layer settings when you save the new view. Saved layer settings are then recalled each time the view is set current. Pick the **OK** button to add the view name to the list in the **View Manager**. Pick the **OK** button in the **View Manager** to finish the new view and exit.

Figure 6-15.
Use the **New View/Shot Properties** dialog box to save the current display as a view, or define a window to set as the view.

Select **Still** for basic 2D views

Pick to create a new view according to the current display in the drawing window

Pick to create a new view by constructing a window around the area to display

Type the view name

Use with sheet sets

Pick to recreate the window boundary

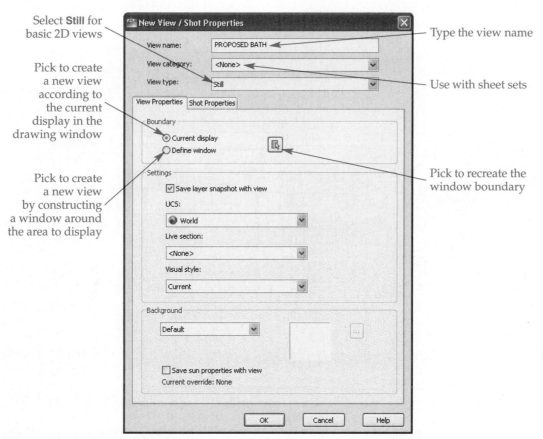

AutoCAD and Its Applications—Advanced describes the other options in the **New View/Shot Properties** dialog box, which are applicable to 3D model animations and the UCS (user coordinate system).

Activating a View

To display a named view in the drawing window without accessing the **View Manager**, select the view from the list in the **Views** panel of the **View** ribbon tab, or the **Custom Model Views** cascading menu of the **View Controls** in-canvas viewport controls selection. To make a named view current from inside the **View Manager**, select the view from the list and pick the **Set Current** button or right-click and choose **Set Current**. The name of the current view appears in the **Current View:** label above the **Views** area. Pick the **Apply** button to display the view, or pick the **OK** button to display the view and exit the dialog box.

Managing Views

Return to the **View Manager** to adjust named views using the appropriate button or shortcut menu option. Apply the **Update Layers** feature to update layers displayed in the view according to changes made to layer states. Use the **Edit Boundaries...** function to define or redefine the view boundary using a window. Use the **Delete** function to remove a named view from the drawing.

Figure 6-16.
Specifying a view boundary using the **Define window** option. This example shows creating a view boundary around the proposed bathroom that corresponds to the PROPOSED BATH named view.

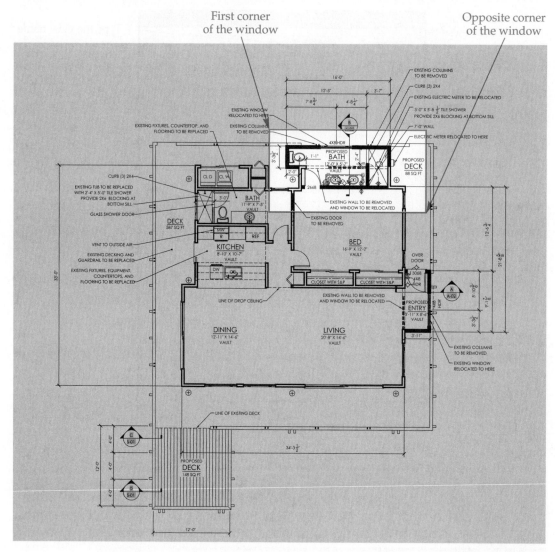

First corner of the window

Opposite corner of the window

Exercise 6-6

Complete the exercise on the companion website.
www.g-wlearning.com/CAD

Tiled Viewports

tiled viewports:
A window or frame within which a drawing is visible in model space.

AutoCAD allows you to divide the model space drawing window into *tiled viewports*. Tiled viewports have both 2D and 3D applications. Use tiled viewports to divide a 2D drawing window into compartments that show different aspects of the drawing project. See **Figure 6-17.** Tiled viewports are most appropriate for viewing large, complex drawings, such as site plans and profiles or floor plans. See *AutoCAD and Its Applications—Advanced* for 3D tiled viewport examples.

The drawing window contains one tiled viewport by default. Additional viewports divide the drawing window into separate tiles that butt against each other. Tiled viewports cannot overlap. Multiple viewports contain different views of the same drawing, displayed at the same time. Only one viewport can be active at any given time. The active viewport has a bold outline, as shown in **Figure 6-17**.

Creating Tiled Viewports

The Viewports dialog box provides one method for creating tiled viewports. **Figure 6-18** shows the **New Viewports** tab of the **Viewports** dialog box. The **Standard viewports:** list contains preset viewport configurations. The configuration name identifies the number of viewports and the arrangement or location of the largest viewport.

Ribbon

View
> Viewports

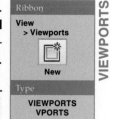

New

VIEWPORTS

Type

VIEWPORTS
VPORTS

Figure 6-17.
A mechanical part drawing viewed using three tiled viewports. Each viewport contains the same drawing, but can present a different display. Tiled viewports are most useful when you are viewing large, complex drawings.

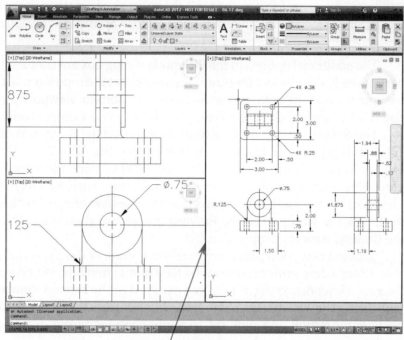

Thick outline around the active viewport

Figure 6-18.
Specify the number and arrangement of tiled viewports in the **New Viewports** tab of the **Viewports** dialog box.

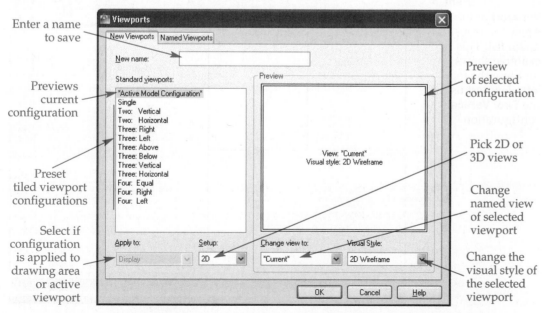

Enter a name to save

Previews current configuration

Preset tiled viewport configurations

Select if configuration is applied to drawing area or active viewport

Preview of selected configuration

Pick 2D or 3D views

Change named view of selected viewport

Change the visual style of the selected viewport

Select a configuration to see a preview of the tiled viewports in the **Preview** area. Select *Active Model Configuration* to preview the current configuration. Pick the **OK** button to divide the drawing window into the selected tiled viewports.

You can activate the same preset viewport configurations available from the **New Viewports** tab of the **Viewport** dialog box using the **Set Viewports** drop-down list in the **Viewports** panel on the **View** ribbon tab, or the **Viewport Configuration List** cascading menu of the Viewport **Controls** in-canvas viewport controls selection. Prompts guide you through the process of creating the correct configuration. The arrangement you choose applies to the active viewport only.

The additional options in the **Viewports** dialog box are useful when two or more viewports already exist. Select **Display** from the **Apply to:** drop-down list to apply the viewport configuration to the entire drawing area. Pick **Current Viewport** from the **Apply to:** drop-down list to apply the new configuration in the active viewport only. See **Figure 6-19**.

The default setting in the **Setup:** drop-down list is 2D, and all viewports show the 2D drawing plane, or top view. If you choose the 3D option, the different viewports display various 3D views of the drawing. At least one viewport shows a 3D isometric view. The other viewports show different views, such as a top view or side view. The viewport configuration is displayed in the **Preview** image. To change a view in a viewport, pick the viewport in the **Preview** image and then select the new viewpoint from the **Change view to:** drop-down list.

Create a new viewport configuration if none of the preset configurations is acceptable. Enter a descriptive name in the **New name:** text box. When you pick the **OK** button, the new viewport configuration is displayed in the **Named Viewports** tab the next time you access the **Viewports** dialog box. See **Figure 6-20**. Select a different named viewport configuration and pick **OK** to apply changes to the drawing area. Named viewport configurations apply to the active viewport only.

Figure 6-19.
You can subdivide a viewport by choosing **Current Viewport** in the **Apply to:** drop-down list. This example shows dividing the top-left viewport using the **Two: Vertical** configuration.

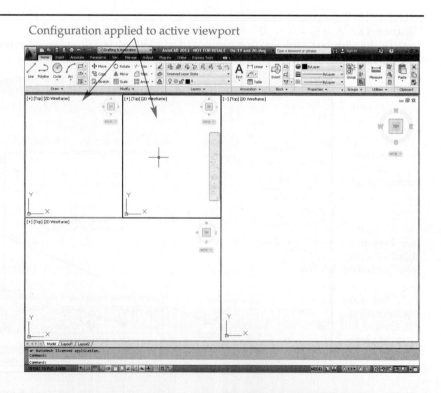

Configuration applied to active viewport

Figure 6-20.
The **Named Viewports** tab displays custom viewports.

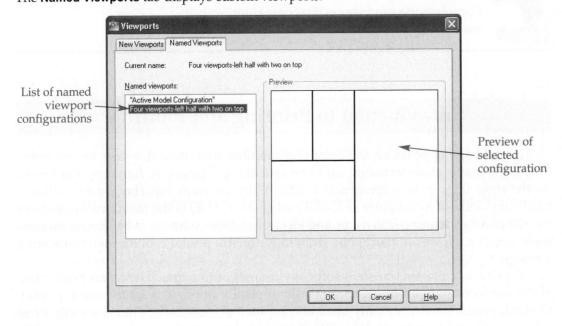

List of named viewport configurations

Preview of selected configuration

Apply the **1 Viewport** option to return the display to the default single viewport.

Working in Tiled Viewports

After you select the viewport configuration and return to the drawing area, move the mouse and notice that only the active viewport contains crosshairs. The cursor is an arrow in the other viewports. To make an inactive viewport active, move the cursor into the inactive viewport and pick.

Depending on the zoom level, as you draw in one viewport, objects appear in other viewports. Try drawing lines and other shapes and notice how drawing affects the viewports. Use a display command, such as **ZOOM**, in the active viewport and notice the results. Only the active viewport reflects the use of the **ZOOM** command.

Joining Tiled Viewports

Use the **Join Viewports** option to join two viewports. Select the dominant viewport, which is the viewport that shows the view you want to display in the joined viewport, or press [Enter] to select the active viewport. Then select the viewport to join with the dominant viewport. AutoCAD joins the two viewports together and retains the dominant view.

Ribbon

View
> Viewports

Join Viewports

Type

VIEWPORTS
VPORTS

VIEWPORTS

The two viewports you join cannot create an L-shape viewport. The adjoining edges of the viewports must be the same size in order to be joined.

Exercise 6-7

Complete the exercise on the companion website.
www.g-wlearning.com/CAD

Introduction to Printing and Plotting

soft copy: The electronic data file of a drawing.

hard copy: A physical drawing produced by a printer or plotter.

A *soft copy* appears on the computer monitor, making a drawing inconvenient to use for many manufacturing and construction purposes. A *hard copy* is useful on the shop floor or at a construction site. A design team can check and redline a hard copy without a computer or CADD software. CADD is the standard throughout the world for generating drawings, and electronic data exchange is becoming increasingly popular. However, hard-copy drawings are still a vital tool for communicating a design.

A printer or plotter transfers soft copy images onto paper. The terms *printer* and *plotter* are interchangeable, although *plotter* typically refers to a large-format printer. Desktop printers are commonly used at computer workstations. They generally print 8 1/2″ × 11″ and sometimes 11″ × 17″ sheets. Desktop printers are most appropriate for printing small drawings and reduced-size test prints. Large-format printers print larger drawings, such as C-size and D-size drawings. The most common types of both desktop and large-format printers are inkjet and laser printers. Traditional pen plotters, which "draw" with actual ink pens, are much less common.

Plotting in Model Space

You typically plot final drawings using a layout in paper space. A layout represents the sheet of paper used to organize and scale, or lay out, and plot or export a drawing or model. However, you can also plot from model space. The following information describes plotting from model space only.

Plotting from model space is common when a layout is unnecessary, to view how model space objects will appear on paper, and to make quick hard copies, such as when submitting basic assignments to your instructor or supervisor. This chapter provides basic plotting information so you can make your first plot. Information about creating and plotting layouts and additional printing and plotting information is provided later in this textbook.

Making a Plot

This section describes one of the many methods for creating a plot from model space. You will explore several additional plot options and settings later in the textbook. Refer to **Figure 6-21** as you read the following plotting procedure.

1. Access the **Plot** dialog box. If the column on the far right of the dialog box shown in **Figure 6-21** is not visible, pick the **More Options** button (>) in the lower-right corner.
2. In the **Printer/plotter** area, select a local or network printer or plotter.
3. In the **Paper size** area, choose a sheet size appropriate for the selected printer or plotter.
4. Select the area to plot from the **Plot area** section. Choose the **Display** option to plot the current screen display, exactly as shown. Pick the **Extents** option to plot the furthest extents of objects in the drawing. The **Limits** option allows you to plot everything inside the specified drawing limits. The **View** option is available

Figure 6-21.
The **Plot** dialog box.

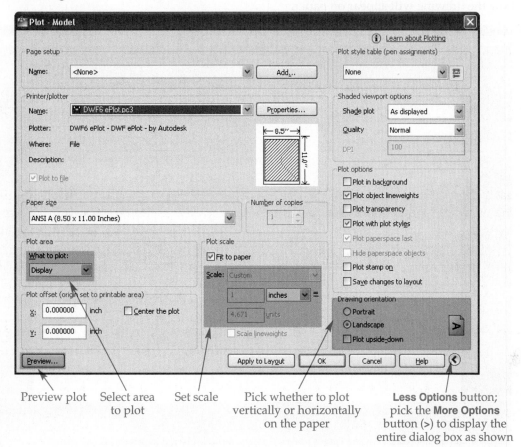

Preview plot   Select area to plot   Set scale   Pick whether to plot vertically or horizontally on the paper   **Less Options** button; pick the **More Options** button (>) to display the entire dialog box as shown

if the file includes named views. Select a named view from the list that appears when you choose the **View** option. When you select the **Window** option, the **Page Setup** dialog box disappears temporarily so you can pick two opposite corners to define a window around the area to plot. Once you create the window, a **Window…** button appears in the **Plot area** section. Pick the button to redefine the opposite corners of a window around the portion of the drawing to plot.

5. Select an option in the **Drawing orientation** area. Choose **Portrait** to orient the drawing vertically (*portrait*) or **Landscape** to orient the drawing horizontally (*landscape*). The **Plot upside-down** option rotates the drawing 180° on the paper.

6. Set the scale in the **Plot scale** area. The scale is a ratio of inches or millimeters to drawing units. Select a scale from the **Scale:** drop-down list or type values in the custom fields. Choose the **Fit to paper** check box to let AutoCAD increase or decrease the plot scale to fill the paper.

7. If necessary, use the **Plot offset (origin set to printable area)** section to set additional left and bottom margins around the plot, or to center the plot.

8. Pick the **Preview…** button to display the sheet as it will look when it plots. See **Figure 6-22**. The cursor appears as a magnifying glass with + and – symbols. Hold down the left mouse button and move the cursor to increase or decrease the displayed image to view more or less detail. Press [Esc] to exit the preview.

9. Pick the **OK** button in the **Plot** dialog box to send the data to the plotting device.

portrait: A vertical paper orientation.

landscape: A horizontal paper orientation.

Figure 6-22.
A plot preview of mechanical part views drawn in model space. The preview shows exactly how the drawing will appear on paper.

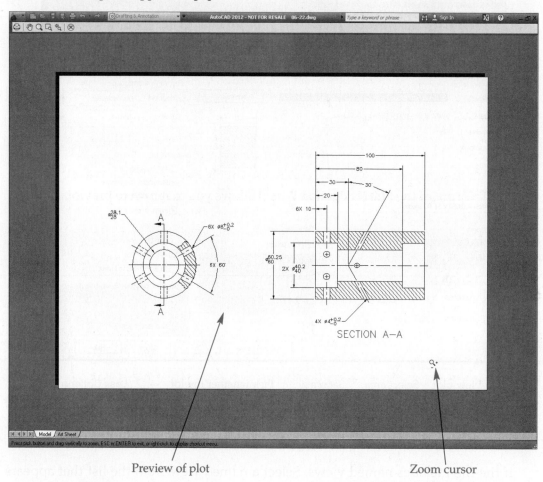

Preview of plot   Zoom cursor

Exercise 6-8

Complete the exercise on the companion website.
www.g-wlearning.com/CAD

Template Development *Zoom All*
Chapter 6 For instructions on zooming to the drawing limits in each template, go to the companion website (www.g-wlearning.com/CAD), select this chapter, and select **Template Development**.

Chapter Review

1. Briefly explain how to use the **Realtime** zoom option.
2. What is the difference between zooming and panning?
3. What two commands allow you to cycle back and forth quickly between prior views that were created by zooming or panning?
4. Which SteeringWheels navigation commands are most appropriate for 2D drafting applications?
5. Explain how to use the **CENTER** command on the **Full Navigation Wheel**.
6. What feature of the **Full Navigation Wheel** allows you to return to previous display settings?
7. Why should you avoid using the ViewCube to rotate the display of a 2D drawing?
8. Provide an example of when regenerating the display is necessary.
9. Which command regenerates all of the viewports?
10. Explain the difference between using layer states and object isolation or object hiding to hide objects.
11. How do you enter a display command transparently at the keyboard?
12. Which command changes the order in which objects are displayed in a drawing?
13. How do you create a named view of the current screen display?
14. How do you display an existing named view?
15. What type of viewport occurs in model space?
16. How can you specify whether a new viewport configuration applies to the entire drawing window or the active viewport?
17. Explain the procedure for joining viewports.
18. Define *hard copy* and *soft copy*.
19. Identify four ways to access the **Plot** dialog box.
20. Describe the difference between the **Display** and **Window** options in the **Plot area** section of the **Plot** dialog box.

Drawing Problems

Start AutoCAD if it is not already started. Start a new drawing from scratch or use an appropriate template of your choice. The template should include layers for drawing the given objects. Add layers as needed. Draw all objects using appropriate layers. Follow the specific instructions for each problem. Use only drawing commands and techniques you have already learned. Do not draw dimensions or text. Use your own judgment and approximate dimensions when necessary.

▼ Basic

1. Draw the surface-mounted fluorescent light fixture shown below. Zoom in and out on the drawing. Pan the screen display. Save the drawing as **P6-1**. Print an 8.5" × 11" copy of the drawing extents, fit to paper, using a portrait orientation.

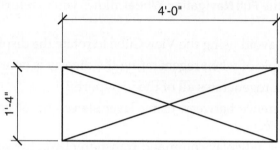

Surface-Mounted Fluorescent Fixture

2. Draw the surface-mounted light fixture shown below. Zoom in and out on the drawing. Pan the screen display. Save the drawing as **P6-2**. Print an 8.5 × 11 copy of the drawing extents, fit to paper, using a portrait orientation.

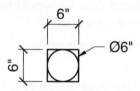

Surface-Mounted Fixture

3. Open the drawing named VW252-03-1200.dwg in the AutoCAD 2012\Sample\Sheet Sets\Manufacturing folder. Pick the **Model** tab below the drawing window to activate model space. Zoom to the extents of the drawing. Zoom in on each drawing view. Use a variety of view commands to view the gears in the full side view, the partial side views, and the isometric view. Close the drawing without saving.

4. Open the drawing named A-02.dwg in the AutoCAD 2012\Sample\Sheet Sets\Architectural folder. Pick the **Model** tab below the drawing window to activate model space. Zoom to the extents of the drawing and print using the **Fit to paper** option. Use a variety of view commands to locate and view the following items:
   A. Front elevation
   B. Doors on the right elevation
   C. Downspouts on the rear elevation
   D. Column on the left elevation
   E. Main entrance
   Mark the location of each item on your hard copy. Close the drawing without saving.

5. Open the drawing named Erosion Control Plan.dwg in the AutoCAD 2012\Sample\Sheet Sets\Civil folder. Pick the **Model** tab below the drawing window to activate model space. Zoom to the extents of the drawing and print using the **Fit to paper** option. Use a variety of view commands to locate and view the following items on the plan:
   A. Coburn Avenue
   B. Proposed track and field
   C. Proposed high school
   D. Proposed tennis courts
   E. Two proposed baseball fields
   F. Two proposed softball fields
   G. Accessible parking near the northeast corner of the high school
   Mark the location of each item on your hard copy. Close the drawing without saving.

▼ **Intermediate**

6. Create a freehand sketch of the default navigation bar that appears in model space. Label each tool button and briefly describe those commands that are most appropriate for 2D drafting applications.

7. Create a freehand sketch of the **Full Navigation Wheel**. Label each wedge and briefly describe those commands that are most appropriate for 2D drafting applications.

8. Draw the gasket shown. Save the drawing as P6-8. Print an 8.5" × 11" copy of the drawing extents, using a 1:1 scale and a landscape orientation.

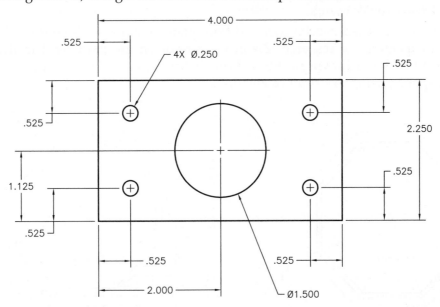

9. Open the drawing named blocks_and_tables_-_imperial.dwg at www.autodesk.com/autocad-samples. Save a copy of the drawing as P6-9. The P6-9 file should be active. Perform the following display functions on the drawing:
   A. Pick the **Model** tab to switch the drawing to model space
   B. Zoom to the drawing extents
   C. Zoom to create a display of the dining room and create a view of this display named Dining Room
   D. Zoom to create a display of the kitchen and create a view of this display named Kitchen
   E. Zoom to create a display of the master bedroom suite and create a view of this display named Master Suite
   F. Display the view named Dining Room
   G. Resave the drawing.

10. Open the drawing named Kitchens.dwg in the AutoCAD 2012\Sample\DesignCenter folder. Save a copy of Kitchen as P6-10. The P6-10 file should be active. Perform the following display functions on the drawing:
    A. Zoom to the drawing extents.
    B. Zoom in on each symbol and research if necessary to identify what the symbol represents.
    C. Create a named view of the extents of each symbol using the **Define Window** option in the **New View** dialog box. Use an appropriate name for each view.
    D. Systematically make each named view current. Edit the boundary if the view does not behave as anticipated.
    E. Resave the drawing.

11. Open the drawing named Pipe Fittings.dwg in the AutoCAD 2012\Sample\DesignCenter folder. Save a copy of Pipe Fittings as P6-11. The P6-11 file should be active. Perform the following display functions on the drawing:
    A. Zoom to the drawing extents.
    B. Zoom in on each symbol and research if necessary to identify what the symbol represents.
    C. Create a named view of the extents of each symbol using the **Define Window** option in the **New View** dialog box. Use an appropriate name for each view.
    D. Systematically make each named view current. Edit the boundary if the view does not behave as anticipated.
    E. Resave the drawing.

12. Draw the enclosed gazebo roof and floor plan shown. Use dimensions based on your experience or research. Zoom in and out on the drawing. Pan the screen display. Save the drawing as P6-12. Print an 8.5″ × 11″ copy of the drawing extents, fit to paper, using a portrait orientation.

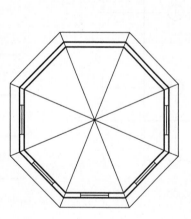
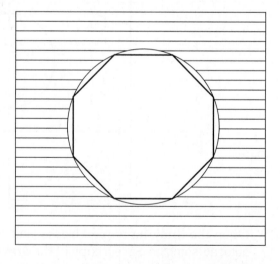

13. Research the design of an existing exercise dumbbell with the following specifications: steel, one-piece hexagon shape, five pounds. Create a dimensioned 2D sketch of the existing design from manufacturer's specifications or from measurements taken from an actual dumbbell. Research the mass properties of steel to design the dumbbell to weigh exactly five pounds. Start a new drawing from scratch or use a decimal template of your choice. Draw the dumbbell from your sketch. Save the drawing as P6-13. Print an 8.5″ × 11″ copy of the drawing extents, using a 1:1 scale and a landscape orientation. Do not dimension the drawing.

AutoCAD Certified Associate Exam Practice

Answer the following questions. Write your answers on a separate sheet of paper.

1. Which of the following actions enters a view command or drawing aid transparently? *Select all that apply.*
   A. picking a button on the status bar
   B. picking a button on the ribbon
   C. picking a button on the navigation bar
   D. typing a colon (:) before the command name
   E. using the mouse wheel

2. How can you display a named view in a drawing? *Select all that apply.*
   A. select the view from the list in the **Views** panel of the **View** ribbon tab
   B. type the view name at the command line
   C. double-click the view name in the **View Manager**
   D. right-click and select the view name

3. When a drawing contains two or more tiled viewports, how can you activate a different viewport? *Select all that apply.*
   A. press the [F3] key
   B. press the space bar
   C. pick anywhere in the viewport to activate
   D. right-click and select the view to activate

AutoCAD Certified Professional Exam Practice

Follow the instructions in each problem. Write your answers on a separate sheet of paper.

1. **Navigate to this chapter on the companion website and open CPE-06zoom.dwg.** Use the view command of your choice to zoom in on the notes in the lower-left corner of the drawing. Which ANSI standard should be used to interpret the graphic symbols for the electrical and electronic elements in this drawing?

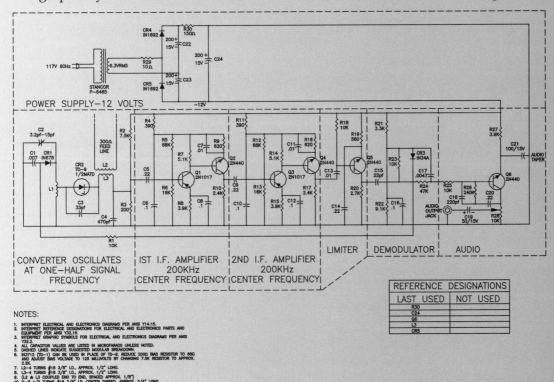

2. **Navigate to this chapter on the companion website and open CPE-06viewport.dwg.**
Create tiled viewports using the **Three: Left** viewport configuration. In the left
viewport, zoom to the drawing extents. In the top-right viewport, zoom to show
the general drawing notes. Answer the following questions.
A. What is the scale of VIEW A?
B. What is the current revision level for this drawing?

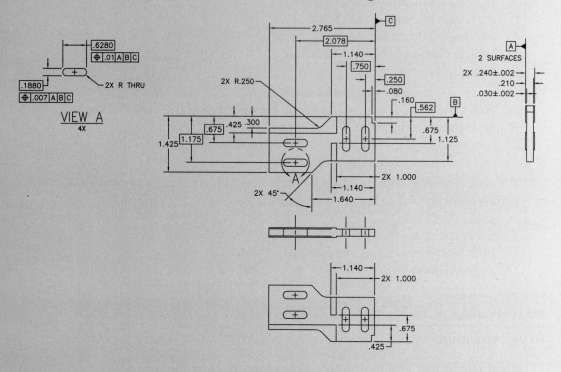

Object Snap and AutoTrack

Learning Objectives

After completing this chapter, you will be able to:

✓ Set running object snap modes for continuous use.
✓ Use object snap overrides for single point selections.
✓ Select appropriate object snaps for various drawing tasks.
✓ Use AutoSnap™ features to speed up point specifications.
✓ Use AutoTrack™ to locate points relative to other points in a drawing.

This chapter explains how to use object snap and AutoTrack tools and options to draw accurately. Object snaps and AutoTrack are very useful and efficient drawing aids. Object snaps and AutoTrack help you create accurate geometric constructions, but they do not constrain, or apply relationships between, objects. Object snaps are also an important aid to parametric drafting. Chapter 22 explains how to use parametric drafting tools to constrain objects.

Object Snap

Object snap increases drafting performance and accuracy through the concept of *snapping*. See **Figure 7-1**. You can use object snap with any command that requires a point selection. Object snap modes identify the object snap point. The AutoSnap feature, which controls object snap, is on by default and displays *markers* while you draw. After a brief pause, a tooltip appears, indicating the object snap mode. See **Figure 7-2**. Refer to the list of standard object snap modes in **Figure 7-3** to identify the appropriate object snap for each drafting task. Object snap use becomes second nature with practice, and greatly increases productivity and accuracy.

object snap: A tool that locates an exact point, such as an endpoint, midpoint, or center point, on or in relation to an existing object.

snapping: Picking a point near the intended position to have the crosshairs "snap" exactly to the specific point.

markers: Visual cues that appear at the snap point to confirm object snap mode and location.

If you cannot see an AutoSnap marker because of the size of the current screen display, you can still confirm the point before picking by reading the tooltip, which indicates if a point is acquired beyond the visible area.

Figure 7-1.
An example of object snaps used to aid construction of specific objects in a part drawing view. Object snaps aid geometric construction for any drawing application.

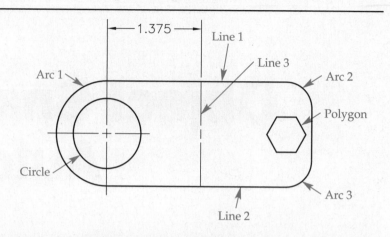

Object Snap	Construction Example
Parallel	Draw Line 2 parallel to Line 1
Endpoint	Begin and end Arc 1 at the endpoints of Line 1 and Line 2
Midpoint	Begin Line 3 at the midpoint of Line 1
Perpendicular	End Line 3 perpendicular to Line 2
Mid Between 2 Points	Locate the center of the Polygon at the midpoint between the center points of Arc 2 and Arc 3

Figure 7-2.
The AutoSnap marker and the related tooltip that appears when you snap to an endpoint and a point of tangency.

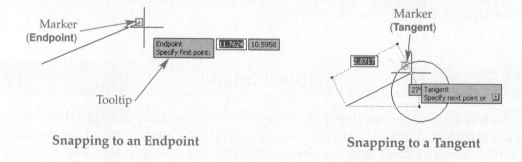

Snapping to an Endpoint

Snapping to a Tangent

Running Object Snaps

Running object snaps are on by default and are often the quickest and most effective way to use object snap. The **Endpoint, Center, Intersection**, and **Extension** running object snap modes are active by default. Right-click on the **Object Snap** or **Object Snap Tracking** button on the status bar and pick the modes to activate or deactivate. See **Figure 7-4A**. You can also set running object snaps using the **Object Snap** tab of the **Drafting Settings** dialog box. See **Figure 7-4B**. A quick way to access the **Object Snap** tab is to right-click on the **Object Snap** or **Object Snap Tracking** button on the status bar and select **Settings...**.

To use most running object snaps, move the crosshairs near the location on an existing object where the object snap should occur. When you see the appropriate

AutoCAD and Its Applications—Basics

Figure 7-3.
Standard object snap modes.

Mode	Marker	Description
Endpoint	□	Locates the nearest endpoint of a line, arc, polyline, elliptical arc, spline, ellipse, ray, solid, or multiline.
Midpoint	△	Finds the point halfway between the endpoints of a line, polyline, circular arc, elliptical arc, polyline arc, spline, or multiline, or finds the start point of an xline.
Center	○	Finds the center point of radial objects, including circles, arcs, ellipses, elliptical arcs, and radial solids.
Node	⊗	Locates a point object drawn with the **POINT**, **DIVIDE**, or **MEASURE** tool, or a dimension definition point.
Quadrant	◇	Locates the closest of the four quadrant points on circles, arcs, elliptical arcs, ellipses, and radial solids. (Some of these objects may not have all four quadrants.)
Intersection	✕	Locates the closest intersection of two objects.
Extension	+	Finds a point along the imaginary extension of an existing line, polyline, arc, polyline arc, elliptical arc, spline, ray, xline, solid, or multiline.
Insertion	⌐	Locates the insertion point of text objects and blocks.
Perpendicular	⌐	Finds a point that is perpendicular to an object from the previously picked point.
Tangent	○	Finds points of tangency between radial and linear objects.
Nearest	⋈	Locates the point on an object that is closest to the crosshairs.
Apparent Intersection	⊠	Locates the intersection between two objects that appear to intersect on-screen in the current view, but may not actually intersect in 3D space. Creating and editing 3D objects is described in *AutoCAD and Its Applications—Advanced*.
Parallel	∥	Finds any point along an imaginary line parallel to an existing line or polyline.
None		Temporarily turns running object snap off during the current selection.

marker and tooltip, pick to locate the point at the exact position on the object. See Figure 7-2.

Toggle running object snaps off and on by picking the **Object Snap** button on the status bar, pressing [F3], right-clicking on the **Object Snap** button on the status bar and selecting the **Enabled** option, or using the **Object Snap On (F3)** check box on the **Object Snap** tab of the **Drafting Settings** dialog box. Turn off running object snaps to locate points without the aid, or to avoid possible confusion of object snap modes. The selected running object snap modes are restored when you reactivate running object snaps.

Figure 7-4.
A—Right-click on the **Object Snap** or **Object Snap Tracking** button on the status bar to activate or deactivate running object snap modes. B—You can also set running object snap modes using the **Drafting Settings** dialog box.

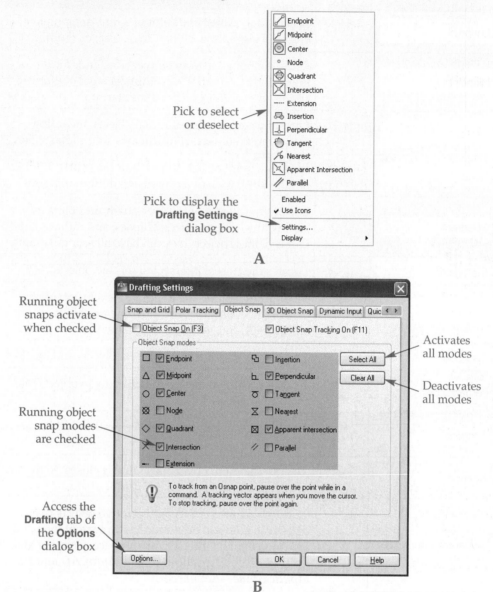

A

B

PROFESSIONAL TIP

Activate only the running object snap modes that you use most often. Too many running object snaps can make it difficult to snap to the appropriate location, especially on detailed drawings with several objects near each other. Use object snap overrides to access object snap modes that you use less often.

By default, a keyboard point entry overrides running object snaps. Use the **Priority for Coordinate Data Entry** area on the **User Preferences** tab of the **Options** dialog box to adjust the default setting.

Object Snap Overrides

Use an *object snap override* to select a specific point if you experience conflicting running object snaps, or to use an object snap that is not running. Running objects snaps return after you make the object snap override selection. All running object snap modes are available as object snap overrides, but some specific object snap options are only available as object snap overrides.

After you access a command and are ready to apply an object snap, hold [Shift] or [Ctrl] and then right-click and choose an object snap override from the shortcut menu shown in **Figure 7-5**. An alternative, when you right-click and AutoCAD does not select the previous point, is to right-click without holding [Shift] or [Ctrl] and select from the **Snap Overrides** cascading menu. Once you activate an object snap override, move the crosshairs near the location on an existing object where the object snap should occur. When you see the corresponding marker and tooltip, pick to locate the point at the exact position on the object.

object snap override: A method of isolating a specific object snap mode while using a drawing or editing command. The selected object snap temporarily overrides the running object snap modes.

You can activate an object snap override by entering the first three letters of the object snap name. For example, enter END to activate the **Endpoint** object snap or CEN to activate the **Center** object snap.

PROFESSIONAL TIP

Remember that object snap modes function with the active command. An error message appears if you try to apply an object snap when no command is active.

Endpoint Object Snap

To snap to an endpoint using the **Endpoint** object snap mode, move the crosshairs near the endpoint of a line, arc, polyline, spline, or ray. When the endpoint marker and tooltip appear, pick to locate the point at the exact endpoint. See **Figure 7-6**.

Figure 7-5.
The **Object Snap** shortcut menu provides quick access to object snap overrides.

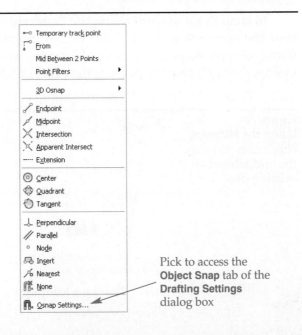

Pick to access the **Object Snap** tab of the **Drafting Settings** dialog box

Figure 7-6.
Using the **Endpoint** object snap to locate the endpoint of an existing line. When using running object snaps, be sure the correct snap marker and tooltip appear before you pick.

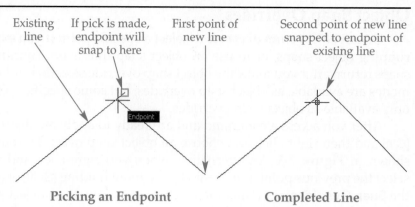

Existing line

If pick is made, endpoint will snap to here

First point of new line

Second point of new line snapped to endpoint of existing line

Picking an Endpoint   **Completed Line**

Midpoint Object Snap

To snap to a midpoint using the **Midpoint** object snap mode, move the crosshairs near the midpoint of a line, arc, or polyline, or the start point of an xline. When the midpoint marker and tooltip appear, pick to locate the point at the exact midpoint. See **Figure 7-7.**

Exercise 7-1

Complete the exercise on the companion website.
www.g-wlearning.com/CAD

Center Object Snap

To snap to a center point using the **Center** object snap mode, move the crosshairs near the *perimeter*, not the center point, of a circle, arc, donut, ellipse, elliptical arc, or polyline arc. You must move the crosshairs near the perimeter of a circular object, especially if the object is large, to acquire the center point. When you see the center marker and tooltip, pick to locate the point at the exact center. See **Figure 7-8.**

Quadrant Object Snap

quadrant: A point on the circumference at the horizontal or vertical quarter of a circle, arc, donut, or ellipse.

To snap to a *quadrant* using the **Quadrant** object snap mode, move the crosshairs near the appropriate 0°, 90°, 180°, or 270° point on the circumference of a circle, arc, donut, ellipse, elliptical arc, or polyline arc. When you see the quadrant marker and tooltip, pick to locate the point at the exact quadrant position. See **Figure 7-9.**

Figure 7-7.
Using the **Midpoint** object snap to locate the midpoint of an existing line.

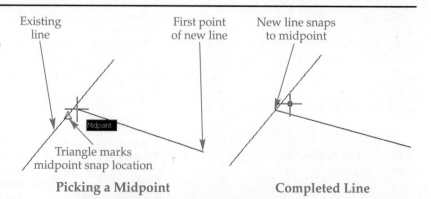

Existing line

First point of new line

New line snaps to midpoint

Triangle marks midpoint snap location

Picking a Midpoint   **Completed Line**

Figure 7-8.
Using the **Center** object snap to locate the center point of existing circles.

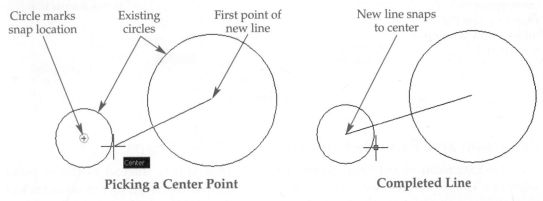

Picking a Center Point | Completed Line

Figure 7-9.
Using the **Quadrant** object snap to locate a quadrant point on the circumference of a circle.

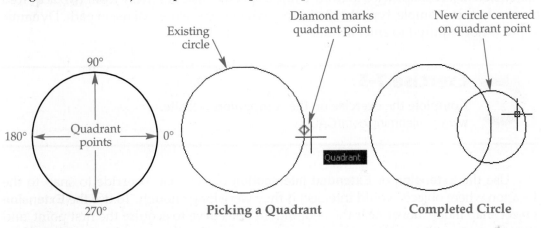

Picking a Quadrant | Completed Circle

The current angle zero direction does not affect quadrant points, but quadrant points always coincide with the angle of the X and Y axes. The quadrant points of circles, arcs, and donuts are at the right (0°), top (90°), left (180°), and bottom (270°), regardless of the rotation of the object. However, the quadrant points of ellipses and elliptical arcs rotate with the object.

Exercise 7-2

Complete the exercise on the companion website.
www.g-wlearning.com/CAD

Intersection Object Snap

To snap to an intersection using the **Intersection** object snap mode, move the crosshairs near the intersection of two or more objects. When you see the intersection marker and tooltip, pick to locate the point at the exact intersection. See **Figure 7-10**.

Figure 7-10.
Using the **Intersection** object snap to locate the intersection of a line and an arc.

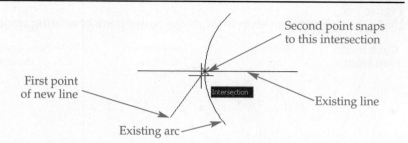

Second point snaps to this intersection

First point of new line

Intersection

Existing line

Existing arc

Extension and Extended Intersection Object Snaps

The **Extension** object snap mode uses *acquired points* instead of direct point selection. To snap to an extension, hover near the endpoint of a curve, but do not select it. A point symbol (+) marks the location when AutoCAD acquires the point. Move the crosshairs away from the acquired point to display an *extension path*. Specify the location of a point along the extension path. **Figure 7-11** shows an example of using an **Extension** object snap twice to draw a line a specific distance away from two acquired points. In this example, type the .8 distance when you see the extension path. Dynamic input is not required to enter a value.

Exercise 7-3

Complete the exercise on the companion website.
www.g-wlearning.com/CAD

Use the **Extension** or **Extended Intersection** object snap override to snap to the location where objects would intersect if they were long enough. To use the **Extension** object snap mode, hover near the endpoint of one curve to acquire the first point, and then hover near the endpoint of another curve to acquire the second point. Then move the crosshairs away from the acquired point, near the location of where the objects would intersect. When you see two extension paths and an intersection icon, pick to locate the point. See **Figure 7-12A**.

To use the **Extended Intersection** object snap override, select objects one at a time using the **Intersection** object snap override. Then move the cursor over one of the objects to display the intersection marker with an ellipsis (...), and pick the object. Then move the cursor over the other object to display the intersection marker at the extended intersection, and pick. See **Figure 7-12B**.

Figure 7-11.
Using the **Extension** object snap to create a line .8 units away from a rectangle.

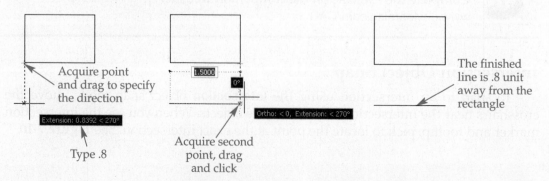

Acquire point and drag to specify the direction

Extension: 0.8392 < 270°

Type .8

1.5000

0°

Ortho: < 0, Extension: < 270°

Acquire second point, drag and click

The finished line is .8 unit away from the rectangle

Figure 7-12.
Locating the center of a circle at the extended intersection of a line and an arc. A—Using the **Extension** object snap. B—Using the **Extended Intersection** object snap.

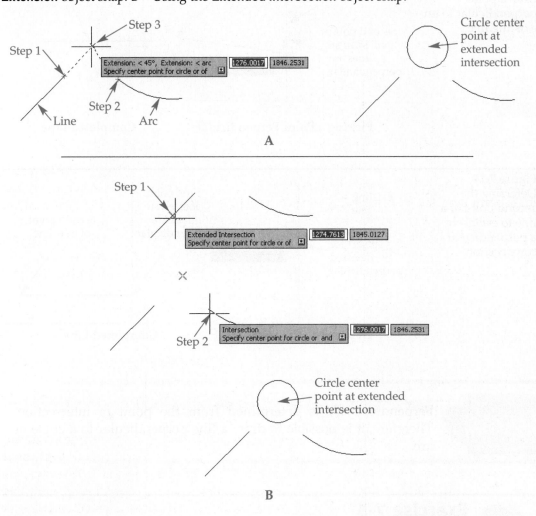

A

B

Exercise 7-4

Complete the exercise on the companion website.
www.g-wlearning.com/CAD

Perpendicular Object Snap

To snap to a perpendicular point using the **Perpendicular** object snap mode, move the crosshairs near the point of perpendicularity on a line, arc, elliptical arc, ellipse, spline, xline, multiline, polyline, or circle, or the endpoint of a line, arc, polyline, or spline. When you see the perpendicular marker and tooltip, pick to locate the point exactly perpendicular to the existing object. See **Figure 7-13**.

Figure 7-14 shows using the **Perpendicular** object snap mode to begin a line perpendicular to an existing object. The tooltip reads Deferred Perpendicular, and the perpendicular marker includes an ellipsis (...). The second endpoint determines the location of the line in a *deferred perpendicular* condition.

deferred perpendicular: A calculation of the perpendicular point that is delayed until you pick another point.

Figure 7-13.
Drawing a line from a point perpendicular to an existing line.

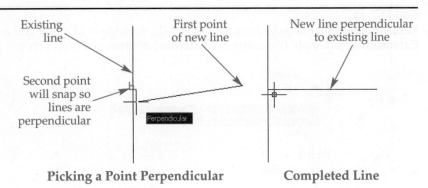

Existing line

First point of new line

New line perpendicular to existing line

Second point will snap so lines are perpendicular

Perpendicular

Picking a Point Perpendicular   **Completed Line**

Figure 7-14.
Deferring the second point of a line to establish a perpendicular construction.

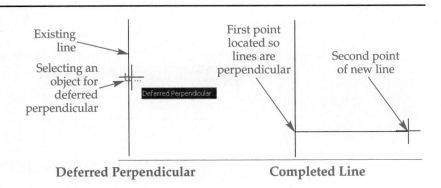

Existing line

Selecting an object for deferred perpendicular

First point located so lines are perpendicular

Second point of new line

Deferred Perpendicular

Deferred Perpendicular   **Completed Line**

Perpendicularity is determined from the point of intersection. Therefore, it is possible to draw a line perpendicular to a circle or arc.

Exercise 7-5

Complete the exercise on the companion website.
www.g-wlearning.com/CAD

Tangent Object Snap

To snap to the point of tangency using the **Tangent** object snap mode, move the crosshairs near an arc, circle, ellipse, elliptical arc, or spline. When you see the tangent marker, pick to locate the point at the exact point of tangency. See **Figure 7-15**.

When drawing an object tangent to two objects, you may need to pick multiple points to fix the point of tangency. Until you identify both endpoints, the object snap specification is for *deferred tangency*. When AutoCAD recognizes both endpoints and calculates the tangency, you can draw the object in the correct location. See **Figure 7-16**.

deferred tangency:
A calculation of the point of tangency that is delayed until you pick both points.

Exercise 7-6

Complete the exercise on the companion website.
www.g-wlearning.com/CAD

Figure 7-15.
Using the **Tangent** object snap to end a line tangent to a circle.

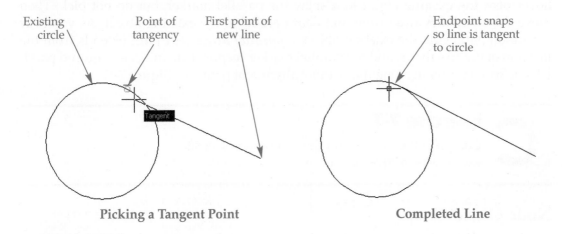

Existing circle

Point of tangency

First point of new line

Endpoint snaps so line is tangent to circle

Picking a Tangent Point

Completed Line

Figure 7-16.
Drawing a line tangent to two circles.

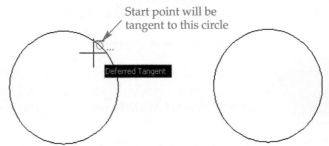

Start point will be tangent to this circle

First Tangent Point Deferred

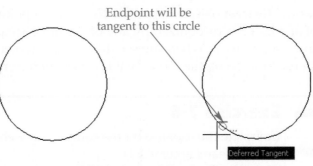

Endpoint will be tangent to this circle

Picking Second Tangent Point

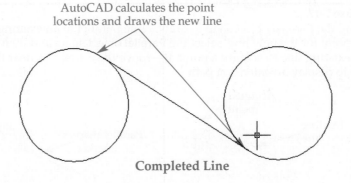

AutoCAD calculates the point locations and draws the new line

Completed Line

Parallel Object Snap

To snap to a point parallel to a line or polyline using the **Parallel** object snap mode, hover near the existing object to display the parallel marker, but do not pick. Then move the crosshairs away from and near parallel to the existing object. As you near a position parallel to the existing object, a *parallel alignment path* extends from the location of the crosshairs, and the parallel marker reappears to indicate acquired parallelism. Specify a point along the parallel alignment path. See **Figure 7-17**.

Exercise 7-7

Complete the exercise on the companion website.
www.g-wlearning.com/CAD

Node Object Snap

Use the **Node** object snap mode to snap to a point drawn using the **POINT**, **DIVIDE**, or **MEASURE** command, or to the origin of an extension line. Move the crosshairs near the node. When you see the node marker and tooltip, pick to locate the point at the exact point.

In order for the **Node** object snap to find a point object, the point must be in a visible display mode. Chapter 8 explains the **POINT**, **DIVIDE**, and **MEASURE** commands and point display mode controls.

Nearest Object Snap

Use the **Nearest** object snap mode to specify a point that is directly on an object, but not at a location recognized by any of the other snap modes. Move the crosshairs near an existing object. When you see the nearest marker and tooltip, pick to locate the point at the location on the object that is closest to the crosshairs.

Exercise 7-8

Complete the exercise on the companion website.
www.g-wlearning.com/CAD

Figure 7-17.
Using the **Parallel** object snap to draw a line parallel to an existing line. A—Select the first endpoint for the new line, select the **Parallel** object snap, and then move the crosshairs near the existing line to acquire a point. B—Move the crosshairs near the location of the parallel line to display an extension path.

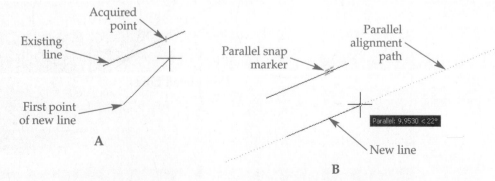

AutoCAD and Its Applications—Basics

Temporary Track Point Snap

The **Temporary track point** snap mode is available only as an object snap override. It allows you to locate a point aligned with or relative to another point. For example, use the **Temporary track point** snap to place the center of a circle at the center of an existing rectangle. At the Specify center point for circle or [3P/2P/Ttr (tan tan radius)]: prompt, select the **Temporary tracking point** snap. Then use the **Midpoint** object snap to pick the midpoint of one of the vertical lines to establish the Y coordinate of the center of the rectangle. See **Figure 7-18A**. When the Specify center point for circle or [3P/2P/Ttr (tan tan radius)]: prompt reappears, reselect the **Temporary tracking point** snap mode. Then use the **Midpoint** object snap mode to pick the midpoint of one of the horizontal lines to establish the X coordinate of the center of the rectangle. See **Figure 7-18B**. Finally, pick to locate the center of the circle where the two tracking vectors intersect, and specify the circle radius. See **Figure 7-18C**.

 The direction in which you move the crosshairs from the temporary tracking point determines the X or Y alignment. Switch between horizontal or vertical tracking as needed.

Snap From

The **From** snap mode is available only as an object snap override and allows you to locate a point using coordinate entry from a specified reference base point. For example, use the **From** snap to place the center of a circle using a polar coordinate entry from the midpoint of an existing line. At the Specify center point for circle or [3P/2P/Ttr (tan tan radius)]: prompt, select the **From** snap mode, and then use the **Midpoint** object snap to pick the midpoint of the line. At the <Offset>: prompt, enter the polar coordinate @2<45 to establish the center of the circle 2 units and at a 45° angle from the midpoint of the line. Specify the radius of the circle to complete the operation. See **Figure 7-19**.

Mid Between 2 Points Snap

The **Mid Between 2 Points** snap function is available only as an object snap override and is very effective for locating a point exactly halfway between two specified points. Use object snaps or coordinate point entry to pick reference points accurately. **Figure 7-20** shows locating the center of a circle between two line endpoints.

Figure 7-18.
Using temporary tracking to locate the center of a rectangle. A—Acquiring the midpoint of the left line. B—Acquiring the midpoint of the bottom line. C—Locating the center point of the circle at the intersection of the tracking vectors.

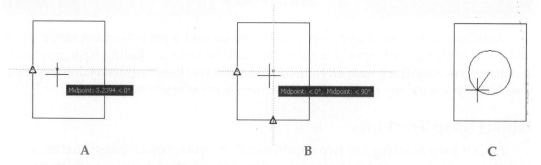

A   B   C

Figure 7-19.
An example of using the **From** point selection mode to locate the center of a circle using the **Midpoint** object snap and polar coordinate entry.

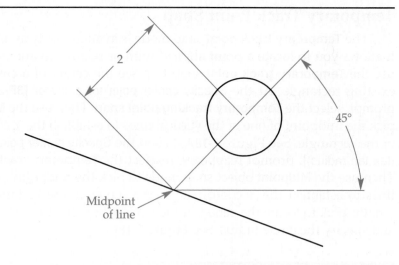

2

45°

Midpoint
of line

Figure 7-20.
Using the **Mid Between 2 Points** option to create a circle with center point located at an exactly equal distance between two line endpoints.

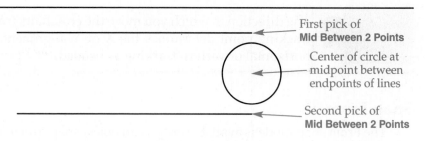

First pick of
Mid Between 2 Points

Center of circle at midpoint between endpoints of lines

Second pick of
Mid Between 2 Points

Object snaps are also important when constructing a parametric drawing, especially for inferring geometric constraints. Chapter 22 explains parametric drafting. 3D object snaps are available for 3D applications, as described in *AutoCAD and Its Applications—Advanced.*

Exercise 7-9

Complete the exercise on the companion website.
www.g-wlearning.com/CAD

alignment paths:
Temporary lines and arcs that coincide with the position of existing objects.

tracking vectors:
Temporary lines that display at specific angles, 0°, 90°, 180°, and 270° by default.

object snap tracking: Mode that provides horizontal and vertical alignment paths for locating points after a point is acquired with object snap.

AutoTrack

AutoTrack offers an object snap tracking mode and a polar tracking mode. Snap and polar tracking are helpful for common drafting tasks, including basic geometric constructions. AutoTrack uses *alignment paths* and *tracking vectors* as drawing aids. Use AutoTrack with any command that requires a point selection.

Object Snap Tracking

Object snap tracking has two requirements: running object snaps must be active, and the crosshairs must hover over the intended selection long enough to acquire the point. Turn object snap tracking on and off by picking the **Object Snap Tracking** button on the status bar, pressing [F11], or checking **Object Snap Tracking On (F11)** in the **Object Snap** tab of the **Drafting Settings** dialog box. Object snap tracking mode works with running object snaps. You must activate object snap tracking, running object snaps, and the appropriate running object snap modes in order for object snap tracking to function properly.

Figure 7-21 shows an example of using object snap tracking with the **Perpendicular** and **Midpoint** running object snaps to draw a line 2 units long, perpendicular to the existing slanted line. Running object snaps, the **Perpendicular** and **Midpoint** running object snap modes, and object snap tracking must be active before you use the **LINE** command to draw the line.

Figure 7-22 shows an example of using object snap tracking with the **Midpoint** running object snap to locate the center point of a circle directly above the midpoint of a horizontal line and to the right of the midpoint of an angled line. Running object snaps, the **Midpoint** running object snap mode, and object snap tracking must be active before you use **Circle, Radius** to draw the circle.

PROFESSIONAL TIP

Use object snap tracking whenever possible to complete tasks that require you to reference locations on existing objects. Often the combination of running object snaps and object snap tracking is the quickest way to construct geometry.

Figure 7-21.
Using object snap tracking to draw a line perpendicular to and at the midpoint of an existing line.

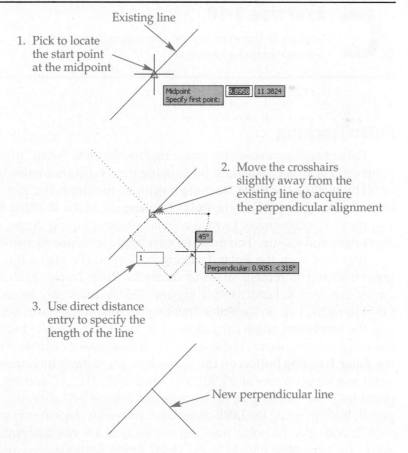

1. Pick to locate the start point at the midpoint

Existing line

Midpoint
Specify first point: 9.8958   11.3824

2. Move the crosshairs slightly away from the existing line to acquire the perpendicular alignment

45°

1

Perpendicular: 0.9051 < 315°

3. Use direct distance entry to specify the length of the line

New perpendicular line

Figure 7-22.
Using object snap tracking to locate the center point of a circle in line with the midpoints of two lines.

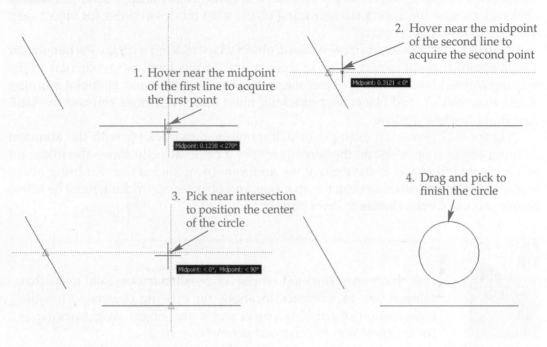

1. Hover near the midpoint of the first line to acquire the first point

2. Hover near the midpoint of the second line to acquire the second point

Midpoint: 0.3121 < 0°

Midpoint: 0.1238 < 270°

3. Pick near intersection to position the center of the circle

4. Drag and pick to finish the circle

Midpoint: < 0°, Midpoint: < 90°

Exercise 7-10

Complete the exercise on the companion website.
www.g-wlearning.com/CAD

Polar Tracking

Polar tracking causes the drawing crosshairs to "snap" to predefined angle increments and is an accurate method of using direct distance entry. Turn polar tracking on or off by picking the **Polar Tracking** button on the status bar, pressing [F10], or checking **Polar Tracking On (F10)** in the **Polar Tracking** tab of the **Drafting Settings** dialog box. As you move the crosshairs toward a polar tracking angle, AutoCAD displays an alignment path and tooltip. The default polar angle increments are 0°, 90°, 180°, and 270°.

Right-click on the **Polar Tracking** button on the status bar and pick an available polar tracking increment angle, or access the **Polar Tracking** tab in the **Drafting Settings** dialog box for total control. See **Figure 7-23**. A quick way to access the **Polar Tracking** tab is to right-click on the **Polar Tracking** button on the status bar and select **Settings...**. Use the **Increment angle** drop-down list to select the angle increments at which polar tracking vectors occur. These are the same angles available when you right-click on the **Polar Tracking** button on the status bar. The default increment is 90, which displays polar tracking vectors at 0°, 90°, 180°, and 270°. The 30° setting shown in **Figure 7-23** provides polar tracking at 30° increments. **Figure 7-24** shows an example of drawing a parallelogram using the **LINE** command and polar tracking set to 30° angle increments.

To add specific polar tracking angles that are not associated with the increment angle, pick the **New** button in the **Polar Angle Settings** area and type an angle in the text box that appears in the **Additional angles** window. The angles you add work with the increment angle setting when you use polar tracking. AutoCAD recognizes only

Figure 7-23.
The **Polar Tracking** tab of the **Drafting Settings** dialog box.

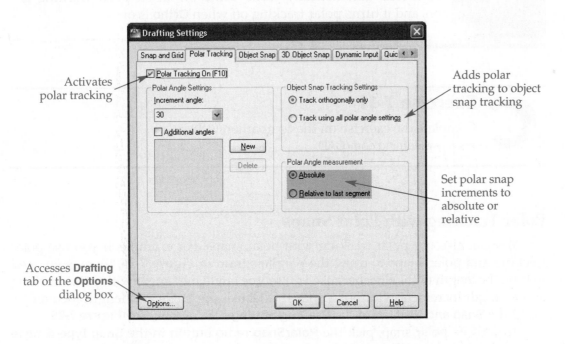

Figure 7-24.
Using polar tracking with 30° angle increments to draw a parallelogram.

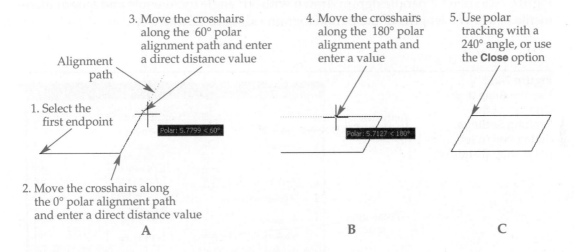

the specific additional angles you enter, not each increment of the angle. Use the **Delete** button to remove angles from the list. Uncheck **Additional angles** to make additional angles inactive.

The **Object Snap Tracking Settings** area sets the angles available with object snap tracking. If you select **Track orthogonally only**, only horizontal and vertical alignment paths are active. If you select **Track using all polar angle settings**, alignment paths are active for all polar snap angles.

The **Polar Angle measurement** setting determines whether the polar snap increments are constant or relative to the previous segment. If you choose **Absolute**, polar snap angles are measured from the base angle of 0° set for the drawing. If you pick **Relative to last segment**, each increment angle is measured from a base angle established by the previously drawn segment.

 AutoCAD automatically turns **Ortho** off when polar tracking is on, and it turns polar tracking off when **Ortho** is on.

 Exercise 7-11

Complete the exercise on the companion website.
www.g-wlearning.com/CAD

Polar Tracking with Polar Snaps

You can also use polar tracking with polar snaps. For example, if you use polar tracking and polar snaps to draw the parallelogram in **Figure 7-24**, there is no need to type the length of the line, because you set the angle increment with polar tracking and a length increment with polar snaps. Establish the angle and length increments using the **Snap and Grid** tab of the **Drafting Settings** dialog box. See **Figure 7-25**.

To activate polar snap, pick the **PolarSnap** radio button in the **Snap type & style** area of the dialog box. This activates the **Polar spacing** area and deactivates the **Snap** area. Set the length of the polar snap increment in the **Polar distance:** text box. If the **Polar distance:** setting is 0, the polar snap distance is the orthogonal snap distance. **Figure 7-26** shows a parallelogram drawn with 30° angle increments and length increments of .75. The lengths of the parallelogram sides are 1.5 and .75.

Figure 7-25.
Use the **Snap and Grid** tab of the **Drafting Settings** dialog box to set polar snap distance.

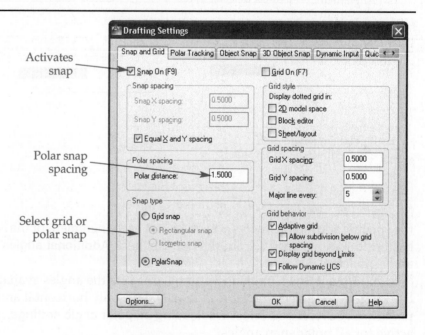

AutoCAD and Its Applications—Basics

Figure 7-26.
Drawing a parallelogram with polar snap. Notice the values that automatically appear in the
input fields.

A B C D

Polar Tracking Overrides

It takes time to set up polar tracking and polar snap options, but it is worth the effort if you intend to draw several objects that can take advantage of these aids. Use polar tracking overrides to define unique polar tracking angles. Polar tracking overrides work for the specified angle whether polar tracking is on or off. To activate a polar tracking override when AutoCAD asks you to specify a point, type a less than symbol (<) followed by the angle. For example, after you access the **LINE** command and pick a first point, enter <30 to set a 30° override. Then move the crosshairs in one of the possible directions and enter a length.

Exercise 7-12

Complete the exercise on the companion website.
www.g-wlearning.com/CAD

AutoSnap and AutoTrack Options

For information about options for controlling the appearance and function of AutoSnap and AutoTrack, go to the companion website (www.g-wlearning.com/CAD), select this chapter, and select **AutoSnap and AutoTrack Options**.

Setting Additional Drawing Aids

For detailed instructions on setting object snaps and polar tracking in your templates to save time and increase efficiency, go to the companion website (www.g-wlearning.com/CAD), select this chapter, and select **Template Development**.

Chapter Review

Answer the following questions. Write your answers on a separate sheet of paper or complete the electronic chapter review on the companion website.
www.g-wlearning.com/CAD

1. Define the term *object snap*.
2. Name the following AutoSnap markers:

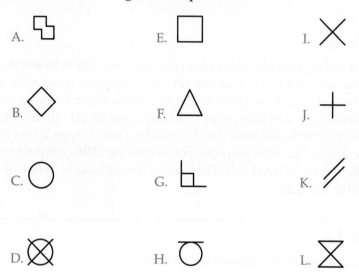

A.   E.   I.

B.   F.   J.

C.   G.   K.

D.   H.   L.

3. Define the term *running object snap*.
4. How do you access the **Drafting Settings** dialog box to change object snap settings?
5. How do you set running object snaps?
6. Why should you activate only the running object snaps you use most often?
7. Describe object snap override.
8. How do you access the **Object Snap** shortcut menu?
9. Where are the four quadrant points on a circle?
10. What does it mean when the tooltip reads Deferred Perpendicular?
11. What is a deferred tangency?
12. Which object snaps depend on acquired points to function?
13. What two display features does AutoTrack use to help you align new objects with existing geometry?
14. What are the two requirements to use object snap tracking?
15. What would you enter to specify a 40° polar tracking override?

Drawing Problems

Start AutoCAD if it is not already started. Start a new drawing from scratch or use an appropriate template of your choice. The template should include layers for drawing the given objects. Add layers as needed. Draw all objects using appropriate layers. Follow the specific instructions for each problem. Use only drawing commands and techniques you have already learned. Use object snap and AutoTrack when possible. Do not draw dimensions or text. Use your own judgment and approximate dimensions when necessary.

▼ Basic

1. Draw the part view shown using object snap modes. Save the drawing as P7-1.

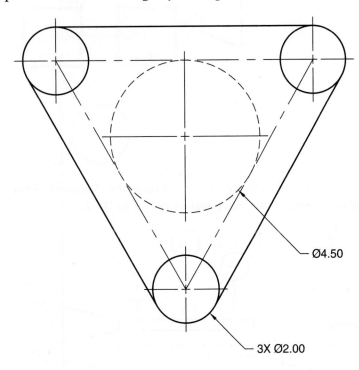

Ø4.50

3X Ø2.00

2. Draw the highlighted objects shown, and then use the object snap modes indicated to draw the remaining objects. Save the drawing as P7-2.

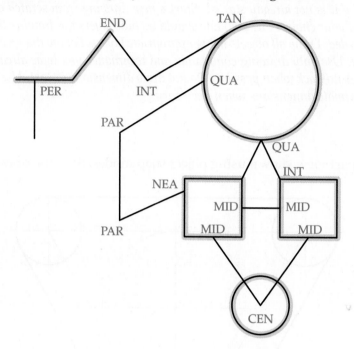

3. Draw the schematic shown using the **Endpoint**, **Tangent**, **Perpendicular**, and **Quadrant** object snap modes. Save the drawing as P7-3.

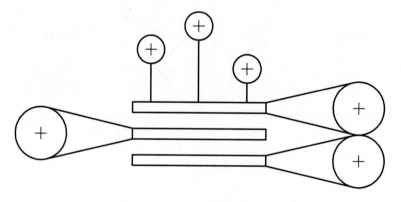

4. Draw the pipe separator shown. Save the drawing as P7-4.

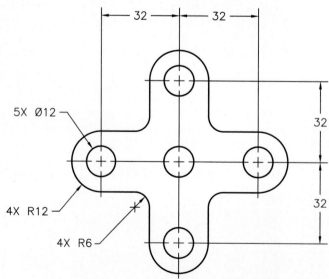

5. Draw the column base detail shown. Save the drawing as P7-5.

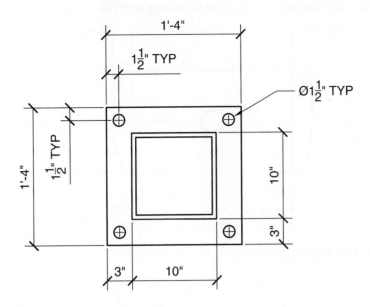

6. Draw the views of the tube hanger shown. Save the drawing as P7-6.

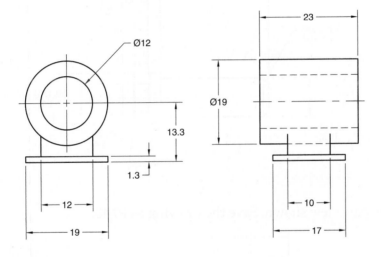

▼ Intermediate

7. Draw the electrical switch schematics shown. Use the **Midpoint**, **Endpoint**, **Tangent**, **Perpendicular**, and **Quadrant** object snap modes. Save the drawing as P7-7.

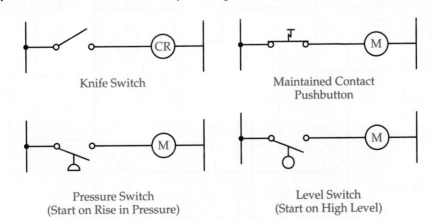

Knife Switch

Maintained Contact
Pushbutton

Pressure Switch
(Start on Rise in Pressure)

Level Switch
(Start on High Level)

8. Draw the pressure cylinder shown. Use the **Arc** option of the **ELLIPSE** command to draw the cylinder ends. Save the drawing as P7-8.

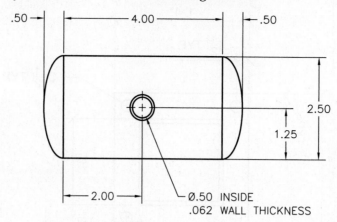

9. Draw the stud shown. Save the drawing as P7-9.

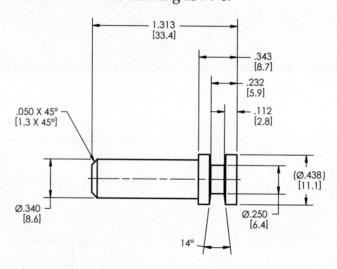

10. Draw the part view shown. Save the drawing as P7-10.

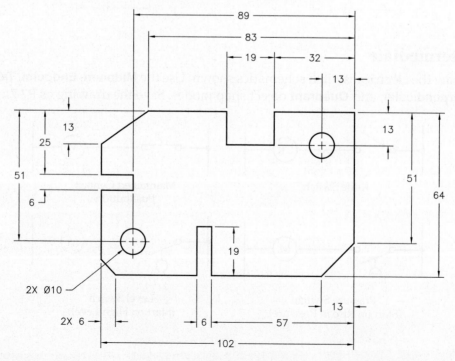

11. Draw the window elevation symbol shown. Save the drawing as P7-11.

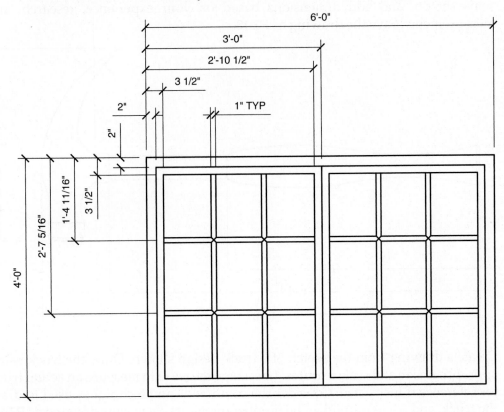

▼ Advanced

12. Use object snap modes to draw the elementary diagram shown. Save the drawing as P7-12.

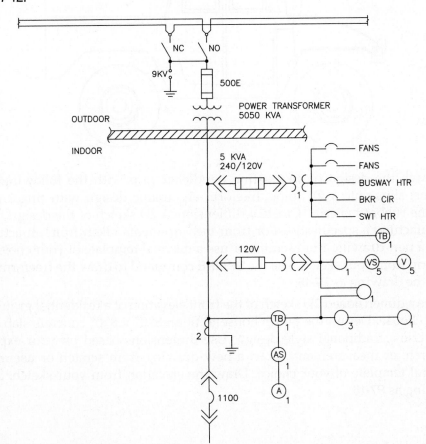

13. Design and draft a hammer similar to the hammer shown. Use the overall dimensions shown, and add dimensions based on your experience, research, and measurements. Save the drawing as P7-13.

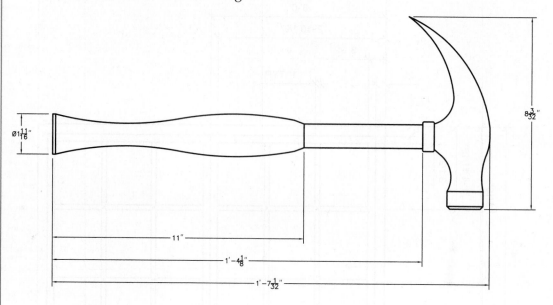

14. Create a drawing from the sketch of a truck design shown. Draw the truck using appropriate size and scale features. Use a tape measure to measure an actual truck for reference if instructed. Consider the commands and techniques used to draw the truck, and try to minimize the number of objects. Save your drawing as P7-14.

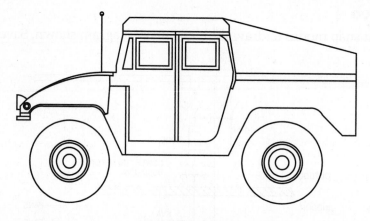
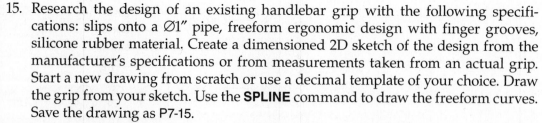

15. Research the design of an existing handlebar grip with the following specifications: slips onto a Ø1" pipe, freeform ergonomic design with finger grooves, silicone rubber material. Create a dimensioned 2D sketch of the design from the manufacturer's specifications or from measurements taken from an actual grip. Start a new drawing from scratch or use a decimal template of your choice. Draw the grip from your sketch. Use the **SPLINE** command to draw the freeform curves. Save the drawing as P7-15.

16. Create a dimensioned 2D sketch of the front elevation of a residential groundwater pump house. Design the pump house to fit an 8'-0" × 8'-0" concrete slab foundation. Use a traditional style design. Use dimensions based on your experience, research, or measurements. Start a new drawing from scratch or use an architectural template of your choice. Draw the elevation from your sketch. Save the drawing as P7-16.

AutoCAD Certified Associate Exam Practice

Answer the following questions. Write your answers on a separate sheet of paper.

1. Which of the labeled points shown can you select using the AutoSnap feature, without entering an object snap override, if the **Midpoint**, **Endpoint**, and **Center** object snaps are set as running object snaps? *Select the one item that best answers the question.*

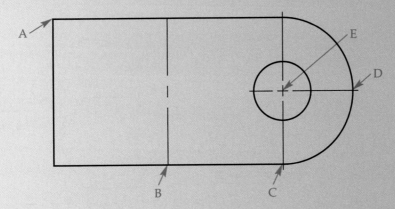

   A. points A, B, and C
   B. points A, C, and D
   C. points A, B, D, and E
   D. points A, B, C, and E
   E. points A and E only

2. Which of the following positions describe quadrant points on a circle? *Select all that apply.*
   A. 0°
   B. 30°
   C. 60°
   D. 90°
   E. 180°

3. Which of the following points can be selected using the **Endpoint** object snap? *Select all that apply.*
   A. a corner of a rectangle
   B. the center of a circle
   C. the end of an arc
   D. a vertex of a polygon
   E. the midpoint of a line

AutoCAD Certified Professional Exam Practice

Follow the instructions in each problem. Write your answers on a separate sheet of paper.

1. **Navigate to this chapter on the companion website and open CPE-07snaps.dwg.**
   Use running object snaps or object snap overrides to create Line A as shown. Restart the **LINE** command and create Line B from the midpoint of Line A, extending 1.5 units upward vertically. What are the coordinates of the upper endpoint of Line B?

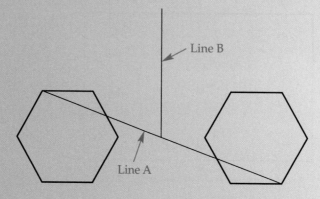

2. **Navigate to this chapter on the Student Web site and open CPE-07polar.dwg.**
   Use the **Drafting Settings** dialog box to create additional polar angles of 20, 40, and 80. Make sure **Polar Angle measurement** is set to **Absolute**. Then use polar tracking to create the following line segments:
   A. A line starting at the right endpoint of the existing line and extending 3 units at 80°
   B. A line starting at the upper endpoint of the previous line and extending 4.5 units at 20°
   C. A line starting at the right endpoint of the previous line and extending 3 units upward vertically
   D. A line starting at the upper endpoint of the previous line and extending 6.3 units at 40°

   What are the coordinates of the upper endpoint of the last line you drew?

Chapter 8

Construction Tools and Multiview Drawings

Learning Objectives

After completing this chapter, you will be able to:

✓ Use the **OFFSET** command to draw parallel and concentric objects.
✓ Place construction points.
✓ Mark points on objects at equal lengths using the **DIVIDE** command.
✓ Mark points on objects at designated increments using the **MEASURE** command.
✓ Create construction lines using the **XLINE** and **RAY** commands.
✓ Create multiview drawings.

This chapter explains how to create parallel offsets, divide objects, place point objects, and use construction lines. You can use these skills and your existing geometric construction skills to create multiview drawings. This chapter describes tools and methods for producing accurate geometric constructions. Construction tools do not constrain, or apply relationships between, objects. Chapter 22 explains how to use parametric drafting tools to constrain objects.

Parallel Offsets

The **OFFSET** command is a common geometric construction tool that is useful for many different drafting tasks. For example, you can offset lines, polylines, or xlines to construct multiviews or form the thickness of architectural floor plan walls. Offset circles, arcs, or other curves to form concentric objects. For example, you can offset a circle to create the wall thickness of a pipe.

Specifying the Offset Distance

Often the best way to use the **OFFSET** command is to enter an offset value at the Specify offset distance or [Through/Erase/Layer] <current>: prompt. For example, to draw a circle concentric to and 1 unit from an existing circle, access the **OFFSET** command and specify an offset distance of 1. Pick the circle to offset, and then pick the side of the circle on which you want the offset to occur. See Figure 8-1. The **OFFSET** command remains active, allowing you to pick another object to offset using the same offset distance. To exit, press [Enter], [Esc], or the space bar, or choose the **Exit** option.

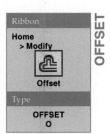

Ribbon

Home
> Modify

Offset

Type

OFFSET
O

OFFSET

Figure 8-1.
Drawing an offset circle using a designated distance.

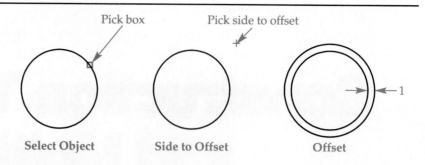

Pick box

Pick side to offset

Select Object Side to Offset Offset

1

Select objects to offset individually. No other selection option, such as window or crossing selection, works for selecting objects to offset.

PROFESSIONAL TIP

When using most commands that prompt you to specify a value, such as distance or height, if you do not know the numeric value, an alternative is to pick two points. The distance between the points sets the value. You typically pick two points on existing objects to specify the appropriate value. Use object snaps, AutoTrack, or coordinate entry to make accurate selections.

Through Option

Another option to specify the offset distance is to pick a point through which the offset occurs. Access the **OFFSET** command and, instead of picking an object to offset, specify the Through option at the Specify offset distance or [Through/Erase/Layer] <current>: prompt. Then pick the object to offset, and pick the point through which the offset occurs. See **Figure 8-2**. The **OFFSET** command remains active, allowing you to pick another object to offset using the **Through** option. Exit the command when you are finished offsetting.

Figure 8-2.
Drawing an offset through a given point.

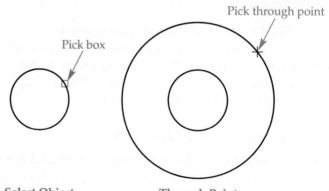

Pick through point

Pick box

Select Object Through Point

Erasing the Source Object

Use the **Erase** option of the **OFFSET** command to erase the source, or original, object during the offset. Start the **OFFSET** command, activate the **Erase** option, and choose **Yes** at the Erase source object after offsetting? prompt. The **Yes** option remains set as the default until you change the setting to **No**. Be sure to change the **Erase** setting back to **No** if the source object should remain the next time you use the **OFFSET** command. Exit the command when you are finished offsetting.

Layer Option

By default, offsets use the same properties as the source object, including layer. Use the **Layer** option of the **OFFSET** command to place the offset object on the current layer, regardless of the layer assigned to the source object. First, make the layer to apply to the offset current. Then start the **OFFSET** command, activate the **Layer** option, and choose **Current** at the **Enter layer option for offset objects:** prompt. The **Current** option remains set as the default until you change the setting to **Source**. Be sure to change the **Layer** setting back to **Source** if the layer assigned to the source object should apply to the offset the next time you use the **OFFSET** command. Exit the command when you are finished offsetting.

Multiple Option

The **Multiple** option is useful for offsetting more than once using the same distance between objects, without having to reselect the object to offset. Access the **OFFSET** command, specify the offset distance, and pick the source object. Then select the **Multiple** option and begin picking to specify the offset direction. See **Figure 8-3**. Exit when you are finished offsetting.

 You can use the **Undo** option, when available, to undo the last offset without exiting the **OFFSET** command.

Figure 8-3.
Use the **Multiple** option to create multiple offsets of the same distance, without having to reselect the source object.

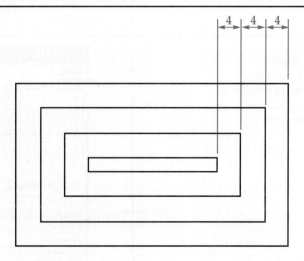

Exercise 8-1

Complete the exercise on the companion website.
www.g-wlearning.com/CAD

Drawing Points

POINT

Ribbon

Home
> Draw

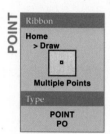

Multiple Points

Type

POINT
PO

Point objects are useful for identifying locations on a drawing and marking positions on objects. You can use the **POINT** command to draw points anywhere in the drawing. Use any appropriate method to specify the location of a point object. To place a single point object and then exit the **POINT** command, enter **POINT** or **PO** at the keyboard. To draw multiple points without exiting the **POINT** command, access the **Multiple Points** function from the ribbon. Press [Esc] to exit the command.

Setting Point Style

DDPTYPE

Ribbon

Home
> Utilities

Point Style

Type

DDPTYPE

Points appear as one-pixel dots by default. The default appearance is functional and does not interfere with objects. However, the one-pixel style is difficult to see and can get lost. Change the point style and size using the **Point Style** dialog box, shown in **Figure 8-4**. Pick the image of the point style you want to use.

Enter a value in the **Point Size:** text box to set the point size. Pick the **Set Size Relative to Screen** button to change the point size relative to the screen magnification, or zoom level. You may need to regenerate the display to view the relative sizes. Pick the **Set Size in Absolute Units** button to make points appear the same size regardless of the screen magnification. See **Figure 8-5**. Pick the **OK** button to exit the **Point Style** dialog box. All existing and new points change to the current style and size.

Exercise 8-2

Complete the exercise on the companion website.
www.g-wlearning.com/CAD

Figure 8-4.
The **Point Style** dialog box provides a quick way to select the point style and change the point size.

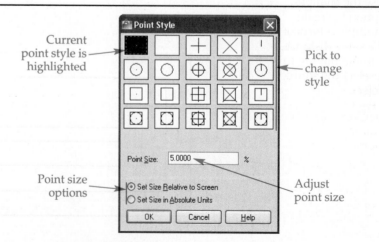

Current point style is highlighted

Pick to change style

Point size options

Adjust point size

Size Setting	Original Point Size	2X Zoom	.5 Zoom
Relative to Screen	⊠	⊠	⊠
Absolute Units	⊠	⊠	⊠

Figure 8-5.
Points sized with the **Set Size Relative to Screen** setting change size as you zoom in and out. Points sized with the **Set Size in Absolute Units** setting remain at a constant size.

Marking an Object at Specified Increments

Use the **DIVIDE** command to place point objects or *blocks* at equally spaced locations on a line, circle, arc, polyline, or spline. AutoCAD calculates the distance between marks based on the number of segments you specify. The **DIVIDE** command does not break an object into an equal number of segments. Access the **DIVIDE** command and select the object to mark. Enter the number of segments, and then exit the command. The point style determines the appearance of the point objects. **Figure 8-6** shows marking seven segments with points.

The **Block** option of the **DIVIDE** command allows you to place a block at each increment, instead of a point object. Select the **Block** option at the Enter the number of segments or [Block]: prompt to insert a block. AutoCAD asks if the block should align with the object, such as rotate around a circle. You will learn about blocks later in this textbook.

block: A symbol or shape saved for repeated use.

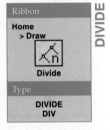

DIVIDE

Ribbon
Home
> Draw

Divide

Type
DIVIDE
DIV

Marking an Object at Specified Distances

Use the **MEASURE** command to place point objects or blocks a specified distance apart on a line, circle, arc, polyline, or spline. In contrast to the **DIVIDE** command, the **MEASURE** command does not break an object at specific lengths. The length of each segment and total length of the object determine the number of segments.

Access the **MEASURE** command and select the object to mark. Measurement begins at the end closest to where you pick the object. Enter the distance between points, and then exit the command. All increments are equal to the specified segment length except the last segment, which may be shorter. The point style determines the appearance of point objects. **Figure 8-7** shows marking .75 unit length segments with points. Use the **Block** option of the **MEASURE** command to place a block at each interval.

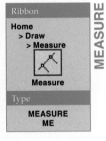

MEASURE

Ribbon
Home
> Draw
> Measure

Measure

Type
MEASURE
ME

Figure 8-6.
Marking seven equal segments with points on a circle and a line using the **DIVIDE** command. An × point style replaces the default point appearance in these examples. Notice that points do not appear at the endpoints of open objects, such as the line.

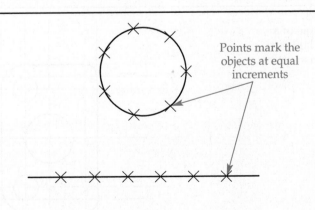

Points mark the objects at equal increments

Figure 8-7.
Using the **MEASURE** command to place point objects on a line at .75 unit intervals. Notice that the last segment may be short, depending on the specified interval and total length of the object.

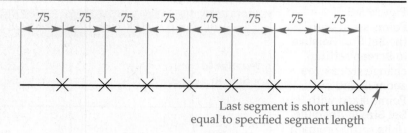

.75 .75 .75 .75 .75 .75 .75 .75

Last segment is short unless / equal to specified segment length

Exercise 8-3

Complete the exercise on the companion website.
www.g-wlearning.com/CAD

Construction Lines and Rays

construction lines: Lines commonly used to lay out a drawing.

The tracking vectors and alignment paths available with the object snap and AutoTrack tools are examples of *construction lines* generated by AutoCAD. Object snap and AutoTrack are very efficient for constructing geometry because vector and alignment lines appear as you draw. Often, however, drawings require construction lines that remain on-screen for reference and future use. You can draw construction geometry using any drawing command, such as **LINE**, **ARC**, or **CIRCLE**. The **XLINE** and **RAY** commands are specifically designed for adding construction lines to help lay out a drawing. See **Figure 8-8**.

PROFESSIONAL TIP

Create construction geometry on a separate construction layer. Use an appropriate layer name, such as **CONST**, **CONSTRUCTION**, or **A-ANNO-NPLT**. Turn off or freeze the construction layer when you do not need it, or you can easily recognize and erase objects drawn on the construction layer if necessary.

Figure 8-8.
An example of a drawing laid out using construction lines.

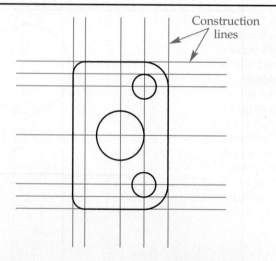

Construction lines

AutoCAD and Its Applications—Basics

Using the XLINE Command

Use the **XLINE** command to draw an infinitely long AutoCAD construction line object, or *xline*. To draw an xline, specify the location of the first point through which the xline passes, or *root point*. Then select a second point through which the xline passes. The **XLINE** command remains active, and the initial root point acts as an axis point. Continue locating additional points from the same root point to create additional xlines. See Figure 8-9. To exit, right-click or press [Enter], [Esc], or the space bar.

Hor and Ver Options

Xline options are available as alternatives to selecting two points. The **Hor** option draws a horizontal xline through a single specified point. The **Ver** option draws a vertical xline through a single specified point. Use the **Hor** or **Ver** option when you know construction lines should be horizontal or vertical. The **XLINE** command remains active with the **Hor** or **Ver** option, allowing you to draw additional horizontal or vertical xlines. Exit when you are finished placing xlines.

Ang Option

Use the **Ang** option to draw an xline at a precise angle through a selected point. Specify an angle by entering a value or picking two points, and then pick a point through which the xline passes. The **Reference** function of the **Ang** option allows you to reference the angle of an existing line object to use as the xline angle. The **Reference** function is useful when you do not know the angle of the xline, but you do know the angle between an existing object and the xline. See Figure 8-10.

XLINE

Ribbon
Home
> Draw

Construction Line

Type

XLINE
XL

xline: A construction line in AutoCAD that is infinite in both directions; helpful for creating accurate geometry and multiviews.

root point: The first point specified to create a construction line or ray.

Figure 8-9.
A—Using the **XLINE** command to draw an infinitely long construction line by specifying two points through which the line passes. B—Constructing multiple xlines that intersect the root point before exiting the command.

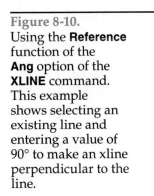
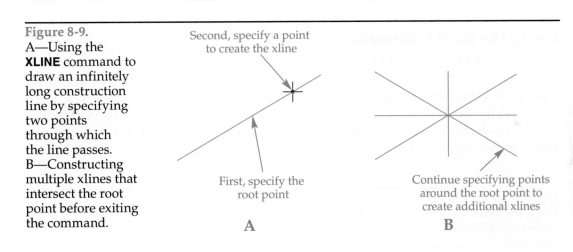

Second, specify a point to create the xline

First, specify the root point

Continue specifying points around the root point to create additional xlines

A

B

Figure 8-10.
Using the **Reference** function of the **Ang** option of the **XLINE** command. This example shows selecting an existing line and entering a value of 90° to make an xline perpendicular to the line.

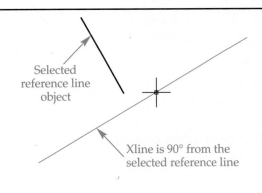

Selected reference line object

Xline is 90° from the selected reference line

Bisect Option

The **Bisect** option draws an xline that bisects a specified angle, using the root point as the vertex. See Figure 8-11. The **Bisect** option is useful for bisecting an angle and similar applications.

Offset Option

The **Offset** option draws an xline parallel to a selected linear object. The **Offset** option of the **XLINE** command functions much like the **OFFSET** command, except that the **Offset** option of the **XLINE** command offsets an xline from the selected object, instead of the object itself. As with the **OFFSET** command, specify an offset distance or use the **Through** option to pick a point through which to draw the xline.

 Although xlines are infinite, they do not change the drawing extents, and they have no effect on zooming operations.

 ## Exercise 8-4

Complete the exercise on the companion website.
www.g-wlearning.com/CAD

Using the RAY Command

Use the **RAY** command to draw an AutoCAD *ray* object. A ray begins at an origin point and extends through another point, similar to the default option of the **XLINE** command. However, a ray is infinite only in the direction of the second point. To draw a ray, specify the location of the root point, followed by a second point through which the ray passes. Continue locating points to create additional rays from the same root point. To exit, right-click or press [Enter], [Esc], or the space bar.

Figure 8-11.
Using the **Bisect** option of the **XLINE** command.

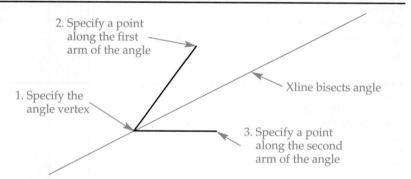

2. Specify a point along the first arm of the angle

Xline bisects angle

1. Specify the angle vertex

3. Specify a point along the second arm of the angle

AutoCAD and Its Applications—Basics

Multiview Drawings

A *multiview drawing* presents views of a product. The general multiview concept applies to most drafting fields. Mechanical drafting uses multiview drawings to represent part and assembly features. Architectural drafting uses plans, elevations, and sections to show building construction and layout. Electronics drafting uses schematic diagrams with electronic symbols to show circuit layout. Civil drafting uses plans and profiles to show the topography of land.

This textbook explains multiview drawings based on the ASME Y14.3 standard, *Multiview and Sectional View Drawings*, and discipline-specific drafting practices. Multiview drawing views are formed through *orthographic projection*. The imaginary *projection plane* is parallel to the object. Thus, the line of sight is perpendicular to the object. The result is 2D views of a 3D object. See **Figure 8-12**.

Six 2D orthographic views are possible: front, right side, left side, top, bottom, and rear. The six orthographic views show the principle sides of an object and are drawn in a standard arrangement for readability. The front view is the central, and typically most important, view. Other views occur around the front view and are directly aligned with, or projected from, the front view when possible. Notice in **Figure 8-13** that the horizontal and vertical edges in the front view align with the corresponding edges in the other views.

multiview drawing:
A presentation of drawing views created through orthographic projection.

orthographic projection:
Projecting object features onto an imaginary plane.

projection plane:
An imaginary projection plane parallel to the object.

Figure 8-12.
Obtaining a front view with orthographic projection.

Line of sight

Projection plane

Three-dimensional object

Two-dimensional drawing

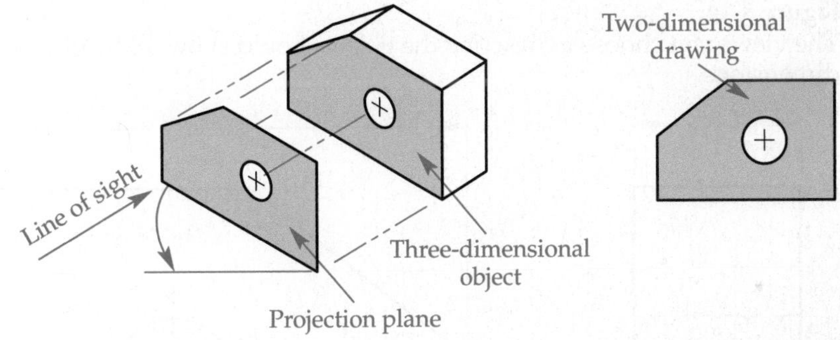

Figure 8-13.
Arrangement of the six orthographic views. The front view is typically central.

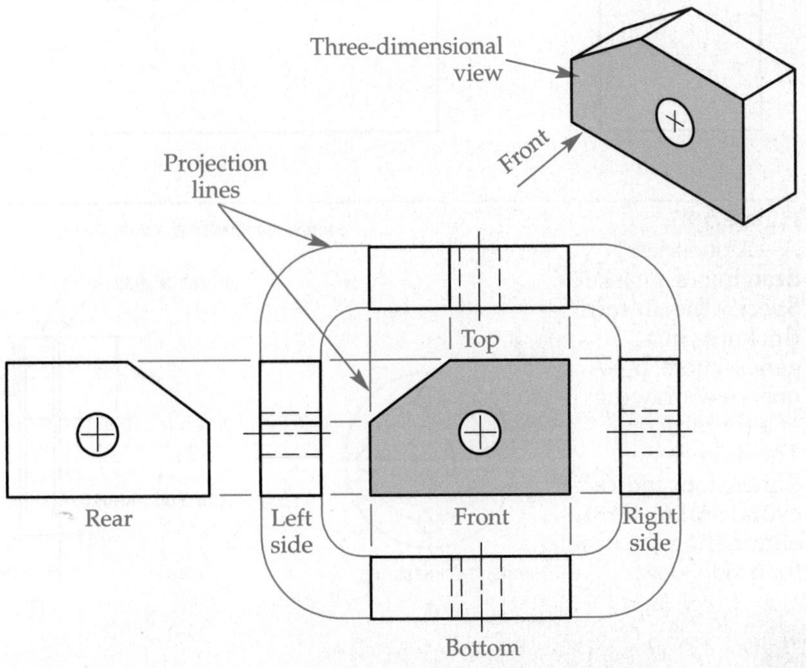

Three-dimensional view

Front

Projection lines

Top

Rear

Left side

Front

Right side

Bottom

Selecting the Front View

The front view is central to most multiview drawings. Consider the following rules when selecting the front view:
- Most descriptive
- Most natural position
- Most stable position
- Provides the longest dimension
- Contains the least number of hidden features

Choosing Additional Views

Select additional views relative to the front view. Few products require all six views. The required number of views depends on the complexity of the object. Use only enough views to describe the object completely. Drawing too many views is time-consuming and can clutter the drawing. **Figure 8-14** shows an example of an object that needs only two views. The two views completely describe the width, height, depth, and features of the object. In some cases, a single view is enough to describe the object. The example shown in **Figure 8-14** could be a one-view drawing with a note specifying the uniform depth. You can often draw a thin part that has a uniform thickness, such as a gasket, with one view. If necessary, provide the thickness or material specification as a general note or in the title block. See **Figure 8-15**.

Figure 8-14.
The views you choose to describe the object should show all height, width, and depth dimensions.

Figure 8-15.
A—A one-view drawing of a gasket. Specify the uniform thickness in a general note. B—A one-view drawing of a thumbscrew. The diameter dimensions indicate cylindrical features, eliminating the need for a side view.

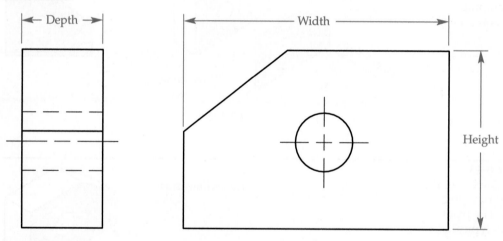

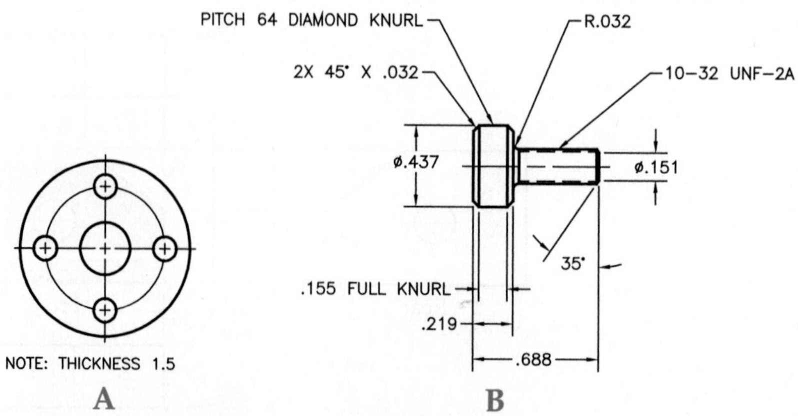

Auxiliary Views

You can sometimes completely describe an object using one or more of the six standard views. However, you must use an *auxiliary view* to describe a surface that appears *foreshortened* in standard orthographic views. Draw an auxiliary view using projection lines perpendicular to a slanted surface. One projection line is sometimes included on the drawing to connect the auxiliary view to the view where the slanted surface appears as a line. The resulting auxiliary view shows the surface at its true size and shape. A *partial auxiliary view* is enough for most applications. See **Figure 8-16**.

Removed Views

Sometimes there is not enough room on a drawing to project directly from one view to another. This requires that you create a *removed view* to locate a view elsewhere on the drawing. An auxiliary view is a common example of a removed view in mechanical drafting, but you can relocate any view if necessary. See **Figure 8-17**. In

auxiliary view: View used to show the true size and shape of a foreshortened surface.

foreshortened: A surface at an angle to the line of sight. Foreshortened surfaces appear shorter than their true size and shape.

partial auxiliary view: An auxiliary view that shows a specific inclined surface of an object, rather than the entire object.

removed view: A view removed from alignment with other views when drawing space is unavailable.

Figure 8-16. Auxiliary views show the true size and shape of an inclined surface. Use a partial auxiliary view to show only the inclined surface, because the other features appear foreshortened and reduce clarity.

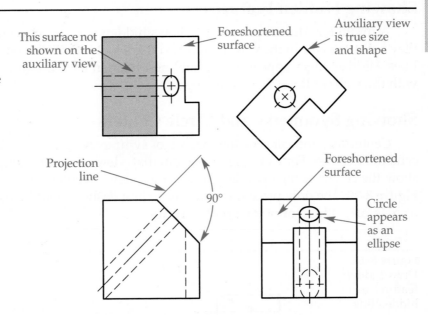

Figure 8-17. A viewing-plane line creates the flexibility to move a view to a location where there is enough space for the view. This process creates a removed view.

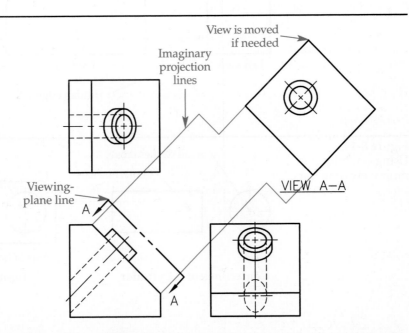

other disciplines, such as architectural drafting, views may not align and often occur on different sheets.

Draw a *viewing-plane line* parallel to the view in which the surface appears as a line. The standard viewing-plane line terminates with bold arrowheads that point toward the surface. A letter labels each end of the viewing-plane line. The letters correlate to the removed view title, such as VIEW A-A, below the removed view to key the viewing plane with the removed view. When you remove more than one view from direct projection, labels continue with B-B through Z-Z, if necessary. Do not use letters I, O, Q, S, X, and Z, because they can be confused with numbers.

> **NOTE** A removed view retains the same angle as if it were projected directly, which is especially important for an auxiliary view.

Showing Hidden Features

A multiview drawing typically shows hidden features of an object, even though they are not visible in the view at which you are looking. Visible edges appear as object lines. Hidden edges appear as hidden lines. Hidden lines are thin to provide contrast with thick object lines. See **Figure 8-18**.

Showing Symmetry and Circle Centers

Centerlines indicate the line or axis of symmetry of symmetrical objects and the centers of circles. For example, in the circular view of a cylinder, centerlines cross to show the center of the cylinder. In the other view, a centerline identifies the axis. See **Figure 8-19**. The only place the small centerline dashes should cross is at the center of a circle, arc, ellipse, or other circular feature.

Figure 8-18.
Draw hidden features using hidden lines.

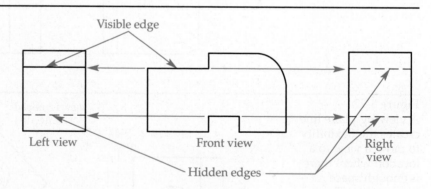

Visible edge

Left view   Front view   Right view

Hidden edges

Figure 8-19.
Using centerlines in multiview drawings.

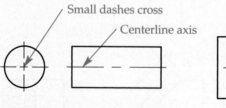

Small dashes cross

Centerline axis

Centerlines of a Cylinder

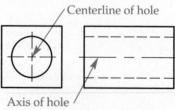

Centerline of hole

Axis of hole

Centerlines of a Hole

Multiview Construction

You can construct a multiview drawing using a variety of techniques, depending on the objects needed, personal working preference, and the information you know about the size and shape of items. Use a combination of construction methods and tools, including coordinate point entry, object snaps, AutoTrack, and construction lines, to produce multiviews.

Orthographic Views

Figure 8-20 shows an example of using object snap tracking and a running **Endpoint** object snap mode to locate points for a left-side view by referencing points on the existing front view. Notice that the AutoTrack alignment path in **Figure 8-20A** provides a temporary construction line. Polar tracking vectors offer a similar temporary construction line. **Figure 8-20B** shows the complete front and left-side views.

Figure 8-21 shows an example of using construction lines to form three views. This example shows offsetting vertical and horizontal xlines to form an xline grid. Use the **Intersection** object snap mode to select the intersecting xlines when drawing line and arc endpoints and the center point of the arc. Notice that a single infinitely long xline can provide construction geometry for multiple views.

Exercise 8-5

Complete the exercise on the companion website.
www.g-wlearning.com/CAD

Figure 8-20.
An example of using object snap tracking to create an additional view. A—Referencing points from the front view to establish the first line of the left-side view. B—The completed front and left-side views.

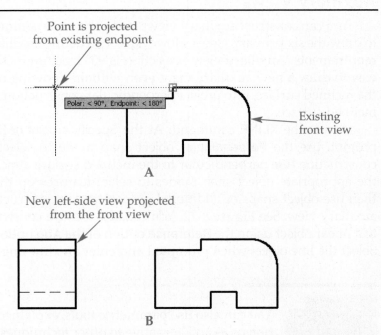

Point is projected from existing endpoint

Polar: < 90°, Endpoint: < 180°

Existing front view

A

New left-side view projected from the front view

B

Figure 8-21.
Using a complete grid of construction lines to form a multiview drawing by "connecting the dots" at the intersections of the construction lines. You can quickly draw the rectangular outlines of the right-side, left-side, and top views using the **RECTANGLE** command.

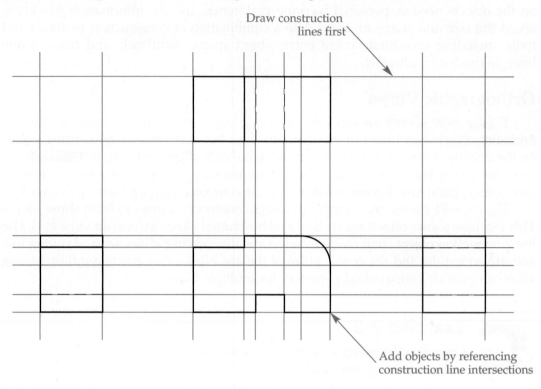

Draw construction
lines first

Add objects by referencing
construction line intersections

Auxiliary Views

You can construct auxiliary views using the same commands and options you use to draw the six primary views. However, constructing auxiliary views presents unique requirements. Auxiliary view projection is 90° from an inclined surface. An effective way to draw a new auxiliary view, even without knowing or calculating the angle of the inclined surface, is to project perpendicular construction lines from features on the inclined surface.

Access the **XLINE** command. At the Specify a point or [Hor/Ver/Ang/Bisect/Offset]: prompt, use the **Perpendicular** object snap mode to select the inclined surface. A construction line perpendicular to the inclined surface attaches to the crosshairs. Use the appropriate object snap modes to select features on the existing view. You can then use object snaps or additional perpendicular construction lines to complete the auxiliary view. See Figure 8-22. You can also make a construction line perpendicular to a linear object using the **Reference** option of the **Ang** option of the **XLINE** command. Select the line object when prompted and enter an xline angle of 90.

You can also use parametric tools, explained in Chapter 22, in addition or as an alternative to other techniques for constructing multiview drawings.

Figure 8-22.
Using construction lines drawn perpendicular to an inclined surface on an existing view to construct an auxiliary view. The completed top and auxiliary views are shown for reference.

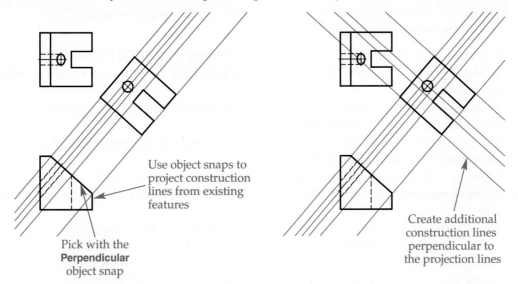

Use object snaps to project construction lines from existing features

Pick with the **Perpendicular** object snap

Create additional construction lines perpendicular to the projection lines

---

Exercise 8-6

Complete the exercise on the companion website.
www.g-wlearning.com/CAD

Template Development Chapter 8

For detailed instructions on choosing a more appropriate point style to use for construction purposes, go to the companion website (www.g-wlearning.com/CAD), select this chapter, and select **Template Development**.

Chapter Test

Answer the following questions. Write your answers on a separate sheet of paper or complete the electronic chapter review on the companion website.
www.g-wlearning.com/CAD

1. List two ways to establish an offset distance using the **OFFSET** command.
2. Which option of the **OFFSET** command allows you to remove the source offset object?
3. How do you draw a single point, and how do you draw multiple points?
4. How do you access the **Point Style** dialog box?
5. If you use the **DIVIDE** command and nothing seems to happen, what should you do to make the points visible?
6. How do you change the point size in the **Point Style** dialog box?
7. What command can you use to place point objects that mark 24 equal segments on a line?
8. What is the difference between the **DIVIDE** and **MEASURE** commands?
9. Why is it a good idea to put construction lines on their own layer?
10. Name the command that allows you to draw infinite construction lines.
11. Name the option that allows you to bisect an angle with a construction line.
12. What is the difference between the construction lines drawn with the command identified in Question 10 and rays drawn with the **RAY** command?
13. What ASME drafting standard applies to multiview drawings?
14. Provide at least four guidelines for selecting the front view of an orthographic multiview drawing.
15. How do you determine how many views of an object are necessary in a multiview drawing?
16. When can you describe a part with only one view?
17. When does a drawing require an auxiliary view?
18. List two methods of aligning the views in a multiview drawing.
19. What is the angle of projection from the slanted surface into the auxiliary view?
20. Describe an effective method of constructing an auxiliary view even if you do not know the angle of the inclined surface.

Drawing Problems

Start AutoCAD if it is not already started. Start a new drawing from scratch or use an appropriate template of your choice. The template should include layers for drawing the given objects. Add layers as needed. Draw all objects using appropriate layers. Follow the specific instructions for each problem. Use only drawing commands and techniques you have already learned. Use object snap and AutoTrack when possible. Do not draw dimensions or text. Use your own judgment and approximate dimensions when necessary.

▼ Basic

1. Draw the front and side views of the offset support shown. Save the drawing as **P8-1**.

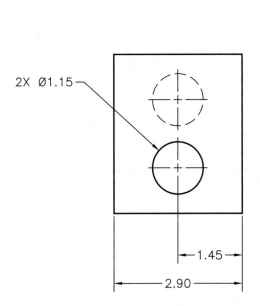
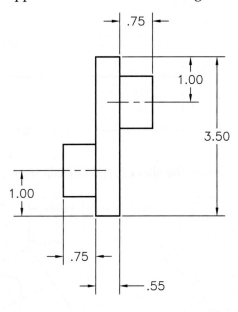

2. Draw the front and top views of the hitch bracket shown. Save the drawing as **P8-2**.

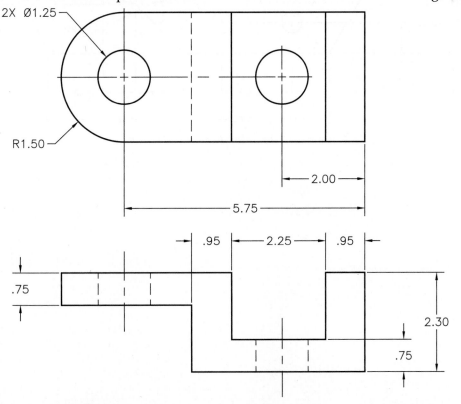

---

3. Draw the spring shown using the **PLINE** command with a width of .024. Save the drawing as P8-3.

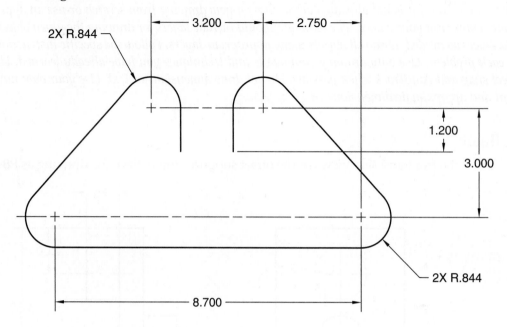

4. Draw the sheet metal chassis shown. Save the drawing as P8-4.

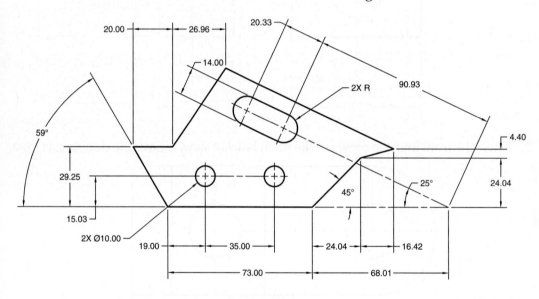

5. Draw the views of the elbow shown. Save the drawing as P8-5.

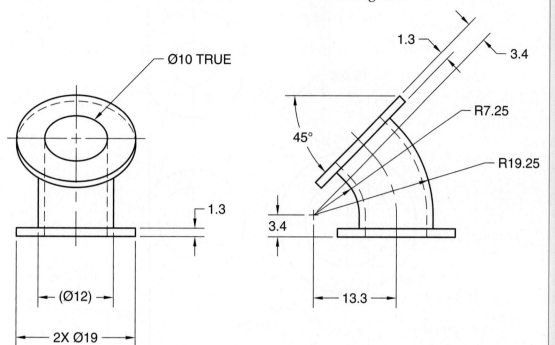

6. Draw the part view shown. Save the drawing as P8-6.

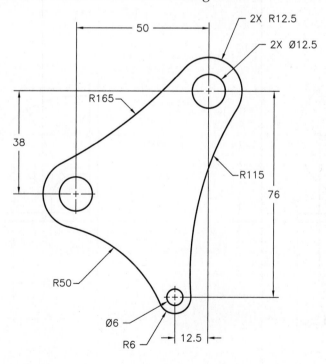

7. Draw the part view shown. Save the drawing as P8-7.

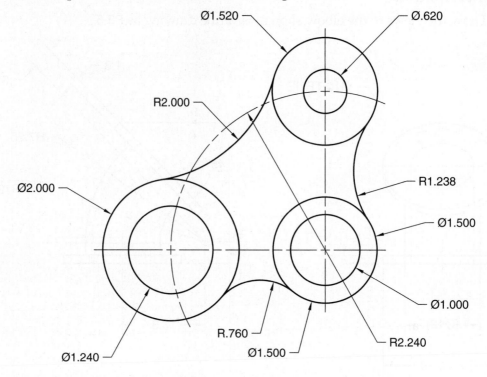

Ø1.520 — Ø.620
R2.000
R1.238
Ø2.000
Ø1.500
Ø1.000
R.760
R2.240
Ø1.240
Ø1.500

8. Draw the views of the elbow shown. Save the drawing as P8-8.

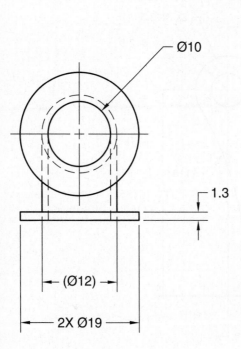

Ø10
1.3
(Ø12)
2X Ø19

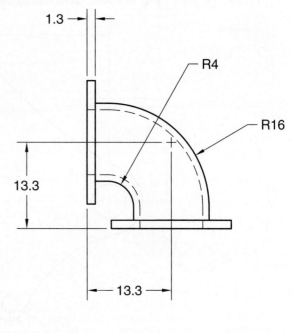

1.3
R4
R16
13.3
13.3

9. Use the **OFFSET** command to draw the elevation of the desk shown. Center 1″ × 4″ rectangular drawer handles 2″ below the top of each drawer. The top of the legs begin 1″ from the edge of the bottom of the desk. Save the drawing as P8-9.

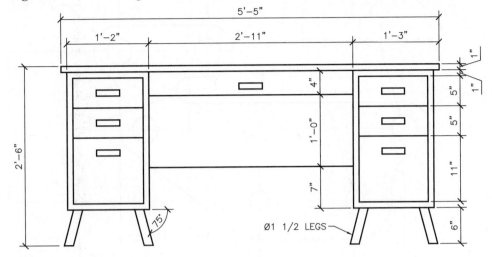

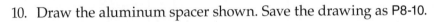

10. Draw the aluminum spacer shown. Save the drawing as P8-10.

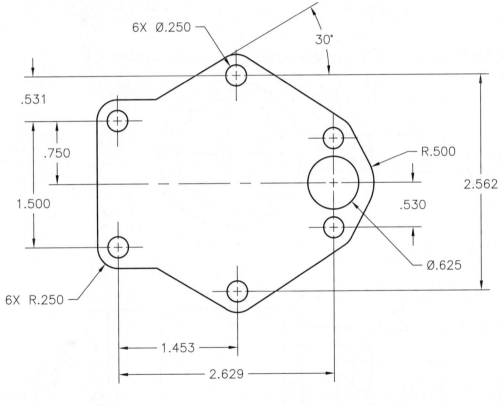

11. Draw the gasket shown. Save the drawing as **P8-11**.

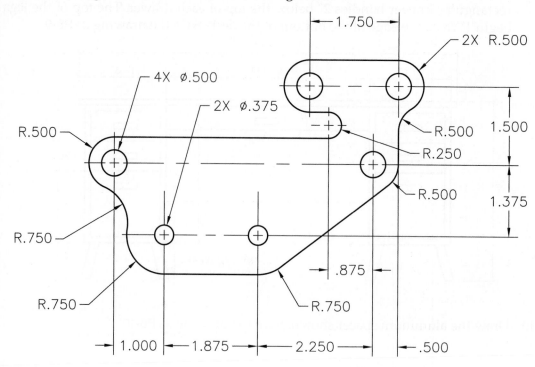

12. Draw the views of the cup shown. Save the drawing as **P8-12**.

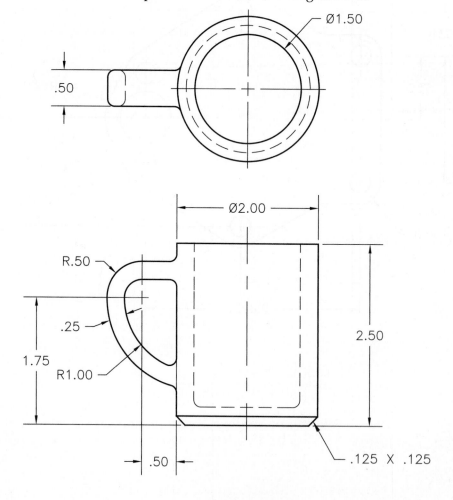

FILLETS AND ROUNDS R.10

13. Draw the views of the bushing shown. Save the drawing as **P8-13**.

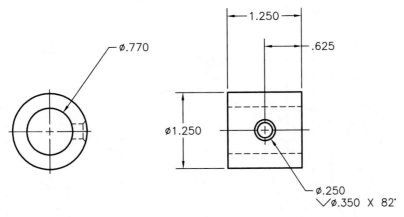

ø.770

1.250

.625

ø1.250

ø.250
∨ø.350 X 82°

14. Draw the views of the wrench shown. Save the drawing as **P8-14**.

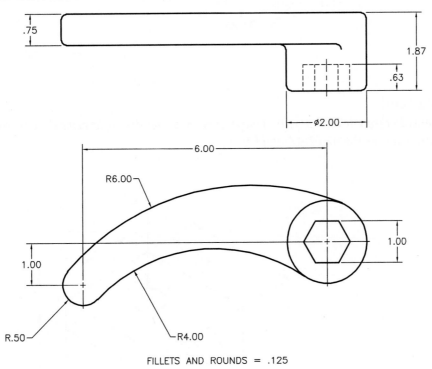

.75

1.87

.63

ø2.00

6.00

R6.00

1.00

1.00

R.50

R4.00

FILLETS AND ROUNDS = .125

15. Draw the views of the support shown. Save the drawing as P8-15.

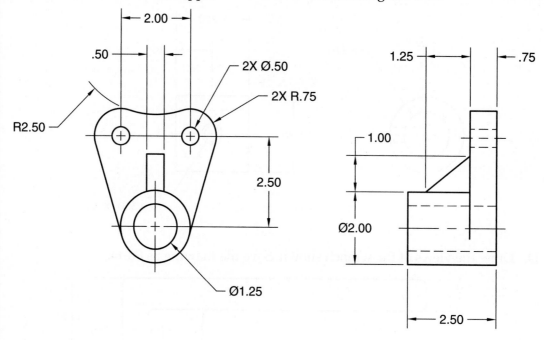

▼ Advanced

For Problems 16 through 18, draw the orthographic views needed to describe the part completely. Save the drawings as P8-16, P8-17, and P8-18.

16.

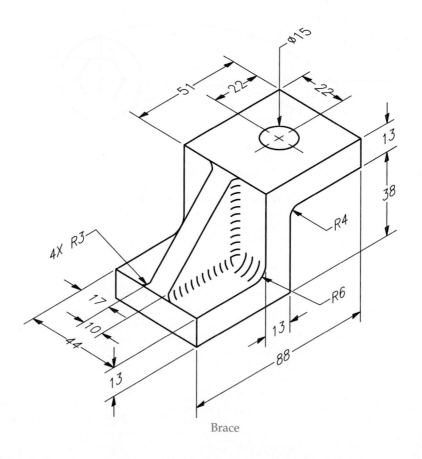

Brace

17.

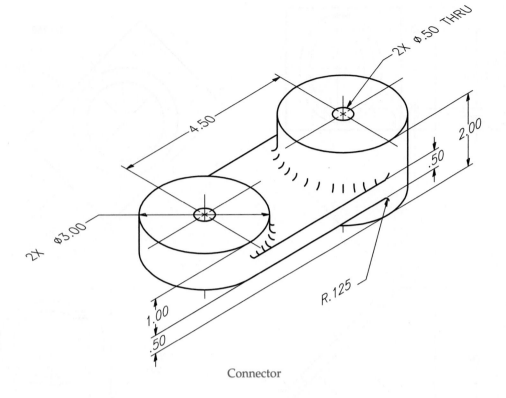

2X Ø .50 THRU

4.50

2.00

.50

2X Ø3.00

R.125

1.00

.50

Connector

18.

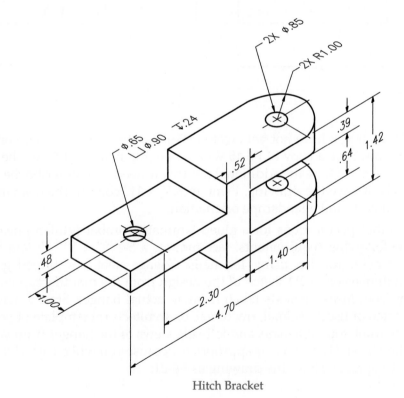

2X Ø.85

2X R1.00

.39

1.42

.64

Ø.65
⌴ Ø.90 ▼.24

.52

.48

1.00

1.40

2.30

4.70

Hitch Bracket

19. Draw all views of the pillow block shown. Save the drawing as P8-19.

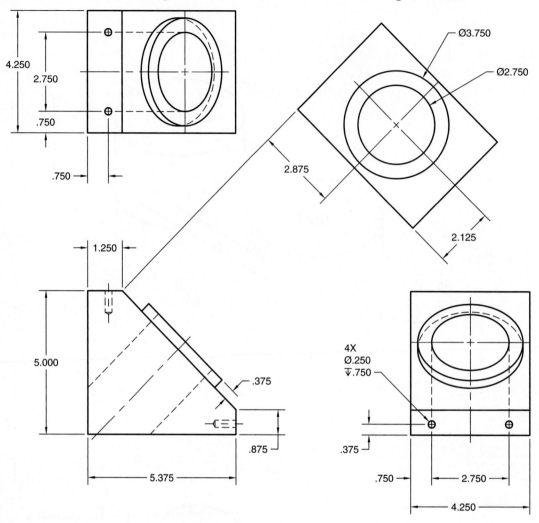

20. Open P5-15. If you have not yet created P5-15, go to Chapter 5 and complete the problem now. Save a copy of P5-15 with a new file name of P8-20. The P8-20 file should be active. Draw the additional multiview needed to describe the nut driver completely. Resave the drawing. Print an 8.5″ × 11″ copy of the drawing extents, using a 1:1 scale and a landscape orientation.

21. Research the specifications for a glued laminated timber (glulam) beam hanger with the following requirements: face-mounts a 5-1/8″ × 10″ glulam beam to a wood member, attaches using 16d nails, 12-gage steel, hot-dipped galvanized. Create a dimensioned 2D sketch of the design from the manufacturer's specifications, or from measurements taken from an actual hanger. Start a new drawing from scratch or use a decimal, fractional, or architectural template of your choice. Draw the front, top, right-side, and left-side views of the hanger from your sketch using the 0 layer. (The 0 layer is appropriate because you will create a block of each view in Chapter 25.) Save the drawing as P8-21.

AutoCAD Certified Associate Exam Practice

Answer the following questions. Write your answers on a separate sheet of paper.

1. Which of the following operations can you complete using the **OFFSET** command? *Select all that apply.*
   A. create an exact copy of a line at a specified distance from the original
   B. create an exact copy of an arc at a specified distance from the original
   C. create a circle concentric with an existing circle
   D. create an arc concentric with an existing arc
   E. create an exact copy of a line through a specified point

2. Which of the following operations can you complete using the **MEASURE** command? *Select all that apply.*
   A. place point objects anywhere in the drawing area
   B. convert point objects into blocks
   C. break a circle into a specified number of pieces
   D. place point objects at specified intervals along a line
   E. place point objects at a specified number of equally spaced locations on an arc

3. Which of the following are primary 2D orthographic views for a mechanical part drawing? *Select all that apply.*
   A. front
   B. rear
   C. auxiliary
   D. central
   E. top

AutoCAD Certified Professional Exam Practice

Follow the instructions in each problem. Write your answers on a separate sheet of paper.

1. **Navigate to this chapter on the companion website and open CPE-08offset.dwg.** Use the **Multiple** option of the **OFFSET** command to offset the existing circle to the outside, using an offset distance of 2, to create Circle 2, Circle 3, and Circle 4. Then offset the outermost circle to the inside using an offset distance of .5 to create Circle 5. What are the coordinates of the top quadrant point of Circle 5?

2. **Navigate to this chapter on the companion website and open CPE-07multiview.dwg.**

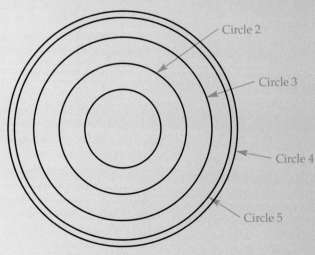

This file contains the front view of a cylindrical spacer, as well as the starting point for the lower-left corner of the right-side view. Change the point display to an × to make it more visible. Then create the right-side view, assuming that the spacer has a total length of 4 units. Use any method of construction described in this chapter to create the view accurately. Use hidden lines to show the inner diameter of the spacer, and include a center line to show the axis. Place the hidden lines and centerlines on the correct layers. What are the coordinates of Point A?

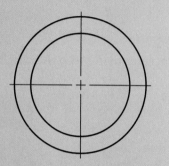 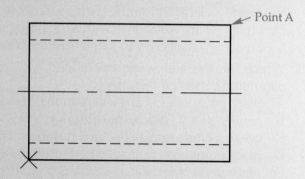

Point A

Chapter

Text Styles and Multiline Text

9

Learning Objectives

After completing this chapter, you will be able to:

✓ Describe and use proper text standards.
✓ Calculate drawing scale and text height.
✓ Develop and use text styles.
✓ Use the **MTEXT** command to create multiline text objects.

Annotation, dimensions, and symbols with *text* provide necessary information about features on a drawing. AutoCAD offers commands and settings to add uniform, easy-to-read text drawn according to drafting standards. This chapter introduces text standards and composition and explains how to use the **MTEXT** command to prepare a text object that can include multiple lines of text, such as paragraphs or a list of general notes. Chapter 10 describes how to create single-line text objects using the **TEXT** command, as well as additional text tools.

annotation: Textual information presented in notes, specifications, comments, and symbols.

text: Lettering on a CADD drawing.

Text Standards and Composition

Industry and company standards dictate how text should appear on a drawing. Consistent text type, format, height, and spacing are critical to legibility and drawing clarity. A drawing should use the same text *font* and format throughout, except for specific cases such as text in traditional architectural title blocks or on maps. Refer to **Figure 9-1** as you read the following information about text type and format.

font: The face design of a letter or number.

The ASME Y14.2 *Line Conventions and Lettering* standard applies to the process of hand lettering each character using one or more single straight or curved elements, in a Gothic font. For example, the letter A has three single-stroke lines. You can achieve the ASME standard with AutoCAD using a font such as Arial, Romans (roman simplex), or Century Gothic. The U.S. National CAD Standard (NCS) recommends a SansSerif font, although some companies prefer a slightly different font such as Arial or Century Gothic. Some architectural companies prefer the Stylus BT, ArchiText, or CountryBlueprint font, because these fonts provide a more traditional architectural hand lettering appearance.

Vertical text is most common. However, inclined text is an approved ASME standard and is used by some companies, most often in structural drafting, and for specific drawing requirements, such as water feature labels on maps. The recommended slant

Figure 9-1.
Examples of common text typefaces and formats.

DIMENSIONS AND TOLERANCES PER ASME Y14.5-2009.
ASME Y14.2 standard: vertical UPPERCASE, Arial font

BEND DOWN 90° R.50
ASME Y14.2 standard: vertical UPPERCASE, Romans font

REMOVE ALL BURRS AND SHARP EDGES.
ASME Y14.2 standard: vertical UPPERCASE, Century Gothic font

SIMPSON LCC5.25-3.5 TYP
U.S. National CAD standard: vertical UPPERCASE, SansSerif font

HOOD W/FAN, VENT TO OUTSIDE AIR
Traditional architecture format: vertical UPPERCASE, Stylus BT font

ALL FRAMING LUMBER TO BE DFL #2 OR BETTER
Traditional architecture format: vertical UPPERCASE, CountryBlueprint font

TYP EACH END TWO FLANGES
ASME Y14.2 standard variation: inclined UPPERCASE, Arial font

Mississippi River
River identification on a map: inclined lowercase, SansSerif font

for inclined text is 68° from horizontal, although some drafters find 75° more appropriate. Uppercase text is standard. However, some companies use lowercase letters for applications such as civil plans or maps.

A drawing displays specific text heights for different purposes. Figure 9-2 lists minimum letter heights based on the ASME Y14.2, *Line Conventions and Lettering* standard. The NCS and many companies, especially those who produce architectural and civil drawings, depart slightly from the ASME standard. The NCS specifies a minimum text height of 3/32" (2.4 mm). Most text is 1/8" (3 mm) high, with titles and similar text 1/4" (3 mm) high.

Figure 9-2.
Minimum letter heights based on the ASME Y14.2, *Line Conventions and Lettering* standard.

Application	Height INCH	Height METRIC (mm)
Most text (dimension values, notes)	.12	3
Drawing title, drawing size, CAGE code, drawing number, revision letter	.24* .12**	6* 3**
Section and view letters	.24	6
Zone letters and numerals in borders	.24	6
Drawing block headings	.10	2.5

*D, E, F, H, J, K, A0, and A1 size sheets

**A, B, C, G, A2, A3, and A4 size sheets

Numbers in dimensions and notes are the same height as standard text. AutoCAD provides several methods for stacking text in fractions. When dimensions contain fractions, the fraction bar should usually appear horizontally between the numerator and denominator. However, many notes have fractions displayed with a diagonal fraction bar (/). In this case, use a dash or space between the whole number and the fraction. **Figure 9-3** shows examples of text for numbers and fractions in different unit formats.

AutoCAD text commands provide great control over text *composition*. You can lay out text horizontally, as is typical when adding notes, or draw text at any angle according to specific requirements. AutoCAD automatically spaces letters and lines of text to help maintain the identity of individual notes. See **Figure 9-4**.

composition: The spacing, layout, and appearance of text.

PROFESSIONAL TIP

Text presentation is important. Refer to appropriate industry, company, or school standards, and consider the following tips when adding text:
- Plan your drawing using rough sketches to allow room for text and notes.
- Arrange text to avoid crowding.
- Place related notes in groups to make the drawing easy to read.
- Place all general notes in a common location. Locate notes in the lower-left corner or above the title block when using ASME standards, in the upper-left corner when using military (MIL) standards, or in the note block when using the NCS.
- Always use the spell checker.

Figure 9-3.
Examples of fractional text for different unit formats.

Decimal Inch	Fractional Inch	Millimeter
2.750 .25	$2\frac{3}{4}$ 2–3/4 2 3/4	2.5 3 0.7

Figure 9-4.
An example of general notes on a part drawing, typed according to the ASME Y14.2, *Line Conventions and Lettering* standard. The standard spacing is also appropriate for use on drawings for other disciplines.

Edges of text should align

Space between words is approximately equal to letter height

Space between numerals with a decimal point between them is a minimum of two-thirds the letter height

NOTES:
1. DRAWING PER IAW MIL-STD-100. CLASSIFICATION PER MIL-T-31000, PARA 3.6.4.
2. DIMENSIONS AND TOLERANCES PER ASME Y14.5-2009.
3. REMOVE ALL BURRS AND SHARP EDGES.
4. BAG ITEM AND IDENTIFY IAW MIL-STD-130, INCLUDE CURRENT REV LEVEL: 64869-XXXXXXXX REV___.

Space between lines of text is half to full height of the letters

Space between letters is approximately equal

Ideally, you should determine drawing scale, scale factors, and text heights before you begin drawing. Incorporate these settings into drawing template files, and make changes when necessary. The drawing scale factor determines how text height appears on-screen and plots.

To understand the concept of drawing scale, look at the portion of a floor plan shown in **Figure 9-5**. You should draw everything in model space at full scale. This means that the bathtub, for example, is actually drawn 5′ long. However, at this scale, text size becomes an issue, because full-scale text that is 1/8″ high is too small compared to the other full-scale objects. See **Figure 9-5A**. As a result, you must adjust the text height according to the drawing scale, as shown in **Figure 9-5B**. You can calculate the scale factor manually and apply it to text height, or you can allow AutoCAD to calculate the scale factor using annotative text.

Scaling Text Manually

text height: The specified height of text, which may be different from the plotting size for text scaled manually.

To adjust *text height* manually according to a specific drawing scale, you must calculate the drawing *scale factor*. **Figure 9-6** provides examples of calculating scale factor. You then multiply the scale factor by the *paper text height* to get the model space text height.

scale factor: The reciprocal of the drawing scale.

For example, a site plan plotted at a 1″ = 60′ scale has a scale factor of 720. Text drawn 1/8″ high is almost invisible, because the drawing is 720 times larger than it is when plotted at the proper scale. Therefore, multiply the scale factor of 720 by the text height of 1/8″ (.125″) to find the 90″ scaled text height for model space. The proper height of 1/8″ text in model space at a 1″ = 60′ scale is 90″.

paper text height: The plotted text height.

Annotative Text

annotative text: Text scaled by AutoCAD according to the specified annotation scale.

AutoCAD scales *annotative text* according to the *annotation scale* you select, which reduces the need for you to calculate the scale factor. Once you choose an annotation scale, AutoCAD applies the corresponding scale factor to annotative text and all other annotative objects. For example, if you manually scale 1/8″ text for a drawing

annotation scale: The drawing scale AutoCAD uses to calculate the height of annotative text.

Figure 9-5.

An example of a portion of a floor plan drawn at full scale in model space. A—Text drawn at full scale (1/8″ high) is too small compared to the large full-scale objects. B—Text scaled (6″ high) to display and plot correctly relative to the size of features on the drawing.

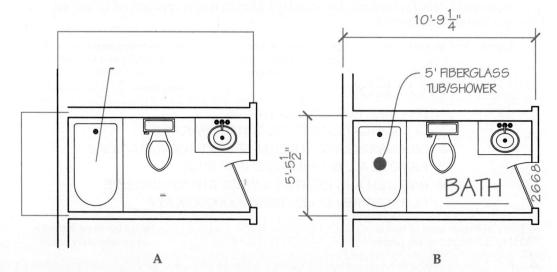

A

B

Figure 9-6.
Examples of calculating scale factor.

Example	Scale	Conversion	Calculation	Scale Factor
Mechanical	1:2	None	2 ÷ 1 = 2	2
Civil	1″ = 60′	1″ = 720″ (60′ = 720″)	720 ÷ 1 = 720	720
Architectural	1/4″ = 1′-0″	1/4″ (.25″) = 12″ (1′ = 12″)	12 ÷ .25 = 48	48
Metric to Inch	1:1	1″ = 25.4 mm	25.4 ÷ 1 = 25.4	25.4
Metric to Inch	1:2	1″ = 25.4 mm × 2 (50.8)	50.8 ÷ 1 = 50.8	50.8

with a 1/4″ = 1′-0″ scale, or a scale factor of 48, you must draw the text using a text height of 6″ (1/8″ × 48 = 6″) in model space. When placing annotative text, using this example, you set an annotation scale of 1/4″ = 1′-0″. Then you draw the text using a paper text height of 1/8″ in model space. The 1/8″ high text is scaled to 6″ automatically according to the preset 1/4″ = 1′-0″ annotation scale.

Annotative text offers several advantages over manually scaled text, including the ability to control text appearance based on the drawing scale and paper text height, while reducing the need to focus on the scale factor. Annotative text is especially effective when the drawing scale changes or when a single sheet includes views at different scales.

PROFESSIONAL TIP

If you anticipate preparing scaled drawings, you should use annotative text and other annotative objects instead of manual scaling. However, scale factor does influence non-annotative items and is still an important value to identify and use throughout the drawing process.

Setting the Annotation Scale

You should usually set the annotation scale before you begin typing text so that the text height is scaled automatically. However, this is not always possible. It may be necessary to adjust the annotation scale throughout the drawing process, especially if you prepare views at different scales on one sheet. This textbook approaches annotation scaling in model space only, using the process of selecting the appropriate annotation scale before typing text. To draw text using another scale, pick the new annotation scale and then type the text.

The **Select Annotation Scale dialog** box appears when you access a text command and an annotative text style is current. This dialog box provides a convenient way to set annotation scale before typing. You will learn about text styles later in this chapter. You can also select the annotation scale from the **Annotation Scale** flyout on the status bar. See **Figure 9-7**. The annotation scale is typically the same as the drawing scale.

This textbook describes many additional annotative object tools. Some of these tools are more appropriate for working with layouts, as explained later in this textbook.

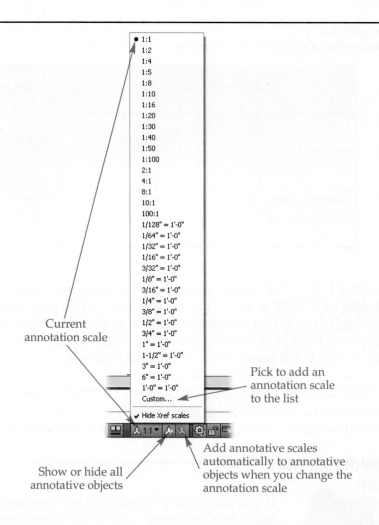

Current annotation scale

Pick to add an annotation scale to the list

Show or hide all annotative objects

Add annotative scales automatically to annotative objects when you change the annotation scale

Editing Annotation Scales

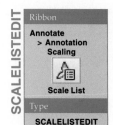

SCALELISTEDIT

Ribbon
Annotate
> Annotation
Scaling

Scale List

Type
SCALELISTEDIT

If a scale is unavailable, or to change an existing scale, pick the **Annotation Scale** flyout on the status bar and choose **Custom...** to access the **Edit Scale List** dialog box. Move the highlighted scale up or down in the list using the **Move Up** or **Move Down** button. To remove the highlighted scale from the list, pick the **Delete** button.

Select the **Edit...** button to open the **Edit Scale** dialog box, where you can change the name of the scale and adjust the scale by entering the paper and drawing units. For example, a scale of 1/4″ = 1′-0″ has a paper units value of .25 or 1 and a drawing units value of 12 or 48.

To create a new annotation scale, pick the **Add...** button to display the **Add Scale** dialog box, which provides the same options as the **Edit Scale** dialog box. Pick the **Reset** button to restore the list to display the default annotation scales. Once you select an annotation scale, you are ready to type annotative text.

Changes you make in the **Edit Scale List** dialog box are stored with the drawing and are specific to the drawing. To make changes to the default scale list saved to the system registry, pick the **Default Scale List...** button in the **User Preferences** tab of the **Options** dialog box to access the **Default Scale List** dialog box. The options are the same as those in the **Edit Scale List** dialog box, but changes are saved as the default for new drawings.

Text Styles

A *text style* presets many text characteristics. Create a text style for each different text appearance or function. For example, use an annotative text style to draw annotative text, and a non-annotative text style to draw non-annotative text. Another example is creating text styles that correspond to a specific text height or other characteristics. Add text styles to your drawing templates for repeated use. Avoid adjusting text format independently of the text style assigned to the text.

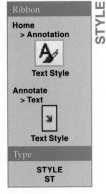

text style: A saved collection of settings for text height, width, oblique angle (slant), and other text effects.

Text Style Dialog Box

Create, modify, and delete text styles using the **Text Style** dialog box. See **Figure 9-8**. The **Styles** list box displays existing text styles. The Annotative text style allows you to create annotative text, as indicated by the icon to the left of the style name. The Standard text style does not use the annotative function.

To make a text style current, double-click the style name, right-click the name and select **Current**, or pick the name and select the **Current** button. Below the **Styles** list box is a drop-down list that you can use to filter the number of text styles displayed in the **Text Style** dialog box. Pick the **All Styles** option to show all text styles in the file, or pick the **Styles in use** option to show only the current style and styles used in the drawing.

Creating New Text Styles

To create a new text style, select an existing text style from the **Styles** list box to use as a base for formatting the new text style. Then pick the **New...** button to open the **New Text Style** dialog box. See **Figure 9-9**. Notice that style1 appears in the **Style Name** text box. Replace the default name with a more descriptive name. For example, the name ARIAL-12 describes a text style that uses the Arial font and characters .12" high.

Figure 9-8.
Use the **Text Style** dialog box to create, rename, delete, and set the characteristics of a text style. The **Big Font:** drop-down list replaces the **Font Style:** drop-down list when you select the **Use Big Font** check box. The **Height** text box replaces the **Paper Text Height** text box when you deselect the **Annotative** check box.

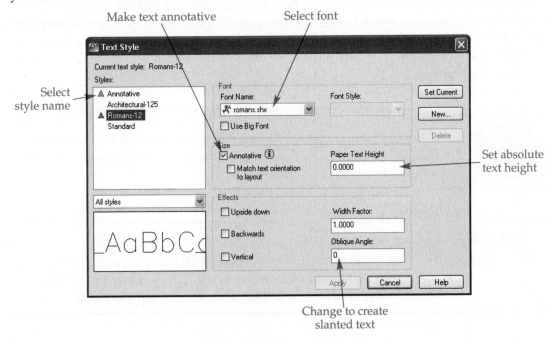

Figure 9-9.
Enter a descriptive
name for the new
text style in the **New
Text Style** dialog
box.

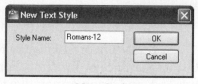

Default Style   New Style

The name ARCHITECTURAL-125 describes a text style that uses the Stylus BT font and characters 1/8″ high.

Text style names can have up to 255 characters, including uppercase and lower-case letters, numbers, dashes (–), underlines (_), and dollar signs ($). After typing the text style name, pick the **OK** button. The new text style appears in the **Styles** list box of the **Text Style** dialog box, and you are ready to adjust text style characteristics. Pick the **Apply** button to apply changes, and pick the **Close** button to exit the **Text Style** dialog box.

PROFESSIONAL TIP

Record the names and details about the text styles you create and keep the information in a log for future reference.

Font Options

Use the **Font Name** drop-down list in the **Font** area of the **Text Style** dialog box to select a font. The list includes TrueType fonts installed on your computer and fonts linked to AutoCAD shape files. TrueType fonts are *scalable fonts* and have an outline. By default, TrueType fonts appear and plot filled. You can recognize fonts from AutoCAD shape files (SHX fonts) by the .shx file extension and the AutoCAD compass icon.

scalable fonts:
Fonts that can be displayed or printed at any size while retaining proportional letter thickness.

The **Font Style** drop-down list is active if the selected font includes options, such as bold or italic. SHX fonts do not provide style options, but some TrueType fonts do. For example, the SansSerif font has Regular, Bold, BoldOblique, and Oblique options. Select a style or combination of styles to change the appearance of the font. The **Use Big Font** check box becomes enabled when you select an SHX font. Pick the check box to replace the **Font Style:** drop-down list with the **Big Font:** drop-down list, from which you can choose a *big font*.

big font: A supplement that provides Asian and other large-format fonts that have characters and symbols not present in other font files.

CAUTION

When a drawing contains a significant amount of text, especially TrueType font text, display changes may be slower and drawing regeneration time may increase.

Reference Material *AutoCAD Fonts*
For a sample of the many fonts available with AutoCAD, go to the **Reference Material** section on the companion website (www.g-wlearning.com/CAD) and select **AutoCAD Fonts**.

Size Options

The **Size** area of the **Text Style** dialog box contains options for defining text style height. Select the **Annotative** check box to set the text style as annotative and display the **Paper Text Height** text box. Deselect the **Annotative** check box to scale text manually and display the **Height** text box.

The default text height is 0. If you set a value other than 0, the text height is fixed for the text style and applies each time you use the text style. As a result, you can only use the specified text height to create single-line text, and you do not have the option of assigning a different text height to objects that reference the text style, such as dimensions.

PROFESSIONAL TIP

Use a text style height of 0 to provide the greatest flexibility when drawing. A prompt asks you to specify a text height when you create single-line text, and you can assign a text height to dimension, multileader, and table styles.

As an alternative, you can use a specific value to limit text height. In this case, the text style fully controls the height applied to single-line text and dimension, multileader, and table styles. This restricts text objects to a certain height, which can increase productivity and accuracy, but requires that you create a text style for each unique height. Dimension, multileader, and table styles are explained later in this textbook.

The **Match text orientation to layout** check box becomes enabled when you pick the **Annotative** check box. Check **Match text orientation to layout** to match the orientation of text in layout viewports with the layout orientation. Layouts are described later in this textbook.

Effects

The **Effects** area of the **Text Style** dialog box offers text format options, including settings for drawing text upside-down, backwards, and vertically. See **Figure 9-10**. The **Vertical** check box is available for SHX fonts. Text on drawings is normally horizontal. Vertical text is appropriate for special effects and graphic designs, and usually works best with a 270° rotation angle.

Figure 9-10.
Special effects for text styles can be set in the **Effects** area of the **Text Style** dialog box.

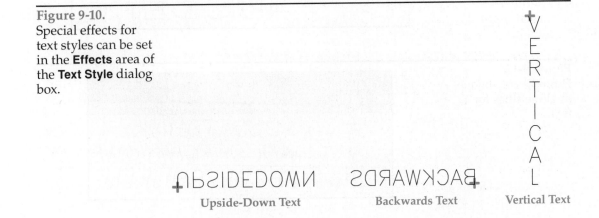

Use the **Width Factor** text box to specify the text character width relative to its height. You can set the width factor between 0.01 and 100. A width factor of 1 is the default and is recommended for most drawing applications. A width factor greater than 1 expands characters, and a factor less than 1 compresses characters. If necessary, reduce the width slightly, and only if you cannot reduce the height. See **Figure 9-11**.

The **Oblique Angle** text box allows you to set the angle at which text inclines. The 0 default draws vertical characters. A value greater than 0 slants characters to the right, and a negative value slants characters to the left. See **Figure 9-12**. Some fonts, such as the italic SHX font, are already slanted.

PROFESSIONAL TIP

AutoCAD text slant is measured from vertical. Use a 22° oblique angle to slant text according to the 68° horizontal incline standard.

The **Preview** area of the **Text Style** dialog box displays an example of the selected font and font effects. This is a convenient way to see how the text appears before using it in a new text style.

Exercise 9-1

Complete the exercise on the companion website.
www.g-wlearning.com/CAD

Figure 9-11.
Examples of width factor settings for text.

Width Factor	Text
1	ABCDEFGHIJKLM
.5	ABCDEFGHIJKLMNOPQRSTUVWXY
1.5	ABCDEFGHI
2	ABCDEFG

Figure 9-12.
Examples of oblique angle settings for text.

Obliquing Angle	Text
0	ABCDEFGHIJKLM
15	ABCDEFGHIJKLM
−15	ABCDEFGHIJKLM

Changing, Renaming, and Deleting Text Styles

Select a text style from the **Styles** list box to edit. Make the necessary changes and pick the **Apply** button to apply the changes. If you make changes to a text style, such as selecting a different font, all existing text objects assigned to the text style are updated. Use a different text style with different characteristics when appropriate.

To rename a text style using the **Text Style** dialog box, slowly double-click on the name or right-click on the name and select **Rename**. To delete a text style using the **Text Style** dialog box, right-click on the name and select **Delete**, or pick the style and select the **Delete** button. You cannot delete a text style that is assigned to text objects. To delete a style that is in use, assign a different style to the text objects that reference the style. You cannot delete or rename the Standard style.

 You can also rename styles using the **Rename** dialog box. Select **Text styles** in the **Named Objects** list to rename the style.

Setting a Text Style Current

Set a text style current using the **Text Style** dialog box by double-clicking the style in the **Styles** list box, right-clicking on the style and selecting **Set current**, or picking the style and selecting the **Set current** button. To set a text style current without opening the **Text Style** dialog box, use the **Text Style** list on the expanded **Annotation** panel of the **Home** ribbon tab or on the **Text** panel of the **Annotate** ribbon tab. See **Figure 9-13**.

**PROFESSIONAL
TIP**

 You can import text styles from existing drawings using **DesignCenter**. See Chapter 5 for more information about using **DesignCenter** to reuse drawing content.

Figure 9-13.
The fastest way to set a style current is to use one of the drop-down lists on the ribbon.

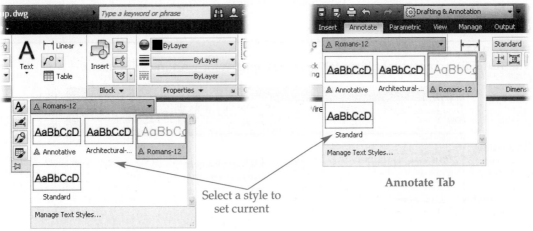

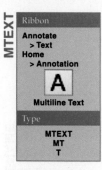

MTEXT

Ribbon

Annotate
> Text
Home
> Annotation

A

Multiline Text

Type

MTEXT
MT
T

text boundary: An imaginary box that sets the location and width for multiline text.

text editor: The area of the multiline or single-line text system where you type text.

Multiline Text

The **MTEXT** command draws a single multiline text, or mtext, object that can include extensive paragraph formatting, lists, symbols, and columns. It is good practice to set the appropriate text style current before you access the **MTEXT** command. Activate the **MTEXT** command to display letters near the crosshairs that indicate the current text style and height. Pick the first corner of the *text boundary*. A prompt then asks you to specify the opposite corner or choose an option. You can use the options to preset text style and height, boundary justification, rotation, width, line spacing, and use of columns. However, it is typically easier to set mtext options while you are typing. Rotation is the only option you cannot adjust while typing. Use the **Rotation** option before selecting the opposite corner to rotate the text boundary, or use an editing command to rotate the mtext object after you complete the **MTEXT** command.

Mtext uses dynamic columns by default to help you organize multiple text columns in an mtext object. Disable columns for typical text requirements without columns. Before selecting the second corner of the text boundary, choose the **Columns** option and then the **No columns** option. An arrow in the boundary shows the direction of text flow and where the boundary will expand as you type, if necessary. Pick the opposite corner of the text boundary to continue. See **Figure 9-14**. You will learn to format columns later in this chapter.

Using the Text Editor

When you select the opposite corner of the text boundary, the **Text Editor** contextual ribbon tab and the *text editor* appear. **Figure 9-15** shows the display with columns

---

Figure 9-14.
The text boundary is the box within which you type. Consider the text boundary the extents of text object, or paragraphs. Long words extend past the boundary in the direction the arrow indicates.

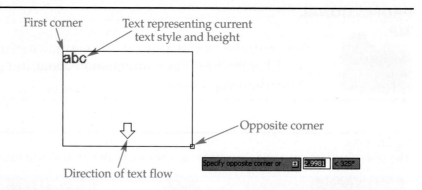

Figure 9-15.
The **Text Editor** contextual ribbon tab provides options for working with mtext.

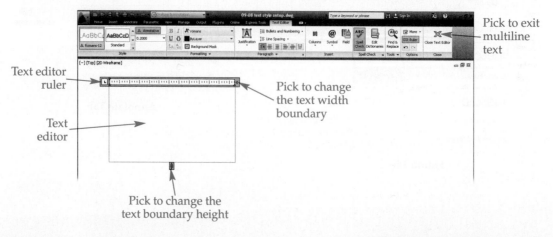

276

AutoCAD and Its Applications—Basics

disabled. Typing mtext is similar to typing using word processing software such as Microsoft® Word. The **Text Editor** ribbon tab provides options for adjusting and formatting text in the text editor. You can access many of the same options found in the **Text Editor** ribbon tab, as well as Windows Clipboard functions, from the shortcut menu that appears when you right-click away from the ribbon. See **Figure 9-16**.

If you close the ribbon, the **Text Formatting** toolbar appears instead of the **Text Editor** ribbon tab. This textbook focuses on using the **Text Editor** ribbon tab to add mtext. The **Text Formatting** toolbar provides the same functions.

The text editor indicates the initial size of the area in which you type. When columns are not active, long words and paragraphs extend past the text editor limits. The text editor is transparent by default so that you can see how the text appears on-screen relative to other objects. Use the **Opaque Background** option to make the text editor opaque. The text editor includes a ruler that displays indent and tab stops and indent and tab markers. Use the **Ruler** option to turn the ruler on or off.

Use the **Undo** and **Redo** commands to undo or redo text editor operations.

To change the width of the text editor when you are not using columns, drag the arrows on the right end of the ruler. An alternative is to right-click on the ruler or the arrows at the bottom of the text editor and choose **Set Mtext Width...** to use the **Set Mtext Width** dialog box. To change the height of the text editor, drag the arrows at the bottom of the text editor, or right-click on the ruler or the arrows at the bottom of the text editor and select **Set Mtext Height...** to use the **Set Mtext Height** dialog box.

Figure 9-16.
Display the text editor shortcut menu by right-clicking away from the ribbon while the text editor is active. Use the shortcut menu as an alternative to the ribbon, or to select options that are not available from the ribbon.

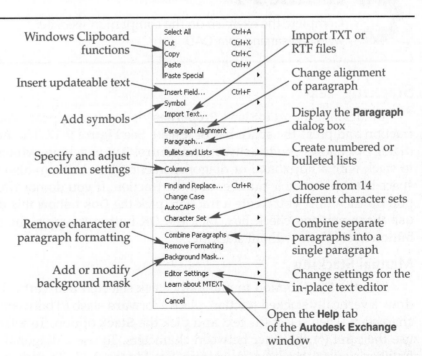

Change the width and height of the text editor to increase or decrease the number of lines of text. Do not press [Enter] to form lines of text, unless you are specifically creating a new paragraph or a new item in a list.

The familiar Windows text editor cursor is displayed at the current text height within the text editor. Begin typing or editing text. The procedure for selecting and editing existing text is the same as in standard Windows text editors. To select all text in the text editor, right-click inside the text editor and choose **Select All**. You can use the **Text Highlight Color...** option to change the selected text highlight color using the **Select Color** dialog box.

When you finish typing, exit the mtext system using the **Close Text Editor** option, or pick outside of the text editor. You can also press [Esc] or right-click and select **Cancel**, but AutoCAD prompts you to save changes to the mtext. The easiest way to reopen the text editor to make changes to text content is to double-click on the mtext object.

Ribbon

Text Editor > Close

Close

The text editor displays text horizontally, right-side up, and forward. Any special effects, such as vertical, backwards, or upside-down, take effect when you exit the text editor.

Reference Material *Shortcut Keys*
For a complete list of keyboard shortcuts for text editing, go to the **Reference Material** section of the companion website (www.g-wlearning.com/CAD) and select **Shortcut Keys**.

Exercise 9-2
Complete the exercise on the companion website.
www.g-wlearning.com/CAD

Stacking Text

Ribbon

Text Editor > Formatting

Stack

By default, the **AutoStack Properties** dialog box appears each time you type a fraction and press the space bar or [Enter]. See **Figure 9-17**. The **AutoStack Properties** dialog box allows you to activate and control AutoStacking, which causes the fraction to stack with a horizontal or diagonal fraction bar. You can also remove the leading space between a whole number and the fraction. If you do not want the dialog box to pop up each time you create a fraction, pick the **Don't show this dialog again; always use these settings** check box. Pick the **OK** button to apply the stack or the **Cancel** button to continue without stacking.

Manual Stacking

Use the **Stack** option to manually stack selected text vertically or diagonally. To draw a vertically stacked fraction, place a forward slash (/) between the top and bottom characters. Then select the text and pick the **Stack** option. To form a *tolerance stack*, use the caret (^) character between characters. To use a diagonal fraction bar, type a number sign (#) between characters. See **Figure 9-18**. To return stacked text to the

tolerance stack:
Text stacked vertically without a fraction bar.

Figure 9-17.
The **AutoStack Properties** dialog box.

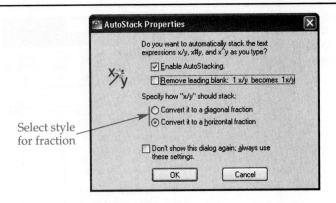

Select style for fraction

Figure 9-18.
Examples of stacked characters. ASME standards recommend that the text height of stacked fraction numerals be the same as the height of other dimension numerals.

	Selected Text	Stacked Text
Vertical Fraction	1/2	$\frac{1}{2}$
Tolerance Stack	1^2	$\frac{1}{2}$
Diagonal Fraction	1#2	½

original unstacked format, select the stacked text and the **Stack** ribbon option, or use the **Unstack** shortcut menu option.

Stack Settings

Adjust stack settings by selecting stacked text, right-clicking, and choosing **Stack Properties** to display the **Stack Properties** dialog box. **Figure 9-19** identifies the **Stack Properties** dialog box features. Select **100%** from the **Text size** drop-down list to conform to ASME standards. This is an appropriate standard to follow for all disciplines.

Adding Symbols

Use the **Symbol** option to insert a common drafting symbol or other unique character not found on a typical keyboard. See **Figure 9-20**. The first two sections in the **Symbol** menu contain common symbols. The third section contains the option to add a *non-breaking space*. Pick a symbol or **Non-breaking Space** to insert at the current location of the text cursor. The **Other...** option opens the **Character Map** dialog box,

Ribbon

**Text Editor
> Insert**

@

Symbol

non-breaking space: A symbol that you insert in place of a space to keep separate words together on one line.

Figure 9-19.
The **Stack Properties** dialog box. Select **100%** from the **Text size** drop-down list to conform to ASME standards.

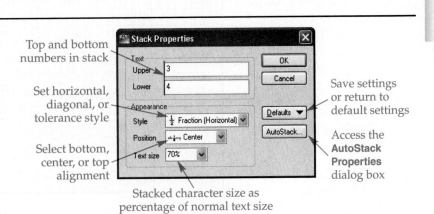

Top and bottom numbers in stack

Set horizontal, diagonal, or tolerance style

Select bottom, center, or top alignment

Stacked character size as percentage of normal text size

Save settings or return to default settings

Access the **AutoStack Properties** dialog box

Figure 9-20.
Access the **Symbol** flyout to add symbols to the text editor.

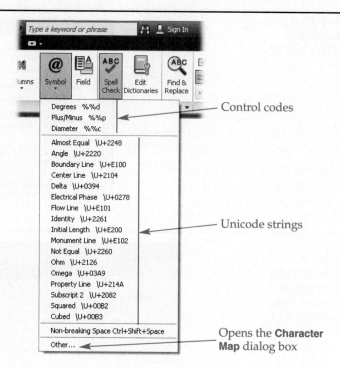

Control codes

Unicode strings

Opens the **Character Map** dialog box

shown in **Figure 9-21**. Use the following steps to insert a symbol from the **Character Map** dialog box:

1. Pick a font from the **Font:** drop-down list to display symbols associated with the font.
2. Locate and pick a symbol, and then pick the **Select** button. The symbol appears in the **Characters to copy:** box. You can copy multiple symbols to the box.
3. Pick the **Copy** button to copy the symbols to the Clipboard.
4. Close the **Character Map** dialog box.
5. In the text editor, place the text cursor at the location where the symbols are to be inserted.
6. Right-click and select **Paste** to paste the symbols at the cursor location.

Figure 9-21.
The **Character Map** dialog box.

Select font from drop-down list

Available symbols

Select to return to the **In-Place Text Editor**

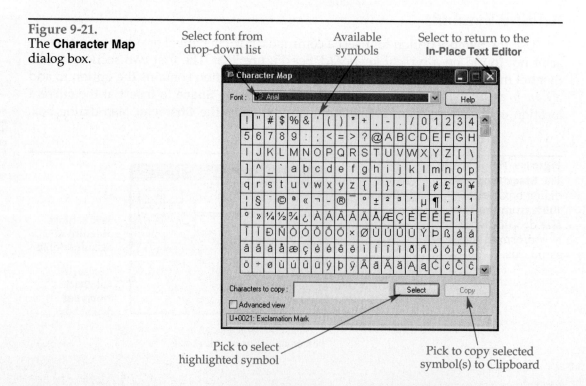

Pick to select highlighted symbol

Pick to copy selected symbol(s) to Clipboard

Exercise 9-3

Complete the exercise on the companion website.
www.g-wlearning.com/CAD

Style Settings

The **Style** panel of the **Text Editor** ribbon tab includes options for changing the text style and for overriding the annotative setting and text height. A single mtext object can use a combination of character formatting and text heights, but it can only be assigned one text style and must be entirely annotative or non-annotative. Use the scroll buttons to the right of the text styles to locate styles, or pick the expansion arrow to display styles in a temporary window. Pick a style different from the current style to apply to all text in the text editor.

Use the **Size** drop-down list to set the text height. The text height for annotative text is the paper text height. The text height for non-annotative text is the text height multiplied by the scale factor. You should usually only change the text height if the current text style uses a 0 height. Otherwise, you override the specified text style height. You can use the **Annotative** button to override the annotative setting of the current text style, but this is typically not appropriate.

Character Formatting

Use the **Formatting** panel of the **Text Editor** ribbon tab to adjust character format. Some of the same settings are also available from the shortcut menu. A single mtext object can use a combination of character formats. Remember, however, that making changes to character formatting overrides specified text style format and preset object properties, such as color. You should usually avoid this practice.

The **Bold** and **Italic** buttons are enabled for some TrueType fonts. Select the appropriate button(s) to make text bold and/or italic. Pick the **Underline** button to underline text. Select the **Overline** button to place a line over text. Use the **Make Uppercase** button to make all selected text uppercase, or use the **Make Lowercase** button to make all selected text lowercase. The **Font** drop-down list allows you to override the text font. The text color is set to ByLayer by default, but you can change the color by picking a color or option from the **Color** drop-down list. Although you should usually define color as **ByLayer**, a single mtext object can have a combination of text colors.

> **NOTE**
> The **AutoCAPS** option turns [Caps Lock] on for typing uppercase text in the mtext editor. [Caps Lock] turns off when you exit the text editor so that text in other programs is not all uppercase.

> Ribbon
>
> **Text Editor**
> **> Tools**
> **> AutoCAPS**

Additional character formatting options are available from the expanded **Formatting** panel. The **Oblique Angle** text box overrides the angle at which text is inclined. The value in the **Tracking** text box determines the amount of space between text characters. The default tracking value is 1, which results in normal spacing. Increase the value to add space between characters, or decrease the value to tighten the spacing between characters. You can enter any value between 0.75 and 4.0. See **Figure 9-22**. The value in the **Width Factor** text box overrides the text character width.

Figure 9-22.
The **Tracking** option for mtext determines the spacing between characters. Normal spacing is typically appropriate for all text on a drawing and adheres to most drafting standards.

AutoCAD tracking
Normal Spacing

AutoCAD tracking
Tracking = 0.75

A u t o C A D   t r a c k i n g
Tracking = 2.0

Using a Background Mask

Sometimes drawings display text over existing objects, such as graphic patterns, making the text difficult to read. A *background mask* can solve this problem. Access the **Background Mask...** option to display the **Background Mask** dialog box. See **Figure 9-23**. To mask the current mtext object, check **Use background mask**. The **Border offset factor:** text box sets the amount of mask, from 1 to 5. The border offset factor works with the text height value according to the formula: border offset factor × text height = total masking distance from the bottom of the text. If you set the border offset factor to 1, the mask occurs directly within the boundary of the text. Use a value greater than 1 to offset the mask beyond the text boundary. See **Figure 9-24**. The **Fill Color** area of the **Background Mask** dialog box allows you to apply color to the mask using the background color or a different color.

Figure 9-23.
The **Background Mask** dialog box controls text mask settings.

Determines how much of the background is masked

Sets mask color same as background color

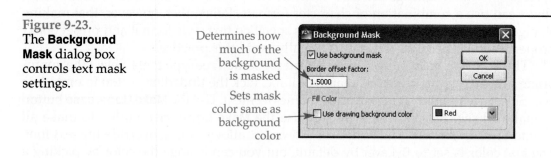

Figure 9-24.
The border offset factor determines the size of the background mask. The text in the figure is 1/8″ with different border offset factors.

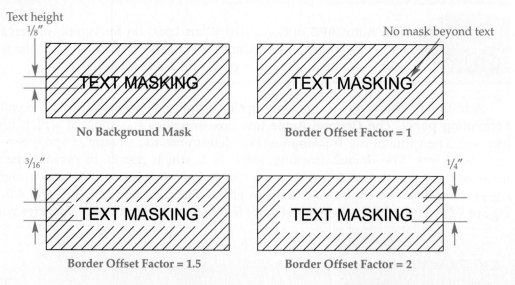

The **Character Set** cascading menu, available from the **More** flyout in the **Options** panel or the shortcut menu, displays a menu of code pages. A code page provides support for character sets used in different languages. Select a code page to apply it to selected text.

Exercise 9-4

Complete the exercise on the companion website.
www.g-wlearning.com/CAD

Paragraph Formatting

Use the **Paragraph** panel of the **Text Editor** ribbon tab to adjust paragraph formatting. Some of the same settings are also available from the shortcut menu. *Justify* the text boundary to control the arrangement and location of text within the text editor. You can also justify the text within the boundary independently of the text boundary justification. This provides flexibility for determining the location and arrangement of text. Justification also determines the direction of text flow. To justify the text boundary, as shown in **Figure 9-25**, select an option from the **Justification** flyout.

Paragraph alignment occurs inside the text boundary. For example, when you apply the **Middle Center** text boundary justification, then set the paragraph alignment to **Left**, the text inside the boundary aligns to the left edge of the text boundary, while

justify: Align the margins or edges of text. For example, left-justified text aligns along an imaginary left border.

paragraph alignment: The alignment of multiline text inside the text boundary.

Figure 9-25.
Options for justifying the mtext boundary.

✕ Insertion point
▪ Grips
– – Text boundary

NOTES:
1. DIMENSIONS AND TOLERANCES PER ASME Y14.5-2009.
2. REMOVE ALL BURRS AND SHARP EDGES.
3. ALL INSIDE BEND RADII TO BE .04.
4. IDENTIFY IAW MIL-STD-130, .12 HIGH GOTHIC STYLE CHARACTERS, CONTRASTING COLOR RUBBER STAMP OR HAND MARK. LOCATE APPROXIMATELY AS SHOWN.
5. TO BE MEASURED IN RESTRAINED CONDITION, 3° MAX DEFLECTION ANGLE (TOTAL).

the text boundary remains positioned according to the **Middle Center** justification. See **Figure 9-26**.

To adjust paragraph alignment, select a paragraph alignment button on the **Paragraph** panel, or pick from the **Paragraph Alignment** cascading menu of the shortcut menu. You can also control paragraph alignment using the **Paragraph** dialog box, shown in **Figure 9-27**. To set paragraph alignment in the **Paragraph** dialog box, pick the **Paragraph Alignment** check box, and then choose the appropriate radio button. **Figure 9-28** shows paragraph alignment options.

The **Paragraph** dialog box also includes tab, indent, paragraph spacing, and paragraph line spacing settings. Use the **Tab** area to set custom tab stops. Pick a tab type radio button, enter a value for the tab in the text box, and then pick the **Add** button to add the tab to the list and ruler. **Figure 9-29** shows and briefly describes each tab option. Add as many custom tabs as necessary. You can also add custom tabs to the ruler by picking the tab button on the far left side of the ruler until the desired tab symbol appears. Then pick a location on the ruler to insert the tab.

The options in the **Left Indent** area set the indentation for the first line of a paragraph of text and the remaining portion of a paragraph. The **First line** indent applies

Ribbon

**Text Editor
> Paragraph**

Paragraph

Figure 9-26.
You can adjust paragraph alignment independently of text boundary justification.

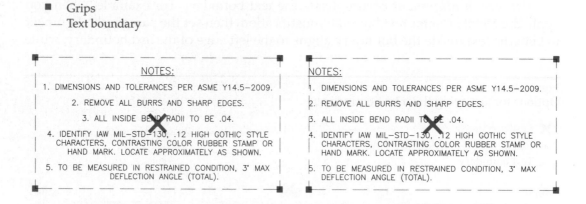

Figure 9-27.
The **Paragraph** dialog box.

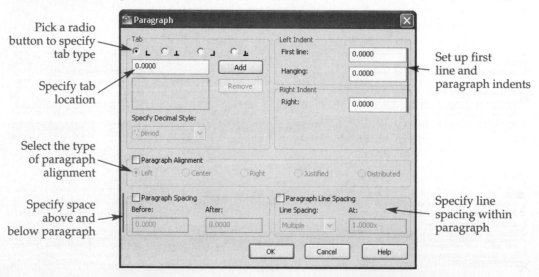

Figure 9-28.

Paragraph alignment options for mtext. In each of these examples, the text boundary justification is set to **Top Left**.

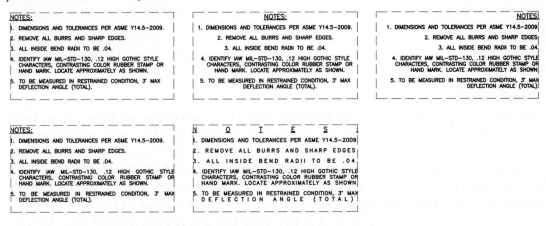

Figure 9-29.

Using custom tabs to position text in the text editor. When you press [Tab], the cursor moves to the tab position. The type of tab determines text behavior.

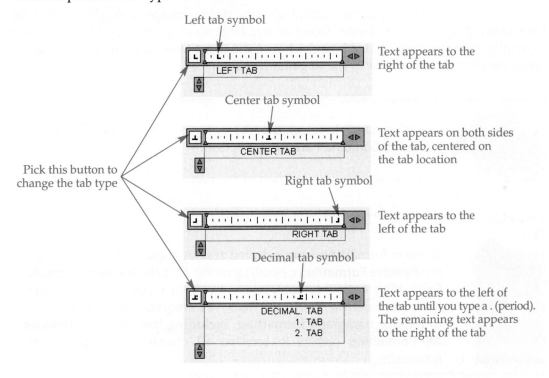

each time you start a new paragraph. As text wraps to the next line, the **Hanging** indent is used. The options in the **Right Indent** area set the indentation for the right side of a paragraph. As you type, the right indent value, not the right edge of the text boundary, determines when the text wraps to the next line.

The options in the **Paragraph Spacing** area define the amount of space before and after paragraphs. To set paragraph *line spacing*, pick the **Paragraph Spacing** check box. Then enter the spacing above a paragraph in the **Before** text box, and the spacing below a paragraph in the **After** text box. **Figure 9-30** shows examples of paragraph spacing settings.

The options in the **Paragraph Line Spacing** area adjust line spacing. Default line spacing for single lines of text is equal to 1.5625 times the text height. To adjust the line

line spacing: The vertical distance from the bottom of one line of text to the bottom of the next line.

Figure 9-30.
Examples of paragraph spacing. Each example uses a text height of .1875" and a first line left indent of .5".

```
┌ ─ ─ ─ ─ ─ ─ ─ ─ ─ ┐
  Paragraph one typed
with no paragpah spacing.
  Paragraph two typed with
no paragraph spacing
└ ─ ─ ─ ─ ─ ─ ─ ─ ─ ┘
      Before: 0"
      After: 0"
```

```
┌ ─ ─ ─ ─ ─ ─ ─ ─ ─ ┐
   Paragraph one typed
with .25 before spacing and
no after spacing.

   Paragraph two typed with
.25 before spacing and no
after spacing.
└ ─ ─ ─ ─ ─ ─ ─ ─ ─ ┘
      Before: .25"
      After: 0"
```

```
┌ ─ ─ ─ ─ ─ ─ ─ ─ ─ ┐
   Paragraph one typed
with .125 before spacing
and .5 after spacing.

   Paragraph two typed with
.125 before spacing and .5
after spacing.
└ ─ ─ ─ ─ ─ ─ ─ ─ ─ ┘
      Before: .125"
      After: .5"
```

spacing, pick the **Paragraph Line Spacing** check box. Select the **Multiple** option from the **Line Spacing** drop-down list to enter a multiple of the text height in the **At** text box. For example, lines with a text height of .12" are spaced .1875" apart. To double-space lines, you could enter a value of 3.125x, making the space between lines of text .375".

To force the line spacing to be the same for all lines of text, select the **Exactly** option from the **Line Spacing** drop-down list and enter a value in the **At** text box. If you enter an exact line spacing that is less than the text height, lines of text stack on top of each other. To add spaces between lines automatically based on the height of the characters in the line, choose the **At Least** option from the **Line Spacing** drop-down list and enter a value in the **At** text box. The result is an equal spacing between lines, even if the text has different heights.

You can also set line spacing without opening the **Paragraph** dialog box, using the **Line Spacing** flyout of the ribbon. Select an available spacing, pick the **More...** button to display the **Paragraph** dialog box, or choose the **Clear Line Spacing** option to apply an automatic spacing similar to the **At Least** function.

Ribbon
Text Editor
> Paragraph
Line Spacing

Ribbon
Text Editor
> Paragraph
> Combine
Paragraphs

Combine multiple selected paragraphs to form a single paragraph using the **Combine Paragraphs** option.

PROFESSIONAL TIP

Remove formatting from selected text by right-clicking and using the **Remove Formatting** cascading menu. Pick the **Remove Character Formatting** option to remove character formatting such as bold, italic, or underline. Select the **Remove Paragraph Formatting** option to remove paragraph formatting, including lists. Pick the **Remove All Formatting** option to remove all character and paragraph formatting.

Exercise 9-5

Complete the exercise on the companion website.
www.g-wlearning.com/CAD

Lists

Drawings often include lists to organize information. Lists provide a way to arrange related items in a logical order and help make lines of text more readable. General notes are usually in list format. Mtext includes systems for automating the

AutoCAD and Its Applications—Basics

process of creating and editing numbered, bulleted, and alphabetical lists. You can create lists as you enter text or apply list formatting to existing text. Lists can contain sublevel items designated with double numbers, letters, or bullets. Default tab settings apply unless you adjust the paragraph options.

List tools are available from the **Numbering** flyout on the **Paragraph** panel of the **Text Editor** ribbon tab or the **Bullets and Lists** cascading menu of the shortcut menu. The **Allow Bullets and Lists** option is active by default and is required to create a list. Unchecking **Allow Bullets and Lists** converts any list items in the text object to plain text characters and disables bullet and list options.

Choose an option from the **Lettered** cascading menu to create an alphabetical list. Use the default **Uppercase** option to use uppercase lettering, or pick **Lowercase** to use lowercase lettering. Choose the **Numbered** option to form a numbered list. Select the **Bulleted** option to create a bulleted list using the default solid circle bullet symbol.

Another method of creating lists is to use the **Allow Auto-list** option. This option, which is active by default, detects characters that frequently start a list and automatically assigns the first list item. For example, if a line of text begins with a number or letter and a period, AutoCAD assumes that you are starting a list and formats additional lines of text to continue the list.

To create a numbered or lettered auto-list, you must include punctuation, such as a period, parenthesis, or colon, and press [Tab] after the number or letter that begins the first item. When you press [Enter] to start a new line of text, the new line has the same formatting as the previous line, and the next consecutive number or letter appears. To end the list, press [Enter] twice. Figure 9-31 shows an example of a numbered list.

When creating a bulleted auto-list, you can use typical keyboard characters, such as a hyphen [-], tilde [~], bracket [>], or asterisk [*], at the beginning of a line. Another option is to insert a symbol at the beginning of a line. Then, to form the list, press [Tab] and type the line of text. When you press [Enter] to start a new line of text, the new line uses the same formatting bullet symbol as the previous line. To end the list, press [Enter] twice. See Figure 9-32.

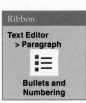

Ribbon

Text Editor > Paragraph

Bullets and Numbering

Figure 9-31.
Framing notes arranged in a numbered list, using the traditional architectural Stylus BT font.

```
FRAMING NOTES:
1.  ALL FRAMING NOTES TO BE DFL #2 OR BETTER.
2.  ALL HEATED WALLS @ HEATED LIVING AREA TO BE 2 X 6 @ 16" OC.
    FRAME ALL EXTERIOR NON-BEARING WALLS W/2 X 6 STUDS @ 24" OC.
3.  USE 2 X 6 NAILER AT THE BOTTOM OF ALL 2-2 X 12 OR 4 X HEADERS
    @ EXTERIOR WALLS, BACK HEADER W/2" RIGID INSULATION.
4.  BLOCK ALL WALLS OVER 10'-0" HIGH AT MID HEIGHT.
```

Figure 9-32.
In addition to the standard solid circle bullet symbol, you can use other keyboard characters or symbols to create a bulleted list.

- An elevation of the beam with end views or sections
- Complete locational dimensions for holes, plates, and angles
- Length dimensions

Bulleted List with Bullet Symbols

~ Connection specifications
~ Cutouts
~ Miscellaneous notes for the fabricator

Bulleted List with Tilde Characters

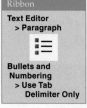

Picking the **Use Tab Delimiter Only** option limits unwanted list formatting by instructing AutoCAD to recognize only tabs to start a list. If the **Use Tab Delimiter Only** option is unchecked, list formatting occurs when a space or tab follows the initial list item character.

You can convert multiple lines of text to a list by selecting the lines of text and picking a list format. AutoCAD detects where you press [Enter] to start a new line of text and lists the lines in sequence. When you create a list in this manner, a tab automatically occurs after the number, letter, or symbol preceding the text. Set tabs and indents to adjust spacing and appearance.

Additional options are available for controlling lists. Use the **Off** option to remove list characters or bulleting from selected text. Pick the **Start** option to renumber or re-letter selected items to create a new list. The numbering or lettering restarts from the beginning, using 1 or A, for example. Choose the **Continue** option to add selected items to a list that exists above the selected line of text. The number of the selected line of text continues from the previous list. Items below the selected line are also renumbered.

Exercise 9-6

Complete the exercise on the companion website.
www.g-wlearning.com/CAD

Columns

Sometimes it is necessary to group text into multiple sections, or columns. A common example is dividing lengthy general notes into columns. See **Figure 9-33**. Mtext with columns is still a single mtext object. This eliminates the need to create multiple text objects to form separate columns of text. You can create columns as you enter text or apply column formatting to existing text.

Earlier in this chapter, you were told to turn columns off before accessing the text editor. This approach is appropriate for typical text requirements without columns, especially as you learn to create mtext. However, mtext is set to form *dynamic columns* using the **Manual height** option by default. Column tools are also available while the text editor is active, from the **Columns** flyout on the **Insert** panel of the **Text Editor** ribbon tab or from the **Columns** cascading menu in the shortcut menu.

dynamic columns:
Columns calculated automatically by AutoCAD according to the amount of text and the specified height and width of the columns.

Dynamic Columns

To form dynamic columns, choose an option from the **Dynamic Columns** cascading menu. Pick the **Auto height** option to produce columns of equal height. **Figure 9-34** shows methods for adjusting dynamic columns using **Auto height**. Increase column width or height to reduce the number of columns, or decrease column width or height to produce more columns. Pick the **Manual height** option to create columns you can adjust individually for height to produce distinct groups of information. Drag the arrows at the bottom of each column to adjust column height. See **Figure 9-35**.

Static Columns

static columns:
Columns in which you divide the text into a specified number of columns.

To form *static columns*, choose the number of columns from the **Static Columns** cascading menu. The display of text in static columns depends on how much text is in the text editor and the height and width of the columns. However, the selected number of columns does not change even if text is not completely filled or extends past a column. **Figure 9-36** shows methods for adjusting static columns. Increasing

Figure 9-33.
An example of general construction notes created as a single mtext object and divided into three columns.

1.01 RELATED WORK

A. REQUIREMENTS: PROVIDE METAL FABRICATION IN ACCORDANCE WITH CONTRACT DOCUMENTS.

1.02 SUBMITTALS

A. SHOP DRAWINGS: INCLUDE PLANS AND ELEVATIONS AT NOT LESS THAN 1" = 1'-0" SCALE, AND INCLUDE DETAILS OF SECTIONS AND CONNECTIONS AT NOT LESS THAN 3" = 1'-0" SCALE. SHOW ANCHORAGE AND ACCESSORY ITEMS. SHOP DRAWINGS FOR ITEMS SPECIFIED BY DESIGN LOAD SHALL INCLUDE ENGINEERING CALCULATIONS AND SHALL BEAR SEAL AND SIGNATURE OF PROFESSIONAL ENGINEER REGISTERED IN STATE IN WHICH PROJECT IS LOCATED.

1.03 DELIVERY, STORAGE AND HANDLING

A. DELIVERY: DELIVER ITEMS, WHICH ARE TO BE BUILT INTO WORK OF OTHER SECTIONS IN TIME TO NOT DELAY WORK.

B. STORAGE: STORE IN UNOPENED CONTAINERS. STORE OFF GROUND AND UNDER COVER, PROTECTED FROM DAMAGE.

C. HANDLING: HANDLE IN MANNER TO PROTECT SURFACES. PREVENT DISTORTION OF, AND OTHER DAMAGE TO, FABRICATED PIECES.

2.01 MATERIALS

A. STRUCTURAL STEEL SHAPES: ASTM A36.

B. STEEL PLATES: ASTM A283 GRADE C FOR BENDING OR FORMING COLD.

C. STEEL TUBING: ASTM A500, GALVANIZED WHERE INDICATED.

D. STEEL BARS AND BAR SHAPES: ASTM A675, GRADE 65; OR ASTM A36.

E. COLD FINISHED STEEL BARS: ASTM A108, GRADE AS SELECTED BY FABRICATOR.

F. STAINLESS STEEL: ASTM A167, TYPE 302 OR 304; NUMBER 4 FINISH.

G. BOLTS AND NUTS: ASTM A307, GRADE A BOLTS.

H. MACHINE SCREWS: F FF-S-92, TYPE COMPATIBLE WITH METALS BEING FASTENED AND FINISHED TO MATCH METAL FINISH.

2.02 FABRICATION, GENERAL

A. INCLUDE SUPPLEMENTARY BARS NECESSARY TO COMPLETE METAL FABRICATION WORK THOUGH NOT DEFINITELY INDICATED.

B. USE MATERIALS OF SIZE AND THICKNESS INDICATED, OR IF NOT INDICATED, OF REQUIRED SIZE AND THICKNESS TO PRODUCE ADEQUATE STRENGTH AND DURABILITY IN FINISHED PRODUCT FOR INTENDED USE.

C. FORM EXPOSED WORK TRUE TO LINE AND LEVEL WITH ACCURATE ANGLES AND SURFACES AND STRAIGHT SHARP EDGES. EASE EXPOSED EDGES TO RADIUS OF APPROXIMATELY 1/32 INCH UNLESS OTHERWISE INDICATED. FORM BENT-METAL CORNERS TO SMALLEST RADIUS WITHOUT CAUSING GRAIN SEPARATION OF OTHERWISE IMPAIRING WORK.

3.01 INSPECTION

A. EXAMINATION: EXAMINE SUBSTRATES, ADJOINING CONSTRUCTION AND CONDITIONS UNDER WHICH WORK IS TO BE INSTALLED. DO NOT PROCEED WITH WORK UNTIL UNSATISFACTORY CONDITIONS HAVE BEEN CORRECTED.

3.02 PREPARATION

A. FIELD MEASUREMENTS: VERIFY DIMENSIONS BEFORE PROCEEDING WITH WORK. OBTAIN FIELD MEASUREMENTS FOR WORK REQUIRED TO BE ACCURATELY FITTED TO OTHER CONSTRUCTION. BE RESPONSIBLE FOR ACCURACY OF SUCH MEASUREMENTS AND PRECISE FITTING AND ASSEMBLY OF FINISHED WORK.

3.03 INSTALLATION

A. INSTALL WORK IN LOCATIONS INDICATED, PLUMB, LEVEL, AND IN LINE WITH ADJACENT MATERIALS WHERE REQUIRED. PROVIDE FASTENINGS INDICATED.

B. FILL SPACE BETWEEN SLEEVES AND POSTS OF RAILINGS WITH SETTING COMPOUND. SLOPE TOP SURFACE TO DRAIN AWAY FROM POSTS.

3.04 ADJUSTING AND CLEANING

A. TOUCH-UP MARRED AND ABRADED SURFACES WITH SPECIFIED PAINT AFTER FIELD ERECTION.

3.05 PROTECTION

A. PROTECT FINISHED SURFACES AGAINST DAMAGE DURING SUBSEQUENT CONSTRUCTION OPERATIONS. REMOVE PROTECTION AT TIME OF SUBSTANTIAL COMPLETION.

Figure 9-34.
Controlling columns using the dynamic column **Auto Height** option. Notice that column text flows automatically from one column to the next.

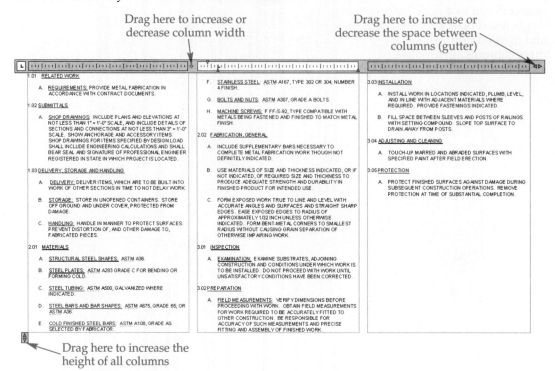

Drag here to increase or decrease column width

Drag here to increase or decrease the space between columns (gutter)

Drag here to increase the height of all columns

Figure 9-35.
Controlling the length of dynamic columns individually, or manually.

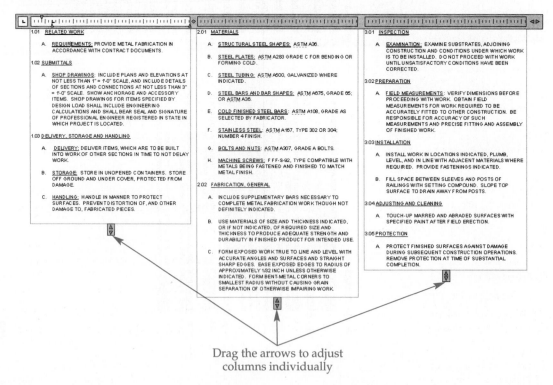

Drag the arrows to adjust columns individually

Figure 9-36.
Controlling static columns.

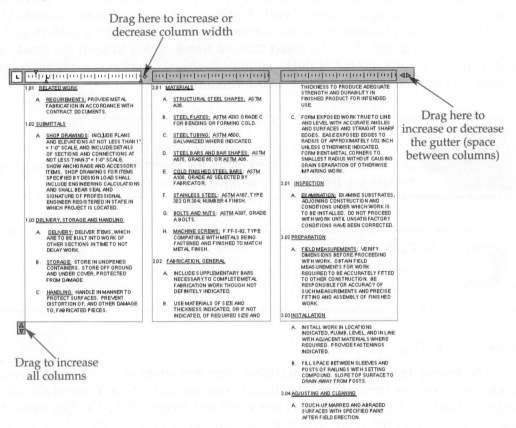

Drag here to increase or decrease column width

Drag here to increase or decrease the gutter (space between columns)

Drag to increase all columns

column width or height rearranges the text in the specified number of columns, but the number of static columns does not change based on column width or height.

To create more than six static columns, pick the **More...** option to access the **Column Settings** dialog box and enter the number of columns in the **Column Number** text box.

Using the Column Settings Dialog Box

You can use the **Column Settings...** dialog box as an alternative method to create columns. To create dynamic columns, select the **Dynamic Columns** radio button, and then pick the **Auto height** or **Manual height** radio button. To create static columns, choose the **Static Columns** radio button and enter the number of static columns in the **Column Number** text box.

Additional controls become available depending on the selected column type radio buttons. The **Height** text box allows you to enter the height for all static or dynamic columns. The **Width** area allows you to set column width and the *gutter*. Enter the column width in the **Column** text box and the gutter width in the **Gutter** text box. The **Total** text box, available only with static columns, allows you to enter the total width of the text editor, which is the sum of the width of all columns and the gutter spacing between columns. To eliminate columns, pick the **No Columns** radio button.

gutter: The space between columns of text.

Controlling Column Breaks

Use the **Insert Column Break** option to specify the line of text at which a new column begins. To assign a column break, first form a dynamic or static column. Then place the cursor at a location in the text editor where a new column is to start, such as the start of a paragraph. Pick the **Insert Column Break** option to form the break. The text shifts to the next column at the location of the break. Continue applying column breaks as needed to separate sections of information.

If you choose to remove columns using the **No Columns** option, any column breaks added using the **Insert Column Break** function remain set. Backspace to remove column breaks.

Exercise 9-7

Complete the exercise on the companion website.
www.g-wlearning.com/CAD

Importing Text

Ribbon

Text Editor
> Tools
> Import Text

The **Import Text** option, also available from the shortcut menu, allows you to import text from an existing text file directly into the text editor. The text file can be either a standard ASCII text file (TXT) or a rich text format (RTF) file. The **Select File** dialog box appears when you access the **Import Text...** option. Select the text file to import and pick the **Open** button. The text is inserted at the location of the text cursor. Imported text becomes part of the mtext object.

PROFESSIONAL TIP

Importing text is useful if someone has already created specifications or notes in a program other than AutoCAD and you want to place the same notes in your drawings.

Template Development Chapter 9

Adding Text Styles

For detailed instructions on adding text styles to each drawing template, go to the companion website (www.g-wlearning.com/CAD), select this chapter, and select **Template Development**.

Chapter Review

Answer the following questions. Write your answers on a separate sheet of paper or complete the electronic chapter review on the companion website.
www.g-wlearning.com/CAD

1. Define *font*.
2. Which ASME standard contains guidelines for lettering?
3. What is text composition?
4. Determine the AutoCAD text height for text to be plotted .188″ high using a half (1:2) scale. Show your calculations.
5. Determine the AutoCAD text height for text to be plotted .188″ high using a scale of 1/4″ = 1′-0″. Show your calculations.
6. Explain the function of annotative text and give an example.
7. What is the relationship between the drawing scale and the annotation scale for annotative text?
8. Define *text style*.
9. Describe how to create a text style that has the name ROMANS-12_15, uses the romans.shx font, has a fixed height of .12, a text width of 1.25, and an oblique angle of 15.
10. What are "big fonts"?
11. When setting text height in the **Text Style** dialog box, what value do you enter so that you can alter the text height when you create single-line text?
12. How would you specify text to display vertically on-screen?
13. What does a width factor of .5 do to text compared to the default width factor of 1?
14. How can you set a text style current without opening the **Text Style** dialog box?
15. Name the command that lets you create multiline text objects.
16. How does the width of the mtext boundary affect what you type?
17. What happens if the mtext you type exceeds the boundary length that you initially establish?
18. What happens by default when you type a fraction in the mtext editor, and what does this allow you to do?
19. How can you draw stacked fractions manually when using the **MTEXT** command?
20. What is the purpose of tracking?
21. What text setting allows you to hide parts of objects behind and around text?
22. What is the difference between text boundary justification and paragraph alignment?
23. Explain how to convert multiple lines of text into a numbered list.
24. Briefly describe the difference between dynamic columns and static columns.
25. From what two file formats can you import text?

Drawing Problems

Start AutoCAD if it is not already started. Start a new drawing for each problem using an appropriate template of your choice. The template should include layers and text styles, when necessary, for drawing the given objects. Add layers and text styles as needed. Draw all objects using appropriate layers and text styles, justification, and format. Follow the specific instructions for each problem. Use only drawing commands and techniques you have already learned. Use your own judgment and approximate dimensions when necessary.

▼ Basic

1. Use the **MTEXT** command to type your name using a text style of your choosing and a text height of 1". Save the drawing as P9-1. Print an 8.5" × 11" copy of the drawing extents, using a 1:1 scale and a landscape orientation. Construct the sheet into a name tag.

2. Use the **MTEXT** command to type the definition of the following terms using a text style with the Arial font and a .12 text height. Save the drawing as P9-2.
   - annotation
   - text
   - composition
   - font
   - justify

3. Use the **MTEXT** command to type the definition of the following terms using a text style with the Romand font and a .12 text height. Save the drawing as P9-3.
   - scale factor
   - annotative text
   - annotation scale
   - text height
   - paper text height

4. Use the **MTEXT** command to type the key notes shown. Use a text style with the Stylus BT font. The heading text height is .25 and the note text height is .125. Save the drawing as P9-4.

   KEY NOTES
   1. SLOPING SURFACE
   2. DIAGONAL SUPPORT STRUT
   3. VENT- PROVIDE NEW CANT FLASHING
   4. BRICK CHIMNEY- REMOVE TO BELOW DECK SURFACE

5. Use the **MTEXT** command to type the general floor plan notes shown. Use a text style with the Century Gothic font. The heading text height is .25 and the note text height is .125. Save the drawing as P9-5.

   GENERAL NOTES
   1. ALL PENETRATIONS IN TOP OR BOTTOM PLATES FOR PLUMBING OR ELECTRICAL RUNS TO BE SEALED. SEE ELECTRICAL PLANS FOR ADDITIONAL SPECIFICATIONS.
   2. PROVIDE 1/2" WATERPROOF GYPSUM BOARD AROUND ALL TUBS, SHOWERS, AND SPAS.
   3. VENT DRYER AND ALL FANS TO OUTSIDE AIR THRU VENT WITH DAMPER.

AutoCAD and Its Applications—Basics

▼ Intermediate

6. Draw the basic organizational chart shown below. Save the drawing as P9-6.

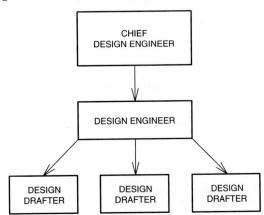

```
┌──────────────────────┐
│        CHIEF          │
│  DESIGN ENGINEER      │
└──────────────────────┘
           │
           ▼
┌──────────────────────┐
│   DESIGN ENGINEER     │
└──────────────────────┘
     │      │      │
     ▼      ▼      ▼
┌────────┐┌────────┐┌────────┐
│ DESIGN ││ DESIGN ││ DESIGN │
│ DRAFTER││ DRAFTER││ DRAFTER│
└────────┘└────────┘└────────┘
```

7. Use the **MTEXT** command to type the general notes for a part drawing shown. Use a text style with the Romans font. The heading text height is 6 mm and the note text height is 3 mm. Save the drawing as P9-7.

NOTES:

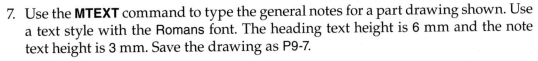

1. DIMENSIONS AND TOLERANCES PER ASME Y14.5–2009.
2. REMOVE ALL BURRS AND SHARP EDGES.

CASTING NOTES UNLESS OTHERWISE SPECIFIED:
1. .31 WALL THICKNESS
2. R.12 FILLETS
3. R.06 CORNERS
4. 1.5°–3.0° DRAFT
5. TOLERANCES
   ± 1° ANGULAR
   ± .03 TWO–PLACE DIMENSIONS
6. PROVIDE .12 THK MACHINING STOCK ON ALL MACHINED SURFACES.

8. Use the **MTEXT** command to type the common framing notes shown. Use a text style with the Stylus BT font. The heading text height is .188 and the note text height is .125. After typing the text exactly as shown, edit the text with the following changes:
   A. Change the \ in item 7 to 1/2.
   B. Change the [in item 8 to 1.
   C. Change the 1/2 in item 8 to 3/4.
   D. Change the ^ in item 10 to a degree symbol.
   Save the drawing as P9-8.

COMMON FRAMING NOTES:
1. ALL FRAMING LUMBER TO BE DFL #2 OR BETTER.
2. ALL HEATED WALLS @ HEATED LIVING AREAS TO BE 2 X 6 @ 24" OC.
3. ALL EXTERIOR HEADERS TO BE 2-2 X 12 UNLESS NOTED, W/ 2" RIGID INSULATION BACKING UNLESS NOTED.
4. ALL SHEAR PANELS TO BE 1/2" CDX PLY W/8d @ 4" OC @ EDGE, HDRS, & BLOCKING AND 8d @ 8" OC @ FIELD UNLESS NOTED.
5. ALL METAL CONNECTORS TO BE SIMPSON CO. OR EQUAL.
6. ALL TRUSSES TO BE 24" OC. SUBMIT TRUSS CALCS TO BUILDING DEPT. PRIOR TO ERECTION.
7. PLYWOOD ROOF SHEATHING TO BE \ STD GRADE 32/16 PLY LAID PERP TO RAFTERS. NAIL W/8d @ 6" OC @ EDGES AND 12" OC @ FIELD.
8. PROVIDE [1/2" STD GRADE T&G PLY FLOOR SHEATHING LAID PERP TO FLOOR JOISTS. NAIL W/10d @ 6" OC @ EDGES AND BLOCKING AND 12" OC @ FIELD.
9. BLOCK ALL WALLS OVER 10'-0" HIGH AT MID.
10. LET-IN BRACES TO BE 1 X 4 DIAG BRACES @ 45^ FOR ALL INTERIOR LOAD-BEARING WALLS.

9. Use the **MTEXT** command to type the caulking notes shown. Use a text style with the SansSerif font. The heading text height is .188 and the note text height is .125. Save the drawing as P9-9.

CAULKING NOTES:

CAULKING REQUIREMENTS BASED ON
OREGON RESIDENTIAL ENERGY CODE

1. SEAL THE EXTERIOR SHEATHING AT CORNERS, JOINTS, DOORS, WINDOWS, AND FOUNDATION SILL WITH SILICONE CAULK.
2. CAULK THE FOLLOWING OPENINGS W/ EXPANDED FOAM, BACKER RODS, OR SIMILAR:
    - ANY SPACE BETWEEN WINDOW AND DOOR FRAMES
    - BETWEEN ALL EXTERIOR WALL SOLE PLATES AND PLY SHEATHING
    - ON TOP OF RIM JOIST PRIOR TO PLYWOOD FLOOR APPLICATION
    - WALL SHEATHING TO TOP PLATE
    - JOINTS BETWEEN WALL AND FOUNDATION
    - JOINTS BETWEEN WALL AND ROOF
    - JOINTS BETWEEN WALL PANELS
    - AROUND OPENINGS

10. Using **MTEXT**, create the electrical notes shown below. Format the notes properly to make them easier to read.
    A. Use the Stylus BT font.
    B. Add the heading "ELECTRICAL NOTES:" on a separate line.
    C. Number each item separately.
    D. Use all capital letters and set the notes in two columns.
    Save the drawing as P9-10.

All garage and exterior plugs and light fixtures to be on GFCI circuit. All kitchen plugs and light fixtures to be on GFCI circuit. Provide a separate circuit for microwave oven. Provide a separate circuit for personal computer. Verify all electrical locations with owner. Exterior spotlights to be on photoelectric cell with timer. All recessed lights in exterior ceilings to be insulation cover rated. Electrical outlet plate gaskets to be insulated on receptacle, switch, and any other boxes in exterior wall. Provide thermostatically controlled fan in attic with manual override; verify location with owner. All fans vent to outside air; all fan ducts to have automatic dampers. Hot water tanks to be insulated to R-11 minimum. Insulate all hot water lines to R-4 minimum; provide alternate bid to insulate all pipes for noise control. Provide 6 sq. ft. of vent for combustion air to outside air for fireplace connected directly to firebox; provide fully closable air inlet. Heating to be electric heat pump; provide bid for single unit near garage or for a unit each floor (in attic). Insulate all heating ducts in unheated areas to R-11; all HVAC ducts to be sealed at joints and corners.

11. Draw the general construction notes exactly as shown in **Figure 9-33** using columns. Save the drawing as P9-11.

12. Draw the flowchart shown. Save the drawing as P9-12.

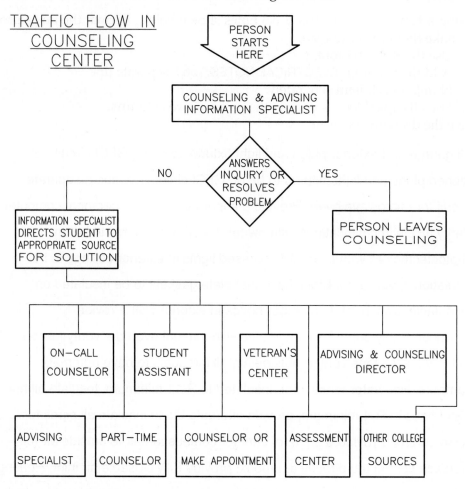

TRAFFIC FLOW IN COUNSELING CENTER

13. Draw the controller schematic shown. Save the drawing as P9-13.

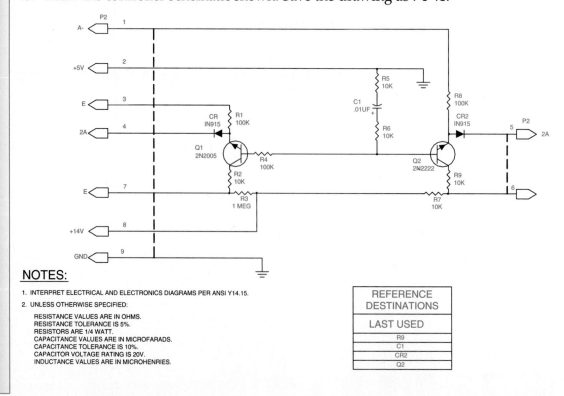

NOTES:

1. INTERPRET ELECTRICAL AND ELECTRONICS DIAGRAMS PER ANSI Y14.15.

2. UNLESS OTHERWISE SPECIFIED:

    RESISTANCE VALUES ARE IN OHMS.
    RESISTANCE TOLERANCE IS 5%.
    RESISTORS ARE 1/4 WATT.
    CAPACITANCE VALUES ARE IN MICROFARADS.
    CAPACITANCE TOLERANCE IS 10%.
    CAPACITOR VOLTAGE RATING IS 20V.
    INDUCTANCE VALUES ARE IN MICROHENRIES.

REFERENCE DESTINATIONS	
LAST USED	
R9	
C1	
CR2	
Q2	

14. Draw the electrical legend shown. Save the drawing as P9-14.

ELECTRICAL LEGEND:

⏀	110 VOLT DUPLEX CONVENIENCE OUTLET
⏀ GFCI	110 VOLT GROUND FAULT CIRCUIT INTERRUPT DUPLEX OUTLET
⏀ GFCI WP	110 VOLT WATERPROOF GFCI DUPLEX OUTLET
⏀	110 VOLT SPLIT WIRED OUTLET
⏀	220 VOLT OUTLET
⟠	JUNCTION BOX
TV	CABLE TELEVISION OUTLET
⟠	CLOCK OUTLET
⏀	DOOR BELL
$	SINGLE POLE SWITCH
$³	THREE-WAY SWITCH
○	CEILING-MOUNTED LIGHT
⟡	WALL-MOUNTED LIGHT
▭	FLUORESCENT LIGHT
⊙	CIRCULAR RECESSED LIGHT
▣	SQUARE RECESSED LIGHT
◖F◗	LIGHT, FAN COMBINATION
◖FH◗	LIGHT, FAN, HEAT COMBINATION
● SD	CEILING-MOUNTED SMOKE DETECTOR
▣SD	WALL-MOUNTED SMOKE DETECTOR

15. Design a poster announcing a party. Design the announcement to fit an 8.5″ × 11″ sheet, using a 1:1 scale and a portrait orientation. Use the **MTEXT** command as needed to provide all relevant information, including the name of the party and a theme statement, time, date, location, and your contact information. Draw non-text objects to enhance and illustrate the announcement. Save the drawing as **P9-15**. Print 8.5″ × 11″ copies of the drawing extents, centered on the sheet, using a 1:1 scale and a portrait orientation. Distribute and post copies of the announcement. Throw the party.

16. Design a greeting card. Design the card to fit an 8.5" × 11" sheet, using a 1:1 scale and a portrait orientation, and using the format shown. Do not draw the dimensions, sheet edges, or fold lines. Use the **MTEXT** command as needed to provide all relevant information, including the occasion, name of the card recipient, and a greeting. Draw non-text objects to enhance and illustrate the card. Save the drawing as P9-16. Print an 8.5" × 11" copy of the drawing extents, centered on the sheet, using a 1:1 scale and a portrait orientation. Fold the card and give it to the recipient.

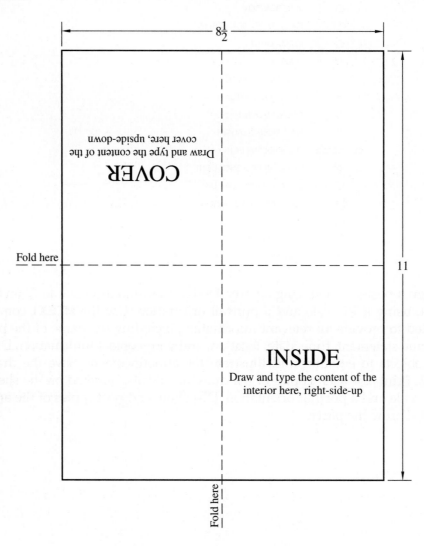

AutoCAD Certified Associate Exam Practice

Answer the following questions. Write your answers on a separate sheet of paper.

1. Which of the following options are available for AutoCAD SHX (shape) fonts? *Select all that apply.*
   - A. backwards
   - B. bold
   - C. italic
   - D. underscore
   - E. vertical

2. Which of the following oblique angles should you specify in AutoCAD to slant text according to the 68° horizontal incline standard? *Select the one item that best answers the question.*
   - A. 0°
   - B. –22°
   - C. 22°
   - D. 68°
   - E. –68°

3. Which of the following symbols can be used to stack selected mtext manually? *Select all that apply.*
   - A. *
   - B. ^
   - C. /
   - D. #
   - E. ~

AutoCAD Certified Professional Exam Practice

Follow the instructions in each problem. Write your answers on a separate sheet of paper.

1. **Navigate to this chapter on the companion website and open CPE-09columns.dwg.** Edit the text to use dynamic columns with the **Auto height** option. Use the arrows on the text editor to create columns with a width of 5.44 units and a height of 2.41 units. How many columns result?

2. **Open a new drawing file and save it as CPE-09mtext.dwg.** Create a text style called SansSerif using the SansSerif font and a text height of .12. Enter the **MTEXT** command and set a text boundary of 3 units. Use the SansSerif text style to insert the legal description shown below into the drawing using all uppercase text. Disable columns. Resave the file. How many lines of text does the legal description occupy with these settings?

   Beginning at a point 20 feet north of center line of Unger Road, north for a length of 655.18', thence S89°33'10"E for a length of 447.75', thence south for a length of 475.53', thence N84°53'42"W for a length of 224.77', thence north for a length of 92.02', thence S81°12'15"W for a length of 196.19', thence south for a length of 256.56', thence S86°55'3"W for a length of 30.23' to the point of beginning.

The Autodesk User Group International (AUGI) organization provides many resources for AutoCAD users. Membership in the organization is free, and the AUGI website allows access to discussion forums, numerous articles and videos, and many other resources.

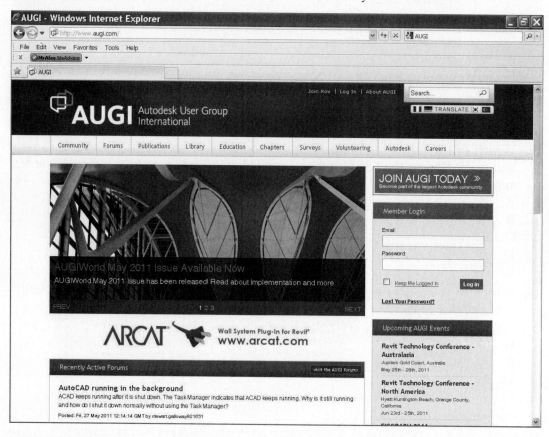

Single-Line Text and Additional Text Tools

Learning Objectives

After completing this chapter, you will be able to:

✓ Use the **TEXT** command to create single-line text.
✓ Insert and use fields.
✓ Check spelling.
✓ Edit existing text.
✓ Search for and replace text automatically.

This chapter describes how to use the **TEXT** command to place a single-line text object. The **TEXT** command is most useful for adding a single character, word, or line of text. Use the **MTEXT** command for all other text requirements. This chapter also presents methods for editing text and other valuable text tools, including fields, spell checking, and ways to find and replace text. You can apply most of the additional text tools described in this chapter to mtext or single-line text objects.

Single-Line Text

It is good practice to set the appropriate text style current before you access the **TEXT** command. Then activate the **TEXT** command to create a single-line text object. Left is the default justification. To use a different justification, choose the **Justify** option at the Specify start point of text [Justify/Style]: prompt before you pick the start point of the text. **Figure 10-1** shows all of the justification options except the **Align** and **Fit** options, which are described later in this chapter. Choose a justification option according to where and how you want text to form.

Pick a point to locate the text according to the specified justification, known as the *justification point*. Next, if the current text style uses a height of 0, enter the text height. If the current text style is annotative, enter the paper text height. If the current text style is not annotative, enter the text height multiplied by the scale factor. The next prompt asks for the text rotation angle. The default value is 0, which draws horizontal text. Specify a rotation angle to pivot the text around the start point in a counterclockwise direction. See **Figure 10-2**.

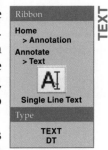

Ribbon

Home
> Annotation
Annotate
> Text

Single Line Text

Type

TEXT
DT

TEXT

justification point:
The point from which text is justified according to the current justification option.

Figure 10-1.
Standard justification options available for drawing single-line text. The difference between some options is evident only when you use values that extend below the baseline, such as the "y" in the word "Justify."

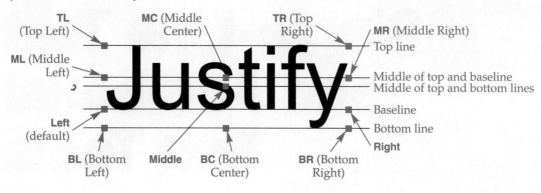

Figure 10-2.
Examples of rotating single-line text. The plus sign indicates the start point.

Changes you make to the default text angle orientation or direction affect the text rotation.

Once you set the justification, height, and rotation angle, a text editor appears on-screen with a cursor equal in height to the text height at the start point. As you type, the text editor increases in size to display the characters. See **Figure 10-3**. Right-click to display a shortcut menu of text options, similar to those available for the **MTEXT** command.

Press [Enter] at the end of each line of text to move the cursor to a start point one line below the preceding line. Each new line of text is a new single-line text object, not a grouped paragraph. When you are finished typing, press [Enter] twice to exit the **TEXT** command. Cancel the **TEXT** command and remove incomplete lines of text by pressing [Esc].

Figure 10-3.
Typing text using the **TEXT** command and default left justification.

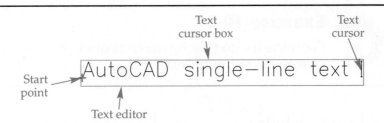

Text cursor box

Text cursor

Start point

AutoCAD single—line text

Text editor

To change selected text to uppercase, right-click and select **UPPERCASE** from the **Change Case** cascading menu. Use the **lowercase** option to change all text to lowercase.

PROFESSIONAL TIP

Set the text style current before accessing the **TEXT** command. A **Style** option is available before you pick the text start point to set the text style, but it is difficult to use.

Align or Fit Justification

Align and **Fit** justification options are available from the **Justify** option at the Specify start point of text [Justify/Style]: prompt, as an alternative to the previous method of creating single-line text. The **Align** option requires you to select the start point and endpoint of the line of text. AutoCAD adjusts the text height to form between the start point and endpoint. The height varies according to the distance between the points and the number of characters. The **Fit** option requires you to select the start point and endpoint of the line of text, and the text height. AutoCAD adjusts character width to fit between the specified points, while keeping text height constant. **Figure 10-4** shows the effects of the **Align** and **Fit** options.

PROFESSIONAL TIP

Avoid using the **Align** and **Fit** options of the **TEXT** command for standard drafting practices, because the text height or width is inconsistent from one line of text to another. In addition, text height adjusted by the **Fit** option can cause one line of text to run into another.

Figure 10-4.
Examples of using the **Align** and **Fit** justification options to create aligned and fit text. The plus signs indicate the start point and endpoint.

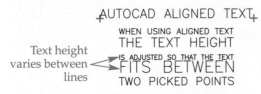

Text height varies between lines

AUTOCAD ALIGNED TEXT
WHEN USING ALIGNED TEXT
THE TEXT HEIGHT
IS ADJUSTED SO THAT THE TEXT
FITS BETWEEN
TWO PICKED POINTS

Align Option

AUTOCAD FIT TEXT
WHEN USING FIT TEXT
THE TEXT WIDTH
IS ADJUSTED SO THAT THE TEXT
HEIGHT REMAINS THE SAME
FOR EACH LINE

Text width varies between lines

Fit Option

Exercise 10-1

Complete the exercise on the companion website.
www.g-wlearning.com/CAD

Adding Symbols

control code sequence: A key sequence beginning with %% that defines symbols in text created with the **TEXT** command.

Type a *control code sequence* to add a symbol with the text command. For example, type %%C2.75 to create the note ∅2.75. In this example, %%C is the code used to add the diameter symbol. Figure 10-5 shows several symbol codes and the symbol the code creates. Add a single percent sign normally. However, when a percent sign must precede another control code sequence, use %%% to force a single percent sign.

Underscoring or Overscoring Text

Type %%U in front of text to underscore and %%O in front of text to overscore. For example, type %%UUNDERSCORING TEXT to create the note <u>UNDERSCORING TEXT</u>.

Figure 10-5.
Common control code sequences used to add symbols to single-line text.

Control Code or Unicode	Type of Symbol	Appearance	Control Code or Unicode	Type of Symbol	Appearance
%%d	Degrees	°	\U+2261	Identity	≡
%%p	Plus/Minus	±	\U+E200	Initial length	
%%c	Diameter	∅	\U+E102	Monument line	
%%%	Percent	%	\U+2260	Not equal	≠
\U+2248	Almost equal	≈	\U+2126	Ohm	Ω
\U+2220	Angle	∠	\U+03A9	Omega	Ω
\U+E100	Boundary line		\U+214A	Property line	⅊
\U+2104	Centerline	℄	\U+2082	Subscript 2	$_2$
\U+0394	Delta	Δ	\U+00B2	Squared	^{2}
\U+0278	Electrical phase	ɸ	\U+00B3	Cubed	^{3}
\U+E101	Flow line				

Use both control code sequences to underscore and overscore text. For example, the control code sequence %%O%%ULINE OF TEXT produces <u>LINE OF TEXT</u>.

The %%U and %%O control codes are toggles that turn underscoring and overscoring on and off. Type %%U before a word or phrase to be underscored, and then type %%U after the word or phrase to turn underscoring off. Text following the second %%U appears without underscoring. For example, enter %%UDETAIL A%%U HUB ASSEMBLY to create <u>DETAIL A</u> HUB ASSEMBLY.

PROFESSIONAL TIP

Underline labels such as SECTION A-A or DETAIL B instead of drawing a line or polyline object under the text. In addition, use the **Middle** or **Center** justification mode. The view labels are automatically underlined and centered under the views or details they identify.

Exercise 10-2

Complete the exercise on the companion website.
www.g-wlearning.com/CAD

Fields

A *field* references drawing properties, AutoCAD functions, and information-related objects, and then displays the content in a text object. For example, insert the **Title** field below a drawing view to display the title of the drawing. You can update fields when changes occur to the reference data. For example, insert the **Date** field into a title block to update the field with the current date as necessary. Fields allow you to create "intelligent" text that uses existing drawing data and displays current information that changes throughout the course of a project.

field: A text object that can display a specific property value, setting, or characteristic.

Inserting Fields

Use the **FIELD** command to add a field to an mtext or text object. To insert a field in an active mtext editor, access the **Field** option, also available by right-clicking and selecting **Insert Field**, or by pressing [Ctrl]+[F]. To insert a field in an active single-line text editor, right-click and select **Insert Field...** or press [Ctrl]+[F].

The **Field** dialog box appears, allowing you to select a field. See **Figure 10-6**. The **Field** dialog box includes many preset fields. You can filter the list of fields by selecting a category from the **Field category** drop-down list. Fields related to the category appear in the **Field names** list box. Pick a field from the list and then choose a format from the **Examples:** list box. If you cannot find an adequate example, create a custom format by typing in the **Data format:** text box.

Pick the **OK** button to insert the field. The field assumes the current text style. By default, field text displays a light gray background. See **Figure 10-7**. This keeps you aware that the text is actually a field and that the value may change. You can deactivate the background in the **Fields** area of the **User Preferences** tab of the **Options** dialog box. See **Figure 10-8**.

Ribbon

Insert
> Data
Text Editor
> Insert

Field

Type

FIELD

FIELD

Figure 10-6.
Select fields using the **Field** dialog box.

Selected field   Current value for field

Select
category
to limit
field list

All available
fields listed

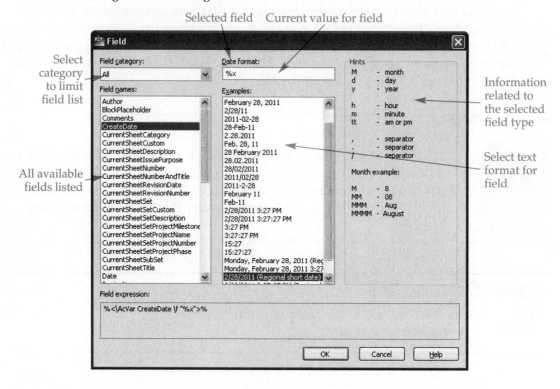

Information
related to
the selected
field type

Select text
format for
field

Figure 10-7.
A date and time field added to an mtext object. The gray background identifies the text as a field.
The date and time field references the computer date and time when you inserted the field.

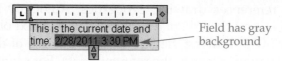

Field has gray
background

Figure 10-8.
Control the
background display
for fields in the **User
Preferences** tab of
the **Options** dialog
box.

Controls
display of field
background

Pick to change
automatic
update settings

Updating Fields

You can *update* fields automatically or manually as changes occur to the referenced data. Examples include updating a **Date** field to correspond to the date today, a **Filename** field to match changes made to the filename, or an **Object** field to match changes made to the properties of an object. Set automatic field updates using the **Field Update Settings** dialog box. To access this dialog box, pick the **Field Update Settings...** button in the **Fields** area of the **User Preferences** tab of the **Options** dialog box. Whenever a selected event, such as saving or regenerating, occurs, all associated fields are automatically updated.

Update fields manually using the **UpdateField** command. After selecting the command, pick the fields to update. Use the **All** selection option to update all fields in a single operation. You can also update a field within the text editor by right-clicking on the field and selecting **Update Field**.

update: The AutoCAD procedure for changing text in a field to reflect the current value of field.

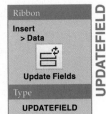

Ribbon
Insert
> Data

Update Fields

Type
UPDATEFIELD

UPDATEFIELD

Editing and Converting Fields

To edit a field, first activate the text object containing the field for editing. A quick way to edit text is to double-click on the text object. Then double-click on the field to display the **Field** dialog box. You can also right-click in the field and pick **Edit Field...**. Use the **Field** dialog box to modify the field settings and pick the **OK** button to apply the changes.

To convert a field to standard text, activate the text object for editing, right-click on the field, and pick **Convert Field To Text**. When you convert a field to text, the current field value becomes text, the association to the field data is lost, and the value can no longer be updated automatically.

You can use fields with many AutoCAD tools, including inquiry commands, drawing properties, attributes, and sheet sets. You will learn specific field applications where appropriate throughout this textbook.

Exercise 10-3

Complete the exercise on the companion website.
www.g-wlearning.com/CAD

Checking Spelling

The quickest way to check for correct spelling in text objects is to use the **Spell Check** function available in the text editors. The **Spell Check** tool is active by default. To toggle **Spell Check** on or off in the mtext or single-line text editor, select the **Check Spelling** option from the **Editor Settings** cascading menu available from the shortcut menu or select **Spell Check** from the **Spell Check** panel on the **Text Editor** ribbon tab. You can also turn **Spell Check** on and off in an active mtext editor by selecting or deselecting the **Spell Check** button on the **Options** panel of the **Text Editor** ribbon tab.

A red dashed line appears under a word that the AutoCAD dictionary does not recognize. Right-click on the underlined word to display options for adjusting the spelling. See **Figure 10-9**. The first section at the top of the shortcut menu provides

Ribbon
Text Editor
> Spell Check

ABC

Spell Check

Figure 10-9.
Checking spelling using the **Spell Check** tool in the mtext editor. The tool functions the same in the single-line text editor.

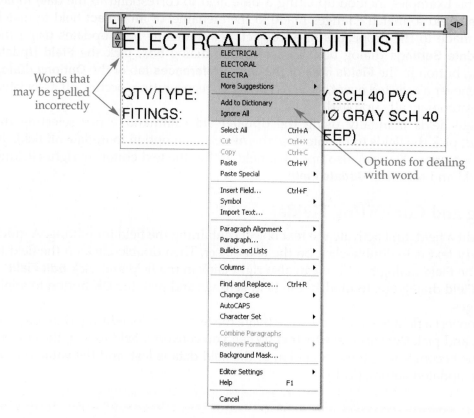

Words that may be spelled incorrectly

Options for dealing with word

suggested replacements for the word. Pick a word to change the spelling in the text editor. If you do not see the correct suggestion, you may be able to find the correct spelling from the **More Suggestions** cascading menu.

If you still cannot find a correct spelling from the suggestions, either the word is spelled correctly but not found in the dictionary, or it is spelled so incorrectly that AutoCAD cannot suggest a replacement. If the word is spelled correctly, pick the **Add to Dictionary** option to add the current word to the custom dictionary. You can add words with up to 63 characters. To use the current spelling without adding the word to the dictionary, pick the **Ignore All** option. Spell checking ignores all words that match the spelling in the active text editor and hides the underline. Add common drafting words and abbreviations to the dictionary, such as the abbreviation for the word SCHEDULE (SCH) in **Figure 10-9**, or ignore the words.

Using the SPELL Command

Ribbon
**Annotate
> Text**

Check Spelling

Type
**SPELL
SP**

The **SPELL** command uses the **Check Spelling** dialog box, shown in **Figure 10-10**, to check the spelling of text objects without activating a text editor. To check spelling, first identify the portion of the drawing to check by selecting an option from the **Where to Check** drop-down list. Pick the **Entire drawing** option to check the spelling of all text objects in the drawing file, including model space and layouts, or choose **Current space/layout** to check spelling only of text objects in the active layout or in model space, if model space is active. You can also pick the **Select text objects** button next to the **Where to Check** drop-down list to enter the drawing window and select specific text objects to check. You do not need to choose the **Selected objects** option from the **Where to Check** drop-down list to check selected objects.

Figure 10-10.
The **Check Spelling** dialog box.

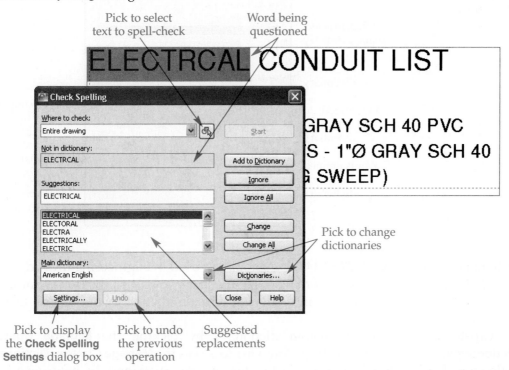

Pick to select text to spell-check

Word being questioned

Pick to change dictionaries

Pick to display the **Check Spelling Settings** dialog box

Pick to undo the previous operation

Suggested replacements

After you define what and where to check, pick the **Start** button to begin spell checking. The first word that may be misspelled becomes highlighted in the drawing window and is active in the **Check Spelling** dialog box. Pick the appropriate button to add, ignore, or change the spelling. Use the **Suggestions:** text box and a change option to type a different spelling that is not available from the list.

PROFESSIONAL TIP

Before you check spelling, you may want to adjust some of the spell-checking preferences provided in the **Check Spelling Settings** dialog box. Access this dialog box by picking the **Settings...** button in the **Check Spelling** dialog box, or select the **Check Spelling Settings...** option from the **Editor Settings** cascading menu available from the text editor shortcut menu. If the mtext editor is already open, pick the small arrow in the lower-right corner of the **Spell Check** panel on the **Text Editor** ribbon tab. The settings apply to spelling checked using the **Spell Check** tool in the active text editor and using the **Check Spelling** dialog box.

Changing Dictionaries

To change the dictionary used to check spelling, pick the **Dictionaries...** button in the **Check Spelling** dialog box. You can also select the **Dictionaries...** option from the **Editor Settings** cascading menu available from the text editor shortcut menu to access the **Dictionaries** dialog box. See **Figure 10-11**.

Use the **Main dictionary** list to select a language dictionary to use as the current main dictionary. You cannot add definitions to the main dictionary. Use the **Custom dictionary** list to select the active custom dictionary. The default custom dictionary is sample.cus. Type a word in the **Content** text box to add or delete from the custom

Ribbon

Text Editor > Spell Check

Edit Dictionaries

Figure 10-11.
The **Dictionaries** dialog box.

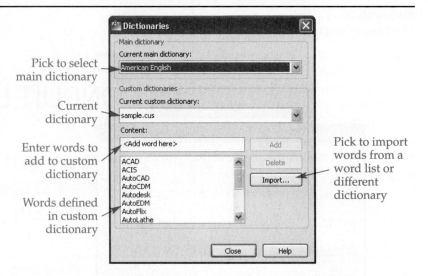

Pick to select main dictionary

Current dictionary

Enter words to add to custom dictionary

Words defined in custom dictionary

Pick to import words from a word list or different dictionary

dictionary. For example, ASME Y14.5 is custom text used in engineering drafting. Pick the **Add** button to accept the custom words in the text box, or pick the **Delete** button to remove the words from the custom dictionary. Custom dictionary entries can use up to 63 characters.

Create and manage a custom dictionary by picking the **Manage Custom Dictionaries…** option from the drop-down list to access the **Manage Custom Dictionary** dialog box. Pick the **New** button to create a new custom dictionary by entering a new file name with a .cus extension. You can add and delete words and combine dictionaries using any standard text editor. If you use a word processor such as Microsoft® Word, be sure to save the file as text only, with no special text formatting or printer codes. Add a custom dictionary by picking the **Add** button, and choose the **Remove** button to delete a custom dictionary from the list. You can also add existing custom dictionaries by picking the **Import…** button from the **Custom dictionary** area.

PROFESSIONAL TIP

Create custom dictionaries for various disciplines. For example, add common abbreviations and brand names for mechanical drawings to a mech.cus file. A separate file named arch.cus might contain common architectural abbreviations and frequently used brand names.

Exercise 10-4

Complete the exercise on the companion website.
www.g-wlearning.com/CAD

Revising Text

The easiest way to reopen the text editor to make changes to text content is to double-click on an mtext or text object. Another technique to re-enter the text editor is to pick the text object to modify and then right-click and select **Mtext Edit…** to revise mtext, or **Edit…** to modify single-line text. A third option is to type DDEDIT to edit mtext or single-line text, or type MTEDIT to edit mtext.

Cutting, Copying, and Pasting Text

Clipboard functions allow you to cut, copy, and paste text from any text-based application, such as Microsoft® Word, into a text editor. You can also cut or copy and paste text from the text editor into other text-based applications. Access clipboard functions from the shortcut menu when an mtext or text editor is active. Text that you paste retains the original text properties.

AutoCAD provides additional paste options for pasting text into the active mtext text editor. Right-click and select an option from the **Paste Special** cascading menu. Pick **Paste without Character Formatting** to paste text without applying preset character formatting such as bold, italic, or underline. Select the **Paste without Paragraph Formatting** option to paste text without applying current paragraph formatting, including lists. Pick the **Paste without Any Formatting** option to paste text without applying any current character and paragraph formatting.

PROFESSIONAL TIP

Importing text is useful if someone has already created specifications or notes in a program other than AutoCAD and you want to place the same notes in your drawings.

Finding and Replacing Text

AutoCAD provides methods for searching for and replacing text with a different word or phrase. You can search for and replace text in an active mtext or single-line text editor, or without opening a text editor.

Using the Find and Replace Tool

Use the **Find and Replace** tool to find and replace text in the active mtext or single-line text editor. Right-click and select **Find and Replace...** to display the **Find and Replace** dialog box. The dialog box is also available in the mtext editor using the **Find & Replace** option. See **Figure 10-12**.

Type the text to search for in the **Find what:** text box. Type the text to substitute in the **Replace with:** text box. Then pick the **Find Next** button to highlight the next instance of the search text. You can then pick the **Replace** or **Replace All** button to replace just the highlighted text or all words that match the search criteria. Check boxes control the characters and words that are recognized.

Ribbon

Text Editor > Tools

Find & Replace

Figure 10-12.
Using the **Find and Replace** dialog box in the mtext editor. The tool functions the same in the single-line text editor.

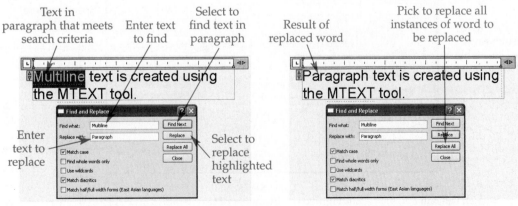

Searching for Text   Replacing Text

Using the FIND Command

FIND

Ribbon
**Annotate
> Text**

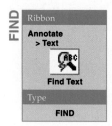

Find Text

Type
FIND

The **FIND** command uses the **Find and Replace** dialog box, shown in **Figure 10-13**, to find and replace text objects without activating a text editor. Another method for accessing the **Find and Replace** dialog box is to enter the text string to find in the **Find Text** text box in the **Text** panel of the **Annotation** ribbon tab and press [Enter]. You can also activate the **FIND** command by right-clicking when no command is active and selecting **Find...**.

To find and replace text, first identify the portion of the drawing to search by selecting an option from the **Where to Check** drop-down list. The options are similar to those in the **Find Where** drop-down list in the **Check Spelling** dialog box. The **Find and Replace** dialog box is much like the dialog box of the same name that appears when you find and replace text within a text editor. However, the **FIND** command version allows you to display the search results in a table within the dialog box and provides more search options. Pick the **More Options** button to display check boxes used to control the characters and words recognized when you are finding and replacing text.

The find and replace strings are saved with the drawing file for future use.

SCALETEXT

Ribbon
**Annotate
> Text**

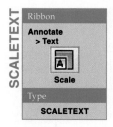

Scale

Type
SCALETEXT

Scaling Text

The **SCALETEXT** command is one option for changing the height of mtext and text objects. Access the **SCALETEXT** command and select the text objects to be scaled. At the prompt, specify the justification for the base point of the scale operation. The justification point determines the point from which the increase or decrease in size occurs.

Figure 10-13.
Using the version of the **Find and Replace** dialog box that appears when you use the **Find** command.

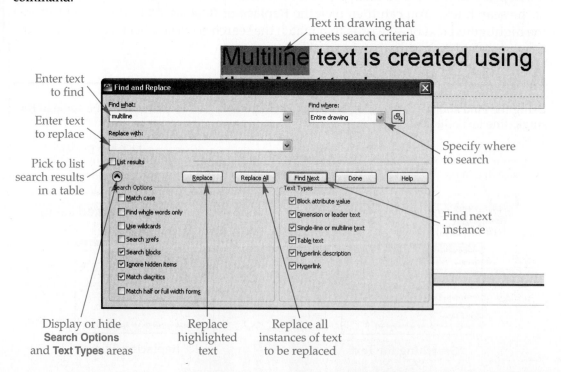

Figure 10-14.
The **Existing** option of the **SCALETEXT** command scales text objects using their individual justification settings.

BL Justification

MC Justification

TR Justification

Original Text

BL Justification

MC Justification

TR Justification

Text Scaled Using
Existing Base Point Option

Figure 10-1 shows the location of all justification points. The **Existing** option scales text objects using their existing justification setting as the base point. See **Figure 10-14**.

After you specify the justification point to use as the base point, AutoCAD prompts for the scaling type. The default **Specify new model height** option allows you to type a new value for the text height of non-annotative objects. If the selected text is annotative, AutoCAD ignores the value you enter. Use the **Paper height** option to type a new paper text height for the annotative objects. If the selected text is non-annotative, AutoCAD ignores the paper height value you enter.

Use **Match object** option to match the height of the text to the height of a different selected text object. Use the **Scale factor** option to scale text objects that have different heights relative to their current heights. For example, a scale factor of 2 scales all selected text objects to twice their current size.

CAUTION

Only use the **SCALETEXT** command to scale non-annotative text.

Changing Justification

Use the **JUSTIFYTEXT** command to change the justification point without moving the text. Pick the text for which you want to change the justification, and then select a new justification option.

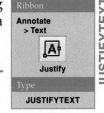

Ribbon

Annotate > Text

Justify

Type

JUSTIFYTEXT

JUSTIFYTEXT

Exercise 10-5

Complete the exercise on the companion website.
www.g-wlearning.com/CAD

Supplemental Material

Isometric Text
For information about constructing text for isometric views, go to the companion website (www.g-wlearning.com/CAD), select this chapter, and select **Isometric Text**.

Express Tools Chapter 10

Text Express Tools
The **Text** panel of the **Express Tools** ribbon tab includes additional text commands. For information about the most useful text express tools, go to the companion website (www.g-wlearning.com/CAD), select this chapter, and select **Text Express Tools**.

Chapter Review

Answer the following questions. Write your answers on a separate sheet of paper or complete the electronic chapter review on the companion website.
www.g-wlearning.com/CAD

1. List two ways to access the **TEXT** command.
2. Write the control code sequence required to draw the following symbols using the **TEXT** command:
   A. 30°
   B. 1.375 ± .005
   C. Ø24
   D. <u>NOT FOR CONSTRUCTION</u>
3. Briefly explain the function and purpose of fields.
4. What is different about the default on-screen display of fields compared to that of text?
5. How can you access the **Field Update Settings** dialog box?
6. Explain how to convert a field to text.
7. What is the quickest way to check your spelling within a current text editor?
8. How do you change the current word if you do not think the word displayed in the **Suggestions:** text box of the **Check Spelling** dialog box is the correct word, but one of the words in the list of suggestions is the correct word?
9. Identify three ways to access the AutoCAD spell checker.
10. How do you change the main dictionary for use in the **Check Spelling** dialog box?
11. Why might you want to create more than one custom dictionary?
12. What happens if you double-click on multiline text?
13. Briefly describe how to find and replace text when an mtext or single-line text editor is open.
14. Name the command that allows you to find text and replace it with different text in a single instance or for every instance in the drawing.
15. When using the **SCALETEXT** command, which base point option would you select to keep the current justification point of the text object?

Drawing Problems

Start AutoCAD if it is not already started. Start a new drawing using an appropriate template of your choice. The template should include layers and text styles for drawing the given objects. Add layers and text styles as needed. Draw all objects using appropriate layers and text styles, justification, and format. Follow the specific instructions for each problem. Use only drawing commands and techniques you have already learned. Use your own judgment and approximate dimensions when necessary.

▼ Basic

1. Create text styles according to the list shown. Use the **TEXT** command and the appropriate text style to type the text shown. Use a .25 unit text height and 0° rotation angle. Save the drawing as P10-1.

ARIAL - A VERY BASIC FONT USED FOR GENERAL-PURPOSE TEXT.

ROMANS — A FONT THAT CLOSELY DUPLICATES THE SINGLE—STROKE LETTERING THAT HAS BEEN THE STANDARD FOR DRAFTING.

ROMANC — A MULTISTROKE DECORATIVE FONT THAT IS GOOD FOR USE IN DRAWING TITLES.

ITALICC — AN ORNAMENTAL FONT THAT IS SLANTED TO THE RIGHT AND HAS THE SAME LETTER DESIGN AS THE COMPLEX FONT.

2. Create text styles according to the list shown. Use the **TEXT** command and the appropriate text style to type the text shown. Use a .25 unit text height. Save the drawing as P10-2.

ARIAL-EXPAND THE WIDTH BY THREE.

MONOTXT-SLANT TO THE LEFT -30°.

ROMANS—SLANT TO THE RIGHT 30°.

ꓤOMAND—BACKWARDS.

ITALICC—UNDERSCORED AND OVERSCORED.

ROMANS—USE 16d NAILS @ 10"OC.

ROMANT—⌀32 (812.8).

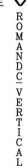

ROMANDC-VERTICAL

3. Open P7-7 and save the file as P10-3. (If you have not yet completed Problem 7-7, work the problem now.) The P10-3 file should be active. Use a text style with the Arial font to add text and titles to the electrical switch schematics as shown in Problem 7-7. Resave the drawing.

4. Open P5-11 and save the file as P10-4. (If you have not yet completed Problem 5-11, work the problem now.) The P10-4 file should be active. Use a text style with the Romans font to add text to the circuit diagram as shown in Problem 5-11. Resave the drawing.

5. Open P7-12 and save the file as P10-5. (If you have not yet completed Problem 7-12, work the problem now.) The P10-5 file should be active. Use a text style with the Romans font to add text to the elementary diagram as shown in Problem 7-12. Resave the drawing.

▼ Intermediate

6. Create text styles with a .375 height and the following fonts: Arial, BankGothic Lt BT, CityBlueprint, Stylus BT, Swis721 BdOul BT, Vineta BT, and Wingdings. Use each text style with the **TEXT** command to type the complete alphabet and numbers 0–10. In addition, type all symbols available on the keyboard and the diameter, degree, and plus/minus symbols. Save the drawing as P10-6.

7. Draw the interior finish schedule shown. Use a text style with the Stylus BT font to add the text. Save the drawing as P10-7. You will learn to create tables using the **TABLE** command in Chapter 21. The purpose of this problem is to practice using the **TEXT** command to place text objects in specific areas, using appropriate justification and format. In general, the **TABLE** command is more appropriate for drawing schedules.

INTERIOR FINISH SCHEDULE											
ROOM	FLOOR					WALLS				CEILING	
	VINYL	CARPET	TILE	HARDWOOD	CONCRETE	PAINT	PAPER	TEXTURE	SPRAY	SMOOTH	BROCADE	PAINT
ENTRY					●							
FOYER			●			●		●				●
KITCHEN			●				●			●		●
DINING				●		●				●	●	●
FAMILY		●				●			●		●	●
LIVING		●				●	●			●	●
MSTR. BATH			●			●				●		●
BATH #2			●			●		●	●		●
MSTR. BED		●				●	●			●	●
BED #2		●				●			●		●	●
BED #3		●				●			●		●	●
UTILITY	●					●			●	●		●

8. Draw the block diagram shown. Use a text style with the Romans font to add the text. Save the drawing as P10-8.

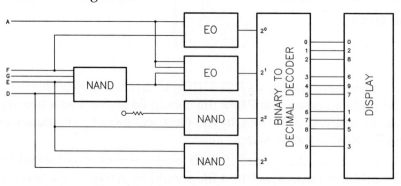

9. Draw the block diagram shown. Use a text style with the Romans font to add the text. Use polylines to create the arrowheads. Save the drawing as P10-9.

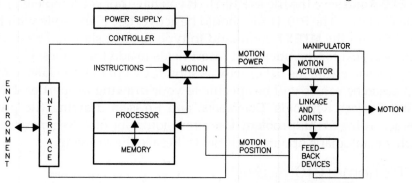

10. Draw the AND/OR schematic shown. Save the drawing as P10-10.

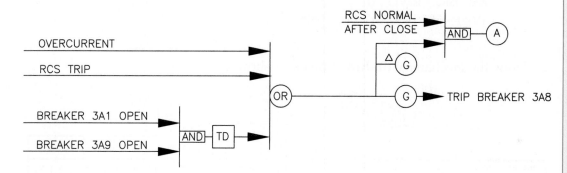

▼ Advanced

11. Open EX2-5 and save the file as P10-11. (If you have not yet completed Exercise 2-5, complete it now.) The P10-11 file should be active. Use a text style with the Century Gothic font and the **MTEXT** command to type the text shown. Use a .25 unit text height. Type text for the titles, and insert fields to add the values. Use appropriate field options to create the list exactly as shown, except that the Author, Approved by, and Checked by values will be specific to your drawing properties. Next, change the Title, Subject, Keywords, Comments, and Revision drawing file properties to values according to this problem. Use a revision value of A. Regenerate the display to apply an automatic update to all fields. Resave the drawing.

> Filename: P10-11.dwg
> Date: 03-01-2011
> Title: EXERCISE 2-5
> Subject: Exercise 2-5
> Author: ABC
> Keywords:Exercise 2-5
> Comments: This is Chapter 2, exercise number five.
> Approved by: ABC
> Checked by: ABC
> Revision level: 0
> Web site: www.g-w.com

12. Draw the mechanical drafting title block shown.

METRIC

R -	CHANGE		DATE	ECN

SPECIFICATIONS

HYSTER COMPANY

THIS PRINT CONTAINS CONFIDENTIAL INFORMATION WHICH IS THE PROPERTY OF HYSTER COMPANY. BY ACCEPTING THIS INFORMATION THE BORROWER AGREES THAT IT WILL NOT BE USED FOR ANY PURPOSE OTHER THAN THAT FOR WHICH IT IS LOANED.

UNLESS OTHERWISE SPECIFIED DIMENSIONS ARE IN ~~INCHES~~ MILLIMETERS AND TOLERANCES FOR:
_____ PLACE DIMS± _____ : _____ PLACE DIMS± _____
ANGLES ± _____ : WHOLE DIMS± _____

DR.		SCALE	DATE
CK. MAT'L.		CK. DESIGN	REL. ON ECN
NAME			

MODEL	DWG. FIRST USED	SIMILAR TO
DEPT.	PROJECT	LIST DIVISION

H | PART NO. | R

13. Draw the mechanical drafting title block and parts list shown. Save the drawing as P10-13.

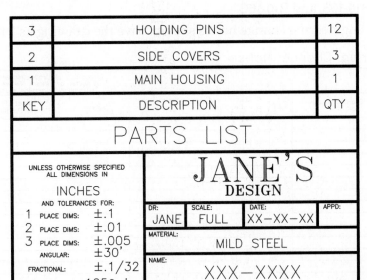

3	HOLDING PINS	12
2	SIDE COVERS	3
1	MAIN HOUSING	1
KEY	DESCRIPTION	QTY

PARTS LIST

UNLESS OTHERWISE SPECIFIED ALL DIMENSIONS IN
INCHES
AND TOLERANCES FOR:
1 PLACE DIMS: ±.1
2 PLACE DIMS: ±.01
3 PLACE DIMS: ±.005
ANGULAR: ±30'
FRACTIONAL: ±.1/32
FINISH: 125? in.

FIRST USED ON: | SIMILAR TO:

JANE'S DESIGN

DR: JANE | SCALE: FULL | DATE: XX−XX−XX | APPD:
MATERIAL: MILD STEEL
NAME: XXX−XXXX

B | PART NO: 123−321 | REV: 0

14. Create a dimensioned 2D sketch of a map providing driving or walking directions from your home to your school or office. Label all features, including roads and distances. Use available measuring devices, such as the odometer in your car, tape measure, or surveying equipment to measure features. Draw the map from your sketch using real-world units. Add text to label features and distances, but do not draw actual dimension objects. Save the drawing as P10-14.

15. Draw the architectural title block shown. Save the drawing as P10-15.

SHT SHTS
SHEET NO.

COMPANY NAME

DATE: PLOTDATE
REVISED:

TITLE1
TITLE2

FILE: PROJNUM

JOB: PROJNUM

BLKID

16. Draw title blocks with borders for electrical, piping, and general drawings. Research sample title blocks to come up with your designs. Save the drawings as P10-16A, P10-16B, and P10-16C.

17. Draw the engineering change notice shown. Save the drawing as P10-17.

Engineering Change Notice

ECN NO.

Disposition of production stock:
A =Alter or rework U=Use in production
T=Transfer to service stock S=Scrap

Qty.	Drawing Size Part No.	R/N	Description	Change	Other Usage in Production	D/S
01						
02						
03						
04						
05						
06						
07						
08						
09						
10						
11						
12						
13						
14						
15						
16						
17						
18						

Reason:

Castings & forgings affected? ☐ Yes ☐ No	Design engineer:	Supervisor approval:	Release date:	Page

AutoCAD Certified Associate Exam Practice

Answer the following questions. Write your answers on a separate sheet of paper.

1. How can you insert a diameter symbol into single-line text? *Select all that apply.*
   A. type %%D
   B. type %%C
   C. type (D)
   D. select it from the **Symbol** menu on the **Text Editor** ribbon tab
   E. right-click and select **Diameter**

2. What is the name of the file in which AutoCAD stores the default custom dictionary? *Select the one item that best answers the question.*
   A. dictionary.smp
   B. sample.cus
   C. sample.dct
   D. sample.dic
   E. sample.dwt

3. Which of the following tasks can you perform using a field? *Select all that apply.*
   A. display the date a drawing was created
   B. display drawing properties in the drawing
   C. change the name of a drawing
   D. change the names of layers
   E. update property displays to reflect their current status

AutoCAD Certified Professional Exam Practice

Follow the instructions in each problem. Write your answers on a separate sheet of paper.

1. **Navigate to this chapter on the companion website and open CPE-10spell.dwg.** Use the default spell checker or the **SPELL** command to check the text in the drawing. What are the first three alternate spellings AutoCAD suggests for "ASME"?

2. **Navigate to this chapter on the companion website and open CPE-10scale.dwg.** Use the **SCALETEXT** command to scale the text using the existing justification point and a scale factor of 2.67. What is the height of the text after the **SCALETEXT** operation?

Students can obtain free AutoCAD software from the Autodesk Education Community website (http://students.autodesk.com). This website also provides discussion forums and other useful resources.

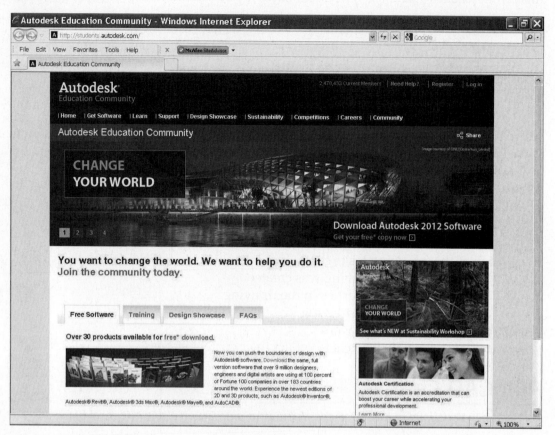

Chapter 11

Modifying Objects

Learning Objectives

After completing this chapter, you will be able to:

✓ Use the **FILLET** command to draw fillets, rounds, and other radii.
✓ Place chamfers and angled corners with the **CHAMFER** command.
✓ Form a spline using the **BLEND** command.
✓ Separate objects using the **BREAK** command and combine objects using the **JOIN** command.
✓ Use the **TRIM** and **EXTEND** commands to edit objects.
✓ Change objects using the **STRETCH** and **LENGTHEN** commands.
✓ Edit the size of objects using the **SCALE** command.
✓ Use the **EXPLODE** command.

AutoCAD includes many commands and options to modify a drawing. Use editing commands to complete common drafting tasks, improve efficiency, and modify a drawing. This chapter explains methods for changing or adding geometry to existing objects using basic editing commands. The approach to editing presented in this chapter is to access a command, such as **STRETCH**, and then follow the prompts to complete the operation. Another technique, presented in Chapter 13, is to select objects first using the crosshairs, then access the editing command, and finally complete the operation.

> **fillet:** A rounded interior corner used to relieve stress or ease the contour of inside corners.
>
> **round:** A rounded exterior corner used to remove sharp edges or ease the contour of exterior corners.

Using the FILLET Command

Design and drafting terminology refers to a rounded interior corner as a *fillet*, and a rounded exterior corner as a *round*. AutoCAD refers to all rounded corners as fillets. The **FILLET** command draws a rounded corner between intersecting or nonintersecting lines, circles, arcs, polylines, and splines. See **Figure 11-1**.

Setting the Fillet Radius

Access the **FILLET** command and then use the **Radius** option to specify the fillet radius. The radius determines the size of a fillet and must be set before you select objects. AutoCAD stores the radius as the new default radius, allowing you to place

Ribbon
Home
> Modify

Fillet

Type
FILLET
F

FILLET

325

Figure 11-1.
Using the **FILLET** command to add fillets and rounds.

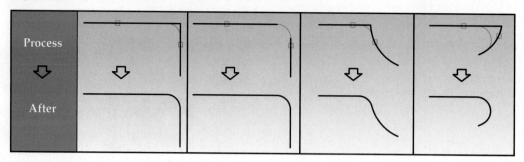

additional fillets of the same size. After you define the radius, select the first object and then hover the pick box over the second object to display a preview of the operation. The **Radius** option, available after you pick the first object, provides another chance to modify the radius before drawing the fillet or round. If the preview looks acceptable, pick the second object to apply the fillet or round.

PROFESSIONAL TIP

Use the **FILLET** command to form sharp corners by specifying a radius of 0, or by holding down [Shift] when you pick the second object.

Only corners large enough to accept the specified fillet radius are eligible for filleting. If the fillet radius is too large, AutoCAD displays the message Distance is too large *Invalid*.

Exercise 11-1

Complete the exercise on the companion website.
www.g-wlearning.com/CAD

Creating Full Rounds

You can use the **FILLET** command to draw a full round between parallel lines. The radius of a fillet between parallel lines is always half the distance between the two lines, regardless of the radius setting. Use this method to create a full round, such as the end radii of a slot.

Polyline Option

Use the **Polyline** option to fillet all corners of a closed polyline. See Figure 11-2. Set the appropriate radius before selecting the **Polyline** option, or use the **Radius** option after you choose the **Polyline** option to modify the radius. If you originally drew the polyline without using the **Close** option, the beginning corner does not fillet, as shown in Figure 11-2.

Figure 11-2.
Using the **Polyline** option of the **FILLET** command.

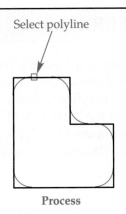
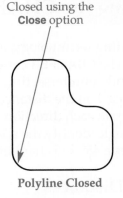
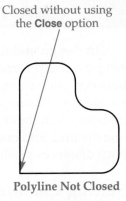

Select polyline

Closed using the **Close** option

Closed without using the **Close** option

Process

Polyline Closed

Polyline Not Closed

Figure 11-3.
Comparison of the **Trim** and **No trim** options of the **FILLET** command.

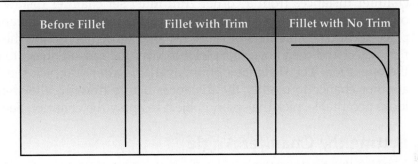

Before Fillet	Fillet with Trim	Fillet with No Trim

Trim Settings

The **Trim** option controls whether the **FILLET** command trims object segments that extend beyond the fillet radius point of tangency. See **Figure 11-3**. Use the default **Trim** setting to trim objects. When you set the **Trim** option to **No trim**, the fillet occurs, but the filleted objects do not change.

PROFESSIONAL TIP

You can fillet objects even when the corners do not meet. If the **Trim** option is set to **Trim**, objects extend as required to generate the fillet and complete the corner. If the **Trim** option is set to **No trim**, objects do not extend to complete the corner.

Multiple Option

Use the **Multiple** option to make several fillets without exiting the **FILLET** command. The prompt for a first object repeats after each fillet is drawn. To exit, press [Enter], the space bar, or [Esc], or right-click and select **Enter**. While using the **Multiple** option, select the **Undo** option to discard the previous fillet if necessary.

Exercise 11-2

Complete the exercise on the companion website.
www.g-wlearning.com/CAD

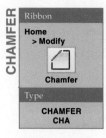

CHAMFER

Ribbon

Home
> Modify

Chamfer

Type

CHAMFER
CHA

chamfer: A small,
angled surface used
to relieve a sharp
corner.

Using the CHAMFER Command

In design and drafting terminology, a *chamfer* is a small, angled surface used to relieve a sharp corner. The **CHAMFER** command allows you to draw an angled corner between intersecting and nonintersecting lines, polylines, xlines, and rays. Determine the size of a chamfer based on the distance from the corner. A 45° chamfer is the same distance from the corner in each direction. See **Figure 11-4**. Typically, two distances or one distance and one angle identify the size of a chamfer. For example, a value of .5 for both distances produces a 45° × .5 chamfer.

Setting Chamfer Distances

Access the **CHAMFER** command and then use the **Distance** option to specify the chamfer distances from a corner. You must set chamfer distances before you select objects. AutoCAD stores the distances as the new default distances, allowing you to place additional chamfers of the same size. Once you define the distances, select the first object and then hover the pick box over the second object to display a preview of the operation. The **Distance** option, available after you pick the first object, provides another chance to modify the distances before drawing the chamfer. If the preview looks acceptable, pick the second object to apply the chamfer. See **Figure 11-5**.

Setting the Chamfer Angle

Use the **Angle** option as an alternative to setting two chamfer distances. Specify the chamfer distance along the first selected curve, followed by the chamfer angle. Then select the objects to chamfer. See **Figure 11-6**. AutoCAD stores the distance and angle, allowing you to place additional chamfers of the same size. The **Angle** option, available after you pick the first object, provides another chance to modify the distance and angle before drawing the chamfer.

Method Option

AutoCAD maintains the specified chamfer distances and the distance and angle until you change the values. You can set the values for each method without affecting the other. Use the **Method** option to toggle between drawing chamfers using the **Distance** and **Angle** options. The **Method** option, available after you pick the first object, provides another chance to modify the method before drawing the chamfer.

Figure 11-4.
Examples of chamfers. An alternative way to dimension a chamfer is to specify the angle by the size, such as 45° × .125 for the 45° chamfer shown.

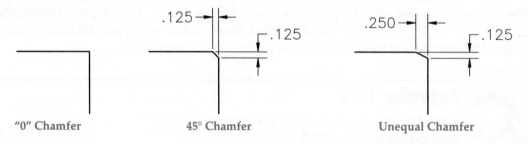

"0" Chamfer   45° Chamfer   Unequal Chamfer

Figure 11-5.
Using the **CHAMFER** command to create chamfered corners.

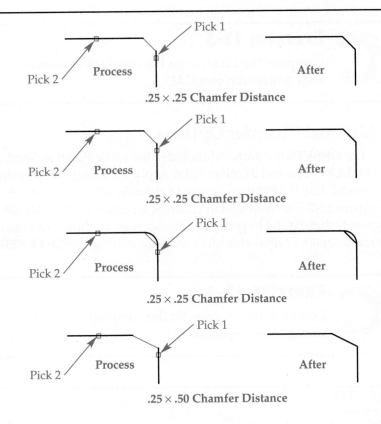

.25 × .25 Chamfer Distance

.25 × .25 Chamfer Distance

.25 × .25 Chamfer Distance

.25 × .50 Chamfer Distance

Figure 11-6.
Using the **Angle** option of the **CHAMFER** command with the chamfer length set at .5 and the angle set at 30°. Notice that the selection order determines where the distance and angle apply.

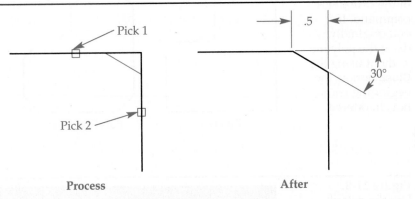

PROFESSIONAL TIP

Use the **CHAMFER** command to form sharp corners by specifying chamfer distances or an angle and distance of 0, or by holding down [Shift] when you pick the second object.

Only corners large enough to accept the specified chamfer size are eligible for chamfering. If the chamfer is too large, AutoCAD displays the message Distance is too large *Invalid*.

Exercise 11-3

Complete the exercise on the companion website.
www.g-wlearning.com/CAD

Additional Chamfer Options

The **CHAMFER** command includes the same **Polyline**, **Trim**, and **Multiple** options as the **FILLET** command. Similar rules apply for using these options with the **CHAMFER** command. Use the **Polyline** option to chamfer all corners of a closed polyline, as shown in **Figure 11-7**. The **Trim** option controls whether the **CHAMFER** command trims object segments that extend beyond the intersection, as shown in **Figure 11-8**. Use the **Multiple** option to make several chamfers without exiting the **CHAMFER** command.

Exercise 11-4

Complete the exercise on the companion website.
www.g-wlearning.com/CAD

Figure 11-7.
Using the **Polyline** option of the **CHAMFER** command. If you originally drew the polyline without using the **Close** option, the beginning corner is not chamfered.

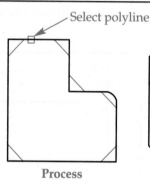

Select polyline

Process   Polyline Closed   Polyline Not Closed

Figure 11-8.
Use the default **Trim** setting to trim objects. When you set the **Trim** option to **No trim**, the chamfer occurs, but chamfered objects do not change.

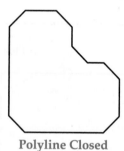
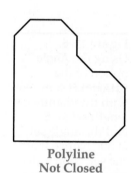

Before Chamfer	Chamfer with Trim	Chamfer with No Trim

Using the BLEND Command

The **BLEND** command allows you to construct a spline between lines, arcs, polylines, splines, elliptical arcs, and helixes. Blending a spline can be effective when an irregular curve is required, typically to fill a break between nonintersecting objects when the curve data is unknown. Access the **BLEND** command and specify the type of continuity using the **CONtinuity** option. The **Tangent** option forms a spline with a mathematical algorithm known as "G1 geometric continuity" and three vertices. The **Smooth** option forms a spline with "G2 geometric continuity" and five vertices. Pick two existing objects to draw the spline. See **Figure 11-9**.

Using the BREAK Command

The **BREAK** command can remove a portion of an object or split an object at a single point, depending on the points and option you select. By default, the point you pick when you select the object to break also locates the first break point. To select a different, possibly more accurate first break point, use the **First point** option at the Specify second break point or [First point]: prompt. See **Figure 11-10**.

If you select the same point for the first and second break points, the **BREAK** command splits the object without removing a portion. At the Specify second break point or [First point]: prompt, use object snaps or coordinate entry, or enter @ to select the same coordinates as the first break point. See **Figure 11-11**. Another option is to use the **Break at Point** option instead of the standard **BREAK** command, which automates the process.

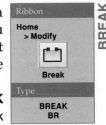

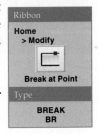

Figure 11-9.
An example of using the **BLEND** command to draw a spline between two lines.

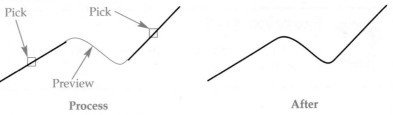

Figure 11-10.
Using the **BREAK** command to break an object. When you use the default method, the first pick selects the object and the first break point. Use the **First point** option to select the object to break and then specify accurate break points using object snaps, such as the **Endpoint** object snaps shown.

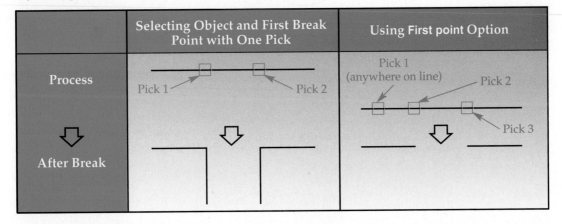

	Selecting Object and First Break Point with One Pick	Using **First point** Option
Process		
After Break		

Figure 11-11.
Using the **BREAK** command to break an object at a single point without removing any of the object. Select the same point for the first and second break points. You can also use the **Break at Point** command to automate the process.

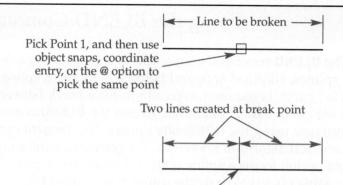

Pick Point 1, and then use object snaps, coordinate entry, or the @ option to pick the same point

Line to be broken

Two lines created at break point

Break point

Select points in a counterclockwise direction when breaking circular objects to ensure that you remove the correct portion of the object. See **Figure 11-12.** Notice in **Figure 11-12** that you can break off the end of an open object by picking the first point on the object and the second point slightly beyond the end to be cut off. AutoCAD selects the endpoint nearest the point you pick.

PROFESSIONAL TIP

Use object snaps to pick a point accurately when using the **First point** option of the **BREAK** command. However, it may be necessary to turn running object snaps off if they conflict with points you try to pick.

Exercise 11-5

Complete the exercise on the companion website.
www.g-wlearning.com/CAD

Figure 11-12.
Select points in a counterclockwise direction when using the **BREAK** command on circular objects, such as the circle shown here.

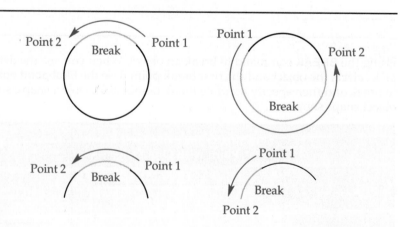

Point 2   Break   Point 1

Point 1   Point 2   Break

Point 2   Break   Point 1

Point 1   Break   Point 2

Using the JOIN Command

While drawing and editing, you sometimes create multiple objects that should be one object. These multiple objects are cumbersome to work with and increase the size of the drawing file. The **JOIN** command is one option for combining lines, polylines, splines, arcs, and elliptical arcs. You can join objects of the same type, or specific combinations of objects.

Joining the Same Object Type

Access the **JOIN** command and select objects of the same type to join. Lines must be collinear, but can share the same endpoint, overlap, or have gaps between segments. See **Figure 11-13**. Polylines must share a common endpoint and cannot have gaps between segments or overlap. See **Figure 11-14**. The same rules apply for joining splines.

Arcs must share the same center point and radius, but can share the same endpoint, overlap, or have gaps between segments. See **Figure 11-15**. Similar rules apply for joining elliptical arcs, although elliptical arcs must share the same axes. Pick arcs or elliptical arcs in a clockwise direction to close the nearest clockwise gap, or pick in a counterclockwise direction to close the nearest counterclockwise gap. Depending on your selections, you may be prompted to convert arcs to a circle. To maintain the original object type, choose the **No** option and reselect the arcs in a counterclockwise direction.

Figure 11-13.
Lines must be collinear to join, but there can be gaps between segments, as shown, or the lines can share the same endpoint or overlap.

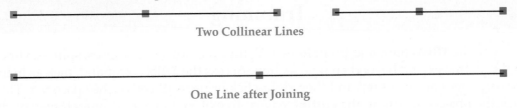

Figure 11-14.
You can join polylines only if they share an endpoint. The same rule applies for joining splines.

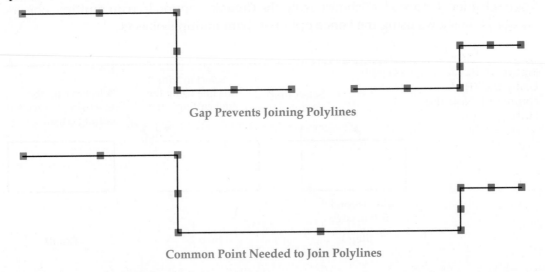

Figure 11-15.
Arcs must share the
same center point
and radius, but there
can be gaps between
segments, as shown,
or the arcs can share
the same endpoint
or overlap.

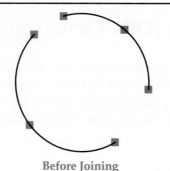
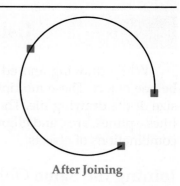

Before Joining   After Joining

After selecting an arc or elliptical arc segment to join, you can use the **cLose** option to form a circle from the arc, or to form an ellipse from an elliptical arc. This option closes the two ends of the selected segment and does not join the segment with any other objects.

Joining Different Object Types

You can join lines, arcs, and polylines to polylines, or join lines, arcs, and polylines to splines. Access the **JOIN** command and select objects of a different type to join. Apply the same rules to join a combination of objects as to join objects of the same type. However, segments cannot have gaps or overlap. The final object becomes the most complex of the original objects.

Trimming

TRIM

cutting edge: An object such as a line, arc, or text that defines the point (edge) at which the object you trim will be cut.

Ribbon

Home
> Modify

Trim

Type

TRIM
TR

Use the **TRIM** command to cut lines, polylines, circles, arcs, ellipses, splines, xlines, and rays that extend beyond an intersection. Access the **TRIM** command, pick as many *cutting edges* as necessary, and then right-click or press [Enter] or the space bar. Then pick the objects to trim to the cutting edges. To exit, right-click or press [Enter] or the space bar. See **Figure 11-16**.

Automatic windowing is often the quickest and most effective method for trimming multiple objects. You can also use crossing, window, or polygon selection. The **TRIM** command offers **Crossing** and **Fence** options that provide standard crossing and fence selection. **Figure 11-17** shows using the **Crossing** option to trim multiple objects. **Figure 11-18** shows using the **Fence** option to trim multiple objects.

Figure 11-16.
Using the **TRIM**
command. Note the
cutting edges.

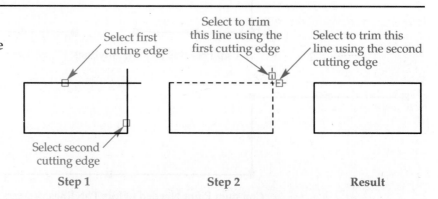

Select first cutting edge

Select to trim this line using the first cutting edge

Select to trim this line using the second cutting edge

Select second cutting edge

Step 1   Step 2   Result

Figure 11-17.
The only objects trimmed with the **Crossing** option are those that cross the edges of the crossing window. Automatic windowing accomplishes the same task.

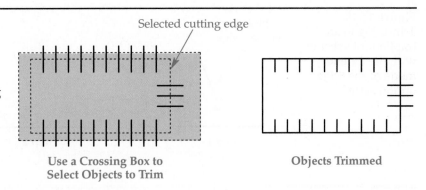

Selected cutting edge

Use a Crossing Box to
Select Objects to Trim

Objects Trimmed

Figure 11-18.
The **Fence** option allows you to select around objects. In this example, the cutting edge consists of the rectangle drawn using the **RECTANGLE** command.

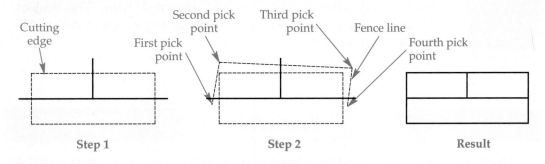

Cutting edge

Second pick point

Third pick point

Fence line

First pick point

Fourth pick point

Step 1

Step 2

Result

To access the **EXTEND** command while using the **TRIM** command, after selecting the cutting edge(s), hold down [Shift] and pick objects to extend to the cutting edge. The **EXTEND** command is described later in this chapter.

Trimming without Selecting a Cutting Edge

To trim objects to the nearest intersection without selecting a cutting edge, access the **TRIM** command and, at the first Select objects or <select all>: prompt, right-click or press [Enter] or the space bar instead of picking a cutting edge. Then pick the objects to trim. You can continue selecting objects to trim without restarting the **TRIM** command. To exit, right-click or press [Enter], the space bar, or [Esc].

Trimming to an Implied Intersection

Use the **Edge** option of the **TRIM** command to trim to an *implied intersection*. The **No extend** mode is active by default, and does not allow you to trim objects that do not intersect. To trim nonintersecting objects, access the **TRIM** command, pick the cutting edges, and select the **Edge** option followed by the **Extend** option. AutoCAD now recognizes implied intersections and allows you to pick objects to trim. See **Figure 11-19**. Adjusting the **Edge** function does not change the selected cutting edges.

implied intersection: The point at which objects would meet if they were extended.

Figure 11-19.
Trimming to an
implied intersection
with the **Extend**
mode of the **Edge**
option active.

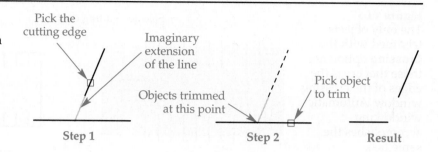

Pick the
cutting edge

Imaginary
extension
of the line

Pick object
to trim

Objects trimmed
at this point

Step 1　　　　　　　　Step 2　　　　　　　Result

Use the **eRase** option of the **TRIM** command to erase objects selected
to trim. Use the **Undo** option to restore previously trimmed objects
without leaving the command. You must activate the **Undo** option
immediately after performing an unwanted trim. The **Project**
option applies to trimming 3D objects, as explained in *AutoCAD
and Its Applications—Advanced.*

Extending

boundary edge:
The edge to which
objects such as
lines, arcs, and
polylines extend.

EXTEND

Ribbon
Home
> Modify

Extend

Type
EXTEND
EX

Use the **EXTEND** command to extend lines, elliptical arcs, rays, open polylines,
arcs, and splines to meet other objects. You cannot extend a closed object because an
unconnected endpoint does not exist. Access the **EXTEND** command, pick as many
boundary edges as necessary, and then right-click or press [Enter] or the space bar.
Now pick the objects to extend to the boundary edges. To exit, right-click or press
[Enter] or the space bar. See **Figure 11-20.**

Automatic windowing is often the quickest and most effective method for
extending multiple objects. You can also use crossing, window, or polygon selection.
The **EXTEND** command offers **Crossing** and **Fence** options that provide standard
crossing and fence selection. **Figure 11-21** shows using the **Crossing** option to extend
multiple objects. **Figure 11-22** shows an example of using the **Fence** option to extend
multiple objects.

Extending without Selecting a Boundary Edge

To extend objects to the nearest intersection without selecting a boundary edge,
access the **EXTEND** command and, at the first Select objects or <select all>: prompt, right-
click or press [Enter] or the space bar instead of picking a boundary edge. Then pick the
objects to extend. You can continue selecting objects to extend without restarting the
EXTEND command. To exit, right-click or press [Enter], the space bar, or [Esc].

Figure 11-20.
Using the **EXTEND**
command. Note the
boundary edge.

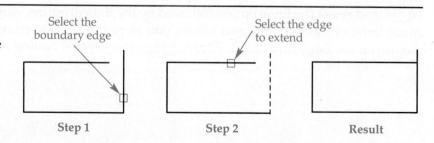

Select the
boundary edge

Select the edge
to extend

Step 1　　　　　　　　Step 2　　　　　　　Result

Figure 11-21.
Selecting objects to extend using the **Crossing** option. Automatic windowing accomplishes the same task.

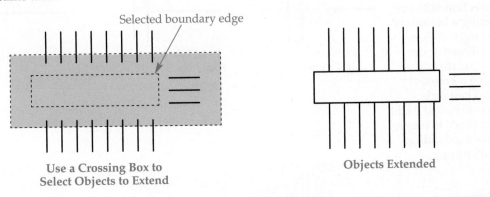

Use a Crossing Box to
Select Objects to Extend

Objects Extended

Figure 11-22.
Extending multiple lines to a boundary edge using the **Fence** option.

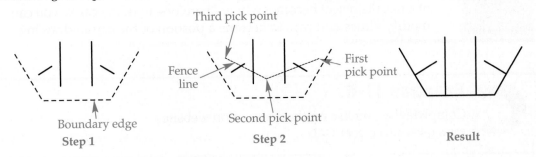

Step 1

Step 2

Result

Additional Extend Options

The **EXTEND** command includes the same **Edge**, **Undo**, and **Project** options as the **TRIM** command, and similar rules apply when using these options with the **EXTEND** command. Use the **Extend** mode of the **Edge** option to extend to an implied intersection, as shown in Figure 11-23. Select the **Undo** option immediately after performing an unwanted extend to restore previous objects without leaving the command. The **Project** option applies to extending 3D objects, as explained in *AutoCAD and Its Applications—Advanced*.

PROFESSIONAL
TIP

Figure 11-24 illustrates how to combine the **EXTEND** and **TRIM** commands, without selecting a boundary edge, to insert a wall in a floor plan. You can apply this process to a variety of applications.

Figure 11-23.
Extending to an implied intersection with the **Extend** mode.

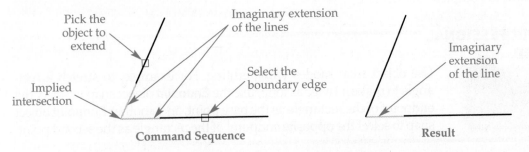

Command Sequence

Result

Figure 11-24.
To extend objects to the nearest intersection without selecting a boundary edge, right-click or press [Enter] or the space bar instead of picking a boundary edge. Then pick objects to extend. Hold down [Shift] to toggle between extending and trimming.

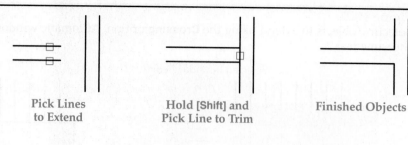

Pick Lines to Extend

Hold [Shift] and Pick Line to Trim

Finished Objects

NOTE When you trim one infinite end of an xline, the object becomes a ray. When you trim both infinite ends of an xline, or the infinite end of a ray, the object becomes a line. Therefore, in many cases, you can modify xlines and rays to become a portion of the actual drawing.

Exercise 11-6

Complete the exercise on the companion website.
www.g-wlearning.com/CAD

Stretching

STRETCH

Ribbon

Home
> Modify

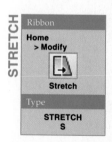

Stretch

Type

STRETCH
S

base point: The initial reference point AutoCAD uses when stretching, moving, copying, and scaling objects.

The **STRETCH** command allows you to modify certain dimensions of an object without changing other dimensions. In mechanical drafting, for example, you can stretch a screw body to create a longer or shorter screw. In architectural design, you can stretch room sizes to increase or decrease square footage.

Once you access the **STRETCH** command, you must use a crossing box or crossing polygon to select only the objects to stretch. This is a very important requirement and is different from object selection for other editing commands. See Figure 11-25. If you select using the pick box or a window, the **STRETCH** command works like the **MOVE** command, described in Chapter 12.

After selecting the objects to stretch, specify the *base point* from which the objects will stretch. Although the position of the base point is often not critical, you may want to select a point on an object, the corner of a view, or the center of a circle, for example. The selection stretches or compresses as you move the crosshairs. Specify a second point to complete the stretch.

PROFESSIONAL TIP

Use object snap modes while editing. For example, to stretch a rectangle to make it twice as long, use the **Endpoint** object snap to select the endpoint of the rectangle as the base point, and another **Endpoint** object snap to select the opposite endpoint of the rectangle as the second point.

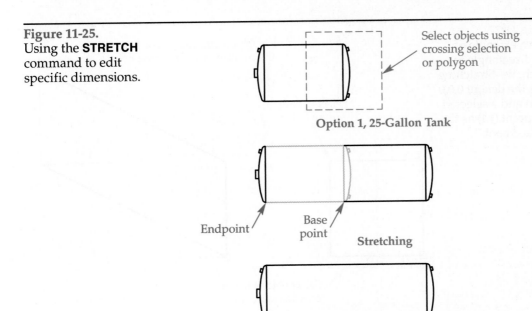

Figure 11-25.
Using the **STRETCH** command to edit specific dimensions.

Select objects using crossing selection or polygon

Option 1, 25-Gallon Tank

Endpoint

Base point

Stretching

Option 2, 50-Gallon Tank

Displacement Option

The **Displacement** option allows you to stretch objects relative to the origin of the UCS (user coordinate system), which is at coordinates 0,0,0 by default. To stretch using a *displacement*, access the **STRETCH** command and use a crossing box or crossing polygon to select only the objects to stretch. Then choose the **Displacement** option instead of defining the base point. At the Specify displacement <0,0,0>: prompt, enter an absolute coordinate to stretch the objects from the origin to the coordinate point. See **Figure 11-26**.

displacement:
The direction and distance in which an object moves.

Using the First Point as Displacement

Another method for stretching an object is to use the first point as the displacement. The coordinates you use to select the base point automatically define the coordinates for the direction and distance for stretching the object. Access the **STRETCH** command and use a crossing box or crossing polygon to select only the objects to stretch. Then specify the base point, and instead of locating the second point, right-click or press [Enter] or the space bar to accept the <use first point as displacement> default. See **Figure 11-27**.

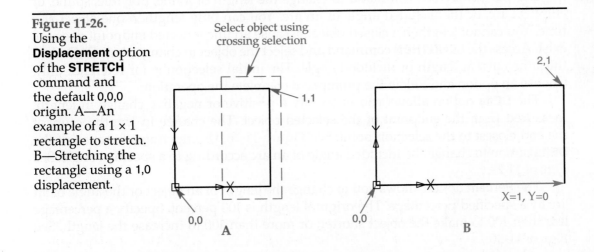

Figure 11-26.
Using the **Displacement** option of the **STRETCH** command and the default 0,0,0 origin. A—An example of a 1 × 1 rectangle to stretch. B—Stretching the rectangle using a 1,0 displacement.

Select object using crossing selection

1,1

0,0

A

2,1

0,0

X=1, Y=0

B

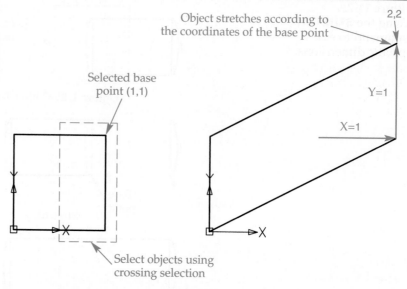

Figure 11-27.
A—An example of a 1 × 1 rectangle to stretch. B—Stretching using the default 0,0,0 origin and a selected base point (1,1) as the displacement.

Object stretches according to the coordinates of the base point

2,2

Selected base point (1,1)

Y=1

X=1

Select objects using crossing selection

PROFESSIONAL TIP

Objects often do not align conveniently for using crossing selection to pick objects to stretch. Consider using a crossing polygon to make selection easier. If the stretch is not as expected, press [Esc] to cancel. The **STRETCH** command and other editing commands work well with polar tracking or **Ortho** mode.

Exercise 11-7

Complete the exercise on the companion website.
www.g-wlearning.com/CAD

Using the LENGTHEN Command

LENGTHEN

Ribbon
Home
> Modify

Lengthen

Type
LENGTHEN
LEN

Use the **LENGTHEN** command to change the length of a line, polyline, spline, or elliptical arc, or the included angle of an arc. You can only lengthen one object at a time. You cannot lengthen a closed object because an unconnected endpoint does not exist. Access the **LENGTHEN** command and select the object to change. AutoCAD identifies the current length or included angle. The initial selection is for reference only. Choose an option and follow the prompts to complete the operation.

The **DElta** option allows you to specify a positive or negative change in length, measured from the endpoint of the selected object. The change in length occurs at the end closest to the selection point. See **Figure 11-28**. Use the **Angle** function of the **DElta** option to change the included angle of an arc according to a specified angle. See **Figure 11-29**.

The **Percent** option allows you to change the length of an object or the angle of an arc by a specified percentage. The original length is 100 percent. Specify a percentage less than 100 to make the object shorter, or more than 100 to increase the length. See **Figure 11-30**.

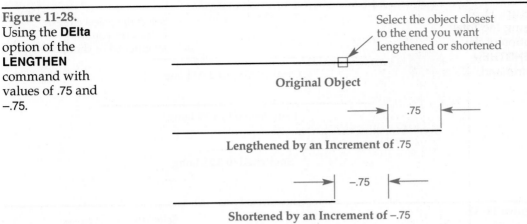

Figure 11-28.
Using the **DElta** option of the **LENGTHEN** command with values of .75 and –.75.

Select the object closest to the end you want lengthened or shortened

Original Object

.75

Lengthened by an Increment of .75

–.75

Shortened by an Increment of –.75

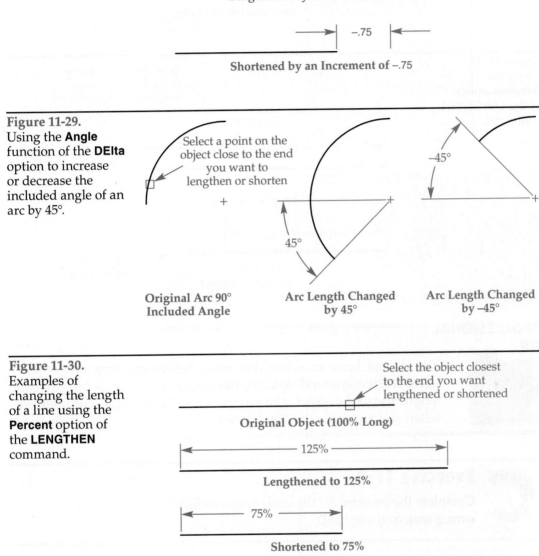

Figure 11-29.
Using the **Angle** function of the **DElta** option to increase or decrease the included angle of an arc by 45°.

Select a point on the object close to the end you want to lengthen or shorten

45°

–45°

Original Arc 90° Included Angle

Arc Length Changed by 45°

Arc Length Changed by –45°

Figure 11-30.
Examples of changing the length of a line using the **Percent** option of the **LENGTHEN** command.

Select the object closest to the end you want lengthened or shortened

Original Object (100% Long)

125%

Lengthened to 125%

75%

Shortened to 75%

The **Total** option allows you to set the total length or angle of the object after the **LENGTHEN** operation. See **Figure 11-31**. The **DYnamic** option lets you drag the endpoint of the object to the desired length or angle using the crosshairs. See **Figure 11-32**. It is helpful to use dynamic input with polar tracking or **Ortho** mode or to have the grid and snap set to usable increments when using the **DYnamic** option.

You can only lengthen lines and arcs dynamically, and you can only decrease the length of a spline.

Figure 11-31.
Using the **Total** option of the **LENGTHEN** command.

Select the object closest to the end you want lengthened or shortened

Original Object 3.00 Long

Lengthened to 3.75 Long

Shortened to 2.25 Long

Figure 11-32.
Using the **DYnamic** option of the **LENGTHEN** command.

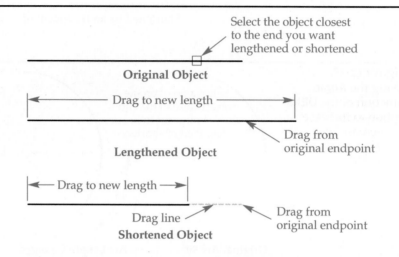

Select the object closest to the end you want lengthened or shortened

Original Object

Drag to new length

Drag from original endpoint

Lengthened Object

Drag to new length

Drag line
Shortened Object

Drag from original endpoint

PROFESSIONAL TIP

You do not have to select the object before entering one of the **LENGTHEN** command options, but doing so indicates the current length and, if the object is an arc, the angle. This is especially helpful when you are using the **Total** option.

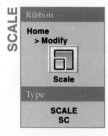

Exercise 11-8

Complete the exercise on the companion website.
www.g-wlearning.com/CAD

Using the SCALE Command

SCALE

Ribbon
Home
> Modify

Scale

Type

SCALE
SC

The **SCALE** command allows you to proportionally enlarge or reduce the size of objects. After you access the **SCALE** command, pick a base point to define where the increase or decrease in size should occur. The change in object size occurs away from (for enlargement) or toward the base point (for reduction) during scaling. The next step is to specify the scale factor. Enter a number to indicate the amount of enlargement or reduction. For example, to make the objects twice the current size, type 2 at the Specify scale factor or [Copy/Reference] <current>: prompt. See Figure 11-33. You can also use coordinate entry or drawing aids to define the scale factor. Figure 11-34 provides examples of scale factors.

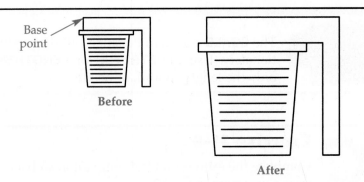

Figure 11-33.
Using the **SCALE** command to make objects twice the previous size. The points associated with the selected objects adjust from the location of the base point.

Base point

Before

After

Figure 11-34.
Scale factors and the resulting sizes.

Scale Factor	Resulting Size
10	10 times bigger
5	5 times bigger
2	2 times bigger
1	Equal to existing size
.75	3/4 of original size
.50	1/2 of original size
.25	1/4 of original size

Reference Option

The **Reference** option is an alternative to entering a scale factor that allows you to specify a new size in relation to an existing dimension. For example, use the **Reference** option to change the size of a part with an overall dimension of 2.50″ to an overall dimension of 3.00″ proportionately. Choose the **Reference** option and specify the current length, 2.5 in the example, at the Specify reference length: prompt. Next, specify the new length for the dimension, 3 in the example. See Figure 11-35.

PROFESSIONAL TIP

Specify the reference length and new length using specific values; or choose points on existing objects. Picking points is especially effective when you do not know the exact reference and new lengths.

Copying While Scaling

The **Copy** option of the **SCALE** command copies and scales the selected objects, leaving the original object unchanged. The copy moves to the specified location.

Figure 11-35.
Using the **Reference** option of the **SCALE** command.

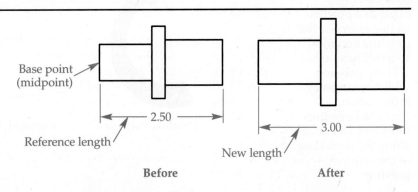

Base point (midpoint)

2.50

Reference length

3.00

New length

Before

After

The **SCALE** command changes all dimensions of an object proportionately. Use the **STRETCH** or **LENGTHEN** command to change only the length, width, or height.

 Exercise 11-9

Complete the exercise on the companion website.
www.g-wlearning.com/CAD

Exploding Objects

Ribbon

**Home
> Modify**

Explode

Type

**EXPLODE
X**

The **EXPLODE** command allows you to change a single object that consists of multiple items into a series of individual objects. For example, you can explode a polyline object into individual line and arc objects. See **Figure 11-36**. Another example is exploding an mtext object to create several text objects. You can explode a variety of other objects, including multilines, regions, dimensions, leaders, and blocks. This textbook explains these objects when appropriate. Access the **EXPLODE** command, pick the object to explode, and right-click or press [Enter] or the space bar to cause the explosion.

 CAUTION

Exploding eliminates the original object properties, characteristics, and associations. The need to explode objects is usually rare. Exploding is only appropriate if no other command or option can produce the desired effect.

 Exercise 11-10

Complete the exercise on the companion website.
www.g-wlearning.com/CAD

Figure 11-36.
Using the **EXPLODE** command to explode a polyline. In this example, the polyline becomes two collinear arcs with no polyline width or tangency information. Exploded polyline lines and arcs occur along the centerline of the original polyline.

Original Polyline   Exploded Polyline

Supplemental Material

Deleting Duplicate Objects
For information about using the **OVERKILL** command to remove duplicate and/or unnecessary objects, go to the companion website (www.g-wlearning.com/CAD), select this chapter, and select **Deleting Duplicate Objects**.

Chapter Review

Answer the following questions. Write your answers on a separate sheet of paper or complete the electronic chapter review on the companion website.
www.g-wlearning.com/CAD

1. How do you specify the size of a fillet?
2. Explain how to set the radius of a fillet to .50.
3. Describe the difference between the **Trim** and **No trim** options of the **CHAMFER** and **FILLET** commands.
4. Which option of the **CHAMFER** command would you use to specify a .125 × .125 chamfer?
5. What is the purpose of the **Method** option of the **CHAMFER** command?
6. Which command allows you to create an irregular curve to fill a break between nonintersecting objects when the curve data is unknown?
7. How can you split an object in two without removing a portion?
8. In what direction should you pick points to break a portion out of a circle or arc?
9. What command can you use to combine two collinear lines into a single line object?
10. What two requirements must be met before you can join two arcs?
11. Name the command that trims an object to a cutting edge.
12. Which command performs the opposite function of the **EXTEND** command?
13. Name the command associated with boundary edges.
14. Name the **TRIM** and **EXTEND** command option that allows you to trim or extend to an implied intersection.
15. List two locations drafters normally choose as the base point when using the **STRETCH** command.
16. Define the term *displacement* as it relates to the **STRETCH** command.
17. Identify the **LENGTHEN** command option that corresponds to each of the following descriptions:
    A. Allows a positive or negative change in length from the endpoint
    B. Changes a length or an arc angle by a percentage of the total
    C. Sets the total length or angle to the value specified
    D. Drags the endpoint of the object to the desired length or angle
18. Write the command aliases for the following commands:
    A. **CHAMFER**
    B. **FILLET**
    C. **BREAK**
    D. **TRIM**
    E. **EXTEND**
    F. **SCALE**
    G. **LENGTHEN**
19. What command would you use to reduce the size of an entire drawing by one-half?
20. Which command removes all width characteristics and tangency information from a polyline?

Drawing Problems

Start AutoCAD if it is not already started. Start a new drawing for each problem using an appropriate template of your choice. The template should include layers and text styles for drawing the given objects. Add layers and text styles as needed. Draw all objects using appropriate layers, text styles, justification, and format. Follow the specific instructions for each problem. Use only drawing and editing commands and techniques you have already learned. Do not draw dimensions. Use your own judgment and approximate dimensions when necessary.

▼ Basic

1. Draw Object A using the **LINE** and **ARC** commands. Make sure the corners overrun and center the arc on the lines, but the arc should not touch the lines. Use the **TRIM**, **EXTEND**, and **STRETCH** commands to make Object B. Save the drawing as P11-1.

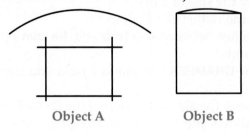

Object A   Object B

2. Open P11-1 and save the file as P11-2. The P11-2 file should be active. Use the **STRETCH** command to convert Object A to Object B. Resave the drawing.

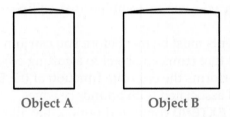

Object A   Object B

3. Refer to Figure 11-25 in this chapter. Draw the object shown in Option 1. Stretch the object to twice its length, as shown in Option 2. Stretch the object again to one and a half times its length. *Hint:* Use endpoint and midpoint object snap modes to stretch accurately. Save the drawing as P11-3.

4. Draw the part view shown. Use the **CHAMFER** command to create the inclined surface. Save the drawing as P11-4.

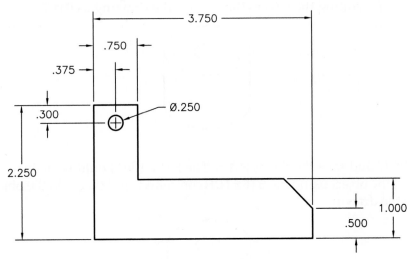

▼ Intermediate

5. Draw the wrench shown. Save the drawing as P11-5.

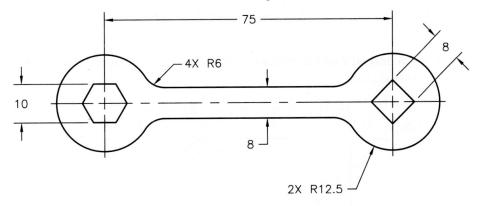

6. Draw the part view shown. Save the drawing as P11-6.

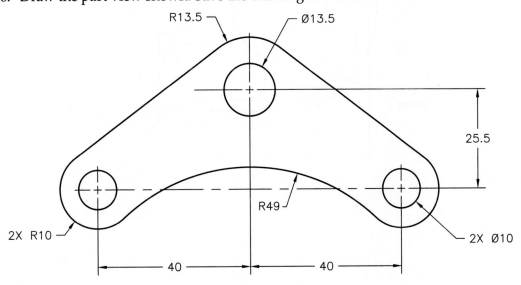

Drawing Problems - Chapter 11

7. Draw the bushing shown using the **ELLIPSE** and **LINE** commands. Draw the ellipses using a 30° rotation angle. Use the **BREAK** or **TRIM** command when drawing and editing the lower ellipse. Save the drawing as P11-7.

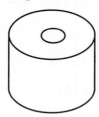

8. Open P11-7 and save the file as P11-8. The P11-8 file should be active. Shorten the height of the object using the **STRETCH** command, and then add the object shown. Resave the drawing.

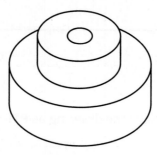

9. Use the **TRIM** and **OFFSET** commands to assist in drawing the part view shown. Save the drawing as P11-9.

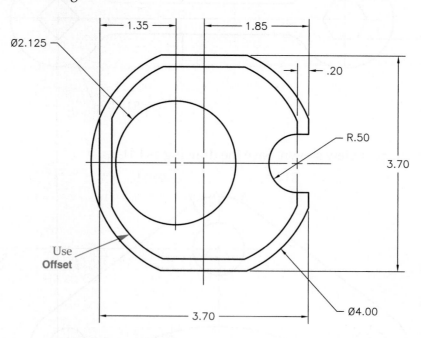

10. Draw the plate shown. Use the **FILLET** command where appropriate. Save the drawing as P11-10.

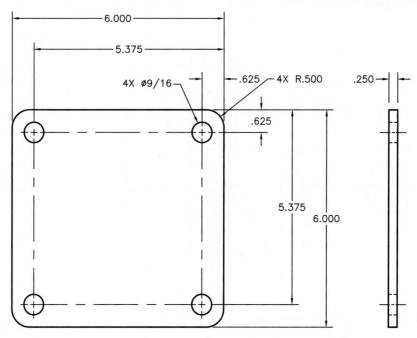

11. Draw the toilet shown. Use dimensions of your choice for undimensioned objects. Save the drawing as P11-11.

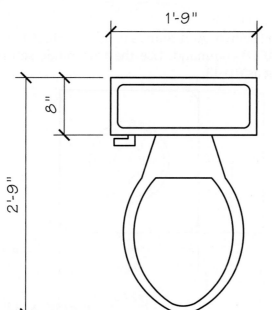

12. Draw the part view shown. Save the drawing as P11-12.

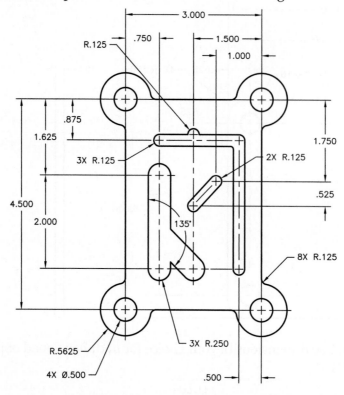

13. Draw the housing shown. Add rounds using the **FILLET** command and chamfers using the **CHAMFER** command. Use the trim mode setting to your advantage. Save the drawing as P11-13.

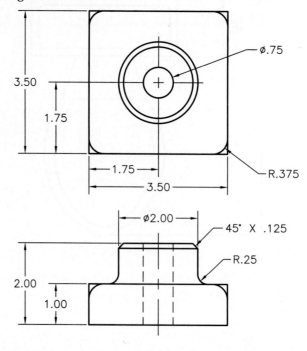

Drawing Problems - Chapter 11

14. Draw the beam wrap detail shown. Save the drawing as P11-14.

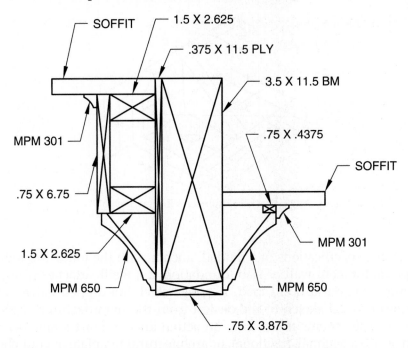

- SOFFIT
- 1.5 X 2.625
- .375 X 11.5 PLY
- 3.5 X 11.5 BM
- MPM 301
- .75 X .4375
- SOFFIT
- .75 X 6.75
- MPM 301
- 1.5 X 2.625
- MPM 650
- MPM 650
- .75 X 3.875

For Problems 15 and 16, draw the orthographic views needed to describe each part completely. Save the drawings as P11-15 and P11-16.

15.

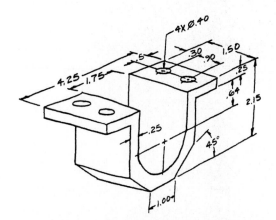

Journal Bracket (Engineer's Rough Sketch)

16.

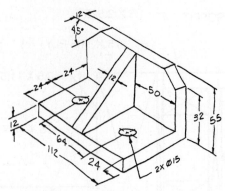

Angle Bracket (Engineer's Rough Sketch)
(Metric)

17. Research the specifications of a mudsill anchor with the following requirements: appropriate for stem wall or slab foundation, 2 × 6 sill, attaches using 10d nails, minimum 700-pound uplift load, 16-gauge steel, hot-dipped galvanized. Create a dimensioned 2D sketch of the design from the manufacturer's specifications, or from measurements taken from an actual anchor. Start a new drawing from scratch or use a decimal, fractional, or architectural template of your choice. Draw the front, top, right-side, and left-side views of the anchor from your sketch using the 0 layer. The 0 layer is appropriate because you will create a block of each view in Chapter 25. Do not draw dimensions. Save the drawing as P11-17.

AutoCAD Certified Associate Exam Practice

Answer the following questions. Write your answers on a separate sheet of paper.

1. Which of the following can you use to extend two line segments to meet exactly at a sharp point? *Select all that apply.*
   A. **CHAMFER** command with a distance of 0
   B. **Extend** mode of the **EXTEND** command
   C. **Extend** mode of the **TRIM** command
   D. **FILLET** command with a radius of 0
   E. **LENGTHEN** command with a delta of 0

2. Which of the following commands can resize a rectangle from 2″ × 4″ to 2″ × 8″ in a single operation? *Select all that apply.*
   A. **EXTEND**
   B. **JOIN**
   C. **LENGTHEN**
   D. **SCALE**
   E. **STRETCH**

3. Which keyboard key or keys can you press to trim an object while the **EXTEND** command is active? *Select the one item that best answers the question.*
   A. [Alt]
   B. [Ctrl]
   C. [F2]
   D. [Shift]
   E. [Shift]+[Ctrl]

AutoCAD Certified Professional Exam Practice

Follow the instructions in each problem. Write your answers on a separate sheet of paper.

1. **Navigate to this chapter on the companion website and open CPE-11fillet.dwg.** Use an appropriate option of the **FILLET** command to round all of the corners of the polyline using a fillet radius of 1.5, as shown in the figure. What are the coordinates of Point 1 (the center of the bottom-right radius)?

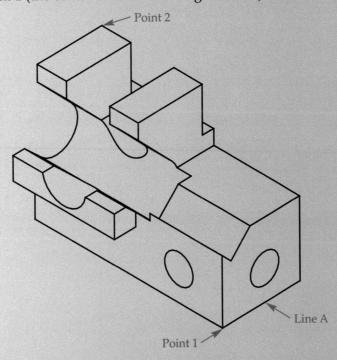

Chapter 11   Modifying Objects

2. **Navigate to this chapter on the companion website and open CPE-11scale.dwg.**
Enlarge the entire object using the **Reference** option of the **SCALE** command. Use
Point 1 as the base point, and use Line A to set the reference length. Specify a new
length of 1.35. Zoom out to see the result. What are the coordinates of Point 2?

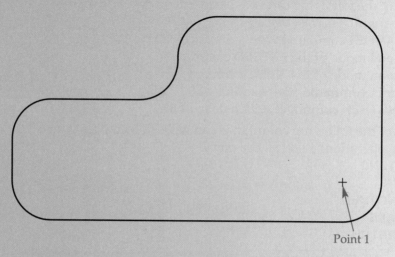

Point 1

Arranging and Patterning Objects

Learning Objectives

After completing this chapter, you will be able to:

✓ Relocate objects using the **MOVE** command.
✓ Change the angular position of objects using the **ROTATE** command.
✓ Use the **ALIGN** command to move and rotate objects at the same time.
✓ Make copies of objects using the **COPY** command.
✓ Draw mirror images of objects using the **MIRROR** command.
✓ Use the **REVERSE** command.
✓ Create patterns of objects using array commands.

This chapter explains methods for arranging and patterning existing objects using basic editing commands. The approach to editing presented in this chapter is to access a command, such as **MOVE**, and then follow the prompts to complete the operation. Another technique, presented in Chapter 13, is to select objects first using the crosshairs, then access the editing command, and finally complete the operation.

Moving Objects

Use the **MOVE** command to move objects to a different location. Access the **MOVE** command and select objects to move. At the next prompt, specify the base point from which the objects will move. Although the position of the base point is often not critical, you may want to select a point on an object, the corner of a view, or the center of a circle, for example. The selection moves as you move the crosshairs. Specify a second point to complete the move. See **Figure 12-1**.

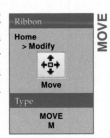

Displacement Option

The **Displacement** option allows you to move objects relative to the origin of the UCS (user coordinate system), which is at coordinates 0,0,0 by default. To move using a displacement, access the **MOVE** command and select objects to move. Then choose the **Displacement** option instead of defining the base point. At the Specify displacement <0,0,0>: prompt, enter an absolute coordinate to move the objects from the origin to the coordinate point. See **Figure 12-2**.

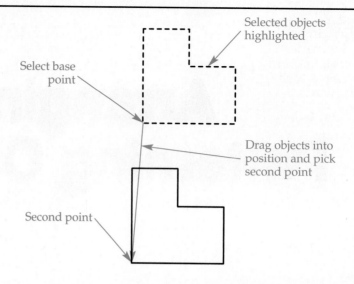

Figure 12-1.
Using the **MOVE** command to relocate objects.

Select base point

Selected objects highlighted

Drag objects into position and pick second point

Second point

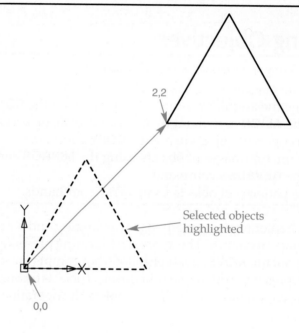

Figure 12-2.
Using the **Displacement** option of the **MOVE** command and the default 0,0,0 origin to move objects. In this example, the origin is the base point and the absolute coordinate point 2,2 is the displacement.

2,2

Selected objects highlighted

Y

X

0,0

Using the First Point as Displacement

Another method for moving an object is to use the first point as the displacement. The coordinates you use for the base point automatically define the coordinates for the direction and the distance to move the object. Access the **MOVE** command and select objects to move. Then specify the base point, and instead of locating the second point, right-click or press [Enter] or the space bar to accept the <use first point as displacement> default. See Figure 12-3.

PROFESSIONAL TIP

Use object snap modes while editing. For example, to move an object to the center of a circle, use the **Center** object snap mode to select the center of the circle.

AutoCAD and Its Applications—Basics

Figure 12-3.
Moving a circle using the selected base point, 1,1 in this example, as the displacement.

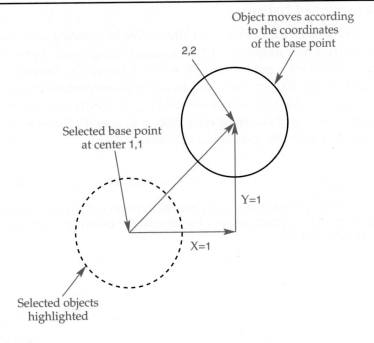

Object moves according to the coordinates of the base point

2,2

Selected base point at center 1,1

Y=1

X=1

Selected objects highlighted

Exercise 12-1

Complete the exercise on the companion website.
www.g-wlearning.com/CAD

Rotating Objects

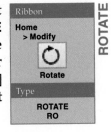

Ribbon

Home
> Modify

Rotate

Type

ROTATE
RO

ROTATE

Use the **ROTATE** command to rotate objects. For example, rotate furniture to adjust an interior design plan, or rotate the north arrow on a site plan. Access the **ROTATE** command and select objects to rotate. Proceed to the next prompt and specify the base point, or axis of rotation, around which the objects rotate. Next, enter a value or specify a point to define a rotation angle at the Specify rotation angle or [Copy/Reference] <current>: prompt. Objects rotate counterclockwise by default. To rotate an object clockwise, use a negative value. See **Figure 12-4.**

Figure 12-4.
Rotating a north arrow on a site plan –30° (330°) and 30°.

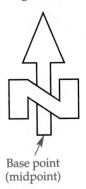

Base point (midpoint)

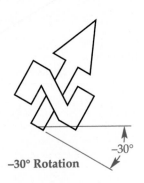

–30°

–30° Rotation

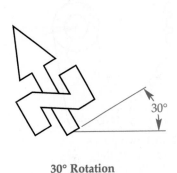

30°

30° Rotation

Reference Option

The **Reference** option is an alternative to entering a rotation angle, and allows you to specify a new angle in relation to an existing angle. For example, use the **Reference** option to rotate a gear from 150° to 90°. Choose the **Reference** option and specify the current angle, 150° in the example, at the Specify reference angle: prompt. Next, specify the angle at which the objects should be, 90° in the example. See **Figure 12-5A**. Use the **Points** function of the **Reference** option to specify the new angle using two points not associated with the selected base point.

Figure 12-5.
Using the **Reference** option of the **ROTATE** command to rotate a gear on an assembly drawing according to the current angle of the objects. A—Entering reference angles. B—Selecting points on a reference line.

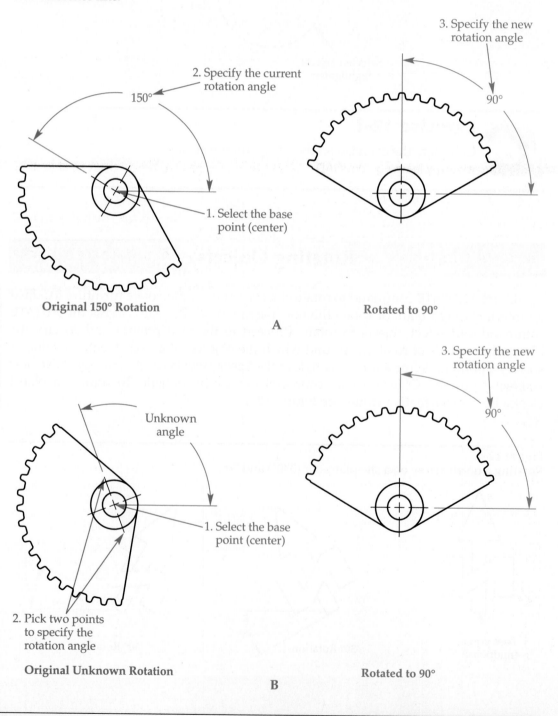

2. Specify the current rotation angle

150°

1. Select the base point (center)

Original 150° Rotation

3. Specify the new rotation angle

90°

Rotated to 90°

A

Unknown angle

1. Select the base point (center)

2. Pick two points to specify the rotation angle

Original Unknown Rotation

3. Specify the new rotation angle

90°

Rotated to 90°

B

Specify the reference and new angles using specific values; or choose points, often on existing objects, as shown in **Figure 12-5B**. Picking points is especially effective when you do not know the exact reference and new angles.

Copying While Rotating

The **Copy** option of the **ROTATE** command copies and rotates the selected objects, leaving the original object unchanged. The copy rotates to the specified angle.

Exercise 12-2

Complete the exercise on the companion website.
www.g-wlearning.com/CAD

Aligning Objects

Use the **ALIGN** command to move and rotate objects in one operation. The **ALIGN** command is primarily meant for 3D applications, but it can be used for 2D drawings. Access the **ALIGN** command and select objects to align. Then specify *source points* and *destination points*. Pick the first source point, followed by the first destination point. Then pick the second source point and the second destination point. Two source and destination points are adequate for aligning 2D objects. Right-click or press [Enter] or the space bar when the prompt requests the third source and destination points. See **Figure 12-6**. The last prompt allows you to change the size of the source objects. Choose **Yes** to scale the source objects if the distance between the source points is different from the distance between the destination points. See **Figure 12-7**.

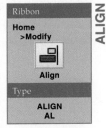

Ribbon
Home
>Modify

Align

Type
ALIGN
AL

ALIGN

source points:
Points to define the original position of an object during an **ALIGN** operation.

destination points:
Points to define the new location of objects during an **ALIGN** operation.

Exercise 12-3

Complete the exercise on the companion website.
www.g-wlearning.com/CAD

Figure 12-6.
Using the **ALIGN** command to move and rotate a kitchen cabinet layout against a wall. Select the **No** choice of the **Scale** option to apply this example.

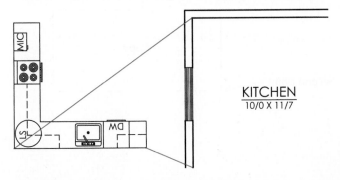

Figure 12-7.
Select the **Yes** choice of the **Scale** option to change the size of an object during the alignment.

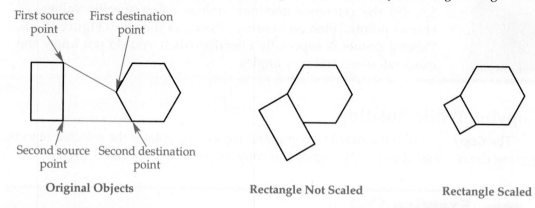

First source point
First destination point
Second source point
Second destination point
Original Objects
Rectangle Not Scaled
Rectangle Scaled

Copying Objects

Ribbon
Home
> Modify

Copy

Type
COPY
CO
CP

COPY

Use the **COPY** command to copy objects. The **COPY** command is similar to the **MOVE** command, except that when you pick a second point, the original objects remain in place and a copy appears. See **Figure 12-8**. Access the **COPY** command, select objects to copy, specify a base point, and pick a location to locate the copy. You can continue creating copies of the selected objects by specifying additional points. Use the **Undo** option to remove copies without exiting the **COPY** command. Press [Enter] or the space bar or right-click and select **Enter** to exit.

The **COPY** command provides the same options as the **MOVE** command, allowing you to specify a base point and a second point, choose a displacement using the **Displacement** option, or define the first point as the displacement.

Figure 12-8.
Using the **COPY** command to duplicate objects.

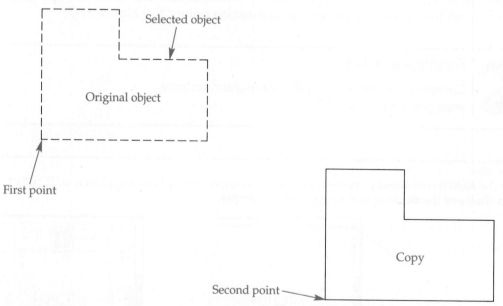

Selected object
Original object
First point
Copy
Second point

The **Multiple** copy mode is active by default and allows you to create several copies of the same object using a single **COPY** operation. To make a single copy and exit the command after placing the copy, use the **mOde** option and activate the **Single** function.

Exercise 12-4

Complete the exercise on the companion website.
www.g-wlearning.com/CAD

Array Option

Use the **Array** option, available after you specify the base point, to create a linear pattern of the selected objects. AutoCAD includes other **ARRAY** commands, described later in this chapter, which are usually more appropriate for patterning objects. However, the **Array** option of the **COPY** command is effective for copying multiple, equally spaced objects quickly. Choose the **Array** option and then enter the total number of copies, including the selected objects, to create. Specify the location of the first copy, which also defines the spacing between copies. An alternative is to use the **Fit** option and then specify the location of the last copy, which divides the number of items equally between the base point and second point. **Figure 12-9** shows an example of developing a pattern using the **Array** option of the **COPY** command.

Exercise 12-5

Complete the exercise on the companion website.
www.g-wlearning.com/CAD

Figure 12-9.
Using the **Array** option of the **COPY** command to draw a linear pattern of rollers along the frame of a conveyer.

Selected objects

Selected base point

Specify second point option—pick here and objects are spaced at the selected distance

Fit option—pick here and objects are spaced evenly within the selected distance

Mirroring Objects

MIRROR

Ribbon
Home
> Modify

Mirror

Type
MIRROR
MI

mirror line: The line
of symmetry across
which objects are
mirrored.

The **MIRROR** command allows you to reflect, or mirror, objects. For example, in mechanical drafting, mirror a part to form the opposite component of a symmetrical assembly. In architectural drafting, mirror a floor plan to create a duplex residence or to accommodate a different site orientation. Access the **MIRROR** command and select the objects to mirror. Then create an imaginary *mirror line* at any angle by specifying two points. After you locate the second mirror line point, you have the option to delete the original objects. See **Figure 12-10**.

The **MIRRTEXT** system variable, which is set to 0 by default, prevents text from reversing during a mirror operation. Change the **MIRRTEXT** value to 1 to mirror text in relation to the original object. See **Figure 12-11**. Backward text is generally not acceptable, except for applications such as reverse imaging.

Exercise 12-6

Complete the exercise on the companion website.
www.g-wlearning.com/CAD

Figure 12-10.
Using the **MIRROR** command to reflect objects over an imaginary mirror line. You have the option of erasing the original objects.

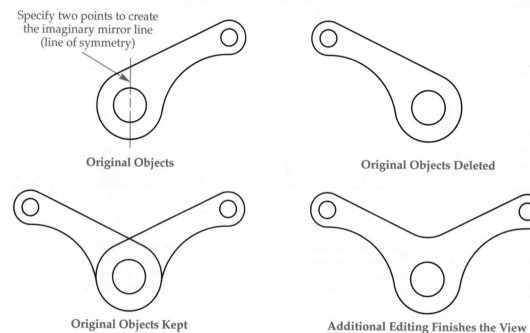

Specify two points to create the imaginary mirror line (line of symmetry)

Original Objects

Original Objects Deleted

Original Objects Kept

Additional Editing Finishes the View

Figure 12-11.
The **MIRRTEXT**
system variable
options.

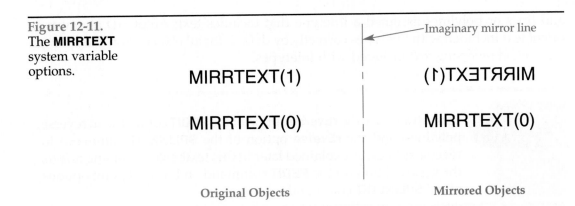

Imaginary mirror line

MIRRTEXT(1)   (1)TXETRRIM

MIRRTEXT(0)   MIRRTEXT(0)

Original Objects   Mirrored Objects

Reversing an Object's Point Calculation

The **REVERSE** command reverses the calculation of points along lines, polylines, splines, and helixes. The previous start point becomes the new endpoint, and the previous endpoint becomes the new start point. As shown in **Figure 12-12**, reversing is apparent when it is applied to specific objects, such as polylines with varying width

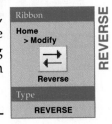

Ribbon
Home
> Modify

Reverse

Type

REVERSE

REVERSE

Figure 12-12.
A—Using the **REVERSE** command to reverse a polyline with varying width. B—Reversing a polyline assigned a linetype that includes text.

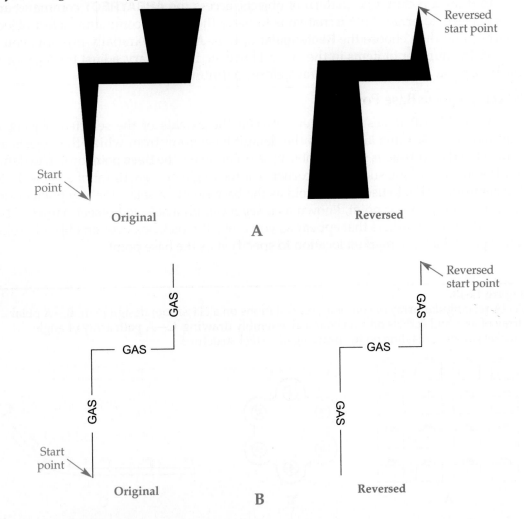

and lines or polylines assigned a linetype that includes text. AutoCAD attempts to orient text included with linetypes correctly by default for all objects. You should typically avoid reversing text included with linetypes.

>
>
> You can also use the **rEverse** option of the **PEDIT** command to reverse polylines, and the **rEverse** option of the **SPLINEDIT** command to reverse splines, as explained later in this textbook. Reversing affects the vertex options of the **PEDIT** command and control point options of the **SPLINEDIT** command.

Arraying Objects

array: Multiple copies of an object arranged in a pattern.

rectangular array: A pattern made up of columns and rows of objects.

polar (circular) array: A circular pattern of objects.

path array: A pattern of objects drawn in reference to another object, or path.

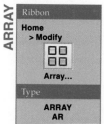

ARRAY

ARRAYRECT

AutoCAD includes methods for creating and modifying a pattern, or *array*, of existing objects. For example, create a *rectangular array* of computer workstations on a classroom design plan, or create a *polar array* (also called a *circular array*) of screws on a mechanical assembly drawing. You can also pattern objects in reference to an existing object using a *path array*. **Figure 12-13** shows examples of arrays.

Rectangular Array

To draw a rectangular pattern of objects, access the **ARRAYRECT** command and select objects to array. An alternative is to issue the **ARRAY** command, select objects to array, and then choose the **Rectangular** option. AutoCAD initially prompts you to specify the number of items in the array. However, if necessary, adjust the location of the base point and angle of the array before continuing.

Specifying the Base Point

AutoCAD calculates the center between the extents of the selected objects, or centroid, and uses this location as the default base point from which the objects are arrayed. If the centroid is not a suitable base point, select the **Base point** option to define a different base point, such as the corner of a rectangle or a quadrant of a circle. Enter the **centroid** option to use the centroid as the base point, or select the **Key point** option to choose a constraint point, known as a *key point*, on a selected object. **Figure 12-14** shows the point markers that appear as you move the pick box over an object to select a key point. Pick the marked location to specify it as the base point.

Figure 12-13.

A—A rectangular array of computer workstations on a classroom design plan. B—A polar array of arcs and screws on a mechanical assembly drawing. C—A path array of angle brackets along a polyline on the drawing of a steel structure.

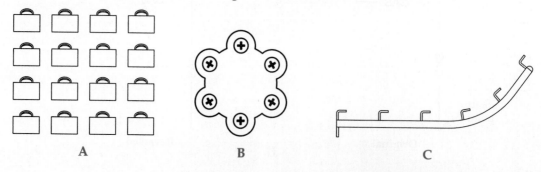

A   B   C

Figure 22-14.
The key points on objects you can select to define a new array base point.

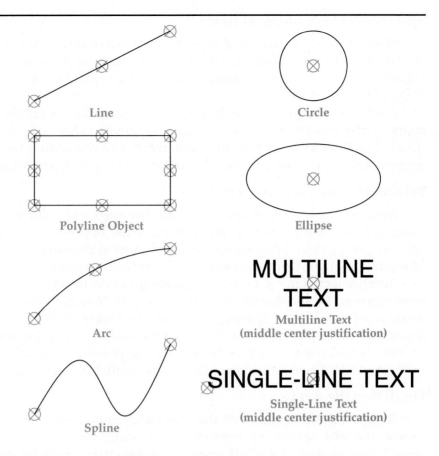

Line

Circle

Polyline Object

Ellipse

Arc

MULTILINE
TEXT

Multiline Text
(middle center justification)

SINGLE-LINE TEXT

Single-Line Text
(middle center justification)

Spline

Setting the Angle

To rotate the array, choose the **Angle** option and then specify the angle of the rows. The alignment of rows and columns rotate, not the objects. See **Figure 12-15.**

Like other objects, arrays form relative to the rotation of the UCS (user coordinate system). For example, the direction of the X axis is horizontal by default. For basic 2D applications, use the **Angle** option to rotate the array, and do not modify the UCS axis direction.

Figure 12-15.
Use the **Angle** option to set the angle of rows in the rectangular array.

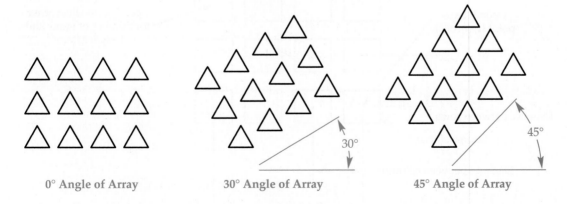

0° Angle of Array

30° Angle of Array

45° Angle of Array

Specifying the Number of Items

Next, specify the number of items in a pattern of rows and columns. The default method is to locate a point at a corner of the array opposite of the base point. See **Figure 12-16**. Move the crosshairs before choosing a point to view a dynamic preview of the number of items.

Another technique is to use the **Count** option to define the number of rows followed by the number of columns. You can enter a numeric value or locate a point to set the count. The **Expression** option, available when you are setting the row and column count, allows you to enter an equation or formula if you do not know the exact count.

Setting the Spacing between Items

Next, set the spacing between items in the rows and columns. The default technique is to locate a point at a corner of the array opposite of the base point, which specifies the distance between rows and columns at the same time. See **Figure 12-16**. Move the crosshairs before choosing a point to view a dynamic preview of the spacing.

Another technique is to use the **Spacing** option to define the distance between rows followed by the distance between columns. You can enter a numeric value or locate a point to set the spacing. As shown in **Figure 12-16**, the offset refers to the distance between a point on an object and the corresponding point in the next row or column. Adjust the direction of the array using positive or negative values, as shown in **Figure 12-17**. The **Expression** option is also available for setting spacing.

Finalizing the Array

To create the array and exit the command, press [Enter], the space bar, or [Esc], choose the **eXit** option, or right-click and select **Enter**. You can also adjust the array before exiting. AutoCAD draws an *associative array* by default. Choose the **ASsociative** option followed by the **No** option to create a *nonassociative array*. Chapter 13 describes working with associative arrays. Use the **EXPLODE** command to remove the associative property from existing arrays.

Use the **Base point** option to redefine the base point, the **Rows** option to change the number of rows, and the **Columns** option to edit the number of columns. The **Total** option associated with the **Rows** and **Columns** options is used to define the total distance from a point on the source objects to the corresponding point on the last row or column. Grips are also available to relocate the array and make changes to the number and spacing of items. Chapter 13 explains modifying objects using grips.

associative array: An adjustable array object; all items are grouped to form a single object that you can modify, such as changing the number of items and spacing between items.

nonassociative array: An array of copied, or static, source objects that do not form a single adjustable array object.

Figure 12-16.
A rectangular array with four columns and three rows. Note that the spacing is the distance between a point on an object and the corresponding point in the next row and column.

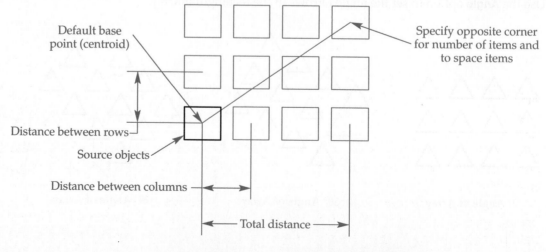

Default base point (centroid)

Specify opposite corner for number of items and to space items

Distance between rows

Source objects

Distance between columns

Total distance

Figure 12-17.
Positive and negative offset distances determine the direction in which an array will grow.

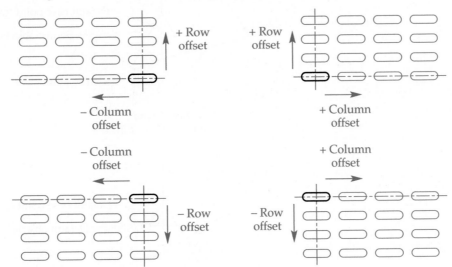

AutoCAD stores the current associative setting as the default for all arrays. Remember to change the associative setting as necessary before finalizing an array.

The **Levels** option allows you to array the source objects along the Z axis to create a 3D array. The Specify the incrementing elevation between rows or [Expression] <0.000>: prompt associated with the **Rows** option is used to step each row a specified value increasingly along the Z axis to create a 3D array. *AutoCAD and Its Applications—Advanced* provides complete information on 3D modeling.

Exercise 12-7

Complete the exercise on the companion website.
www.g-wlearning.com/CAD

Polar Array

To draw a circular pattern of objects, access the **ARRAYPOLAR** command and select objects to array. An alternative is to issue the **ARRAY** command, select objects to array, and then choose the **POlar** option. AutoCAD initially prompts you to specify the center point of the array. The **Base point** option is available and operates the same as when you are creating a rectangular array. However, the purpose of relocating the base point for a polar array is apparent only when you edit an associative polar array, as explained in Chapter 13.

Next, specify the center point around which the objects will be arrayed. The array forms around the Z axis of the UCS. Finish constructing the array based in the information you know about the pattern. **Figure 12-18** shows examples of polar arrays.

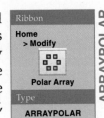

Ribbon
Home
> Modify

Polar Array

Type

ARRAYPOLAR

ARRAYPOLAR

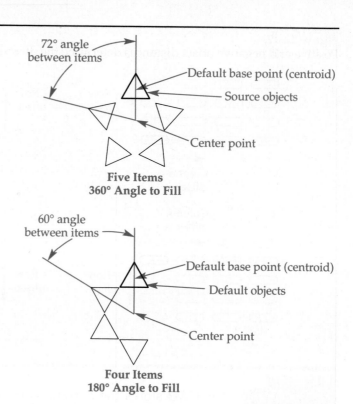

Figure 12-18.
Examples of polar arrays.

72° angle between items

Default base point (centroid)

Source objects

Center point

Five Items
360° Angle to Fill

60° angle between items

Default base point (centroid)

Default objects

Center point

Four Items
180° Angle to Fill

Specifying the Total Number of Items and Angle to Fill

By default, AutoCAD prompts you to enter the number of items to array. This method is effective if you know the total number of items and the total angle to fill with the items in the array. For most applications, enter a numeric value for the total number of items, including the source objects. The **Expression** option allows you to enter an equation or formula if you do not know the exact number of items. An alternative is to move the crosshairs around the center point to view a dynamic preview of the array and pick when you see the desired number of items.

Then enter a positive angle to array objects in a counterclockwise direction, or enter a negative angle to array objects in a clockwise direction. Enter 360 to create a complete circular array. If you prefer, move the crosshairs around the center point to view a dynamic preview of the array and specify a point location to define the angle to fill.

Specifying the Total Number of Items and Angle between Items

After you select objects to array and locate the center point, choose the **Angle between** option to specify the total number of items and the angle between adjacent objects in the array. Enter a numeric value or locate a point to set the angle between items. Then specify the total number of items as previously described.

Specifying Angle between Items and Angle to Fill

After you select objects to array and locate the center point, choose the **Angle between** option to specify the angle between adjacent objects and the total angle to fill with the items in the array. Enter a numeric value or locate a point to set the angle between items. Then set the total angle to fill as previously described.

Finalizing the Array

To create the array and exit the command, press [Enter], the space bar, or [Esc], choose the **eXit** option, or right-click and select **Enter**. You can also adjust the array before exiting. Use the **ASsociative** option to create an associative or nonassociative array. Use the **Base point** option to redefine the base point, the **Items** option to change the number of items, the **Angle between** option to edit the angle between items, and the **Fill angle** option to edit the angle to fill.

The **ROWs** option allows you to form multiple rows during the array, as shown in Figure 12-19. Enter the total number of rows, including the source row. The **Expression** option allows you to enter an equation or formula. Then enter the distance between rows or choose the **Total** option to define the total distance from a point on the source objects to the corresponding point on the last row. The **Expression** option is also available. Respond to the last prompt with a 0 for 2D applications.

The **ROTate items** option is used to control whether items are rotated as they are arrayed. AutoCAD rotates items perpendicular to the center point by default. Choose the **ROTate items** option followed by the **No** option to position items in the same orientation as the source objects. See Figure 12-20. Grips are also available to relocate the array and make changes to the number and arrangement of items. Chapter 13 explains modifying objects using grips.

The **Axis of rotation** option, available after you select objects to array, allows you to array the source objects along a specified axis to create a 3D array. The **Levels** option allows you to array the source objects along the Z axis to create a 3D array. The Specify the incrementing elevation between rows or [Expression] <0.000>: prompt associated with the **ROWs** option is used to step each row a specified value increasingly along the Z axis to create a 3D array. *AutoCAD and Its Applications—Advanced* provides complete information on 3D modeling.

Figure 12-19.
Use the **ROWs** option to form multiple rows during the array. This example shows a nozzle design drawn using a polar array with 10 items, 360° angle to fill, and three rows.

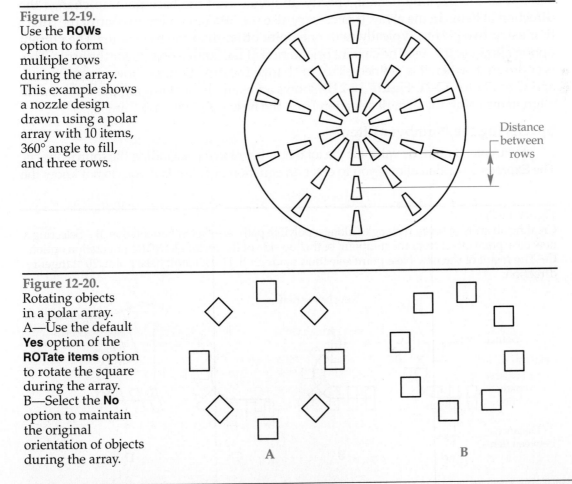

Distance between rows

Figure 12-20.
Rotating objects in a polar array. A—Use the default **Yes** option of the **ROTate items** option to rotate the square during the array. B—Select the **No** option to maintain the original orientation of objects during the array.

A

B

Exercise 12-8

Complete the exercise on the companion website.
www.g-wlearning.com/CAD

Path Array

To array objects along a path, access the **ARRAYPATH** command and select objects to array. Right-click or press [Enter] or the space bar to continue. Then pick a line, circle, arc, ellipse, polyline, spline, or helix to use as the path along which the source objects will be arrayed. Pick the object near where you want the array to begin, because the default base point is the endpoint of the path closest to where you select the path. An alternative is to issue the **ARRAY** command, select objects to array, choose the **PAth** option, and then pick the path. Once you select the objects to array and the path, AutoCAD prompts you to specify the number of items in the array. However, if necessary, you can adjust the orientation of items in the array before continuing.

Specifying the Orientation

Select the **Orientation** option to adjust the location of the base point and the rotation of items in the array. The default base point is the endpoint of the path closest to where you select the path, not a point on the source objects. See **Figure 12-21A**. Enter the **end of path curve** option to use the default endpoint as the base point, or specify a different location to use as the base point. The **Key point** option is available for choosing a key point for the base point, as explained for creating a rectangular array. Parts B and C of **Figure 12-21** show an example of changing the default base point for a path array.

The next prompt after you specify the base point requests a direction to align items with the path. By default, specify an angle from the base point to change the direction of items in the array. If necessary, use the **2Points** option to control the direction using two points, typically with one point other than the base point. The **current** option allows you to use the current orientation of the source objects for the array. This is appropriate when it is not necessary to change the direction, as shown in parts A, B, and C of **Figure 12-21**. **Figure 12-21D** shows an example of changing the direction to align items using the default selection from the base point and a 20° angle.

Specifying the Number of Items

Next, enter a numeric value for the total number of items, including the source objects. The **Expression** option allows you to enter an equation or formula if you do not know the

---

Figure 12-21.
Creating an array of seven rectangles along a polyline path. A—Default orientation. B—Selecting a new base point offset from the midpoint of the top side of the rectangle, **NONe** orientation option. C—The result of the new base point selection made in B. D—Default base point, 20° tangent direction.

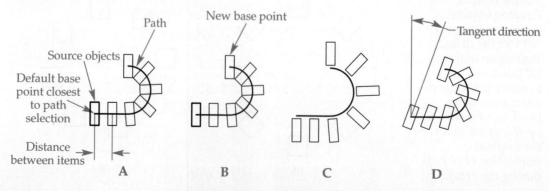

exact number of items. An alternative is to move the crosshairs along the path to view a dynamic preview of the array and pick when you see the desired number of items.

Setting the Spacing between Items

AutoCAD then prompts you to specify the distance between items in the array. The **Expression** option is available if necessary. Another common technique is to choose the **Divide** option to place items at equally spaced locations on the path. See Figure 12-21. AutoCAD calculates the distance between items based on the number of items you specify. The third method is to choose the **Total** option to define the total distance from a point on the source objects to the corresponding point on the last item.

Finalizing the Array

To create the array and exit the command, press [Enter], the space bar, or [Esc], choose the **eXit** option, or right-click and select **Enter**. You can also adjust the array before exiting. Use the **ASsociative** option to create an associative or nonassociative array. Use the **Base point** option to redefine the base point and the **Items** option to change the number of items. The **Rows** option allows you to form multiple rows during the array, similar to the **ROWs** option available for creating a polar array.

Use the **Align items** option to control the alignment of items along the path. Items align with the path by default. See Figure 12-22A. Choose the **Align** items option, followed by the **No** option, to apply the alignment of the source objects to all items along the path. See Figure 12-22B.

The **NORmal** orientation option allows you to array items normal (often perpendicular) to the path to create a 3D array. The **Levels** option, available before exiting, allows you to array the source objects along the Z axis to create a 3D array. The **Z direction** option, also available before exiting, is used to maintain the orientation or items along a 3D path. *AutoCAD and Its Applications—Advanced* provides complete information on 3D modeling.

Exercise 12-9

Complete the exercise on the companion website.
www.g-wlearning.com/CAD

Figure 12-22.
Creating an array of four palm trees along a spline path for a landscape elevation. Use the **Align items** option to set alignment of items along the path.

Items Aligned to Path   Items Not Aligned to Path

Chapter Review

1. How would you rotate an object 45° clockwise?
2. Briefly describe the two methods of using the **Reference** option of the **ROTATE** command.
3. Name the command that you can use to move and rotate an object at the same time.
4. How many points must you select to align an object in a 2D drawing?
5. Explain the difference between the **MOVE** and **COPY** commands.
6. Which command allows you to draw a reflected image of an existing object?
7. What is the purpose of the **REVERSE** command?
8. What is the difference between polar and rectangular arrays?
9. What does AutoCAD use as the default base point for a rectangular array?
10. Suppose an object is 1.5″ (38 mm) wide and you want to create a rectangular array with .75″ (19 mm) spacing between objects. What should you specify for the distance between columns?
11. Define *associative array.*
12. How do you specify a clockwise circular array rotation?
13. What value should you specify for the angle to fill to create a complete circular array?
14. List the objects you can use as a path for a path array.
15. Describe the spacing between items that occurs when you use the **Divide** option of the **ARRAYPATH** command.

Drawing Problems

Start AutoCAD if it is not already started. Start a new drawing for each problem using an appropriate template of your choice. The template should include layers and text styles for drawing the given objects. Add layers and text styles as needed. Draw all objects using appropriate layers, text styles, justification, and format. Follow the specific instructions for each problem. Use only drawing and editing commands and techniques you have already learned. Do not draw dimensions. Use your own judgment and approximate dimensions when necessary.

▼ Basic

1. Open P11-1 and save the file as P12-1. (If you have not yet completed problem 11-1, complete it now.) The P12-1 file should be active. Rotate the object 90° to the right and mirror the object to the left. Use the vertical base of the object as the mirror line. The final drawing should look like the example below. Resave the drawing.

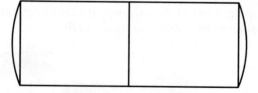

2. Open P11-2 and save the file as P12-2. (If you have not yet completed problem 11-2, complete it now.) The P12-2 file should be active. Make two copies of the object to the right of the original object. Scale the first copy 1.5 times the size of the original object. Scale the second copy 2 times the size of the original object. Move the objects so they are approximately centered in the drawing area. Move the objects as needed to align the bases of all objects and provide an equal amount of space between the objects. The final drawing should look like the example below. Resave the drawing.

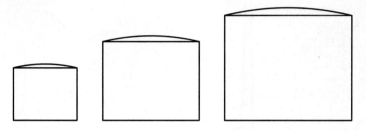

3. Draw Objects A, B, and C. Make a copy of Object A two units up. Make four copies of Object B three units up, center to center. Make three copies of Object C three units up, center to center. Save the drawing as P12-3.

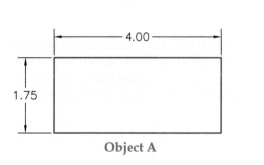

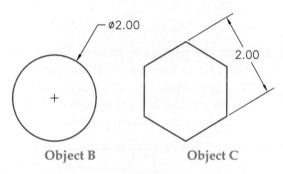

Object A   Object B   Object C

4. Open P11-4 and save the file as P12-4. (If you have not yet completed problem 11-4, complete it now.) The P12-4 file should be active. Draw a mirror image as Object B. Then remove the original view and move the new view so that Point 2 is at the original Point 1 location. Resave the drawing.

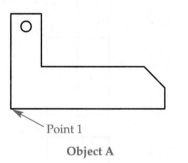

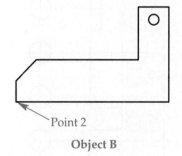

Point 1   Point 2

Object A   Object B

5. Draw the palm trees along the spline for the portion of the landscape elevation shown. Use the **ARRAYPATH** command as needed. Save the drawing as P12-5.

6. Draw the part view shown. The object is symmetrical; therefore, draw only one half. Mirror the other half into place. Use the **CHAMFER** and **FILLET** commands to your best advantage. All fillets and rounds are .125. Use the **JOIN** command where necessary. Use the **Array** option of the **COPY** command to array the row of ⌀.500 holes. Save the drawing as P12-6.

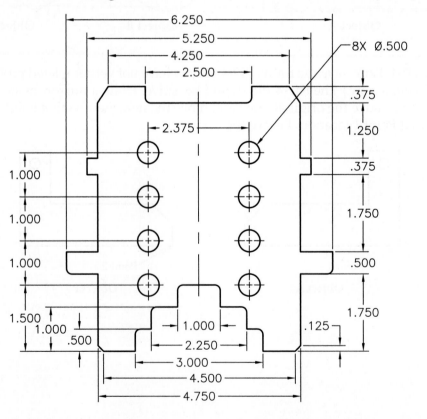

7. Draw the portion of the part view shown. Mirror the right half into place. Use the **CHAMFER** and **FILLET** commands to your best advantage. Save the drawing as P12-7.

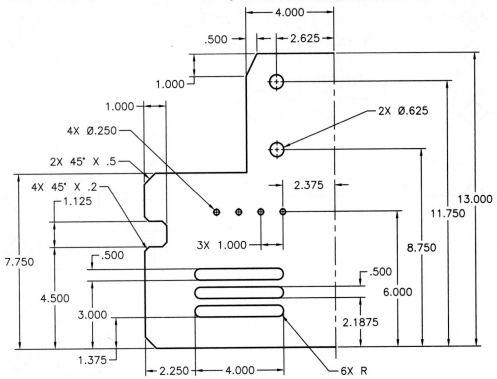

8. Draw the electronic schematic symbols shown. Mirror the drawing, but make sure the text remains readable. Delete the original image during the mirroring process. Save the drawing as P12-8.

2b1

TRANSFER

5a2 4a1 8a1

LTS. HTRS. FANS

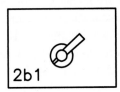

2b1

1b1

RESET

11b1

BYPASS

9. Draw the timer schematic shown. Save the drawing as P12-9.

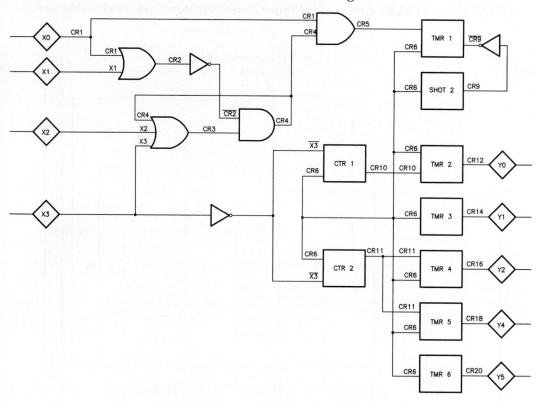

10. Use tracking and object snaps to draw the board shown, based on the following instructions:
    A. Draw the outline first, followed by the ten ∅.500 holes (A).
    B. The holes labeled B are located vertically halfway between the centers of the holes labeled A. They have a diameter one-quarter the size of the holes labeled A.
    C. The holes labeled C are located vertically halfway between the holes labeled A and B. Their diameter is three-quarters of the diameter of the holes labeled B.
    D. The holes labeled D are located horizontally halfway between the centers of the holes labeled A. These holes have the same diameter as the holes labeled B.
    E. Draw the rectangles around the circles as shown.
    F. Do not draw dimensions, notes, or labels.
    G. Save the drawing as P12-10.

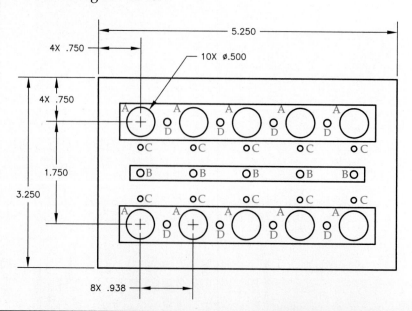

AutoCAD and Its Applications—Basics

11. Draw the portion of the gasket shown on the left. Use the **MIRROR** command to complete the gasket as shown on the right. Save the drawing as P12-11.

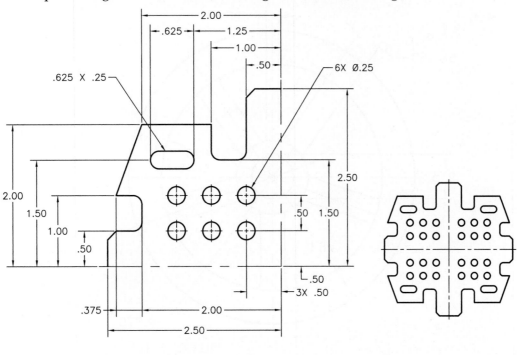

ALL FILLETS AND ROUNDS R.125.
CHAMFERS 45° X .125

12. Draw the padded bench shown. Use the **COPY** and **ARRAY** commands as needed. Save the drawing as P12-12.

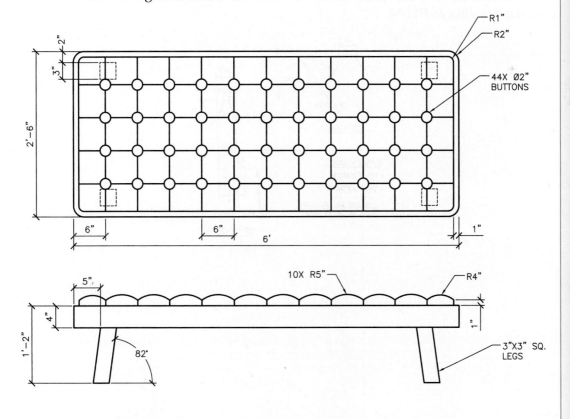

13. Draw the hand wheel shown. Use the **ARRAYPOLAR** command to draw the spokes. Save the drawing as P12-13.

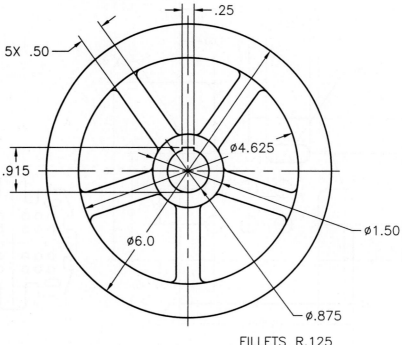

FILLETS R.125

▼ Advanced

14. Use the engineer's sketch and notes shown to draw the sprocket. Create a front and side view of the sprocket. Use the **ARRAYPOLAR** command as needed. Save the drawing as P12-14.

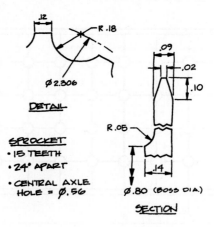

15. Draw the views of the sprocket shown. Use **ARRAYPOLAR** to construct the hole and tooth arrangements. Save the drawing as P12-15.

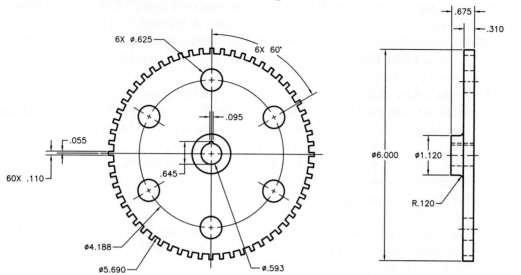

16. Draw the refrigeration system schematic shown. Save the drawing as P12-16.

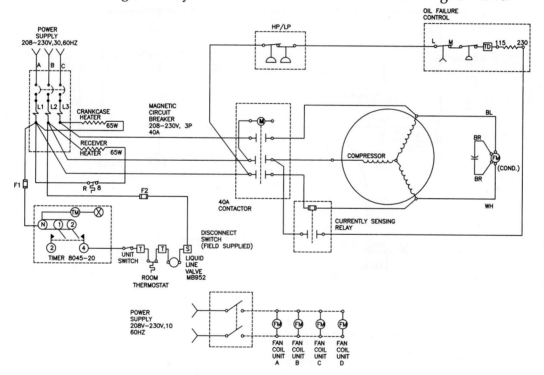

Drawing Problems - Chapter 12

17. The structural sketch shown is a steel column arrangement on a concrete floor slab for a new building. The I-shaped symbols represent the steel columns. The columns are arranged in "bay lines" and "column lines." The column lines are numbered 1, 2, and 3. The bay lines are labeled A through G. The width of a bay is 24'-0". Line balloons, or tags, identify the bay and column lines. Draw the arrangement, using **ARRAYRECT** for the steel column symbols and the tags. The following guidelines will help:

A. Begin a new drawing using an architectural template.
B. Select architectural units and set up the drawing to print on a 36 × 24 sheet size. Determine the scale required for the floor plan to fit on this sheet size and specify the drawing limits accordingly.
C. Draw the steel column symbol to the dimensions given.
D. Set the grid spacing at 2'-0" (24").
E. Set the snap spacing at 12".
F. Draw all other objects.
G. Place text inside the balloon tags. Set the running object snap mode to **Center** and justify the text to **Middle**. Make the text height 6".
H. Save the drawing as P12-17.

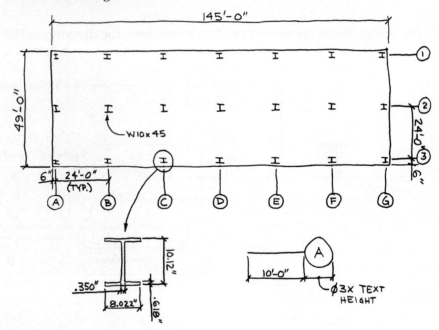

18. The sketch shown is a proposed classroom layout of desks and chairs. One desk is shown with the layout of a chair, keyboard, monitor, and tower-mounted computer (drawn with dotted lines). All of the desk workstations should have the same configuration. The exact sizes and locations of the doors and windows are not important for this problem. Use the following guidelines to complete this problem:

A. Begin a new drawing.
B. Choose architectural units.
C. Set up the drawing to print on a C-size sheet, and be sure to create the drawing in model space.
D. Use the appropriate drawing and editing commands to complete this problem quickly and efficiently.
E. Draw the desk and computer hardware to the dimensions given.
F. Do not dimension the drawing.
G. Save the drawing as P12-18.

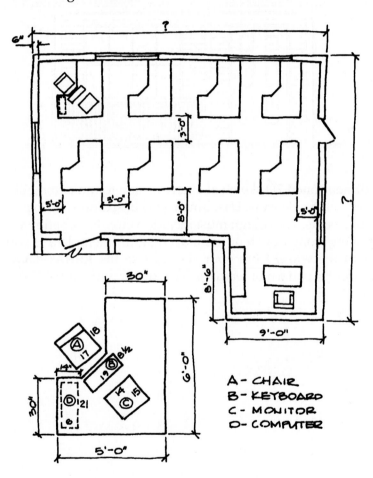

A - CHAIR
B - KEYBOARD
C - MONITOR
D - COMPUTER

19. Draw the front elevation of this house. Create the features proportional to the given drawing. Use the **ARRAYRECT** and **TRIM** commands to place the siding and porch rails evenly. Save the drawing as P12-19.

20. Create a dimensioned 2D sketch of a new design for an automobile wheel. Sketch a front view and a side view. Use dimensions based on your experience, research, and measurements. The design must include a circular repetition of features. Start a new drawing from scratch or use a decimal-unit template of your choice. Draw the views of the wheel from your sketch. Use the **ARRAYPOLAR** command to draw the circular pattern of features. Save the drawing as P12-20.

AutoCAD Certified Associate Exam Practice

Answer the following questions. Write your answers on a separate sheet of paper.

1. Which of the following can you do using the **MOVE** command? *Select all that apply.*
   A. move objects relative to the origin (0,0,0)
   B. move objects from a base point to a second specified point
   C. rotate objects during the move operation
   D. use the first point you pick as the point of displacement
   E. use a scale factor

2. In which order do you pick the source and destination points when using the **ALIGN** command? *Select the one item that best answers the question.*
   A. destination point 1, destination point 2, source point 1, source point 2
   B. destination point 1, source point 1, destination point 2, source point 2
   C. destination point 2, source point 2, destination point 1, source point 1
   D. source point 1, destination point 1, source point 2, destination point 2
   E. source point 1, source point 2, destination point 1, destination point 2

3. In the array shown below, what would you enter for the row and column offsets? *Select the one item that best answers the question.*
   A. row offset 1.00, column offset 2.00
   B. row offset 2.00, column offset 1.00
   C. row offset 2.00, column offset 4.00
   D. row offset 4.00, column offset 2.00
   E. row offset 5.00, column offset 6.00
   F. row offset 6.00, column offset 5.00

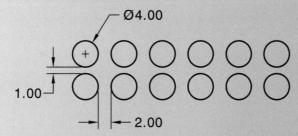

AutoCAD Certified Professional Exam Practice

Follow the instructions in each problem. Write your answers on a separate sheet of paper.

1. **Navigate to this chapter on the companion website and open CPE-12array.dwg.** Use the **ARRAYPOLAR** command to finish the view of a fan plate as shown. Analyze the drawing and use the most appropriate options for the polar array. What are the coordinates of Point 1?

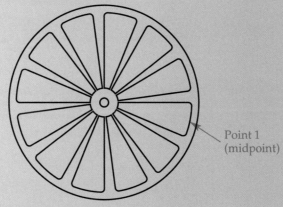

Point 1
(midpoint)

2. **Navigate to this chapter on the companion website and open CPE-12mirror.dwg.**
   Create a mirror line starting at absolute coordinates 13′,8′ and extending 5′ at 120°.
   Mirror the couch across this line. What are the coordinates of Point 1?

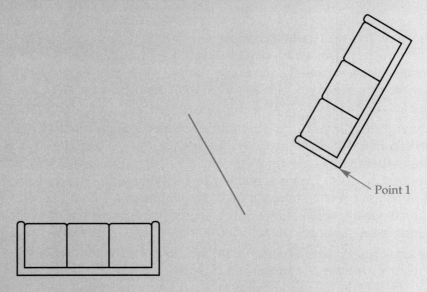

Point 1

Grips, Properties, and Additional Selection Techniques

Learning Objectives

After completing this chapter, you will be able to:

✓ Use grips to stretch, move, rotate, scale, mirror, and copy objects.
✓ Edit associative arrays.
✓ Adjust object properties using the **Quick Properties** panel and the **Properties** palette.
✓ Use the **MATCHPROP** command to match object properties.
✓ Edit between drawings.
✓ Use the **ADDSELECTED** command to draw an object based on an existing object.
✓ Create selection sets using the **SELECTSIMILAR** and **Quick Select** commands.

One approach to editing is to access a command, such as **ERASE**, **FILLET**, **MOVE**, or **COPY**, select the objects to modify, and follow prompts to complete the operation. This chapter explains the alternative approach of selecting objects first and then using editing commands or object properties to make changes. This chapter also describes additional selection options, selection set filters, and related tools.

Grips

Use the crosshairs to select objects and display *grips*. See **Figure 13-1**. Selected objects become highlighted and grips initially appear as *unselected grips*. Unselected grips are blue (Color 140) by default. Grips are specific to object type. Most objects include the standard filled-square grips at critical and editable points on the object. Several objects, including elliptical arcs, mtext, polylines, splines, associative arrays, tables, hatches, and blocks, also have specialized grips. For example, elliptical arcs include filled-arrow grips for adjusting the length of the elliptical arc. This textbook explains the grips specific to various object types when applicable.

Move the crosshairs over an unselected grip to snap to the grip. Then pause to change the color of the grip to pink (Color 11). Hovering over an unselected grip and allowing it to change color helps you select the correct grip, especially when multiple grips are close together. A tooltip or options may appear, depending on the object and grip.

grips: Small boxes that appear at strategic points on a selected object, allowing you to edit the object directly.

unselected grips: Grips that you have not yet picked to perform an operation.

Figure 13-1.
Grips appear at specific locations on objects when you select the objects while no drawing or editing command is active.

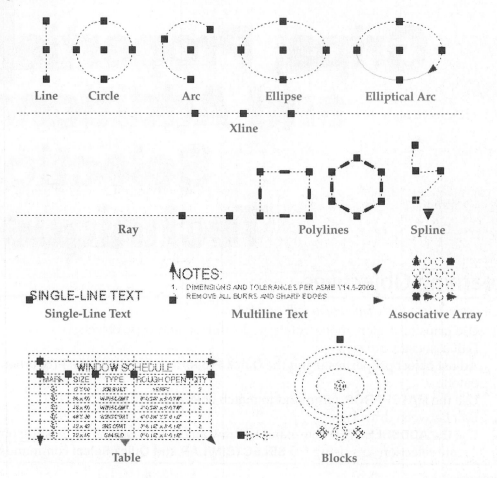

Line   Circle   Arc   Ellipse   Elliptical Arc

Xline

Ray   Polylines   Spline

Single-Line Text   Multiline Text   Associative Array

Table   Blocks

Pick a grip to perform an editing operation at the location of the grip. A *selected grip* appears red (Color 12) by default. If you select grips on more than one object, what you do with the selected grips affects all of the selected objects. Objects that have both unselected and selected grips become part of the current selection set.

To remove objects from a selection set, hold down [Shift] and pick the objects to deselect. Select additional objects, without pressing [Shift], to add to the grip selection set. [Shift] also allows you to select multiple grips. Hold down [Shift] and then select each grip. Remember not to release [Shift] until you pick all the grips you want to activate. While still holding down [Shift], you can pick a selected (red) grip to return it to the unselected (blue) state. **Figure 13-2** shows an example of modifying two circles at the same time using selected grips.

Figure 13-2.
You can modify multiple objects at the same time by pressing [Shift] to select additional grips. In this example, if you only make one grip hot, you can only edit one circle.

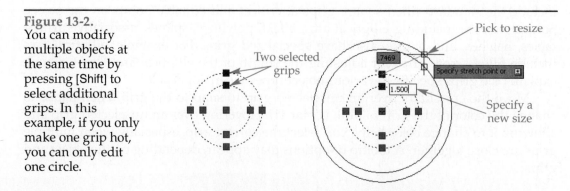

Two selected grips

Pick to resize

.7469

Specify stretch point or

1.500

Specify a new size

Press [Esc] to deactivate the current grip operation. Press [Esc] again to deselect all objects and hide the grips. You can also right-click and pick **Deselect All** to deselect objects and hide all grips.

 Use the options in the **Grip Size** and **Grips** areas of the **Selection** tab in the **Options** dialog box to control grip size and color.

noun/verb selection: Performing tasks in AutoCAD by selecting the objects before activating a command.

verb/noun selection: Performing tasks in AutoCAD by activating a command before selecting objects.

PROFESSIONAL TIP

 You can perform some conventional operations by selecting objects before you access a command. For example, you can select objects to erase and then activate the **ERASE** command or press [Delete]. This technique is available by default and is controlled by the **Noun/verb selection** check box in the **Selection Modes** area of the **Selection** tab in the **Options** dialog box.

Standard Grip Commands

Standard grip boxes provide access to the **STRETCH, MOVE, ROTATE, SCALE,** and **MIRROR** commands. In addition, the **Copy** option of the **MOVE** command and sometimes, depending on the selected grip, the **STRETCH** command imitate the **COPY** command. Select grips to display options at the dynamic input cursor and at the command line. Do not attempt to use conventional means of command access, such as the ribbon. The first command is **STRETCH**, as indicated by the ** STRETCH ** Specify stretch point or [Base point/Copy/Undo/eXit]: prompt. Use the **STRETCH** command, or press [Enter] or the space bar or right-click and select **Enter** to cycle through the **MOVE, ROTATE, SCALE,** and **MIRROR** commands.

An alternative to cycling through commands is to select grips, right-click, and select an option from the shortcut menu. A third method to activate a command is to enter the first two characters of the command name. Type MO for **MOVE**, MI for **MIRROR**, RO for **ROTATE**, SC for **SCALE**, or ST for **STRETCH**.

Stretching

Stretching using grips is similar to stretching using the **STRETCH** command, except that the selected grip acts as the stretch base point. In addition, depending on the selected grip and type of object, stretching using a grip can result in a move, rotate, or scale operation. See **Figure 13-3**. Stretch individual grips, or select multiple grips as needed depending on the desired result. See **Figure 13-4**.

Use the **Base point** option to specify a base point instead of using the selected grip as the base point. Select the **Undo** option to undo the previous operation. Choose the **eXit** option or press [Esc] to exit without completing the stretch. When you finish stretching, the selected grips return to the unselected state. Press [Esc] to hide the grips.

Dynamic input and other drawing aids, such as polar tracking, are very useful for grip editing. Dynamic input is especially effective with the **STRETCH** grip command. **Figure 13-5** shows an example of using dimensional input to modify the size of a circle or offset the circle a specific distance. In this example, enter the new radius of the circle in the distance input field, or press [Tab] to enter an offset in the other distance input field. Another example is modifying the length of an ellipse axis by selecting the appropriate quadrant grip and using dimensional input to edit the value. These are

Figure 13-3.
Examples of using the **STRETCH** grip command with dynamic input active. Note the selected grip in each case and the relevant dynamic input fields.

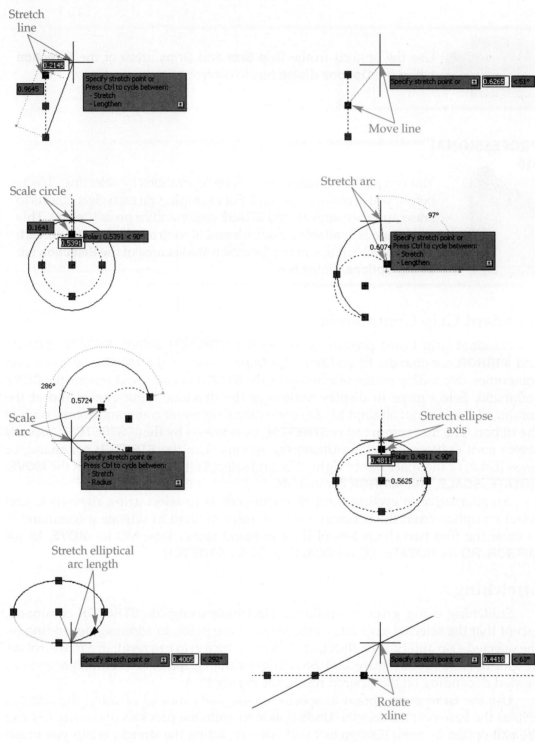

Figure 13-4.
Stretching a drawing consisting of three lines and an arc. A—Select a corner to stretch.
B—Hold down [Shift] to select multiple grips to stretch.

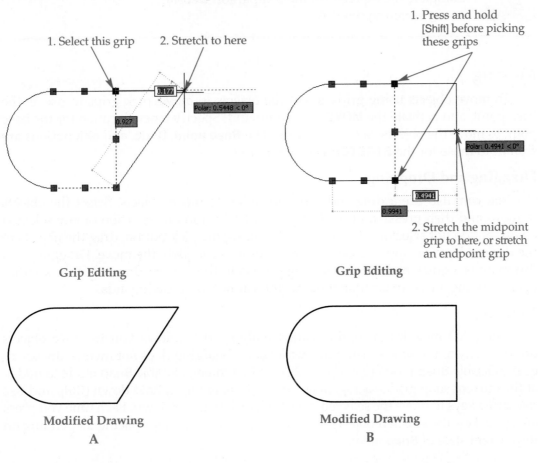

Grip Editing

Grip Editing

Modified Drawing
A

Modified Drawing
B

Figure 13-5.
Using the dimensional input feature of dynamic input with the **STRETCH** grip command.

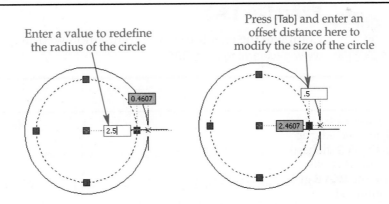

just two examples of using dynamic input with grips. You can apply similar processes to edit most objects.

PROFESSIONAL TIP

Use grid snaps, coordinate entry, polar tracking, object snaps, and object snap tracking with any grip editing command to improve accuracy. When dynamic input is active, depending on the selected grip and type of objects, pressing [Tab] activates a different set of inputs. Press [Tab] until you receive the desired input fields.

Exercise 13-1

Complete the exercise on the companion website.
www.g-wlearning.com/CAD

Moving

To move objects using grips, select the objects to move, pick grips to use as the base point, and activate the **MOVE** grip command. Specify a new location for the base point to move the objects. See **Figure 13-6**. The **Base point**, **Undo**, and **eXit** options are similar to those for the **STRETCH** grip command.

Dragging and Dropping

You can also use a drag-and-drop operation to move objects. Select the objects to move and then press and hold down the pick button on a portion of any selected object, but do not select a grip. While still holding the pick button, drag the objects to the desired location and release the pick button to complete the move. Dragging and dropping is a quick method for moving objects in the current drawing or to another open drawing, but is inaccurate because you cannot use drawing aids.

NEW

nudging: Moving objects orthogonally by selecting the objects and using the arrow keys on the keyboard.

Nudging

AutoCAD includes an option called *nudging* that allows you to move objects orthogonally to the screen using the arrow keys. Nudging does not involve the use of grips. Disable **Snap** mode to nudge at 2-pixel increments. Enable **Snap** mode to nudge at the current snap grid spacing. Select the objects to move, hold down [Ctrl], and use the arrow keys to move the selected objects right, left, up, or down. Each time you press an arrow key, the selected objects move two pixels or one snap spacing, depending on the current state of **Snap** mode.

Exercise 13-2

Complete the exercise on the companion website.
www.g-wlearning.com/CAD

Figure 13-6.
The selected grip is the default base point when you use the **MOVE** grip command.

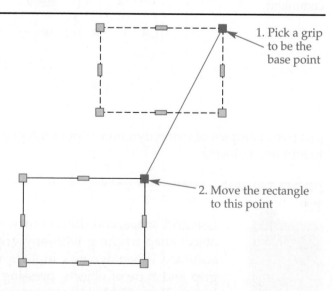

1. Pick a grip to be the base point

2. Move the rectangle to this point

Rotating

To rotate objects using grips, select the objects to rotate, pick a grip to use as the base point, and activate the **ROTATE** grip command. Specify a rotation angle to rotate the objects. The **Base point**, **Undo**, and **eXit** options are similar to those for the **STRETCH** grip command.

Use the **Reference** option to specify a new angle in relation to an existing angle. The reference angle is often the current angle of the objects. If you know the value of the current angle, enter the value at the prompt. Otherwise, pick two points to identify the angle. Enter a value for the new angle or pick a point. **Figure 13-7** shows **ROTATE** grip command options.

Exercise 13-3

Complete the exercise on the companion website.
www.g-wlearning.com/CAD

Scaling

To scale objects using grips, select the objects to scale, pick a grip to use as the base point, and activate the **SCALE** grip command. Enter a scale factor or pick a point to increase or decrease the size of the objects. The **Base point**, **Undo**, and **eXit** options are similar to those for the **STRETCH** grip command.

Use the **Reference** option to specify a new size in relation to an existing size. The reference size is often the current length, width, or height of the objects. If you know the current size, enter the value at the prompt. Otherwise, pick two points to identify the size. Enter a value for the new size or pick a point. **Figure 13-8** shows **SCALE** grip command options.

Exercise 13-4

Complete the exercise on the companion website.
www.g-wlearning.com/CAD

Figure 13-7.
Using the **ROTATE** grip command with the default rotation angle option and with the **Reference** option.

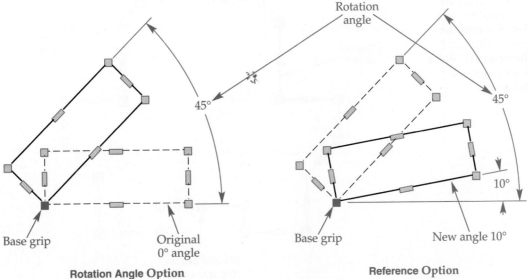

Figure 13-8.
When using the **SCALE** command with grips, enter a scale factor or use the **Reference** option.

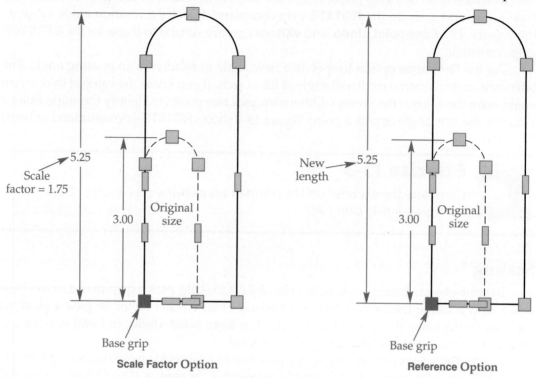

Scale factor = 1.75

5.25

3.00

Original size

Base grip

Scale Factor Option

New length

5.25

3.00

Original size

Base grip

Reference Option

Mirroring

To mirror objects using grips, select the objects to mirror, pick a grip to use as the first point of the mirror line, and activate the **MIRROR** grip command. Then pick another grip or any point on-screen to locate the second point of the mirror line. See **Figure 13-9.** Unlike the non-grip **MIRROR** command, the grip version does not give you the immediate option to delete the old objects. Old objects are deleted automatically.

Figure 13-9.
When you use grips to access the **MIRROR** command, the selected grip becomes the first point of the mirror line, and the original object is automatically deleted.

Selected grip, the first point of the mirror line

Original object

Mirrored object

Second point of the mirror line

To keep the original objects, use the **Copy** option of the **MIRROR** grip command. The **Base point**, **Undo**, and **eXit** options are similar to those for the **STRETCH** command.

Exercise 13-5

Complete the exercise on the companion website.
www.g-wlearning.com/CAD

Copying

Each standard grip editing command includes the **Copy** option. The effect of using the **Copy** option depends on the selected objects, grip, and command. The original selected objects remain unchanged, and the copy stretches when the **STRETCH** grip command is active, rotates when the **ROTATE** grip command is active, or scales when the **SCALE** grip command is active. The **Copy** option of the **MOVE** grip command is the true copy operation, allowing you to copy from the selected grip. The selected grip acts as the copy base point. Create as many copies of the selected object as needed, and then exit the command.

Exercise 13-6

Complete the exercise on the companion website.
www.g-wlearning.com/CAD

Object-Specific Grip Options

Several objects have specialized grips or additional grip options. Context-sensitive commands are available at the endpoint grips of a line or arc, the midpoint grip of an arc, and the endpoint grip of an elliptical arc when the endpoint is at a quadrant. You can access and apply the same context-sensitive object-specific grip commands in three different ways. **Figure 13-10** illustrates each technique. **Figure 13-11** illustrates

Figure 13-10.
Apply one of the following methods to access context-sensitive grip commands at the endpoint grips of a line or arc, the midpoint grip of an arc, and the endpoint grip of an elliptical arc when the endpoint is at a quadrant. A—Hover over an unselected grip. B—Pick a grip and press [Ctrl] to cycle through options. C—Pick a grip and then right-click to display options.

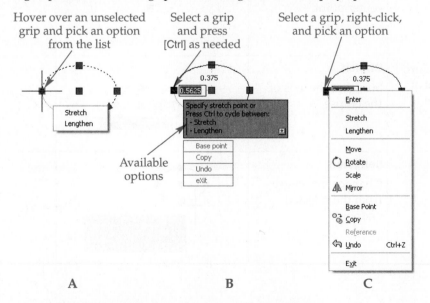

A   B   C

Figure 13-11.

Examples of options available for modifying lines, arcs, and elliptical arcs using grips.

Grip	Option	Process	Result
Line endpoint	**Stretch** Stretches the line at the selected endpoint	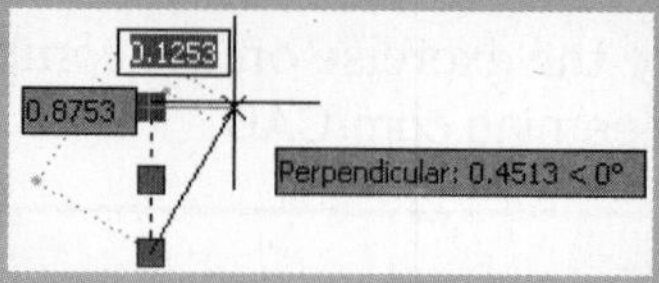	
	Lengthen Changes the length from the selected endpoint	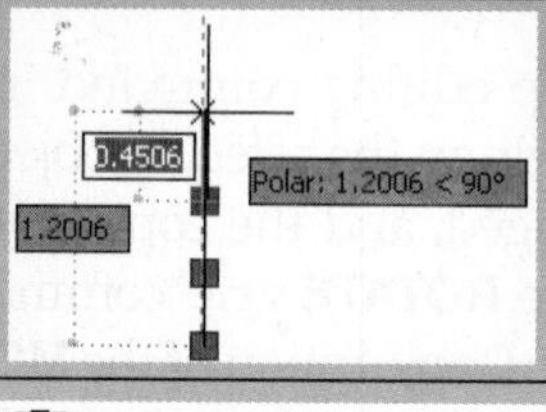	
Arc endpoint	**Stretch** Stretches the arc at the selected endpoint	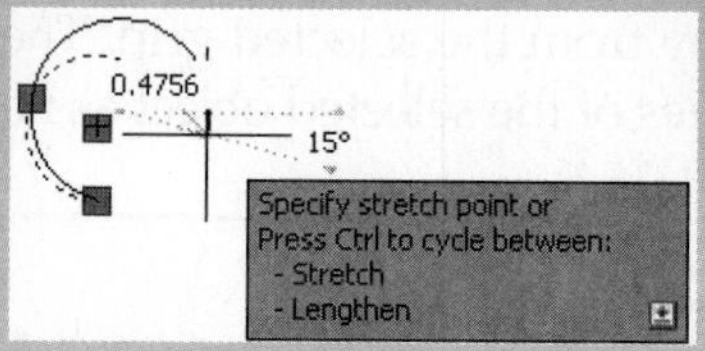	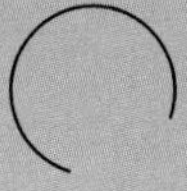
	Lengthen Changes the length from the selected endpoint	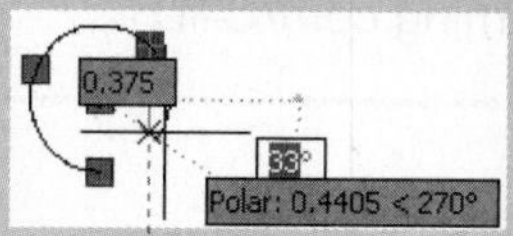	
Arc midpoint	**Stretch** Stretches the radius and center point from the endpoints	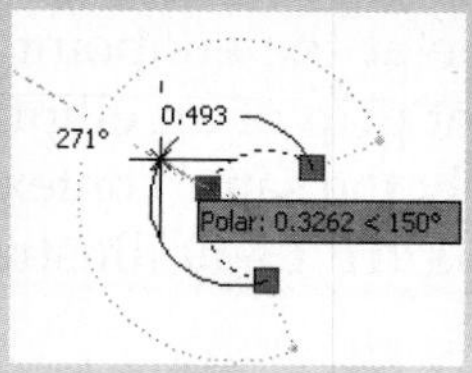	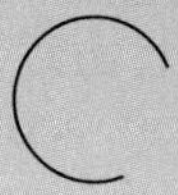
	Radius Changes the radius and location of the endpoints from the center point	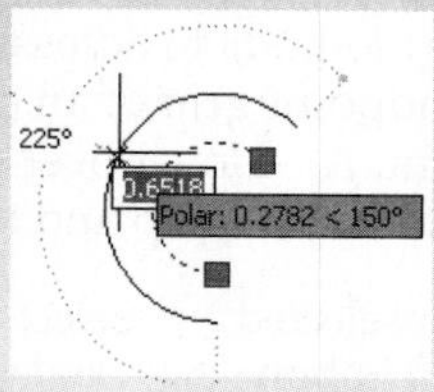	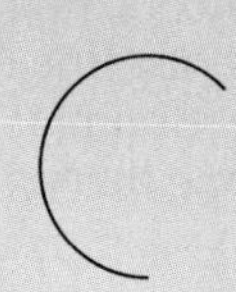
Elliptical arc endpoint at quadrant	**Stretch** Stretches the axis of the ellipse	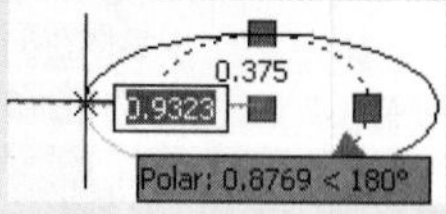	
	Lengthen Changes the length from the selected endpoint	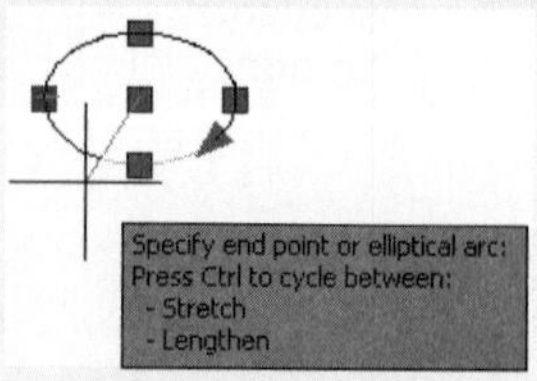	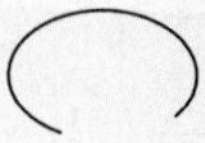

and explains the process of using object-specific grip commands to edit lines, arcs, and elliptical arcs. Use dynamic input when possible to complete the operation.

Select an mtext object to display a standard grip at the justification point, grips for modifying the mtext boundary width and height, and grips for adjusting columns. See **Figure 13-12.** Using grips is an alternative to re-entering the text editor to make changes to mtext layout, as described in Chapters 9 and 10. This textbook explains specialized grips related to other objects when applicable.

Figure 13-12.
Use grips to apply standard editing commands to the justification base point, and to adjust the mtext boundary or columns.

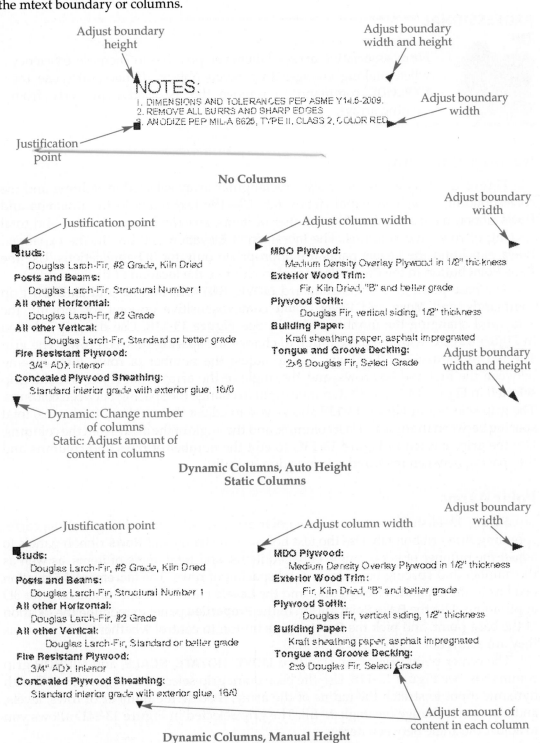

ARRAYEDIT

Ribbon
**Home
> Modify**

Edit Array

Type

ARRAYEDIT

Editing Associative Arrays

AutoCAD creates associative rectangular, polar, and path arrays by default. To create a nonassociative array, select **No** at the **ASsociative** option before finalizing an array, as described in Chapter 12. An associative array is a single array object. The major advantage of using associative arrays is the ability to modify the parameters of the array, and to edit or replace the source objects without recreating the array. The easiest way to edit an associative array is to select the array to display grips and the context-sensitive **Array** ribbon tab.

PROFESSIONAL TIP

Form associative arrays whenever possible to increase efficiency when making changes to arrayed objects. If necessary, use the **EXPLODE** command to remove the associative property from existing arrays.

Rectangular Array

Figure 13-13 shows an associative rectangular array selected for editing and the corresponding **Array** contextual ribbon tab. Use the text boxes in the **Columns** and **Rows** ribbon panels to adjust the number of items, spacing between items, and total spacing of rows and columns. The **Incremental Elevation** text box in the expanded **Rows** panel and the **Levels** panel are appropriate only for 3D applications. Use the **Base Point** button in the **Properties** panel to redefine the location of the base point.

The base point grip offers standard **MOVE**, **ROTATE**, **SCALE**, and **MIRROR** grip commands, and **Move** and **Level Count** context-sensitive options for moving the array and changing the number of levels. See **Figure 13-13B**. Use the grip selected in **Figure 13-13C** with dynamic input, to change the spacing between rows. The grip selected in **Figure 13-13D** allows you to adjust the number of rows, total spacing between the first and last rows, and the angle of the array from rows. Use the grip selected in **Figure 13-13E** with dynamic input to change the spacing between columns. The grip selected in **Figure 13-13F** allows you to adjust the number of columns, total spacing between the first and last columns, and the angle of the array from the columns. Use the grip selected in **Figure 13-13G** to edit the number of rows and columns and the spacing between rows and columns dynamically.

Polar Array

Figure 13-14 shows an associative polar array selected for editing and the corresponding **Array** ribbon tab. Use the text boxes in the **Items** and **Rows** ribbon panels to adjust the number of items, angle between items, and total angle of items, as well as the number and spacing between or total spacing of rows. The **Incremental Elevation** text box in the expanded **Rows** panel and the **Levels** panel are appropriate only for 3D applications. Use the **Base Point** button in the **Properties** panel to redefine the location of the base point, and pick the **Rotate items** button to control whether items rotate as they are arrayed.

The center point grip offers standard **MOVE**, **ROTATE**, **SCALE**, and **MIRROR** grip commands. See **Figure 13-14B**. Use the base point grip selected in **Figure 13-14C** with dynamic input to stretch the radius of the array; modify the number of rows, levels, and items; and change the angle to fill. The grip selected in **Figure 13-14D** allows you to adjust the angle between items.

Figure 13-13.
Editing an associative rectangular array of a pattern of slots.

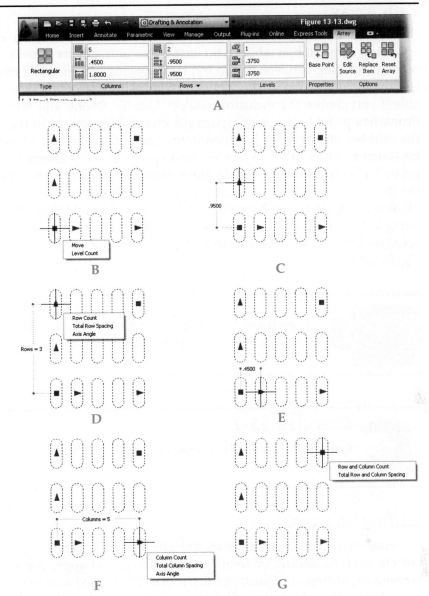

Figure 13-14.
Editing an associative polar array of teeth on a spur gear.

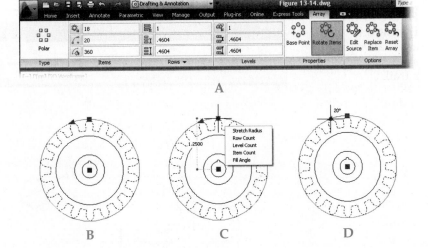

Path Array

Figure 13-15 shows an associative path array selected for editing and the corresponding **Array** ribbon tab. Use the text boxes in the **Items** panel to adjust the number of items, spacing between items, and total spacing of items. The **Item Spacing** text box is available if you choose the divide method. The **Total Item Distance** text box is available if you choose the measure method. Use the **Divide** and **Measure** buttons in the **Properties** panel to toggle the spacing method. The options in the **Rows** panel control the number and spacing between items and total spacing of rows. The **Incremental Elevation** text box in the expanded **Rows** panel and the **Levels** panel are only appropriate for 3D applications. Use the **Base Point** button in the **Properties** panel to redefine the location of the base point, and pick the **Align Items** button to control whether items align with the path. The **Z direction** button is used to maintain the orientation of items along a 3D path. The base point grip offers **Move**, **Row Count**, and **Level Count** context-sensitive options to move the array and change the number of rows and levels. See **Figure 13-15B**.

When using the **Move** option to edit a path array, pick the **Continue** button when you see the alert box to continue with the move operation.

 Exercise 13-7

Complete the exercise on the companion website.
www.g-wlearning.com/CAD

Editing Source Objects

Ribbon

Array
> Options

Edit Source

Associative rectangular, polar, and path arrays include an option to edit the source objects, such as adding or removing geometry, and apply the changes to the array without exploding or recreating the pattern. The easiest way to edit the source objects of an associative array is to select the array to display grips and the context-sensitive **Array** ribbon tab. Then pick the **Edit Source** button from the **Options** panel on the

Figure 13-15.
Editing an associative path array of trees along a trail.

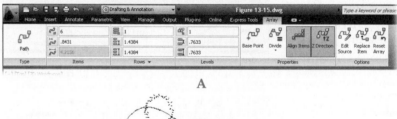

A

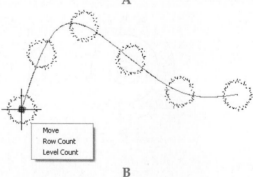

B

ribbon. Pick any item in the selected array. You do not have to pick the original source objects. Pick the **OK** button when you see the alert box to enter the array editing state.

Use drawing and editing commands to modify the selected item. **Figure 13-16** shows a basic change made using the **STRETCH** command to modify the items in a rectangular array. When you finish editing, pick the **Save Changes** button in the **Edit Array** panel on the **Home** ribbon tab to save changes and exit the array editing state, or pick the **Discard Changes** button to exit without saving.

 The layer you assign to source objects applies to all items in the array. When working with an associative array, you must edit and then assign a different layer to the source objects in order to change the layer on which the array is drawn.

Replacing Items

Associative rectangular, polar, and path arrays also offer the option to replace specific items with different objects. The easiest way to replace items in an associative array is to select the array to display grips and the context-sensitive **Array** ribbon tab. Then pick the **Replace Item** button from the **Options** panel of the **Array** ribbon tab. Pick objects not related to the array to use as the replacement for items in the array. Right-click or press [Enter] or the space bar to continue. A rubberband line attaches to the base point of the array for reference. Select an appropriate base point for the replacement objects. Use the **centroid** and **Key Point** options if necessary.

Next, pick items in the array to replace with the selected replacement objects, and then right-click or press [Enter] or the space bar to continue. An alternative is to select the **Source Objects** option to replace all items in the array. To finalize the array and exit the command, press [Enter], the space bar, or [Esc], choose the **eXit** option, or right-click and select **Enter**. AutoCAD updates the array and erases the replacement objects. **Figure 13-17** shows an example of replacing specific items in a path array.

 The easiest way to return an associative array to its original state, without replacements, is to select the array to display grips and the context-sensitive **Array** ribbon tab. Then pick the **Reset Array** button from the **Options** panel. However, you cannot reset an array to the original state if you use the **Source Objects** option and then exit array editing.

Figure 13-16.
Editing source objects to modify an associative array. This example shows changing the size of slots in a rectangular array.

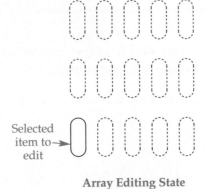

Selected item to edit

Array Editing State   Updated Array

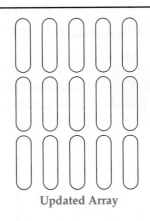

Figure 13-17.
Replacing items in an associative array. This example shows replacing specific trees along a trail in a path array.

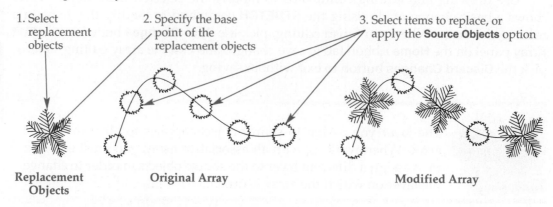

1. Select replacement objects

2. Specify the base point of the replacement objects

3. Select items to replace, or apply the **Source Objects** option

Replacement Objects

Original Array

Modified Array

Exercise 13-8

Complete the exercise on the companion website.
www.g-wlearning.com/CAD

Object Properties

Every object has specific properties. Properties include geometry characteristics, such as the coordinates of the endpoints of a line in X,Y,Z space, the diameter of a circle, or the area of a rectangle. Layer is another property associated with all objects. The layer on which you draw an object defines other properties, including color, linetype, and lineweight. Most objects also include object-specific properties. For example, multiline text has a variety of text properties, and an associative rectangular array has column, row, and other properties that define the array.

AutoCAD provides many options for adjusting object properties, depending on the object and the properties assigned to it. One method is to use grip editing or editing commands, such as **STRETCH** or **ROTATE**, to make changes. Another method is to adjust layer characteristics using layer tools. You can also use the multiline text editor to adjust existing multiline text properties. A different technique to view and make changes to the properties of any object is to use the **Quick Properties** panel or the **Properties** palette. These tools are especially effective for modifying a particular property or set of properties for multiple objects at once.

PROFESSIONAL TIP

View object, color, layer, and linetypes properties by hovering over an object. This is a quick way to reference basic object information. See **Figure 13-18**.

Figure 13-18.
Hover over an object to view its color, layer, and linetype properties.

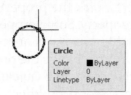

Circle
Color   ■ ByLayer
Layer   0
Linetype   ByLayer

Using the Quick Properties Panel

The **Quick Properties** panel, shown in **Figure 13-19**, appears by default when you double-click on certain objects. For example, the **Quick Properties** panel appears by default when you double-click on a line, but the text editor opens when you double-click on an mtext or text object. You can also set the **Quick Properties** panel to display with a single-click on an object. A quick way to enable or disable this function is to pick the **Quick Properties** button on the status bar or type [Ctrl]+[Shift]+[P].

The **Quick Properties** panel floats by default above and to the right of the crosshairs. The drop-down list at the top of the **Quick Properties** panel identifies the selected object. Properties associated with the selected object are displayed below the drop-down list in rows. For example, if you pick a circle, the **Quick Properties** panel lists rows of circle properties.

The **Quick Properties** panel lists common properties associated with the selected objects by default. You should recognize most of the properties included in the **Quick Properties** panel. You can pick multiple objects by using a window to select them, then right-clicking and selecting **Quick Properties**, or by double-clicking sequentially on more than one object. When you pick multiple objects, use the **Quick Properties** panel to modify all of the objects, or pick a specific object type from the drop-down list to modify. See **Figure 13-20**. Select **All (n)** to change the properties of all selected objects. Only properties shared by all selected objects appear when you choose **All (n)**. Select the appropriate object type to modify a single type of object.

Figure 13-19.
Use the **Quick Properties** panel to display and modify certain object properties, such as the basic properties of the selected line shown.

Type of object selected   Customize button   Turn off **Quick Properties** panel

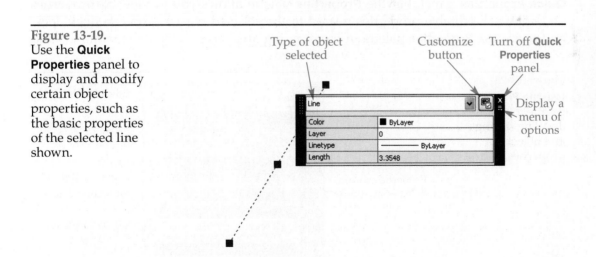

Display a menu of options

Figure 13-20.
The **Quick Properties** panel with three objects selected. You can edit the objects individually or select **All (3)** to edit all of the objects together.

Total number of objects selected

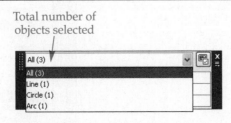

To change a property, pick the property or current value. The way you change a value depends on the property. Some properties, such as the **Radius** property of an arc or circle or the **Text height** property of an mtext or text object, display a text box. Enter a new value in the text box to change the property. Most text boxes display a calculator icon on the right side that opens the **QuickCalc** tool for calculating values. Chapter 15 covers using **QuickCalc**. Other properties, such as the **Layer** property, display a drop-down list of selections. A pick button is available for geometric properties, such as the **Center X** and **Center Y** properties of a circle. Select the pick button to specify a new coordinate. Choose an **...** (ellipsis) button to open a dialog box related to the property. Press [Esc] or pick the **Close** button in the upper-right corner of the panel to hide the **Quick Properties** panel.

Right-click on a **Quick Properties** side bar or pick the **Options** button on the **Quick Properties** panel to access options for adjusting the display and function of the **Quick Properties** panel. The **Quick Properties** tab of the **Drafting Settings** dialog box includes many of the same settings, as well as additional options.

Exercise 13-9

Complete the exercise on the companion website.
www.g-wlearning.com/CAD

Using the Properties Palette

The **Properties** palette, shown in **Figure 13-21**, provides the same function as the **Quick Properties** panel, but the **Properties** palette allows you to view *all* properties and adjust all editable properties related to the selected objects. You can dock, lock, and resize the **Properties** palette in the drawing area. You can access commands and

PROPERTIES

Ribbon
View
> Palettes

Properties
Home
> Properties

Properties

Type
PROPERTIES
PROPS
CH
MO
[Ctrl]+[1]

Figure 13-21.
Use the **Properties** palette to modify drawing settings and object properties.

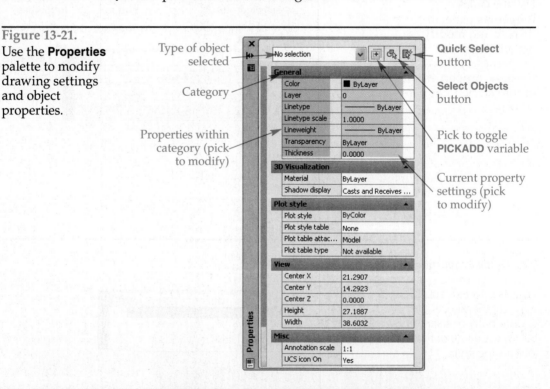

Type of object selected

Category

Properties within category (pick to modify)

Quick Select button

Select Objects button

Pick to toggle **PICKADD** variable

Current property settings (pick to modify)

continue to work while displaying the **Properties** palette. To close the palette, pick the **X** in the top-left corner, select **Close** from the shortcut menu, or press [Ctrl]+[1].

If you already selected an object, you can access the **Properties** palette by right-clicking and selecting **Properties**. You can also double-click some objects to select them and open the **Properties** palette automatically.

Working with properties in the **Properties** palette is similar to working with properties in the **Quick Properties** panel. The drop-down list at the top of the **Properties** palette identifies the selected object. Properties associated with the selected object display below the drop-down list in categories and property rows. The categories and rows update to display properties associated with your selections. If you do not select an object, the **General**, **3D Visualization**, **Plot style**, **View**, and **Misc** categories list the current drawing settings.

When you pick multiple objects, use the **Properties** palette to modify all of the objects, or pick a specific object type from the drop-down list to modify. See **Figure 13-22**. Select **All (*n*)** to change the properties of all selected objects. Only properties shared by all selected objects appear when you choose **All (*n*)**.

You should recognize most of the properties listed in the **Properties** palette. Do not adjust properties that you do not recognize. For example, the **3D Visualization** category and any properties related to the Z axis are for use in 3D applications.

To change a property, pick the property or current value. The way you change a value depends on the property, just as it does in the **Quick Properties** panel. Use the appropriate text box, drop-down list, or button to modify the value. After you make changes to the objects, press [Esc] to clear grips and remove the objects from the **Properties** palette. Close the **Properties** palette when you are finished.

The upper-right portion of the **Properties** palette contains three buttons. The left button toggles the value of the **PICKADD** system variable, which determines whether you need to hold down [Shift] when adding objects to a selection set. Pick the **Select Objects** button in the middle to deselect the currently selected objects and change the crosshairs to a pick box, allowing you to select other objects. Pick the **Quick Select** button on the right to access the **Quick Select** dialog box from which you can create a selection set, as described later in this chapter.

General Properties

The **General** category of the **Properties** palette allows you to modify general object properties such as color, layer, linetype, linetype scale, plot style, lineweight, transparency, and thickness. See **Figure 13-23**. The **Quick Properties** panel also lists some general properties.

Figure 13-22.
The **Properties** palette with four objects selected. You can edit the objects individually or select **All (4)** to edit all of the objects together.

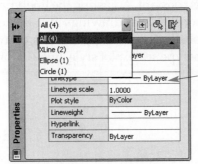

When All is selected, only properties common to all selected objects appear

Figure 13-23.
A—The **Properties** palette with a line object selected. Line objects have options in only three property categories. B—The **Properties** palette with a circle object selected.

Type of object selected

General properties

Start point and endpoint coordinates

These values cannot be directly modified, but change if endpoints are modified

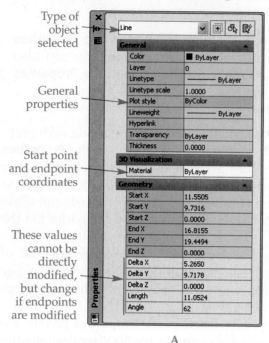

A

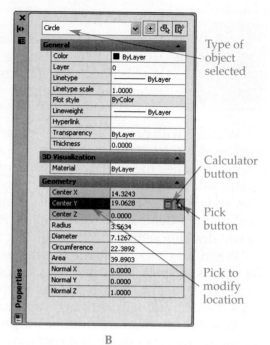

Type of object selected

Calculator button

Pick button

Pick to modify location

B

You can change the layer of a selected object by choosing a layer from the **Layer Control** drop-down list in the **Layers** panel on the **Home** ribbon tab. You can override color, linetype, lineweight, and plot style by choosing from the appropriate drop-down list in the **Properties** panel on the **Home** ribbon tab. Override transparency by selecting from the flyout and using the slider, also in the **Properties** panel on the **Home** ribbon tab.

CAUTION

absolute value:
In property settings, a value set directly instead of referenced by layer or a block. An absolute value ignores the current layer settings.

Color, linetype, lineweight, and transparency should typically be set as **ByLayer**. Changing color, linetype, lineweight, or transparency to a value other than **ByLayer** overrides logical properties, making the property an *absolute value*. Therefore, if the color of an object is set to red, for example, it appears red regardless of the color assigned to the layer on which you draw the object.

Linetype scale should usually be set globally so the linetype scale of all objects is constant. Adjusting the linetype scale of individual objects can create nonstandard drawings and make it difficult to adjust linetype scale globally. For most applications, you should not override color, linetype, linetype scale, plot style, lineweight, transparency, or thickness.

Geometry Properties

The **Geometry** category of the **Properties** palette allows you to modify object coordinates and dimensions. Refer again to **Figure 13-23**. The properties in the **Geometry** category vary depending on the selection. **Figure 13-23A** highlights the X, Y, and Z coordinates that you can use to relocate the start point and endpoint of a line. **Figure 13-23B** highlights the X, Y, and Z coordinates that you can use to relocate the center of a circle. Enter a value, select the calculator to calculate a value, or use the pick button to specify a point on-screen. The **Quick Properties** panel also lists some geometry properties.

PROFESSIONAL TIP

The **Radius**, **Diameter**, **Circumference**, and **Area** properties are especially useful for modifying the dimension of a circle. Circle properties are an example of the useful information and instinctive adjustments often available using the **Properties** palette. The **Properties** palette provides editable properties for most objects.

Exercise 13-10

Complete the exercise on the companion website.
www.g-wlearning.com/CAD

Text Properties

The **Text** category appears when you select an mtext or single-line text object. **Figure 13-24** shows text properties associated with mtext. The **Properties** palette provides a convenient way to modify a variety of text properties without re-entering the text editor. The **Properties** palette is especially effective to adjust a particular property for multiple selected text objects. For example, change the annotative setting of all text

Figure 13-24.
The **Properties** palette shows the properties of selected multiline text. The properties of single-line text are slightly different from those of mtext.

Text category

Properties specific to text objects

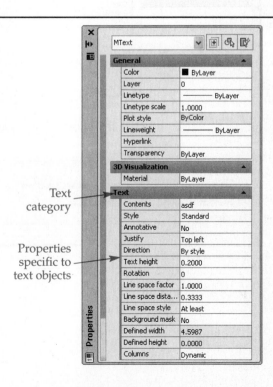

in the drawing using the **Annotative** property row, or reset the height of multiple mtext or text objects using the **Height** property row. The **Quick Properties** panel also lists some text-specific properties.

PROFESSIONAL TIP

Fields provide an effective way to display drawing information within mtext and text objects. You can use fields with many AutoCAD tools, including inquiry commands, drawing properties, attributes, and sheet sets. You can update fields when changes occur to the reference data. Chapter 10 explains using fields. **Figure 13-25** shows acquiring object properties to display in fields.

Exercises 13-11 and 13-12

Complete the exercises on the companion website.
www.g-wlearning.com/CAD

Figure 13-25.
Pick the **Object** field to add a property for a specific object to a field. Pick the **Select object** button to select the object. Properties specific to the object type appear. Select the property and format for the field.

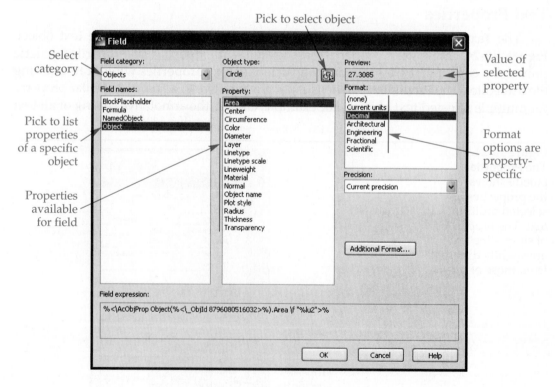

AutoCAD and Its Applications—Basics

Matching Properties

The **MATCHPROP** command allows you to match, or "paint," properties from one object to other objects. You can match properties in the same drawing or between drawings. When you first access the **MATCHPROP** command, AutoCAD prompts you for the *source object*. After you select the source object, AutoCAD displays the properties it will paint. The next prompt allows you to pick the *destination objects*.

To change the paint properties, select the **Settings** option before picking the destination objects. The **Property Settings** dialog box appears, showing the properties to be painted. See **Figure 13-26**. Properties are replaced in the destination objects if the corresponding **Property Settings** dialog box check boxes are active. For example, to paint only the layer property and text style of one text object to another text object, uncheck all boxes except the **Layer** and **Text** property check boxes. Pick the **OK** button to select destination objects.

Ribbon
Home
> Clipboard

Match Properties

Type
MATCHPROP
PAINTER
MA

MATCHPROP

source object:
When matching properties, the object with the properties you want to copy to other objects.

destination object:
When matching properties, the object that receives the properties of the source object.

Exercise 13-13

Complete the exercise on the companion website.
www.g-wlearning.com/CAD

Figure 13-26.
In the **Property Settings** dialog box for the **MATCHPROP** command, select the properties to paint to the destination objects.

Properties painted to the destination objects

Properties particular to specific objects

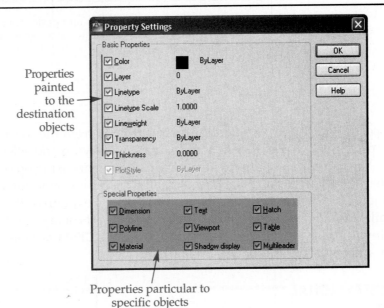

Editing between Drawings

You can edit in more than one drawing at a time and edit between open drawings. For example, you can copy objects from one drawing to another. You can also refer to a drawing to obtain information, such as a distance, while working in a different drawing.

Figure 13-27 shows two drawings, each of a different section for the same home remodel project, tiled vertically. The Windows *copy and paste* function allows you to copy objects from one drawing to another. For example, copy the rafters and exterior studs from the Proposed Entry Section A drawing to paste and reuse them in the

copy and paste: A Windows function that allows you to copy an object from one location and paste it into another.

Figure 13-27.
Tile multiple drawings to make editing between drawings easier.

Active drawing

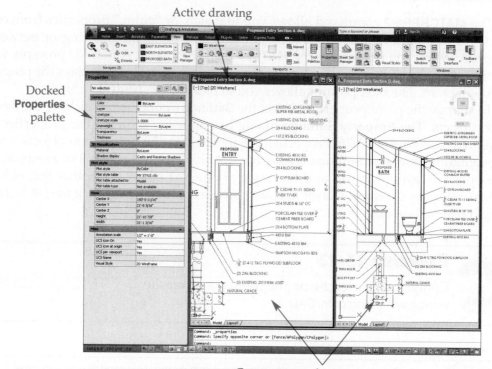

Docked
Properties
palette

Cut or copy and paste, or use the
MATCHPROP tool to work between open drawings

Proposed Bath Section B. Select the objects to copy from the source drawing and choose a copy option. Then switch to the destination drawing and select a paste option.

You can cut, copy, and paste between documents using options from the **Clipboard** panel on the **Home** ribbon tab, the **Clipboard** cascading shortcut menu, or the Windows-standard keyboard shortcuts. **Figure 13-28** briefly explains cut, copy, and paste options available when you right-click after selecting objects or after cutting or copying objects. Many of the same options are available from the **Clipboard** panel on the **Home** ribbon tab, and by typing, as shown.

**PROFESSIONAL
TIP**

You may find it more convenient to use the **MATCHPROP** command to match properties between drawings. To use the **MATCHPROP** command between drawings, select the source object from one drawing and the destination object from another.

Exercise 13-14

Complete the exercise on the companion website.
www.g-wlearning.com/CAD

Figure 13-28.
Right-click options available for cutting, copying, and pasting objects in the active drawing or between drawings.

Cutting or Copying		
Option	Keyboard Shortcut	Description
Cut	[Ctrl]+[X]	Erases selected objects from the drawing and places the objects in the Clipboard.
Copy	[Ctrl]+[C]	Copies selected objects to the Clipboard.
Copy with Base Point	[Ctrl]+[Shift]+[C]	Copies selected objects to the Clipboard using a specific base point to position the copied objects for pasting. When prompted, select a logical base point, such as a corner or center point of an object.
Pasting		
Option	Keyboard Shortcut	Description
Paste	[Ctrl]+[V]	Pastes the objects on the Clipboard to the drawing. If you used the **Copy with Base Point** option, the objects attach to the crosshairs at the specified base point.
Paste as Block	[Ctrl]+[Shift]+[V]	Pastes and "joins" all objects on the Clipboard to the drawing as a block. The pasted objects act as a single object. Blocks are covered later in this textbook. Use the **EXPLODE** tool to break up the block.
Paste to Original Coordinates		Pastes the objects on the Clipboard to the same coordinates at which they were located in the original drawing.

Add Selected

The **ADDSELECTED** command allows you to draw a new object using the properties of an existing object, without locating and selecting the object command or presetting the layer or other properties. An easy way to access the **ADDSELECTED** command is to select the object to replicate and then right-click and choose **Add Selected**. AutoCAD initiates the drawing command and assigns properties corresponding to the selected object. For example, pick a circle, right-click and choose **Add Selected**, and draw a circle as if you had accessed the **CIRCLE** command. AutoCAD applies the properties of the selected circle to the new circle, regardless of the current settings.

Type
ADDSELECTED

Select Similar

The **SELECTSIMILAR** command provides another method of creating a selection set. AutoCAD selects all objects in the drawing that match certain properties of objects

Type
SELECTSIMILAR

Figure 13-29.
Use the **Quick Select** dialog box to create a specific selection set.

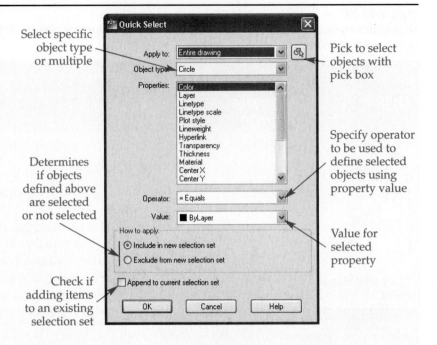

Select specific object type or multiple

Pick to select objects with pick box

Determines if objects defined above are selected or not selected

Specify operator to be used to define selected objects using property value

Value for selected property

Check if adding items to an existing selection set

you select. An easy way to access the **SELECTSIMILAR** command is to pick the object type you want to select throughout the drawing and then right-click and choose **Select Similar**. By default, all objects of the same type and layer are selected. For example, pick a polyline object drawn on a HIDDEN layer, and then right-click and choose **Select Similar** to create a selection set of all polylines drawn on the HIDDEN layer.

To specify the properties for selecting similar objects, you must type **SELECTSIMILAR** before picking objects and choose the **SEttings** option to display the **Select Similar Settings** dialog box. Use the check boxes to filter the properties that must match the objects you pick in order for other objects to select. The more boxes you check, the more properties must match in order for AutoCAD to select objects.

You can pick multiple objects with different properties to select all similar objects. For example, pick a line, arc, and spline, each assigned to a different layer, to select all lines, arcs, and splines with matching layers.

Quick Select

QSELECT

Ribbon
Home
> Utilities

Quick Select

Type
QSELECT

The **QSELECT**, or **Quick Select**, command is similar to the **SELECTSIMILAR** command, but it provides additional filters. For example, use the **SELECTSIMILAR** command to pick all circles in a drawing, but use the **Quick Select** command to pick Ø2" circles. Access the **Quick Select** command to display the **Quick Select** dialog box shown in **Figure 13-29**. The **Quick Select** dialog box provides options for specifying the exact objects to include in or exclude from the selection set.

Begin the process by selecting the **Entire drawing** option from the **Apply to:** drop-down list to have access to all object types in the drawing for creating a selection set. An alternative is to pick specific objects to create an initial filter of just the selected objects. Pick objects before accessing the **Quick Select** command, or choose the **Select Objects** button to return to the drawing window temporarily to pick objects. The **Apply to:**

drop-down list then displays **Current selection**. To return to the entire drawing format, choose the **Entire drawing** option or check **Append to current selection set** at the bottom of the dialog box.

Pick a specific object type from the **Object type:** drop-down list to create a selection set according to the object type. The **Multiple** option displays properties common to different objects in the entire drawing or the selected objects. Pick a property from the **Properties:** list to narrow the selection set. The items in the **Properties:** list vary depending on the specified object type or **Multiple** option.

The **Value:** text box or drop-down list allows you to specify a property value to narrow the selection set. Use the **Operator:** drop-down list to assign a *relative operator* to control which objects are selected according to the specified property value. For example, to select all ∅2″ circles in the drawing, select the **Entire drawing** option, a **Circle** object type, the **Diameter** property, and the **= Equals** operator, and type a value of 2.

> **relative operators:** In math, functions that determine the relationship between data items.

The **Include in new selection set** radio button creates a selection set according to the quick select settings, as previously described. The **Exclude from new selection set** radio button reverses the selection set to select all objects except those you specify. Using the previous example, all objects except ∅2″ circles would be selected. Pick the **OK** button to create the selection set.

You can also access the **Quick Select** dialog box by right-clicking in the drawing area and choosing **Quick Select...** or by picking the **Quick Select** button on the **Properties** palette.

PROFESSIONAL TIP

Creating a selection set according to specific properties can be very useful. For example, suppose you design a sheet metal part with many different size holes (circle objects) at different locations, and 20 of the holes accept 1/8″ screws. A design change occurs, and the holes must accept 3/16″ screws instead. You could select and modify each circle individually, but it is more efficient to create a selection set of the 20 circles of the same size and modify them at the same time. Use the **Quick Properties** panel or the **Properties** palette to adjust properties of objects selected using the **SELECTSIMILAR** and **Quick Select** commands. You can also use parametric tools, explained in Chapter 22, to make all the circles equal in size. Then, when you change the diameter of one circle, all circles change to the new value.

Exercise 13-15

Complete the exercise on the companion website.
www.g-wlearning.com/CAD

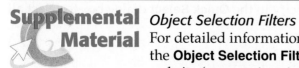

Supplemental Material

Object Selection Filters

For detailed information about selecting multiple objects using the **Object Selection Filters** dialog box, go to the companion website (www.g-wlearning.com/CAD), select this chapter, and select **Object Selection Filters**.

Express Tools

Chapter 13

Selection Express Tools

The **Modify** panel of the **Express Tools** ribbon tab includes additional editing commands. For information about the most useful selection set express tools, go to the companion website (www.g-wlearning.com/CAD), select this chapter, and select **Selection Express Tools**.

Chapter Review

Answer the following questions. Write your answers on a separate sheet of paper or complete the electronic chapter review on the companion website.
www.g-wlearning.com/CAD

1. Name the editing commands that are available using standard grips.
2. How can you select a grip command other than the default **STRETCH**?
3. What is the purpose of the **Base Point** option in the grip commands?
4. Explain the function of the **Undo** option in the grip commands.
5. Which **ROTATE** grip option would you use to rotate an object from an existing 60° angle to a new 25° angle?
6. Identify the advantages of using an associative array.
7. Briefly explain how to change the source objects in an array.
8. Explain how to replace one or more items in an associative array.
9. Describe the options for editing object properties.
10. Describe what happens when you double-click on a line.
11. Identify at least two ways to access the **Properties** palette.
12. How can you change the linetype of an object using the **Properties** palette?
13. For most applications, what value should you use for the color, linetype, and lineweight of objects?
14. Explain how to change the radius of a circle from 1.375 to 1.875 using the **Properties** palette.
15. What command changes the properties of existing objects to match the properties of a different object?
16. Briefly explain how the Windows copy and paste function works to copy an object from one drawing to another.
17. Name the paste option that joins a group of objects as a block when pasted.
18. What is the purpose of the **ADDSELECTED** command? Provide an example.
19. What is the purpose of the **SELECTSIMILAR** command? Provide an example.
20. List the information you would specify in the **Quick Select** dialog box to select all Ø6" circles in a drawing.

Drawing Problems

Start AutoCAD if it is not already started. Start a new drawing for each problem using an appropriate template of your choice. The template should include layers and text styles for drawing the given objects. Add layers and text styles as needed. Draw all objects using appropriate layers, text styles, justification, and format. Follow the specific instructions for each problem. Use only drawing and editing commands and techniques you have already learned. Do not draw dimensions. Use your own judgment and approximate dimensions when necessary.

▼ Basic

1. Draw the objects labeled A. Then use the **STRETCH** grip command to make the objects look like the objects labeled B. Save the drawing as P13-1.

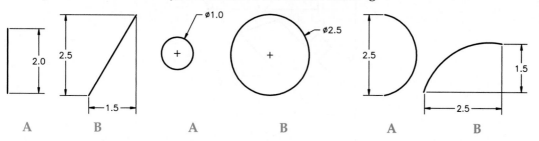

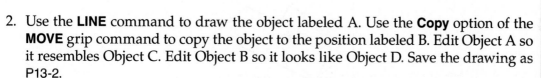

2. Use the **LINE** command to draw the object labeled A. Use the **Copy** option of the **MOVE** grip command to copy the object to the position labeled B. Edit Object A so it resembles Object C. Edit Object B so it looks like Object D. Save the drawing as P13-2.

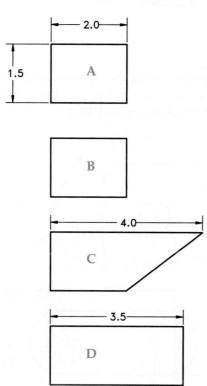

3. Use the **LINE** command to draw the view labeled A. Copy the object, without rotating it, to a position below, as indicated by the dashed lines. Rotate the object 45°. Copy the rotated object labeled B to a position below, as indicated by the dashed lines. Use the **Reference** option to rotate the object labeled C to 25°, as shown. Save the drawing as P13-3.

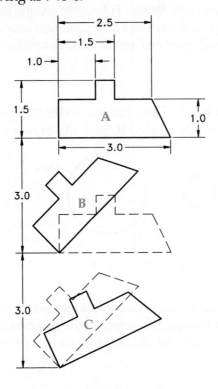

▼ **Intermediate**

4. Draw the individual objects (vertical line, horizontal line, circle, arc, and three-line shape) in A using the dimensions given. Use these objects and grips to create the view shown in B. Save the drawing as P13-4.

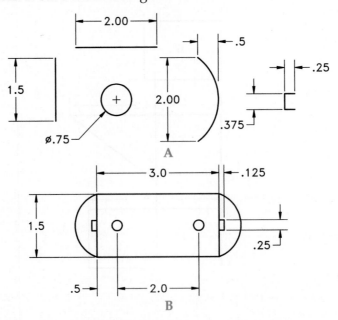

5. Open P13-4 and save the file as P13-5. The P13-5 file should be active. Copy the view two times to positions B and C. Use the **SCALE** grip command to scale the view in position B to 50% of its original size. Use the **Reference** option of the **SCALE** grip command to enlarge the view in position C from the existing 3.0 length to 4.5, as shown in C. Resave the drawing.

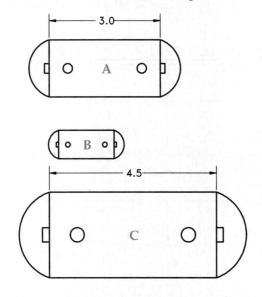

6. Draw the portion of the gasket shown in A. Use an associative rectangular array to pattern the ∅.25 holes. Use the **MIRROR** grip command to complete the gasket as shown in B. Save the drawing as P13-6.

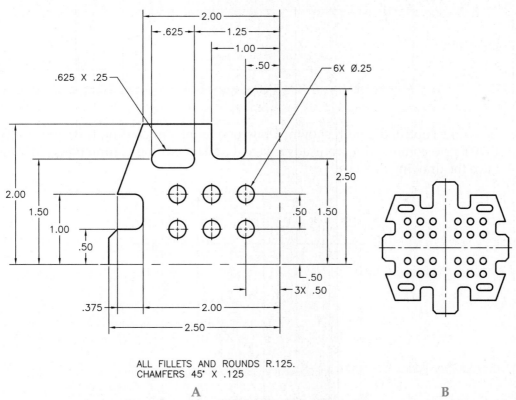

ALL FILLETS AND ROUNDS R.125.
CHAMFERS 45° X .125

A   B

7. Open P13-6 and save the file as P13-7. The P13-7 file should be active. Use the **Properties** palette to change the diameters of the circles from .25 to .125. Change the layer assigned to the slots to a layer that uses the PHANTOM linetype. Be sure the linetype scale allows the linetypes to display correctly. Resave the drawing.

8. Draw an assembly view similar to the view shown within the boundaries of the given dimensions. All other dimensions are flexible. Save the drawing as P13-8.

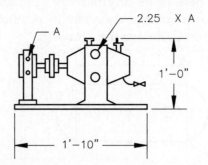

9. Draw the half of the gasket shown. Mirror the drawing to complete the other half of the gasket. Save the drawing as P13-9.

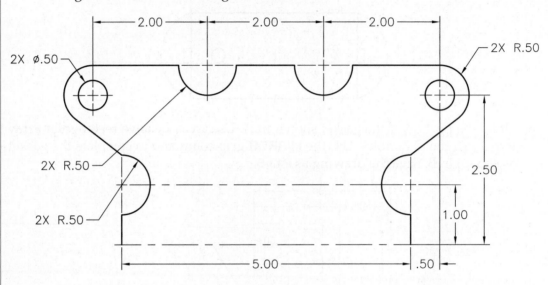

10. Draw the control diagram shown. Draw one branch (including text) and use the **COPY** grip command to your advantage. Use text editing commands as needed. Save the drawing as P13-10.

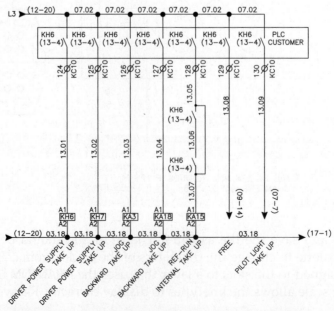

(Design and drawing by EC Company, Portland, Oregon)

▼ Advanced

11. Draw the folded and flat pattern views of the sheet metal bracket shown. The part material is 18-gauge steel. Save the drawing as P13-11.

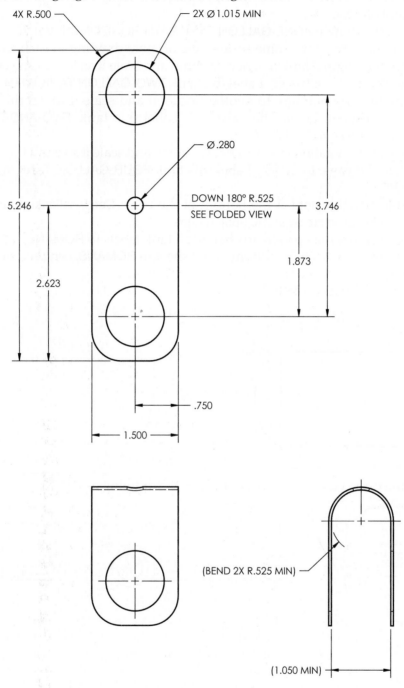

12. Draw a tank similar to the one shown within the boundaries of the given dimensions. All other dimensions are flexible. After drawing the tank, create a page for a vendor catalog, as follows:
    - All labels should be ROMAND text, centered directly below the view. Use a text height of .125".
    - Label the drawing ONE-GALLON TANK WITH HORIZONTAL VALVE.
    - Keep the valve at the same scale as the original drawing in each copy.
    - Copy the original tank to a new location and scale it so it is 2 times its original size. Rotate the valve 45°. Label this tank TWO-GALLON TANK WITH 45° VALVE.
    - Copy the original tank to another location and scale it to 2.5 times the size of the original. Rotate the valve 90°. Label this tank TWO-AND-ONE-HALF-GALLON TANK WITH 90° VALVE.
    - Copy the two-gallon tank to a new position and scale it so it is 2 times this size. Rotate the valve to 22°30". Label this tank FOUR-GALLON TANK WITH 22°30" VALVE.
    - Left-justify this note at the bottom of the page: Combinations of tank size and valve orientation are available upon request.
    - Use the **Properties** palette to change all tank labels to ROMANC, .25" high.
    - Change the note at the bottom of the sheet to ROMANS, centered on the sheet, using uppercase letters.
    - Save the drawing as P13-12.

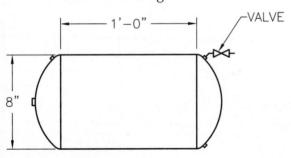

13. Draw the three views of a sports car. Save the drawing as P13-13.

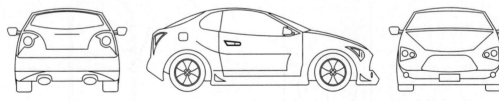

14. Draw the views of the sailboat shown. Save the drawing as P13-14.

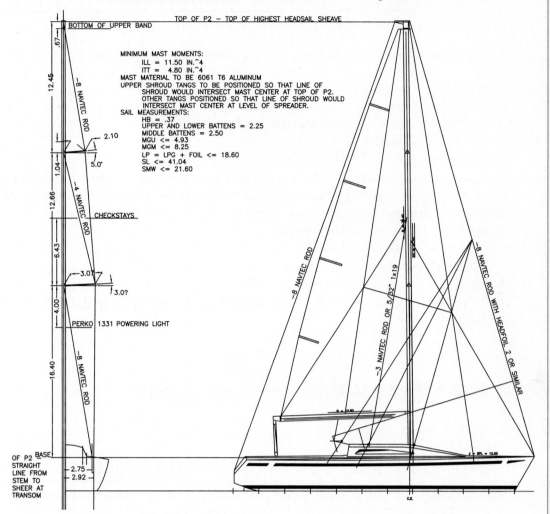

15. Create a dimensioned 2D sketch of a patio or deck plan. Include an outdoor kitchen with a grill, single-burner cooktop, and refrigerator. Add ample seating areas, a table with chairs, and a hot tub. Use dimensions based on your experience, research, and measurements. Start a new drawing from scratch or use an architectural template of your choice. Draw the patio or deck from your sketch. Save the drawing as P13-15.

AutoCAD Certified Associate Exam Practice

Answer the following questions. Write your answers on a separate sheet of paper.

1. Which of the following operations can be performed using grips? *Select all that apply.*
   A. exploding a polygon into individual lines
   B. reflecting an object to create a mirror image
   C. reversing the order of point calculation for an object
   D. rotating an object by 17°
   E. scaling an object to half its current size

2. If you do not use the **Base point** option of a grip command, which point does AutoCAD automatically select as the base point for the grip editing operation? *Select the one item that best answers the question.*
   A. the lower-left corner of the object
   B. the origin (0,0,0)
   C. a point you specify
   D. the selected grip

3. Which of the following relative operators are available for filtering data? *Select all that apply.*
   A. *
   B. /
   C. =
   D. <
   E. >
   F. >=

AutoCAD Certified Professional Exam Practice

Follow the instructions in each problem. Write your answers on a separate sheet of paper.

1. **Navigate to this chapter on the companion website and open CPE-13grips.dwg.** Use grips to rotate the metal plate 23.5°, as shown. Select the appropriate grip or use the **Base point** option to select the base point for the rotation. What are the coordinates of the center of the hole?

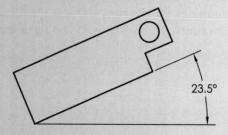

2. **Navigate to this chapter on the companion website and open CPE-13select.dwg.** Enter the **SELECTSIMILAR** command using the command line or dynamic input and specify the **SEttings** option. Disable all of the **Similar Based On** check boxes except **Color**, **Layer**, and **Name**. Select the port in the lower-left corner of the connector system and press [Enter]. How many ports are selected based on your settings?

Chapter 14

Polyline and Spline Editing Tools

Learning Objectives

After completing this chapter, you will be able to:

✓ Edit polylines with the **PEDIT** command.
✓ Use context-sensitive polyline grip commands.
✓ Create polyline boundaries.
✓ Edit splines with the **SPLINEDIT** command.
✓ Use context-sensitive spline grip commands.

You can modify polyline and spline objects using standard editing commands such as **ERASE**, **STRETCH**, and **SCALE**, but AutoCAD also provides specific commands to edit polylines and splines. Use the **PEDIT** command or polyline grip commands to modify polylines. Use the **SPLINEDIT** command or spline grip commands to modify splines. This chapter also describes how to create polyline boundaries and explores additional options for converting polylines and splines.

Using the PEDIT Command

Access the **PEDIT** command and select a polyline to edit, or choose the **Multiple** option to select multiple polylines to edit. To select a wide polyline, pick the edge of a polyline segment rather than the center. Choose an option to activate the appropriate editing function.

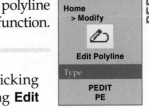

Ribbon
Home
> Modify
Edit Polyline

Type
PEDIT
PE

PEDIT

You can also access **PEDIT** command options by double-clicking on a polyline, or by right-clicking on a polyline and choosing **Edit Polyline** from the **Polyline** cascading menu.

PROFESSIONAL TIP

Select the **Undo** option immediately after performing an unwanted edit to restore previous objects without leaving the command. Use the **Undo** option more than once to step back through each operation.

Converting Objects to a Polyline

You can use the **PEDIT** command to convert a line, arc, or spline to a polyline. Access the **PEDIT** command and select the object to convert. A prompt asks if you want to turn the object into a polyline. Select the **Yes** option to make the conversion and continue using the **PEDIT** command.

Opening and Closing a Polyline

Use the **Open** option to open a closed polyline and the **Close** option to close an open polyline. See Figure 14-1. The **Close** option is available if you originally closed the polyline by drawing the final segment manually, or if you have used the **Open** option of the **PEDIT** command. The **Open** option is available if you have used the **Close** option of the **PLINE** or **PEDIT** command.

Joining Polylines

The **Join** option of the **PEDIT** command provides the same function as the **JOIN** command, allowing you to create a single polyline object from connected polylines or from a polyline connected to lines and arcs. As when you use the **JOIN** command, the objects to be joined must share a common endpoint and cannot overlap or have gaps between segments. Choose the **Join** option and pick the objects to join. You can include the original polyline in the selection set, but it is not necessary. See Figure 14-2.

PROFESSIONAL TIP

After you join objects into a continuous polyline, use the **Close** option to close the polyline if necessary.

Figure 14-1.
Open and closed polylines.

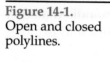

Open Polyline   Closed Polyline

Figure 14-2.
Joining a polyline to other connected lines and an arc.

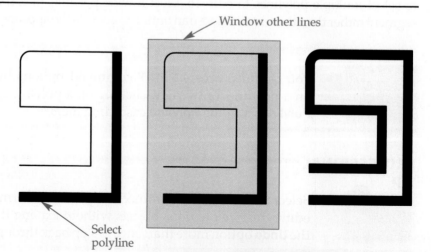

Window other lines

Select polyline

Changing Polyline Width

Choose the **Width** option of the **PEDIT** command to specify a new width to assign to the polyline. See **Figure 14-3**. The width of the original polyline can be constant, or it can vary, but all segments change to the new, constant width.

Exercise 14-1

Complete the exercise on the companion website.
www.g-wlearning.com/CAD

Editing a Polyline Vertex

The **Edit vertex** option of the **PEDIT** command allows you to perform several operations at a *polyline vertex*. When you enter the **Edit vertex** option, an "X" marker appears on-screen at the first polyline vertex. The **Edit vertex** option is more difficult to use than other editing commands. It includes **Break**, **Insert**, **Move**, and **Straighten** functions that you can perform more easily using context-sensitive polyline grips or standard editing commands such as **BREAK** and **STRETCH**. However, the **Tangent** and **Width** functions are unique to the **Edit vertex** option and are sometimes useful. The **Tangent** option is most appropriate when used with the **Fit** option, as explained later in this chapter.

polyline vertex:
The point at which two polyline segments meet.

If an **Edit vertex** option does not appear to take effect, use the **Regen** option to regenerate the polyline. Use the **eXit** option to return to the **PEDIT** prompt.

Use the **Width** function to change the starting and ending widths of a polyline segment. Enter the **Edit vertex** option and use the **Next** and **Previous** options to move the "X" marker to the first vertex where you want to change the polyline width. Activate the **Width** function and specify the starting and ending width of the polyline segment. See **Figure 14-4**.

Figure 14-3.
Changing the width of a polyline.

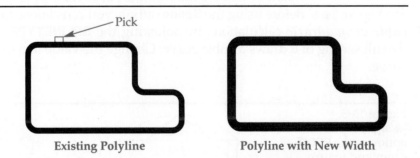

Existing Polyline   Polyline with New Width

Figure 14-4.
Changing the width of a polyline segment with the **Width** vertex editing function. Use the **Regen** option to display the change if the width does not appear to change.

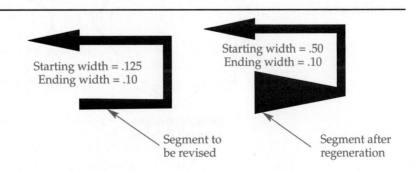

Starting width = .125
Ending width = .10

Starting width = .50
Ending width = .10

Segment to be revised   Segment after regeneration

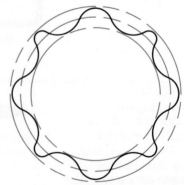

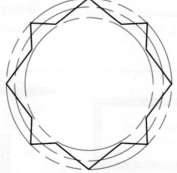

Fit Option

curve fitting: Converting a polyline into a series of smooth curves.

The **Fit** option of the **PEDIT** command uses *curve fitting* to convert straight polyline segments to a series of smooth curves. For example, you could use curve fitting to draw lines on a graph, add an underlayment symbol to a structural section, or construct a freeform shape by picking accurate points and drawing straight polyline segments. Then use the **Fit** option to smooth the segments. The **Fit** option creates a *fit curve* by constructing pairs of arcs that pass through control points. You can use the vertices of the polyline as the control points or specify different control points.

fit curve: A curve that passes through all of its fit points.

Prior to curve fitting, you can use the **Tangent** function of the **Edit vertex** option to assign each vertex a tangent direction. AutoCAD then fits the curve based on the preset tangent directions. Specifying tangent directions is optional and provides a way to edit vertices when the **Fit** option of the **PEDIT** command does not produce the best results. Enter the **PEDIT** command, select the **Edit vertex** option, and use the **Next** and **Previous** options to move the "X" marker to the first vertex to be changed. Choose the **Tangent** option and specify a tangent direction in degrees or pick a point in the expected direction. An arrow at the vertex indicates the direction. Continue to move the marker to other vertices and use the **Tangent** option as necessary.

When all of the vertices you want to change include a specified tangent direction, enter the **Fit** option to create the curve. You can also enter the **PEDIT** command, select a polyline, and enter the **Fit** option without adjusting tangencies. **Figure 14-5** shows a polyline formed into a smooth curve using the **Fit** option. If the resulting curve is not correct, use the **Edit vertex** option to make changes as necessary.

Spline Option

spline curve: A curve that passes through the first and last fit points and is influenced by the other fit points.

cubic curve: A very smooth curve created by the **PEDIT Spline** option with SPLINETYPE set at 6.

quadratic curve: A curve created by the **PEDIT Spline** option with **SPLINETYPE** set at 5. The curve is tangent to the polyline segments between the intermediate control points.

When you edit a polyline with the **Fit** option, the resulting curve passes through each polyline vertex. The **Spline** option also smoothes the corners of a straight-segment polyline, but the **Spline** option creates a *spline curve* that approximates a true B-spline. See **Figure 14-6.** Before using the **Spline** option, you can choose to create a spline using *cubic* or *quadratic* calculations by adjusting the **SPLINETYPE** system variable. The default setting of 6 draws a cubic curve. Change the value to 5 to generate a quadratic curve.

Figure 14-5.
Using the **Fit** option of the **PEDIT** command to turn a polyline into a smooth curve.

Existing Straight-Segment Polyline

Polyline after Curve Fitting

Figure 14-6.
A comparison of polylines edited with the **Fit** and **Spline** options of the **PEDIT** command. The **SPLINETYPE** system variable controls whether the **Spline** option creates a quadratic or cubic curve.

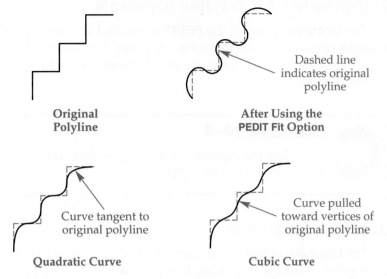

Original Polyline

After Using the PEDIT Fit Option

Dashed line indicates original polyline

Curve tangent to original polyline

Quadratic Curve

Curve pulled toward vertices of original polyline

Cubic Curve

After Using the PEDIT Spline Option

Set the number of line segments used to construct spline curves by entering a value in the **Segments in a polyline curve** text box in the **Display resolution** area of the **Display** tab in the **Options** dialog box. After changing the value, reissue the **Spline** option of the **PEDIT** command to apply the setting. The default value is 8, which creates a smooth spline curve with moderate regeneration time. Increasing the **Segments in a polyline curve** value creates a smoother spline curve, but increases regeneration time and drawing file size. See **Figure 14-7**.

The **Fit** and **Spline** options of the **PEDIT** command create approximations of a B-spline curve. Use the **SPLINE** command to create a true B-spline curve.

Exercise 14-3

Complete the exercise on the companion website.
www.g-wlearning.com/CAD

Figure 14-7.
A comparison of curves drawn with different display resolution settings.

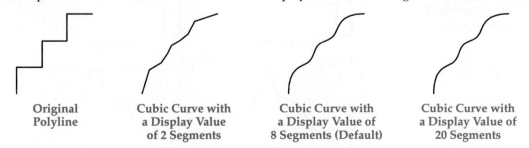

Original Polyline

Cubic Curve with a Display Value of 2 Segments

Cubic Curve with a Display Value of 8 Segments (Default)

Cubic Curve with a Display Value of 20 Segments

Straightening All Polyline Segments

The **Decurve** option of the **PEDIT** command returns a polyline edited with the **Fit** or **Spline** option to its original form. Specified tangent directions remain, however, for future reference. You can also use the **Decurve** option to straighten arc segments of a polyline. See **Figure 14-8**.

Exercise 14-4

Complete the exercise on the companion website.
www.g-wlearning.com/CAD

Linetype Generation

The **Ltype gen** option determines how linetypes other than Continuous generate in relation to polyline vertices. For example, if you use a Center linetype and disable the **Ltype gen** option, the polyline has a long dash at each vertex. When you activate the **Ltype gen** option, the polyline has a constant pattern in relation to the polyline as a whole. See **Figure 14-9**.

The **rEverse** option of the **PEDIT** command applies the same operation as the **REVERSE** command. Reversing affects the vertex options of the **PEDIT** command.

Figure 14-8.
The **Decurve** option straightens all curved segments of a polyline.

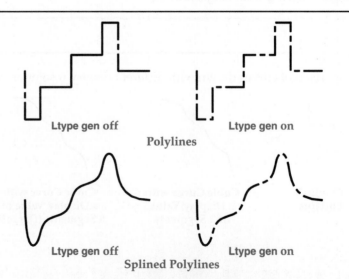

Original Polyline

Polyline after Using the Decurve Option

Figure 14-9.
A comparison of polylines and splined polylines with the **Ltype gen** option of the **PEDIT** command on and off.

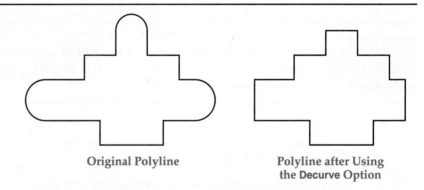

Ltype gen off   Ltype gen on

Polylines

Ltype gen off   Ltype gen on

Splined Polylines

Polyline Grip Commands

Pick a polyline object to display primary grips at each vertex and secondary grips at the midpoint of each segment. See **Figure 14-10**. Vertex grips have the standard filled-square grip appearance, and the midpoint grips appear as filled rectangles. Use any of the grips to apply standard grip editing commands, including **STRETCH**, **MOVE**, **ROTATE**, **SCALE**, **MIRROR**, and **Copy**. Polyline grips also offer commands specific to working with polylines. Using these context-sensitive commands is often the best way to add or remove a vertex, convert a straight segment to an arc, or convert an arc to a straight segment.

You can access and apply the same context-sensitive polyline grip commands in three different ways. **Figure 14-10** explains each technique. Different options are

Figure 14-10.
Pick a polyline object to display grips at each vertex and at the midpoint of each segment. Apply one of the following methods to access context-sensitive polyline grip commands. A—Hover over an unselected grip to display a menu of options, and then pick an option from the list. B—Pick a grip and then right-click and select an option. C—Pick a grip and then press [Ctrl] to cycle though options.

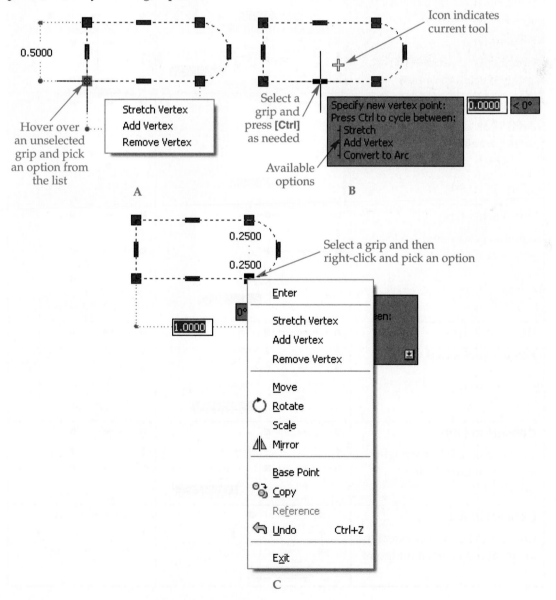

available depending on whether you hover over or select a vertex or a midpoint grip, and whether the segment is straight or an arc. **Figure 14-11** illustrates the process of using polyline grip commands to edit polylines. Specify a point if necessary to complete an operation. **Base point**, **Copy**, **Undo**, and **eXit** options are available for stretching a vertex. These options behave the same as the standard **STRETCH** grip edit command.

Figure 14-11.
Examples of options available for modifying polylines using grips.

Option	Process	Result
Stretch Vertex Stretches the polyline at the selected vertex	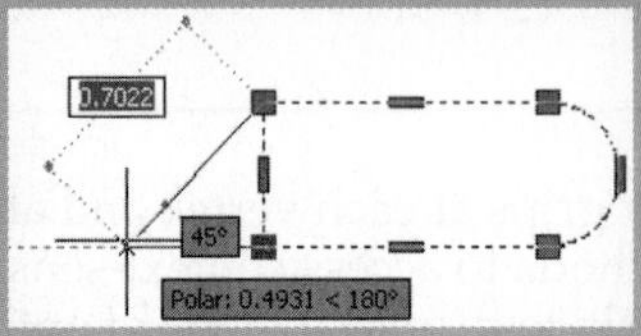	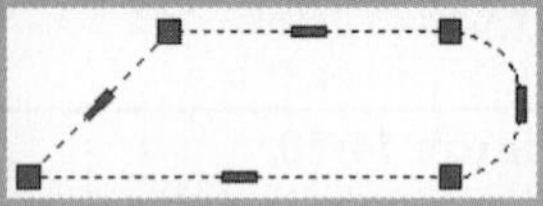
Stretch (straight segment) Stretches the polyline at the selected midpoint	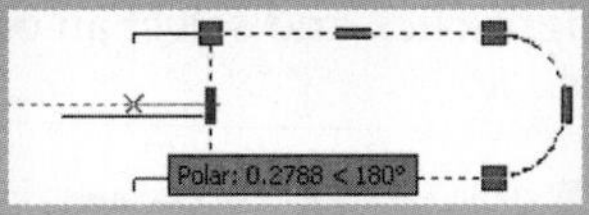	
Stretch (arc) Changes the radius of the arc at the selected midpoint	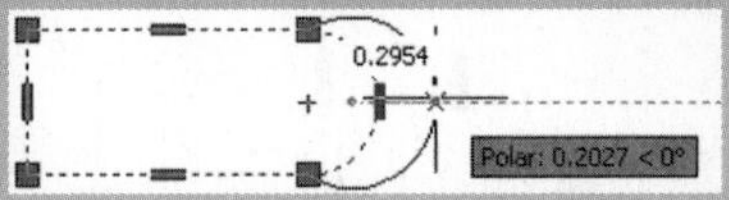	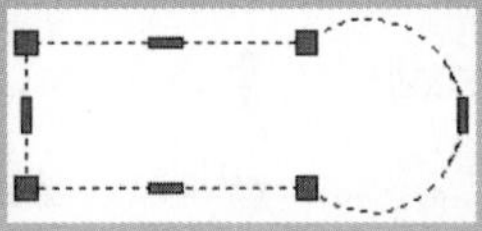
Add Vertex (straight segment) Adds a vertex using a selected vertex or midpoint	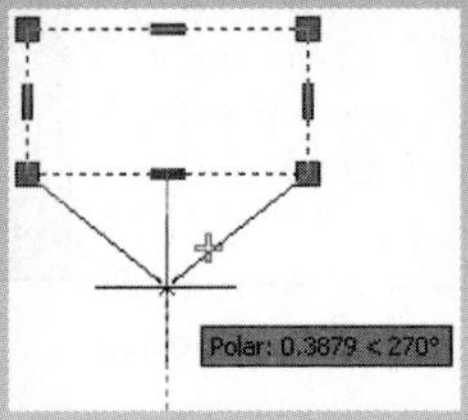	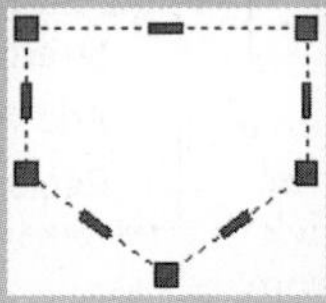
Add Vertex (arc) Adds a vertex using a selected vertex or midpoint	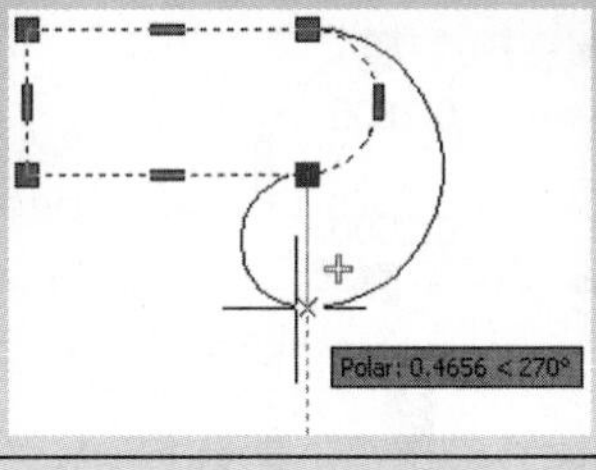	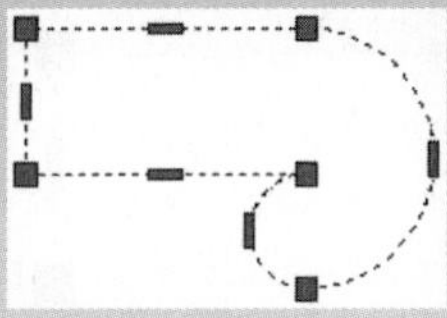
Remove Vertex Removes a selected vertex	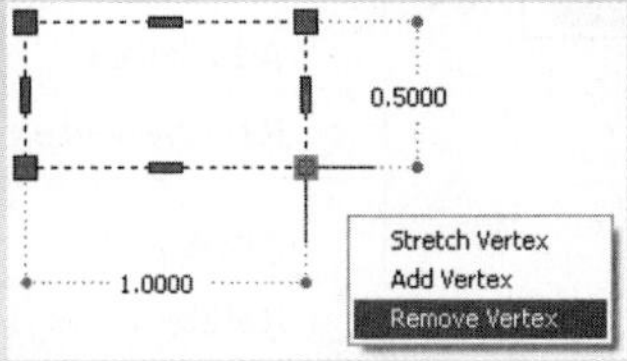	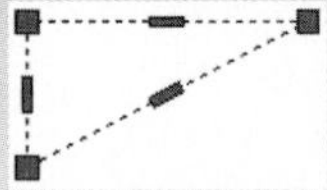
Convert to Line Converts an arc to a straight segment at the selected midpoint	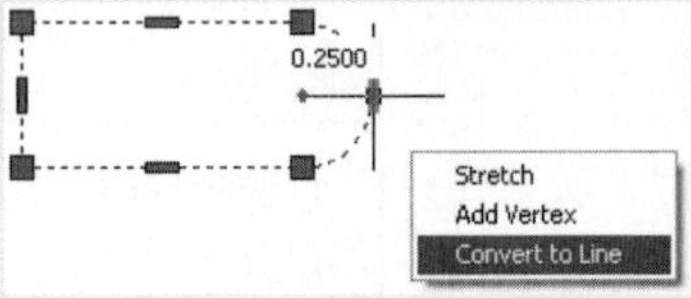	
Convert to Arc Converts a straight segment to an arc at the selected midpoint	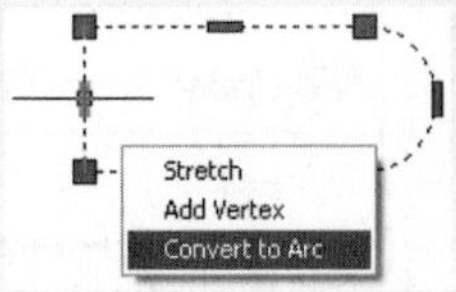	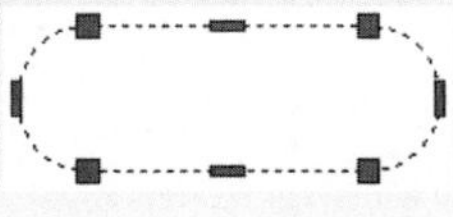

To select specific segments of a polyline object, instead of all segments, hold down [Ctrl] while picking. The polyline still is edited as one continuous object.

PROFESSIONAL
TIP

Use object snap tracking, polar tracking, object snaps, coordinate entry, or grid snaps with any grip editing command to improve accuracy.

Exercise 14-5

Complete the exercise on the companion website.
www.g-wlearning.com/CAD

Creating a Polyline Boundary

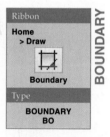

Access the **BOUNDARY** command to create a polyline boundary from linear objects that form a closed area. The **Boundary Creation** dialog box appears as shown in **Figure 14-12**. Select the default **Polyline** option from the **Object type:** drop-down list to create a polyline around the specified area. Select **Region** to create a *region* that you can use for area calculations, shading, extruding a solid model, and other purposes.

The **Current viewport** setting is active by default in the **Boundary set** drop-down list. This setting defines the *boundary set* from everything in the current viewport, even if it is not in the current display. An alternative is to pick the **New** button to return to the drawing window and select objects to use to create a boundary set. When you are finished selecting objects, right-click or press [Enter] or the space bar. The **Boundary Creation** dialog box returns with **Existing set active** in the **Boundary set** drop-down list to indicate that the boundary set references the selected objects.

The **Island detection** setting specifies whether *islands* within the boundary are treated as boundary objects. See **Figure 14-13**. Check **Island detection** to form separate boundaries from islands within a boundary.

region: A closed 2D area that can have physical properties such as a centroid and product of inertia.

boundary set: The part of the drawing AutoCAD evaluates to define a boundary.

island: A closed area inside a boundary.

Figure 14-12.
The **Boundary Creation** dialog box.

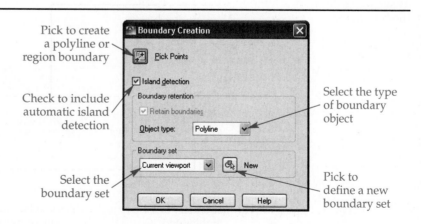

Pick to create a polyline or region boundary

Check to include automatic island detection

Select the boundary set

Select the type of boundary object

Pick to define a new boundary set

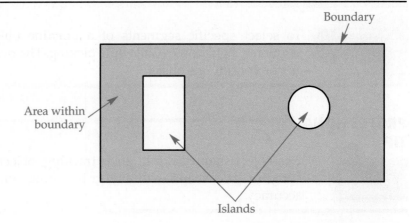

Figure 14-13.
You can include or exclude islands when defining a boundary set.

Boundary

Area within boundary

Islands

Select the **Pick Points** button, located in the upper-left corner of the **Boundary Creation** dialog box, to create the boundaries. Pick a point in each boundary area. If a point you pick is inside a closed area, the boundary becomes highlighted, as shown in **Figure 14-14**. The **Boundary Definition Error** alert box appears if the point you pick is not within a closed polygon. Pick **OK** and try again.

Unlike an object created with the **JOIN** command or the **Join** option of the **PEDIT** command, a polyline boundary created with the **BOUNDARY** command does not replace the original objects. The polyline traces over the defining objects with a polyline. The separate objects still exist underneath the newly created boundary. To avoid duplicate geometry, move the boundary to another location on-screen, erase the original objects, and then move the boundary back to its original position.

PROFESSIONAL TIP

You can simplify area calculations by using the **BOUNDARY** command or by joining objects with the **JOIN** command or the **Join** option of the **PEDIT** command before using the **AREA** command. To retain the original objects, explode the polyline after the area calculation if you used the **JOIN** command or the **Join** option of the **PEDIT** command. Erase the polyline boundary after the calculation if you used the **BOUNDARY** command. Chapter 15 covers the **AREA** command.

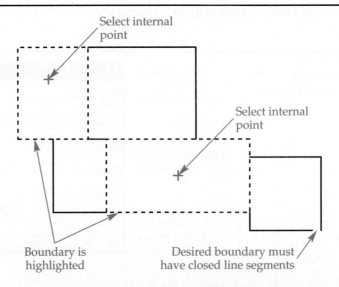

Figure 14-14.
Pick inside closed areas to highlight the boundary.

Select internal point

Select internal point

Boundary is highlighted

Desired boundary must have closed line segments

Using the SPLINEDIT Command

Ribbon
Home
> Modify

Edit Spline

Type
SPLINEDIT
SPE

Splines are complex curves that you can adjust and calculate using a variety of methods. This chapter introduces specialized spline editing commands. Access the **SPLINEDIT** command and select a spline to edit. Fit point grips locate *fit points* and identify a fit point spline. Control vertex grips locate *control vertices* and identify a control vertex spline. See **Figure 14-15**. Choose an option to activate the appropriate editing function. To access several spline editing options without first issuing the **SPLINEDIT** command, pick a spline to edit and then right-click and choose an option from the **Spline** cascading menu.

fit points: Points through which a spline passes that determine the shape of the spline.

control vertices: Points that are used to define the curve shape and change the curve design. Adding control points typically increases the complexity of the curve.

You can also access the **SPLINEDIT** command by double-clicking on a spline.

PROFESSIONAL TIP

Select the **Undo** option immediately after performing an unwanted edit to restore the previous spline without leaving the command. Use the **Undo** option more than once to step back through each operation.

Opening and Closing a Spline

Use the **Open** option to open a closed spline and the **Close** option to close an open spline. See **Figure 14-16**. The **Open** option is available if you originally closed the

Figure 14-15.
Access the **SPLINEDIT** command to display fit point or control vertex grips and a list of options for editing splines.

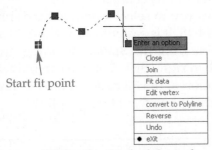

Start fit point

Fit Creation Method Control

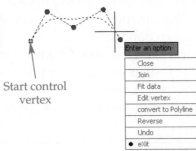

Start control vertex

Control Vertices Creation Method

Figure 14-16.
Closing an open spline. Note the significant difference between closing a spline created using the **Fit** method and a spline created using the **CV** method, even though the open spline is the same shape.

Original
Open Spline

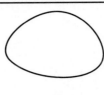

Closed Fit
Points Spline

Closed Control
Vertices Spline

by drawing the final segment manually or if you have used the **Close** option of ~~LINE~~ or **SPLINEDIT** command. The effect of closing a spline varies depending on ~~spline~~ creation method and associated options.

> **NOTE**
> The **Fit data** option includes **Open** and **Close** functions that behave the same as the **Open** and **Close** options previously described.

Joining Splines

The **Join** option provides the same function as the **JOIN** command, allowing you to create a single spline object from connected splines or from a spline connected to polylines, lines, or arcs. As when you use the **JOIN** command, the objects to be joined must share a common endpoint and cannot overlap or have gaps between segments. Choose the **Join** option and pick the objects to join. You can include the original spline in the selection set, but it is not necessary. See **Figure 14-17**.

> **PROFESSIONAL TIP**
> After you join objects into a continuous spline, use the **Close** option to close the spline if necessary.

Fit Data Option

The **Fit data** option allows you to perform several operations at a spline fit point. If you apply the **Fit data** option to a spline showing control vertices, the spline converts to show fit points. The **Fit data** option is more difficult to use than other editing commands. It includes **Add, Close, Delete, Move, Purge**, and **Tangents** functions that you can perform more easily using standard editing commands or context-sensitive spline grips. You will learn to grip-edit splines later in this chapter.

The **Kink** function of the **Fit data** option allows you to select a point on the spline to add a sharp point known as a *kink*. See **Figure 14-18**. It is easier to add a kink to a fit point spline using the **Add Fit Kink** option available from the **Spline** cascading menu that appears when you pick a spline to edit and then right-click, but you can only add a kink to a control vertex spline using the **Kink** option of the **Fit data** option. However, adding a kink to a control vertex spline changes the spline to a fit point spline.

Figure 14-17.
Joining a spline to other connected objects. The single object is now a complex spline that you can further edit to create a freeform shape.

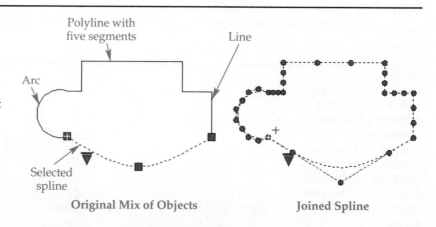

Polyline with five segments

Line

Arc

Selected spline

Original Mix of Objects

Joined Spline

Figure 14-18.
Adding a kink to a spline using the **Kink** function of the **Fit data** option.

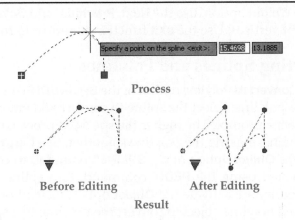

The **toLerance** option is unique to the **Fit data** option and allows you to adjust fit tolerance values. The results are immediate, so you can adjust the fit tolerance as necessary to produce different results. Use the **eXit** function to return to the **SPLINEDIT** prompt.

Edit Vertex Option

The **Edit vertex** option allows you to perform several operations at a spline control vertex. If you apply a specific **Edit vertex** function to a spline showing fit points, the spline converts to show control vertices. Like the **Fit data** option, the **Edit vertex** option is more difficult to use than other editing commands. It includes **Add**, **Delete**, and **Move** functions that you can perform more easily using standard editing commands or context-sensitive spline grips.

The **Elevate order** and **Weight** options are unique to the **Edit vertex** option and are sometimes useful. **Figure 14-19** briefly explains and illustrates these functions. When

Figure 14-19.
Spline editing functions that are unique to the **Edit vertex** option of the **SPLINEDIT** command.

Option	Process	Result
Elevate order Adds control points for greater control of spline shape and spline *order* refinement. Specify a value between the current number of vertices and 26.	3 Original Control Vertices	Order Elevated to 10
Weight Adjusts the spline weight at the selected vertex. The larger the weight, the closer the spline pulls toward the vertex. Specify a value greater than or equal to 1.	Original Weight of 1 Applied to the Second Vertex	Modified Weight of 5 Applied to the Second Vertex

order: In a spline, the degree of the spline polynomial + 1.

using the **Weight** option, use the **Next**, **Previous**, and **Select point** functions to navigate to different vertices. Use the **eXit** function to return to the **SPLINEDIT** prompt.

Converting Splines and Polylines

The **Convert to Polyline** option of the **SPLINEDIT** command allows you to convert a spline to a polyline. Select the spline to convert and enter a value at the Specify a precision <current>: prompt. The higher the specified precision, the more vertices are added to the polyline, making the polyline smoother. See **Figure 14-20**.

Use the **Object** option of the **SPLINE** command to convert a spline-fitted polyline object, created using the **PEDIT** command, to a spline object. When you access the **SPLINE** command, activate the **Object** option instead of defining points. Then pick a spline-fitted polyline object to convert the polyline to a spline.

The **rEverse** option of the **SPLINEDIT** command applies the same operation as the **REVERSE** command. Reversing affects the fit point and vertex options of the **SPLINEDIT** command.

PROFESSIONAL TIP

A spline created by fitting a spline curve to a polyline is a linear approximation of a true spline and is not as accurate as a spline drawn using the **SPLINE** command. An additional advantage of spline objects over smoothed polylines is that splines use less disk space.

Exercise 14-6

Complete the exercise on the companion website.
www.g-wlearning.com/CAD

Figure 14-20.
The effects of changing precision when converting a spline to a polyline.

| Original Spline | Converted with Default Precision of 10 | Converted with Precision of 1 |

AutoCAD and Its Applications—Basics

Spline Grip Commands

Pick a spline object to display grips at each fit point or vertex, depending on the creation method. See **Figure 14-21**. A parameter grip appears near the start point of the spline. Pick the grip and then select the **Show Fit Points** option to change the spline to use fit points, or pick the **Show Control Vertices** option to change spline to use control vertices. These options are also available when you select a spline and right-click.

Fit point grips have the standard filled-square grip appearance, and vertex grips display as filled circles. A "+" mark through the grip indicates the start point of the spline. Use any of the grips to apply standard grip editing commands, including **STRETCH, MOVE, ROTATE, SCALE, MIRROR**, and **Copy**. Spline grips also offer commands specific to working with splines. Using these context-sensitive commands is often the best way to move, add, or remove a fit point, vertex or kink, adjust tangent direction, and refine vertices.

There are three different ways to access and apply the same context-sensitive spline grip commands. **Figure 14-21** explains each technique. An icon identifies the current operation. Different options are available depending on whether you hover over or select a fit point or a vertex, and whether the point is at the end of the spline. **Figure 14-22** provides examples of using spline grip commands to edit splines. **Base point**, **Copy**, **Undo**, and **eXit** options are available for stretching a fit point or vertex. They behave the same as in the standard **STRETCH** grip edit command.

Figure 14-21.
Pick a spline object to display grips at each fit point or vertex and a parameter grip near the start point. Apply one of the following methods to access context-sensitive spline grip commands. A—Hover over an unselected grip to display a menu of options, and then pick an option from the list. B—Pick a grip and then press [Ctrl] to cycle through options. C—Pick a grip, right-click, and select an option.

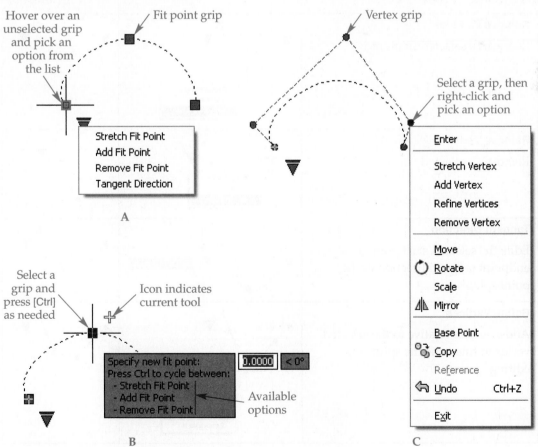

Figure 14-22.
Examples of options available for modifying splines using grips.

Option	Process	Result
Stretch Fit Point Stretches the spline at the selected fit point	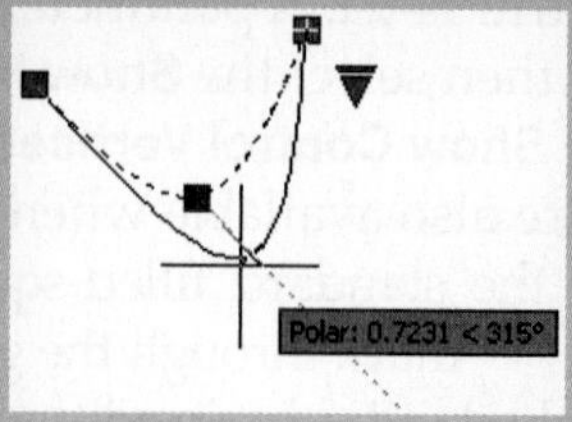	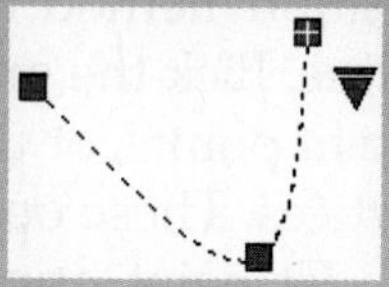
Stretch Vertex Stretches the spline at the selected vertex	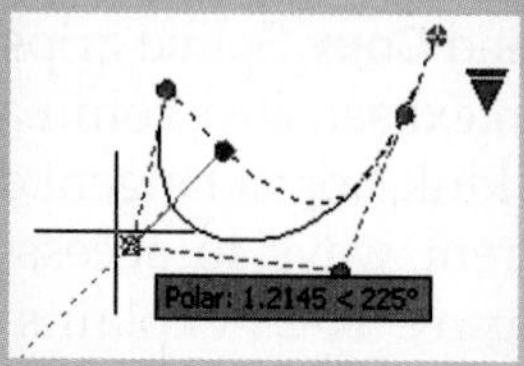	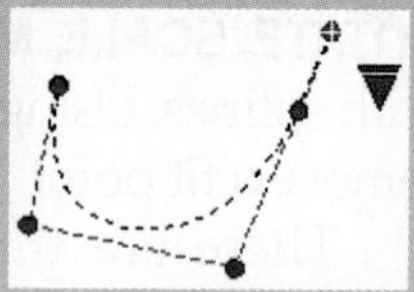
Add Fit Point Adds a fit point to the left of the selected fit point	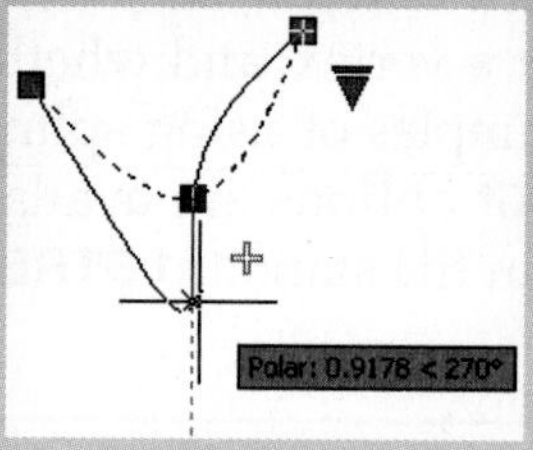	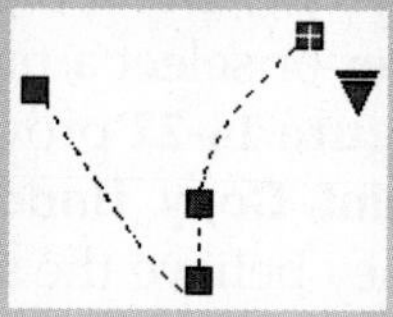
Add Vertex Adds a vertex to the left of the selected vertex	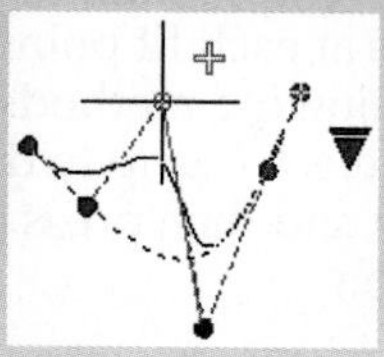	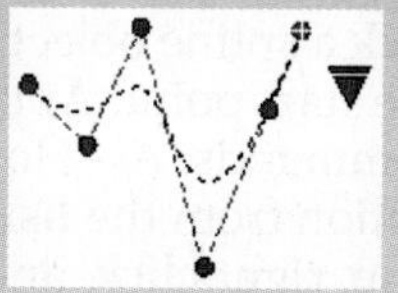
Remove Fit Point Removes a selected fit point	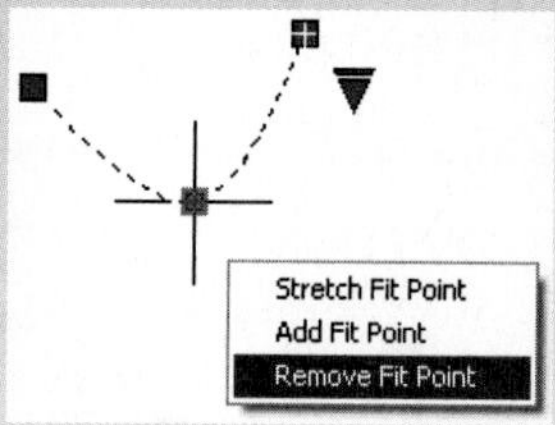	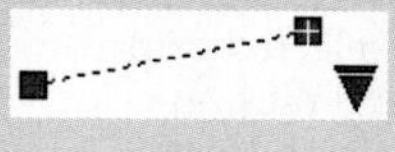
Remove Vertex Removes a selected vertex	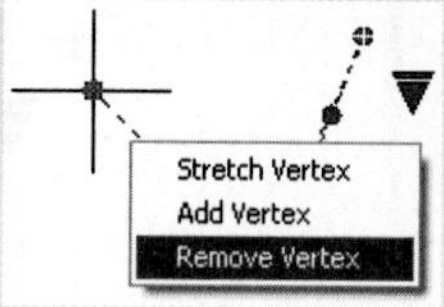	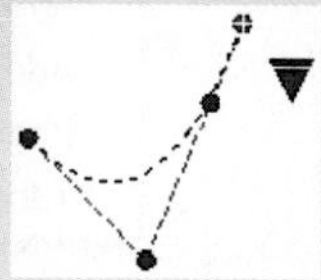
Tangent Direction Edits the selected start point or endpoint tangent direction for fit point splines	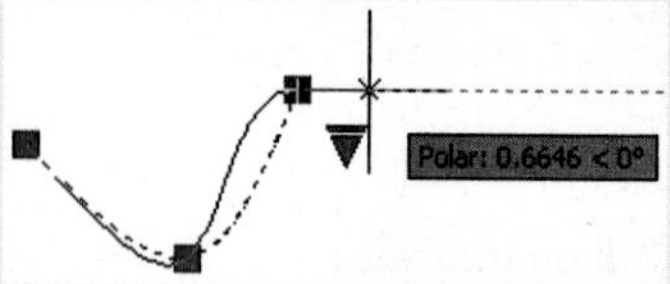	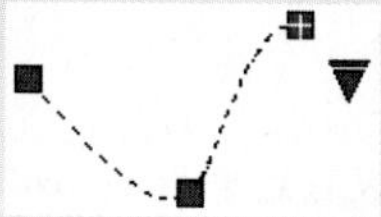
Refine Vertices Adds vertices relative to the selected vertex to fine-tune the spline for editing	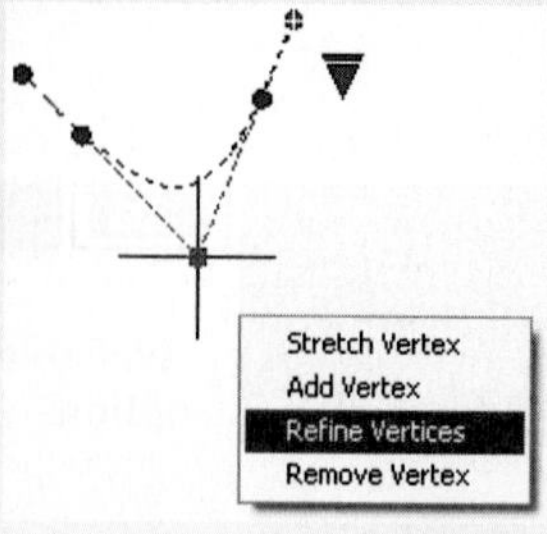	

PROFESSIONAL
TIP

Use object snap tracking, polar tracking, object snaps, coordinate entry, or grid snaps with any grip editing command to improve accuracy.

Exercise 14-7

Complete the exercise on the companion website.
www.g-wlearning.com/CAD

Chapter Review

Answer the following questions. Write your answers on a separate sheet of paper or complete the electronic chapter review on the companion website.
www.g-wlearning.com/CAD

1. Name the command and option required to turn three connected lines into a single polyline.
2. When you enter the **Edit vertex** option of the **PEDIT** command, where does AutoCAD place the "X" marker?
3. Why might it appear that nothing happens when you change the starting and ending widths of a polyline?
4. Which **PEDIT** command option and function allow you to change the starting and ending widths of a polyline?
5. Name the **PEDIT** command option and function used for curve fitting.
6. How do you move the "X" marker to edit a different polyline vertex?
7. Explain the difference between a fit curve and a spline curve.
8. Compare a quadratic curve, cubic curve, and fit curve.
9. Which **SPLINETYPE** system variable setting allows you to draw a quadratic curve?
10. Explain how you can adjust the way polyline linetypes are generated using the **PEDIT** command.
11. Describe how to use grips to stretch a straight polyline segment at the midpoint.
12. Describe how to use grips to straighten a polyline arc.
13. Name the command used to create a polyline boundary.
14. Which **Fit data** function of the **SPLINEDIT** command allows you to add a sharp point to a spline?
15. Identify the **Fit data** function of the **SPLINEDIT** command that lets you increase, but not decrease, the number of control points appearing on a spline curve.
16. Name **Fit data** function of the **SPLINEDIT** command that controls the pull exerted by a control point on a spline.
17. Name the **SPLINE** command option that allows you to turn a spline-fitted polyline into a true spline.
18. Describe how to use grips to add a fit point to a control vertex spline.
19. Describe how to use grips to remove a vertex from a fit point spline.
20. Which spline grip option allows you to add vertices relative to a selected vertex to fine-tune the spline?

Chapter 14   Polyline and Spline Editing Tools

Drawing Problems

Start AutoCAD if it is not already started. Start a new drawing for each problem using an appropriate template of your choice. The template should include layers and text styles for drawing the given objects. Add layers and text styles as needed. Draw all objects using appropriate layers, text styles, justification, and format. Follow the specific instructions for each problem. Use only drawing and editing commands and techniques you have already learned. Do not draw dimensions. Use your own judgment and approximate dimensions when necessary.

▼ Basic

1. Use the **LINE** command to draw two connected lines. Use the **PEDIT** command to convert one of the lines to a polyline, and then use the **Join** option to convert the line and polyline into a single polyline object. Use the **LINE** command to draw a rectangle. Use the **PEDIT** command to convert one of the lines to a polyline, and then use the **Join** option to convert the three remaining lines and polyline into a single polyline object. Save the drawing as P14-1.

2. Use the **SPLINE** command to draw a spline of your own design with at least four fit points. Use the **Convert to Polyline** option of the **SPLINEDIT** command to convert the spline to a polyline. Save the drawing as P14-2.

3. Use the **RECTANGLE** command to draw a 50 mm × 25 mm rectangle. Use grip editing to convert the straight 25 mm segments to R25 mm arcs, creating a full-round slot. Save the drawing as P14-3.

4. Use the **POLYGON** command to draw the hexagon shown in A. Use the **Remove Vertex** and **Stretch** polyline grip edit commands to edit the hexagon as shown in B. Save the drawing as P14-4.

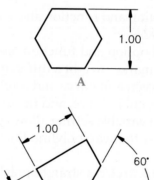

▼ Intermediate

5. Open P4-15 and save the file as P14-5. If you have not yet completed problem 4-15, do so now. The P14-5 file should be active. Use polyline editing to change the original objects to a rectangle as shown. Use the **Decurve** and **Width** options of the **PEDIT** command and polyline grip editing. Resave the drawing.

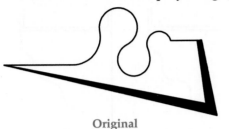

Original

After Editing

6. Open P4-16 and save the file as P14-6. If you have not yet completed problem 4-16, do so now. The P14-6 file should be active. Make the following changes.
   A. Combine the two polylines using the **Join** option of the **PEDIT** command.
   B. Change the beginning width of the left arrow to 1.0 and the ending width to .2.
   C. Draw a polyline .062 wide, similar to Line A, as shown.
   Resave the drawing.

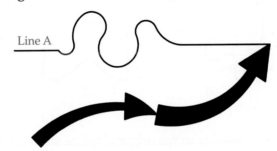

Line A

7. Draw four polylines .032 wide, using the following absolute coordinates for all four objects.

Point	Coordinates	Point	Coordinates	Point	Coordinates
1	1,1	5	3,3	9	5,5
2	2,1	6	4,3	10	6,5
3	2,2	7	4,4	11	6,6
4	3,2	8	5,4	12	7,6

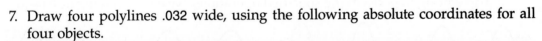

Leave the first polyline as drawn. Use the **Fit** option of the **PEDIT** command to smooth the second polyline. Use the **Spline** option of the **PEDIT** command to turn the third polyline into a quadratic curve. Make the fourth polyline into a cubic curve. Use the **Decurve** option of the **PEDIT** command to return one of the three edited polylines to its original form. Save the drawing as P14-7.

8. Use the **PLINE** command to draw four copies of a patio plan similar to the plan shown in A. Draw the house walls 6″ wide. Leave the first plan as drawn. Use the **PEDIT** command to create the designs shown. Use the **Fit** option for B, a quadratic spline for C, and a cubic spline for D. Change the **SPLINETYPE** system variable as required. Save the drawing as P14-8.

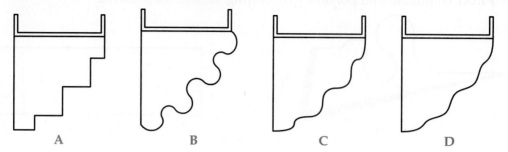

A   B   C   D

9. Open P14-8 and save the file as P14-9. The P14-9 file should be active. Use grip editing to create four new patio designs similar to A, B, C, and D below. Resave the drawing.

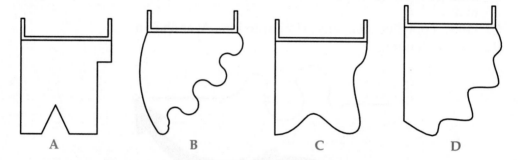

A   B   C   D

10. Draw a fit point spline similar to the original spline shown. Copy the spline seven times to create a layout similar to the layout shown. Use the **SPLINEDIT** command or grip editing to perform the specified operations. Save the drawing as P14-10.

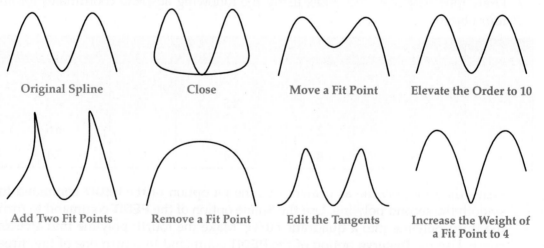

Original Spline   Close   Move a Fit Point   Elevate the Order to 10

Add Two Fit Points   Remove a Fit Point   Edit the Tangents   Increase the Weight of a Fit Point to 4

▼ Advanced

11. Draw the exterior door elevation shown. Use polylines and the **OFFSET** command to draw the features. Save the drawing as P14-11.

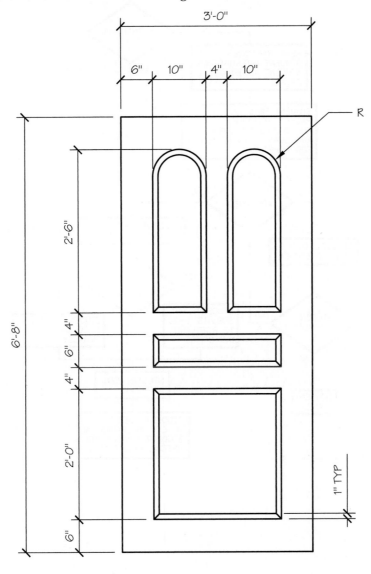

12. Draw the flow chart shown. Use polylines to draw the connecting lines, arrows, and diamonds. Save the drawing as P14-12.

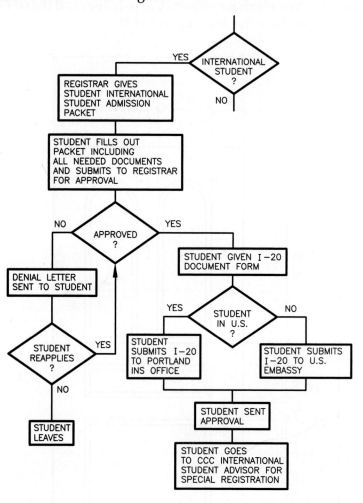

13. Draw the flow chart shown. Use polylines to draw the connecting lines, arrows, and diamonds. Save the drawing as P14-13.

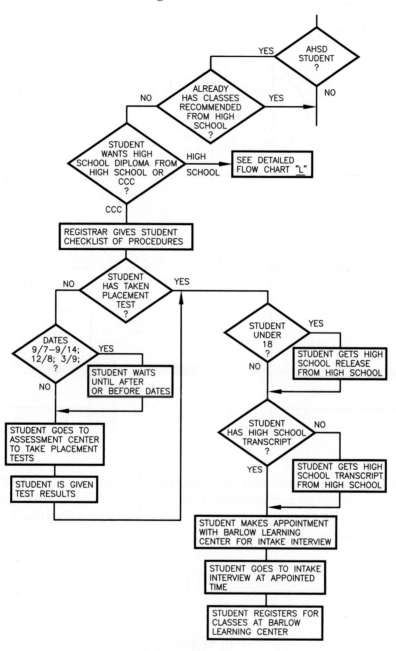

14. Draw the part views shown. Save the drawing as P14-14.

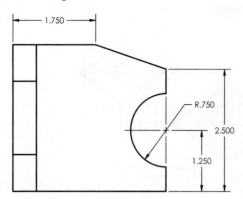

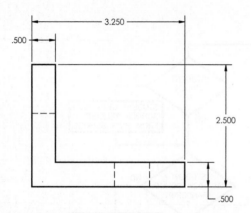

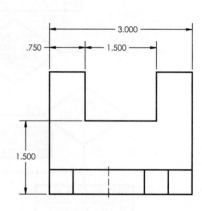

15. Use the **SPLINE** command and other commands, such as **ELLIPSE**, **MIRROR**, and **OFFSET**, to design an architectural door knocker similar to the knocker shown. Use an appropriate text command and font to place your initials in the center. Save the drawing as P14-15.

16. Draw the roof plan shown. Save the drawing as P14-16.

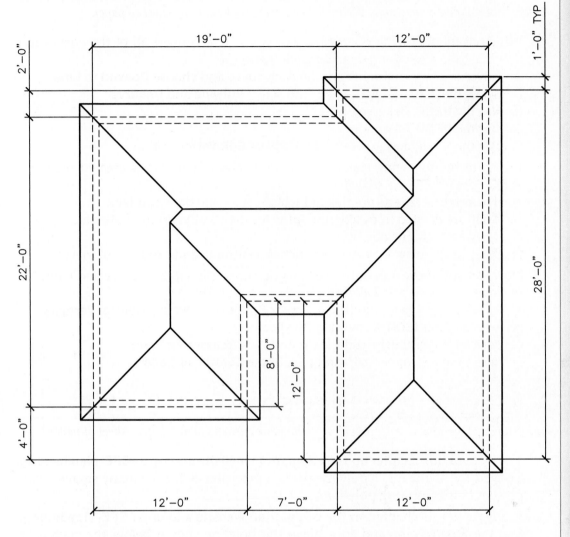

17. Find a tree appropriate for residential landscaping. Create a 2D sketch of the tree elevation. Use available measuring devices, such as a tape measure, to dimension the size and location of features accurately. Start a new drawing from scratch or use an architectural template of your choice. Draw the tree elevation from your sketch. Save the drawing as P14-17.

18. Create a dimensioned 2D sketch of a view of a new design for a cell phone. Include circles, ellipses, polylines, and freeform ergonomic edges in the design. Use available measuring devices, such as a tape measure and caliper, to measure an actual cell phone for reference if necessary. Start a new drawing from scratch or use a decimal template of your choice. Draw all of the views needed to describe the cell phone completely from your sketch. Use the **SPLINE** command to draw the freeform curves. Save the drawing as P14-18.

AutoCAD Certified Associate Exam Practice

Answer the following questions. Write your answers on a separate sheet of paper.

1. Which of the following operations can you use to convert all of the curves in a polyline to straight segments? *Select all that apply.*
   A. hover over the midpoint grip on each curve and choose **Convert to Line**
   B. select the polyline, right-click, and select **Polyline** and **Decurve**
   C. use the **PEDIT Decurve** option
   D. use the **PEDIT Fit** option
   E. use the **Straighten** function of the **PEDIT Edit Vertex** option

2. Which of the following statements are true about boundaries created using the default options? *Select all that apply.*
   A. The original objects are deleted when you create the boundary.
   B. They are created in exactly the same location as the original objects.
   C. They are polyline objects.
   D. They have physical properties such as centroids and products of inertia.

3. How can you show the control vertex grips on a spline that was created using the fit points method? *Select all that apply.*
   A. select the spline, right-click, and pick **Spline** and **Display Control Vertices**
   B. use the **SPLINEDIT Convert to Polyline** option
   C. select a fit point grip, right-click, and pick **Remove Fit Point**
   D. select the parameter grip and pick **Show Control Vertices**

AutoCAD Certified Professional Exam Practice

Follow the instructions in each problem. Write your answers on a separate sheet of paper.

1. **Navigate to this chapter on the companion website and open CPE-14spline.dwg.** Convert the spline into a polyline using a precision of 2. How many segments are present in the resulting polyline?

2. **Navigate to this chapter on the companion website and open CPE-14pedit.dwg.** Use the **PEDIT** command to achieve the polyline shown below the centerline. Mirror the result using the centerline as the mirror line. What are the coordinates of Point 1?

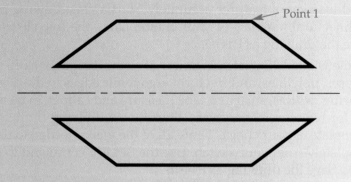
Point 1

Chapter

15

Obtaining Drawing Information

Learning Objectives

After completing this chapter, you will be able to:

✓ Measure distance, radius, diameter, angles, and area.
✓ List data related to a single point, an object, a group of objects, or an entire drawing.
✓ Determine the drawing status.
✓ Determine the amount of time spent in a drawing session.
✓ Perform calculations using the **QuickCalc** calculator.

This chapter describes tools that allow you to retrieve geometric values such as point coordinates, distance, angle, and area. You will learn methods for referencing object data, determining drawing status, and calculating time spent working on a drawing. You will also learn how to use the **QuickCalc** calculator to calculate values while you work.

Taking Measurements

Taking measurements from a drawing is common during the designing and drafting processes. Mechanical drafting examples include measuring a circle to confirm the size of a hole, measuring the angle between two surfaces, and calculating the volume of a part. Architectural drafting examples include checking the dimensions of a room, measuring the location of features within a room, and calculating square footage.

Using Grips

Grips and dynamic input provide one way to view basic object dimensions. To identify the location of a point that corresponds to a grip, confirm that the coordinate display field in the status bar is on. Then pick the object to activate grips and hover over a grip. The coordinates of the point appear in the coordinate display field of the status bar. Use grips and dynamic input to view dimensions between grips. Pick the object to activate grips and hover over a grip to display dimensions. The information that appears varies depending on the object type and the selected grip. See **Figure 15-1**.

Figure 15-1.
Examples of hovering over grips to display dimensions at the dynamic input cursor. Dynamic input must be active to view dimensions, but it does not have to be active to list the coordinates of a grip in the status bar.

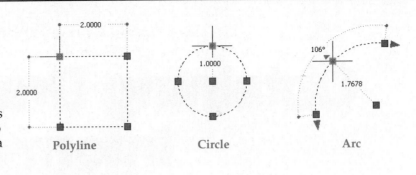

Polyline   Circle   Arc

Exercise 15-1

Complete the exercise on the companion website.
www.g-wlearning.com/CAD

Using the MEASUREGEOM Command

The **MEASUREGEOM** command allows you to measure distance, radius, angle, area, perimeter, and volume. The ribbon is an effective way to access **MEASUREGEOM** command options directly. See **Figure 15-2**. When you access the **MEASUREGEOM** command by typing, you must activate a measurement option before you begin. The **MEASUREGEOM** command remains active after you take measurements, allowing you to continue measuring without reselecting the command. Select the **eXit** option or press [Esc] to exit the command.

Measuring Distance

Use the **Distance** option of the **MEASUREGEOM** command to find the distance between points. Specify the first point, followed by the second point. The linear distance between the points, angle in the XY plane, and delta (change in) X and Y values appear on-screen, as shown in **Figure 15-3**, and at the command line. In a 2D drawing, the angle from the XY plane and delta Z values are always 0, as indicated at the command line. The first point you specify defines the vertex of the angular dimension, as shown in **Figure 15-3**.

After you specify the first point, you can choose the **Multiple points** function to measure the distance between multiple points. AutoCAD calculates the distance between each point and displays the value at the command line during selection. Several options are available for picking multiple points, as described in **Figure 15-4**. When you finish measuring, use the **Total** or **Close** option to display the total distance between all points. **Figure 15-5** shows an example of using the **Multiple points** function to calculate the perimeter of a shape.

Figure 15-2.
Use the flyout in the **Utilities** panel of the **Home** ribbon tab to access specific **MEASUREGEOM** command options.

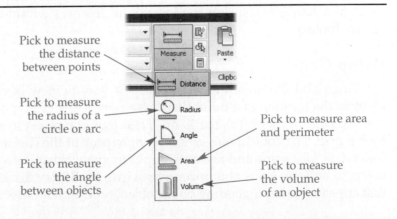

Pick to measure the distance between points

Pick to measure the radius of a circle or arc

Pick to measure the angle between objects

Pick to measure area and perimeter

Pick to measure the volume of an object

 Use coordinate entry, object snap modes, and other drawing aids to pick points when using the **MEASUREGEOM** command.

 You should typically use the **Multiple points** function of the **Distance** option to calculate the total distance between points of an open shape. The **Area** option of the **MEASUREGEOM** command, described later in this chapter, is more suited to calculating the perimeter of a closed shape.

Figure 15-3.
The data provided by the **Distance** option. Notice that the first point defines the vertex of the angular value.

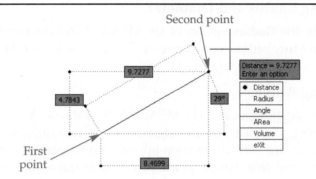

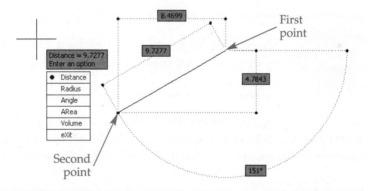

Figure 15-4.
Options available for the **Multiple points** function of the **Distance** option.

Option	Description
Arc	Measures the length of an arc; includes the same functions available for drawing arcs. Choose the **Line** function to return to measuring the distance between linear points.
Length	Measures the specified length of a line.
Undo	Cancels the effects of an unwanted selection, returning to the previous measurement point.
Total	Finishes multiple point selection and calculates the total distance between points.
Close	Connects the current point to the first point; finishes multiple point selection and calculates the total distance between points.

Figure 15-5.
An example of using the **Multiple points** function of the **Distance** option to calculate the total distance between several points along lines and an arc.

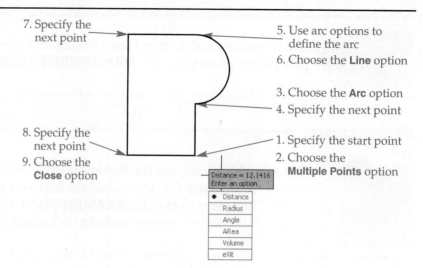

7. Specify the next point

5. Use arc options to define the arc

6. Choose the **Line** option

3. Choose the **Arc** option

4. Specify the next point

8. Specify the next point

9. Choose the **Close** option

1. Specify the start point

2. Choose the **Multiple Points** option

Measuring Radius and Diameter

Specify the **Radius** option of the **MEASUREGEOM** command and pick an arc or circle to measure radius and diameter. The dimensions appear on-screen, as shown in **Figure 15-6**, and at the command line.

Measuring Angles

Use the **Angle** option of the **MEASUREGEOM** command to find the angle between lines or points. To measure the angle between two lines, select the first line followed by the second line. The dimension appears on-screen, as shown in **Figure 15-7A**, and at the command line. Select an arc to measure the angle between arc endpoints. See **Figure 15-7B**.

Measuring an angle by selecting a circle and a point is typically most suitable for measuring 3D objects, such as a cylinder.

Choose the **Specify vertex** option when it is more appropriate to measure the angle between points, instead of selecting objects. Then select the angle vertex, followed by the first angle endpoint, and finally the second angle endpoint. See **Figure 15-8**.

Figure 15-6.
The data provided by the **Radius** option when you measure a circle or arc.

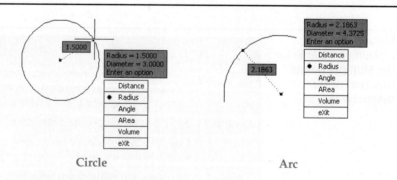

Circle

Arc

Figure 15-7.
The data provided by the **Angle** option. A—Selecting two lines. B—Picking an arc.

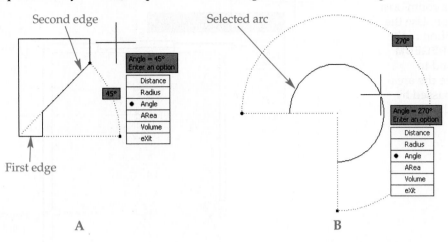

A   B

Figure 15-8.
Examples of when it is more appropriate to specify a vertex to measure an angle.

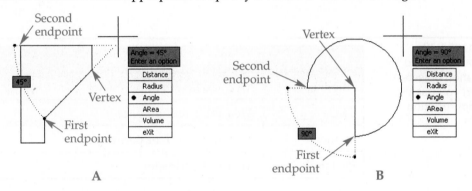

A   B

Exercise 15-2

Complete the exercise on the companion website.
www.g-wlearning.com/CAD

Measuring Area

Use the **Area** option of the **MEASUREGEOM** command to find the area between selected points or the area of an object. Specify the first point (corner) of the area to measure, followed by all other points (corners). The **Arc**, **Length**, and **Undo** options work the same as they do for measuring distance using the **Multiple points** function of the **Distance** option. A green (color 100) background fills the area to help you visualize the area. When you finish specifying perimeter corners, use the **Total** or **CLose** option to display the area between the points and the perimeter. The values appear on-screen, as shown in Figure 15-9, and at the command line.

Use the **Object** function of the **Area** option to find the area of a polyline object, circle, or spline without picking corners. Access the **Area** option, activate the **Object** function instead of picking vertices, and then select the object to display values. AutoCAD displays the area of the object and a second value. The second value returned by the **Object** function varies, depending on the object type, as shown in Figure 15-10.

Figure 15-9.
Measuring the area of a room on a floor plan. Use the **Area** option of the **MEASUREGEOM** command to calculate the area encompassed by selected corners.

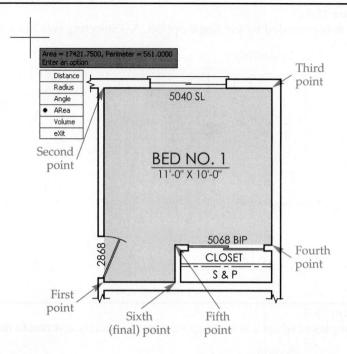

Figure 15-10.
Values returned for common objects when you use the **Object** function.

Object	Values Returned
Polyline	Area and length or perimeter
Circle	Area and circumference
Spline	Area and length or perimeter
Rectangle	Area and perimeter

You can calculate the area of an open polyline or spline object, but only if the endpoints show an apparent closure.

The **Area** option includes functions that allow you to calculate the sum of multiple areas during a single operation. Before selecting corners or an object, activate the **Add area** option and define the first area by picking corners or using the **Object** function. Continue adding areas as needed, or use the **Subtract area** option to remove areas from the selection set. AutoCAD calculates a running total of the area as you add or remove areas. The **Area** option remains in effect until you end the command.

Figure 15-11 shows an example of using the **Add area** and **Subtract area** functions of the **Area** option in the same operation. In this example, select the **Add area** option, and then select the **Object** option and pick the rectangle (polyline). Right-click or press [Enter] or the space bar at the (ADD mode) select objects: prompt to continue. The area and perimeter of the rectangle and a total area appear.

Then choose the **Subtract area** option to enter **SUBTRACT** mode. Select the **Object** option and pick the circles to subtract the area of the circles from the area of the rectangle. The area and circumference of each circle appear, along with the total area of the rectangle minus the total area of the subtracted circles. Right-click or press [Enter] or the space bar at the (SUBTRACT mode) select objects: prompt to continue. Press [Enter]

Figure 15-11.
Measuring the area of a part view. To calculate the area of a rectangle drawn with the **RECTANGLE** command, first select the outer boundary of the rectangle using the **Add area** function of the **Area** option. Select the inner circle boundaries using the **Subtract area** option. The total calculation appears.

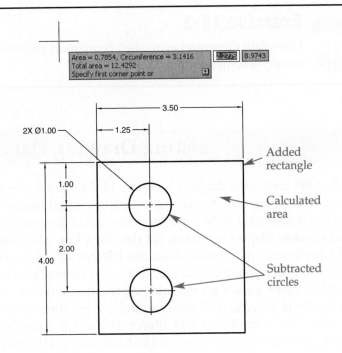

or the space bar or right-click and choose **Enter** to display the total area and return to the **MEASUREGEOM** command prompt.

PROFESSIONAL TIP

Calculating the area, circumference, and perimeter of shapes drawn with the **LINE** command can be time-consuming, because you must specify each vertex. Use the **PLINE** command when possible to construct a single object, or join existing objects. Use the **Object** function of the **Area** option to add or subtract objects.

You can use the **Volume** option of the **MEASUREGEOM** command to measure the volume of a basic drawing in 2D multiview format, but the option is most appropriate for measuring 3D objects. The **Region/Mass Properties** option provides data related to the properties of a 2D region or 3D solid. *AutoCAD and Its Applications—Advanced* describes the **Volume** and **Region/Mass Properties** options of the **MEASUREGEOM** command.

Traditional inquiry commands are available by typing command names. The **ID** command displays the coordinates of a selected point. The **DIST** command finds the distance between points, and the **AREA** command calculates area. These commands function much like the options available with the **MEASUREGEOM** command, but they are not as interactive.

Exercise 15-3

Complete the exercise on the companion website.
www.g-wlearning.com/CAD

Listing Drawing Data

LIST

Ribbon

Home
> Properties

List

Type

LIST
LI
LS

DBLIST

Type

DBLIST

The **LIST** command displays a variety of data about selected objects. Access the **LIST** command, select the objects, and right-click or press [Enter] or the space bar. The data for each object is displayed at the command line and in the text window. **Figure 15-12A** shows the text window display when you list data for a line. The Delta X and Delta Y values indicate the horizontal and vertical distances between the *from point* and *to point* of the line. **Figure 15-12B** identifies the measurements of a line provided by the **LIST** command.

Figure 15-13 shows an example of the text window listing data for a selection set of multiple objects, including a circle, a polyline drawn using the **RECTANGLE** command, and a multiline text object. If the list data does not fit in the window, AutoCAD prompts you to press [Enter] to display additional data.

The **DBLIST** (database list) command lists all data for every object in the current drawing. The information appears in the same format used by the **LIST** command, although the text window does not appear automatically.

Figure 15-12.
A—The text window displayed when you use the **LIST** command to list the properties of a line.
B—The various data and measurements of a line provided by the **LIST** command.

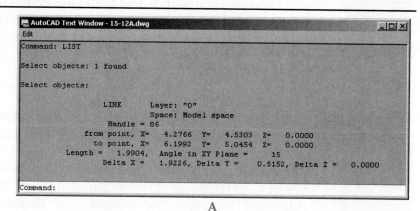

```
AutoCAD Text Window - 15-12A.dwg
Edit

Command: LIST

Select objects: 1 found

Select objects:

            LINE       Layer: "0"
                       Space: Model space
              Handle = 86
      from point, X=    4.2766  Y=    4.5303   Z=   0.0000
        to point, X=    6.1992  Y=    5.0454   Z=   0.0000
      Length =   1.9904,   Angle in XY Plane =      15
              Delta X =   1.9226, Delta Y =    0.5152, Delta Z =   0.0000

Command:
```

A

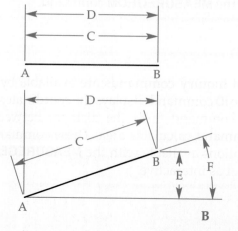

A = XY location
B = XY location
C = length
D = delta X
E = delta Y
F = angle

B

Figure 15-13.
Using the **LIST** command to list the properties of multiline text, a circle, and a rectangle.

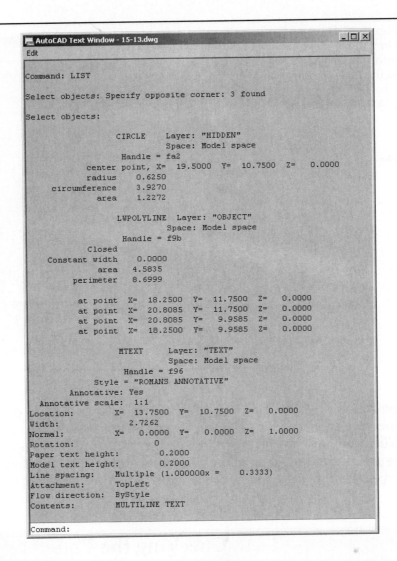

```
AutoCAD Text Window - 15-13.dwg                                    _ □ X
Edit

Command: LIST

Select objects: Specify opposite corner: 3 found

Select objects:

                        CIRCLE      Layer: "HIDDEN"
                                    Space: Model space
                    Handle = fa2
              center point, X=  19.5000  Y=  10.7500  Z=   0.0000
                  radius     0.6250
          circumference     3.9270
                  area       1.2272


                        LWPOLYLINE  Layer: "OBJECT"
                                    Space: Model space
                    Handle = f9b
                Closed
        Constant width     0.0000
                  area       4.5835
              perimeter     8.6999


              at point  X=  18.2500  Y=  11.7500  Z=   0.0000
              at point  X=  20.8085  Y=  11.7500  Z=   0.0000
              at point  X=  20.8085  Y=   9.9585  Z=   0.0000
              at point  X=  18.2500  Y=   9.9585  Z=   0.0000


                        MTEXT       Layer: "TEXT"
                                    Space: Model space
                    Handle = f96
                Style = "ROMANS ANNOTATIVE"
           Annotative: Yes
        Annotative scale: 1:1
  Location:        X=  13.7500  Y=  10.7500  Z=   0.0000
  Width:              2.7262
  Normal:          X=   0.0000  Y=   0.0000  Z=   1.0000
  Rotation:                0
  Paper text height:        0.2000
  Model text height:        0.2000
  Line spacing:     Multiple (1.000000x =    0.3333)
  Attachment:       TopLeft
  Flow direction:   ByStyle
  Contents:         MULTILINE TEXT

Command:
```

PROFESSIONAL TIP

The **LIST** command provides most of the information about an object, including the area and perimeter of polylines. The **LIST** command also reports object color and linetype, unless both are set to BYLAYER.

Exercise 15-4

Complete the exercise on the companion website.
www.g-wlearning.com/CAD

Reviewing the Drawing Status

The **STATUS** command provides a method to display a variety of drawing information at the command line and in the text window, as shown in **Figure 15-14**. The number of objects in a drawing refers to the total number of objects—erased and existing. Free dwg disk (C:) space: represents the space left on the drive that contains

Type
STATUS

STATUS

Figure 15-14.
Drawing information shown in the text window when you use the **STATUS** command.

```
AutoCAD Text Window - 15-14.dwg                                      _ □ ×
Edit
Command: STATUS
503 objects in C:\Documents and Settings\bhumecki\Desktop\Working DWGs\15-14.dwg
Undo file size:      48457 bytes
Model space limits are X:      0.0000   Y:      0.0000   (Off)
                       X:     34.0000   Y:     22.0000
Model space uses       X:     13.7500   Y:      9.9585
                       X:     20.8085   Y:     11.7500
Display shows          X:      7.5787   Y:      7.4953
                       X:     41.7712   Y:     27.5019
Insertion base is      X:      0.0000   Y:      0.0000   Z:      0.0000
Snap resolution is     X:      0.1250   Y:      0.1250
Grid spacing is        X:      0.2500   Y:      0.2500

Current space:         Model space
Current layout:        Model
Current layer:         "HIDDEN"
Current color:         BYLAYER -- 3 (green)
Current linetype:      BYLAYER -- "HIDDEN"
Current material:      BYLAYER -- "Global"
Current lineweight:    BYLAYER
Current elevation:      0.0000  thickness:    0.0000
Fill on  Grid off  Ortho off  Qtext off  Snap off  Tablet off
Object snap modes:     Center, Endpoint, Intersection, Extension
Free dwg disk (C:) space: 80539.2 MBytes
Free temp disk (C:) space: 80539.2 MBytes
Free physical memory: 6341.4 Mbytes (out of 8175.0M).
Press ENTER to continue:
Free swap file space: 8527.6 Mbytes (out of 9765.0M).

Command:
```

the drawing file. Drawing aid settings appear, along with the current settings for layer, linetype, and color. Press [Enter] if necessary to proceed to additional information. When you finish reviewing the status, press [F2] to close the text window. You can also switch to the drawing window without closing the text window by picking anywhere inside the drawing window or using the Windows [Alt]+[Tab] feature.

Checking the Time

The **TIME** command displays the current time and time related to the current drawing session. **Figure 15-15** shows an example of drawing time data displayed in the text window. The drawing creation time starts when you begin a new drawing, not when you first save a new drawing. The **SAVE** command affects the Last updated: time. However, all drawing session time is lost if you exit AutoCAD without saving.

The Windows operating system maintains the date and time settings for the computer. You can change these settings in the Windows Control Panel.

PROFESSIONAL TIP

field: A text object that displays a property, setting, or value for an object, drawing, or computer system.

Fields are an effective way to display drawing information within mtext and text objects. You can use fields with many AutoCAD tools, including inquiry commands, drawing properties, attributes, and sheet sets. You can update fields when changes occur to the reference data. Chapter 10 explains using fields.

Figure 15-15.
Information displayed in the text window when you use the **TIME** command.

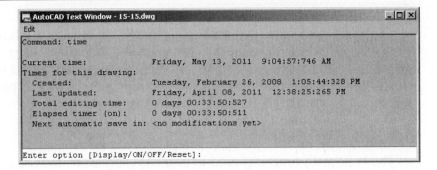

```
AutoCAD Text Window - 15-15.dwg                              _ | □ | X |
Edit
Command: time

Current time:           Friday, May 13, 2011  9:04:57:746 AM
Times for this drawing:
  Created:              Tuesday, February 26, 2008  1:05:44:328 PM
  Last updated:         Friday, April 08, 2011  12:38:25:265 PM
  Total editing time:   0 days 00:33:50:527
  Elapsed timer (on):   0 days 00:33:50:511
  Next automatic save in: <no modifications yet>


Enter option [Display/ON/OFF/Reset]:
```

Exercise 15-5

Complete the exercise on the companion website.
www.g-wlearning.com/CAD

Using QuickCalc

Most drafting projects require you to make calculations. For example, you may need to double-check dimensions or, when working from a sketch with missing dimensions, you may need to calculate a distance or angle. Drafters often make calculations using a handheld calculator. An alternative is to use **QuickCalc**, which is a palette containing a basic calculator, scientific calculator, units converter, and variables feature. See **Figure 15-16**. Use **QuickCalc** as you would a handheld calculator. You can also use **QuickCalc** while a command is active to paste calculations when a prompt asks for a specific value.

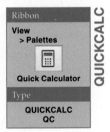

Ribbon

View
> Palettes

Quick Calculator

Type

QUICKCALC
QC

QUICKCALC

Figure 15-16.
The **QuickCalc** palette allows you to perform a variety of calculations.

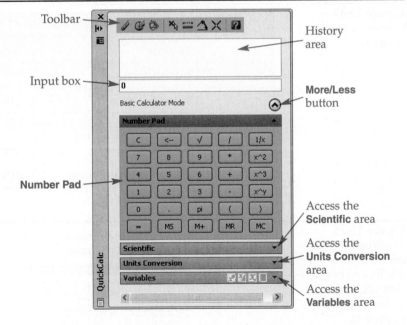

Toolbar

History area

Input box

More/Less button

Basic Calculator Mode

Number Pad

Access the **Scientific** area

Access the **Units Conversion** area

Access the **Variables** area

QuickCalc

You can also access the **QuickCalc** palette by right-clicking in the drawing window and selecting **QuickCalc**, or from specific areas such as the button that sometimes appears next to a text box in a dialog box, the **Quick Properties** panel, or the **Properties** palette.

Entering Expressions

The basic math functions used in numeric expressions include addition, subtraction, multiplication, division, and exponential notation. You can enter grouped expressions by using parentheses to break up the expressions to be calculated separately. For example, to calculate 6 + 2 and then multiply the sum by 4, enter (6+2)*4. The result will be wrong if you do not add the parentheses. **Figure 15-17** shows the symbols used for basic math operators.

Add expressions in the input box by picking buttons on the **Number Pad** or using keyboard keys. After creating the expression, pick the equal (**=**) button on the number pad or press [Enter] to evaluate the expression. **Figure 15-18** shows options found on the **Number Pad** that are not available from the keyboard.

The result of an evaluated expression appears in the input box, and the expression moves to the history area. **Figure 15-19** shows the **QuickCalc** palette before and after calculating 96.27 + 23.58. When you are using only the input box in **QuickCalc**, pick the **More/Less** button below the input box to hide the additional sections, saving valuable drawing space. When all areas of the **QuickCalc** palette are displayed, the button is an up arrow and its tooltip reads Less. Pick the button again to display hidden areas.

If you move the cursor outside of the **QuickCalc** palette, the drawing area is automatically activated. Pick anywhere inside the **QuickCalc** palette to reactivate **QuickCalc**.

Figure 15-17.
Common math operators.

Symbol	Function	Example
+	Addition	3+26
−	Subtraction	270–15.3
*	Multiplication	4*156
/	Division	265/16
^	Exponent	22.6^3
( )	Grouped expressions	2*(16+2^3)

Figure 15-18.
The **Number Pad** area contains additional options that you cannot access using the keyboard.

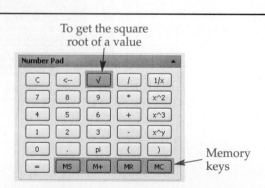

AutoCAD and Its Applications—Basics

Figure 15-19.
A—Add an expression to the input box. B—After creating the expression, press [Enter] to evaluate the expression. The history area stores the expression and result.

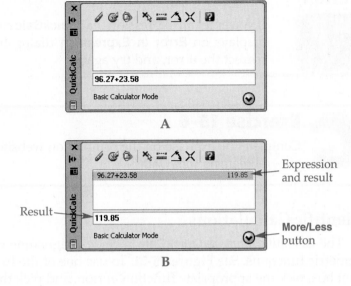

96.27+23.58

Basic Calculator Mode

A

Expression and result

96.27+23.58   119.85

Result

119.85

Basic Calculator Mode

More/Less button

B

Clearing the Input and History Areas

After picking the equal (=) button or pressing [Enter] to evaluate an expression, you can create a new expression without clearing the last result. AutoCAD automatically starts a new expression. You can clear the input box manually by placing the cursor in the input box and pressing [Backspace] or [Delete], or by picking the **Clear** button from the **QuickCalc** toolbar. To clear the history area, pick the **Clear History** button. See **Figure 15-20**.

Figure 15-20.
Use the buttons on the **QuickCalc** toolbar to clear the input box and history areas.

Pick to clear history area

Pick to clear the input box

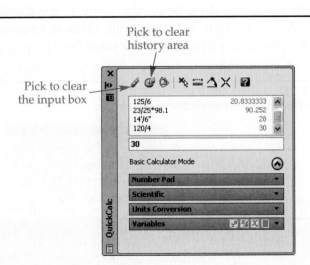

125/6	20.8333333
23/25*98.1	90.252
14'/6"	28
120/4	30

30

Basic Calculator Mode

Number Pad
Scientific
Units Conversion
Variables

Exercise 15-6

Complete the exercise on the companion website.
www.g-wlearning.com/CAD

Scientific Calculations

The **Scientific** area of **QuickCalc** includes trigonometric, exponential, and some geometric functions. See **Figure 15-21**. To use one of the functions, add a value to the input box, pick the appropriate function button, and pick the equal (**=**) button or press [Enter]. When you pick a function button, the input box value appears in parentheses after the expression. For example, to get the sine of 14, clear the input box, type 14 in the input box, and pick the **sin** button. The input box now reads sin(14). Pick the equal (**=**) button or press [Enter] to view the result.

You can pick a function button first, but doing so puts a default value of 0 in parentheses. Place the cursor in the input box to type a different number in the parentheses if needed.

Converting Units

The **Units Conversion** area allows you to convert one unit type to another. The available unit types are **Length**, **Area**, **Volume**, and **Angular**. For example, to use the unit converter to convert 23 centimeters to inches, pick in the **Units type** field to display the drop-down list. See **Figure 15-22**. Pick the drop-down list button to display the different unit types and select **Length**. Activate the **Convert from** field and select **Centimeters** from the drop-down list. Activate the **Convert to** field and select **Inches** from the drop-down list. Type 23 in the **Value to convert** field and pick the equal (**=**) button or press [Enter]. The **Converted value** field displays the converted units.

Figure 15-21.
The scientific functions available in **QuickCalc**.

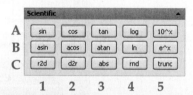

	1	2	3	4	5
A	Sine	Cosine	Tangent	Base–10 Log	Base–10 Exponent
B	Arcsine	Arccosine	Arctangent	Natural Log	Natural Exponent
C	Convert Radians to Degrees	Convert Degrees to Radians	Absolute Value	Round	Truncate

Figure 15-22.
Picking the current unit type activates the field and displays the drop-down list button.

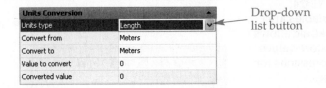

Drop-down list button

To pass the converted value to the input box for use in an expression, pick the **Return Conversion to Calculator Input Area** button. See **Figure 15-23**. If the button is not visible, pick once on the converted units in the **Converted value** field.

Exercise 15-7

Complete the exercise on the companion website.
www.g-wlearning.com/CAD

Using Variables

If you use an expression or value frequently, you can save it as a *variable*. Use the **Variables** area, shown in **Figure 15-24**, to create, edit, and delete variables, and to pass variables to the input box. The **Variables** area of **QuickCalc** includes predefined *constant* and *function* variables.

To create a new variable, select the **New Variable...** button to open the **Variable Definition** dialog box shown in **Figure 15-25**. Type a name for the variable in the **Name:** field. Select a group to contain the variable using the **Group with:** field. Type the value or the expression for the variable in the **Value or expression:** field. Give a description for the variable in the **Description** field. Pick the **OK** button to save the variable and display it in the **Variables** area.

To edit a variable, pick the variable to modify and select the **Edit Variable** button to reopen the **Variable Definition** dialog box. Pick the **Delete** button to delete the selected variable. Pick the **Return Variable to Input Area** button or double-click the variable name to pass the selected variable to the input box.

You can also access variables by right-clicking in the **Variable** area to display a shortcut menu. Two additional options are available from the menu. Pick the **New Category** option to create a new category for saved variables. Select the **Rename** option or slowly double-click on a variable to rename the selected variable.

variable: A text item that represents another value and is available for future reference.

constant: An expression or value that stays the same.

function: An expression or value that asks for user input to get values to pass to the expression.

Figure 15-23.
After you convert a value, pass the value to the input box for use in an expression.

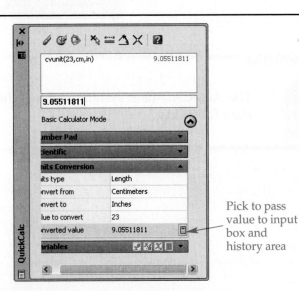

Pick to pass value to input box and history area

Figure 15-24.
The **Variables** area of **QuickCalc** allows you to store values and expressions for later use.

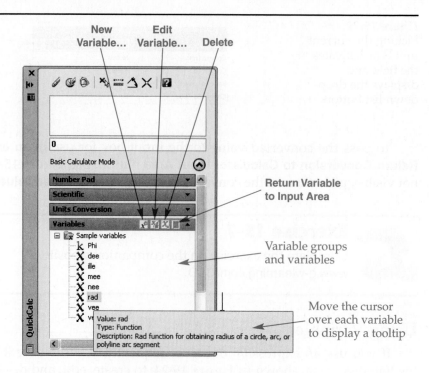

New Variable...
Edit Variable...
Delete

Return Variable to Input Area

Variable groups and variables

Move the cursor over each variable to display a tooltip

Figure 15-25.
Use the **Variable Definition** dialog box to define new variables.

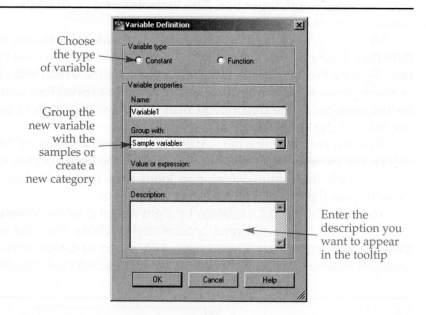

Choose the type of variable

Group the new variable with the samples or create a new category

Enter the description you want to appear in the tooltip

PROFESSIONAL TIP

The AutoCAD help system explains the predefined variables and their functions.

Using Drawing Values

Tools available from the **QuickCalc** toolbar allow you to pass values from the drawing to the **QuickCalc** input box, and from the input box to the drawing. When you select any of the buttons shown in **Figure 15-26**, the **QuickCalc** palette temporarily disappears so that you can select points from the drawing window.

Pick the **Get Coordinates** button to select a point from the drawing window and display the X,Y,Z coordinates in the input box. Pick the **Distance Between Two Points** button and select two points in the drawing area to display the distance between the points in the input box. Select the **Angle of Line Defined by Two Points** button and pick two points on a line to calculate the angle of the line and display the angle in the input box. Select the **Intersection of Two Lines Defined by Four Points** button to find the intersection of two lines by picking points on the two lines. The X,Y,Z coordinates of the intersection appear in the input box.

Using QuickCalc with Commands

The previous information focuses on using the **QuickCalc** palette to calculate unknown values while you are drafting, much like using a handheld calculator or the Windows Calculator. You can also use **QuickCalc** while a command is active to pass a calculated value to the command line as a response to a prompt. AutoCAD provides a few alternatives for using **QuickCalc** while a command is active.

If the **QuickCalc** palette is active when you access a command, when the prompt requesting an unknown value appears, calculate the value using the **QuickCalc** palette and then press the **Paste value to command line** button to pass the value as a response to the prompt. For example, to draw a line a distance of 14'8" + 26'3" horizontally from a start point, activate the **QuickCalc** palette, access the **LINE** command, and pick a start point. Then use polar tracking or **Ortho** mode to move the crosshairs to the right or left of the start point so the line is at a 0° or 180° angle. At the Specify next point or [Undo]: prompt, enter 14'8" + 26'3" in the **QuickCalc** palette input box and pick the equal (=) button or press [Enter]. The result in the input box is 40'11". Pick the **Paste value to command line** button to respond to the prompt with a 40'11" value. Press [Enter] or the space bar or right-click and select **Enter** to draw the 40'11" line.

If the **QuickCalc** palette is not active while you are using a command, you can still calculate and use a value. When the prompt requesting an unknown value appears, access **QuickCalc** by right-clicking and selecting **QuickCalc** or typing 'QC. A **QuickCalc** window, which is not the same as the **QuickCalc** palette, opens in command calculation mode. See **Figure 15-27**. Use the necessary tools to evaluate an expression. Then pick the **Apply** button to pass the value back to the command line, and close the **QuickCalc** window.

Figure 15-26.
Use buttons on the **QuickCalc** toolbar to pass values from **QuickCalc** to AutoCAD and retrieve values from AutoCAD to pass to **QuickCalc**.

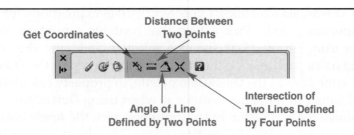

Get Coordinates

Distance Between Two Points

Angle of Line Defined by Two Points

Intersection of Two Lines Defined by Four Points

Figure 15-27.
The **QuickCalc** window appears when you access **QuickCalc** while a command is active. The **Apply** and **Close** buttons are available at the bottom of the window.

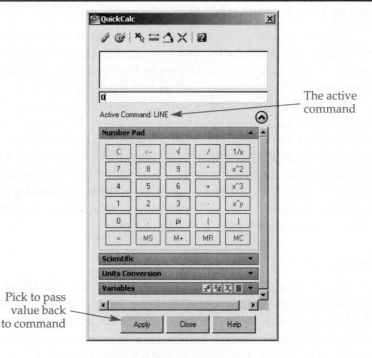

The active command

Pick to pass value back to command

Exercise 15-8

Complete the exercise on the companion website.
www.g-wlearning.com/CAD

Using QuickCalc with Object Properties

QuickCalc also allows you to calculate expressions for an object while using the **Properties** palette. Pick a field that contains a numeric value to display the calculator icon. **Figure 15-28** shows a selected circle and the active **Diameter** field in the **Properties** palette. Pick the calculator icon to open the **QuickCalc** window, again not the same item as the **QuickCalc** palette, in property calculation mode. Use expressions and values in the same manner as when using **QuickCalc** at any other time. Once you evaluate the expression in the input box, pick the **Apply** button to pass the value to the property field in the **Properties** palette. The object automatically updates based on the new value.

Figure 15-28.
The calculator
icon appears
when you select a
numeric field in the
Properties palette.

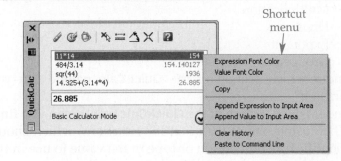

Selected object

Pick to access **QuickCalc**

Active field

Additional QuickCalc Options

The history area contains settings and features that you can access only from a shortcut menu, shown in Figure 15-29. Right-click anywhere in the history area to access the menu with the following options:

- **Expression Font Color**—changes the color of the expression font.
- **Value Font Color**—changes the color of the value font.
- **Copy**—copies the expression and value to the Windows clipboard.
- **Append Expression to Input Area**—passes the expression to the input box.
- **Append Value to Input Area**—passes the value to the input box.
- **Clear History**—clears the history area.
- **Paste to Command Line**—passes the value to the Command: prompt.

Picking the **Properties** button on the **QuickCalc** palette displays a shortcut menu. The settings allow you to change the palette appearance, including its ability to dock, hide, or appear transparent.

Figure 15-29.
The history area
shortcut menu
contains additional
QuickCalc functions
and settings.

Shortcut menu

Chapter Review

Answer the following questions. Write your answers on a separate sheet of paper or complete the electronic chapter review on the companion website.
www.g-wlearning.com/CAD

1. Explain how to use grips to identify the location of a point and the dimensions of an object.
2. What types of information does the **Distance** option of the **MEASUREGEOM** command provide?
3. What information does the **Area** option of the **MEASUREGEOM** command provide?
4. To add the areas of several objects when using the **Area** option of the **MEASUREGEOM** command, when do you select the **Add area** function?
5. Explain how measuring the area of a polyline using the **Area** option of the **MEASUREGEOM** command is different from measuring the area of an object drawn with the **LINE** command.
6. What is the purpose of the **LIST** command?
7. Describe the meaning of *delta X* and *delta Y*.
8. What command, other than **MEASUREGEOM** and **AREA**, provides the area and perimeter of an object?
9. What is the function of the **DBLIST** command?
10. Which command allows you to list drawing aid settings for the current drawing?
11. What information does the **TIME** command provide?
12. When does the drawing creation time start?
13. What term describes a text object that displays a set property, setting, or value for an object?
14. List three ways to open the **QuickCalc** palette.
15. Name the four sections of the **QuickCalc** palette.
16. Give the proper symbol to use for the following math functions:
    A. Addition
    B. Subtraction
    C. Multiplication
    D. Division
    E. Exponent
    F. Grouped expressions
17. Under which section of the **QuickCalc** palette can you find the square root function?
18. Under which section of the **QuickCalc** palette can you find the arccosine function?
19. When using one of the scientific functions, which should you do first: pick the scientific function button or type in the value to use in the input box? Why?
20. Name the four types of units that you can convert using **QuickCalc**.
21. What term describes a text item that represents another value and can be accessed later as needed?
22. Which **QuickCalc** button passes the value in the **QuickCalc** input box to respond to a prompt?
23. How can you start **QuickCalc** while a command is active?
24. When using **QuickCalc** while a command is active, how do you pass the value to respond to a prompt?
25. When the **Properties** palette is open, what do you need to do first to see the calculator icon to use **QuickCalc**?

Drawing Problems

Start AutoCAD if it is not already started. Start a new drawing for each problem using an appropriate template of your choice. The template should include layers and text styles, when necessary, for drawing the given objects. Add layers and text styles as needed. Draw all objects using appropriate layers, text styles, justification, and format. Follow the specific instructions for each problem. Use only drawing and editing commands and techniques you have already learned. Do not draw dimensions. Use your own judgment and approximate dimensions when necessary.

▼ Basic

1. Use **QuickCalc** to calculate the result of the following equations.
    A. 27.375 + 15.875
    B. 16.0625 − 7.1250
    C. 5 × 17′-8″
    D. 48′-0″ ÷ 16
    E. (12.625 + 3.063) + (18.250 − 4.375) − (2.625 − 1.188)
    F. 7.252

2. Use **QuickCalc** to convert 4.625″ to millimeters.

3. Use **QuickCalc** to convert 26 mm to inches.

4. Use **QuickCalc** to convert 65 miles to kilometers.

5. Use **QuickCalc** to convert 5 gallons to liters.

6. Use **QuickCalc** to find the square root of 360.

7. Use **QuickCalc** to calculate 3.25 squared.

▼ Intermediate

8. Show the calculation and answer used with the **LINE** command to make an 8″ line .006 in./in. longer in a pattern to allow for shrinkage in the final casting. Show only the expression and answer.

9. Solve for the deflection of a structural member. The formula is written as *PL3/48EI*, where *P* = pounds of force, *L* = length of beam, *E* = modulus of elasticity, and *I* = moment of inertia. The values to be used are *P* = 4000 lbs, *L* = 240″, and *E* = 1,000,000 lbs/in². The value for *I* is the result of the beam (width × height³)/12, where width = 6.75″ and height = 13.5″.

10. Calculate the coordinate located at 4,4,0 + 3<30.

11. Calculate the coordinate located at (3 + 5,1 + 1.25,0) + (2.375,1.625,0).

12. Draw the part view shown. Check the time when you start the drawing. Draw all the features using the **PLINE** and **CIRCLE** commands. Use the **Area** option of the **MEASUREGEOM** command and the **Object, Add area**, and **Subtract area** functions to calculate the following:
A. The area and perimeter of Object A.
B. The area and perimeter of area B. The slot ends are full radius.
C. The area and circumference of one of the circles (C).
D. The area of Object A, minus the area of Object B.
E. The area of Object A, minus the areas of the other three features.
Enter the **TIME** command and note the editing time spent on your drawing. Save the drawing as P15-12.

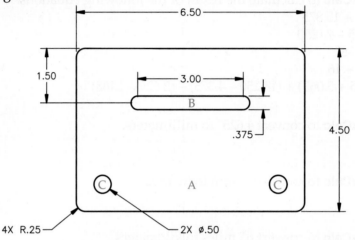

13. Draw the deck shown using the **PLINE** command. Use the **POLYGON** command to draw the hot tub. Use the following guidelines to complete this problem:
A. Specify architectural units for your drawing. Use 1/2" fractions and decimal degrees. Leave the remaining settings for the drawing units at the default values.
B. Set the limits to 100′,80′ and use the **All** option of the **ZOOM** command.
C. Set the grid spacing to 2′ and the snap spacing to 1′.
D. Calculate the measurements listed below.
    a. The area and perimeter of the deck (A).
    b. The area and perimeter of the hot tub (B).
    c. The area of the deck minus the area of the hot tub.
    d. The distance between Point C and Point D.
    e. The distance between Point E and Point C.
    f. The coordinates of Points C, D, and F.
E. Enter the **DBLIST** command and check the information listed for the drawing.
F. Enter the **TIME** command and note the total editing time spent on the drawing.
G. Save the drawing as P15-13.

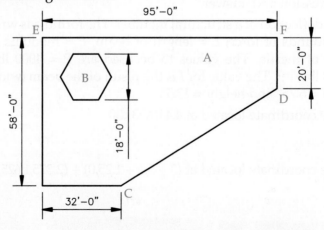

AutoCAD and Its Applications—Basics

14. The drawing shown is a side view of a pyramid. The pyramid has four sides. Create an auxiliary view showing the true size of a pyramid face. Save the drawing as P15-14. Calculate the following:
    A. The area of one side.
    B. The perimeter of one side.
    C. The area of all four sides.
    D. The area of the base.
    E. The true length (distance) from the midpoint of the base on one side to the apex.

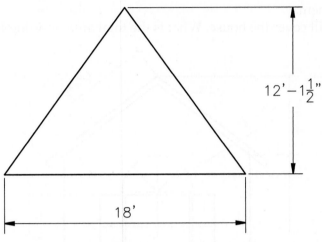

$12'-1\frac{1}{2}"$

$18'$

▼ **Advanced**

15. Make the following calculations given the right triangle shown.
    A. Length of side c (hypotenuse).
    B. Sine of angle A.
    C. Sine of angle B.
    D. Cosine of angle A.
    E. Tangent of angle A.
    F. Tangent of angle B.

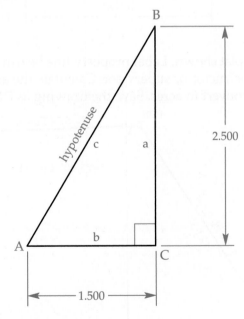

16. Draw the house elevation shown. Draw the windows as single lines only (the location of the windows is not critical). The spacing between each of the second-floor windows is 3″. The width of this end of the house is 16′-6″. The length of the roof is 40′. Use the **PLINE** command to assist in creating the specific shapes in this drawing, except as previously noted. Save the drawing as P15-16. Calculate the following:
   A. The total area of the roof.
   B. The diagonal distance from one corner of the roof to the other.
   C. The area of the first-floor window.
   D. The total area of all second-floor windows, including the 3″ spaces between each of them.
   E. Siding will cover the house. What is the total area of siding for this end?

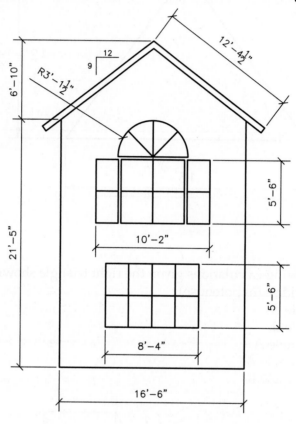

17. Draw the property plat shown. Label property line bearings and distances only if required by your instructor or supervisor. Calculate the area of the property plat in square feet and convert to acres. Save the drawing as P15-17.

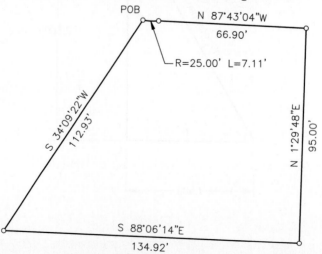

AutoCAD and Its Applications—Basics

18. Draw the subdivision plat. Label the drawing as shown. Calculate the acreage of each lot and record each value as a label inside the corresponding lot (for example, .249 AC). Save the drawing as P15-18.

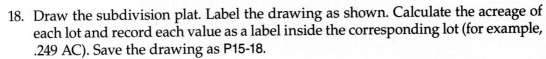
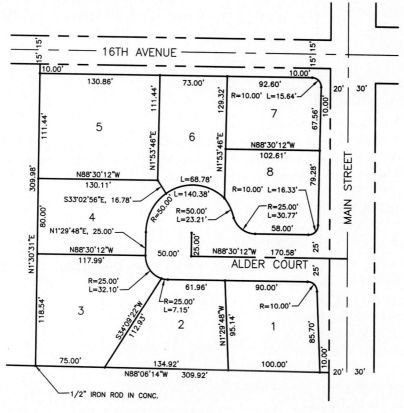
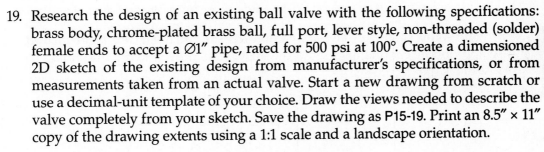

19. Research the design of an existing ball valve with the following specifications: brass body, chrome-plated brass ball, full port, lever style, non-threaded (solder) female ends to accept a Ø1" pipe, rated for 500 psi at 100°. Create a dimensioned 2D sketch of the existing design from manufacturer's specifications, or from measurements taken from an actual valve. Start a new drawing from scratch or use a decimal-unit template of your choice. Draw the views needed to describe the valve completely from your sketch. Save the drawing as P15-19. Print an 8.5" × 11" copy of the drawing extents using a 1:1 scale and a landscape orientation.

AutoCAD Certified Associate Exam Practice

Answer the following questions. Write your answers on a separate sheet of paper.

1. Which of the following commands can you use to find the length of a line? *Select all that apply.*
   A. **DBLIST**
   B. **DIST**
   C. **LIST**
   D. **MEASUREGEOM**
   E. **TIME**

2. Which of the following expressions in **QuickCalc** adds 3.7, 4.2, and 8.5, then divides the result by the sum of 9.8 and 17.4? *Select the one item that best answers the question.*
   A. 3.7+4.2+8.5÷9.8+17.4
   B. (3.7+4.2+8.5)÷(9.8+17.4)
   C. 3.7+4.2+8.5/9.8+17.4
   D. (3.7+4.2+8.5)/(9.8+17.4)
   E. =sum(3.7,4.2,8.5)/sum(9.8,17.4)

3. Which of the following conversions are possible in **QuickCalc**? *Select all that apply.*
   A. cubic feet to cubic meters
   B. degrees Fahrenheit to degrees Celsius
   C. degrees to radians
   D. miles to kilometers
   E. ounces to milligrams

AutoCAD Certified Professional Exam Practice

Follow the instructions in each problem. Write your answers on a separate sheet of paper.

1. **Navigate to this chapter on the companion website and open CPE-15circum.dwg.** Use appropriate commands to find the answers to the following questions:
   A. What is the circumference of Circle 1?
   B. What is the total area of Circles 3, 4, 5, 6, 7, and 8?

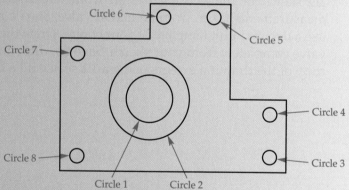

2. **Navigate to this chapter on the companion website and open CPE-15perimeter.dwg.** What is the perimeter of the gear?

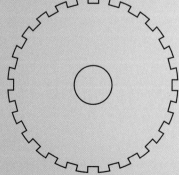

16

Dimension Standards and Styles

Learning Objectives

After completing this chapter, you will be able to:

✓ Describe common dimension standards and practices.

✓ Set drawing scale using manual and annotative techniques.

✓ Create and manage dimension styles.

A *dimension* typically includes numerical values, lines, symbols, and notes. **Figure 16-1** shows typical dimension elements and dimensioning applications. Use dimension commands to dimension the size and location of features and objects. Dimension styles control the initial appearance of dimension elements. Dimensional constraint tools, explained in Chapter 22, allow you to use dimensions to control object size and location.

dimension: A description of the size, shape, or location of features on an object or structure.

Dimension Standards and Practices

Dimensions help communicate drawing information. Each drafting field uses different dimensioning practices. Dimensioning practices depend on product requirements, manufacturing or construction accuracy, standards, and tradition. It is important for you to draw dimensions according to industry and company standards. Dimension standards help to ensure that product manufacturing or construction is accurate. In addition, consistent dimension formatting is critical to legibility and drawing clarity. A drawing should use the same general dimension format throughout when possible.

This textbook presents mechanical drafting dimensioning standards according to ASME Y14.5-2009, *Dimensioning and Tolerancing*, published by the American Society of Mechanical Engineers (ASME). When appropriate, this textbook also references International Standards Organization (ISO) standards and discipline-specific standards, including the United States National CAD Standard® (NCS) and American Welding Society (AWS) standards.

Figure 16-1.
Dimensions describe the size and location of objects and features. Follow accepted conventions when dimensioning.

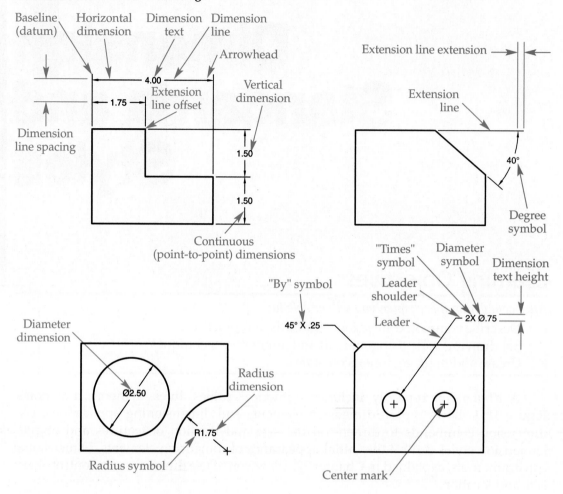

Unidirectional Dimensioning

unidirectional dimensioning:
A dimensioning system in which all dimension values are displayed horizontally on the drawing.

Unidirectional dimensioning is common in mechanical drafting. The term *unidirectional* means "in one direction." Unidirectional dimensioning allows you to read all dimensions from the bottom of the sheet. Unidirectional dimensions often have arrowheads at the ends of dimension lines. The dimension value usually appears in a break near the center of the dimension line. See **Figure 16-2.**

Figure 16-2.
Basic mechanical drafting views using unidirectional dimensions. All dimension values and notes read horizontally on the drawing.

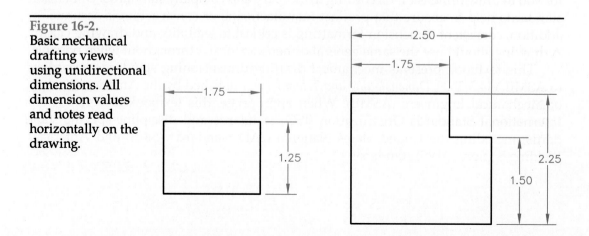

Aligned Dimensioning

Aligned dimensions are most common on architectural and structural drawings. Dimension values read at the same angle as the dimension line. Horizontal values read horizontally, and vertical values are rotated 90° to read from the right side of the sheet. Notes usually read horizontally. Tick marks, dots, or arrowheads commonly terminate aligned dimension lines. In architectural drafting, you generally place the dimension number above the dimension line and use tick mark terminators. See Figure 16-3.

aligned dimensioning: A dimensioning system in which dimension values align with dimension lines.

Size and Location Dimensions

Size dimensions provide the size of physical geometric *features*. See Figure 16-4. *Location dimensions* locate features. See Figure 16-5. Dimension to the center of circular features, such as holes and arcs, in the view in which they appear circular. Dimension to the edges of rectangular features. The *rectangular coordinate system* and the *polar coordinate system* are the two basic systems for applying location dimensions. See Figure 16-6. An example of location dimensions used in architectural drafting is dimensioning to the center of windows and doors on a floor plan.

size dimensions: Dimensions that provide the size of physical features.

feature: Any physical portion of a part or object, such as a surface, hole, window, or door.

location dimensions: Dimensions that locate features on an object without specifying the size of the feature.

rectangular coordinate system: A system for locating dimensions from surfaces, centerlines, or center planes using linear dimensions.

polar coordinate system: A coordinate system in which angular dimensions locate features from surfaces, centerlines, or center planes.

Figure 16-3.
An example of aligned dimensioning in architectural drafting. Notice the tick marks used instead of arrowheads and the placement of the dimensions above the dimension line.

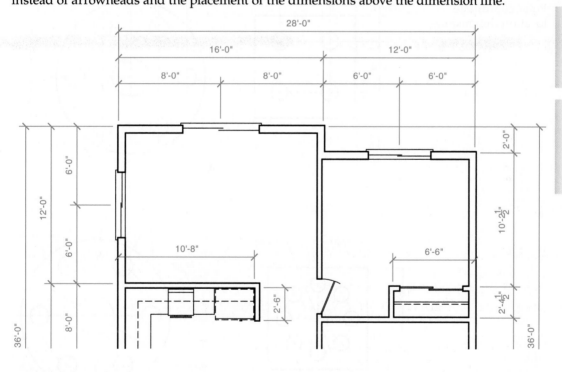

Figure 16-4.
Size dimensions describe the size of features, such as those in this part view.

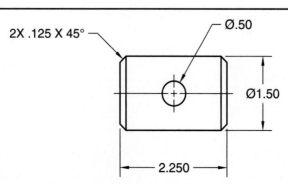

Figure 16-5.
Using location dimensions to locate circular and rectangular features on part drawings.

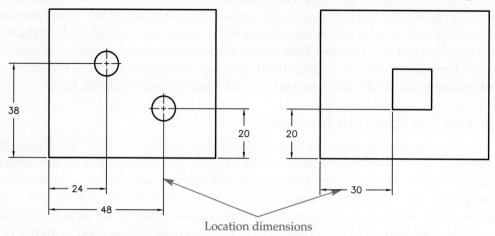

Location dimensions

Figure 16-6.
A—Rectangular coordinate location dimensions. B—Polar coordinate location dimensions.

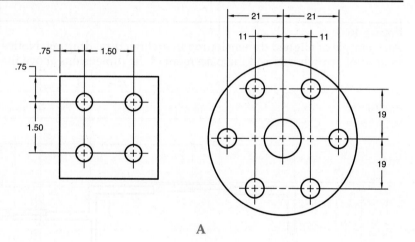

A

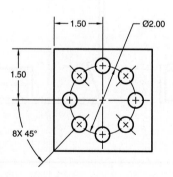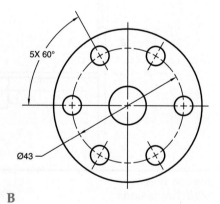

B

Figure 16-7.
A—A specific note added to a mechanical part drawing. B—A specific note added to an architectural roof plan. C—General notes on a mechanical part drawing. D—General notes on an architectural floor plan.

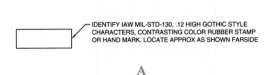

IDENTIFY IAW MIL-STD-130, .12 HIGH GOTHIC STYLE CHARACTERS, CONTRASTING COLOR RUBBER STAMP OR HAND MARK. LOCATE APPROX AS SHOWN FARSIDE

A

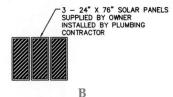

3 – 24" X 76" SOLAR PANELS SUPPLIED BY OWNER INSTALLED BY PLUMBING CONTRACTOR

B

NOTES:
1. DIMENSIONS AND TOLERANCES PER ASME Y14.5-2009.
2. REMOVE ALL BURRS AND SHARP EDGES R.020 MAX UNLESS OTHERWISE SPECIFIED.
3. HEAT TREAT H900 TO Rc 42-44.

C

GENERAL NOTES:
1. PROVIDE SCREENED VENTS @ EA. 3RD. JOIST SPACE @ ALL ATTIC EAVES.
2. PROVIDE SCREENED ROOF VENTS @ 10'-0" O.C. (1/300 VENT TO ATTIC SPACE).
3. USE 1/2" CCX PLY. @ ALL EXPOSED EAVES.
4. USE 300# COMPOSITION SHINGLES OVER 15# FELT.

D

Notes

Specific notes and *general notes* provide another way to describe feature size, location, or additional information. See Figure 16-7. Specific notes are attached to the dimensioned feature using a leader line. Place general notes in the lower-left corner, upper-left corner, or above or next to the title block, depending on sheet size and industry, company, or school practice.

specific notes: Notes that relate to individual or specific features on the drawing.

general notes: Notes that apply to the entire drawing.

Dimensioning Features and Objects

In mechanical drafting, you dimension flat surfaces using measurements for each feature. If you provide an overall dimension, you should omit one dimension, because the overall dimension controls the omitted value. See Figure 16-8A. In architectural drafting, it is common to place all dimensions without omitting any when possible to help make construction easier. See Figure 16-8B.

Dimensioning Cylindrical Shapes

You typically dimension the diameter and the length of a cylindrical shape in the view in which the cylinder appears rectangular. See Figure 16-9. The diameter symbol next to the dimension indicates that the part is a cylinder. This allows you to omit the view in which the cylinder appears as a circle.

Dimensioning Square and Rectangular Features

You usually dimension square and rectangular features in the views that show the length and height. If appropriate, add a square symbol preceding the dimension for a square feature to eliminate the need for an additional view. See Figure 16-10.

Dimensioning Cones and Regular Polygons

One method to dimension a conical shape is to dimension the length and the diameters at both ends. An alternative is to dimension the taper angle and the length. Regular polygons that have an even number of sides are usually dimensioned by giving the distance across the flats and the length. Figure 16-11 shows examples of dimensioning cones and regular polygons.

Figure 16-8.
A—Dimensioning flat surfaces.
B—Dimensioning architectural features.

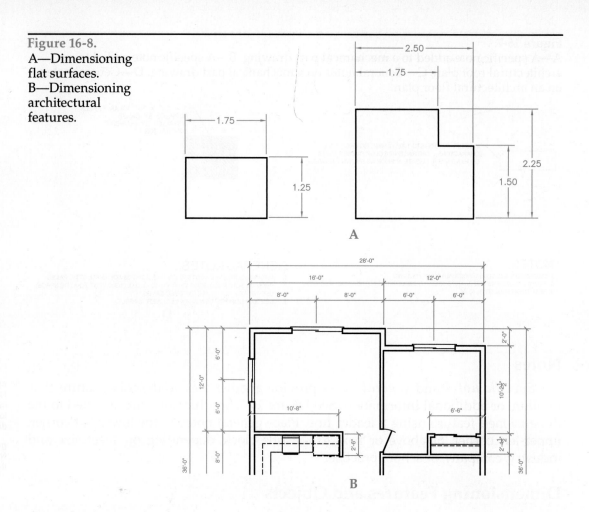

A

B

Figure 16-9.
Dimensioning cylindrical shapes.

Circular view can be omitted

Ø32   56

Figure 16-10.
Dimensioning square and rectangular features.

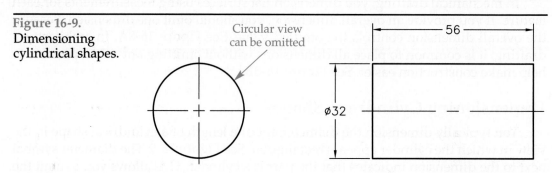

Square symbol

□28   50   28   38

50

Square view can be omitted

Figure 16-11.
Dimensioning cones and hexagonal cylinders.

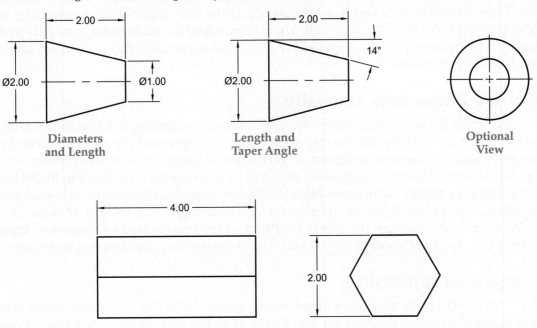

Drawing Scale and Dimensions

Ideally, you should determine drawing scale, scale factors, and dimension size characteristics before you begin drawing. Incorporate these settings into your drawing template files, and make changes when necessary. The drawing scale factor determines how dimensions plot and appear on-screen.

To help understand the concept of drawing scale, look at the portion of a floor plan shown in **Figure 16-12**. You should draw everything in model space at full scale. This means that the bathtub, for example, is actually drawn 5' long. However, at this

Figure 16-12.
An example of a portion of a floor plan drawn at full scale in model space. A—Dimensions drawn at full scale (1/8" text) is too small compared to the large, full scale objects. B—Dimensions scaled (6" text) to display and plot correctly relative to the size of features on the drawing.

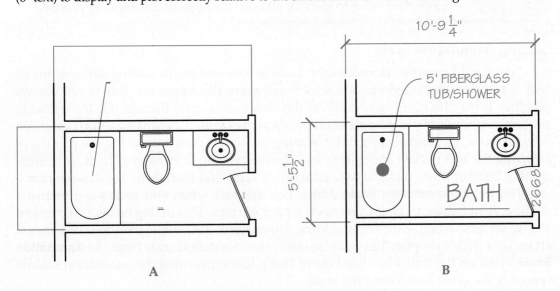

scale, dimension appearance becomes an issue, because full-scale dimension elements, such as 1/8″ dimension text, are too small compared to the other full-scale objects. See Figure 16-12A. As a result, you must adjust the size of dimensions according to the drawing scale. See Figure 16-12B. You can calculate the scale factor manually and apply it to dimensions, or you can allow AutoCAD to calculate the scale factor using annotative dimensions.

Scaling Dimensions Manually

To adjust the size of dimension elements manually according to a specific drawing scale, you must calculate the drawing scale factor. You then multiply the scale factor by the plotted size of dimension elements to get the model space size of dimension elements. Apply the scale factor to all dimension elements by entering the scale factor in the **Fit** tab of the **New** (or **Modify**) **Dimension Style** dialog box, described later in this chapter. For example, if you manually scale dimensions for a drawing with a 1/4″ = 1′-0″ scale, or a scale factor of 48, you must enter 48 in the **Fit** tab of the **New** (or **Modify**) **Dimension Style** dialog box. Refer to Chapter 9 for information on determining the drawing scale factor.

Annotative Dimensions

AutoCAD scales annotative dimensions according to the annotation scale you select, which reduces the need for you to calculate the scale factor. Once you choose an annotation scale, AutoCAD applies the corresponding scale factor to annotative dimensions and all other annotative objects. When you place annotative dimensions using the previous example, you set an annotation scale of 1/4″ = 1′-0″. Then when you draw annotative dimensions, AutoCAD scales dimension elements automatically according to the preset 1/4″ = 1′-0″ annotation scale.

Annotative dimensions offer several advantages over manually scaled dimensions, including the ability to control dimension appearance based on the drawing scale and plotted size of dimension elements, while reducing the need to focus on the scale factor. Annotative dimensions are especially effective when the drawing scale changes or when a single sheet includes views at different scales.

PROFESSIONAL TIP

If you anticipate preparing scaled drawings, you should use annotative dimensions and other annotative objects instead of manual scaling. However, scale factor does influence non-annotative items and is still an important value to identify and use throughout the drawing process.

Setting Annotation Scale

You should usually set annotation scale before you begin adding dimensions so that dimension characteristics are scaled automatically. However, this is not always possible. It may be necessary to adjust the annotation scale throughout the drawing process, especially if you prepare multiple drawings with different scales on one sheet. This textbook approaches annotation scaling in model space only, using the process of selecting the appropriate annotation scale before placing dimensions. To draw dimensions at another scale, pick the new annotation scale and then draw the dimensions.

The **Select Annotation Scale** dialog box appears when you access a dimension command and an annotative dimension style is current. This dialog box is a convenient way to set annotation scale before adding dimensions. You will learn about dimension styles later in this chapter. You can also select the annotation scale from the **Annotation Scale** flyout on the status bar. See Figure 16-13. Remember that the annotation scale is typically the same as the drawing scale.

Figure 16-13.
Pick the **Annotation Scale** flyout on the status bar to activate an annotation scale.

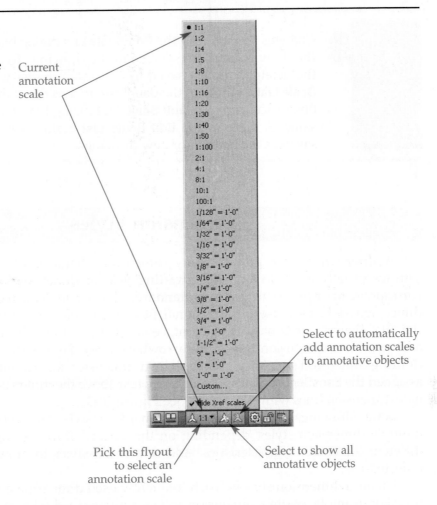

Current annotation scale

Select to automatically add annotation scales to annotative objects

Pick this flyout to select an annotation scale

Select to show all annotative objects

This textbook describes many additional annotative object tools. Some of these tools are more appropriate for working with layouts, as explained later in this textbook.

Editing Annotation Scales

To add a scale that is unavailable, or to change an existing scale, pick the **Annotation Scale** flyout on the status bar and choose **Custom...** to access the **Edit Scale List** dialog box. Move the highlighted scale up or down in the list using the **Move Up** or **Move Down** button. To remove the highlighted scale from the list, pick the **Delete** button.

Select the **Edit...** button to open the **Edit Scale** dialog box, where you can change the name of the scale and adjust the scale by entering the paper and drawing units. For example, a scale of 1/4" = 1'-0" uses a paper units value of .25 or 1 and a drawing units value of 12 or 48.

To create a new annotation scale, pick the **Add...** button to display the **Add Scale** dialog box, which provides the same options as the **Edit Scale** dialog box. Pick the **Reset** button to restore the list to display the default annotation scales. After you select an annotation scale, you are ready to place annotative dimensions.

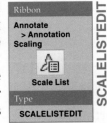

Ribbon

Annotate > Annotation Scaling

Scale List

Type

SCALELISTEDIT

SCALELISTEDIT

Changes you make in the **Edit Scale List** dialog box are stored with the drawing and are specific to the drawing. To make changes to the default scale list saved to the system registry, pick the **Default Scale List...** button in the **User Preferences** tab of the **Options** dialog box to access the **Default Scale List** dialog box. The options are the same as those in the **Edit Scale List** dialog box, but changes are saved as the default for new drawings.

Dimension Styles

dimension style: A saved configuration of dimension appearance settings.

A *dimension style* presets many dimension characteristics. Dimension style settings usually apply to a specific drafting field or dimensioning application and correspond to appropriate drafting standards. For example, a mechanical drafting dimension style may use unidirectional placement, reference a text style assigned the Arial, Romans, or Century Gothic font, center text in a break in the dimension line, and terminate dimension lines with arrowheads. See **Figure 16-2**. An architectural drafting dimension style may use an aligned dimension format, reference a text style assigned the SansSerif or Stylus BT font, place text above the dimension line, and terminate dimension lines with tick marks. See **Figure 16-3**.

Some drawings require only one dimension style. However, you may need multiple dimension styles, depending on the variety of dimensions you apply and different dimension characteristics. Add dimension styles to drawing templates for repeated use.

Create a dimension style for each frequently used dimension appearance or function. For example, create a dimension style for unspecified tolerances and a different dimension style for a common specified tolerance. Another example is developing a separate dimension style to add dimensions with a common prefix or suffix.

PROFESSIONAL TIP

AutoCAD provides the flexibility to control the appearance of dimensions for various dimensioning requirements without using a separate dimension style. For example, you can change the precision of or add a prefix to a limited number of special-case dimensions, instead of creating separate dimension styles.

DIMSTYLE

Ribbon
Home
> Annotation

Dimension Style

Annotate
> Dimensions

Dimension Style

Type
DIMSTYLE
DIMSTY
DDIM
DST
D

Dimension Style Manager

Create, modify, and delete dimension styles using the **Dimension Style Manager** dialog box. See **Figure 16-14**. The **Styles:** list box displays existing dimension styles. The Annotative dimension style allows you to create annotative dimensions, as indicated by the icon to the left of the style name. The Standard dimension style does not use the annotative function.

To make a dimension style current, double-click the style name, right-click on the name and select **Set current**, or pick the name and select the **Current** button. Use the **List:** drop-down list to filter the number of dimension styles displayed in the **Styles:** list box. Pick the **All Styles** option to show all dimension styles in the file, or pick the **Styles in use** option to show only the current style and styles used in the drawing. If the current drawing contains external references (xrefs), use the **Don't list styles in**

Figure 16-14.
The **Dimension Style Manager** dialog box.

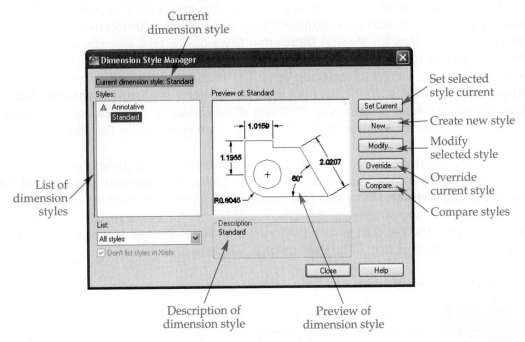

Current dimension style

Set selected style current

Create new style

Modify selected style

Override current style

Compare styles

List of dimension styles

Description of dimension style

Preview of dimension style

Xrefs box to eliminate xref-dependent dimension styles from the **Styles:** list box. This setting is valuable because you cannot set xref dimension styles current or use them to create new dimensions. You will learn about external references later in this textbook.

The **Description** area provides information about the selected dimension style. The **Preview of:** image displays a representation of the dimension style and changes according to the selections you make. If you change dimension settings without creating a new dimension style, the changes are automatically stored as a dimension style override.

Creating New Dimension Styles

To create a new dimension style, select an existing dimension style from the **Styles:** list box to use as a base for formatting the new dimension style. Then pick the **New...** button to open the **Create New Dimension Style** dialog box. See **Figure 16-15.** You can base the new dimension style on the formatting of a different dimension style by selecting from the **Start With** drop-down list. Notice that Copy of followed by the name of the existing style appears in the **New Style Name** text box. Replace the default name with a more descriptive name, such as Mechanical or Architectural. Dimension

Figure 16-15.
The **Create New Dimension Style** dialog box.

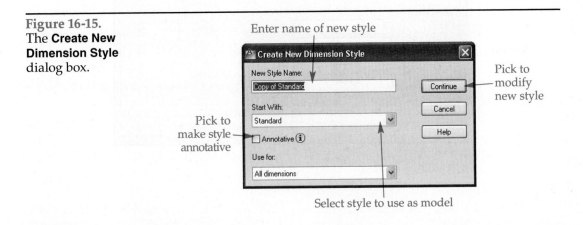

Enter name of new style

Pick to modify new style

Pick to make style annotative

Select style to use as model

style names can have up to 255 characters, including uppercase or lowercase letters, numbers, dashes (–), underlines (_), and dollar signs ($).

Pick the **Annotative** check box to make the dimension style annotative. You can also make the dimension style annotative by selecting the **Annotative** check box in the **Fit** tab of the **New** (or **Modify**) **Dimension Style** dialog box, described later in this chapter. The **Use for** drop-down list specifies the type of dimensions to which the new style applies. Use the **All dimensions** option to create a new dimension style for all types of dimensions. If you select the **Linear dimensions**, **Angular dimensions**, **Radius dimensions**, **Diameter dimensions**, **Ordinate dimensions**, or **Leaders and Tolerances** option, you create a sub-style of the dimension style specified in the **Start With:** text box.

Pick the **Continue** button to open the **New Dimension Style** dialog box and adjust dimension style settings. See **Figure 16-16**. The **Lines**, **Symbols and Arrows**, **Text**, **Fit**, **Primary Units**, **Alternate Units**, and **Tolerances** tabs display groups of settings for specifying dimension appearance, as described in this chapter. After completing the style definition, pick the **OK** button to return to the **Dimension Style Manager** dialog box. Pick the **Close** button to exit the **Dimension Style Manager** dialog box.

The preview image shown in the upper-right corner of each tab of the **New** (or **Modify**) **Dimension Style** dialog box displays a representation of the dimension style and changes according to the selections you make.

Figure 16-16.
The **Lines** tab of the **New** (or **Modify**) **Dimension Style** dialog box.

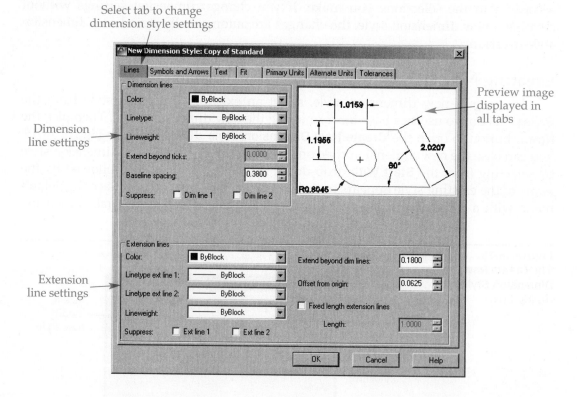

AutoCAD stores dimension style settings as *dimension variables*. Dimension variables have limited practical uses and are more likely to apply to advanced applications such as scripting and customizing. Changing dimension variables by typing the variable name is not a recommended method for changing dimension style settings. Changes made in this manner can introduce inconsistencies with other dimensions. You should make changes to dimensions by redefining styles or performing style overrides.

dimension variables: System variables that store the values of dimension style settings.

Reference Material

Dimension Variables
For a list of dimension variables, go to the **Reference Material** section of the companion website (www.g-wlearning.com/CAD) and select **Dimension Variables**.

Lines Tab

Figure 16-16 shows the **Lines** tab of the **New** (or **Modify**) **Dimension Style** dialog box. The **Lines** tab controls the display of dimension and extension lines. A dimension style presets the appearance of dimension and extension lines for common applications. You can edit specific dimensions when necessary without using a separate dimension style.

Dimension Line Settings

The **Dimension lines** area of the **Lines** tab allows you to set dimension line format. **Color**, **Linetype**, and **Lineweight** drop-down lists are available for changing the dimension line color, linetype, and lineweight. *Associative dimensions* are block objects. Chapter 24 explains blocks and provides complete information on using ByBlock, ByLayer, or absolute color, lineweight, and linetype with blocks. For now, as long as you assign a specific layer to a dimension object and do not change the dimension properties to absolute values, the default ByBlock properties are acceptable.

associative dimension: A dimension in which all elements are linked to, or associated with, the dimensioned object; updates when the associated object changes.

The **Extend beyond ticks** text box is inactive unless you select oblique or architectural tick terminators from the **Symbols and Arrows** tab of the **New** (or **Modify**) **Dimension Style** dialog box. Architectural tick marks or oblique arrowheads are common dimension line terminators on architectural drawings. In architectural dimensioning formats, dimension lines sometimes extend past extension lines, as shown in **Figure 16-17**. The 0.00 default draws dimension lines that do not extend past extension lines.

Figure 16-17.
Using the **Extend beyond ticks** setting to allow the dimension line to extend past the extension line. With the default value of 0, the dimension line ends at the extension line.

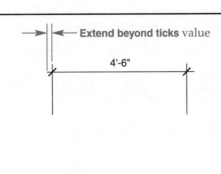

Use the **Baseline spacing** text box to change the spacing between the dimension lines of baseline dimensions created with the **DIMBASELINE** command. The default spacing is too close for most drawings, as shown in Figure 16-18. ASME standards recommend a minimum spacing of .375″ (10 mm) from a drawing feature to the first dimension line and a minimum spacing of .25″ (6 mm) between dimension lines. A minimum spacing of 3/8″ is common for architectural drawings. These minimum recommendations are generally less than the spacing required by actual company or school standards. A value of .5″ (12 mm) or .75″ (19 mm) is usually more appropriate.

Check the **Suppress** boxes to hide the first, second, or both sides of dimension lines and dimension line terminators for dimension lines broken by a value. The **Dim line 1** and **Dim line 2** check boxes refer to the first and second points you pick when drawing a dimension. Both dimension lines appear by default. Figure 16-19 shows the result of using dimension line suppression options.

Extension Line Settings

The **Extension lines** area of the **Lines** tab allows you to set extension line format. **Color, Linetype ext line 1, Linetype ext line 2**, and **Lineweight** drop-down lists are available for changing the extension line color, linetype, and lineweight from the default ByBlock setting, if necessary. You can use the **Linetype ext line 1** and **Linetype ext line 2** drop-down lists to override the linetype applied to each extension line. Extension lines 1 and 2 correspond to the first and second points you pick when drawing a dimension.

Use the **Extend beyond dim lines** option to set the distance the extension line runs past the dimension line. See Figure 16-20. ASME standards recommend a .125″ (3 mm)

Figure 16-18.
The **Baseline spacing** setting controls the spacing between dimension lines when you use the **DIMBASELINE** command.

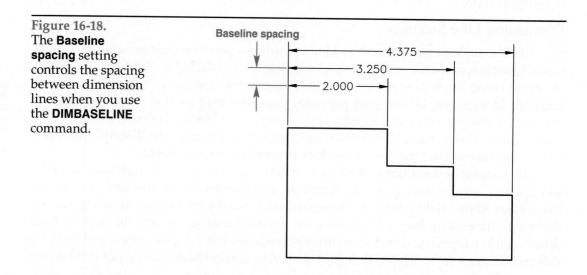

Figure 16-19.
Using the **Dim line 1** and **Dim line 2** dimensioning settings. "Off" is equivalent to an unchecked **Suppress** check box in the **Lines** tab.

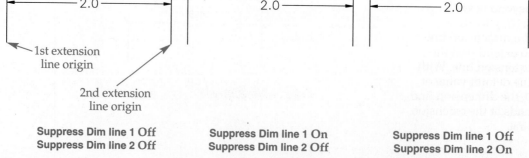

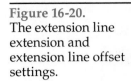

Figure 16-20.
The extension line
extension and
extension line offset
settings.

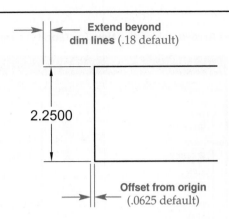

extension line extension. The **Offset from origin** option specifies the distance between the object and the beginning of the extension line. Most applications require this small offset. ASME standards recommend a .063″ (1.5 mm) extension line offset. When an extension line meets a centerline, however, use a setting of 0.0 to prevent a gap.

The **Fixed length extension lines** check box sets a given length for extension lines. Check this box to activate the **Length** text box. The value in the **Length** text box sets a restricted length for extension lines, measured from the dimension line toward the extension line origin.

Check the **Suppress** boxes to hide the first, second, or both extension lines. The **Ext line 1** and **Ext line 2** check boxes refer to the first and second points you pick when drawing a dimension. Both extension lines appear by default. **Figure 16-21** shows a common example of suppressing extension lines when they coincide with object lines.

Symbols and Arrows Tab

Figure 16-22 shows the **Symbols and Arrows** tab of the **New** (or **Modify**) **Dimension Style** dialog box. The **Symbols and Arrows** tab controls the appearance of dimension line and leader terminators, center marks, and other symbol components of dimensions. A dimension style presets the appearance of symbols and arrows for common applications. You can edit specific dimensions when necessary without using a separate dimension style.

Arrowhead Settings

Use the appropriate drop-down list in the **Arrowheads** area to select the arrowhead to use for the first, second, and leader arrowheads. The default arrowhead is **Closed filled**, which is recommended by ASME standards, although **Closed blank**, **Closed**, or **Open** arrowheads are sometimes used. A leader pointing to a surface in a view where the surface appears as a plane terminates with a small dot. **Figure 16-23** shows arrowhead styles. If you pick a new arrowhead in the **First:** drop-down list, AutoCAD automatically makes the same selection for the **Second:** drop-down list. When you select the **Oblique** or **Architectural tick** arrowhead, the **Extend beyond ticks:** text box in the **Lines** tab becomes activated.

Figure 16-21.
Examples of when
it is appropriate to
suppress extension
lines. Place dimension
lines away from
objects, when
possible, to display
both extension lines.

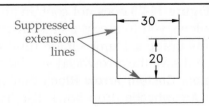

Figure 16-22.
The **Symbols and Arrows** tab of the **New** (or **Modify**) **Dimension Style** dialog box.

Select tab to specify arrow style

Arrowhead properties

Center mark properties

Dimension break size

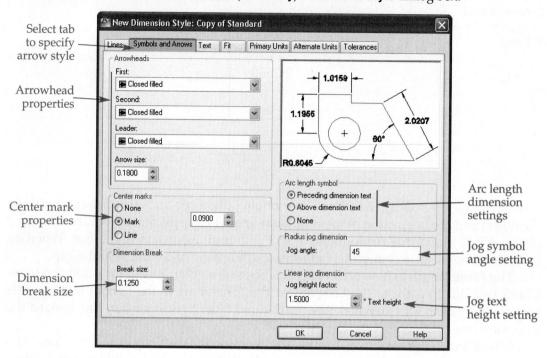

Arc length dimension settings

Jog symbol angle setting

Jog text height setting

Figure 16-23.
Examples of dimensions drawn using the options in the **Arrowheads** drop-down lists.

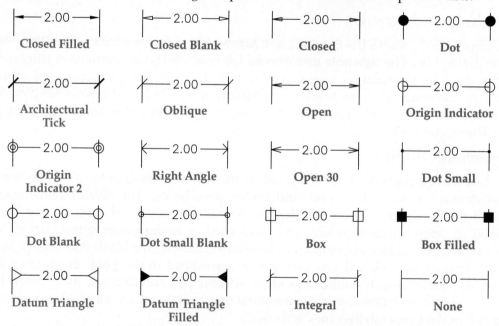

Notice that **Figure 16-23** does not contain an example of a user arrow. The **User Arrow...** option allows you to access a custom arrowhead. You must first design an arrowhead that fits inside a 1 unit square (unit block) with a dimension line "tail" 1 unit in length, and save the arrowhead as a block. Blocks are described later in this textbook. The **Select Custom Arrow Block** dialog box appears when you pick **User Arrow...** from an **Arrowheads** drop-down list. Type the name of the custom arrow block in the **Select from Drawing Blocks:** text box or pick a block from the drop-down list and then pick **OK** to apply the arrowhead to the style.

AutoCAD and Its Applications—Basics

Use the **Arrow size:** text box to change arrowhead size. **Figure 16-24** shows the measurement you specify to set the arrowhead size. A .125″ (3 mm) arrowhead size is most common, especially on mechanical drawings.

Center Mark Settings

The **Center marks** area allows you to select the way center dashes and centerlines appear in circles and arcs when you use circular feature dimensioning commands. The ASME standard refers to AutoCAD center marks as the center dashes of centerlines. Center marks are typically applied to circular objects that are too small to receive centerlines. The **None** option results in no center marks or centerlines in circles and arcs. Fillets and rounds generally have no center marks. The **Mark** option places center dashes. The **Line** option places centerlines.

Use the **Size:** text box to change the size of the center mark and centerline. The size defines half the length of a centerline dash and the distance that the centerline extends past the object. A value of .0625″ (1.5 mm) is appropriate for the centerline dash half-length, but does not provide for the preferred .125″ (3 mm) extension past the object. **Figure 16-25** shows the results of specifying center marks and centerlines.

Adjusting Break Size

The **Dimension Break** area controls the amount of extension line that is hidden when you use the **DIMBREAK** command. Specify a value in the **Break size:** text box to set the total length of the break. **Figure 16-26** shows an example of a .125″ (3 mm) extension line break. The default size is .125″ (3 mm). ASME standards do not recommend breaking extension lines.

Adding an Arc Length Symbol

The **Arc Length Symbol** area controls the placement of the arc length symbol when you use the **DIMARC** command. The default **Preceding dimension text** option places the symbol in front of the dimension value. Select the **Above dimension text** radio button to place the arc length symbol over the length value. See **Figure 16-27**. Pick the **None** radio button to hide the symbol.

Figure 16-24.
ASME standards specify an arrowhead size of .125″ (3 mm). The **Closed filled, Closed blank,** and **Closed** arrowhead styles adhere to the standard 3:1 ratio of length to width.

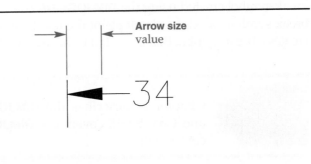

Figure 16-25.
Arcs and circles displayed with center marks and centerlines. Use a size of .0625″ (1.5 mm) to achieve ASME standards for center marks, but not the extension past objects.

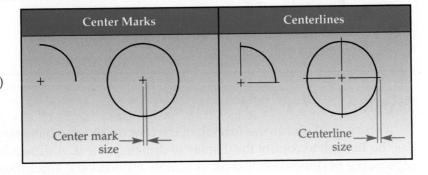

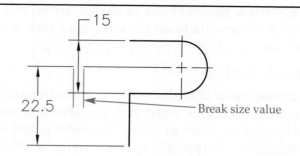

Figure 16-26.
Use the **Break size** setting to specify the length of the break created using the **DIMBREAK** command.

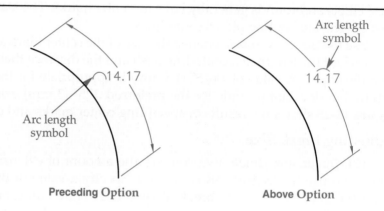

Figure 16-27.
You can place the arc length symbol in front of or above the arc dimension text.

Arc length symbol

Arc length symbol

Preceding Option

Above Option

Adjusting Jog Angle

Use the **Jog angle** setting in the **Radius jog dimension** area to set the appearance of the break line applied to the jog symbol when you use the **DIMJOGGED** command. This value sets the incline formed by the line connecting the extension line and dimension line. The default angle is 45°.

Setting Jog Height

The **Jog height factor** setting in the **Linear jog dimension** area controls the size of the break symbol created using the **DIMJOGLINE** command. This value sets the height of the break symbol based on a multiple of the text height. For example, the default value of 1.5 creates a break symbol that is .18" tall if the text height is .12". The default angle is 45°.

Chapter 17 describes the **DIMJOGLINE** command in more detail, and Chapter 18 covers the **DIMJOGGED**, **DIMARC**, and **DIMBREAK** commands.

Exercise 16-1

Complete the exercise on the companion website.
www.g-wlearning.com/CAD

Text Tab

Figure 16-28 shows the **Text** tab of the **New** (or **Modify**) **Dimension Style** dialog box. Use the **Text** tab to control the display of dimension text. A dimension style presets the appearance of dimension text for common applications. You can edit specific dimensions when necessary without using a separate dimension style.

Select tab to set up dimension text

Set appearance of the text

Set location of text relative to dimension line

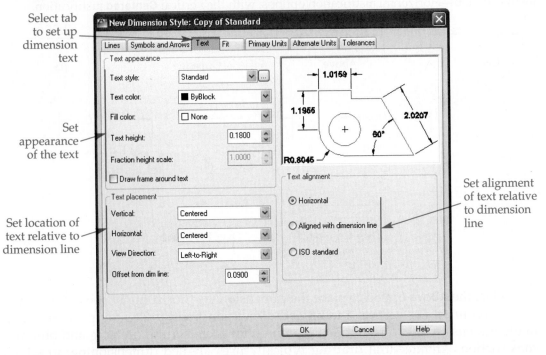

Set alignment of text relative to dimension line

Text Appearance Settings

Use the **Text appearance** area to set the dimension text style, color, height, and frame. A dimension style references a text style for the appearance of dimension values. Pick an existing text style from the **Text style** drop-down list. To create or modify a text style, pick the ellipsis (**...**) button next to the drop-down list to launch the **Text Style** dialog box. Use the **Text color** drop-down list to specify the appropriate text color, which should be ByBlock for typical applications.

Use the **Text height** text box to specify the dimension text height. Dimension text height is commonly the same as the text height used for most other drawing text, except for titles, which are often larger. The default dimension text height of .18″ (2.5 mm) is an acceptable standard. Many companies use a text height of .125″ (3 mm). The ASME standard recommends a .12″ (3 mm) text height. The ASME standard text height for titles and labels is .24″ (6 mm).

The **Fraction height scale** setting controls the height of fractions for architectural and fractional unit dimensions. The value in the **Fraction height scale** box is multiplied by the text height value to determine the height of the fraction. A value of 1.0 creates fractions that are the same text height as regular (nonfractional) text, which is an accepted standard. A value less than 1.0 makes the fraction smaller than the regular text height.

Select the **Draw frame around text** check box to create a box around the dimension text. A *basic dimension* is a common application for framed text, as described later in this textbook. The setting for the **Offset from dim line** value, explained later in this chapter, determines the distance between the text and the frame.

basic dimension:
A theoretically perfect dimension used to describe the exact size, profile, orientation, and location of a feature.

Text Placement Settings

The **Text placement** area controls text placement relative to the dimension line. See **Figure 16-29.** The **Vertical:** drop-down list provides vertical justification options. Use the default **Centered** option to place dimension text centered in a gap in the dimension line. This is the most common dimensioning practice in mechanical drafting and many other fields.

Figure 16-29.
Dimension text justification options. A—Vertical justification options, with the horizontal **Centered** justification. B—Horizontal justification options, with the vertical **Centered** justification.

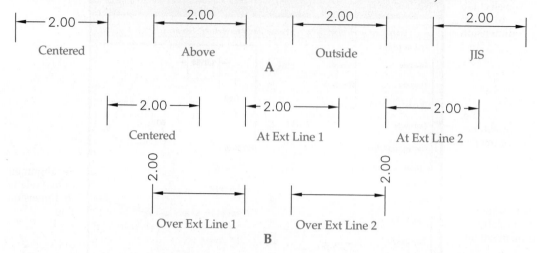

Select the **Above** option to place the dimension text horizontally above horizontal dimension lines. For vertical and angled dimension lines, the text appears in a gap in the dimension line. This option is common for architectural drafting and building construction. Architectural drafting typically uses aligned dimensioning, in which the dimension text aligns with the dimension lines and all text reads from either the bottom or the right side of the sheet.

Pick the **Outside** option to place the dimension text outside the dimension line and either above or below a horizontal dimension line or to the right or left of a vertical dimension line. The direction in which you move the cursor determines the above/ below and left/right placement. Choose the JIS option to align the text according to the Japanese Industrial Standards (JIS).

The **Horizontal:** drop-down list provides options for controlling the horizontal placement of dimension text. Pick the default **Centered** option to center dimension text between the extension lines. Select the **At Ext Line 1** option to locate the text next to the extension line you place first, or choose **At Ext Line 2** to locate the text next to the extension line you place second. Pick **Over Ext Line 1** to place the text aligned with and over the first extension line, or select **Over Ext Line 2** to place the text aligned with and over the second extension line. Placing text aligned with and over an extension line is not common practice.

The **View Direction:** drop-down list determines the reading direction of dimension text. Use the default **Left-to-Right** option to make text readable from left to right or from bottom to top, depending on the text placement and alignment. Choose the **Right-to-Left** option to flip dimension text. Text may appear inverted and reads from right to left or from top to bottom, depending on the text placement and alignment. Changing text view direction to right-to-left is not common practice.

The **Offset from dim line:** text box sets the gap between the dimension line and dimension text, the distance between the leader shoulder and text, and the space between text and a frame. The gap should be half the text height for most applications. **Figure 16-30** shows the gap in linear dimensions and dimensions that use a leader or frame.

Text Alignment Settings

Use the **Text alignment** area to specify unidirectional or aligned dimensions. The **Horizontal** option draws the unidirectional dimensions that are commonly used in mechanical drafting. The **Aligned with dimension line** option creates aligned dimensions, typical for architectural drafting. The **ISO Standard** option creates aligned

Figure 16-30.
The **Offset from dim line** value controls the space between dimension text and the dimension line, leader shoulder, and frame.

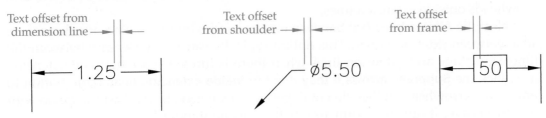

dimensions when the text falls between the extension lines, and horizontal dimensions when the text falls outside the extension lines.

Fit Tab

Figure 16-31 shows the **Fit** tab of the **New** (or **Modify**) **Dimension Style** dialog box. The settings in the **Fit** tab establish dimension *fit format*. A dimension style presets fit format for common applications. You can edit specific dimensions when necessary without using a separate dimension style.

fit format: The arrangement of dimension text and arrowheads on a drawing.

Fit Options

The **Fit options** area controls how text, dimension lines, and arrows behave when there is not enough room between the extension lines to accommodate all of the items. The dimension style settings, such as the height of dimension text, offset, and arrowheads, influence fit performance. All fit options place text and dimension lines with arrowheads inside the extension lines if space is available. All fit options except the **Always keep text between ext lines** option place arrowheads, dimension lines, and text outside the extension lines when space is limited.

Choose the default **Either text or arrows (best fit)** radio button to move either the dimension value or the arrows outside extension lines first. Pick the **Arrows** radio

Figure 16-31.
The **Fit** tab of the **New** (or **Modify**) **Dimension Style** dialog box.

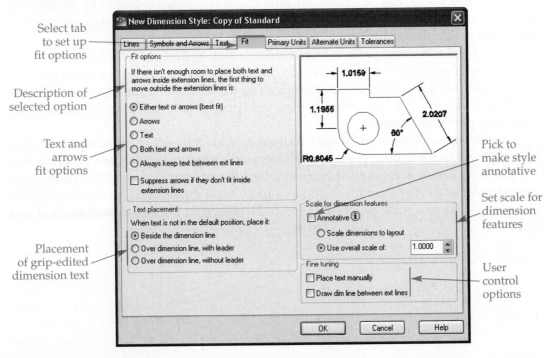

button to attempt to place arrowheads outside extension lines first, followed by text. Pick the **Text** radio button to attempt to place text outside extension lines first, followed by arrowheads. Choose the **Both text and arrows** radio button to move both text and arrowheads outside extension lines.

Select the **Always keep text between ext lines** radio button to place the dimension value between extension lines. This option typically causes interference between the dimension value and extension lines when there is limited space between extension lines. Pick the **Suppress arrows if they don't fit inside extension lines** radio button to remove the arrowheads if they do not fit between extension lines. Use this option with caution, because it can create dimensions that violate standards.

Text Placement Settings

Sometimes it becomes necessary to move the dimension text from its default position. You can stretch the dimension text independently of the dimension. The selected option in the **Text placement** area presets the effect of stretching dimension text.

Select the **Beside the dimension line** radio button to restrict dimension text movement. You can stretch the text with the dimension line, but only within the same plane as the dimension line. Pick the **Over dimension line, with leader** radio button to stretch the text in any direction away from the dimension line. A leader line forms connecting the text to the dimension line. Choose the **Over dimension line, without leader** radio button to stretch the text in any direction away from the dimension line without including a leader.

PROFESSIONAL TIP

To return the dimension text to its default position, select the dimension and use the **Reset Text Position** context-sensitive grip option or right-click and select **Home text** from the **Dim Text position** cascading menu.

Text Scale Options

Use the **Scale for dimension features** area to set the dimension scale factor. Select the **Annotative** check box to create an annotative dimension style. The **Annotative** check box is already set when you modify the Annotative dimension style or pick the **Annotative** check box in the **Create New Dimension Style** dialog box.

Select the **Scale dimensions to layout** radio button to dimension in a floating viewport in a paper space layout. You must add dimensions to the model in a floating viewport in order for this option to function. Scaling dimensions to the layout adjusts the overall scale according to the active floating viewport by setting the overall scale equal to the viewport scale factor.

Pick the **Use overall scale of** radio button to scale a drawing manually. Enter the drawing scale factor to be applied to all dimension elements in the corresponding text box. The scale factor is multiplied by the plotted dimension size to get the dimension size in model space. For example, if you manually scale dimensions for a drawing with a 1:2 (half) scale, or a scale factor of 2 ($2 \div 1 = 2$), enter 2 in the **Use overall scale of** text box.

You can draw dimensions in model space and layout space. Set the model space dimension scale according to the drawing scale factor to achieve the correct dimension appearance. Associative paper space dimensions automatically adjust to model modifications and do not require scaling. In addition, if you dimension in paper space, you can dimension the model differently in two viewports. However, paper space dimensions are not visible when you work in model space, so you must be careful not to move a model space object into a paper space dimension. Avoid using non-associative paper space dimensions.

Fine Tuning Settings

The **Fine tuning** area provides additional options for controlling the placement of dimension text. Select the **Place text manually** check box to increase flexibility when placing dimensions, allowing you to locate text to the side within extension lines or outside of extension lines. However, the **Place text manually** feature can make it more cumbersome to offset dimension lines equally, and it is not necessary for standard dimensioning practices.

The **Draw dim line between ext lines** option forces AutoCAD to place the dimension line inside the extension lines, even when the text and arrowheads are outside. The default application is to place the dimension line and arrowheads outside the extension lines. See **Figure 16-32**. Although some companies prefer the appearance, forcing the dimension line inside the extension lines is not an ASME standard.

Exercise 16-2

Complete the exercise on the companion website.
www.g-wlearning.com/CAD

Primary Units Tab

Figure 16-33 shows the **Primary Units** tab of the **New** (or **Modify**) **Dimension Style** dialog box. The **Primary Units** tab controls linear and angular dimension units. A dimension style presets units for typical dimensioning requirements. You can edit specific dimensions when necessary without using a separate dimension style.

Linear Dimension Settings

The **Linear dimensions** area allows you to specify settings for primary linear dimensions. The options in the **Unit format** drop-down list are the same as those in the **Length** area of the **Drawing Units** dialog box. Typically, primary linear dimension unit format is the same as the corresponding drawing units.

Figure 16-32.
The effect of the **Draw dim line between ext lines** option in the **Fine tuning** area of the **Fit** tab.

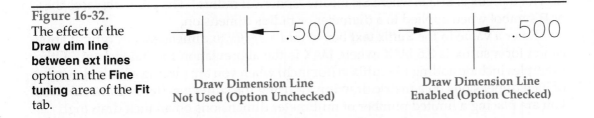

Draw Dimension Line
Not Used (Option Unchecked)

Draw Dimension Line
Enabled (Option Checked)

Figure 16-33.
The **Primary Units** tab of the **New** (or **Modify**) **Dimension Style** dialog box.

Settings for linear units

Select tab to set up primary dimension units

Settings for angular units

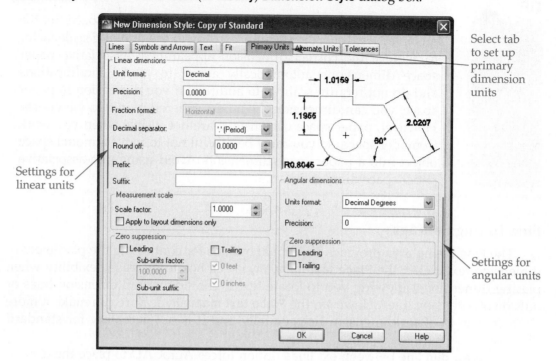

The **Precision** drop-down list sets the precision applied to dimensions, which may be the same as the related drawing units precision. A drawing often includes different precisions applied to specific dimensions. With decimal units, precision determines the number of zeros that follow the decimal place.

Precision settings in mechanical drafting depend on the accuracy required to manufacture specific features. Some features require greater precision, generally due to fits between mating parts. For example, a precision setting of 0.00 represents less exactness than a setting of 0.0000. Chapter 19 further explains this concept. Common precisions in mechanical drafting include 0.0000, 0.000, and 0.00.

When you specify fractional units, precision values identify the smallest desired fractional denominator. Precisions of 1/16 to 1/64 are common, but you can choose a precision from 1/2 to 1/256 . A precision of 0 displays no fractional values.

Use the **Decimal separator** drop-down list to specify commas, periods, or spaces as separators for decimal numbers. The '.' **(Period)** option is the default and is appropriate for typical applications. The **Fraction format** drop-down list is available if the unit format is **Architectural** or **Fractional**. The options for controlling the display of fractions are **Diagonal**, **Horizontal**, and **Not Stacked**.

The **Round off** text box specifies the accuracy of rounding for dimension numbers. The default is 0, which means that no rounding takes place and all associated dimensions specify the value exactly as measured. If you enter a value of .1, all dimensions are rounded to the closest .1 unit. For example, an actual measurement of 1.188 is rounded to 1.2. Rounding is inappropriate for most applications.

Enter a value in the **Prefix text** box to add a *prefix* to dimensions. A typical application for a prefix is SR3.5, where SR means "spherical radius." The prefix replaces the ∅ or R symbol when applied to a diameter or radius dimension.

Enter a value in the **Suffix text** box to add a *suffix* to dimensions. A typical application for a suffix is 3.5 MAX, where MAX is the abbreviation for "maximum." Other examples include adding the suffix in (for inch) when you are placing a limited number of inch dimensions on a metric drawing, or using the suffix mm (for millimeter) when you are placing a limited number of millimeter dimensions on an inch drawing.

prefix: A special note or application placed before the dimension value.

suffix: A special note or application placed after the dimension value.

A prefix or suffix is normally a special specification, used in only a few cases on a drawing. As a result, it is often most effective to enter a prefix or suffix using the **MText** or **Text** option of the related dimensioning command, or edit an existing dimension value to add a prefix or suffix.

Set the scale factor of linear dimensions in the **Scale factor:** text box of the **Measurement scale** area. If you set a scale factor of 1, dimension values display their measured value. If the scale factor is 2, dimension values display two times the measured value. For example, an actual measurement of 2 inches displays as 2 with a scale factor of 1, but the same measurement displays as 4 when the scale factor is 2. Place a check in the **Apply to layout dimensions only** check box to make the linear scale factor active only for dimensions created in paper space.

Zero Suppression Options

The **Zero suppression** area provides options for suppressing primary unit leading and trailing zeros and for controlling function *sub-units*. Uncheck **Leading** to leave a zero on decimal units less than 1, such as 0.5. This option is suitable for creating metric dimensions as recommended by the ASME standard. Check **Leading** to remove the 0 on decimal units less than 1, such as .5. Apply this option to create inch dimensions as recommended by the ASME standard. The **Leading** check box is not available for architectural units.

Uncheck **Trailing** to leave zeros after the decimal point based on the precision. This setting is usually suitable for decimal inch dimensioning because trailing zeros control tolerances for manufacturing processes. Check **Trailing** for metric dimensions to conform to the ASME standard. The **Trailing** check box is not available for architectural units.

The **0 feet** and **0 inches** check boxes are available for architectural and engineering units. Check **0 feet** to remove the zero in dimensions given in feet and inches when there are zero feet. For example, check **0 feet** to display a dimension as 11″, or uncheck **0 feet** to display the same dimension as 0'-11″.

Check **0 inches** to remove the zero when the inch portion of dimensions displayed in feet and inches is less than one inch, such as 12'-7/8″. If **0 inches** is unchecked, the same dimension reads 12'-0 7/8″. In addition, this option removes the zero from a dimension with no inch value; for example, 12' appears instead of 12'-0″.

The **Sub-units factor** and **Sub-unit suffix** text boxes become available when you use decimal units and select the **Leading** check box. Most drawings use a single format for all dimension values. For example, all dimensions on a decimal inch drawing are measured in inches, or decimals of an inch. Sub-units allow you to apply a different unit format to dimensions that are smaller than the primary unit format, without using decimals. For example, if you use meters to dimension most objects on a metric civil engineering drawing, you can use a **Sub-units factor** value of 100 (100 cm/m) and a **Sub-unit suffix** of cm to dimension objects smaller than one meter using centimeters instead of decimals of a meter. Then, when you dimension an object that is 0.5 meters, the dimension reads 500 cm.

sub-units: Unit formats smaller than the primary unit format. For example, centimeters can be defined as a sub-unit of meters.

For drawings that do not require sub-units but do suppress leading zeros, specify no sub-unit suffix. As long as you do not add a suffix, there is no need to change the sub-unit factor, although a factor of 0 also disables sub-units.

Angular Dimension Settings

The **Angular dimensions** area allows you to specify settings for primary angular dimensions. The options in the **Units format** drop-down list are the same as those in the **Angle** area of the **Drawing Units** dialog box. Typically, the primary angular dimension unit format is the same as the corresponding drawing units. Use the **Precision** drop-down list to set the appropriate angular dimension value precision. The **Zero suppression** area has check boxes for suppressing leading and trailing zeros on angular dimensions. Zero suppression for angular units is usually the same as applied to linear dimensions.

Alternate Units Tab

Figure 16-34 shows the **Alternate Units** tab of the **New** (or **Modify**) **Dimension Style** dialog box. Use the **Alternate Units** tab to set *alternate units*, or *dual dimensioning units*. Dual dimensioning practices are no longer a recommended ASME standard. ASME recommends that drawings be dimensioned using inch units or metric units only. However, other applications do use alternate units. Some companies who use manufacturers and vendors both in the United States and internationally require dual dimensioning.

Select the **Display alternate units** check box to enable alternate units. The **Alternate Units** tab includes most of the same settings found in the **Primary Units** tab. The value in the **Multiplier for alt units** text box is multiplied by the primary unit to establish the value for the alternate unit. A value of 25.4 allows you to use millimeters as alternate units on an inch-unit drawing. The **Placement** area controls the location of the alternate-unit dimension value. You can place alternate units either after or below the primary value.

> **alternate units (dual dimensioning units):** Dimensions in which measurements in one system, such as inches, are followed by bracketed measurements in another system, such as millimeters.

Figure 16-34.

The **Alternate Units** tab of the **New** (or **Modify**) **Dimension Style** dialog box.

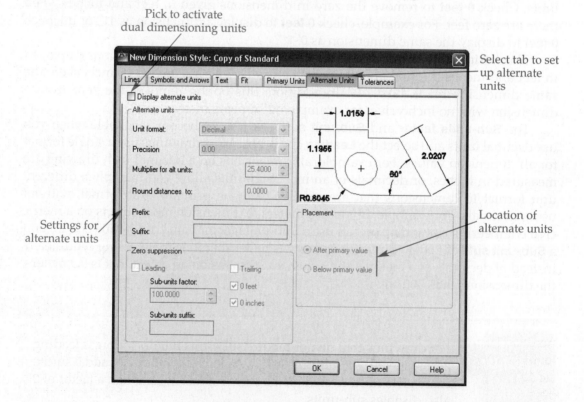

Chapter 19 describes the **Tolerances** tab found in the **New** (or **Modify**) **Dimension Style** dialog box.

Exercise 16-3

Complete the exercise on the companion website.
www.g-wlearning.com/CAD

Developing Dimension Styles

Creating and using dimension styles is an important element of drafting with AutoCAD. Carefully evaluate the characteristics of dimensions, and check school, company, and national standards to verify the accuracy of dimension settings. Figure 16-35 provides possible settings for three common dimension styles. Use the AutoCAD default values for settings not listed.

PROFESSIONAL TIP

To save valuable drafting time, add dimension styles to template drawings.

Exercise 16-4

Complete the exercise on the companion website.
www.g-wlearning.com/CAD

Changing Dimension Styles

Use the **Dimension Style Manager** to change the characteristics of an existing dimension style. Pick the **Modify** button to open the **Modify Dimension Style** dialog box. When you make changes to a dimension style, such as selecting a different text style or linear precision, all existing dimensions assigned the modified dimension style are updated. Use a different dimension style with different characteristics when appropriate to prevent updating existing dimensions.

To *override* a dimension style, pick the **Override** button in the **Dimension Style Manager** to open the **Override Current Style** dialog box. An example of an override is including a text prefix for a few of the dimensions in a drawing. The **Override** button is available only for the current style. Once you create an override, it is current and appears as a branch, called the *child*, of the *parent* style. Override settings are lost when any other style, including the parent, is set current.

Sometimes it is useful to view the details of two styles to determine their differences. Select the **Compare...** button in the **Dimension Style Manager** to display the **Compare Dimension Styles** dialog box. Compare two styles by selecting the name of one style from the **Compare:** drop-down list and the name of the other from the **With:** drop-down list. The differences between the selected styles display in the dialog box.

override: A temporary change to the current style settings; the process of changing a current style temporarily.

child: A style override.

parent: The dimension style from which a style override is formed.

Figure 16-35.
This chart shows dimension settings for typical mechanical and architectural drawings.

Setting	Mechanical—Inch	Mechanical—Metric (mm)	Architectural—U.S. Customary
Baseline spacing	.5	12	1/2″
Extend beyond dimension lines	.125	3	1/8″
Offset from origin	.063	1.5	1/16″ or 3/32″
Arrowhead options	Closed filled, Closed, or Open	Closed filled, Closed, or Open	Architectural tick, Dot, Closed filled, Oblique, or Right angle
Arrow size	.125	3	1/8″
Center marks	Line	Line	Mark
Center mark size	.0625	1.5	1/16″
Text style	Romans, Arial, or Century Gothic	Romans, Arial, or Century Gothic	SansSerif, Arial, Century Gothic, Stylus BT, or ArchiText
Text height	.12	3	1/8″
Vertical and horizontal text placement	Centered	Centered	Vertical: Above Horizontal: Centered
View direction	Left-to-right	Left-to-right	Left-to-right
Offset from dimension line	.063	1.5	1/16″
Text alignment	Horizontal	Horizontal	Aligned with dimension line
Linear unit format	Decimal	Decimal	Architectural
Linear precision	0.0000	0.00	1/16″
Linear zero suppression	Suppress only the leading zero	Suppress only the trailing zero	Suppress only the 0 feet zero
Sub-units factor	0	Disabled	Disabled
Sub-units suffix	None	Disabled	Disabled
Angular unit format	Decimal degrees	Decimal degrees	Decimal degrees
Angular precision	0	0	0
Angular zero suppression	Suppress only the leading angular dimension zero	Suppress only the trailing angular dimension zero	Suppress only the leading angular dimension zero
Alternate units	Do not display	Do not display	Do not display
Tolerances	By application	By application	None

The **New Dimension Style**, **Modify Dimension Style**, and **Override Current Style** dialog boxes have the same tabs.

Renaming and Deleting Dimension Styles

To rename a dimension style using the **Dimension Style Manager**, slowly double-click on the name or right-click on the name and select **Rename**. To delete a dimension style using the **Dimension Style Manager**, right-click on the name and select **Delete**. You cannot delete a dimension style assigned to dimension objects. To delete a style that is in use, first assign a different style to the dimension objects that reference the style to be deleted.

You can also rename styles using the **Rename** dialog box. Select **Dimension styles** in the **Named Objects** list to rename a dimension style.

Type
RENAME

Setting a Dimension Style Current

Set a dimension style current using the **Dimension Style Manager** by double-clicking the style in the **Styles** list box, right-clicking on the name and selecting **Set current**, or picking the style and selecting the **Set current** button. To set a dimension style current without opening the **Dimension Style Manager**, use the **Dimension Style** list in the expanded **Annotation** panel of the **Home** ribbon tab or on the **Dimensions** panel of the **Annotate** ribbon tab.

PROFESSIONAL TIP

You can import dimension styles from existing drawings using **DesignCenter**. See Chapter 5 for more information about using **DesignCenter** to import file content.

Exercise 16-5

Complete the exercise on the companion website.
www.g-wlearning.com/CAD

The **Dimension** panel of the **Express Tools** ribbon tab includes a **DIMEX** command that allows you to export a dimension style as a separate .dim file, and a **DIMIM** command that allows you to import a .dim file as a dimension style into the current drawing. For most applications, use **DesignCenter** to reuse dimension styles from existing drawings or templates, without creating a separate file.

Template Development

Chapter 16

Adding Dimension Styles
For instructions on adding dimension styles to each drawing template, go to the companion website (www.g-wlearning.com/CAD), select this chapter, and select **Template Development**.

Chapter Review

Answer the following questions. Write your answers on a separate sheet of paper or complete the electronic chapter review on the companion website.
www.g-wlearning.com/CAD

1. List at least three factors that influence a company's dimensioning practices.
2. Name the current ASME document that specifies mechanical drafting dimensioning practices.
3. Name two basic coordinate systems used to create location dimensions.
4. Define the term *general notes*.
5. Briefly explain the difference between placing specific and general notes on a drawing.
6. Explain how to dimension a cylinder using a single view.
7. Describe two ways to dimension a cone.
8. When is the best time to determine the drawing scale and scale factors for a drawing?
9. Explain how to add a scale to the **Annotation Scale** flyout in the status bar.
10. Define *dimension style*.
11. Name the dialog box used to create dimension styles.
12. Identify two ways to access the dialog box identified in Question 11.
13. Name the dialog box tab used to control the appearance of dimension lines and extension lines.
14. Name at least four arrowhead types that are available in the **Symbols and Arrows** tab for common use on architectural drawings.
15. Name the dialog box tab used to control the settings that display the dimension text.
16. What has to happen before you can assign a text style to a dimension style?
17. What is the ASME-recommended height for dimension numbers and notes on drawings?
18. Name the dialog box tab used to control settings that adjust the location of dimension lines, dimension text, arrowheads, and leader lines.
19. How can you delete a dimension style from a drawing?
20. How do you set a dimension style current?

Drawing Problems

Start AutoCAD if it is not already started. Follow the specific instructions for each problem.

▼ Basic

1. Start a new drawing from a template and create a RomanS text style using the romans font. Create the Mechanical (Inch) dimension style shown in Figure 16-35. Use the default AutoCAD settings for the dimension style settings not listed. Save the drawing as P16-1.

2. Start a new drawing from a template and create a RomanS text style using the romans font. Create the Mechanical (Metric) dimension style shown in Figure 16-35. Use the default AutoCAD settings for the dimension style settings not listed. Save the drawing as P16-2.

3. Start a new drawing from a template and create a Stylus BT text style using the Stylus BT font. Create the Architectural dimension style shown in Figure 16-35. Use the default AutoCAD settings for the dimension style settings not listed. Save the drawing as P16-3.

4. Write a short report explaining the difference between unidirectional and aligned dimensioning. Use a word processor and include sketches giving examples of each method.

5. Write a short report explaining the difference between size and location dimensions. Use a word processor and include sketches giving examples of each method.

6. Write a short report describing the difference between dimensioning for mechanical drafting (drafting for manufacturing) and architectural drafting. Use a word processor and include sketches giving examples of each method.

7. Make sketches showing the standard practice for dimensioning a cylindrical object, a square object, and a conical object.

8. Make sketches showing the standard practice for dimensioning angles. Make one sketch showing coordinate dimensioning and another showing angular dimensioning.

▼ Intermediate

9. Find a copy of the ASME Y14.5-2009, *Dimensioning and Tolerancing*, standard and write a report of approximately 350 words explaining the importance and basic content of this standard.

10. Interview your drafting instructor or supervisor and determine what dimension standards exist at your school or company. Write them down and keep them with you as you learn AutoCAD. Make notes as you progress through this textbook on how you use these standards. Also, note how you could change the standards to match the capabilities of AutoCAD.

▼ Advanced

11. Create a freehand sketch of Figure 16-1. Label each dimension item. To the side of the sketch, write a short description of each item.

12. Research civil drafting and create a template establishing the dimension styles for a civil drawing.

13. Visit a local manufacturing company at which design drafting work is part of the business. Write a report with sketched examples identifying the standards used at the company.

14. Visit a local architect or architectural designer. Write a report with sketched examples identifying the standards used at the company.

15. Visit a local civil engineering company at which design drafting work is part of the business. Write a report with sketched examples identifying the standards used at the company.

AutoCAD Certified Associate Exam Practice

Answer the following questions. Write your answers on a separate sheet of paper.

1. Which of the following items are common components of a dimension? *Select all that apply.*
   A. boundaries
   B. lines
   C. numerical values
   D. symbols
   E. xlines

2. What is the term for a dimension style in which all text reads from the bottom of the sheet? *Select the one item that best answers the question.*
   A. aligned dimensioning
   B. polar dimensioning
   C. rectangular dimensioning
   D. unidirectional dimensioning

3. What is the smallest number of views that can be used to dimension a cylindrical shape? *Select the one item that best answers the question.*
   A. one
   B. two
   C. three
   D. four

AutoCAD Certified Professional Exam Practice

Follow the instructions in each problem. Write your answers on a separate sheet of paper.

1. **Navigate to this chapter on the companion website and open CPE-16annoscale.dwg.** What annotation scale is currently assigned to this drawing?

2. **Navigate to this chapter on the companion website and open CPE-16dimstyle.dwg.** What text style is assigned to the Mechanical dimension style? What font is assigned to the text style?

Chapter 17

Linear and Angular Dimensioning

Learning Objectives

After completing this chapter, you will be able to:

✓ Add linear dimensions to a drawing.
✓ Add angular dimensions to a drawing.
✓ Draw datum and chain dimensions.
✓ Dimension multiple objects using the **QDIM** command.

Drawings often require a variety of dimensions to describe the size and shape of features. Linear and angular dimensions are two of the most common types of dimensions. This chapter covers the process of adding linear and angular dimensions to a drawing using several dimensioning commands. It also covers adding a break symbol to a dimension line and using the **QDIM** command.

Linear Dimensions

Linear dimensions usually measure straight distances, such as distances between horizontal, vertical, or slanted surfaces. Use the **DIMLINEAR** command to draw a single linear dimension. Dimension commands reference the current dimension style and the points or objects you select to create a dimension object. When you use the **DIMLINEAR** command, you create a dimension object that includes all related dimension style characteristics, dimension and extension lines, arrowheads, and a dimension value that describes the distance between selected points.

Access the **DIMLINEAR** command, pick a point to locate the origin of the first extension line, and then pick a point to locate the origin of the second extension line. See **Figure 17-1**. Use object snap modes and other drawing aids to pick the exact origin of extension lines. Once you establish the extension line origins, use the options that appear at the Specify dimension line location or [Mtext/Text/Angle/Horizontal/Vertical/Rotated]: prompt as needed. To apply the default settings and create a linear dimension, specify a point to locate the dimension line. See **Figure 17-2**.

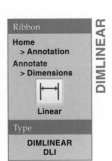

Ribbon

Home
> Annotation
Annotate
> Dimensions

⊢⊣

Linear

Type

DIMLINEAR
DLI

DIMLINEAR

Figure 17-1.
Establishing extension line origins. The **Endpoint** and **Intersection** object snap modes are useful for locating origins accurately.

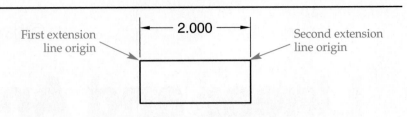

First extension line origin   2.000   Second extension line origin

Figure 17-2.
Establishing the location of a dimension line. ASME standards recommend a minimum spacing of .375″ (10 mm) from a drawing feature to the first dimension line and a minimum spacing of .25″ (6 mm) between dimension lines. A value of .5″ (12 mm) or .75″ (19 mm) is usually more appropriate. Maintain consistent spacing throughout a drawing.

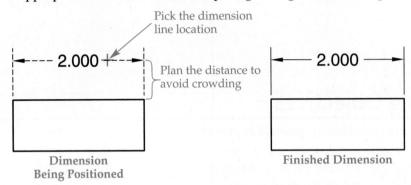

Pick the dimension line location

2.000

Plan the distance to avoid crowding

Dimension Being Positioned

2.000

Finished Dimension

When you dimension objects with AutoCAD, the objects are measured exactly as drawn if you had set the scale factor of dimensions to 1 in the **Scale factor:** text box of the **Measurement scale** area of the **New** (or **Modify**) **Dimension Style** dialog box. This makes it important for you to draw objects and features accurately and to select the origins of extension lines accurately.

PROFESSIONAL TIP

Use preliminary planning sheets and sketches to help determine proper dimension line location and distances between dimension lines to avoid crowding. Create offset construction geometry to pick with object snap modes when placing dimension lines, or use drawing aids such as AutoSnap and AutoTrack or **Grid** and **Snap** modes.

Exercise 17-1

Complete the exercise on the companion website.
www.g-wlearning.com/CAD

Selecting an Object to Dimension

An alternative method to locate extension line origins is to pick a line, polyline segment, circle, or arc to dimension. You can use this option whenever you see the Specify first extension line origin or <select object>: prompt. Press [Enter] or the space bar or right-click and then pick the object to dimension. When you select a line, polyline segment, or arc, extension lines originate from the endpoints. When you pick a circle, extension lines originate from the quadrant closest to the selection and the opposite quadrant. See **Figure 17-3**.

Adjusting Dimension Text

The value attached to the dimension corresponds to the distance between the origins of extension lines. Use the **Mtext** option after you establish the extension line origins to adjust the dimension value using the multiline text editor. See **Figure 17-4**. The highlighted value represents the current dimension value. Add to or modify the dimension text and then close the text editor. The **DIMLINEAR** command continues, allowing you to pick the dimension line location.

The **Text** option, also available after you establish the extension line origins, uses the single-line text editor to change dimension text, even though the final dimension value is mtext. The current dimension value appears in brackets. Add to or modify the value as necessary, and then press [Enter] to exit the option. The **DIMLINEAR** command continues, allowing you to pick the dimension line location.

Dimension values are horizontal or aligned with the dimension line, according to the current dimension style format. The **Angle** option has limited applications, but allows you to rotate the dimension text. Enter the desired angle at the Specify angle of dimension text: prompt to use this option.

PROFESSIONAL TIP

If you forget to use the **Mtext** or **Text** option before placing a dimension, or to modify an existing dimension value, double-click on the dimension value to display the multiline text editor. Chapter 20 provides more information on editing dimensions.

Figure 17-3.
AutoCAD can determine extension line origins automatically when you use the **Select object** option to pick a line, polyline segment, arc, or circle.

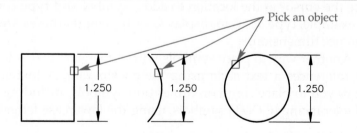

Figure 17-4.
When you use the **Mtext** option, the **Text Editor** ribbon tab appears with the dimension value calculated by AutoCAD in a text box for editing.

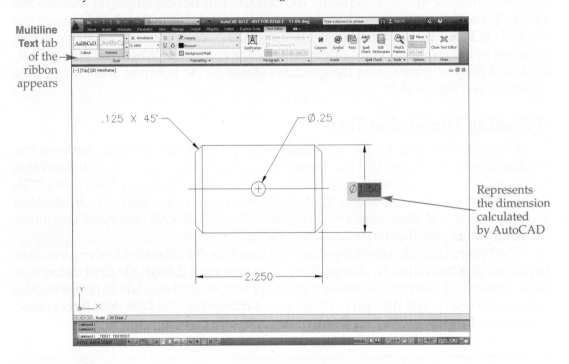

Multiline **Text** tab of the ribbon appears

Represents the dimension calculated by AutoCAD

Including Symbols with Dimension Text

Some AutoCAD dimension commands automatically place appropriate symbols with the dimension value. For example, when you dimension an arc using the **DIMRADIUS** command, an R appears before the dimension value. When you dimension a circle using the **DIMDIAMETER** command, a ∅ symbol appears before the dimension value. The ASME standard recommends these symbols. However, dimension commands such as **DIMLINEAR** do not automatically place certain symbols or add necessary characters.

One option is to use the **Mtext** option to activate the multiline text editor. Place the cursor at the location to add a symbol and then use options from the **Symbol** flyout or cascading submenu, or type characters. For example, pick the **Diameter** symbol from the **Symbol** flyout to add ∅ before the dimension value. Another example is enclosing a reference dimension in parentheses, as recommended by the ASME standard. To create a reference dimension, type open and close parentheses around the highlighted value.

You can also add content using the **Text** option and the single-line text editor. Place the cursor at the location to add a symbol and type control codes or characters. For example, type %%C to display ∅ or type parentheses around the value to create a reference dimension.

Another way to place symbols with dimension text is to create a dimension style that references a text style using the gdt.shx font. A text style with the gdt.shx font allows you to place common dimension symbols, including geometric dimensioning and tolerancing (GD&T) symbols, using the lowercase letter keys.

PROFESSIONAL TIP

Although you can add a prefix and suffix to a dimension style, it is usually more appropriate to adjust the limited number of dimensions that require a prefix or suffix.

Reference Material *Drafting Symbols*
For more information about common drafting symbols and the gdt.shx font, go to the **Reference Material** section of the companion website (www.g-wlearning.com/CAD) and select **Drafting Symbols** in the list.

Exercise 17-2

Complete the exercise on the companion website.
www.g-wlearning.com/CAD

Controlling the Dimension Line Angle

The **Horizontal** and **Vertical** options, available after you establish the extension line origins, are helpful when it is difficult to produce the appropriate horizontal or vertical dimension line, such as when you are dimensioning the horizontal or vertical distance of a slanted surface. The **Horizontal** option restricts the **DIMLINEAR** command to dimension only a horizontal distance. The **Vertical** option restricts the **DIMLINEAR** command to dimension only a vertical distance. The **Mtext**, **Text**, and **Angle** options are available to change the dimension text value if necessary.

The **Rotated** option, also available after you establish the extension line origins, allows you to specify a dimension line angle. Practical applications include dimensioning angled surfaces and auxiliary views. This technique is different from other dimensioning commands because you provide a dimension line angle. See **Figure 17-5**. At the Specify angle of dimension line <0>: prompt, enter a value or specify two points on the line.

PROFESSIONAL TIP

AutoCAD dimensioning should be accurate and neat. You can achieve consistent, professional results by using the following guidelines:
- Never round off decimal values when entering locations, distances, or angles. For example, enter .4375 for 7/16, rather than .44.
- Set the precision to the most common precision level in the drawing before adding dimensions. Adjust the precision as needed for each dimension.
- Always use drawing aids, such as object snaps, to ensure the accuracy of dimensions.
- Never type a different dimension value from what appears highlighted or in <> brackets. To change a dimension, revise the drawing or dimension settings. Only adjust dimension text when it is necessary to add prefixes and suffixes, or to use a different text format.

Exercise 17-3

Complete the exercise on the companion website.
www.g-wlearning.com/CAD

Figure 17-5.
Rotating a
dimension for an
angled view.

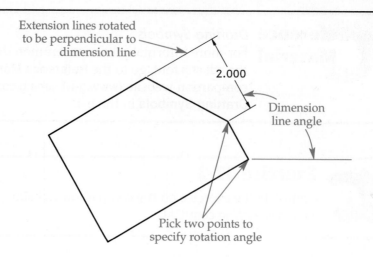

Extension lines rotated
to be perpendicular to
dimension line

2.000

Dimension
line angle

Pick two points to
specify rotation angle

Dimensioning Angled Surfaces and Auxiliary Views

When you dimension a surface drawn at an angle, such as a feature shown in an auxiliary view, it is often necessary to align the dimension line with the surface, so that the extension lines are perpendicular to the surface. To dimension these features properly, use the **DIMALIGNED** command or the **Rotated** option of the **DIMLINEAR** command.

Figure 17-6 shows the results of using the **DIMALIGNED** command. Notice the difference between the aligned dimension in this figure and the rotated dimension in **Figure 17-5**. You can usually use the **DIMALIGNED** command when the length of the extension lines is equal. The **Rotated** option of the **DIMLINEAR** command is often necessary when extension lines are unequal.

DIMALIGNED

Ribbon
Home
> Annotation
Annotate
> Dimensions
Aligned
Type
DIMALIGNED
DAL

Exercises 17-4 and 17-5

Complete the exercises on the companion website.
www.g-wlearning.com/CAD

Dimensioning Long Objects

When you create a drawing of a long part that has a constant shape, the view may not fit on the sheet, or it may look strange compared to the rest of the drawing. To overcome this problem, use a *conventional break* (or *break*) to shorten the view. **Figure 17-7** shows examples of standard break lines. For many long parts, a conventional break is required to display views or increase view scale without increasing sheet size. Dimensions added to conventional breaks describe the actual length of the product in its unbroken form. The dimension line includes a break symbol to indicate

conventional break (break): Removal of a portion of a long, constant-shaped object to make the object fit better on the sheet.

Figure 17-6.
The **DIMALIGNED**
command allows
you to place
dimension lines
parallel to angled
features.

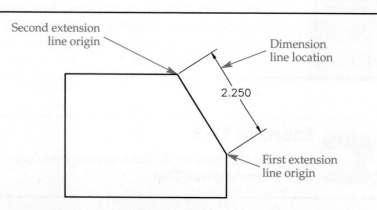

Second extension
line origin

Dimension
line location

2.250

First extension
line origin

AutoCAD and Its Applications—Basics

Figure 17-7.
Standard break lines.

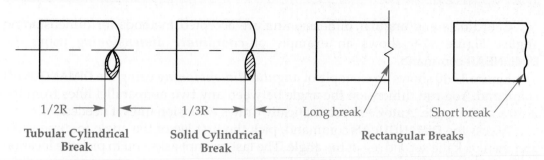

Tubular Cylindrical Break   Solid Cylindrical Break   Rectangular Breaks

1/2R   1/3R   Long break   Short break

that the drawing view is broken and that the feature is longer than it appears. See **Figure 17-8**.

Use the **DIMJOGLINE** command to add a break symbol to dimension lines created using the **DIMLINEAR** or **DIMALIGNED** command. Access the **DIMJOGLINE** command and pick a linear or aligned dimension line. Then pick a location on the dimension line to place the break symbol, as shown in **Figure 17-8**. An alternative to selecting the location of the break symbol is to press [Enter] to accept the default location. You can move the break later using grip editing or by reusing the **DIMJOGLINE** command to select a different location. To remove the break symbol, access the **DIMJOGLINE** command and select the **Remove** option.

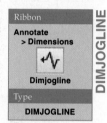

Ribbon

Annotate
> Dimensions

Dimjogline

Type

DIMJOGLINE

DIMJOGLINE

Use the **Mtext** or **Text** option or edit the dimension value to specify the actual length. You can add only one break symbol to a dimension line.

Exercise 17-6

Complete the exercise on the companion website.
www.g-wlearning.com/CAD

Figure 17-8.
Using the **DIMJOGLINE** command to place a break symbol on a dimension line.

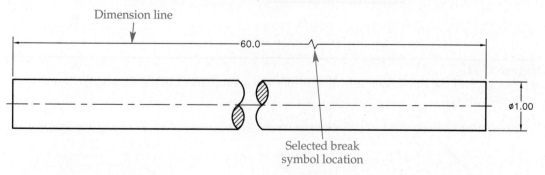

Dimension line

60.0

Ø1.00

Selected break symbol location

Coordinate and angular dimensioning are accepted methods for dimensioning angles. **Figure 17-9** shows an example of *coordinate dimensioning* using the **DIMLINEAR** command.

Figure 17-10 shows an example of *angular dimensioning* using the **DIMANGULAR** command. You can dimension the angle between any two nonparallel lines from the *vertex* of the angle. AutoCAD automatically draws extension lines if needed.

Access the **DIMANGULAR** command, pick the first leg of the angle to dimension, and then pick the second leg of the angle. The last prompt asks you to pick the location of the dimension line arc. **Figure 17-11** shows examples of angular dimensions and the effect that limited space may have on dimension fit and placement. Fit characteristics apply to most dimensions.

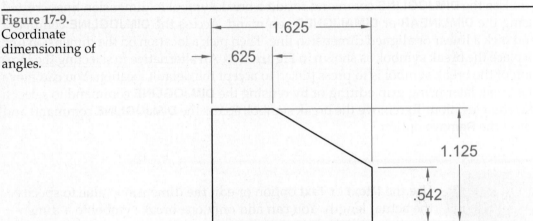

Figure 17-9. Coordinate dimensioning of angles.

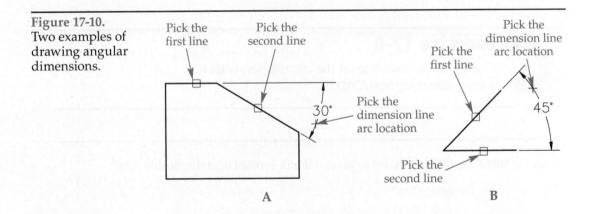

Figure 17-10. Two examples of drawing angular dimensions.

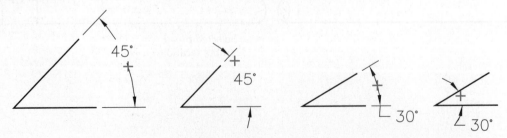

Figure 17-11. The location of the dimension line determines the arrangement of the dimension line arc, text, and arrows.

PROFESSIONAL TIP

You can create four different dimensions (two different angles) with an angular dimension. To preview these options before selecting the dimension line location, use the cursor to move the dimension around an imaginary circle. You can use the **Quadrant** option to isolate a specific quadrant of the imaginary circle and force the dimension to produce the value found in the selected quadrant.

Dimensioning Angles on Arcs and Circles

Use the **DIMANGULAR** command to dimension the included angle of an arc or a portion of a circle. When you dimension an arc, the center point becomes the angle vertex, and the two arc endpoints establish the extension line origins. See Figure 17-12.

When you dimension a circle using **DIMANGULAR**, the center point becomes the angle vertex, and two specified points locate the extension line origins. See Figure 17-13. The point you pick to select the circle locates the origin of the first extension line. You then select the second angle endpoint, which locates the origin of the second extension line.

PROFESSIONAL TIP

Using angular dimensioning for circles increases the number of possible solutions for a given dimensioning requirement, but the actual uses are limited. One application is dimensioning an angle from a quadrant point to a particular feature without first drawing a line to dimension. Another benefit of this option is the ability to specify angles that exceed 180°.

Angular Dimensioning through Three Points

You can also use the **DIMANGULAR** command to establish an angular dimension according to the vertex and two angle line endpoints. See Figure 17-14. To apply this technique, press [Enter] or the space bar, or right-click after the first prompt. Then pick

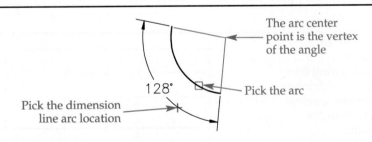

Figure 17-12.
Placing angular dimensions on arcs.

The arc center point is the vertex of the angle

128°

Pick the dimension line arc location

Pick the arc

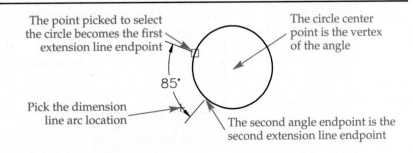

Figure 17-13.
Placing angular dimensions on circles.

The point picked to select the circle becomes the first extension line endpoint

The circle center point is the vertex of the angle

85°

Pick the dimension line arc location

The second angle endpoint is the second extension line endpoint

Figure 17-14.
Placing angular
dimensions using
three points.

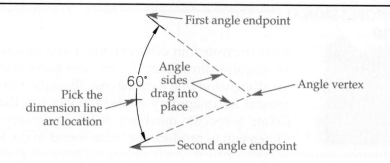

First angle endpoint

Angle
sides
drag into
place

60°

Pick the
dimension line
arc location

Angle vertex

Second angle endpoint

the vertex, followed by the two points. You can also use this method to dimension angles over 180°.

Exercise 17-7

Complete the exercise on the companion website.
www.g-wlearning.com/CAD

Baseline, Chain, and Direct Dimensioning

baseline dimensioning:
A method of dimensioning in which several dimensions originate from a common surface, centerline, or center plane.

datum:
Theoretically perfect surface, plane, point, or axis from which measurements can be taken.

chain dimensioning:
A method of dimensioning in which dimensions appear in a line from one feature to the next.

Mechanical drafting often requires *baseline dimensioning*, in which each dimension references a *datum* and is independent of the others. This achieves more accuracy in manufacturing. **Figure 17-15** shows an object dimensioned with baseline dimensioning and surface datums.

Mechanical drafting sometimes uses *chain dimensioning*. However, this method provides less accuracy than baseline dimensioning because each dimension is dependent on other dimensions in the chain, resulting in tolerance buildup. In mechanical drafting, it is common to leave one dimension out and provide an overall dimension that controls the missing value. See **Figure 17-16**. Architectural drafting uses chain dimensioning in most applications to reduce the need to calculate or find dimension values during construction. Architectural drafting practices usually show dimensions for all features, plus an overall dimension.

Figure 17-15.
Baseline dimensioning on a part view. Maintain consistent spacing throughout the drawing.

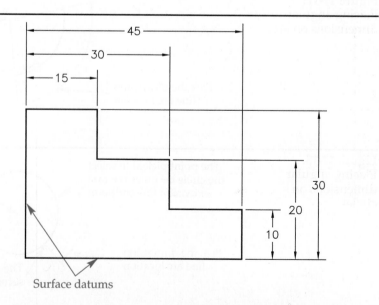

Surface datums

Figure 17-16.
Chain dimensioning on a part view. The example on the left is more common, depending on the design and dimensioning requirements. The example on the right includes an optional reference dimension.

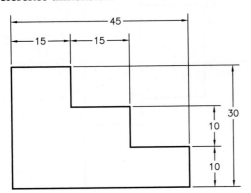
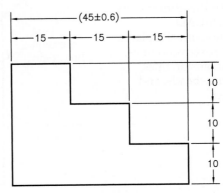

Direct dimensioning controls the specific size or location of one or more specific features, often resulting in the least tolerance buildup. Figure 17-17 shows an example of direct dimensioning on a part view. Baseline or chain dimensioning could be used to dimension the view, but would create more tolerance accumulation.

direct dimensioning: A type of dimensioning applied to control the specific size or location of one or more specific features.

Baseline Dimensioning

The **DIMBASELINE** command controls baseline dimensioning and allows you to select several points to define a series of baseline dimensions. You can create baseline dimensions with linear, angular, and ordinate dimensions. Chapter 18 describes ordinate dimensions.

The **DIMBASELINE** command continues from an existing dimension. Therefore, you must create the first dimension using a suitable dimensioning command. For example, for linear dimensions, use the **DIMLINEAR** command. The first point you select when drawing the linear dimension defines the datum. Then access the **DIMBASELINE** command and pick the next second extension line origin. Continue picking extension line origins until you have dimensioned all the features. Press [Enter] or the space bar or right-click twice, once at the Specify a second extension line origin or: prompt, and again at the Select base dimension: prompt, to create the dimensions and exit the command. Notice that as you pick additional extension line origins, AutoCAD automatically places the dimension text; you do not specify a location. Figure 17-18 shows an example of using the **DIMBASELINE** command to pick two additional extension line origins to add to an existing linear dimension. Use the **Undo** option to undo previously drawn dimensions.

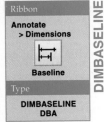

Ribbon
Annotate
> Dimensions
Baseline
Type
DIMBASELINE
DBA

DIMBASELINE

Figure 17-17.
An example of direct dimensioning between surfaces X and Y. Baseline dimensioning locates the other surfaces.

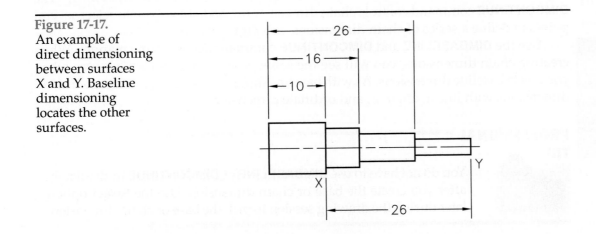

Figure 17-18.
Using the **DIMBASELINE** command to add baseline dimensions. AutoCAD automatically places the extension lines, dimension lines, arrowheads, and text.

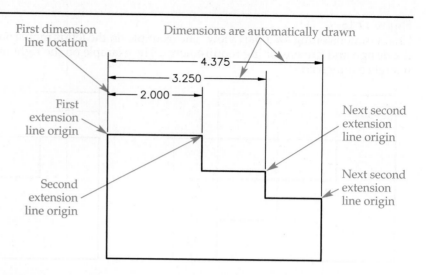

First dimension line location

Dimensions are automatically drawn

4.375

3.250

2.000

First extension line origin

Second extension line origin

Next second extension line origin

Next second extension line origin

AutoCAD automatically selects the most recent dimension as the base dimension unless you specify a different dimension. To add baseline dimensions to an existing dimension other than the most recent dimension, use the **Select** option by pressing [Enter] or the space bar, or by right-clicking and selecting **Enter** at the first prompt. At the Select base dimension: prompt, pick the dimension to serve as the base. The extension line nearest the point where you select the dimension establishes the datum. Then select the new second extension line origins as described.

You can also draw baseline dimensions to angular features. First, draw an angular dimension. Then enter the **DIMBASELINE** command. Figure 17-19 shows angular baseline dimensions. As with linear dimensions, you can pick an existing angular dimension other than the most recent dimension.

PROFESSIONAL TIP

Specify the distance between dimension lines in the **Lines** tab of the **New** (or **Modify**) **Dimension Style** dialog box. The ASME minimum distance between dimension lines is .375″ (10 mm), but this is usually too close. A distance of .75″ (19 mm) from the object to the first dimension line and .5″ (12 mm) between other dimension lines is often ideal.

Chain Dimensioning

continued dimensioning: The AutoCAD term for chain dimensioning.

AutoCAD refers to chain dimensioning as *continued dimensioning*. The **DIMCONTINUE** command controls chain dimensioning and allows you to select several points to define a series of chain dimensions. See Figure 17-20.

Use the **DIMBASELINE** and **DIMCONTINUE** commands in the same manner. When creating chain dimensions, you will see the same prompts and options you see when you create baseline dimensions. As with baseline dimensions, you can create continued dimensions with linear, angular, and ordinate dimensions.

DIMCONTINUE

Ribbon

Annotate > Dimensions

Continue

Type

DIMCONTINUE DCO

PROFESSIONAL TIP

You do not have to use **DIMBASELINE** or **DIMCONTINUE** immediately after you create the base or chain dimension. Use the **Select** option later during the drawing session to pick the base or chain dimension.

Figure 17-19.
Using the
DIMBASELINE
command to add
baseline dimensions
to angular features.

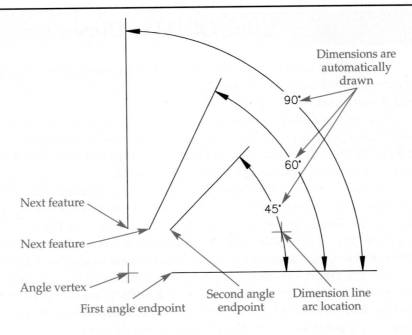

Dimensions are
automatically
drawn

90°

60°

45°

Next feature

Next feature

Angle vertex

First angle endpoint

Second angle
endpoint

Dimension line
arc location

Figure 17-20.
Using the
DIMCONTINUE
command to create
chain dimensions.

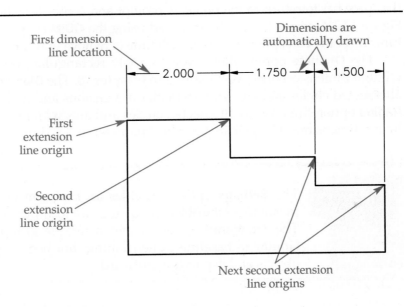

First dimension
line location

Dimensions are
automatically drawn

2.000   1.750   1.500

First
extension
line origin

Second
extension
line origin

Next second extension
line origins

The **Baseline Dimension** option available for grip-editing linear, and angular dimensions, and the **Continue Dimension** option, available for grip-editing linear, angular, and ordinate dimensions, provide an effective alternative to the **DIMBASELINE** and **DIMCONTINUE** commands. Chapter 20 provides more information on editing dimensions.

Exercise 17-8

Complete the exercise on the companion website.
www.g-wlearning.com/CAD

Using QDIM to Dimension

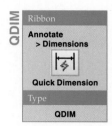

QDIM

Ribbon
Annotate
> Dimensions

Quick Dimension

Type
QDIM

The **QDIM** command automates the dimensioning processing by creating a group of dimensions based on the objects you select. You can also use the **QDIM** command to edit dimensions, as explained in Chapter 20.

Access the **QDIM** command and select the objects to dimension. Right-click or press [Enter] or the space bar to display a preview of the dimensions attached to the cursor. By default, AutoCAD establishes an extension line from every line, polyline, and arc endpoint, and circle and arc center point.

To apply chain dimensioning, use the default **Continuous** option and specify the location of the dimension lines. See Figure 17-21A. To apply baseline dimensioning, choose the **Baseline** option and specify the location of the first dimension line. See Figure 17-21B. If the **QDIM** command does not reference the appropriate datum, use the **datum Point** option before locating the dimensions to specify a different datum point.

Use the **Edit** option before locating the dimensions to add dimensions to, or remove dimensions from, the current set. Marks indicate the points acquired by the **QDIM** command. Use the **Add** function to specify a point to add a dimension, or use the **Remove** function to specify a point to remove the corresponding dimension. Right-click or press [Enter] or the space bar to return to the previous prompt and continue using the **QDIM** command. Figure 17-21C shows a view dimensioned using the **QDIM** command twice. The **Remove** function of the **Edit** option was used each time to remove unwanted dimensions.

The **Ordinate** option allows you to apply rectangular coordinate dimensioning without dimension lines, as explained in Chapter 18. The **Diameter** option dimensions all selected circles and arcs with diameter dimensions and no linear dimensions. The **Radius** option dimensions all selected circles and arcs with radius dimensions and no linear dimensions. Chapter 18 describes diameter and radius dimensions.

The **Settings** option provides an **Endpoint** or **Intersection** toggle meant to set the object snap mode for locating extension line origins. The **Staggered** option creates a unique grouping of dimensions, similar to baseline dimensioning, but beginning from the center objects and expanding outward.

Figure 17-21.
The **QDIM** command can dimension multiple selected objects, without you having to pick extension line origins. This example shows selecting all objects initially, and then using the **Remove** function of the **Edit** option to deselect unwanted points.

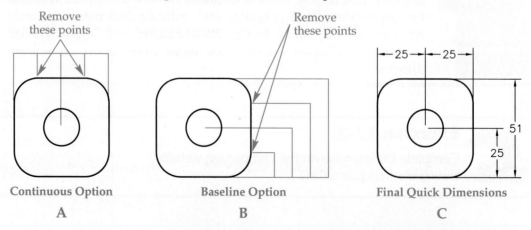

Continuous Option
A

Baseline Option
B

Final Quick Dimensions
C

Exercise 17-9

Complete the exercise on the companion website.
www.g-wlearning.com/CAD

Chapter Review

Answer the following questions. Write your answers on a separate sheet of paper or complete the electronic chapter review on the companion website.
www.g-wlearning.com/CAD

1. Name the two **DIMLINEAR** options that allow you to change dimension text.
2. Give two examples of symbols that automatically appear with some dimensions.
3. What is one purpose of the AutoCAD gdt.shx font?
4. Name the two dimensioning commands that provide linear dimensions for angled surfaces.
5. Which command allows you to place a break symbol in a dimension line?
6. Name the command used to dimension angles in degrees.
7. Describe a way to specify an angle in degrees if the angle is greater than 180°.
8. Which type of dimensioning is generally preferred for manufacturing because of its accuracy?
9. Define *direct dimensioning*.
10. How do you place a baseline dimension from the origin of the previously drawn dimension?
11. How do you place a baseline dimension from the origin of a dimension drawn during a previous drawing session?
12. What is the conventional term for the type of dimensioning AutoCAD refers to as continuous dimensioning?
13. Which command other than **DIMBASELINE** creates baseline dimensions?
14. Explain how to remove dimensions from the current set when you use the **QDIM** command.
15. Name at least three options for dimensioning that are available through the **QDIM** command.

Drawing Problems

Start AutoCAD if it is not already started. Start a new drawing for each problem using an appropriate template of your choice. The template should include layers and text, and dimension styles, when necessary, for drawing the given objects. Add layers, text, and dimension styles as needed. Draw all objects using appropriate layers and text and dimension styles, justification, and format. Follow the specific instructions for each problem. Use only drawing and editing commands and techniques you have already learned. Use your own judgment and approximate dimensions when necessary. Apply dimensions accurately using ASME or appropriate industry standards.

Note: Some of the problems in this chapter are built on problems from previous chapters. If you have not yet completed those problems, complete them now.

▼ Basic

1. Start a new drawing from scratch or use a fractional unit template of your choice. Save the file as P17-1. Open P3-7 and copy one instance of Object A and Object B to the P17-1 drawing. The P17-1 file should be active. Dimension the views as shown.

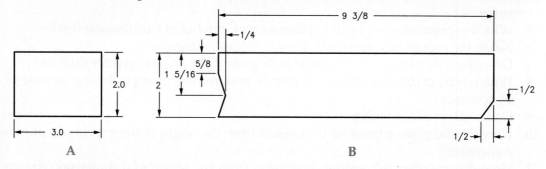

2. Open P3-5 and save the file as P17-2. The P17-2 file should be active. Dimension the drawing as shown.

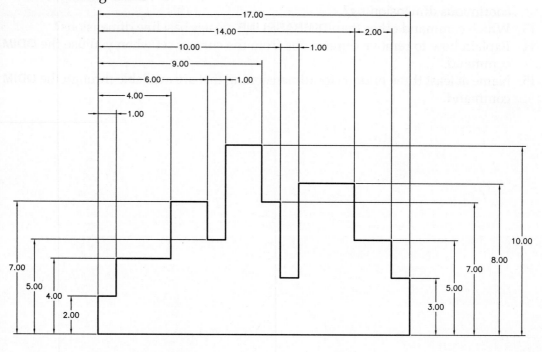

3. Open P3-10 and save the file as P17-3. The P17-3 file should be active. Dimension the views as shown.

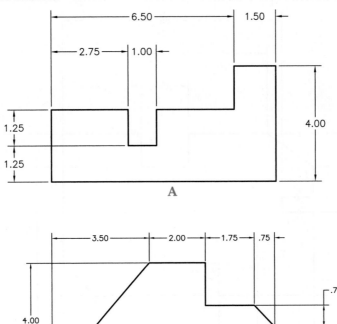

A

B

4. Open P3-13 and save the file as P17-4. The P17-4 file should be active. Dimension the view as shown.

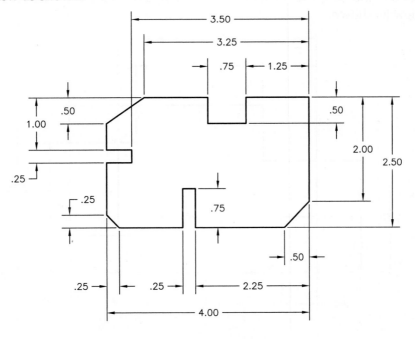

5. Open P3-14 and save the file as P17-5. The P17-5 file should be active. Dimension the views as shown. Note that this is a metric drawing.

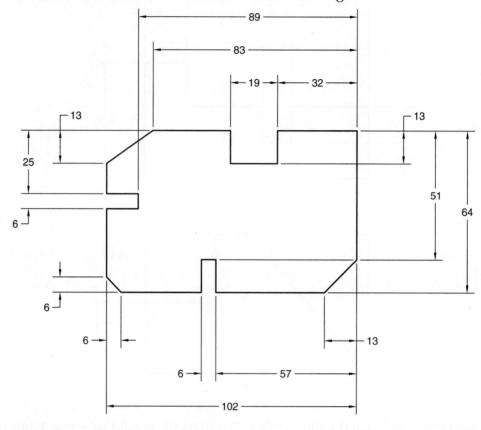

6. Open P3-8 and save the file as P17-6. The P17-6 file should be active. Dimension the view as shown.

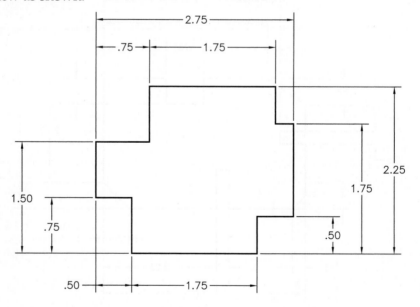

AutoCAD and Its Applications—Basics

7. Open P3-9 and save the file as P17-7. The P17-7 file should be active. Dimension the view as shown.

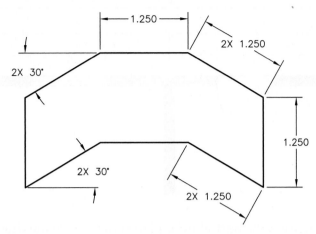

8. Write a report explaining the difference between baseline and chain dimensioning. Use a word processor and include sketches giving examples of each method.

▼ Intermediate

9. Draw and dimension the views of the shaft shown. Save the drawing as P17-9.

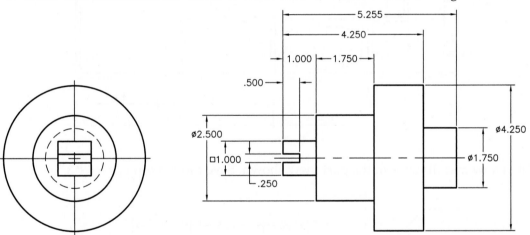

10. Open P3-15 and save the file as P17-10. The P17-10 file should be active. Dimension the view as shown.

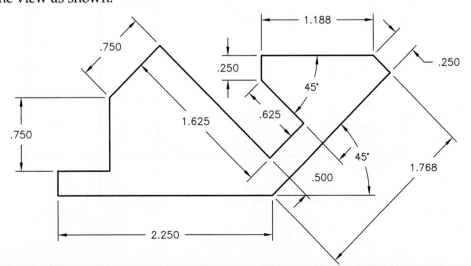

11. Draw and dimension the partial floor plan shown. Save the drawing as P17-11.

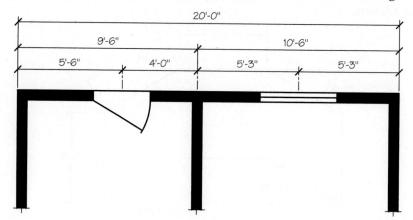

12. Draw and dimension the partial floor plan shown. Save the drawing as P17-12.

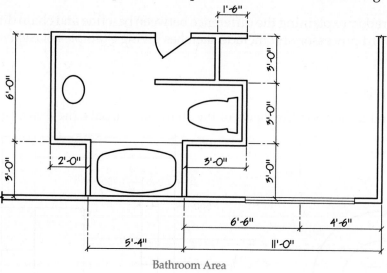

Bathroom Area

13. Draw and dimension the part view shown. Save the drawing as P17-13.

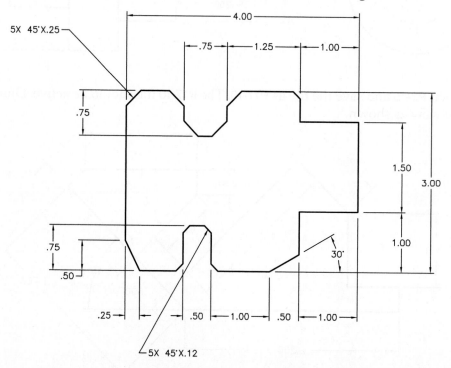

14. Draw and dimension the part view shown. Save the drawing as P17-14.

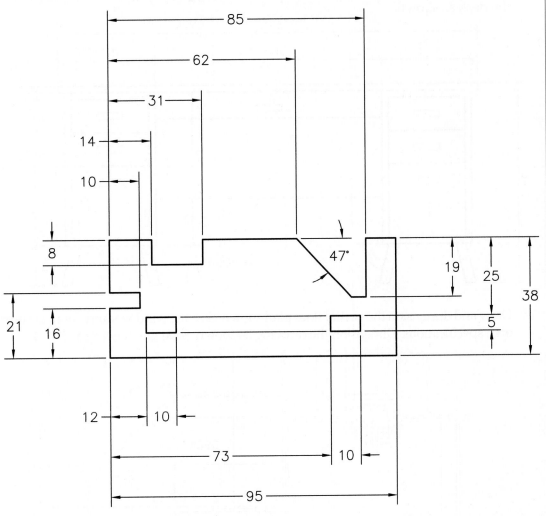

▼ **Advanced**

15. Open P3-3 shown and save the file as P17-15. The P17-15 file should be active. Make one copy of the view at a new location to the right of the original view. Dimension the view on the left using baseline dimensioning. Dimension the view on the right using chain dimensioning.

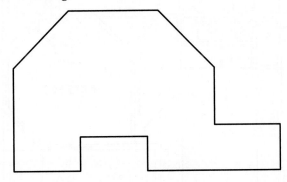

16. Open P8-9 and save the file as P17-16. The P17-16 file should be active. Dimension the desk as shown.

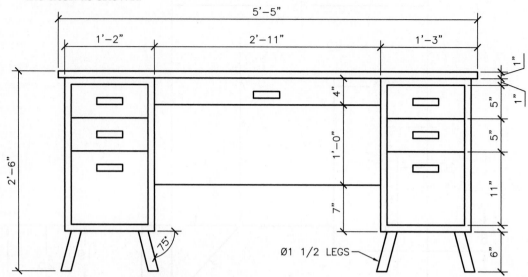

17. Draw and dimension the partial floor plan shown. Size the windows and doors to your own specifications. Dimension the drawing as shown. Save the drawing as P17-17.

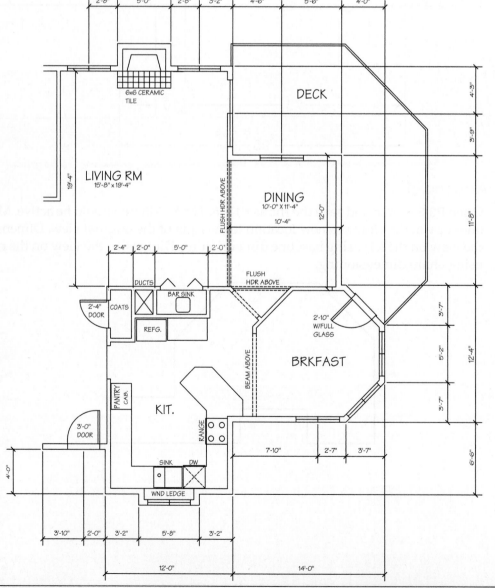

AutoCAD Certified Associate Exam Practice

Answer the following questions. Write your answers on a separate sheet of paper.

1. Which of the following is the minimum ASME-recommended spacing from the object to the first dimension line? *Select all that apply.*
   A. .375"
   B. .5"
   C. .75"
   D. 10 mm
   E. 19 mm
   F. 12 mm

2. How can you add a symbol to dimension text? *Select all that apply.*
   A. apply control codes or characters
   B. create a dimension style that references the gdt.shx font
   C. hold [Ctrl] and press an appropriate letter key
   D. press the appropriate function key
   E. use the **Mtext** option of the dimensioning command
   F. use the **Text** option of the dimensioning command

3. Which of the following commands continues dimensioning from a previously placed dimension? *Select all that apply.*
   A. **DIMANGULAR**
   B. **DIMBASELINE**
   C. **DIMCONTINUE**
   D. **DIMJOGLINE**
   E. **DIMLINEAR**

AutoCAD Certified Professional Exam Practice

Follow the instructions in each problem. Write your answers on a separate sheet of paper.

1. **Navigate to this chapter on the companion website and open CPE-17angles.dwg.**
   What are the measured values of ANGLE A and ANGLE B? Use the appropriate dimensioning command(s) to find the measurements.

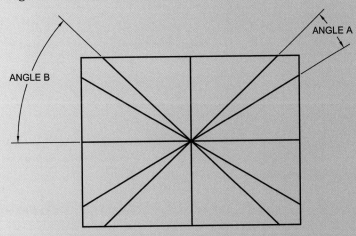

2. **Navigate to this chapter on the companion website and open CPE-17dimstyle.dwg.**
   Use the **Dimension Style Manager** to change the baseline spacing in the Standard
   dimension style to .5, and leave all other settings as they are currently set. Use
   DIMBASELINE to finish the dimensions as shown. What are the coordinates of
   Point 1?

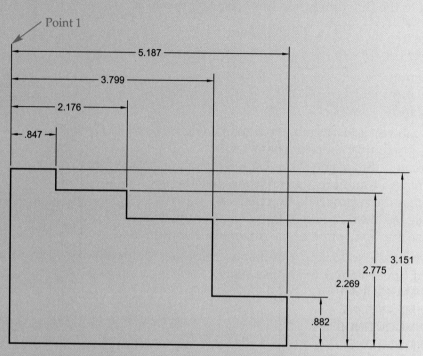

Chapter 18

Dimensioning Features and Alternate Practices

Learning Objectives

After completing this chapter, you will be able to:

✓ Dimension circles, arcs, and other curves.
✓ Create and use multileader styles.
✓ Draw leaders using the **MLEADER** command.
✓ Apply alternate dimensioning practices.
✓ Dimension using the **DIMORDINATE** command.
✓ Mark up a drawing using the **REVCLOUD** and **WIPEOUT** commands.

A drawing must describe the size and location of all features for manufacturing or construction. This chapter explains options for dimensioning object features and introduces alternate mechanical drafting dimensioning practices. This chapter also introduces basic redlining techniques using the **REVCLOUD** and **WIPEOUT** commands.

Dimensioning Circles, Arcs, and Other Curves

AutoCAD includes several commands for dimensioning circular and noncircular curves. Circles are generally dimensioned according to their diameter, and arcs are dimensioned by their radius. Curves that do not have a constant radius require other methods, as described in this chapter.

Dimensioning Circles

The **DIMDIAMETER** command allows you to dimension a circle or arc with a diameter. However, diameter is usually used to describe the size of circles. The ASME standard for dimensioning arcs is to identify the radius. Access the **DIMDIAMETER** command and select a circle or arc to display a leader line and a diameter dimension value attached to the cross-hairs. Specify a point to locate the dimension value. See **Figure 18-1**.

Like other dimension commands, the **DIMDIAMETER** command references the current dimension style and the object you select to create a dimension object. The dimension includes all related dimension style characteristics, centerlines, leader, arrowheads, and a dimension value associated with the diameter. The leader points to the center of the circle or arc, as recommended by the ASME standard.

Ribbon

Home
> Annotation
Annotate
> Dimensions

Diameter

Type

DIMDIAMETER
DDI

DIMDIAMETER

Figure 18-1.
Using the **DIMDIAMETER** command to dimension a circle.

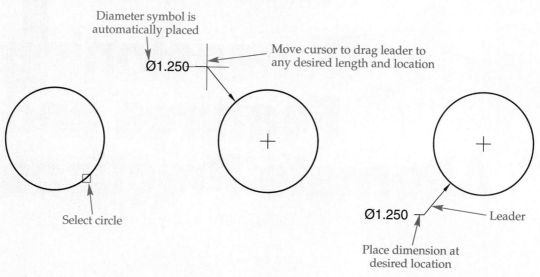

Diameter symbol is
automatically placed

Ø1.250

Move cursor to drag leader to
any desired length and location

Select circle

Ø1.250 — Leader

Place dimension at
desired location

The **DIMDIAMETER** command includes the **Mtext**, **Text**, and **Angle** options described in Chapter 17. Use the **Mtext** or **Text** option to add information to or change the dimension value. The **Angle** option changes the dimension text angle, although this practice is not common.

Exercise 18-1

Complete the exercise on the companion website.
www.g-wlearning.com/CAD

Dimensioning Holes

Dimension holes in the view in which they appear as circles. Give location dimensions to the center and a leader showing the diameter. The **DIMDIAMETER** command is effective for dimensioning holes. To note multiple holes of the same size, dimension the size of one hole using the **DIMDIAMETER** command and the **Mtext** or **Text** option. Precede the diameter with the number of holes followed by X and then a space. See the 2X Ø.50 dimension in **Figure 18-2**.

PROFESSIONAL TIP

The ASME standard recommends a small space between the object and the extension line. To specify the space, adjust the **Offset from origin** setting in the dimension style to an appropriate positive value, such as .063 (1.5 mm). This is very useful *except* when you dimension to centerlines to locate circular features. When you pick the endpoint of the centerline, a positive value leaves an unacceptable space between the centerline and the beginning of the extension line. Change the **Offset from origin** setting to 0 to remove the gap. Be sure to change back to the positive setting before dimensioning other objects.

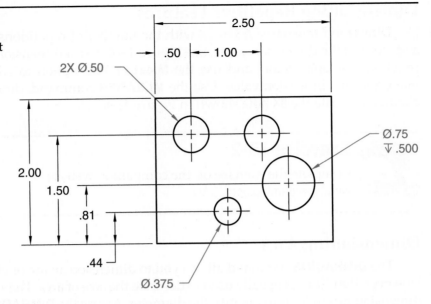

Figure 18-2.
Dimensioning holes on a mechanical part drawing view.

Dimensioning for Manufacturing Processes

Counterbore, spotface, and *countersink* manufacturing processes are examples of hole features that are dimensioned using symbols. Dimension these manufacturing processes in the view in which holes appear as circles, with a leader providing machining information in a note. See **Figure 18-3**. The **Mtext** option of the **DIMDIAMETER** command provides a convenient method for dimensioning manufacturing processes. Many symbols are available in the gdt.shx font. You can also create custom symbols as *blocks*. Blocks are described later in this textbook.

counterbore: A larger-diameter hole machined at one end of a smaller hole that provides a place for the screw head.

spotface: A larger-diameter hole machined at one end of a smaller hole that provides a smooth, recessed surface for a washer; similar to a counterbore, but not as deep.

countersink: A cone-shaped recess at one end of a hole that provides a mating surface for a screw head of the same shape.

block: A symbol previously created and saved for reuse.

Reference Material *Drafting Symbols*
For more information about common drafting symbols and the gdt.shx font, go to the **Reference Material** section of the companion website (www.g-wlearning.com/CAD) and select **Drafting Symbols** in the list.

Figure 18-3.
Dimension notes for machining processes. You can insert symbols as blocks or use lowercase letters in the gdt.shx font.

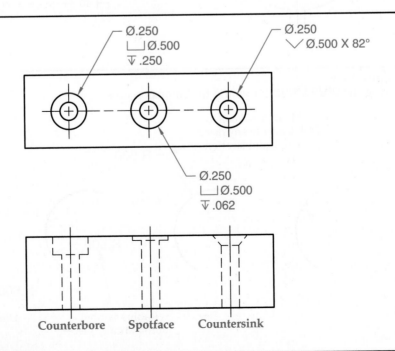

Dimensioning Repetitive Features

repetitive features: Many features having the same shape and size.

Dimension *repetitive features* with the number of repetitions followed by an X, a space, and the dimension value. See **Figure 18-4**. Use a dimension command appropriate for the application, and use the **Mtext** or **Text** option to add repetitive information to the dimension value. Use the **MLEADER** command, described later in this chapter, to create the 8X note shown in **Figure 18-4**.

Exercise 18-2

Complete the exercise on the companion website.
www.g-wlearning.com/CAD

Dimensioning Arcs

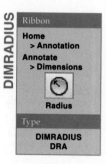

The **DIMRADIUS** command allows you to dimension an arc or circle with a radius. However, the radius is usually used to describe the size of arcs. The ASME standard for dimensioning circles is to identify the diameter. Access the **DIMRADIUS** command and select an arc or circle to display a leader line and a radius dimension value attached to the crosshairs. Specify a point to locate the dimension value. See **Figure 18-5**.

Figure 18-4.
Dimensioning repetitive features (shown in color) on a drawing view of different mechanical parts.

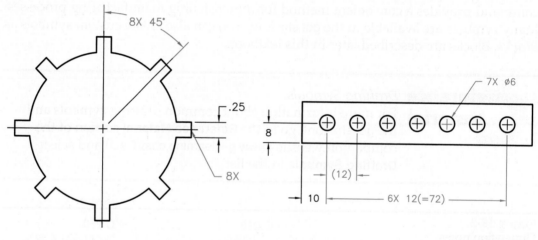

Figure 18-5.
Using the **DIMRADIUS** command to dimension arcs.

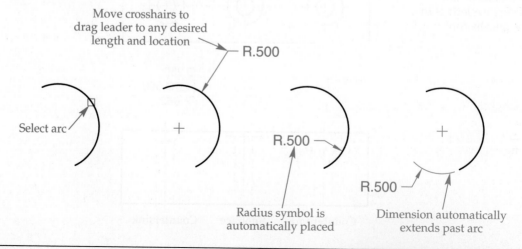

AutoCAD and Its Applications—Basics

Like other dimension commands, the **DIMRADIUS** command references the current dimension style and the object you select to create a dimension object. The dimension includes all related dimension style characteristics, centerlines, leader, arrowheads, and a dimension value associated with the radius. The leader points to the center of the arc or circle, as recommended by the ASME standard.

 The **DIMRADIUS** command includes the **Mtext**, **Text**, and **Angle** options described in Chapter 17. Use the **Mtext** or **Text** option to add information to or change the dimension value. The **Angle** option changes the dimension text angle, although this practice is not common.

Dimensioning Arc Length

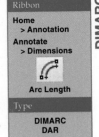

Access the **DIMARC** command to dimension the length of an arc. Select an arc or polyline arc segment to display the arc length symbol and dimension value attached to the crosshairs. Specify a point to locate the dimension value. By default, the arc length symbol occurs before the text. The ASME standard recommends placing the symbol over the text, as shown in **Figure 18-6**. The dimension style controls the symbol placement.

Before placing the arc length dimension, you can add information to or change the dimension value using the **Mtext** or **Text** option. The **Angle** option is available to change the text angle. Use the **Partial** option to dimension a portion of the arc length. Select two points on the arc to dimension the length between the points. The **Leader** option, which is available when the arc is greater than 90°, allows you to add a leader pointing to the arc from the dimension value.

Dimensioning Large Arcs

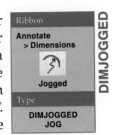

Use the **DIMJOGGED** command to dimension an arc that is so large that the center point cannot appear on the layout. Jogging a dimension line is most appropriate for large arcs, but you can select a circle. Access the **DIMJOGGED** command and pick an arc or circle. Then pick a location for the origin of the center. This point represents the center of the arc or circle. The associated radius value does not change. Select a location for the dimension line and then pick a location for the break symbol. See **Figure 18-7**. You can move the components of the dimension by grip editing after you place the dimension.

Dimensioning Fillets and Rounds

You can dimension *fillets* and *rounds* individually as arcs, using the **DIMRADIUS** command, or collectively in a general note. See **Figure 18-8**. On mechanical drawings, it is common to include a general note such as ALL FILLETS AND ROUNDS R.125 UNLESS OTHERWISE SPECIFIED on the drawing.

fillets: Small inside arcs designed to strengthen inside corners.

rounds: Small arcs on outside corners used to relieve sharp corners.

Figure 18-6.
Using the **DIMARC** command to dimension the length of an arc.

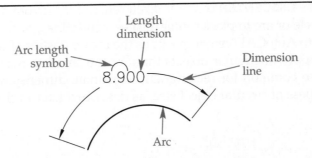

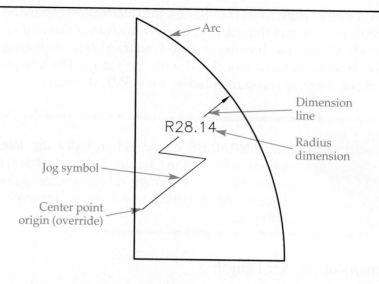

Figure 18-7.
Using the **DIMJOGGED** command to place a radius dimension for a large arc.

Arc

Dimension line

Radius dimension

R28.14

Jog symbol

Center point origin (override)

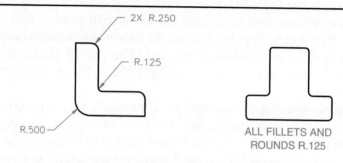

Figure 18-8.
Dimensioning fillets and rounds.

2X R.250

R.125

R.500

ALL FILLETS AND ROUNDS R.125

Exercise 18-3

Complete the exercise on the companion website.
www.g-wlearning.com/CAD

Dimensioning Other Curves

Dimension curves as arcs when possible. When an arc does not have a constant radius, dimension to points along the curve using the **DIMLINEAR** command. See **Figure 18-9**.

Adding Center Dashes and Centerlines

DIMCENTER

Ribbon
Annotate
> Dimensions

Center Mark

Type
DIMCENTER
DCE

Depending on the current dimension style setting, when you use the **DIMDIAMETER** and **DIMRADIUS** commands, small circles and arcs automatically receive center dashes. Large circles and arcs display center dashes or centerlines, or no symbol appears. Use the **DIMCENTER** command to add center dashes or centerlines to objects that are not dimensioned using the **DIMDIAMETER** or **DIMRADIUS** commands. Access the **DIMCENTER** command and pick a circle or an arc to display center marks.

The **DIMCENTER** command references the current dimension style and the size of the circle or arc to place center dashes, centerlines, or no symbol. The ASME standard refers to AutoCAD center marks as the center dashes of centerlines. Center marks typically apply to circular objects that are too small to receive centerlines. Center marks are also common for rectangular coordinate dimensioning without dimension lines, regardless of circular object size, as described later in this chapter.

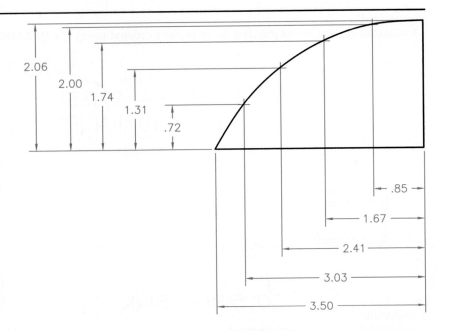

Figure 18-9.
Dimensioning curves that do not have a constant radius.

2.06
2.00
1.74
1.31
.72
.85
1.67
2.41
3.03
3.50

Drawing Leader Lines

The **DIMDIAMETER** and **DIMRADIUS** commands automatically place *leader lines* when you dimension circles and arcs. AutoCAD multileaders created using the **MLEADER** command allow you to add leader lines for other applications, such as specific notes. Multileaders consist of single or multiple lines of *annotations*, including symbols, and leaders. Multileader styles control multileader characteristics, such as leader format, annotation style, and arrowhead size. You can create multi-segment leaders and align and group separate leaders. Chapter 20 describes adding and removing multiple leader lines and aligning leaders.

leader line: A line that connects a note or symbol to a specific feature or location on a drawing.

annotation: Textual information presented in notes, specifications, comments, and symbols.

QLEADER and **LEADER** commands are available to create leader lines, but they provide fewer options than the **MLEADER** command.

multileader styles: Saved configurations for the appearance of leaders.

shoulder: A short horizontal line usually added to the end of straight leader lines.

Multileader Styles

A *multileader style* presets many multileader characteristics. Multileader style settings correspond to appropriate drafting standards and usually apply to a specific drafting field or dimensioning application. In mechanical drafting, properly drawn leaders have one straight segment extending from the feature to a horizontal *shoulder* that is 1/8"–1/4" (3 mm–6 mm) long. While most other fields also use straight leaders, AutoCAD provides the option of drawing curved leaders, which are common in architectural drafting. See **Figure 18-10**.

Multileader Style Manager

Create, modify, and delete multileader styles using the **Multileader Style Manager** dialog box. See **Figure 18-11**. The **Styles:** list box displays existing multileader styles. The Annotative multileader style allows you to create annotative leaders, as indicated by the icon

Ribbon

Home
> Annotation

Multileader Style

Annotate
> Leaders

Multileader Style

Type

MLEADERSTYLE
MLS

MLEADERSTYLE

Figure 18-10.
Multileader styles control the display of leaders created using the **MLEADER** command.

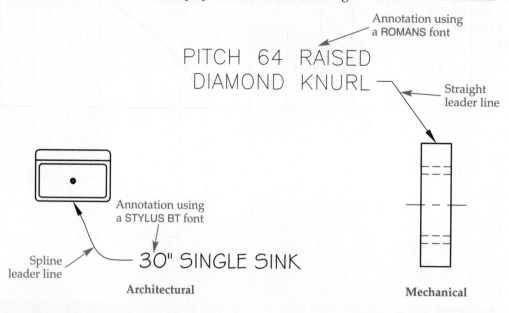

Annotation using
a ROMANS font

PITCH 64 RAISED
DIAMOND KNURL

Straight
leader line

Annotation using
a STYLUS BT font

Spline
leader line

30" SINGLE SINK

Architectural

Mechanical

Figure 18-11.
The **Multileader Style Manager** dialog box.

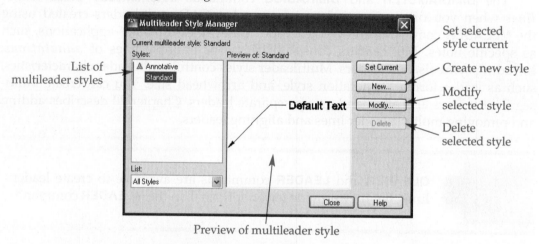

List of
multileader styles

Set selected
style current

Create new style

Modify
selected style

Delete
selected style

Preview of multileader style

to the left of the style name. The Standard multileader style does not use the annotative function. To make a multileader style current, double-click the style name, right-click on the name and select **Set current**, or pick the name and select the **Current** button. Use the **List:** drop-down list to filter the multileader styles displayed in the **Styles:** list box.

The **Preview of:** image displays a representation of the multileader style and changes according to the selections you make.

Creating New Multileader Styles

To create a new multileader style, select an existing multileader style from the **Styles:** list box to use as a base for formatting the new multileader style. Then pick the **New...** button to open the **Create New Multileader Style** dialog box. See **Figure 18-12.**

Figure 18-12.
The **Create New
Multileader Style**
dialog box.

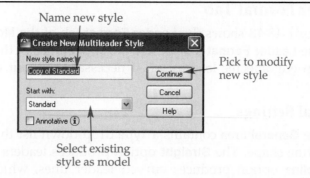

Name new style

Pick to modify
new style

Select existing
style as model

You can base the new multileader style on the formatting of a different multileader style by selecting from the **Start With** drop-down list. Notice that Copy of followed by the name of the existing style appears in the **New Style Name** text box. Replace the default name with a more descriptive name, such as Mechanical, Architectural, Straight, or Spline. Multileader style names can have up to 255 characters, including uppercase or lowercase letters, numbers, dashes (–), underlines (_), and dollar signs ($).

Pick the **Annotative** check box to make the multileader style annotative. Pick the **Continue** button to access the **Modify Multileader Style** dialog box, shown in **Figure 18-13**. The **Leader Format**, **Leader Structure**, and **Content** tabs display groups of settings for specifying leader appearance, as described in this chapter. After completing the style definition, pick the **OK** button to return to the **Multileader Style Manager** dialog box.

The preview image in the upper-right corner of each **Modify Multileader Style** dialog box tab displays a representation of the multileader style and changes according to the selections you make.

Figure 18-13.
The **Leader Format** tab of the **Modify Multileader Style** dialog box.

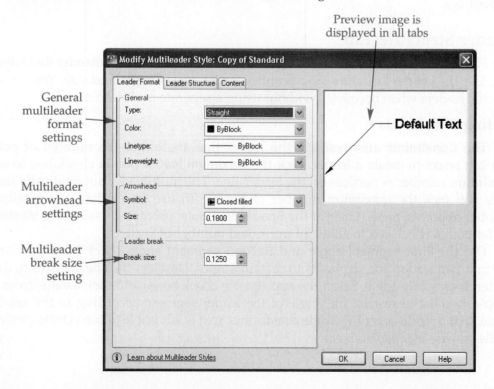

Preview image is
displayed in all tabs

General
multileader
format
settings

Multileader
arrowhead
settings

Multileader
break size
setting

Default Text

Leader Format Tab

Figure 18-13 shows the **Leader Format** tab of the **Modify Multileader Style** dialog box. The **Leader Format** tab presets the appearance of the leader line and arrowhead. You can edit specific leaders when necessary without using a separate multileader style.

General Settings

The **General** area contains a **Type** drop-down list that you can use to specify the leader line shape. The **Straight** option produces leaders with straight line segments. The **Spline** option produces curved leader lines, which are common in architectural drafting. Pick the **None** option to create a multileader style that does not use a leader line. Use the **None** option to create a leader that you can associate with other leaders using the **MLEADERALIGN** and **MLEADERCOLLECT** commands, described in Chapter 20.

Color, **Linetype**, and **Lineweight** drop-down lists are available for changing the dimension line color, linetype, and lineweight. Multileaders are block objects. Chapter 24 explains blocks and provides complete information on using ByBlock, ByLayer, or absolute color, lineweight, and linetype with blocks. For now, as long as you assign a specific layer to a multileader and do not change the multileader properties to absolute values, the default ByBlock properties are acceptable.

Arrowhead Settings

The **Arrowhead** area sets the leader arrowhead style and size. Select the arrowhead style from the **Symbol:** drop-down list. The arrowhead symbol options are the same as those for dimension style arrowheads. Set the arrowhead size using the **Size:** text box. A .125″ (3 mm) arrowhead size is most common, especially on mechanical drawings. Leader arrowheads are typically the same size as dimension arrowheads.

Adjusting Break Size

The **Leader Break** area controls the amount of leader line hidden by the **DIMBREAK** command. Specify a value in the **Break size:** text box to set the total length of the break. The default size is .125″ (3 mm). ASME standards do not recommend breaking leader lines.

Leader Structure Tab

Figure 18-14 shows the **Leader Structure** tab of the **Modify Multileader Style** dialog box. Use the **Leader Structure** tab to control leader construction and size. You can edit specific leaders when necessary without using a separate multileader style.

Setting Constraints

The **Constraints** area restricts the leader line angle and the number of points you can select to create a leader. Pick the **Maximum leader points** check box to set a maximum number of vertices on the leader line. The multileader automatically forms once you pick the maximum number of points. To use fewer than the maximum number of points, press [Enter] at the Specify next point: prompt. Deselect the **Maximum leader points** check box to allow an unlimited number of vertices.

Use the **First segment angle** and **Second segment angle** check boxes to restrict the first two leader line segments to certain angles. Deselect the check boxes to draw leader lines at any angle. Select the appropriate check boxes and pick a value from the drop-down list to restrict the angle of the leader segment according to the selected value. **Ortho** mode overrides angle constraints, so it is advisable to turn **Ortho** mode off while you are placing leaders.

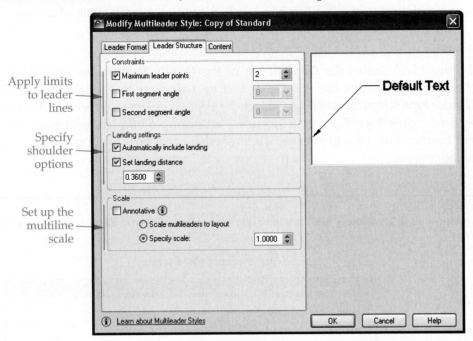

Apply limits to leader lines

Specify shoulder options

Set up the multiline scale

PROFESSIONAL TIP

The ASME standard for leaders recommends that leader lines have angles not less than 15° and not greater than 75° from horizontal. Use the **First segment angle** and **Second segment angle** settings to help maintain this standard.

Landing Settings

The **Landing settings** area controls the display and size of the *landing* and is available only for straight-line multileader styles. Select the **Automatically include landing** check box to display a shoulder automatically when you select the second leader line point. This is the preferred method for creating straight leader lines. Deselect the check box to create leaders without shoulders, or to pick a third point to draw the shoulder manually. When you check **Automatically include landing**, the **Set landing distance** check box becomes enabled. Check **Set landing distance** to define a specific shoulder length, typically 1/8″–1/4″ (3 mm–6 mm), in the text box. If you deselect the text box, a prompt asks for the shoulder length when you place a leader.

> **landing:** The AutoCAD term for a leader shoulder.

Scale Options

The **Scale** area sets the multileader scale factor. Select the **Annotative** check box to create an annotative multileader style. The **Annotative** check box is already set when you modify the Annotative multileader style or pick the **Annotative** check box in the **Create New Multileader Style** dialog box.

Select the **Scale dimensions to layout** radio button to add leaders in a floating viewport in a paper space layout. You must add leaders to the model in a floating viewport in order for this option to function. Scaling leaders to the layout allows the overall scale to adjust according to the active floating viewport by setting the overall scale equal to the viewport scale factor.

Pick the **Use overall scale of** radio button to scale a drawing manually. Enter the drawing scale factor to apply to all leader elements in the corresponding text box. The scale factor is multiplied by the plotted leader size to get the size of the leader in model space. For example, if you manually scale multileaders for a drawing with a 1:2 (half) scale, or a scale factor of 2 (2 ÷ 1 = 2), enter 2 in the **Use overall scale of** text box.

Content Tab

Figure 18-15 shows the **Content** tab of the **Modify Multileader Style** dialog box. The **Content** tab controls the display of text or a block with the leader line. Use the **Multileader type:** drop-down list to select the type of object to attach to the end of the leader line or shoulder. Figure 18-16 shows an example of a leader drawn with each content option. You can edit specific leaders when necessary, without using a separate multileader style.

Figure 18-15.
The **Content** tab of the **Modify Multileader Style** dialog box with the **Mtext** multileader type selected.

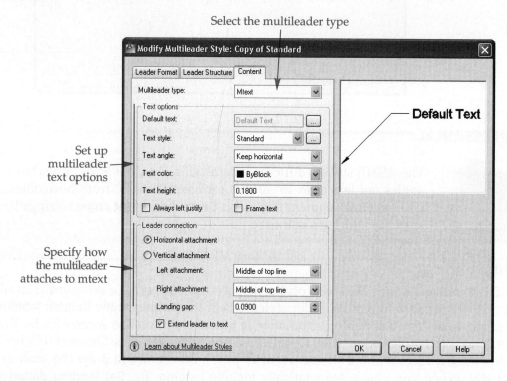

Figure 18-16.
Examples of each multileader content type.

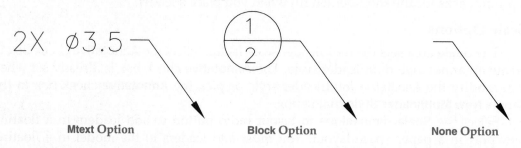

AutoCAD and Its Applications—Basics

Attaching Mtext

Pick the **Mtext** option from the **Multileader type:** drop-down list, as shown in **Figure 18-15**, to attach multiline text to the leader. The **Text options** and **Leader connection** areas of the **Content** tab appear when you select the **Mtext** content type. The **Default** text option allows you to specify a value to attach leaders during leader placement. This is useful when the same note or symbol is required throughout a drawing. Pick the ellipsis (…) button to return to the drawing window and use the multiline text editor to enter the default text value. Close the text editor to return to the **Modify Multileader Style** dialog box.

A multileader style references a text style for the appearance of leader text. Pick an existing text style from the **Text style** drop-down list. To create or modify a text style, pick the ellipsis (…) button next to the drop-down list to launch the **Text Style** dialog box.

Select an option from the **Text angle** drop-down list to control the angle at which text appears in reference to the angle of the leader line or shoulder. See **Figure 18-17**. Use the **Text color** drop-down list to specify the text color, which should be ByBlock for typical applications. Use the **Text height** text box to specify the leader text height. Leader text height is usually the same as dimension text height.

The **Always left justify** option forces leader text to left-justify, regardless of the leader line direction. Pick the **Frame text** check box to create a box around the leader text. A basic dimension is a common application for framed text, as described later in this textbook.

The **Leader connection** area contains options that determine how mtext is positioned relative to the endpoint of the leader line or shoulder. Most drawings require leaders that use the **Horizontal** attachment option. Use the **Left attachment:** drop-down list to define how multiple lines of text are positioned when the leader is on the left side of the text. Use the **Right attachment:** drop-down list to define how multiple lines of text are positioned when the leader is on the right side of the text. **Figure 18-18** shows typical selections.

Select the **Extend leader to text** check box to lengthen the shoulder to text when multiple lines of text do not align according to the leader attachment. **Figure 18-19** shows examples of using multiline text justification and the **Extend leader to text**

Figure 18-17.
Text angle options available for mtext.

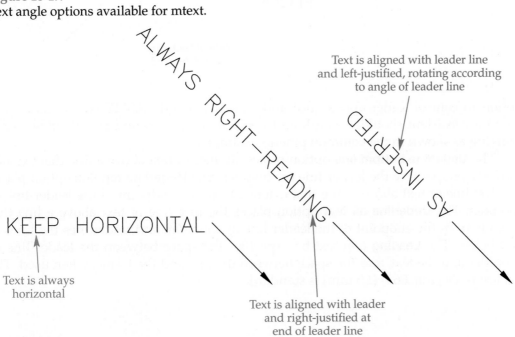

Figure 18-18.
Horizontal alignment options. The shaded examples are the recommended ASME standards.

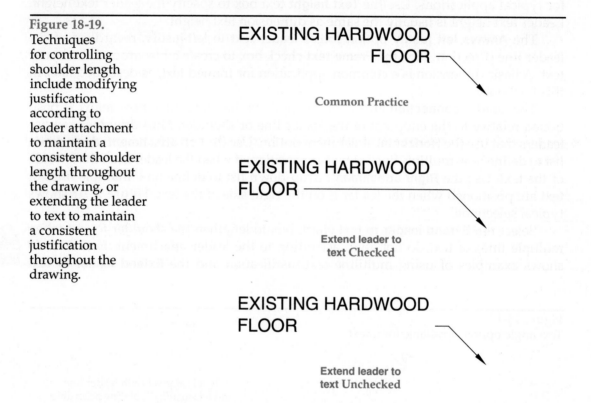

Figure 18-19.
Techniques for controlling shoulder length include modifying justification according to leader attachment to maintain a consistent shoulder length throughout the drawing, or extending the leader to text to maintain a consistent justification throughout the drawing.

EXISTING HARDWOOD
FLOOR

Common Practice

EXISTING HARDWOOD
FLOOR

Extend leader to
text Checked

EXISTING HARDWOOD
FLOOR

Extend leader to
text Unchecked

option to control leader appearance and shoulder length. ASME standards recommend a consistent shoulder length and appropriate justification for all leaders on a drawing as shown in the common practice example.

The **Underline bottom line** option places all lines of text above a line that extends from the endpoint of the leader line or shoulder. The **Underline top line** option places the first line of text above a line that extends from the endpoint of the leader line or shoulder. The **Underline all text** option places the first line of text above a line that extends from the endpoint of the leader line or shoulder, and underlines the lines of text below. The **Leading gap** text box specifies the space between the leader line or shoulder and the text, and the space between the text and the frame, when used. The default is .09″, but .063″ (1.5 mm) is standard.

Common drafting practice is to use the **Middle of bottom line** option for left attachment and the **Middle of top line** option for right attachment. The **Middle of text** option is sometimes necessary to display a single line of text that includes a horizontally stacked fraction.

Selecting the **Vertical attachment** radio button is uncommon. This option eliminates the possible use of a shoulder and connects the leader endpoint to the top center or bottom center of the text, depending on the leader line position. Use the **Top attachment:** drop-down list to define how text is positioned when the leader is above the text. Use the **Bottom attachment:** drop-down list to define how text is positioned when the leader is below the text. **Figure 18-20** shows each option.

Attaching a Symbol

Pick the **Block** option from the **Multileader type:** drop-down list, as shown in **Figure 18-21**, to attach a block to the leader. Blocks are described later in this textbook. Several blocks are available by default from the **Source block:** drop-down list. Pick the **User Block...** option to select a custom block. The **Select Custom Content Block** dialog box appears when you pick the **User Block...** option. Pick a block in the current drawing from the **Select from Drawing Blocks:** drop-down list and then pick the **OK** button.

Use the **Attachment:** drop-down list to specify how to attach the block to the leader. Pick the **Insertion point** option to attach the block to the leader according to the block insertion point, or base point. Choose the **Center Extents** option to attach the block directly to the leader, aligned to the center of the block, even if the block insertion point is not on the block itself. See **Figure 18-22**.

Use the **Color:** drop-down list to specify the block color, which should be ByBlock for typical applications. Use the **Scale:** text box to proportionately increase or decrease the block size. Scale does not affect the appearance of the leader line, arrowhead, or shoulder, or the scale applied to the multileader object.

Using No Content

Select the **None** option from the **Multileader type:** drop-down list to end the leader without annotation. Use the **None** option when you need to create only a leader, without text or a symbol attached to the leader line or shoulder.

Figure 18-20.
Vertical alignment options.

Center	Underline and Center	Center	Overline and Center

Figure 18-21.
The **Content** tab of the **Modify Multileader Style** dialog box with the **Block** multileader type selected.

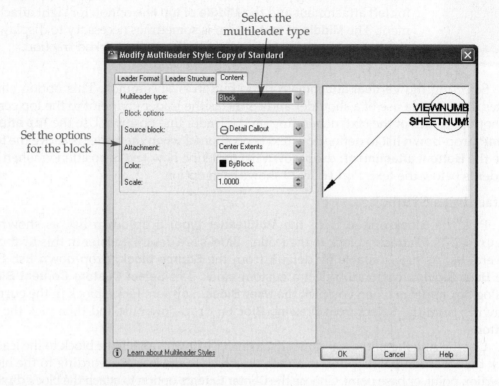

Select the multileader type

Set the options for the block

Figure 18-22.
Adjusting multileader block attachment.

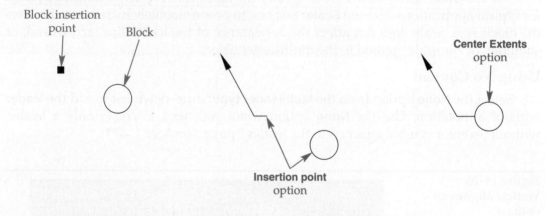

Block insertion point

Block

Center Extents option

Insertion point option

PROFESSIONAL TIP

Add leaders to existing multileaders using the **Add Leader** tool. This eliminates the need to create a separate multileader style that uses the **None** multileader content type for most applications.

Exercise 18-4

Complete the exercise on the companion website.
www.g-wlearning.com/CAD

Changing Multileader Styles

Use the **Multileader Style Manager** to change the characteristics of an existing multileader style. Pick the **Modify** button to open the **Modify Multileader Style** dialog box. When you make changes to a multileader style, such as selecting a different text or arrowhead style, all existing leaders assigned the modified multileader style are updated. Use a different multileader style with different characteristics when appropriate.

Renaming and Deleting Multileader Styles

To rename a multileader style using the **Multileader Style Manager**, slowly double-click on the name or right-click on the name and select **Rename**. To delete a multileader style using the **Multileader Style Manager**, right-click the name and select **Delete**. You cannot delete a multileader style that is assigned to leaders. To delete a style that is in use, assign a different style to the leaders that reference the style to be deleted.

 You can also rename styles using the **Rename** dialog box. Select **Multileader styles** in the **Named Objects** list to rename a multileader style.

Setting a Multileader Style Current

Set a multileader style current using the **Multileader Style Manager** by double-clicking the style in the **Styles:** list box, right-clicking on the name and selecting **Set current**, or picking the style and selecting the **Set Current** button. To set a multileader style current without opening the **Multileader Style Manager**, use the **Multileader Style** drop-down list located in the expanded **Annotation** panel on the **Home** ribbon tab or the **Leaders** panel on the **Annotate** ribbon tab.

PROFESSIONAL TIP

 You can import multileader styles from existing drawings using **DesignCenter**. See Chapter 5 for more information about using **DesignCenter** to import file content.

Inserting Multileaders

Use the **MLEADER** command to insert a leader. How you insert a leader depends on the current multileader style settings and the option you choose to construct the leader. In general, there are three methods for inserting a multileader, depending on what portion of the leader you locate first. Review the components of a leader, shown in **Figure 18-23**, before reading the options for creating a multileader.

To use the default **Specify leader arrowhead location** option, first specify the point at which the arrowhead touches. Then specify a point to locate where the leader ends and the shoulder begins. If the **Mtext** option is active, enter leader text using the multi-line text editor.

Choose the **leader Landing first** option to specify a point to locate where the leader ends and the shoulder begins first. Then specify the point at which the arrowhead touches. If the **Mtext** option is active, enter leader text using the multiline text editor.

Ribbon

Home
> Annotation
Annotate
> Leaders

Multileader

Type

MLEADER
MLD

MLEADER

Figure 18-23.
Examples of leaders created using the **MLEADER** command. A—An architectural leader created using a spline leader line, the **Specify leader arrowhead location** option, and three leader points. B—A mechanical leader created using a straight leader line, the **leader Landing first** option, and two leader points.

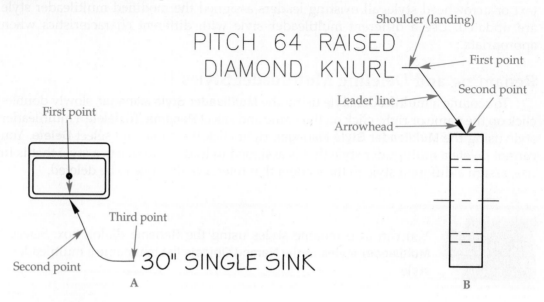

Use the **Content first** option to define the leader content first. The **Mtext** option allows you to type text using the multiline text editor. Then specify the point at which the arrowhead touches.

Select **Options** to access a list of options to override the current multileader style characteristics. The options are the same as those found in the **Modify Multileader Style** dialog box.

 Additional tools are available for adding and removing multiple leader lines, arranging multiple leaders, and combining leader content, as explained in Chapter 20.

 Exercise 18-5

Complete the exercise on the companion website.
www.g-wlearning.com/CAD

Dimensioning Chamfers

chamfer: An angled surface used to relieve sharp corners.

Dimension *chamfers* of 45° with a leader giving the angle and linear dimension, or with two linear values. See **Figure 18-24**. Place the leader using the **MLEADER** command. Chamfers other than 45° include either the angle and a linear dimension or two linear dimensions. See **Figure 18-25**. Use the **DIMLINEAR** and **DIMANGULAR** commands for this purpose.

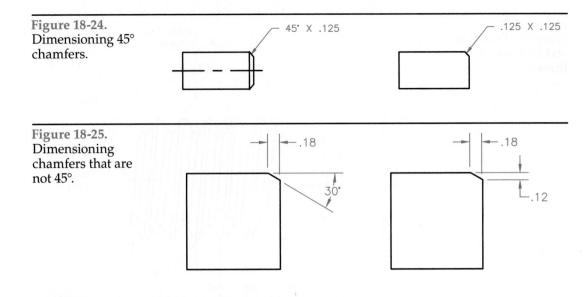

Figure 18-24. Dimensioning 45° chamfers.

45° X .125

.125 X .125

Figure 18-25. Dimensioning chamfers that are not 45°.

.18

30°

.18

.12

Exercise 18-6

Complete the exercise on the companion website.
www.g-wlearning.com/CAD

Thread Drawings and Notes

Figure 18-26 shows the elements of external and internal screw threads. However, threads commonly appear on a drawing as a simplified representation in which a hidden line indicates thread depth. See Figure 18-27. Threaded parts often include a chamfer to help engage the mating thread.

Thread representations show the reader that a thread exists, but the thread note gives the exact specifications. The thread note typically connects to the thread with a leader. See Figure 18-28. The most common thread forms are the Unified and metric screw threads, but a variety of other thread forms are required for specific applications.

The following format specifies the thread note for Unified screw threads:

3/4-10 UNC-2 A
(1)   (2)   (3)   (4)(5)

(1) Major diameter of thread, given as a fraction or number
(2) Number of threads per inch
(3) Thread series: UNC = Unified National Coarse, UNF = Unified National Fine
(4) Class of fit: 1 = large tolerance, 2 = general-purpose tolerance, 3 = tight tolerance
(5) Thread type: A = external thread. B = internal thread

The following format specifies the thread note for metric threads:

M14X2
(1) (2)   (3)

(1) M = metric thread.
(2) Major diameter in millimeters.
(3) Pitch in millimeters.

There are too many screw threads to describe in detail in this textbook. Refer to the *Machinery's Handbook*, published by Industrial Press Inc., or a comprehensive mechanical drafting text for more information.

Figure 18-26.
Features of external
and internal screw
threads.

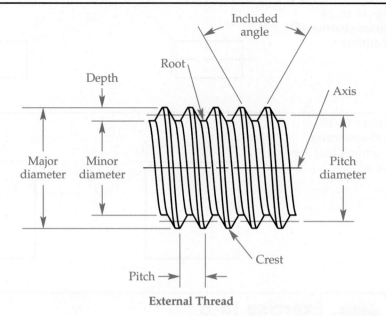

External Thread

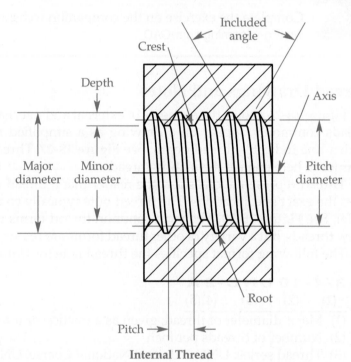

Internal Thread

The **Dimension** panel of the **Express Tools** ribbon tab includes **QLATTACH**, **QLATTACHSET**, and **QLDETACHSET** commands that allow you to attach an annotation object to and detach it from leaders created using the **QLEADER** and **LEADER** commands. Use multileaders instead of these leader commands and express tools.

Exercise 18-7
Complete the exercise on the companion website.
www.g-wlearning.com/CAD

Figure 18-27.
Simplified thread representations.

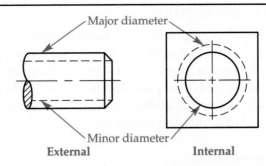

Major diameter

Minor diameter

External   Internal

Figure 18-28.
Displaying the thread note with a leader.

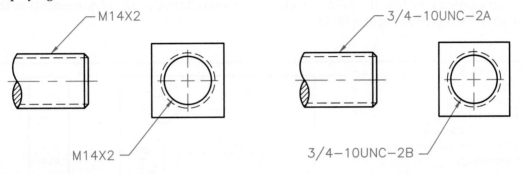

M14X2

M14X2

3/4−10UNC−2A

3/4−10UNC−2B

Alternate Dimensioning Practices

Omitting dimension lines is common for drawings in industries that use computer-controlled machining processes, and for drawings in which unconventional dimensioning practices are required because of product features. Rectangular coordinate dimensioning without dimension lines, tabular dimensioning, and chart dimensioning are examples of dimensioning methods that omit dimension lines.

Rectangular Coordinate Dimensioning without Dimension Lines

Rectangular coordinate dimensioning without dimension lines is popular in mechanical drafting for specific applications such as precision sheet metal part drawings and electronics drafting, especially for chassis layout. Each dimension represents a measurement originating from a *datum*. See **Figure 18-29.** Identification letters label holes or similar features. Often a table, keyed to the identification letters, indicates feature size or specifications.

rectangular coordinate dimensioning without dimension lines: A type of dimensioning that includes only extension lines and text aligned with the extension lines.

datum: The 0 dimension, baseline, or common point from which all measurements are made while dimensioning.

Figure 18-29.
Rectangular coordinate dimensioning without dimension lines.

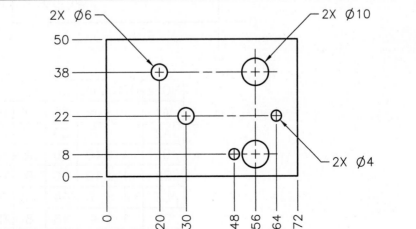

2X ⌀6   2X ⌀10

2X ⌀4

Tabular Dimensioning

In *tabular dimensioning*, each feature receives a label with a letter or number that correlates to a table. See **Figure 18-30**. Some companies take this practice one step further and display the location and size of features in the table from an X and a Y axis. The depth of features is also provided from the Z axis where appropriate. Each feature is labeled with a letter or number that correlates to the table, as shown in **Figure 18-31**. Chapter 21 explains drawing tables.

Chart Dimensioning

Chart dimensioning may take the form of unidirectional, aligned, or tabular dimensioning or rectangular coordinate dimensioning without dimension lines. Chart dimensioning provides flexibility when dimensions change as the requirements of the product change. See **Figure 18-32**.

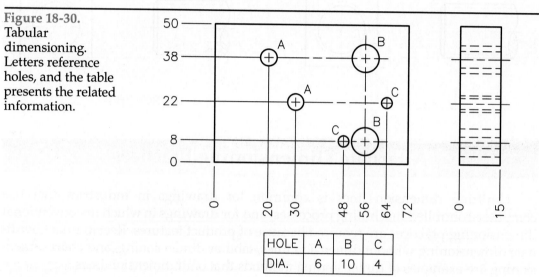

Figure 18-30. Tabular dimensioning. Letters reference holes, and the table presents the related information.

HOLE	A	B	C
DIA.	6	10	4

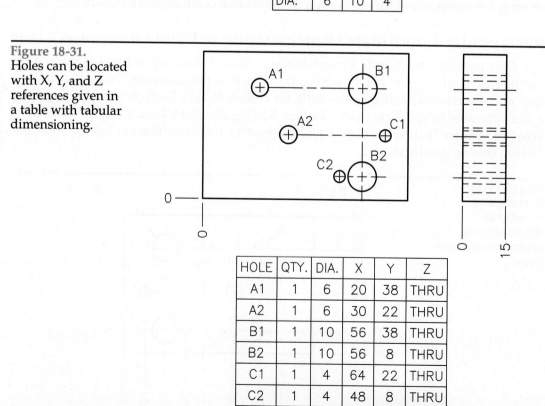

Figure 18-31. Holes can be located with X, Y, and Z references given in a table with tabular dimensioning.

HOLE	QTY.	DIA.	X	Y	Z
A1	1	6	20	38	THRU
A2	1	6	30	22	THRU
B1	1	10	56	38	THRU
B2	1	10	56	8	THRU
C1	1	4	64	22	THRU
C2	1	4	48	8	THRU

Figure 18-32.
Chart dimensioning.

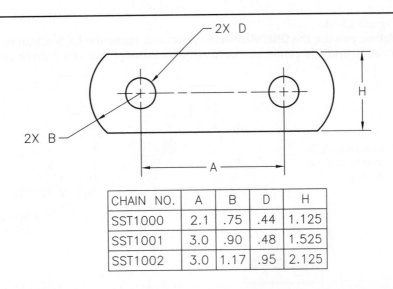

CHAIN NO.	A	B	D	H
SST1000	2.1	.75	.44	1.125
SST1001	3.0	.90	.48	1.525
SST1002	3.0	1.17	.95	2.125

Creating Ordinate Dimension Objects

AutoCAD refers to rectangular coordinate dimensioning without dimension lines as *ordinate dimensioning*. In order to create ordinate dimension objects accurately, you must move the default origin (0,0,0 coordinate) to the object datum. This involves understanding the AutoCAD world coordinate system and user coordinate systems. Once you establish the datum by temporarily moving the origin, use the **DIMORDINATE** command to place ordinate dimension objects.

ordinate dimensioning: The AutoCAD term for rectangular coordinate dimensioning without dimension lines.

Introduction to WCS and UCS

The origin of the *world coordinate system (WCS)* has been at the 0,0,0 point for the drawings you have created throughout this textbook. In most cases, this is appropriate. However, when you apply rectangular coordinate dimensioning without dimension lines, it is best to originate dimensions from a primary datum, which is often a corner of the object. Depending on how you draw the object, this point may or may not align with the WCS origin.

world coordinate system (WCS): The AutoCAD rectangular coordinate system. In 2D drafting, the WCS contains four quadrants, separated by the X and Y axes.

The WCS is fixed, but the origin of a *user coordinate system (UCS)* can move to any point. The UCS is described in detail in *AutoCAD and Its Applications—Advanced*. In general, a UCS allows you to set your own coordinate system and origin. Measurements made with the **DIMORDINATE** command originate from the current UCS origin. By default, this is the 2D 0,0 origin.

user coordinate system (UCS): A temporary override of the WCS in which the origin (0,0,0) is moved to a location specified by the user.

A quick method for relocating the origin is to pick the UCS icon in the lower-left corner of the model space drawing window to display grips. Use the origin grip box and the **Move Origin Only** option and move the UCS icon to specify a new origin point, such as the corner of an object or another appropriate datum. Press [Esc] to deselect the UCS icon and hide grips. See **Figure 18-33**.

NEW

When you finish drawing dimensions from a datum, you can leave the UCS origin at the datum or move it back to the WCS origin. To return to the WCS, grip-edit the UCS icon and apply the **World** option, or select the **WCS** option from the **WCS** menu of the ViewCube.

Figure 18-33.
Before you use the **DIMORDINATE** command, move the UCS origin to the appropriate datum location and add center marks to circular features that you plan to dimension.

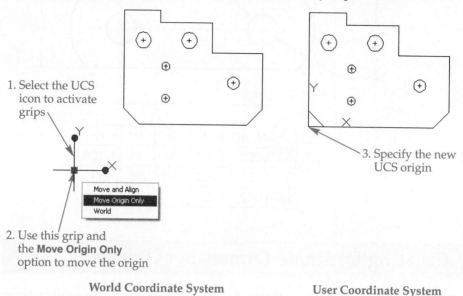

1. Select the UCS icon to activate grips

2. Use this grip and the **Move Origin Only** option to move the origin

3. Specify the new UCS origin

Move and Align
Move Origin Only
World

World Coordinate System   User Coordinate System

Using the DIMORDINATE Command

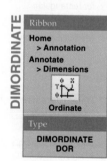

DIMORDINATE

Ribbon
Home
> Annotation
Annotate
> Dimensions

Ordinate

Type
DIMORDINATE
DOR

The **DIMORDINATE** command automatically places an extension line and a dimension at the point you specify. The dimension measures an X or Y coordinate distance from the UCS origin. Since you are working in the XY plane, you may want to set vertical and horizontal polar tracking or turn on **Ortho** mode before using the **DIMORDINATE** command. In addition, if the drawing includes circular features, use the **DIMCENTER** command to place center dashes as shown in Figure 18-33. This conforms to ASME standards and provides something to pick when you dimension circular features.

Access the **DIMORDINATE** command, and when the Specify feature location: prompt appears, pick a point to locate the origin of the extension line. If the feature is the corner of the object, pick an endpoint or other appropriate position. If the feature is a circle, pick the end of the center dash, as shown in Figure 18-34, not the center of the object. This leaves the required space between the center mark and the extension line. Zoom in if needed and use object snap modes when necessary. The next prompt asks for the leader endpoint, which refers to the extension line endpoint. Specify the endpoint of the extension line.

If the X axis or Y axis distance between the feature and the extension line endpoint is large, the axis AutoCAD uses for the dimension by default may not be correct. When

Figure 18-34.
Pick the endpoints of center marks to establish the correct offset, or develop a specific dimension style with an extension line origin offset of 0 for placing dimensions from center marks.

Pick this endpoint when using a dimension style that has an extension line offset

Pick this endpoint when using a dimension style that has no extension line offset

this happens, use the **Xdatum** or **Ydatum** option to specify the axis from which the dimension originates. The **Mtext**, **Text**, and **Angle** options are identical to the options available with other dimensioning commands. Pick the extension line endpoint to complete the process.

Figure 18-35A shows ordinate dimensions placed on the object. Notice that the dimension text aligns with the extension lines. Aligned dimensioning is standard with ordinate dimensioning. Add missing lines, such as centerlines or fold lines, to complete the drawing. Identify the holes with letters and create a correlated dimensioning table if appropriate. See Figure 18-35B.

Figure 18-35.
A—Placing ordinate dimensions. B—Completing the drawing.

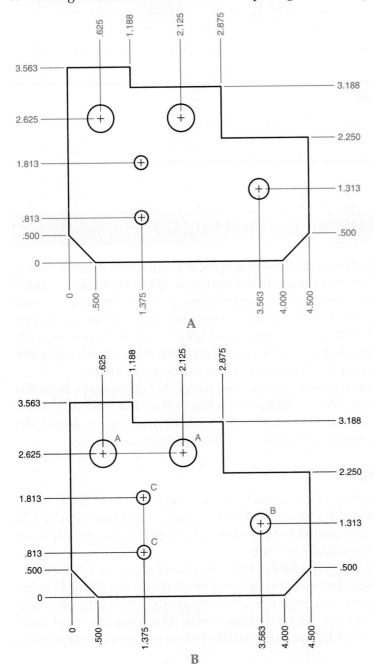

HOLE DATA		
HOLE	QTY.	DIAMETER
A	2	.500
B	1	.375
C	2	.250

Most ordinate dimensioning tasks work best with polar tracking or **Ortho** mode on. However, when the extension line is too close to an adjacent dimension, it is best to stagger the extension line as shown in the following illustration. With **Ortho** mode off, the extension line is automatically staggered when you pick the second extension line point, as shown.

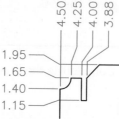

Exercise 18-8

Complete the exercise on the companion website.
www.g-wlearning.com/CAD

Marking Up Drawings

marking up (redlining): The process of reviewing a drawing and marking required changes.

Marking up, or *redlining*, is not a dimensioning practice, but it is similar to dimensioning in that it helps to explain information to the reader. Redlines are typically added directly to a final drawing by someone who reviews the drawing for accuracy and design changes. You might experience this process with your instructor or supervisor. Common mark-up techniques include redlining a plot with a red pen, using separate mark-up software to review exported drawings, or redlining directly in the drawing file. AutoCAD redlines become part of the drawing to document revision history.

You can redline a drawing with any appropriate AutoCAD commands, typically using a separate layer. Redlining often includes basic objects, text, and leaders. In some cases, you may add redline dimensions and even an entire drawing or detail. The **REVCLOUD** and **WIPEOUT** commands are also common mark-up commands.

Creating Revision Clouds

revision cloud: A polyline of sequential arcs used to form a cloud shape around changes in a drawing.

A *revision cloud* is a polyline of sequential arcs forming a cloud-shaped object. **Figure 18-36** shows both styles of revision clouds with a leader and note attached. A revision cloud points the drafter and other members of a design team to a specific portion of the drawing that may require an edit.

Drawing a revision cloud using the **REVCLOUD** command is somewhat different than drawing most other objects, because a single pick is all that is required. To draw a revision cloud, pick a start point, and then move the crosshairs around the objects to be enclosed until you return close to the start point. AutoCAD closes the cloud automatically and exits the command. Options are available before you pick the start point.

Defining Arc Length

Use the **Arc length** option to specify the size of revision cloud arcs. The value measures the length of an arc from the arc start point to the arc endpoint. AutoCAD

REVCLOUD

Ribbon
Home
> Draw
Annotate
> Markup

Revision Cloud

Type
REVCLOUD

Figure 18-36.
Examples of revision clouds identifying areas of a drawing to be modified. Notice the leaders describing the changes. You can create revision clouds using the Normal style or the Calligraphy style.

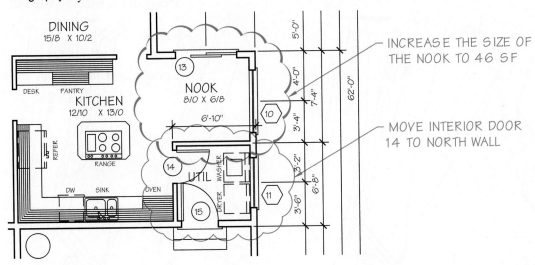

prompts for the minimum arc length and then for the maximum arc length. Specifying different minimum and maximum values causes the revision cloud to have an uneven, hand-drawn appearance.

Converting Objects to Revision Clouds

Use the **Object** option to convert a circle, closed polyline, ellipse, polygon, or rectangle to a revision cloud. Pick the object to convert to a revision cloud. Use the **No** option at the Reverse direction: prompt, or use the **Yes** option to reverse the direction of the cloud arcs.

Changing Revision Cloud Style

Use the **Style** option to change the revision cloud style. The default style is Normal, which displays arcs with a consistent width. When you specify the Calligraphy style, the start and end widths of the individual arcs are different, creating a more stylized revision cloud. See **Figure 18-36**.

Exercise 18-9

Complete the exercise on the companion website.
www.g-wlearning.com/CAD

Using the WIPEOUT Command

The **WIPEOUT** command allows you to clear a portion of the drawing without erasing objects. The command is sometimes appropriate for applications similar to those for **REVCLOUD**, most often redlining. **Figure 18-37** shows an example of a wipeout used to lay out the location of a proposed building site on a site plan.

Specify the first corner of the wipeout, followed by all other perimeter corners. Use the **Undo** option as needed to reverse the effects of an incorrect selection. Use the **Close** option, press [Enter] or the space bar, or right-click and choose **Enter** to create the wipeout.

Ribbon

Home
> Draw

Annotate
> Markup

Wipeout

Type

WIPEOUT

WIPEOUT

Figure 18-37.
Using the **WIPEOUT** command to clear a proposed building site on a site plan. Objects below the wipeout still exist. Further information has been added to the wipeout in this example.

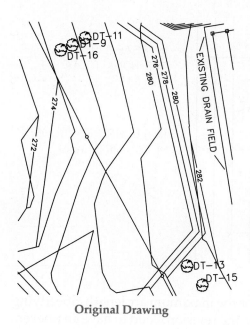

Original Drawing

Pick points to create the wipeout

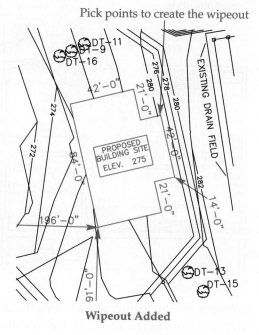

Wipeout Added

An alternative to picking points is to use the **Polyline** option and select a closed polyline object to convert to a wipeout. Use the **Frames** option to turn the display of all wipeout boundaries on or off. You may need to regenerate the display to observe the effects of changing the frame setting. To reveal objects hidden by a wipeout, freeze or turn off the wipeout layer, use draw order options, or erase the wipeout if it is no longer needed.

Template Development Chapter 18

Adding Multileader Styles
For detailed instructions on adding multileader styles to each drawing template, go to the companion website (www.g-wlearning.com/CAD), select this chapter, and select **Template Development**.

Chapter Review

Answer the following questions. Write your answers on a separate sheet of paper or complete the electronic chapter review on the companion website.
www.g-wlearning.com/CAD

1. Which command provides diameter dimensions for circles?
2. Which command provides radius dimensions for arcs?
3. Explain how to add a center mark to a circle without using the **DIMDIAMETER** or **DIMRADIUS** command.
4. What is the most common size for leader arrowheads?
5. What angle constraints should you use for leaders to maintain the ASME standard?
6. What is the usual length for the shoulder of a leader in mechanical drafting?
7. Describe two ways to dimension a 45° chamfer.
8. Identify the elements of this Unified screw thread note: 1/2-13UNC-2B.
    A. 1/2
    B. 13
    C. UNC
    D. 2
    E. B
9. Identify the elements of this metric screw thread note: M14 X 2.
    A. M
    B. 14
    C. 2
10. Define the term *rectangular coordinate dimensioning without dimension lines*.
11. What term does AutoCAD use to refer to rectangular coordinate dimensioning without dimension lines?
12. Explain the importance of the user coordinate system (UCS) for drawing ordinate dimension objects.
13. What is the purpose of a revision cloud?
14. How do you close a revision cloud?
15. What is the purpose of the **WIPEOUT** command?

Drawing Problems

Start AutoCAD if it is not already started. Start a new drawing for each problem using an appropriate template of your choice. The template should include layers and text, dimension, and multileader styles, when necessary, for drawing the given objects. Add layers and text, dimension, and multileader styles as needed. Draw all objects using appropriate layers and text, dimension, and multileader styles, justification, and format. Follow the specific instructions for each problem. Use only drawing and editing commands and techniques you have already learned. Use your own judgment and approximate dimensions when necessary. Apply dimensions accurately using ASME or appropriate industry standards.

▼ Basic

1. Draw and dimension the part view shown. Save the drawing as P18-1.

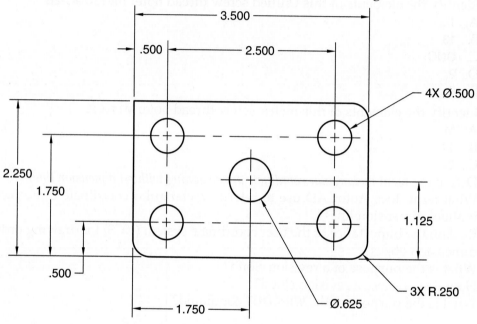

2. Draw and dimension the part views shown. Save the drawing as P18-2.

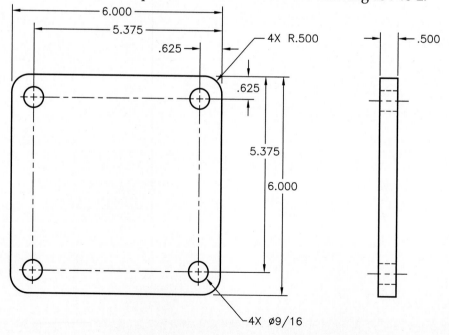

3. Draw and dimension the part view shown. Save the drawing as P18-3.

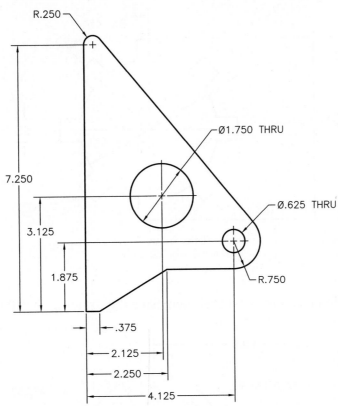

4. Draw and dimension the pin shown. Save the drawing as P18-4.

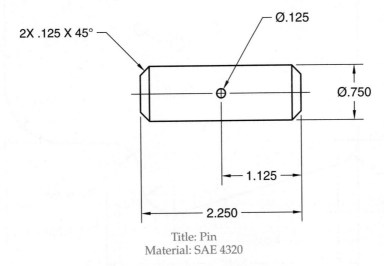

Title: Pin
Material: SAE 4320

5. Draw and dimension the spline shown. Save the drawing as P18-5.

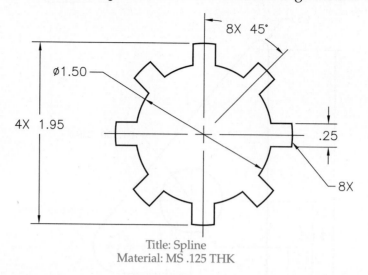

Title: Spline
Material: MS .125 THK

6. Draw and dimension the gasket shown. Save the drawing as P18-6.

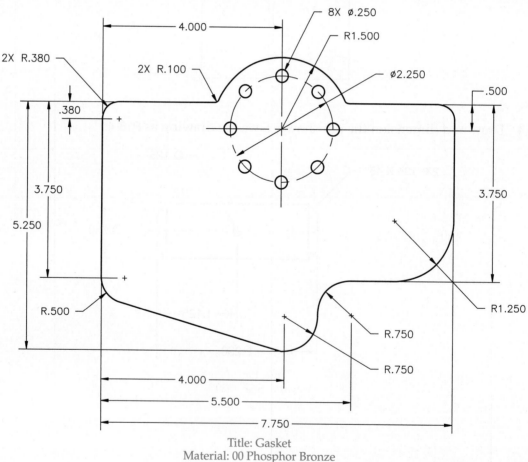

Title: Gasket
Material: 00 Phosphor Bronze

7. Draw and dimension the SST1001 chain link shown. Do not draw the table. Save the drawing as P18-7.

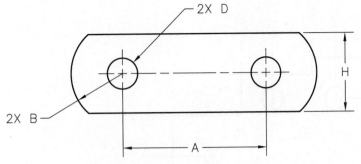

CHAIN NO.	A	B	D	H
SST1000	2.1	.75	.44	1.125
SST1001	3.0	.90	.48	1.525
SST1002	3.0	1.17	.95	2.125

8. Draw and dimension the part view shown. Save the drawing as P18-8.

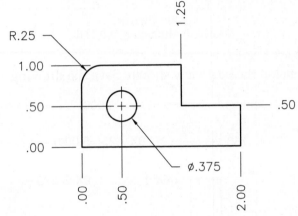

9. Draw and dimension the chassis spacer shown. Do not draw the table. Save the drawing as P18-9.

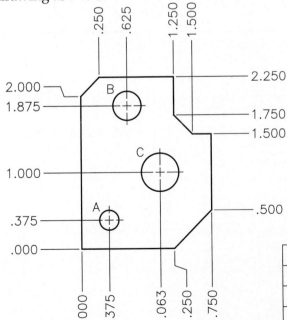

KEY	DIAMETER	DEPTH
A	.25	THRU
B	.38	THRU
C	.50	THRU

Title: Chassis Spacer
Material: .008 Aluminum

10. Draw and dimension the chassis shown. Do not draw the table. Save the drawing as P18-10.

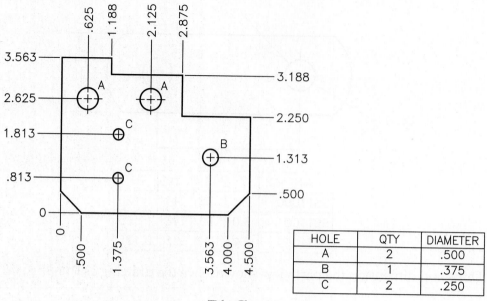

HOLE	QTY	DIAMETER
A	2	.500
B	1	.375
C	2	.250

Title: Chassis
Material: Aluminum .100 THK

11. Draw and dimension the part view shown. Save the drawing as P18-11.

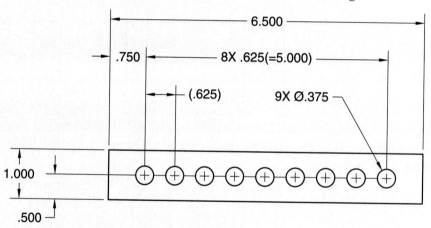

▼ Intermediate

12. Draw and dimension the thumb screw shown. Save the drawing as **P18-12**. Print an 8.5″ × 11″ copy of the drawing extents, using a 1:1 scale and landscape orientation.

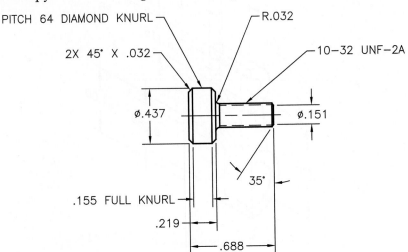

PITCH 64 DIAMOND KNURL

R.032

2X 45° X .032

10–32 UNF–2A

ø.437

ø.151

35°

.155 FULL KNURL

.219

.688

For Problems 13 through 15, draw and dimension the orthographic views needed to describe the part completely. Save the drawings as **P18-13**, **P18-14**, *and* **P18-15**.

13.

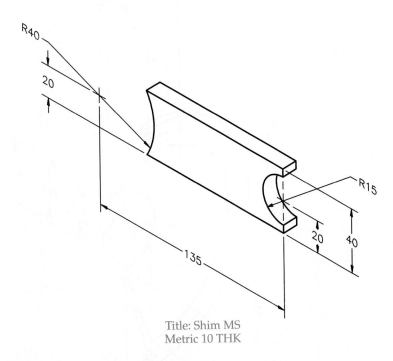

R40

20

R15

135

20

40

Title: Shim MS
Metric 10 THK

Chapter 18 Dimensioning Features and Alternate Practices

Drawing Problems – Chapter 18

14. Half of the drawing is removed for clarity. Draw the entire part.

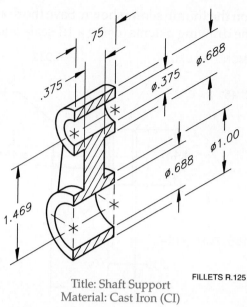

FILLETS R.125

Title: Shaft Support
Material: Cast Iron (CI)

15. Half of the drawing is removed for clarity. Draw the entire part.

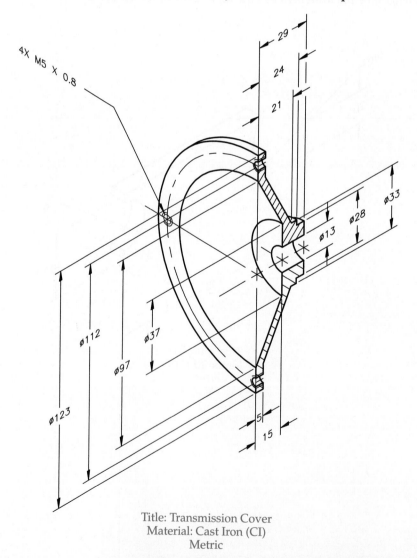

Title: Transmission Cover
Material: Cast Iron (CI)
Metric

16. Draw and dimension the part views shown. Save the drawing as P18-16.

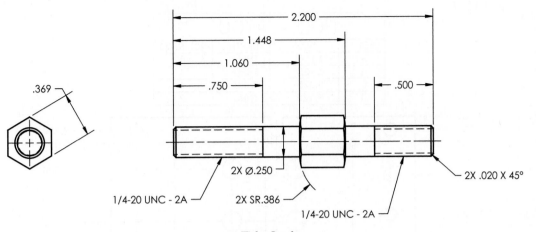

Title: Stud
Material: Stainless Steel

17. Draw and dimension the part view shown. Do not draw the table. Save the drawing as P18-17.

HOLE LAYOUT			
KEY	SIZE	DEPTH	NO. REQD
A	ø.250	THRU	6
B	ø.125	THRU	4
C	ø.375	THRU	4
D	R.125	THRU	2

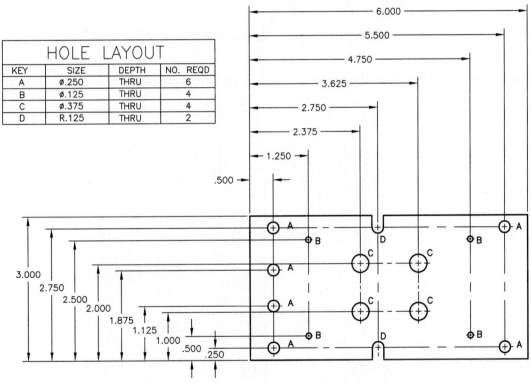

Title: Chassis Base (datum dimensioning)
Material: 12 gage Aluminum

18. Draw and dimension the part view shown. Do not draw the table. Save the drawing as P18-18.

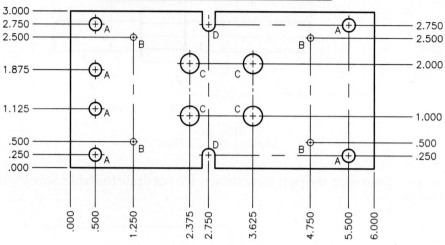

Title: Chassis Base (arrowless dimensioning)
Material: 12 gage Aluminum

19. Draw and dimension the part view shown. Do not draw the table. Save the drawing as P18-19.

HOLE LAYOUT				
KEY	X	Y	SIZE	TOL
A1	.500	2.750	⌀.250	±.002
A2	.500	1.875	⌀.250	±.002
A3	.500	1.125	⌀.250	±.002
A4	.500	.250	⌀.250	±.002
A5	5.500	2.750	⌀.250	±.002
A6	5.500	.250	⌀.250	±.002
B1	1.250	2.500	⌀.125	±.001
B2	1.250	.500	⌀.125	±.001
B3	4.750	2.500	⌀.125	±.001
B4	4.750	.500	⌀.125	±.001
C1	2.375	2.000	⌀.375	±.005
C2	2.375	1.000	⌀.375	±.005
C3	3.625	2.000	⌀.375	±.005
C4	3.625	1.000	⌀.375	±.005
D1	2.750	2.750	R.125	±.002
D2	2.750	.250	R.125	±.002

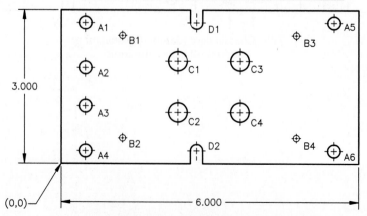

Title: Chassis Base (arrowless tabular dimensioning)
Material: 12 gage Aluminum

20. Draw and dimension the part views shown. Do not draw the table. Save the drawing as P18-20.

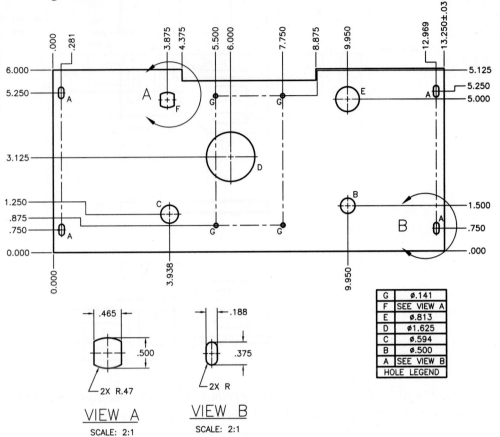

	HOLE LEGEND	
G	⌀.141	
F	SEE VIEW A	
E	⌀.813	
D	⌀1.625	
C	⌀.594	
B	⌀.500	
A	SEE VIEW B	

VIEW A
SCALE: 2:1

VIEW B
SCALE: 2:1

21. Draw and dimension the part views shown. Save the drawing as P18-21.

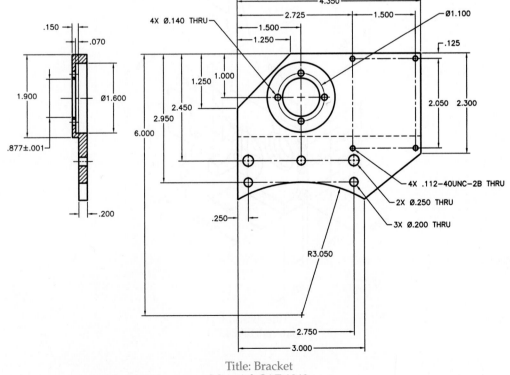

Title: Bracket
Material: SAE 1040

22. Draw and dimension the part views shown. Save the drawing as P18-22.

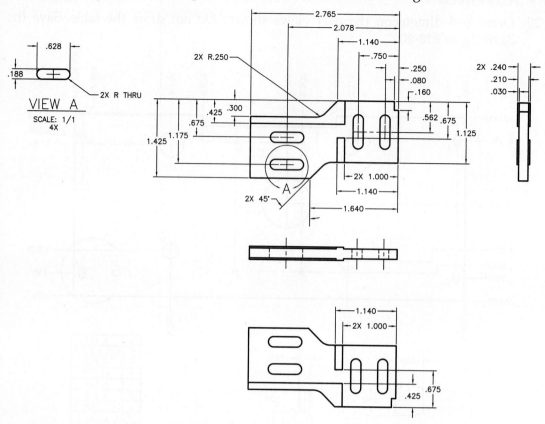

VIEW A

SCALE: 1/1
4X

2X R THRU

For Problems 23 and 24, draw and dimension the orthographic views needed to describe the part completely. Save the drawings as P18-23 and P18-24.

23.

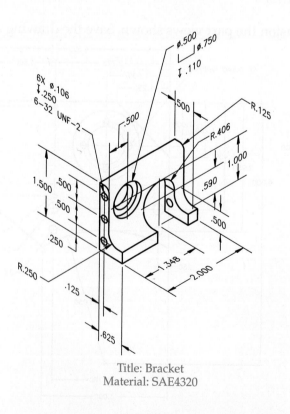

Title: Bracket
Material: SAE4320

24. Open P17-17 and save the file as P18-24. (If you have not yet completed problem 17-17, complete it now.) The P18-24 file should be active. Add the client-requested redlines to the floor plan as shown.

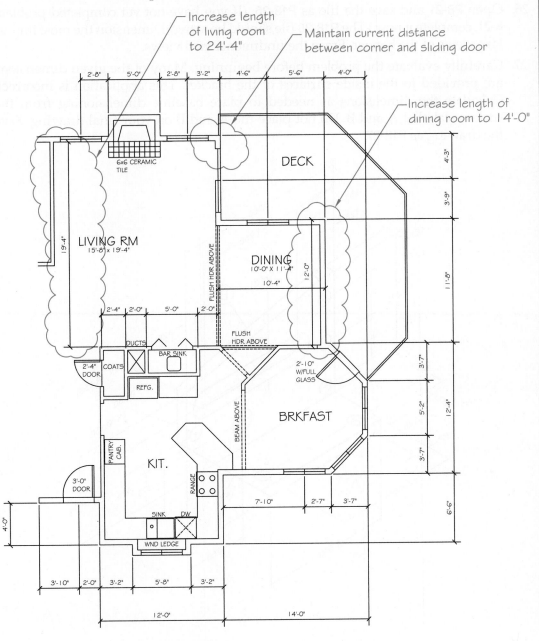

25. Open P5-16 and save the file as P18-25. (If you have not yet completed problem 5-16, complete it now.) The P18-25 file should be active. Dimension the drawing of the kitchen.

26. Open P8-21 and save the file as P18-26. (If you have not yet completed problem 8-21, complete it now.) The P18-26 file should be active. Dimension the most important views of the hanger. Erase the undimensioned views.

27. Carefully evaluate the problem before beginning. Many of the given dimensions are provided to the inside surfaces of the bracket. This application is incorrect. Calculate the dimensions as needed to place baseline dimensioning from the surfaces labeled A and B. Do not place the A and B on your final drawing. Save the drawing as P18-27.

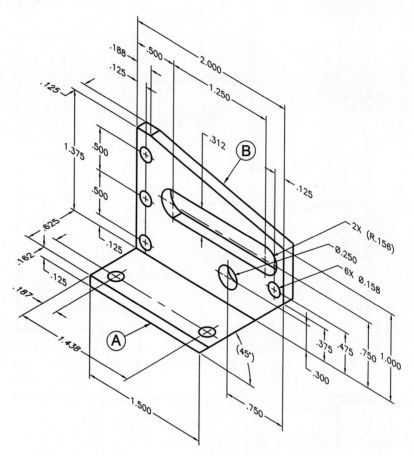

AutoCAD Certified Associate Exam Practice

Answer the following questions. Write your answers on a separate sheet of paper.

1. To comply with the ASME standard, what should the **Offset from origin** setting be when you dimension to a centerline? *Select all that apply.*
   - A. 0
   - B. .012
   - C. .063
   - D. .12 mm
   - E. 1 mm
   - F. 1.5 mm

2. What is the correct term for the shaded area in the hole shown below? *Select the one item that best answers the question.*
   - A. counterbore
   - B. countersink
   - C. fillet
   - D. landing
   - E. round
   - F. spotface

3. Which of the following commands can be used to dimension an arc? *Select all that apply.*
   - A. **DIMARC**
   - B. **DIMCENTER**
   - C. **DIMJOGGED**
   - D. **DIMJOGLINE**
   - E. **DIMRADIUS**

AutoCAD Certified Professional Exam Practice

Follow the instructions in each problem. Write your answers on a separate sheet of paper.

1. **Navigate to this chapter on the companion website and open CPE-18ordinate.dwg.** Use rectangular coordinate dimensioning without dimension lines to dimension the drawing as shown. What are the values of A, B, and C?

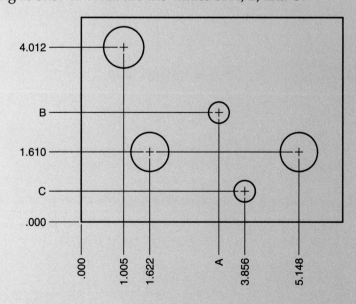

2. **Navigate to this chapter on the companion website and open CPE-18chart.dwg.** Create the drawing shown using the dimensions for item DRI203. Use the **Node** object snap to start the lower-right corner of the object at the point object provided in the drawing file. What are the coordinates of Point 1?

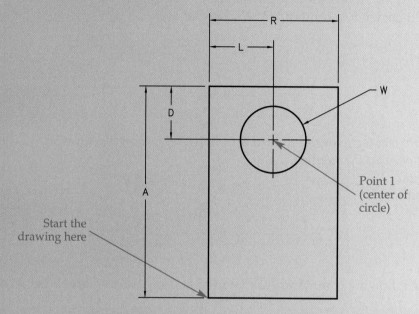

ITEM NO.	A	D	L	R	W
DRI201	4.000	1.000	1.250	2.500	1.261
DRI202	6.000	1.500	1.875	3.750	1.892
DRI203	9.500	2.375	2.969	5.938	2.995

Chapter

19

Dimensioning with Tolerances

Learning Objectives

After completing this chapter, you will be able to:

✓ Define and use dimensioning and tolerancing terminology.
✓ Set the precision for dimensions and tolerances.
✓ Set up the primary units for use with inch or metric dimensions.
✓ Specify an appropriate tolerance method.
✓ Create and use specified tolerance dimension styles.
✓ Explain the purpose of geometric dimensioning and tolerancing (GD&T).

This chapter introduces general tolerancing as applied to *conventional dimensioning* and explains how to create dimensions with specified tolerances for mechanical drawings common in the manufacturing industry. This chapter also introduces geometric dimensioning and tolerancing (GD&T) symbols and offers information on how you can learn more about GD&T.

conventional dimensioning: Dimensioning without the use of geometric tolerancing.

Tolerancing Fundamentals

Every dimension has a *tolerance*, except dimensions specifically identified as reference, maximum, minimum, or stock. The tolerancing practice depends on specific engineering and manufacturing applications, interrelated features, and industry and company preference. You can apply tolerance to dimensions indirectly by placing the information in the title block or a general note. See **Figure 19-1**. Any dimension that requires a tolerance that is different from the general tolerances given in the title block or general note must have the specific tolerance applied directly to the dimension on the drawing. See **Figure 19-2**.

The dimension stated as 12.50±0.25 in **Figure 19-2A** is in a style known as *plus-minus dimensioning*. The tolerance of this dimension is the difference between the maximum and minimum *limits*. The plus-minus tolerance style applies when the variance is the same in the positive and negative directions. In this case, the upper limit is 12.75 (12.50 + 0.25 = 12.75), and the lower limit is 12.25 (12.50 − 0.25 = 12.25). To find the tolerance, subtract the lower limit from the upper limit. The tolerance in this example is 0.50 (12.75 − 12.25 = 0.50). The *specified dimension* of the feature shown in **Figure 19-2** is 12.50.

tolerance: The total amount by which a specific dimension is permitted to vary.

plus-minus dimensioning: A tolerance style in which the positive and negative variance is equal and is preceded by a ± symbol.

limits: The largest and smallest numerical values the feature can have.

specified dimension: The part of the dimension from which the limits are calculated.

Figure 19-1.
The title block or a general note provides indirect tolerance specifications. A—An unspecified tolerance on an inch drawing. B—An unspecified tolerance on a metric drawing. Metric tolerancing is generally controlled by the ISO 2768—*General Tolerances* standard developed by the International Organization for Standardization (ISO).

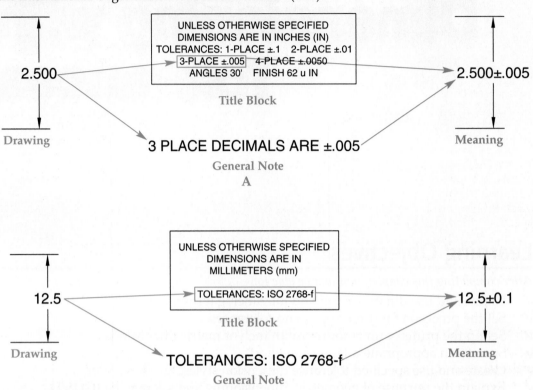

UNLESS OTHERWISE SPECIFIED
DIMENSIONS ARE IN INCHES (IN)
TOLERANCES: 1-PLACE ±.1   2-PLACE ±.01
3-PLACE ±.005   4-PLACE ±.0050
ANGLES 30'   FINISH 62 u IN

Title Block

2.500

Drawing

3 PLACE DECIMALS ARE ±.005

General Note

2.500±.005

Meaning

A

UNLESS OTHERWISE SPECIFIED
DIMENSIONS ARE IN
MILLIMETERS (mm)

TOLERANCES: ISO 2768-f

Title Block

12.5

Drawing

TOLERANCES: ISO 2768-f

General Note

12.5±0.1

Meaning

B

Figure 19-2.
A—Plus-minus dimensioning. B—Limit dimensioning.

12.50±0.25

Plus-Minus Dimensioning
A

12.75
12.25

Limit Dimensioning
B

limit dimensioning:
Method in which the upper and lower limits are given, instead of the specified dimension and tolerance.

bilateral tolerance:
A tolerance style that permits variance in both the positive and negative directions from the specified dimension.

Limit dimensioning, shown in Figure 19-2B, is an alternative method of showing and calculating tolerance. Limit dimensioning is most common for defining fits between mating parts, such as a sliding fit between a hole and shaft or a press fit between a hole and bearing. Some companies, or departments within a company, such as an inspection department, prefer limits dimensioning because it does not require calculating limits. However, the actual dimension of the object in the drawing is unknown.

Plus-minus dimensioning uses a bilateral or unilateral tolerance format, depending on the application. Figure 19-3 shows examples of equal and unequal *bilateral tolerances*. Bilateral tolerancing is the most common tolerancing method. Manufacturers typically prefer equal bilateral tolerancing because they attempt to manufacture features as close to the specified dimension as possible. Figure 19-4 shows examples of *unilateral*

Figure 19-3.
Examples of plus-
minus dimensioning
values using
bilateral tolerances.

24 ± 0.1
Metric

$.750\pm.005$
Inch

$45°30'\pm0°5'$
Angular

Equal Bilateral Tolerance

$24^{+0.08}_{-0.20}$
Metric

$.750^{+.002}_{-.003}$
Inch

$45.5°^{+0.2°}_{-0.5°}$
Angular

Unequal Bilateral Tolerance

Figure 19-4.
A unilateral
tolerance allows
variation in only one
direction from the
specified dimension.

$24^{0}_{-0.2}$

$.625^{+.000}_{-.004}$

$25.5°^{0}_{-0.5°}$

$24^{+0.2}_{0}$
Metric

$.625^{+.004}_{-.000}$
Inch

$25.5°^{+0.5°}_{0}$
Angular

tolerances. Some companies use unilateral tolerances to define fits between mating parts. However, manufacturers who use the drawing to program computerized numerically controlled (CNC) machining equipment often avoid unilateral tolerancing.

Basic dimensions establish true position from datums and between interrelated features, and define true profile. A rectangle around the dimension value distinguishes a basic dimension from other types of dimensions. *Single limits* are sometimes applied to various features, such as chamfers, fillets, rounds, hole depths, and thread lengths. The abbreviation for minimum (MIN) or maximum (MAX) follows the dimension value to describe a single limit application. The design determines the unspecified limit.

unilateral tolerance: A tolerance style that permits a variation in only one direction from the specified dimension.

basic dimension: A theoretically exact dimension used in geometric dimensioning and tolerancing.

single limits: Limit dimensions used when the specified dimension cannot be any more than the maximum or less than the minimum given value.

Dimensioning Units

The ASME Y14.5-2009 *Dimensioning and Tolerancing* standard has separate recommendations for the display of inch, metric, and angular dimensions. **Figure 19-5**, **Figure 19-6**, and **Figure 19-7** briefly explain the rules for each type of dimension. The U.S. unit of measure commonly used on engineering drawings is the inch. The SI unit of measure commonly used on engineering drawings is the millimeter. Company or school policy and product requirements determine the actual units used on engineering drawings.

Place the general note UNLESS OTHERWISE SPECIFIED, ALL DIMENSIONS ARE IN INCHES (or MILLIMETERS) on the drawing when all dimensions are in inches or millimeters. Follow millimeter dimensions on an inch drawing with the abbreviation MM. Follow inch dimensions on a metric drawing with the abbreviation IN. **Figure 19-2**, **Figure 19-3**, and **Figure 19-4** show examples of displaying inch, metric, and angular dimensions with specified tolerances.

Figure 19-5.
Dimensioning rules for inch dimensions.

Rules for Inch Dimensions	Examples
A zero does not precede a decimal inch that is less than one.	.5
Express a specified dimension to the same number of decimal places as its tolerance. Add zeros to the right of the decimal point if needed.	.250±.005 (additional zero added to .25)
Fractional inches generally indicate a larger tolerance, or give nominal sizes, such as in a thread callout.	Dimension value: 2 1/2±1/32 Thread: 1/2-13UNC-2B
Plus and minus values of an inch tolerance have the same number of decimal places.	$.250^{+.005}_{-.010}$ .255 .240
Unilateral tolerances use the + or – symbol, and the 0 value has the same number of decimal places as the value that is greater or less than 0.	$.250^{+.005}_{-.000}$ $.250^{+.000}_{-.005}$
Inch limit tolerance values have the same number of decimal points. When displaying limit tolerance values on one line, the lower value precedes the higher value, and a dash separates the values. When displaying stacked limit tolerance values, place the higher value above the lower value.	One line: 1.000–1.062 Stacked: $\begin{array}{c}1.062\\1.000\end{array}$
Basic dimension values have the same number of decimal places as their associated tolerance.	2.000

Setting Primary Units

A dimension style controls the appearance of dimensions, including dimension values and tolerance. The initial phase of dimensioning with tolerances involves setting the appropriate values for the primary units of the dimension style. Use the **Primary Units** tab of the **New** (or **Modify**) **Dimension Style** dialog box, shown in **Figure 19-8**, to set the dimension units and precision.

Use the **Precision:** drop-down list in the **Linear dimensions** area to set the number of zeros displayed after the decimal point of the specified dimension. The **Zero suppression** settings control the display of zeros before and after the decimal point. For inch dimensions, the **Leading** options should be on, and the **Trailing** options should be off. For typical metric dimensions, without using subunits, the **Leading** options should be off, and the **Trailing** options should be on.

Figure 19-6.
Dimensioning rules for metric dimensions.

Rules for Metric Dimensions	Examples
Omit the decimal point and zero when the dimension is a whole number.	12
A zero precedes a decimal millimeter that is less than one.	0.5
When the dimension is greater than a whole number by a fraction of a millimeter, follow the last digit to the right of the decimal point with a zero. This rule is true unless the dimension displays tolerance values.	12.5
Plus and minus values of a metric tolerance have the same number of decimal places. Add zeros to fill in where needed.	$24^{+0.25}_{-0.10}$ 24.25 24.00
Metric limit tolerance values have the same number of decimal points. When displaying limit tolerance values on one line, the lower value precedes the higher value, and a dash separates the values. When displaying stacked limit tolerance values, place the higher value above the lower value. Examples in ASME Y14.5 show no zeros after the specified dimension to match the tolerance.	One line: 7.5–7.6 Stacked: $\frac{7.6}{7.5}$ 24±0.25 24.5±0.25
When applying unilateral tolerances, use a single 0 without a + or – sign for the 0 part of the value.	$24\ {}^{0}_{-0.2}$ $24\ {}^{+0.25}_{0}$
Basic dimension values follow the same display rules as stated for other metric numbers.	$\boxed{24}$ $\boxed{24.5}$

Figure 19-7.
Dimensioning rules for angular dimensions.

Rules for Angular Dimensions	Examples
Establish angular dimensions in degrees (°) and decimal degrees (30.5°), or in degrees (°), minutes ('), and seconds (").	24°15'30"
The plus and minus tolerance values and the angle have the same number of decimal places.	30.0°±0.5° (not 30°±0.5°)
Where only specifying minutes or seconds, precede the number of minutes or seconds with 0° or 0°0', as applicable.	0°45'30" 0°0'45"

Figure 19-8.
The **Primary Units** tab of the **New** (or **Modify**) **Dimension Style** dialog box sets the unit format and precision of linear dimensions.

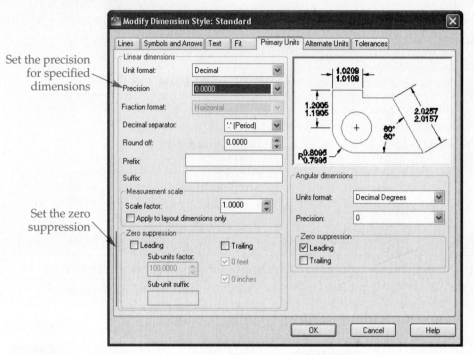

Set the precision for specified dimensions

Set the zero suppression

Setting Tolerance Methods

The **Tolerances** tab of the **New** (or **Modify**) **Dimension Style** dialog box, shown in **Figure 19-9**, allows you to create a specified tolerance dimension style. The default option in the **Method:** drop-down list is **None**. This means dimensions use an unspecified tolerance format. As a result, most of the options in the **Tolerances** tab are disabled. When you pick a different tolerance method from the drop-down list, appropriate options become enabled, and the preview image reflects the selected method. **Figure 19-10** shows the drop-down list options.

The following information describes tolerance methods and the settings unique to each. You will learn about general settings, including tolerance precision, height, vertical position, alignment, and zero suppression, later in this chapter.

Symmetrical Tolerance Method

symmetrical tolerance: The AutoCAD term for an equal bilateral tolerance.

Select the **Symmetrical** option from the **Method:** drop-down list to create a *symmetrical tolerance*. Use the **Symmetrical** option to draw dimensions that display an equal bilateral tolerance in the plus-minus format. See **Figure 19-11**. Enter a tolerance value in the **Upper value:** text box. Although the **Lower value:** text box is disabled, you can see that the value in the **Lower value:** text box matches the value in the **Upper value:** text box.

Figure 19-9.
The **Tolerances** tab of the **New** (or **Modify**) **Dimension Style** dialog box contains formatting settings for tolerance dimensions.

Select a tolerance method

Set the precision for tolerance dimensions

Settings should match the **Zero suppression** linear dimension settings in the **Primary Units** tab

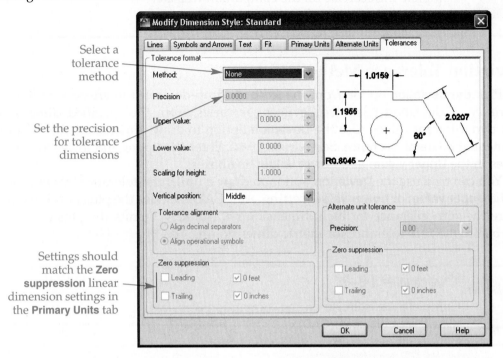

Figure 19-10.
Select a tolerance method from the **Method:** drop-down list in the **Tolerance format** area.

Select a tolerance method

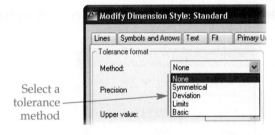

Figure 19-11.
Setting the **Symmetrical** tolerance method option current, with an equal bilateral tolerance value of .005.

Specified tolerance method

Equal bilateral tolerance value

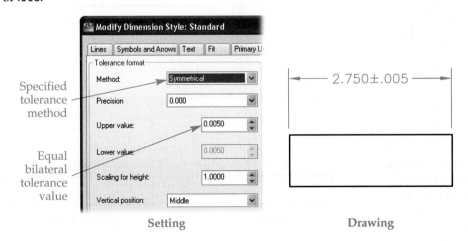

Setting   Drawing

Exercise 19-1

Complete the exercise on the companion website.
www.g-wlearning.com/CAD

Deviation Tolerance Method

deviation tolerance: The AutoCAD term for an unequal bilateral tolerance.

Pick the **Deviation** option from the **Method:** drop-down list to create a *deviation tolerance*. A deviation tolerance deviates, or varies, from the specified dimension with two different values. Use the **Deviation** option to draw dimensions that display an unequal bilateral tolerance. See Figure 19-12. Enter the upper and lower tolerance values in the **Upper value:** and **Lower value:** text boxes.

You can also use the **Deviation** option to draw a unilateral tolerance by entering 0 for the **Upper value:** or **Lower value:** setting. AutoCAD includes the plus or minus sign before the zero tolerance for inch dimensioning. AutoCAD omits the plus or minus sign before the zero tolerance for metric dimensioning. See Figure 19-13.

Exercise 19-2

Complete the exercise on the companion website.
www.g-wlearning.com/CAD

Figure 19-12.
Setting the **Deviation** tolerance method option current, with unequal bilateral tolerance values.

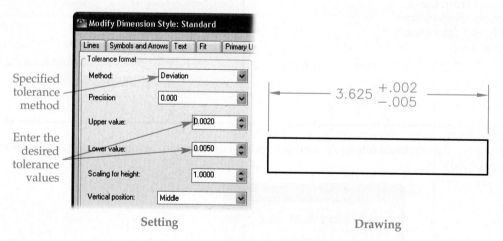

Specified tolerance method

Enter the desired tolerance values

Setting

Drawing

Figure 19-13.
When you form a unilateral tolerance, AutoCAD automatically places the plus or minus symbol in front of the zero tolerance when using inch units. AutoCAD omits the plus symbol with metric units.

Inch Units

Metric Units

Limits Tolerance Method

Select the **Limits** option from the **Method:** drop-down list to apply limits dimensioning. See **Figure 19-14**. Use the **Upper value:** and **Lower value:** text boxes to enter the upper and lower tolerance values to add and subtract from the specified dimension. The upper and lower values can be equal or different.

Exercise 19-3

Complete the exercise on the companion website.
www.g-wlearning.com/CAD

Basic Tolerance Method

Pick the **Basic** option from the **Method:** drop-down list to draw basic dimensions. See **Figure 19-15**. Few options are enabled in the dialog box because a basic dimension has no tolerance. A rectangle around the dimension value distinguishes a basic dimension from other dimensions.

Checking **Draw frame around text** in the **Text** tab of the **New** (or **Modify**) **Dimension Style** dialog box also activates the basic tolerance method.

PROFESSIONAL TIP

Choose a tolerance method based on the characteristics of the specified tolerance. If the upper and lower values are equal, choose the **Symmetrical** option to create an equal bilateral tolerance. If the upper and lower values vary, use the **Deviation** option. Use the **Limits** option to show only the minimum and maximum allowed values.

Figure 19-14.
Selecting the **Limits** tolerance method and setting limit values.

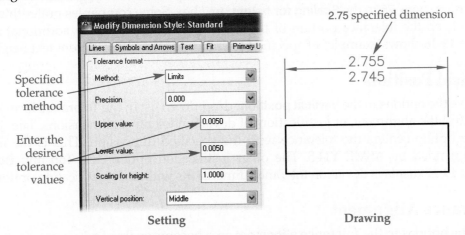

Setting   Drawing

Figure 19-15.
Use the **Basic** tolerance method for basic dimensioning. The dimension text for a basic dimension appears inside a frame.

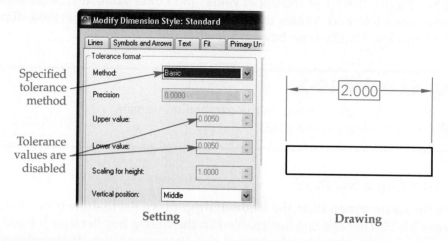

Specified tolerance method

Tolerance values are disabled

Setting

Drawing

Specifying Tolerance Settings

After you choose the tolerance method, adjust the tolerance settings to achieve the required tolerances. You can set the tolerance precision, height, position, alignment, and zero suppression characteristics of tolerances.

Tolerance Precision

Adjust the tolerance precision after you choose the tolerance method. AutoCAD automatically makes the tolerance precision in the **Tolerances** tab the same precision you set in the **Primary Units** tab. If the tolerance precision does not reflect the correct level of precision, change the precision using the **Precision:** drop-down list in the **Tolerance format** area.

Tolerance Height

Use the **Scaling for height:** text box in the **Tolerance format** area to set the text height of tolerance values in relation to the text height of the specified dimension. The default of 1.0000 makes the tolerance values same height as the specified dimension text. This is the format recommended by ASME Y14.5.

To make the height of tolerance values three-quarters the height of the specified dimension, type .75 in the **Scaling for height:** text box. Some companies prefer this practice to keep the tolerance portion of the dimension from taking up additional space. Figure 19-16 shows examples of specified tolerance values with different text heights.

Vertical Position

Use the options in the **Vertical position:** drop-down list in the **Tolerance** format area to control the alignment, or justification, of deviation tolerance dimensions. The default **Middle** option centers the tolerance with the specified dimension. This is the format recommended by ASME Y14.5. The other justification options are **Top** and **Bottom**. Figure 19-17 displays deviation tolerance dimensions with each justification option.

Tolerance Alignment

The options in the **Tolerance alignment** area become enabled when you use a deviation or limits tolerance method. Tolerance alignment controls the left and right tolerance

Figure 19-16.
Using different scale settings for the text height of tolerance dimensions. Use a scale of 1 to adhere to ASME standards.

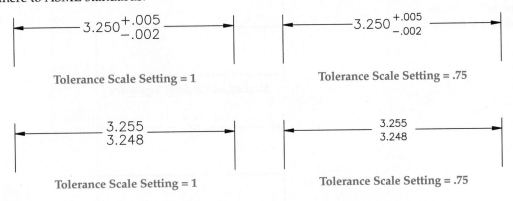

Tolerance Scale Setting = 1   Tolerance Scale Setting = .75

Tolerance Scale Setting = 1   Tolerance Scale Setting = .75

Figure 19-17.
Examples of the tolerance justification options for deviation tolerance dimensions. Use the **Middle** option to adhere to ASME standards.

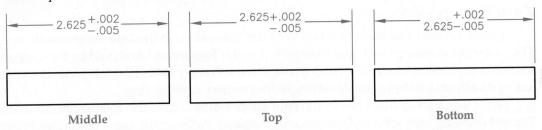

Middle   Top   Bottom

justification. When using a deviation tolerance method, pick the **Align decimal separators** radio button to align the upper and lower tolerance value decimal points vertically. Select the **Align operational symbols** radio button to align the upper and lower tolerance plus and minus symbols vertically. See **Figure 19-18**. When using the limits tolerance method, pick the **Align decimal separators** radio button to align the upper and lower limit decimal points vertically. Select the **Align operational symbols** radio button to left-justify the upper and lower limits. See **Figure 19-19**.

Figure 19-18.
Changing tolerance alignment for use with a deviation tolerance method.

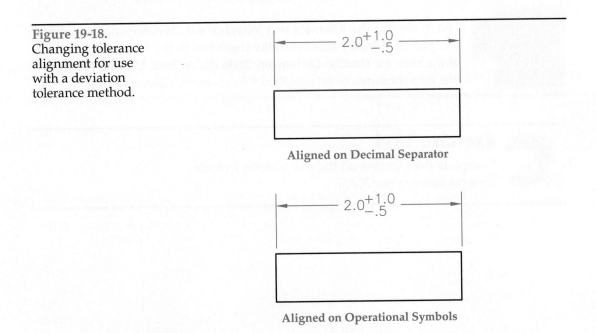

Aligned on Decimal Separator

Aligned on Operational Symbols

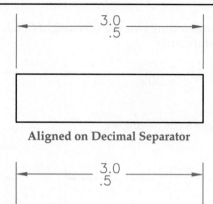

3.0
.5

Aligned on Decimal Separator

3.0
.5

Aligned on Operational Symbols

Zero Suppression

You must select a tolerance method to enable the options in the **Zero suppression** area. The suppression settings for linear dimensions in the **Tolerances** tab should be the same as the zero suppression tolerance format settings in the **Primary Units** tab. AutoCAD does not automatically match the tolerance setting to the primary units setting.

Select the **Leading** check box in the **Zero suppression** area of the **Tolerances** tab when you are drawing inch specified tolerance dimensions. Activate the same option for linear dimensions in the **Primary Units** tab. You can then draw inch-specified tolerance dimensions without placing the zero before the decimal point, as recommended by ASME standards. These settings allow you to create a tolerance dimension such as .625±.005.

Deselect the **Leading** check box in the **Zero suppression** area of the **Tolerances** tab when you are drawing metric specified tolerance dimensions. Deactivate the same option for linear dimensions in the **Primary Units** tab. This allows you to place a metric specified tolerance dimension with the zero before the decimal point, such as 12±0.2, as recommended by ASME standards.

The options in the **Alternate unit** tolerance area become enabled when you pick the **Display alternate units** check box in the **Alternate Units** tab of the **New** (or **Modify**) **Dimension Style** dialog box. Use the **Alternate unit tolerance** area to set specified tolerances for alternate units.

Exercise 19-4

Complete the exercise on the companion website.
www.g-wlearning.com/CAD

Introduction to GD&T

Geometric dimensioning and tolerancing (GD&T) is the dimensioning and tolerancing of individual features of a part where the permissible variations relate to characteristics of form, profile, orientation, runout, or the relationship between features. For complete coverage of GD&T, refer to *Geometric Dimensioning and Tolerancing* by David A. Madsen and David P. Madsen, published by Goodheart-Willcox Company, Inc.

geometric dimensioning and tolerancing (GD&T): The dimensioning and tolerancing of individual features of a part where the permissible variations relate to characteristics of form, profile, orientation, runout, or the relationship between features.

Reference Material — *Drafting Symbols*

For the names and examples of GD&T symbols and symbol applications, go to the **Reference Material** section of the companion website (www.g-wlearning.com/CAD) and select **Drafting Symbols** in the list.

Supplemental Material — *GD&T with AutoCAD*

For information about creating GD&T symbols using AutoCAD, go to the companion website (www.g-wlearning.com/CAD), select this chapter, and select **Using GD&T Tools in AutoCAD**.

Chapter Review

Answer the following questions. Write your answers on a separate sheet of paper or complete the electronic chapter review on the companion website.
www.g-wlearning.com/CAD

1. Define the term *tolerance*.
2. What are the limits of the tolerance dimension 3.625±.005?
3. Give an example of an equal bilateral tolerance in inches and in metric units.
4. Give an example of an unequal bilateral tolerance in inches and in metric units.
5. Give an example of a unilateral tolerance in inches and in metric units.
6. What is the purpose of the **Symmetrical** tolerance method option?
7. What is the purpose of the **Deviation** tolerance method option?
8. What is the purpose of the **Limits** tolerance method option?
9. How do you set the number of zeros displayed after the decimal point for a tolerance dimension?
10. Explain the result of setting the **Scaling for height:** option to 1 in the **Tolerances** tab.
11. What setting should you use for the **Scaling for height:** option if you want the tolerance dimension height to be three-quarters of the specified dimension height?
12. Name the tolerance dimension justification option recommended by the ASME standards.
13. Which **Zero suppression** settings should you choose for linear and tolerance dimensions when using inch units?
14. Which **Zero suppression** settings should you choose for linear and tolerance dimensions when using metric units?
15. What is the purpose of geometric dimensioning and tolerancing?

Drawing Problems

Start AutoCAD if it is not already started. Start a new drawing for each problem using an appropriate template of your choice. The template should include layers and text, dimension, and multileader styles, when necessary, for drawing the given objects. Add layers and text, dimension, and multileader styles as needed. Draw all objects using appropriate layers and text, dimension, and multileader styles, justification, and format. Follow the specific instructions for each problem. Use only drawing and editing commands and techniques you have already learned. Use your own judgment and approximate dimensions when necessary. Apply dimensions accurately using ASME or appropriate industry standards.

▼ Basic

1. Draw and dimension the part view shown. Save the drawing as P19-1.

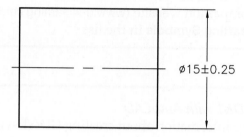

ø15±0.25

2. Draw and dimension the part view shown. Save the drawing as P19-2.

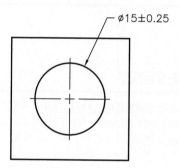

ø15±0.25

For Problems 3 through 6, draw and dimension the orthographic views needed to describe the part completely. Save the drawings as P19-3, P19-4, P19-5, and P19-6.

3.

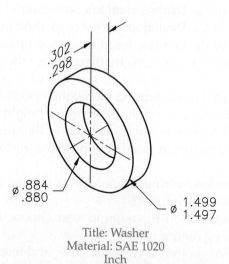

.302
.298

ø .884
.880

ø 1.499
1.497

Title: Washer
Material: SAE 1020
Inch

4.

SØ.562   Ø.375 FLAT

Ø.249 $^{+.000}_{-.001}$

▽.400

Title: Handle
Material: Bronze
Inch

5.

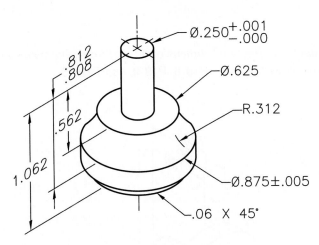

Ø.250 $^{+.001}_{-.000}$

.812
.808

.562

1.062

Ø.625

R.312

Ø.875±.005

.06 X 45°

ALL OTHER THREE PLACE DECIMALS ±.010

6.

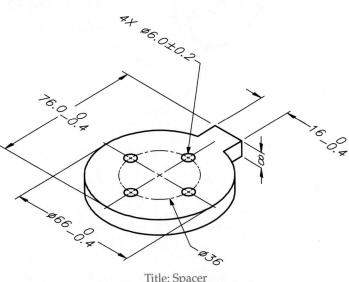

4X Ø6.0±0.2

76.0 $^{0}_{-0.4}$

16 $^{0}_{-0.4}$

8

Ø66 $^{0}_{-0.4}$

Ø36

Title: Spacer
Material: Cold Rolled Steel
Metric

▼ **Intermediate**

7. Draw and dimension the threaded stud part views shown. Save the drawing as P19-7.

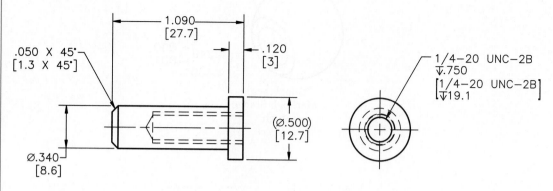

For Problems 8 through 10, draw and dimension the orthographic views needed to describe the part completely. Save the drawings as P19-8, P19-9, *and* P19-10.

8.

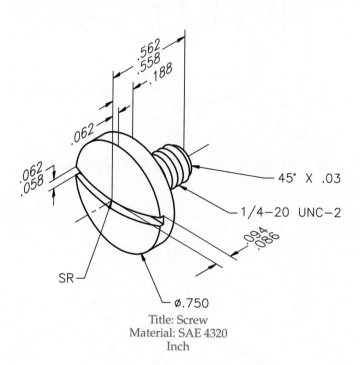

45° X .03

1/4—20 UNC—2

SR

ø.750

Title: Screw
Material: SAE 4320
Inch

9. A portion of the drawing is removed for clarity. Draw the entire part.

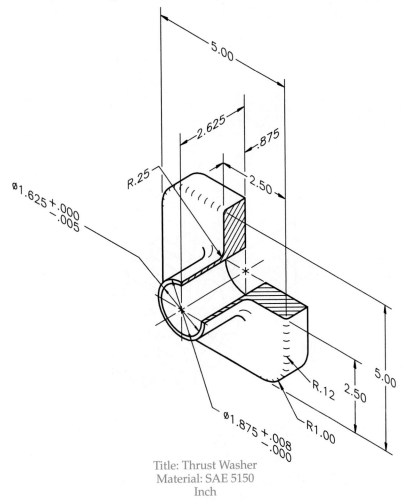

5.00

2.625

.875

R.25

2.50

Ø1.625 +.000
        -.005

*

Ø1.875 +.008
        -.000

R.12

R1.00

2.50

5.00

Title: Thrust Washer
Material: SAE 5150
Inch

10.

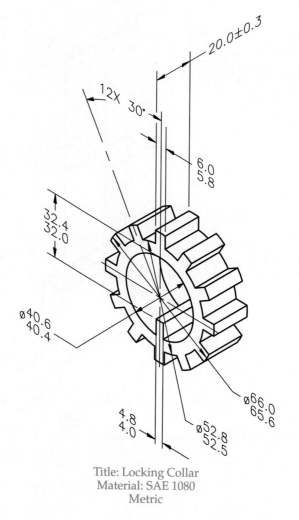

20.0±0.3

12X 30°

6.0
5.8

32.4
32.0

ø40.6
40.4

ø66.0
65.6

4.8
4.0

ø52.8
52.5

Title: Locking Collar
Material: SAE 1080
Metric

11. Draw and dimension the vise clamp part views shown. Save the drawing as **P19-11**.

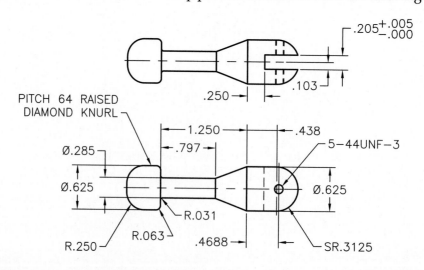

.205+.005
−.000

.250

.103

PITCH 64 RAISED
DIAMOND KNURL

1.250

.797

.438

5−44UNF−3

Ø.285

Ø.625

Ø.625

R.031

R.063

R.250

.4688

SR.3125

Drawing Problems - Chapter 19

For Problems 12 through 16, draw and dimension the orthographic views needed to describe the part completely. Save the drawings as **P19-12, P19-13, P19-14, P19-15,** *and* **P19-16.** *Use the GD&T commands and practices described in the "Using GD&T Tools in AutoCAD" supplement available in the Supplemental Material for this chapter on the companion website.*

12. *Note:* Untoleranced dimensions are ±0.3.

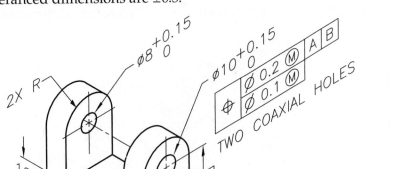

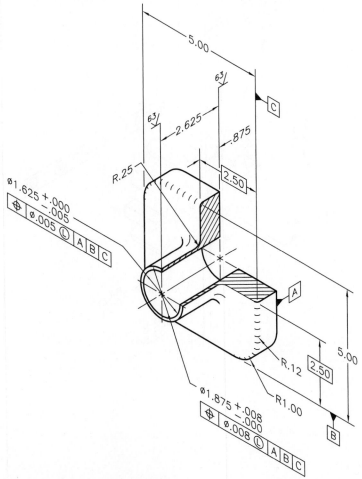

13. Open **P19-9** and save the file as **P19-13**. The P19-13 file should be active. Add the geometric tolerancing applications shown. Untoleranced dimensions are ±.02 for two-place decimal precision and ±.005 for three-place decimal precision.

14. Open P19-6 and save the file as P19-14. The P19-14 file should be active. Add the geometric tolerancing applications shown. Untoleranced dimensions are ±0.5.

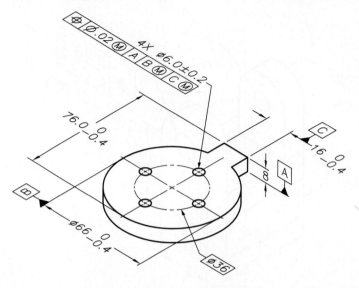

15. Open P19-10 and save the file as P19-15. The P19-15 file should be active. Add the geometric tolerancing applications shown.

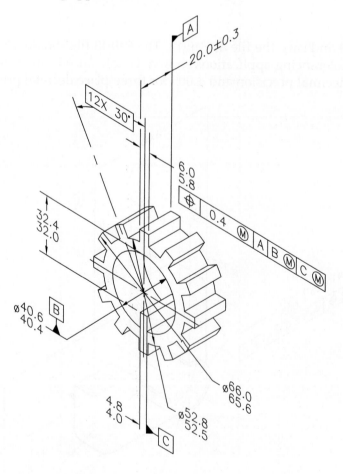

16. *Note:* Half of the drawing is removed for clarity. Draw the entire part. Untoleranced dimensions are ±.010.

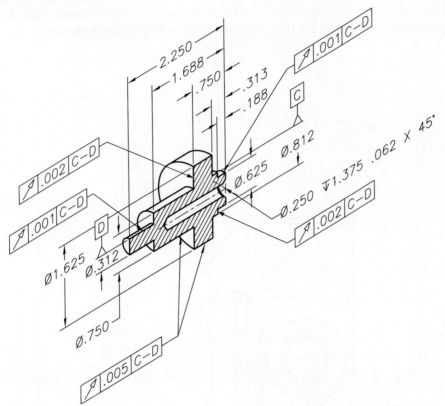

17. Open P18-21and save the file as P19-17. (If you have not yet done problem 18-21, do it now.) The P19-17 file should be active. Use the GD&T commands and practices described in the "Using GD&T Tools in AutoCAD" supplement available in the Supplemental Material for this chapter on the companion website.

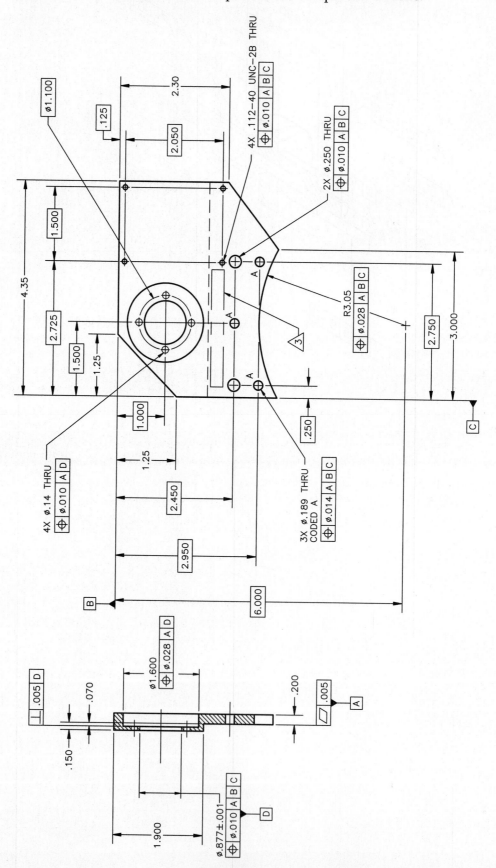

18. Draw and dimension the orthographic views needed to describe the part completely. Half of the drawing is removed for clarity. Draw the entire part. Untoleranced dimensions are ±.010. Use the GD&T commands and practices described in the "Using GD&T Tools in AutoCAD" supplement available in the Supplemental Material for this chapter on the companion website. Save the drawing as P19-18.

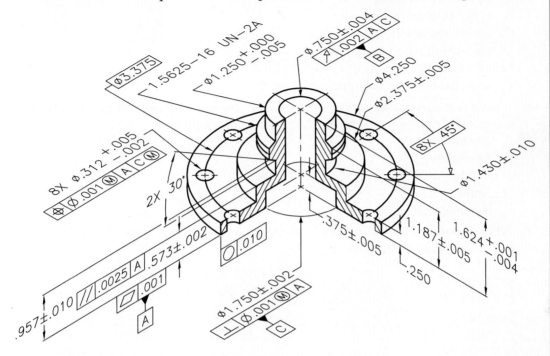

19. Research the design of an existing shaft collar with the following specifications: two-piece clamp-on, 1" bore, 1/4-28UNF screw threads. Create a dimensioned 2D sketch of the existing design from manufacturer's specifications, or from measurements taken from an actual shaft collar. Start a new drawing from scratch or use a decimal template of your choice. Draw and dimension each part of the collar from your sketch. Save the drawing as P19-19.

AutoCAD Certified Associate Exam Practice

Answer the following questions. Write your answers on a separate sheet of paper.

1. Which of the following terms describes the dimension 18.75±.25? *Select all that apply.*
   A. deviation tolerance
   B. equal bilateral tolerance
   C. limit dimensioning
   D. plus-minus dimensioning
   E. symmetrical tolerance
   F. unequal bilateral tolerance

2. Which AutoCAD tolerancing method can you use to create an unequal bilateral tolerance? *Select the one item that best answers the question.*
   A. **Basic**
   B. **Deviation**
   C. **Limits**
   D. **None**
   E. **Symmetrical**

3. What is the specified dimension in the tolerance shown here? *Select the one item that best answers the question.*
   A. 10.17
   B. 10.50
   C. 10.60
   D. 10.62
   E. 10.67

AutoCAD Certified Professional Exam Practice

Follow the instructions in each problem. Write your answers on a separate sheet of paper.

1. **Navigate to this chapter on the companion website and open CPE-19limits.dwg.** Create a new dimension style named Limits and select the **Limits** tolerancing method. Do not change any other settings. Use the Limits dimension style to create the two dimensions shown. What are the limits of dimensions A and B?

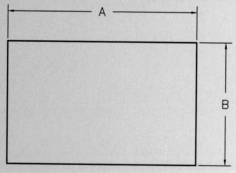

2. **Navigate to this chapter on the companion website and open CPE-19unilateral.dwg.** Create a new dimension style named Unilateral and select the appropriate tolerancing method to create a unilateral tolerance. Set an upper limit of 0 and a lower limit of –.021. Use the Unilateral dimension style to create the two dimensions shown. What are the limits of dimensions C and D?

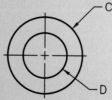

Editing Dimensions

Learning Objectives

After completing this chapter, you will be able to:

✓ Describe and control associative dimensions.
✓ Control the appearance of existing dimensions and dimension text.
✓ Override dimension style settings and match dimension properties.
✓ Copy properties from an existing object to other objects.
✓ Change dimension line spacing and alignment.
✓ Break dimension, extension, and leader lines.
✓ Create inspection dimensions.
✓ Edit existing multileaders.

You can modify dimensions using standard editing commands such as **ERASE** and **STRETCH**. AutoCAD also provides specific commands and options to adjust dimensions. This chapter describes techniques for editing dimension placement, value, and appearance.

Associative Dimensioning

A dimension is a group of elements treated as a single object. For example, you can access the **ERASE** command and pick any portion of a dimension to erase the entire dimension object. Additionally, dimensions reference objects or points. When you edit dimensioned objects with commands such as **STRETCH**, **MOVE**, **ROTATE**, and **SCALE**, dimensions change accordingly. See **Figure 20-1**.

An *associative dimension* forms by default when you select objects or pick points using object snaps. For example, if you dimension the Ø1.0 circle in **Figure 20-1** using the **DIMDIAMETER** command, and then change size of the circle to Ø1.50, the diameter dimension adapts to show the correct size of the modified circle. Create associative dimensions when possible and practical by selecting objects or using object snaps. Associative dimensions are related directly to object size and make revisions easier.

associative dimension: A dimension associated with an object. The dimension value updates automatically when the object changes.

Figure 20-1.
An example of a revised drawing. Dimensions adjust to the modified geometry, and the dimension values update to reflect the size and location of the modified geometry.

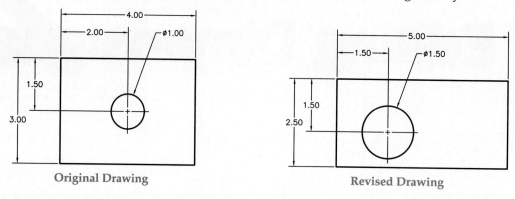

Original Drawing

Revised Drawing

Figure 20-2.
Select a dimension and then right-click to access this shortcut menu with options for adjusting individual dimensions.

Dimension editing options

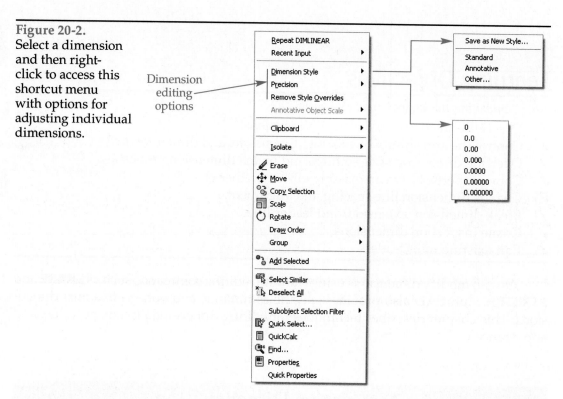

non-associative dimension: A dimension linked to point locations, not an object; does not update when the object changes.

A *non-associative dimension* forms when you select points without using object snaps. A non-associative dimension is still a single object that updates when you make changes to the dimension, such as stretching the extension line origin. Non-associative dimensions are appropriate when associative dimensions would result in dimensioning difficulty or unacceptable standards. When using non-associative dimensions, remember to edit the dimension with the object it dimensions, or adjust the dimension after the object changes.

Refer to the **Associative** property in the **General** area of the **Properties** palette to determine whether a dimension is associative.

PROFESSIONAL TIP

Dimension commands allow you to dimension a drawing, but they do not control object size and location. Chapter 22 explains how to use dimensional constraints to control object size and location. If you anticipate creating a drawing with features that will require significant or constant change, you may want to use dimensional constraints instead of, or in addition to, traditional dimensioning practices.

Associating Dimensions with Objects

Dimensions are associated with objects by default when you select objects or pick points using object snaps. To deactivate associative dimensioning for new objects, access the **Options** dialog box and deselect the **Make new dimensions associative** check box in the **Associative Dimensioning** area of the **User Preferences** tab.

Often the easiest way to convert a non-associative dimension to an associative dimension is to select the dimension to display grips. Then stretch the appropriate grip, such as the grip at the origin of a linear dimension extension line, to the corresponding object snap point using the appropriate object snap mode.

You can also convert dimensions using the **DIMREASSOCIATE** command. Access the command and select the dimension to associate with an object. An X marker appears at a dimension origin, such as the origin of a linear dimension extension line or the center of a radial dimension. Select a point on an object to associate with the marker location. Repeat the process to locate the second object point for the first extension line, if required. Use the **Next** option to advance to the next definition point. Use the **Select object** option to select an object to associate with the dimension. The extension line endpoints automatically associate with the object endpoints.

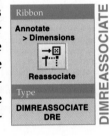

Ribbon
Annotate
> Dimensions

Reassociate

Type
DIMREASSOCIATE
DRE

DIMREASSOCIATE

To disassociate a dimension from an object, grip-edit the dimension to stretch an appropriate grip point away from the associated object, or use the **DIMDISASSOCIATE** command.

Type
DIMDISASSOCIATE

The **Dimension** panel of the **Express Tools** ribbon tab includes a **DIMREASSOC** command, not to be confused with the **DIMREASSOCIATE** command, that allows you to change the overridden value of an associated dimension back to the actual associated dimension value. Access the **DIMREASSOC** command and select associative dimensions to change. The **DIMREASSOCIATE** command creates an associative dimension, but does not change an overridden dimension value.

Definition Points

Definition points, or *defpoints*, form automatically when you create a dimension. Use the **Node** object snap to snap to a definition point. If you select an object to edit and want to include dimensions in the edit, you must include the definition points in the selection set. AutoCAD automatically creates a Defpoints layer and places definition points on the layer. By default, the Defpoints layer does not plot. You can only plot definition points if you rename the Defpoints layer and then set the renamed layer to plot. Definition points are displayed even if you turn off or freeze the Defpoints layer.

definition points (defpoints): The points used to specify the dimension location and the center point of the dimension text.

Exercise 20-1

Complete the exercise on the companion website.
www.g-wlearning.com/CAD

PROFESSIONAL TIP

You can edit individual dimension properties without exploding a dimension using dimension shortcut menu options, grip editing, or the **Properties** palette to create a *dimension style override*.

dimension style override: A temporary alteration of dimension style settings that does not actually modify the style.

Dimension Editing Tools

As the drawing process evolves and design changes occur, you will find it necessary to make changes to dimensioned objects and dimensions. AutoCAD includes dimension-specific editing tools and techniques to help you adjust dimensions as necessary.

Assigning a Different Dimension Style

To assign a different dimension style to existing dimensions, pick the dimensions to change and select a different dimension style from the **Dimension Style** drop-down list on the **Home** or **Annotation** ribbon tab. A second technique is to select the dimensions to change and choose a different dimension style from the **Quick Properties** panel or the **Properties** palette.

A third method is to select the dimensions to edit and then right-click to display the shortcut menu shown in **Figure 20-2**. Use the **Dimension Style** cascading menu to assign a different dimension style to the dimension or to save a new dimension style based on the properties of the selected dimension.

The **Update** dimension tool provides another way to change the dimension style assigned to existing dimensions. Before you access the **Update** tool, set the dimension style to be assigned to existing dimensions current. Then access the **Update** tool and pick the dimensions to change them to the current style.

Ribbon

Annotate > Dimensions

Update

Editing the Dimension Value

AutoCAD provides the flexibility to control the appearance of dimensions for various dimensioning requirements without using a separate dimension style. For example, you can add a prefix or suffix to the dimension value or adjust the dimension text format for a limited number of special-case dimensions, instead of creating separate dimension styles. **Figure 20-3** shows adding a diameter symbol to a linear diameter dimension if you forget to use the **Mtext** or **Text** option of the **DIMLINEAR** command.

Figure 20-3.
Adding a diameter
symbol to an
existing dimension.

Original   Diameter Symbol Added

The easiest method to modify an existing dimension value is to double-click on the dimension value to activate the multiline text editor. The highlighted value represents the current dimension value. Add to or modify the dimension text and then close the text editor.

 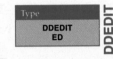

NOTE

You can also use the **DDEDIT** command to edit a dimension value.

CAUTION

You can replace the highlighted dimension value, but this action disassociates the dimension value with the object or points it dimensions. Therefore, leave the default value intact whenever possible.

Exercise 20-2

Complete the exercise on the companion website.
www.g-wlearning.com/CAD

Dimension Grip Commands

A drawing typically includes certain dimensions that require special treatment or modification. For example, proper dimensioning practice requires dimensions that are clear and easy to read. This sometimes involves moving the text of adjacent dimensions to separate the text elements. See **Figure 20-4**. AutoCAD offers a variety of commands and options for adjusting dimensions, but often the quickest method for making basic changes is to use grips. Select a dimension to display grips at editable locations on the dimension.

Most dimension grips offer the standard **STRETCH**, **MOVE**, **ROTATE**, **SCALE**, **MIRROR**, and **Copy** grip commands. Remember when using these commands that the

Figure 20-4.
Staggering
dimension text
for improved
readability.

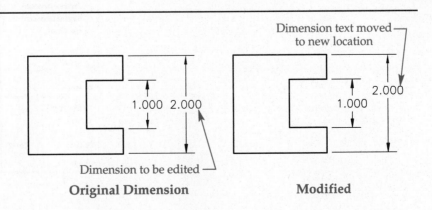

Dimension text moved
to new location

Dimension to be edited

Original Dimension   **Modified**

dimension is a single object with variables controlled by the dimension style. Certain dimension grips also provide access to context-sensitive options. You can access and apply the same context-sensitive grip commands in three different ways. **Figure 20-5** briefly explains and illustrates each technique.

Figure 20-6 illustrates the process of using the dimension text grip commands to edit the display of a dimension value. **Figure 20-7** shows the process of using the grip commands available at the ends of dimension lines or extension lines, depending on the type of dimension. The options also vary depending on the type of dimension. Use drawing aids and construction geometry when necessary to complete the operation accurately.

PROFESSIONAL TIP

Activate the **Place text manually** check box in the **Fit** tab of the **New** (or **Modify**) **Dimension Style** dialog box to provide greater flexibility for the initial placement of dimensions when necessary.

Select a dimension and then right-click to display the shortcut menu shown in **Figure 20-2**. Use the **Dimension Style** cascading menu, previously described, to assign a different dimension style to the dimension or to save a new dimension style based on the properties of the selected dimension. The **Precision** cascading menu includes options to adjust the number of decimal places displayed with a dimension value.

Figure 20-5.
Using grips to make basic changes to dimensions. A—Hover over an unselected grip to display a menu of options, and then pick an option from the list. B—Pick a grip, right-click, and select an option. C—Pick a grip and press [Ctrl] to cycle through options.

Hover over an unselected grip and pick

1.250

Stretch
Move with Dim Line
Move Text Only
Move with Leader
Above Dim Line
Center Vertically
Reset Text Position

A

Select a grip and press [Ctrl] as needed

.375

Stretch
Continue Dimension
Baseline Dimension
Flip Arrow

B

Select a grip, right-click, and pick an option

Enter

Stretch
Continue Dimension
Baseline Dimension
Flip Arrow

Move
Rotate
Scale
Mirror

Base Point
Copy
Reference
Undo   Ctrl+Z

Exit

C

The **Precision** cascading menu often provides the easiest way to specify an alternative tolerance. The **Remove Style Overrides** option clears dimension style overrides assigned to the selected dimension. You will explore dimension style overrides later in this chapter.

Exercise 20-3

Complete the exercise on the companion website.
www.g-wlearning.com/CAD

Using the DIMTEDIT Command

The **DIMTEDIT** command provides another method to change the placement and orientation of existing dimension text. Access the **DIMTEDIT** command and select the dimension to alter. Specify a new point to stretch the text and automatically reestablish the break in the dimension line.

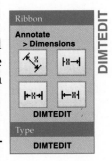

Figure 20-6.
Examples of options available for modifying dimension text using grips.

Option	Process	Result
Stretch Stretches the text with or without the dimension line		
Move with Dimension Line Always stretches the text with the dimension line		
Move Text Only Moves text independently of the dimension, which is usually not appropriate		
Move with Leader Moves text independently of the dimension and adds a leader line		
Above Dim Line Places text above the dimension line		
Center Vertically Returns text above the dimension line to centered between the dimension line		
Reset Text Position Returns text to its original position		

Figure 20-7.
Examples of options available for modifying or creating dimensions using grips.

Option	Process	Result
Stretch Stretches the text with or without the dimension line		
Continue Dimension Provides an effective alternative to using the **DIMCONTINUE** command		
Baseline Dimension Provides an effective alternative to using the **DIMBASELINE** command		
Flip Arrow Flips the direction of a dimension arrowhead to the opposite side of the extension line or object that the arrow touches		

The **DIMTEDIT** command also provides options to relocate dimension text to a specific position and rotate the text. However, it is usually quicker to select the appropriate button from the expanded **Dimensions** panel of the **Annotation** ribbon tab. Select the **Left** option to move horizontal text to the left and vertical text down. Choose the **Right** option to move horizontal text to the right and vertical text up. Use the **Center** option to center the text on the dimension line. Select the **Home** option to move the text back to its original position. Use the **Angle** option to rotate the dimension text. **Figure 20-8** shows the result of using each **DIMTEDIT** command option.

Figure 20-8.
A comparison of the **DIMTEDIT** command options.

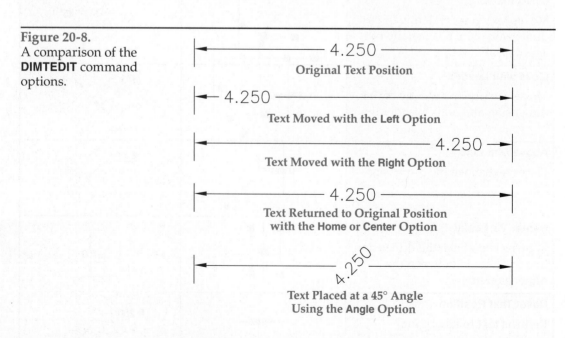

Exercise 20-4

Complete the exercise on the companion website.
www.g-wlearning.com/CAD

Using the DIMEDIT Command

The **DIMEDIT** command, not to be confused with the **DIMTEDIT** command, provides **Home** and **Rotate** options that function the same as the **Home** and **Angle** options of the **DIMTEDIT** command. The **New** option is similar to using the **DDEDIT** command to edit dimension values. When you activate the **New** option, the multiline text editor appears with the associated dimension value highlighted. Add to or modify the dimension text and then close the text editor.

The **Oblique** option is unique to the **DIMEDIT** command and allows you to change the extension line angle without affecting the associated dimension value. Figure 20-9A shows an example of adjusting the placement of dimensions when space is limited by changing existing linear dimensions to use oblique extension lines. Figure 20-9B shows an example of using oblique extension lines to orient extension lines properly with the angle of the stairs in a stair section. Notice that the associated values and orientation of the dimension lines in these examples do not change.

Ribbon
Annotate > Dimensions
Oblique
Type
DIMEDIT

Figure 20-9.
Drawing dimensions with oblique extension lines.

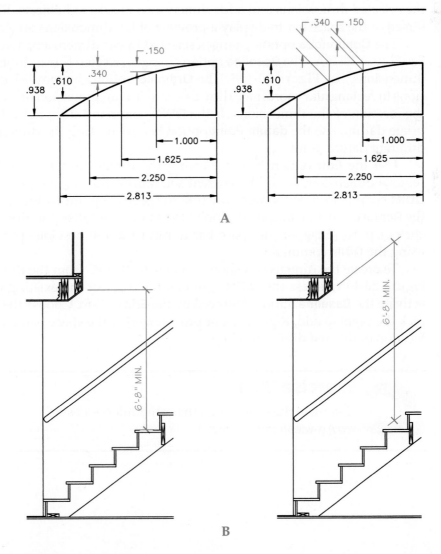

To create oblique extension lines, dimension the object using the **DIMLINEAR** and **DIMALIGNED** commands as appropriate, even if the dimensions are crowded or overlap. Then access the **Oblique** option of the **DIMEDIT** command. The quickest way to access the **Oblique** option is to pick the corresponding button from the expanded **Dimensions** panel of the **Annotation** ribbon tab. Then pick the linear and aligned dimensions to be redrawn at an oblique angle and specify the obliquing angle. Plan carefully to make sure you enter the correct obliquing angle. Obliquing angles originate from 0° East and revolve counterclockwise. Enter a specific value or pick two points to define the obliquing angle.

Exercise 20-5

Complete the exercise on the companion website.
www.g-wlearning.com/CAD

Editing Dimensions with the QDIM Command

QDIM

Ribbon
Annotate > Dimensions
Quick Dimension

Type
QDIM

The **QDIM** command provides options for replacing, adding, removing, and rearranging existing linear or ordinate dimensions. The **QDIM** command does not change diameter or radius dimensions. Access the **QDIM** command and select the dimensions to modify, and any other objects to be dimensioned. The **QDIM** command replaces the selected dimensions and adds dimensions to selected objects. Right-click or press [Enter] or the space bar to display a preview of the dimensions attached to the cursor.

The **Continuous** option changes selected linear dimensions to chain dimensions. See Figure 20-10A. The **Baseline** option changes selected linear dimensions to baseline dimensions. See Figure 20-10B. The **Ordinate** option changes selected linear dimensions to rectangular coordinate dimensions without dimension lines. You must reselect the location of the dimensions. If the **QDIM** command does not reference the appropriate datum, use the **datum Point** option before locating the dimensions to specify a different datum point.

Use the **Edit** option before locating the dimensions to add dimensions to, or remove dimensions from, the current set. Marks indicate the points acquired by the **QDIM** command. Use the **Add** function to specify a point to add a dimension, or use the **Remove** function to specify a point to remove the corresponding dimension. Right-click or press [Enter] or the space bar to return to the previous prompt and continue using the **QDIM** command.

To create the dimensions shown in Figure 20-10C from the dimensions shown in Figure 20-10A, access the **QDIM** command and select the existing dimensions. Then activate the **Baseline** option, followed by the **Edit** option. Choose the **Add** function and pick the point to add. Right-click or press [Enter] or the space bar, and then specify the location of the first dimension line.

Exercise 20-6

Complete the exercise on the companion website.
www.g-wlearning.com/CAD

Figure 20-10.
Use the **QDIM** command to change the arrangements or type of existing dimensions and to add or remove dimensions.

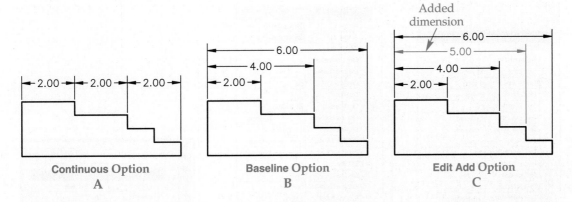

Continuous Option
A

Baseline Option
B

Edit Add Option
C

Overriding Dimension Style

Drawings often include dimensions that require settings slightly different from the assigned dimension style. These dimensions may be too few to merit creating a new style. Perform a dimension style override for these situations. For example, use a dimension style with an **Offset from origin** value of .063 to conform to ASME standards for most dimensions. Apply a dimension style override with an **Offset from origin** value of 0 to three dimensions that should not display an extension line offset.

Existing Dimensions

The **Properties** palette is an effective tool for overriding the dimension style assigned to existing dimensions. The **Properties** palette divides dimension properties into several categories. See Figure 20-11. To change a property, access the proper category, pick the property to highlight, and adjust the corresponding value. Most changes made using the **Properties** palette override the dimension style assigned to the selected dimension. The changes do not alter the original dimension style and do not apply to new dimensions.

The **Quick Properties** panel provides a limited number of dimension properties and style overrides.

New Dimensions

Use the **Dimension Style Manager** to override the dimension style assigned to dimensions you are about to create. An example of an override is including a text prefix for a few dimensions. Select the dimension style to override from the **Styles** list and then pick the **Override** button to open the **Override Current Style** dialog box. The **Override** button is available only for the current style. The **Override Current Style** dialog box includes the same tabs as the **New Dimension Style** and **Modify Dimension Style** dialog boxes. Make the necessary changes and pick the **OK** button. The override becomes current and appears as a branch under the original style labeled <style overrides>. Close the **Dimension Style Manager** and draw the dimensions.

Figure 20-11.
The **Properties** palette allows you to edit dimension properties and create dimension style overrides.

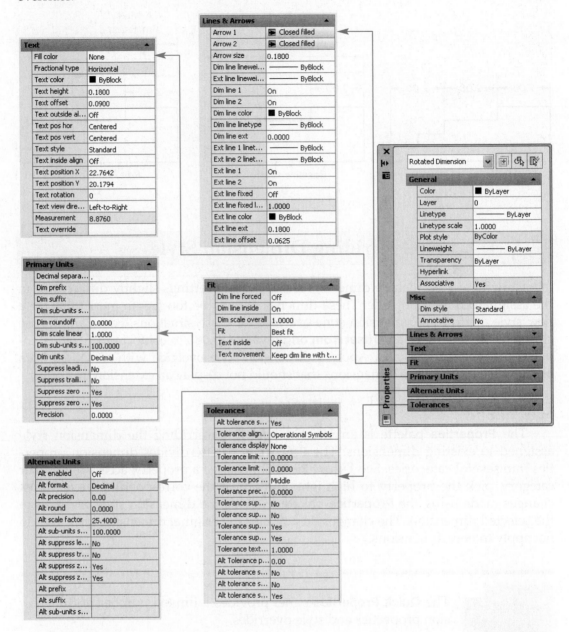

To clear style overrides, return to the **Dimension Style Manager** and set a different style current. The override settings are lost when you set a different style, including the parent style, current. To incorporate the overrides into the overridden style, right-click on the <style overrides> name and select **Save to current style**. To save the changes to a new style, pick the **New...** button. Then select <style overrides> in the **Start With** drop-down list in the **Create New Dimension Style** dialog box. In the **New Dimension Style** dialog box, pick **OK** to save the overrides as a new style.

You can also select a dimension and then right-click and choose the **Remove Style Overrides** option to clear dimension style overrides assigned to the selected dimension.

Carefully evaluate the dimensioning requirements in a drawing before performing a style override. It may be better to create a new style. Consider generating a new dimension style if several dimensions require the same overrides. If only one or two dimensions need the same changes, an override is usually more productive.

Exercise 20-7

Complete the exercise on the companion website.
www.g-wlearning.com/CAD

Using the MATCHPROP Command

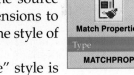

The **MATCHPROP** command allows you to copy, or "paint," properties from one object to other objects, including dimensions. You can match properties in the same drawing or between drawings. Access the **MATCHPROP** command, pick the source dimension with the desired properties, and then pick the destination dimensions to change. Press [Enter] or the space bar, or right-click and select **Enter** to exit. The style of the source dimension is applied to destination dimensions.

If you override the dimension style of the source dimension, the "base" style is applied along with the dimension style override. Reapplying the "base" style removes the overrides.

The **Property Settings** dialog box, available by selecting the **Settings** option before picking the destination objects, includes a **Dimension** check box. AutoCAD checks the box by default, allowing you to match dimensions.

Exercise 20-8

Complete the exercise on the companion website.
www.g-wlearning.com/CAD

Using the DIMSPACE Command

The amount of space between a drawing view and the first dimension line and the space between dimension lines vary depending on the drawing and industry or company standards. ASME standards recommend a minimum spacing of .375" (10 mm) from a drawing feature to the first dimension line and a minimum spacing of .25" (6 mm) between dimension lines. A minimum spacing of 3/8" is common for architectural drawings. These minimum recommendations are generally less than the spacing required by actual company or school standards. A value of .5″ (12 mm) or .75″ (19 mm) is usually more appropriate.

Typically, the spacing between dimension lines is equal, and chain dimensions align. See **Figure 20-12**. You generally determine the correct location and spacing of dimension lines before and while dimensioning. However, you can adjust dimension line spacing and alignment after you place dimensions. This is a common requirement when there is a need to increase or decrease the space between dimension lines, such as when the drawing scale changes, or when dimensions are unequally spaced or misaligned.

Grips or the **STRETCH**, **DIMTEDIT**, and **QDIM** commands are common methods for adjusting the location and alignment of dimension lines. However, you must determine the exact location or amount of stretch applied to each dimension line before using these commands. An alternative is to use the **DIMSPACE** command, which allows you to adjust the space equally between dimension lines or to align dimension lines.

Access the **DIMSPACE** command and select the *base dimension*, followed by each dimension to be spaced. Right-click or press [Enter] or the space bar to display the Enter value or [Auto]: prompt. Enter a value to space the dimension lines equally. For example, enter .5 to space the selected dimension lines .5″ apart. Enter a value of 0 to align the dimensions. See **Figure 20-13**. Use the **Auto** option to space dimension lines using a value that is twice the height of the dimension text.

DIMSPACE

Ribbon
**Annotate
> Dimensions**

Adjust Space

Type
DIMSPACE

base dimension:
The dimension line that remains in the same location, with which other dimension lines align or spaced.

NOTE   Use the **DIMSPACE** command to space and align linear and angular dimensions.

Figure 20-12.
Correct drafting practice requires equal space and alignment between dimension lines for readability. The correct example uses a spacing of .75″ (19 mm) from a drawing feature to the first dimension line and a spacing of .5″ (12 mm) between dimension lines.

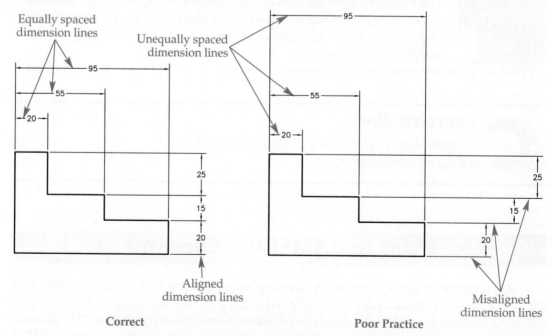

Figure 20-13.
Using the **DIMSPACE** command to space and align dimension lines correctly.

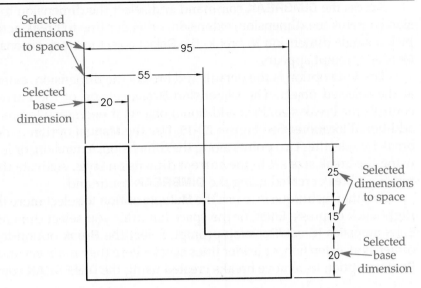

Selected dimensions to space

Selected base dimension

Selected dimensions to space

Selected base dimension

Exercise 20-9

Complete the exercise on the companion website.
www.g-wlearning.com/CAD

Using the DIMBREAK Command

ASME and many other drafting discipline standards state that when dimension, extension, or leader lines cross a drawing feature or another dimension, neither line is broken at the intersection. See **Figure 20-14**. However, you can use the **DIMBREAK** command to create breaks if desired.

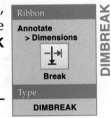

Ribbon
Annotate > Dimensions
Break
Type
DIMBREAK

Figure 20-14.
ASME and many other drafting discipline standards state that when dimension, extension, or leader lines cross a drawing feature or another dimension, the line is not broken at the intersection.

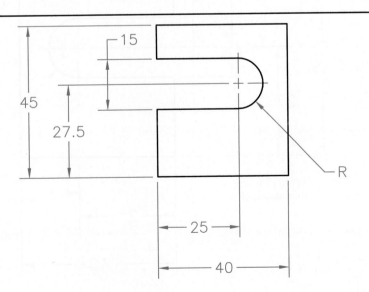

Access the **DIMBREAK** command and select the dimension to break. This dimension contains the dimension, extension, or leader line to break across an object. If you pick a single dimension to break, the Select object to break dimension or [Auto/Restore/Manual]: prompt appears.

The **Auto** option is the default and breaks the dimension, extension, or leader line at the selected object. The **Dimension Break** setting of the current dimension style controls the break size. Pick additional objects if necessary to break the dimension at additional locations. See Figure 20-15. Use the **Manual** option to define the size of the break by selecting two points along the dimension, extension, or leader line, instead of using the break size set in the current dimension style. Activate the **Restore** option to remove a break created using the **DIMBREAK** command.

Another technique is to use the **Multiple** option to select more than one dimension. Right-click or press [Enter] or the space bar after you select dimensions to display the Enter an option [Break/Restore]: prompt. Select the **Break** option to break the selected dimension, extension, or leader lines everywhere they intersect another object. Use the **Restore** option to remove breaks created using the **DIMBREAK** command.

Exercise 20-10

Complete the exercise on the companion website.
www.g-wlearning.com/CAD

Figure 20-15.
Use the **DIMBREAK** command to break dimension, extension, or leader lines when they cross an object. *Caution:* This example created violates ASME standards and is for reference only. Extension and leader lines do not break over object lines, but some drafters prefer to break an extension line when it crosses a dimension line.

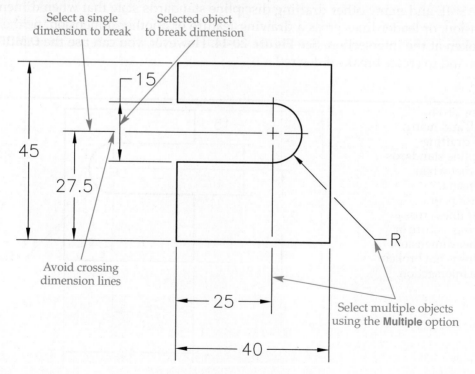

Creating Inspection Dimensions

Inspections and tests occur throughout the design and manufacturing of a product. Tests help ensure the correct size and location of product features. In some cases, size and location dimensions include information about how frequently a test on the dimension occurs for consistency and tolerance during the manufacturing process. See **Figure 20-16**. Use the **DIMINSPECT** command to add inspection information to most types of existing dimensions.

Access the **DIMINSPECT** command to display the **Inspection Dimension** dialog box, shown in **Figure 20-17**. Pick the **Select dimensions** button and choose the dimensions to which you want to apply inspection information. You can select multiple dimensions, although the same inspection specifications apply to each. Pick the appropriate radio button in the **Shape** area to define the shape of the inspection dimension frame. The inspection dimension contains the inspection label, the dimension value, and the inspection rate. Select the **None** option to omit frames around values.

Pick the **Label** check box to include a label, and type the label in the text box. The label appears on the left side of the inspection dimension and identifies the dimension. The inspection dimension shown in **Figure 20-16** is labeled A. The dimension frame houses the dimension value specified when you created the dimension. The length of the part shown in **Figure 20-16** is 2.500, as created using the **DIMLINEAR** command. The **Inspection rate** check box is active by default. Enter a value in the text box to indicate how often to test the dimension. The inspection rate for the dimension shown in **Figure 20-16** is 100%. This rate has different meanings depending on the application. In this example, the inspection rate of 100% means that the manufacturer must check the length of the part for tolerance every time the part is added to an assembly.

To remove an inspection dimension, access the **DIMINSPECT** command, pick the **Select dimensions** button in the **Inspection Dimension** dialog box, and choose the dimensions from which you want to remove inspection information. Right-click or press [Enter] or the space bar to return to the **Inspection Dimension** dialog box, and pick the **Remove Inspection** button to return the dimension to its condition prior to adding the inspection content.

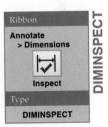

Exercise 20-11

Complete the exercise on the companion website.
www.g-wlearning.com/CAD

Figure 20-16.
An inspection dimension added to a part drawing. This example shows an angular shape with a label, dimension, and inspection rate frame.

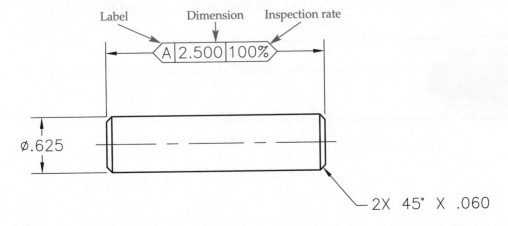

Figure 20-17.
The **Inspection
Dimension** dialog
box allows you
to add inspection
information to
existing dimensions.

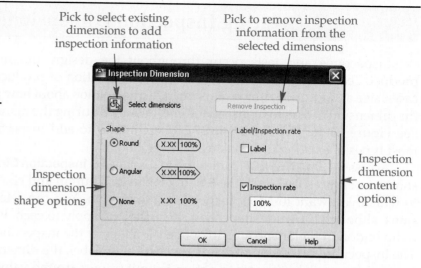

Pick to select existing
dimensions to add
inspection information

Pick to remove inspection
information from the
selected dimensions

Inspection
dimension
shape options

Inspection
dimension
content
options

Multileader Editing Tools

AutoCAD provides commands and options to control the appearance of multileaders for various dimensioning and annotation requirements. Edit multileaders using methods similar to those you use to edit dimensions. Use the **Properties** palette or **Quick Properties** panel to override specific multileader properties. You can also use the **MATCHPROP** command. In addition to these general editing techniques, specific tools allow you to add and remove leader lines and space, align, and group multileader objects.

Assigning a Different Multileader Style

To assign a different multileader style to existing multileaders, pick the multileaders to change and select a different multileader style from the **Multileader Style** drop-down list on the **Home** or **Annotation** ribbon tab. A second technique is to select the multileaders to change and choose a different multileader style from the **Quick Properties** panel or the **Properties** palette.

A third method is to select the multileaders to edit and then right-click to display the shortcut menu shown in Figure 20-18. Use the **Multileader Style** cascading menu to assign a different multileader style to the multileader or to save a new multileader style based on the properties of the selected multileader.

Editing Multileader Mtext

The easiest method to modify multileader text created using the **Mtext** content option is to double-click on the multileader text to activate the multiline text editor. Add to or modify the multileader text and then close the text editor.

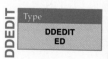

You can also use the **DDEDIT** command to edit multileader mtext.

Figure 20-18.
Select a multileader and then right-click to access this shortcut menu with options for adjusting individual multileaders.

Multileader options

- <u>R</u>epeat HELP
- Recent Input ▸
- Annotative Object Scale ▸
- Clipboard ▸
- <u>I</u>solate ▸
- Erase
- <u>M</u>ove
- Cop<u>y</u> Selection
- Sca<u>l</u>e
- R<u>o</u>tate
- Dra<u>w</u> Order ▸
- Group ▸
- Add Leader
- Remove Leader
- Multileader Style ▸
- A<u>d</u>d Selected
- Select Similar
- Deselect <u>A</u>ll
- Subobject Selection Filter ▸
- <u>Q</u>uick Select...
- QuickCalc
- <u>F</u>ind...
- Properties
- Quick Properties

- Save as New Multileader Style...
- Standard
- Inch
- Struc
- Weld
- Other...

Multileader Grip Commands

You can use standard editing and multileader-specific commands to adjust multileaders, but often the quickest method for making basic changes and adding and removing leaders and vertices is to use grips. Select a multileader to display grips at editable locations on the multileader. See **Figure 20-19**. Most multileader grips offer the standard **STRETCH**, **MOVE**, **ROTATE**, **SCALE**, **MIRROR**, and **Copy** grip commands. Remember that the multileader is a single object with variables controlled by the multileader style.

Context-Sensitive Options

Certain multileader grips also provide access to context-sensitive options. Access and apply context-sensitive grip commands using one of three methods, similar to grip-editing dimensions as shown in **Figure 20-5**. **Figure 20-19** explains the process of using multileader grip commands. Use drawing aids and construction geometry when possible to complete the operation accurately.

Multiple leaders are not a recommended ASME standard. For example, do not use two leaders from the same value to dimension two chamfers of the same size. However, multiple leaders are appropriate for some applications, such as welding symbols. Multiple leaders are also appropriate for some architectural drawings.

Shortcut Menu Options

Select a multileader and then right-click to display the shortcut menu shown in **Figure 20-18**. Use the **Add Leader** option to add a leader line to the multileader object, or use the **Remove Leader** option to remove a leader line. The **Multileader Style** cascading menu, described previously, allows you to assign a different multileader style to the multileader or to save a new multileader style based on the properties of the selected multileader.

Figure 20-19.
Use grips to make basic changes to multileaders, and add and remove leader lines and vertices.

Grip	Option	Process	Result
Landing endpoint	Lengthens the landing from the endpoint closest to the content	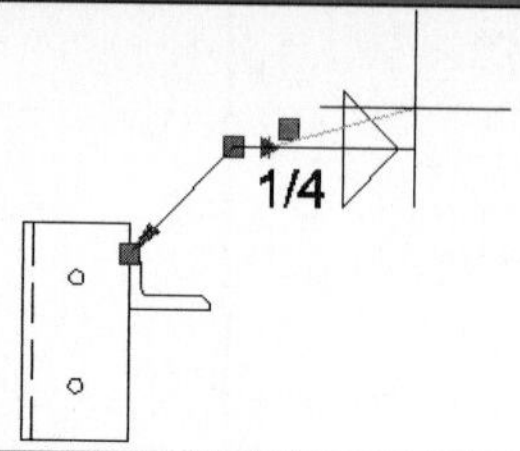	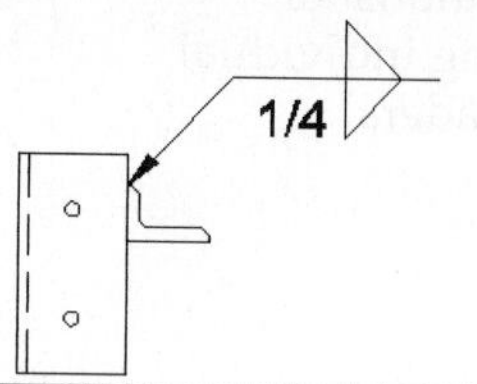
Landing endpoint	**Stretch** Stretches the leader at the selected endpoint	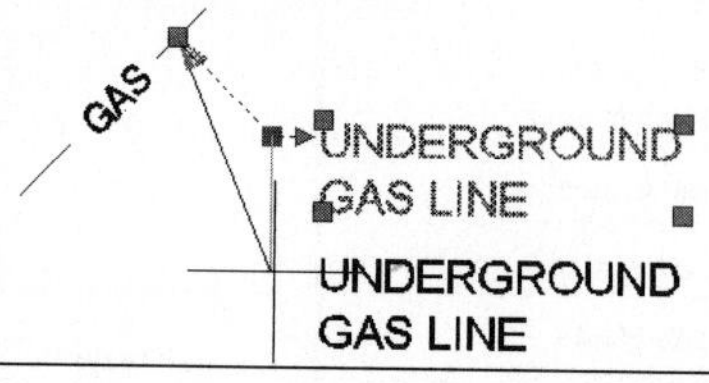	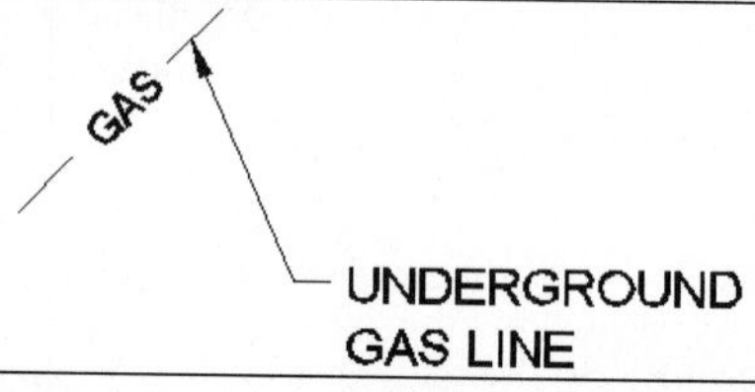
	Lengthen Landing Lengthens the landing from the selected endpoint	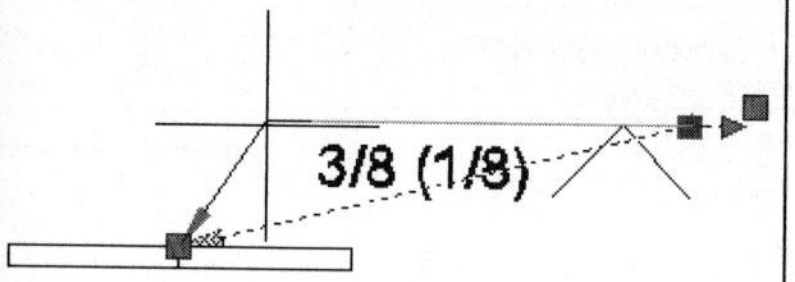	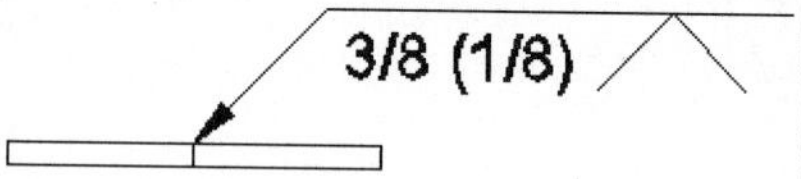
	Add Leader Adds leaders to the content	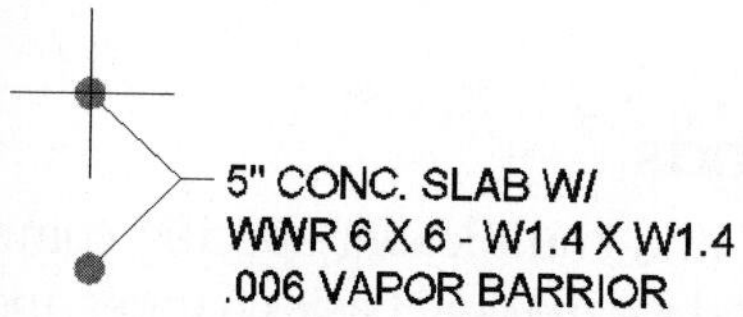	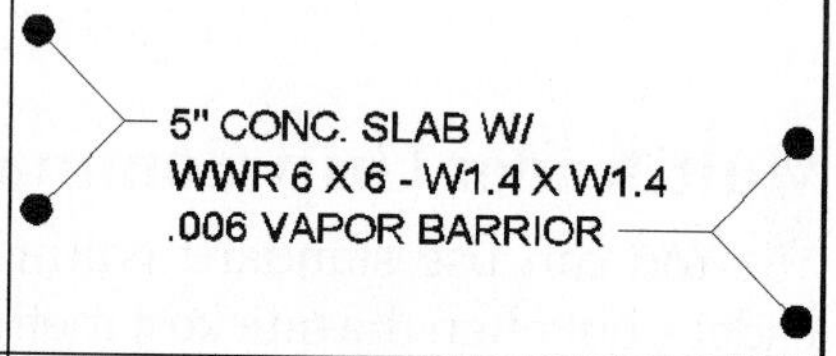
Leader line endpoint or vertex	**Stretch** Stretches the leader line at the selected endpoint or vertex	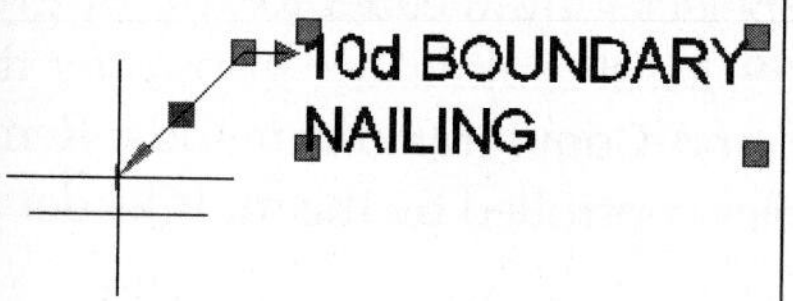	
	Add Vertex Adds a leader point	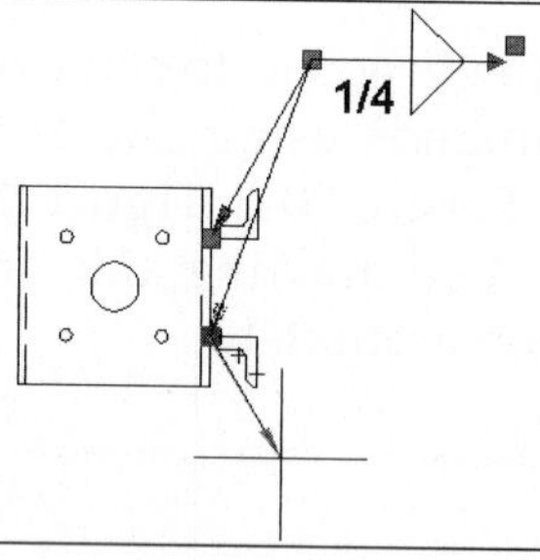	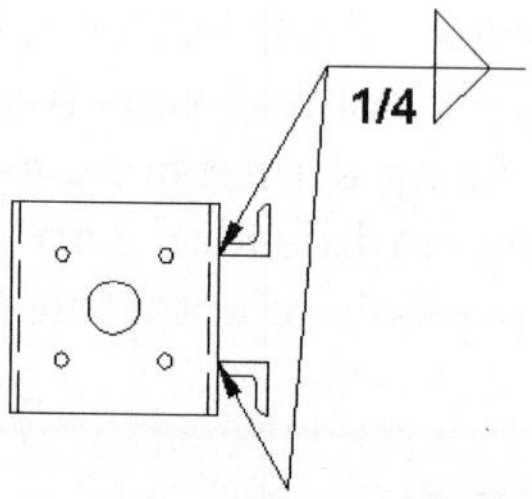
Leader line endpoint	**Remove Leader** Removes a selected leader line	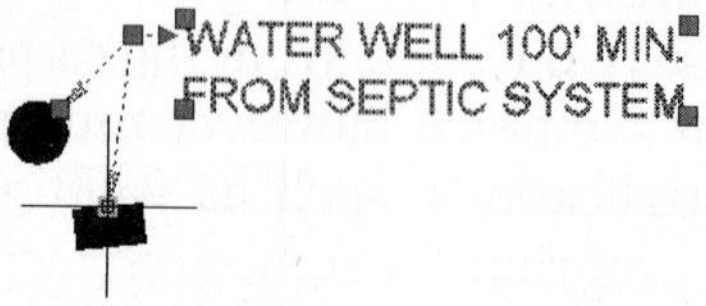	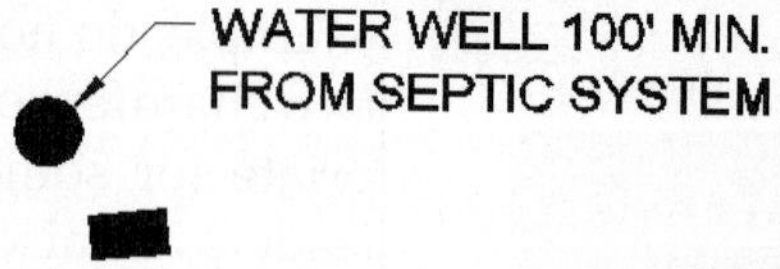
Vertex	**Remove Vertex** Removes a selected vertex	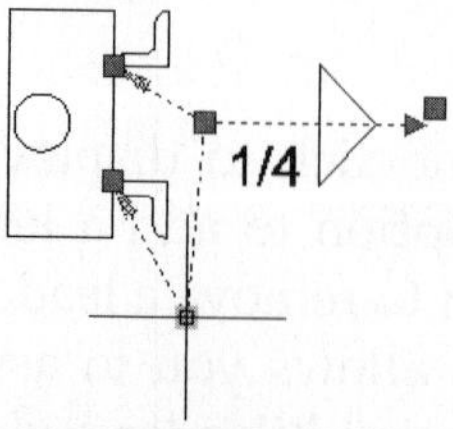	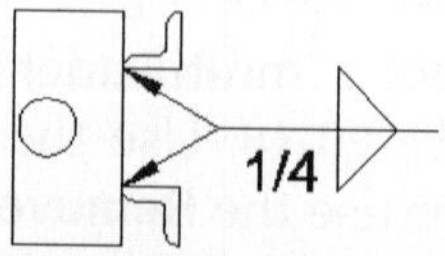

Using the MLEADEREDIT Command

The **MLEADEREDIT** command is an alternative to using grips to add leader lines to, and remove them from, an existing multileader object. To add a leader line to a multileader object, pick the **Add Leader** button from the ribbon and select the multileader to receive the additional leader line. Pick a location for the additional leader line arrowhead. You can place as many additional leader lines as needed without accessing the command again. When you are finished, press [Enter], [Esc], or the space bar or right-click and select **Enter**. All leader lines are grouped to form a single multileader object.

To remove an unneeded leader line, pick the **Remove Leader** button from the ribbon and select the multileader object that includes the leader to remove. You can also select the multileader, right-click, and choose **Remove Leader**. Select the leader lines to remove and press [Enter], [Esc], or the space bar or right-click and select **Enter**.

Ribbon

Home
> Annotation

Annotate
> Multileaders

Add Leader

Type

MLEADEREDIT

Ribbon

Home
> Annotation

Annotate
> Multileaders

Remove Leader

Type

MLEADEREDIT

> If you type **MLEADEREDIT** to access the command, you must activate the **Remove leaders** option to remove leader lines.

PROFESSIONAL TIP

> To adjust the properties of a specific leader line in a group of leaders attached to the same content, hold down [Ctrl] and pick the leader to modify. Then access the **Properties** palette. Options specific to the selected leader appear, and all other properties are filtered out.

Exercise 20-12

Complete the exercise on the companion website.
www.g-wlearning.com/CAD

Aligning Multileaders

An advantage of using multileaders is the ability to space and align leaders in an easy-to-read pattern. You typically determine the correct location and spacing of leaders before and while dimensioning. However, you can adjust leader spacing and alignment after you place multileaders. This is a common requirement when there is a need to increase or decrease the space between leaders, such as when the drawing scale changes, or when leaders are unequally spaced or misaligned. See **Figure 20-20**.

The **STRETCH** command and grips are commonly used to adjust the location and alignment of leaders. However, you must determine the exact location of or amount of stretch applied to each leader before using these commands. An alternative is to use the **MLEADERALIGN** command, which allows you to align and adjust the space between leaders.

Access the **MLEADERALIGN** command and select the leaders to space and align. You can use the **MLEADERALIGN** command to adjust the location of a single leader in reference to another leader, but for most applications, you should select several leaders. Select each leader to space or align and right-click or press [Enter] or the space bar. When a prompt asks you to select the multileader to align to, activate **Options** to change the multileader alignment.

Ribbon

Home
> Annotation

Annotate
> Multileaders

Align

Type

MLEADERALIGN

MLEADERALIGN

Figure 20-20.
Leaders that are equally spaced and aligned improve drawing readability.

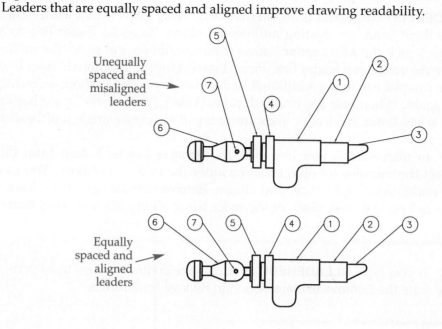

Unequally spaced and misaligned leaders

Equally spaced and aligned leaders

Figure 20-21.
Applying the **Use current spacing** option to align and equally space leaders.

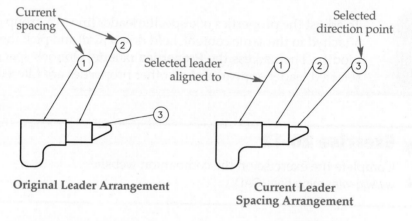

Current spacing

Selected direction point

Selected leader aligned to

Original Leader Arrangement

Current Leader Spacing Arrangement

Use Current Spacing Option

Apply the **Use current spacing** option to align and space the selected leaders equally according to the distance between one of the selected leaders and the next closest leader. Select the multileader with which all other leaders should be aligned and spaced. Then specify the direction of the leader arrangement by entering or picking a point. The space between leaders is maintained if possible, depending on the selected direction. See **Figure 20-21**.

Distribute Option

Select the **Distribute** option to align and distribute the leaders, or to place them at equally spaced locations between two points. The first point you specify identifies the location of one of the leaders and determines where distribution begins. The second point you specify identifies the location of the last leader. All other leaders are distributed equally between the two points. Leaders align with the first point. See **Figure 20-22**.

Make Leader Segments Parallel Option

Use the **make leader segments Parallel** option to make all the selected leader lines parallel to one of the selected leader lines. Select an existing leader to keep in the same

Figure 20-22.
Using the **Distribute** option to align and equally space leaders. This example uses horizontally aligned points.

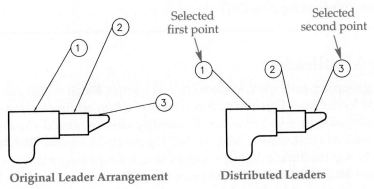

Original Leader Arrangement   Distributed Leaders

location and at the same angle. All other leaders form parallel to the selection. The length of each leader line, except for the leader aligned to, increases or decreases in order to become parallel with the first leader. See **Figure 20-23**.

Specify Spacing Option

Choose the **Specify spacing** option to align and equally space the selected leaders according to the distance, or clear space, between the extents of the content of each leader. Select the multileader with which all other leaders should be aligned and spaced. Specify the direction of the leader arrangement by entering or picking a point. See **Figure 20-24**.

Figure 20-23.
Using the **make leader segments Parallel** option to make leader lines parallel to each other.

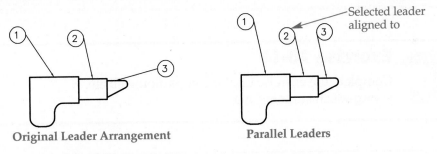

Original Leader Arrangement   Parallel Leaders

Figure 20-24.
Using the **Specify spacing** option to align and equally space leaders.

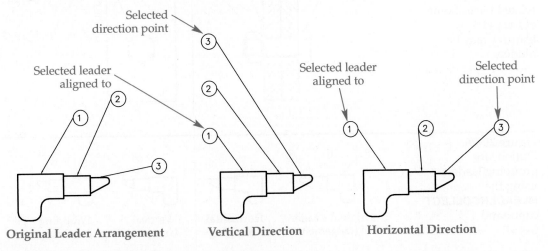

Original Leader Arrangement   Vertical Direction   Horizontal Direction

Exercise 20-13

Complete the exercise on the companion website.
www.g-wlearning.com/CAD

Grouping Multileaders

Ribbon

Home
> Annotation
Annotate
> Multileaders

Collect

Type

MLEADERCOLLECT

You can group separate multileaders created using a **Block** multileader content style to use a single leader line. This practice is common when adding *balloons* to assembly drawings. *Grouped balloons* allow you to identify closely related clusters of assembly components, such as a bolt, washer, and nut. See **Figure 20-25**. Use the **MLEADERCOLLECT** command to group multiple existing leaders using a single leader line.

Access the **MLEADERCOLLECT** command and select the leaders to group. The order in which you select leaders determines how they are grouped. Select leaders in a sequential order, ending with the leader line you want to keep.

The options illustrated in **Figure 20-26** are available after you select the leaders. Select the **Horizontal** option to align grouped content horizontally, or the **Vertical** option to align grouped content vertically. Pick a point to locate the grouped leader. Select the **Wrap** option to wrap grouped content to additional lines as needed when the number of items exceeds a specified width or quantity. Enter the width at the Specify width: prompt, or use the **Number** option to enter a quantity not to exceed before the grouped leaders wrap. Then pick a point to locate the grouped leader.

balloons: Circles that contain a number or letter to identify the assembly component and correlate the component to a parts list or bill of materials. Balloons connect to a component with a leader line.

grouped balloons: Balloons that share the same leader, which typically connects to the most obviously displayed component.

You can only use the **MLEADERCOLLECT** command to group symbols attached to leaders created using the **Block** content style.

Exercise 20-14

Complete the exercise on the companion website.
www.g-wlearning.com/CAD

Figure 20-25.
An example of grouped balloons identifying closely related parts. Some of the parts or features may be hidden.

Figure 20-26.
Options for grouping leaders using the **MLEADERCOLLECT** command

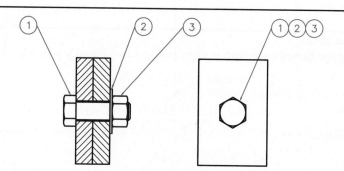

Original Leader Arrangement   Horizontal Grouping   Vertical Grouping   Wrapping the Group

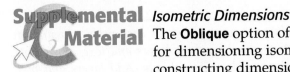

Supplemental Material

Isometric Dimensions

The **Oblique** option of the **DIMEDIT** command is one option for dimensioning isometric drawings. For information about constructing dimensions for isometric views, go to the companion website (www.g-wlearning.com/CAD), select this chapter, and select **Isometric Dimensions**.

Chapter Review

Answer the following questions. Write your answers on a separate sheet of paper or complete the electronic chapter review on the companion website.
www.g-wlearning.com/CAD

1. Define *associative dimension*.
2. Why is it important to have associative dimensions for editing objects?
3. Which **Options** dialog box setting controls associative dimensioning?
4. Which command allows you to convert non-associative dimensions to associative dimensions?
5. Which command allows you to convert associative dimensions to non-associative dimensions?
6. What are definition points?
7. List the context-sensitive options available for grip-editing a dimension value.
8. Name three methods of changing the dimension style of a dimension.
9. How does the **Dimension Update** tool affect selected dimensions?
10. Explain how to add a diameter symbol to a dimension text value using the **DDEDIT** command.
11. Name the command that allows you to control the placement and orientation of an existing associative dimension text value.
12. Name two applications in which you might need to create oblique extension lines.
13. Which command and option can you use to add a new baseline dimension to an existing set of baseline dimensions?
14. When you use the **Properties** palette to edit a dimension, what is the effect on the dimension style?
15. How do you access the **Property Settings** dialog box?
16. Which command can you use to adjust the space equally between dimension lines or align dimension lines without having to determine the exact location or amount of stretch needed?
17. What two options are available when you use the **Multiple** option of the **DIMBREAK** command?
18. What command allows you to add information about how frequently the manufacturer should test a dimension for consistency and tolerance during the manufacturing of a product?
19. Name an application in which leaders with multiple leader lines are common.
20. Identify the four options available to change multileader alignment.

Drawing Problems

Start AutoCAD if it is not already started. Start a new drawing for each problem using an appropriate template of your choice. The template should include layers and text, dimension, and multileader styles, when necessary, for drawing the given objects. Add layers and text, dimension, and multileader styles as needed. Draw all objects using appropriate layers and text, dimension, and multileader styles, justification, and format. Follow the specific instructions for each problem. Use only drawing and editing commands and techniques you have already learned. Use your own judgment and approximate dimensions when necessary. Apply dimensions accurately using ASME or appropriate industry standards.

Note: Some of the problems in this chapter are built on problems from previous chapters. If you have not yet completed those problems, complete them now.

▼ Basic

1. Open P17-9 and save the file as P20-1. The P20-1 file should be active. Edit the drawing as follows:
   A. Erase the front (circular) view.
   B. Stretch the vertical dimensions to provide more space between dimension lines. Be sure the space you create is the same between all vertical dimensions.
   C. Stagger the existing vertical dimension text numbers if they are not staggered as shown in the original problem.
   D. Erase the 1.750 horizontal dimension and then stretch the 5.255 and 4.250 dimensions to make room for a new baseline dimension from the baseline to where the 1.750 dimension was located. This should result in a new baseline dimension that equals 2.750. Be sure all horizontal dimension lines are equally spaced.
   E. Resave the file.

2. Open P18-1 and save the file as P20-2. The P20-2 file should be active. Edit the drawing as follows:
   A. Stretch the total length from 3.500 to 4.000, leaving the holes the same distance from the edges.
   B. Fillet the upper-left corner. Modify the 3X R.250 dimension accordingly.
   C. Resave the drawing.

3. Open P17-12 and save the file as P20-3. The P20-3 file should be active. Edit the drawing as follows: Make the bathroom 8′-0″ wide by stretching the walls and vanity that are currently 6′-0″ wide to 8′-0″. Do this without increasing the size of the water closet compartment. Provide two equally spaced oval sinks where there is currently one. Resave the drawing.

4. Open P18-19 and save the file as P20-4. The P20-4 file should be active. Edit the drawing as follows:
   A. Lengthen the part .250 on each side for a new overall dimension of 6.500.
   B. Change the width of the part from 3.000 to 3.500 by widening an equal amount on each side.
   C. Resave the drawing.

5. Open P18-17 and save the file as P20-5. The P20-5 file should be active. Edit the drawing as follows:
   A. Shorten the .75 thread on the left side to .50.
   B. Shorten the .388 hexagon length to .300.
   C. Resave the drawing.

6. Draw the shim shown at A. Then edit the .150 and .340 values using oblique dimensions as shown at B. Save the drawing as P20-6.

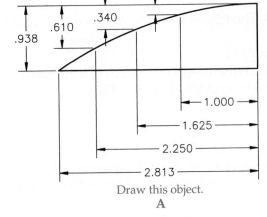

Draw this object.
A

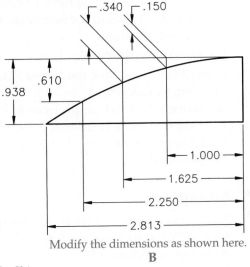

Modify the dimensions as shown here.
B

Title: Shim

▼ Intermediate

7. Draw and dimension the swivel screw shown. Save the drawing as P20-7.

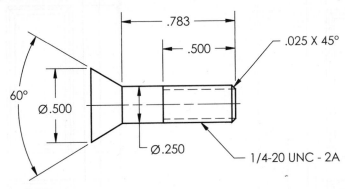

8. Open **P18-4** and save the file as **P20-8**. The **P20-8** file should be active. Edit the drawing as follows:
   A. Use the existing drawing as the model and make four copies.
   B. Leave the original drawing as it is and edit the other four pins in the following manner, keeping the Ø.125 hole exactly in the center of each pin.
   C. Give one pin a total length of 1.500.
   D. Create the next pin with a total length of 2.000.
   E. Edit the third pin to a length of 2.500.
   F. Change the last pin to a length of 3.000.
   G. Organize the pins on your drawing in a vertical row ranging in length from the smallest to the largest. You may need to change the drawing limits.
   H. Resave the drawing.

Drawing Problems – Chapter 20

9. Open P18-5 and save the file as P20-9. The P20-9 file should be active. Edit the drawing as follows:
   A. Modify the spline to have twelve projections, rather than eight.
   B. Change the angular dimension, linear dimension, and 8X dimension to reflect the modification.
   C. Resave the drawing.

10. Open P18-11 and save the file as P20-10. The P20-10 file should be active. Edit the drawing as follows:
    A. Stretch the total length from 6.500 to 7.750.
    B. Add two more holes that continue the equally spaced pattern of .625 apart.
    C. Change the 8X .625(=5.00) dimension to read 10X .625(=6.250).
    D. Resave the drawing.

▼ Advanced

11. Draw and dimension the door elevation shown at A. Save the drawing as P20-7A. Open P20-7A and save the file as P20-7B. The P20-7B file should be active. Edit the drawing as shown at B.

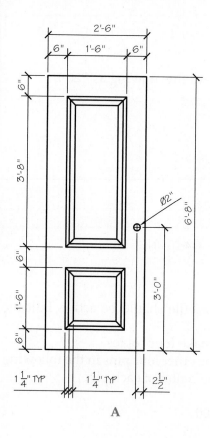

A

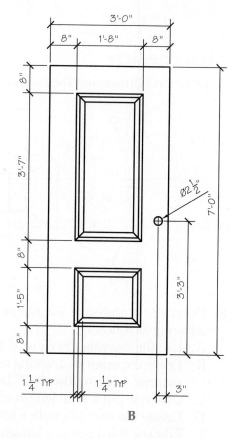

B

12. Open P17-17 and save the file as P20-12. The P20-12 file should be active. Make the client-requested revisions to the floor plan as shown. Make sure the dimensions reflect the changes.

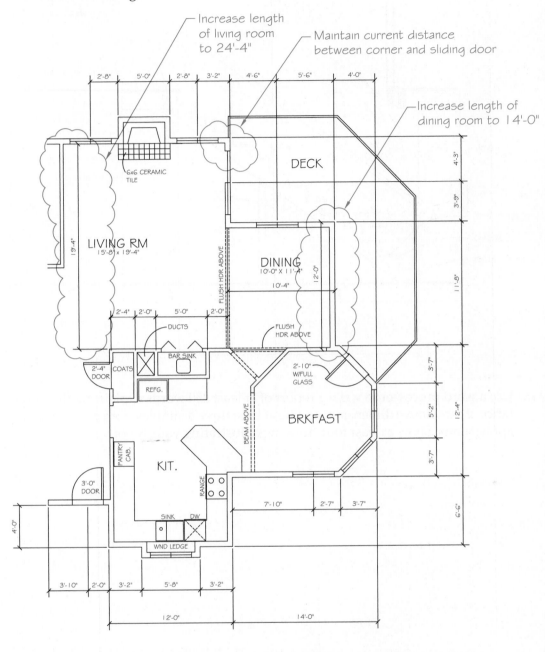

Increase length of living room to 24'-4"

Maintain current distance between corner and sliding door

Increase length of dining room to 14'-0"

DECK

6x6 CERAMIC TILE

LIVING RM
15'-8" x 19'-4"

FLUSH HDR ABOVE

DINING
10'-0" X 11'-4"

10'-4"

DUCTS

BAR SINK

FLUSH HDR ABOVE

COATS

2'-4" DOOR

REFG.

2'-10" W/FULL GLASS

BEAM ABOVE

BRKFAST

PANTRY CAB.

KIT.

3'-0" DOOR

RANGE

SINK   DW

WND LEDGE

13. Design and draw a vice clamp similar to the vice clamp shown in **Figure 20-20**. Add balloons and a parts list to the drawing. Save the drawing as P20-13.

14. Open P8-20 and save the file as P20-14. The P20-14 file should be active. Add balloons and a parts list to the drawing of the nut driver.

15. Open P11-17 and save the file as P20-15. The P20-15 file should be active. Dimension the most important views of the anchor. Erase the undimensioned views.

Drawing Problems - Chapter 20

16. Draw and dimension the stairs cross section shown. Use oblique dimensions where necessary. Save the drawing as P20-16.

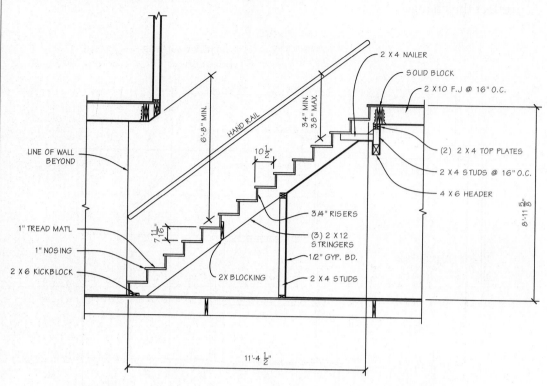

17. Use a word processor to write a report of at least 250 words explaining the importance of associative dimensioning. Site at least three examples from actual industry applications. Show at least four drawings illustrating your report.

AutoCAD Certified Associate Exam Practice

Answer the following questions. Write your answers on a separate sheet of paper.

1. Which of the following commands can you use to add a prefix to an existing linear dimension value? *Select all that apply.*
   A. **DDEDIT**
   B. **DIMEDIT New** option
   C. **DIMLINEAR Mtext** option
   D. **DIMTEDIT**
   E. **QDIM Edit** option

2. Which of the following commands allows you to space dimensions equally? *Select all that apply.*
   A. **DIMBASELINE**
   B. **DIMBREAK**
   C. **DIMORDINATE**
   D. **DIMSPACE**
   E. **MATCHPROP**

3. Which of the following commands allow you to adjust the location and alignment of dimension lines? *Select all that apply.*
   A. **DDEDIT**
   B. **DIMSPACE**
   C. **DIMTEDIT**
   D. **QDIM**
   E. **STRETCH**

AutoCAD Certified Professional Exam Practice

Follow the instructions in each problem. Write your answers on a separate sheet of paper.

1. **Navigate to this chapter on the companion website and open CPE-20align.dwg.** Use the appropriate command to align leaders 1 and 3 horizontally with leader 2, as shown. Use **Ortho** to ensure that the balloon alignment is exactly horizontal. What are the coordinates of the balloon grip on leader 1?

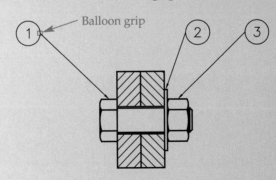

Chapter 20   Editing Dimensions

2. **Navigate to this chapter on the companion website and open CPE-20distribute.dwg.** Use the appropriate command to distribute the four leaders equally between Point A and Point B. What are the coordinates of the balloon grip of leader 4?

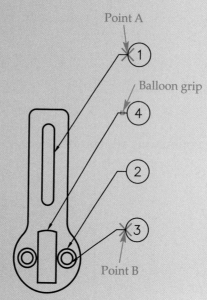

Chapter 21
Tables

Learning Objectives

After completing this chapter, you will be able to:

✓ Create and modify table styles.
✓ Insert tables into a drawing.
✓ Edit tables.
✓ Create formulas in table cells to perform calculations.

This chapter describes how to create *tables*, which are common elements on technical drawings. Examples of table applications include bills of materials, parts lists, schedules, legends, tabular and chart dimensioning, revision history and status blocks, and other *associated lists*. **Figure 21-1** shows an example of a schedule and a parts list, and highlights the features of a table.

> **table:** An arrangement of rows and columns that organize data to make it easier to read.

> **associated list:** The ASME term describing tables added or related to engineering drawings.

Table Styles

A *table style* presets many table characteristics. Create a table style for each different table appearance or function. For example, to draw a parts list, use a table style preset to the standard format of a parts list. To draw a wire list, use a different table style preset to the standard format of a wire list. Add table styles to drawing templates for repeated use. Avoid adjusting table formats independently of the table style assigned to the table.

> **table style:** A saved collection of table settings, including direction, text appearance, and margin spacing.

Table Style Dialog Box

Create, modify, and delete table styles using the **Table Style** dialog box. See **Figure 21-2**. The **Styles** list box displays existing table styles. The Standard table style is the default. To make a table style current, double-click the style name; right-click the name and select **Set current**, or pick the name and select the **Set current** button. Below the **Styles** list box is a drop-down list that you can use to filter the number of table styles displayed in the **Table Style** dialog box. Pick the **All Styles** option to show all table styles in the file, or pick the **Styles** in use option to show only the current style and styles used in the drawing.

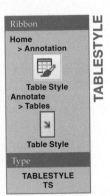

Ribbon

Home
> Annotation

Table Style

Annotate
> Tables

Table Style

Type

TABLESTYLE
TS

TABLESTYLE

629

Figure 21-1.
A—An example of a window schedule added to an architectural floor plan to identify windows and related information. This table uses a down direction. B—An example of a parts list added to a mechanical assembly drawing to identify assembly components. This table uses an up direction.

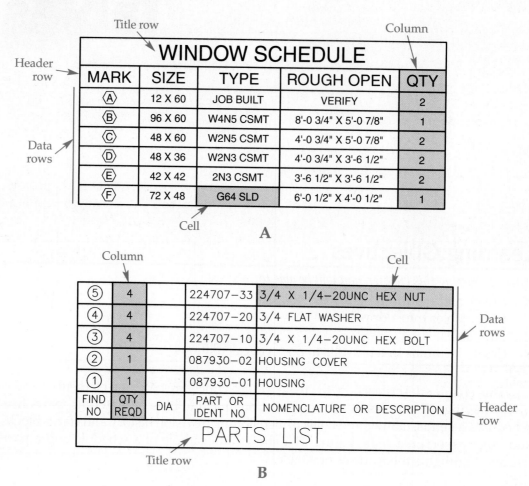

Figure 21-2.
The **Table Style** dialog box.

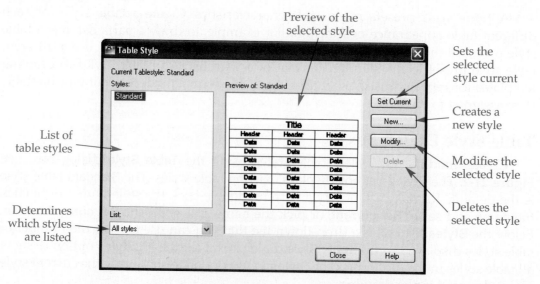

NOTE: You can also open the **Table Style** dialog box from the **Insert Table** dialog box, described later in this chapter, by picking the **Launch the Table Style dialog** button.

Creating New Table Styles

To create a new table style, select an existing table style from the **Styles** list box to use as a base for formatting the new table style. Then pick the **New...** button to open the **Create New Table Style** dialog box. See **Figure 21-3**. You can base the new table style on the formatting of a different table style by selecting from the **Start With** drop-down list. Notice that Copy of followed by the name of the existing style appears in the **New Style Name** text box. Replace the default name with a more descriptive name, such as Parts List, Parts List No Heading, or Door Schedule.

Table style names can have up to 255 characters, including uppercase and lowercase letters, numbers, dashes (–), underlines (_), and dollar signs ($). After typing the table style name, pick the **Continue** button to open the **New Table Style** dialog box and adjust table style settings. See **Figure 21-4**. Pick the **OK** button to apply changes and close the **New Table Style** dialog box. Pick the **Close** button to exit the **Table Style** dialog box.

Figure 21-3.
In the **Create New Table Style** dialog box, specify the name of the new table style and the existing style to copy as a basis for the new style.

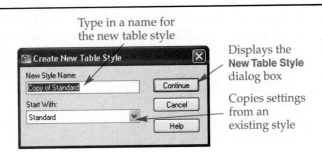

Figure 21-4.
Use the **New Table Style** dialog box to specify the formatting properties of a new table style. This figure shows adjusting the **Data** cell style.

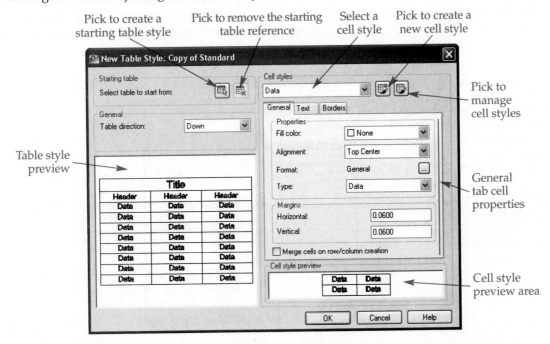

Record the names and details about the table styles you create and keep this information in a log for future reference.

Table Direction

The **Table direction** setting in the **General** area of the **New Table Style** dialog box determines the placement of title and header rows and the order of data rows. Select **Down** from the drop-down list to place data rows below the title and header rows. Select **Up** to place data rows above the title and header rows. See **Figure 21-1**.

Cell Styles

cell styles: Styles that allow you to assign specific formatting to data, header, and title row cells.

Use the **Cell Styles** area of the **New Table Style** dialog box to control *cell styles*. Pick the **Data**, **Header**, or **Title** option from the **Cell Styles** drop-down list to display the properties corresponding to the selected row. **Figure 21-4** shows the **Data** cell style selected. Set cell formatting properties using the **General**, **Text**, and **Borders** tabs. The options in these tabs are the same for adjusting data, header, and title cell styles.

General Tab Settings

The **General** tab, shown in **Figure 21-4**, allows you to set general table characteristics. The **Fill color** drop-down provides options for filling cells. The default **None** setting does not fill cells with a color, and is appropriate for most drafting applications. The drawing window color determines the on-screen table display. Pick a color from the drop-down list to fill cells with the color. Fill cells with color to highlight or organize table information.

The **Alignment** drop-down list specifies text justification within the cell. The **Format** setting shows the current cell format, which is General by default. Pick the ellipsis (...) button to access the **Table Cell Format** dialog box. See **Figure 21-5**. The **Data Type** area lists options for formatting the selected table cell. Pick the appropriate format, such as **Text** or **Currency**, to access options for adjusting the format characteristics. Different options are available depending on the selected format.

Use the **Type** drop-down list of the **New Table Style** dialog box to specify the cell data type. Pick the **Data** option to define a data cell type. Choose the **Label** option if the cell is a label type, such as a column heading or the table title. The **Margins** area provides a **Horizontal** and **Vertical** text box for controlling the horizontal and vertical

Figure 21-5.
Many different data types are available to format a table cell. For example, select the **Currency** data type to enable the cell to recognize values as currency and format values appropriately.

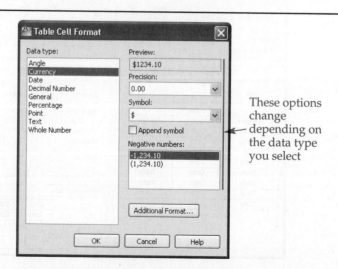

These options change depending on the data type you select

space between cell content and borders. The default varies depending on the current units. For example, for decimal units the default is .06" (1.5 mm).

Pick the **Merge cells on row/column creation** check box to merge the row of cells together to form a single cell. This box is checked by default for the Title cell style. The title cell provides an example of when it is suitable to merge cells. The title applies to the entire table, or to each column.

Text Tab Settings

The **Text** tab, shown in **Figure 21-6**, allows you to set text characteristics for the selected cell style. The **Text style** drop-down list displays the text styles found in the current drawing. Select a style or pick the ellipsis (...) button to the right of the drop-down list to open the **Text Style** dialog box to create or modify a text style.

Use the **Text height** text box to specify the text height. The default varies depending on the current units and cell style, such as .18" (4.5 mm) for data and header and .25" (6.3 mm) for title when the drawing specifies decimal units. The **Text height** text box is inactive if you assign a text height other than 0 to the text style. Use the **Text color** drop-down list to set the text color. The **Text angle** text box controls the rotation angle of text within the table cell. **Figure 21-7** shows an example of a 90° text angle applied to the **Header** cell style.

Borders Tab Settings

The **Borders** tab, shown in **Figure 21-8**, allows you to control the border display and characteristics for the selected cell style. Use the **Lineweight** drop-down list to assign a unique lineweight to cell borders. Use the **Linetype** drop-down list to assign a unique linetype to cell borders. As when creating layers, you must load linetypes before you can apply them to borders. Use the **Color** drop-down list to set the cell border color.

Pick the **Double line** check box to add another line around the default single-line border style. The **Spacing** edit box is available when you check **Double line**, allowing you to enter the distance between the double lines. The default spacing varies depending on the current units, such as .045 (1.125 mm) when the drawing uses decimal units.

Figure 21-6.
The **Text** tab in the **New Table Style** dialog box allows you to set text properties.

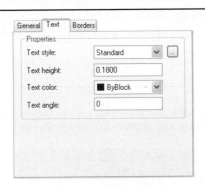

Figure 21-7.
In this table, a 90° text angle has been applied to the header cell style. This table is an example of a room schedule added to an architectural floor plan to list the characteristics of floor areas.

ROOM SCHEDULE					
NUMBER	NAME	LENGTH	WIDTH	HEIGHT	AREA
1	BEDROOM 1	11'-0"	10'-0"	9'-0"	110 SQ. FT.
2	BEDROOM 2	10'-0"	11'-0"	9'-0"	110 SQ. FT.
3	MASTER BEDROOM	12'-0"	14'-0"	9'-0"	168 SQ. FT.
4	LIVING ROOM	12'-0"	16'-0"	9'-0"	192 SQ. FT.
5	DINING ROOM	11'-0"	12'-0"	9'-0"	132 SQ. FT.
6	KITCHEN	11'-0"	10'-0"	9'-0"	110 SQ. FT.

Figure 21-8.
The **Borders** tab in
the **New Table Style**
dialog box allows
you to set cell border
properties.

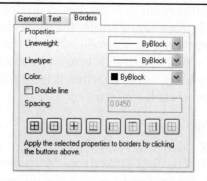

The **Border** buttons control how the **Lineweight, Linetype, Color**, and **Double line** properties apply to cell borders. From right to left, the options are **All Borders, Outside Borders, Inside Borders, Bottom Border, Left Border, Top Border, Right Border**, and **No Borders**. After you set border properties, select or deselect the buttons as needed. See **Figure 21-9**.

AutoCAD treats a table like a block. Chapter 24 explains blocks and provides complete information on using ByBlock, ByLayer, or absolute color, lineweight, and linetype with blocks. For now, as long as you assign a specific layer to a table, and do not change the table properties to absolute values, the default **ByBlock** cell properties are acceptable.

Creating Cell Styles

The default **Data, Header**, and **Title** cell styles are adequate for typical table applications. However, you can develop additional cell styles to increase the flexibility and options for creating tables. For example, you could create a cell style called **Data Yellow** that is the same as the **Data** cell style but fills cells with a yellow color. Then when you draw a table, you can choose the **Data** or the **Data Yellow** cell style, depending on the application.

Figure 21-9.
Border options available for table cells. This figure shows applying border options to data cell borders. The thin lines are shown for reference only.

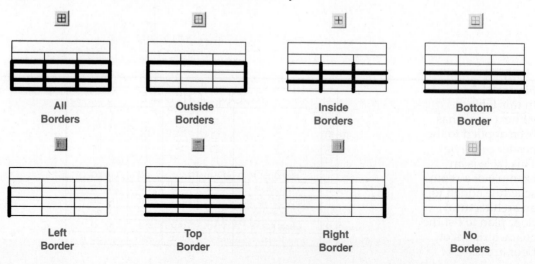

AutoCAD and Its Applications—Basics

To create a new cell style, select an existing cell style from the **Cell Styles** area drop-down list to use as a base for formatting the new cell style. Then pick the **Create new cell style...** button from the **Cell Styles** area, or select **Create new cell style...** from the **Cell Styles** area drop-down list to display the **Create New Cell Style** dialog box. Type a name for the new cell style in the **New Style Name** text box. You can base the new cell style on the formatting of a different cell style by selecting from the **Start With** drop-down list.

Use the **Manage Cell Styles** dialog box, shown in Figure 21-10, to create, rename, and delete cell styles. To access this dialog box, pick the **Manage Cell Style dialog...** button from the **Cell Styles** area, or select **Manage cell styles...** from the **Cell Styles** area drop-down list.

> The **Preview** and **Cell style preview** areas of the **New Table Style** dialog box allow you to see how the selected table style characteristics appear in a table. This provides a convenient way to observe changes made to a table style, without creating a table.

Exercise 21-1

Complete the exercise on the companion website.
www.g-wlearning.com/CAD

Starting Table Styles

One technique for creating a table is to use a starting table style to base a new table on an existing table. You can consider a starting table style to be a table template that includes preset table properties and specific rows, columns, and data entries. Using a starting table style is much like copying a complete table and editing the table as needed. A starting table style can save time if you often prepare similar tables. For example, use a starting table style of a standard parts list to create a parts lists quickly. Another example is using a starting table style of a finished door schedule to add a similar door schedule to a plan that contains most of the same doors.

A starting table style references the characteristics of an existing table, including the number of columns and rows and the table direction. Other table style characteristics, such as text style, are set according to the selected base table style. As a result, it is usually most appropriate to create a new starting table style using a base table style

Figure 21-10.
Use the **Manage Cell Styles** dialog box to create new cell styles and to rename and delete existing cell styles.

Right-click to access **New**, **Rename**, and **Delete** options from the shortcut menu

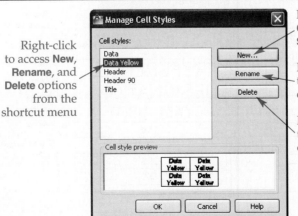

Pick to display the **Create New Cell Style** dialog box

Pick to rename the selected cell style

Pick to delete the selected cell style

that is the same as that used to draw the reference table. For example, if you create a door schedule using a table style named Door Schedule, you should base the new starting table style on the Door Schedule table style.

To create a starting table style, pick the **Select table to start from** button in the **Starting table** area. Then pick a border line of the existing table to reference. The preview displays the selected table and the table style settings of the base table style. See **Figure 21-11**. Modify the table direction and cell style options using the **General** and **Cell Styles** areas. Pick the **Remove Table** button to remove the table reference from the table style. Pick the **Start from Table style** insertion option, described later in this chapter, to add a table to a drawing using a starting table style.

Changing, Renaming, and Deleting Table Styles

Select a table style from the **Styles** list box to edit. Then pick the **Modify** button to access the **Modify Table Style** dialog box, which is the same as the **New Table Style** dialog box. If you make changes to a table style, such as merging cells, all existing table objects assigned the modified table style are updated. Use a different table style with different characteristics when appropriate.

To rename a table style using the **Table Style** dialog box, slowly double-click on the name or right-click on the name and select **Rename**. To delete a table style using the **Table Style** dialog box, right-click on the name and choose **Delete**, or pick the style and select the **Delete** button. You cannot delete a table style that is assigned to table objects. To delete a style that is in use, assign a different style to the tables that reference the style. You cannot delete or rename the Standard style.

RENAME

Type

RENAME

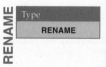 You can also rename styles using the **Rename** dialog box. Select **Table styles** in the **Named Objects** list to rename the style.

Figure 21-11.
Creating a starting table style that references an existing table. The table in this example is a list of reference designations, identifying last used components on an electrical schematic.

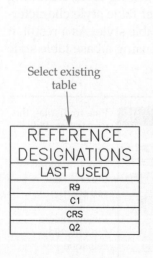

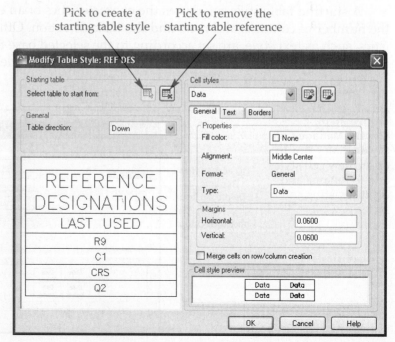

Select existing table

Pick to create a starting table style

Pick to remove the starting table reference

Setting a Table Style Current

Set a table style current using the **Table Style** dialog box by double-clicking the style in the **Styles** list box, right-clicking on the style and selecting **Set current**, or picking the style and selecting the **Set current** button. To set a table style current without opening the **Table Style** dialog box, use the **Table Style** flyout on the expanded **Annotation** panel of the **Home** ribbon tab, or the **Tables** panel of the **Annotate** ribbon tab. See Figure 21-12.

PROFESSIONAL TIP

You can import table styles from existing drawings using **DesignCenter**. See Chapter 5 for more information about using **DesignCenter** to reuse drawing content.

Inserting Tables

The **TABLE** command allows you to insert an empty table with a specified number of rows and columns. After you insert the table, you can type text and insert content into the table cells. The **TABLE** command also provides other methods for inserting tables, such as beginning a table using a starting table style, forming a table from data in an existing Microsoft® Excel spreadsheet or CSV (comma-separated) file, and creating a table by referencing AutoCAD data. Access the **TABLE** command to display the **Insert Table** dialog box. See Figure 21-13.

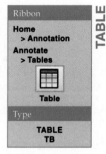

Ribbon
Home
> Annotation
Annotate
> Tables

Table

Type
TABLE
TB

Placing an Empty Table

To place an empty table, select a table style from the **Table Style** drop-down list, or pick the ellipsis (...) button to create or modify a table style. Next, pick the **Start from empty table** radio button in the **Insert options** area to create an empty table. The preview area shows a representation of a table using the current table style, but it does not adjust to column and row settings. Pick the **Specify insertion point** radio button in the **Insertion Behavior** area to create a table using the values in the **Column & row settings** area, and then select a single point to place the table in the drawing.

Figure 21-12.
The fastest way to set a style current is to use one of the drop-down lists on the ribbon.

Access from the
Home Ribbon Tab

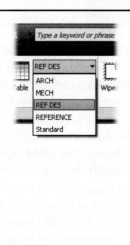

Access from the
Annotate Ribbon Tab

Figure 21-13.
The **Insert Table** dialog box, shown with the **Start from empty table** insert option selected.

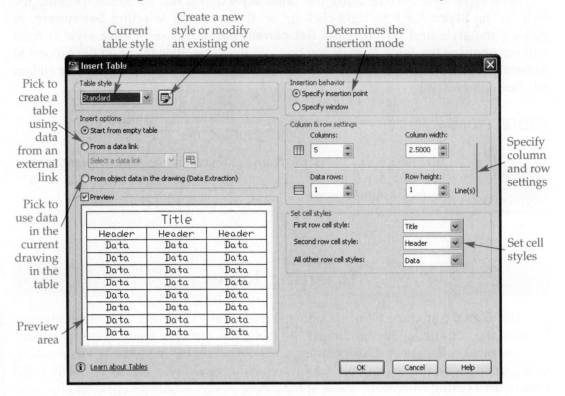

Create a new
Current   style or modify   Determines the
table style   an existing one   insertion mode

Pick to create a table using data from an external link

Pick to use data in the current drawing in the table

Preview area

Specify column and row settings

Set cell styles

Use the **Columns** text box to specify the total number of table columns. Choose a **Column width** value to establish the initial width of each column. To help avoid initial crowding, enter a column width larger than necessary and then resize the columns later. Use the **Data rows** text box to specify the total number of data rows. Choose a **Row height** value to establish the initial height of each row based on the number of lines typed and the margin settings assigned to the table style. When you pick the **OK** button, AutoCAD prompts you to specify the table insertion point. See **Figure 21-14A**.

When you use the **Specify insertion point** option to place a table, the cursor attaches to the table based on the table style direction.

Pick the **Specify window** radio button in the **Insertion Behavior** area to create a table that fits within a rectangular area you create. The radio buttons in the **Column & row settings** area control which column and row settings are active. To set a fixed number of columns, choose the **Columns** radio button. The selected table width determines the width of each column. The alternative is to pick the **Column width** radio button to set a fixed column width. The selected table width determines the total number of columns.

Choose the **Data rows** radio button to set a fixed number of rows. The selected table height determines the height of each data row. The alternative is to pick the **Row height** radio button to set a fixed row height. The selected table height determines the total number of rows. When you pick the **OK** button, AutoCAD prompts you to select the upper-left and lower-right corners of the table. AutoCAD uses the fixed **Column & row settings** values to adjust the table to fit the window. See **Figure 21-14B**.

Figure 21-14.
Two ways to insert an empty table. A—Using the **Specify insertion point** radio button to select a single insertion point to create a table with three fixed columns and five fixed data rows. B—Using the **Specify window** radio button to specify an area using two pick points, creating a table with three columns and five data rows that adjust according to the window size.

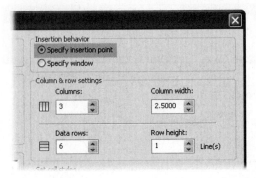

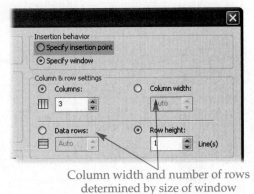

Column width and number of rows determined by size of window

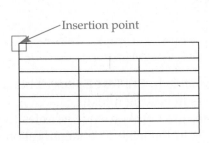

Insertion point

First point of window

Second point of window

A

B

You specify the total number of data rows when constructing a table. By default, the table will also include a header row and title row, depending on cell style settings. The current table style text height and cell margin settings determine the default value for row height. For example, enter a row height of 1 if you plan to have a single line of text in each cell.

PROFESSIONAL TIP

You can add, delete, and fully adjust rows and columns as needed. Therefore, it is not critical that you enter the exact number and size of columns and rows before inserting a table.

Exercise 21-2

Complete the exercise on the companion website.
www.g-wlearning.com/CAD

Adding Cell Content

When you insert a table, the Text Editor contextual ribbon tab appears, with the text editor cursor in the title cell ready for typing. See **Figure 21-15**. Typing in a cell is

Figure 21-15.
Use the **Text Editor** ribbon tab to add and modify table cell text. A blinking cursor, dashed border, and light gray background indicate the active cell.

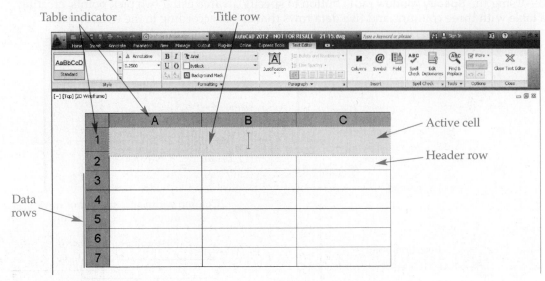

like typing mtext. The options and settings available in the **Text Editor** ribbon tab and shortcut menu function the same in table cells as in editing mtext.

A dashed line around the border and a light gray background indicates the active cell. The *table indicator* identifies individual cells in the table. The identification system helps you to assign formulas to table cells for calculation purposes, as described later in this chapter. Before adding content to a cell, adjust the text settings in the **Text Editor** ribbon, if needed. Remember, however, that making changes to some text characteristics overrides the settings specified in the text style or table style, which is often not appropriate.

Hold [Alt] and press [Enter] to insert a return within the cell. When you finish entering text in the active cell, press [Tab] to move to the next cell. Hold [Shift] and press [Tab] to move the cursor backward (to the left or up) and make the previous cell active. Press [Enter] to make the cell directly below the current cell active, or exit the text editor if the cursor is at the last cell. You can also use the arrow keys to navigate table cells.

When you finish typing, exit the text editor system using the **Close Text Editor** option, or pick outside of the table editor. You can also press [Esc] twice. The easiest way to reopen the text editor to make changes to text in a cell is to double-click in the cell. Figure 21-16 shows a finished table.

table indicator:
The grid of letters and numbers that identify individual cells in a table.

Ribbon
Text Editor > Close

Close Text Editor

Exercise 21-3

Complete the exercise on the companion website.
www.g-wlearning.com/CAD

Using a Starting Table Style

To place a table using a starting table style, select a starting table style from the **Table Style** drop-down list or select the ellipsis (...) button to create or modify a starting table style. The **Start from Table Style** radio button becomes activated in the **Insert** options area. See Figure 21-17. The preview area shows a preview of the parent table with the current table style settings and table options.

The **Specify insertion point** option is the only method for inserting a table using a starting table style. However, you can add columns and rows to the table using the **Additional columns** and **Additional rows** text boxes. You can also select the items from

Figure 21-16.
A completed parts
list table added
to a mechanical
assembly drawing
to identify assembly
components.

PARTS LIST

FIND NO	QTY REQD	DIA	PART OR IDENT NO	NOMENCLATURE OR DESCRIPTION
1	1		100-TBL-001	TABLE
2	2		100-CLJA-45	CLAMP JAWS
3	2		202-PIV-32	PIVOT ARM
4	1		340-HAND-06	LOCKING HANDLE
5	3		38009561	1/4-28UNF NYLOCK NUT
6	2		567-ADJR-98	ADJUSTING ROD HANDLE
7	2		786-SPG-64	1/8 SPRING PIN

Figure 21-17.
You can use the **Insert Table** dialog box to create a new table using a starting table style.

Pick to
create a new
table using
a starting
table style

**Start from
Table Style**
becomes
active

Preview of
parent table

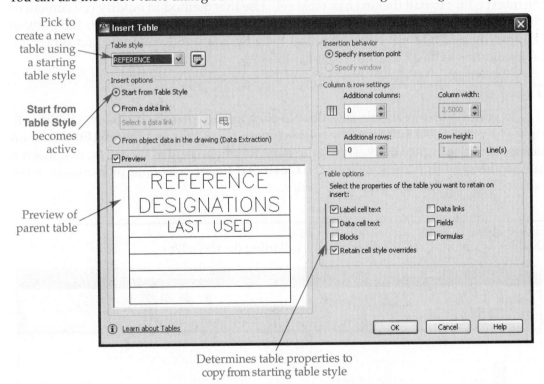

Determines table properties to
copy from starting table style

the parent table to include in the new table using the check boxes in the **Table options** area. For example, pick the **Data cell text** check box to create a new table that contains all the text added to the data cells of the parent table.

Pick the **OK** button and specify the insertion point of the table. The **Text Editor** ribbon tab appears with the text editor cursor in the title cell, ready for adding new content or editing existing values. Exit the text editor when you are finished. Figure 21-18 shows a table created by referencing an existing starting table style, with two additional rows.

Exercise 21-4

Complete the exercise on the companion website.
www.g-wlearning.com/CAD

Figure 21-18.
Use a starting table style to create a new or modified table quickly. This example shows creating a new table of electronic reference designations using the starting table style created in Figure 21-11.

Existing table style used to form a starting table style
REFERENCE DESIGNATIONS
LAST USED
R9
C1
CRS
Q2

In the new table, all table options are retained and two rows are added
REFERENCE DESIGNATIONS
LAST USED
R9
C1
CRS
Q2
T4
R6

Editing Tables

AutoCAD provides several options to edit existing tables. One option is to re-enter the mtext editor to edit the text in a table cell. Use this method to modify cell content or change text format. Another option is to make changes to the table layout. Table layout changes include adding, removing, and resizing rows and columns, and wrapping table columns to break a large table into sections.

Text Editor

To edit the text in a table cell, double-click inside the cell or pick inside the cell, right-click, and select **Edit Text**. The cell becomes a text editor and the **Text Editor** ribbon tab appears. See **Figure 21-19**. This is the same format presented when you first insert a table. When you finish editing, use the **Close Text Editor** option, or pick outside of the table editor. You can also press [Esc] twice.

Figure 21-19.
Double-click inside a cell to edit text in the cell using the text editor.

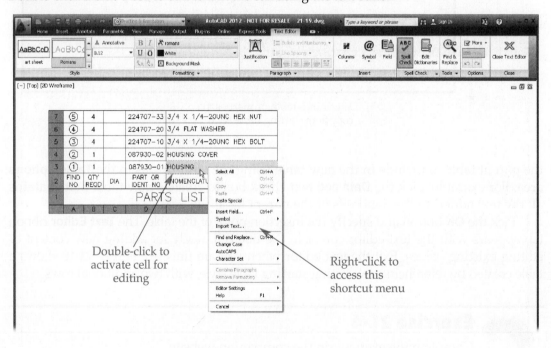

Double-click to activate cell for editing

Right-click to access this shortcut menu

Exercise 21-5

Complete the exercise on the companion website.
www.g-wlearning.com/CAD

Table Cell Editor

You can access several table layout settings by picking (single-clicking) inside a cell to make the cell active and display the **Table Cell** contextual ribbon tab. The highlighted cell includes grips. The **Table Cell** ribbon tab contains options for adjusting table and individual cell layout. You can access many of the same options in **Table Cell** ribbon tab and Windows Clipboard functions from the shortcut menu that appears when you right-click away from the ribbon. See **Figure 21-20**. Use an option shown in **Figure 21-21** to select multiple cells and apply changes to all the cells at once.

Make sure you pick completely inside of the cell. If you accidentally select one of the cell borders, the entire table becomes the selected object. Editing table layout by selecting a cell border is described later in this chapter.

Auto-Fill

The most effective method for copying the content of one cell to multiple cells is to use the *auto-fill* function. **Figure 21-22A** shows the basic steps to use auto-fill. Pick inside the cell that contains the content to copy. Select the diamond-shaped auto-fill grip, and then right-click to choose an auto-fill option, if necessary. Finally, move the cursor to the last cell to fill and pick inside the cell.

auto-fill: A table function that fills selected cells based on the contents of another cell.

Figure 21-20.
Pick inside a cell to access several options for modifying the table layout.

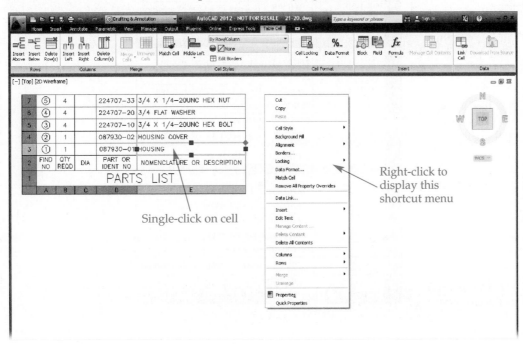

Figure 21-21.
Selecting multiple cells in a table to edit. A—Using the pick-and-drag method. B—Picking a range of cells using [Shift]. C—Selecting a row, column, or the entire table.

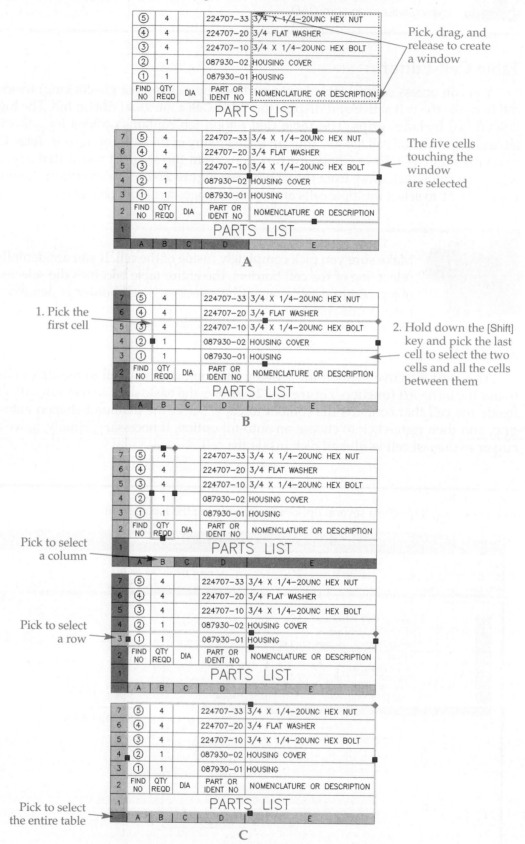

Pick, drag, and release to create a window

The five cells touching the window are selected

A

1. Pick the first cell

2. Hold down the [Shift] key and pick the last cell to select the two cells and all the cells between them

B

Pick to select a column

Pick to select a row

Pick to select the entire table

C

The **Fill Series** option fills cells with the content of the selected cell and applies format overrides. This option also automatically increases or decreases values of certain data types, such as dates and whole numbers, as the fill occurs. See **Figure 21-22B**. The **Fill Series Without Formatting** option fills cells with the content of the selected cell, but does not include format overrides.

Figure 21-22.
A—Using the auto-fill function to copy cell content to multiple cells. B—Using the **Fill Series** option to fill sequential whole number data. C—Using the **Copy Cells** option to copy data that would fill sequentially when using the **Fill Series** option. D—The completed door schedule uses auto-fill seven times to fill cells quickly. A door schedule typically appears on an architectural floor plan to identify doors and related information.

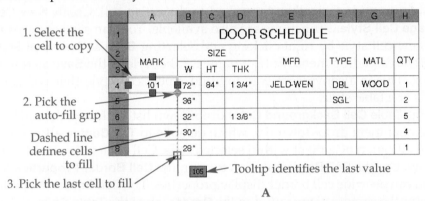

1. Select the cell to copy
2. Pick the auto-fill grip
Dashed line defines cells to fill
Tooltip identifies the last value
3. Pick the last cell to fill

A

Reference cell
Using **Fill Series** to increase the value of whole number data as cells fill

B

Reference cell
Using **Copy Cells** creates a copy of the reference cell without applying a sequence

C

DOOR SCHEDULE							
MARK	SIZE			MFR	TYPE	MATL	QTY
	W	HT	THK				
101	72"	84"	1 3/4"	JELD-WEN	DBL	WOOD	1
102	36"	84"	1 3/4"	JELD-WEN	SGL	WOOD	2
103	32"	84"	1 3/8"	JELD-WEN	SGL	WOOD	5
104	30"	84"	1 3/8"	JELD-WEN	SGL	WOOD	4
105	28"	84"	1 3/8"	JELD-WEN	SGL	WOOD	1

D

The **Copy Cells** option copies the content of the selected cell and applies format overrides, but creates a static cell copy that does not adjust data values. See **Figure 21-22C**. The **Copy Cells Without Formatting** option copies the content of the selected cell, but does not include any format overrides. The **Fill Formatting Only** option fills the cells only with format overrides applied to the selected cell, allowing you to enter cell content manually. **Figure 21-22D** shows a table finished using multiple auto-fills.

Modifying Cell Style

Ribbon
Table Cell
> Cell Styles
> By Row/
Column

The table style and the text style assigned to the table style determine the appearance of most cell properties. However, you can override the cell style assigned to specific cells if necessary. Cell style options are available from the **Cell Styles** panel of the **Table Cell** ribbon tab and from the shortcut menu. Use the **Table Cell Style** drop-down list to override the cell style applied to the active cell. Create **New Cell Style...** and **Manage Cell Style...** functions are also available. You can save changes made to a cell as a new cell style by right-clicking and selecting **Save as New Cell Style...** from the **Cell Style** cascading menu. Enter a name for the style in the **Save as New Cell Style** dialog box. This is a convenient way to build a new cell style that you can apply to other cells or a table style.

Ribbon
Table Cell
> Cell Styles
Background Color

Use the **Table Cell Background Color** drop-down list to change the cell background color. The alignment drop-down list, which defaults to **Top Center**, allows you to override the justification of content within selected cells. Cell content is placed in relation to cell borders. Use the **Cell Borders** option to open the **Cell Border Properties** dialog box, where you can override cell border display properties. The **Cell Border Properties** dialog box contains the same options found in the **Border** tab of the **Table Style** dialog box.

Ribbon
Table Cell
> Cell Styles
Top Center

To copy format settings from one cell to another, select the cell with the settings to copy, and then access the **Match Cell** option. Pick a destination cell, which receives the properties. Select another cell to match, or right-click to exit.

Adjusting Cell Format

Ribbon
Table Cell
> Cell Styles
Edit Borders

Cell format options are available from the **Cell Format** panel of the ribbon and from the shortcut menu. The **Cell Locking** feature provides options for locking cells to protect data from unintended or inappropriate changes. The locked icon appears when you move the cursor over a locked cell. The **Unlocked** option unlocks the cell so you can make changes to cell content and format. The **Content Locked** option locks only the content of the cell, allowing you to make changes to cell format. The **Format Locked** option locks only the cell format, allowing you to make changes to cell content. The **Content and Format Locked** option locks the cell against changes in content and format.

Ribbon
Table Cell
> Cell Styles
Match Cell

Override the data format of cells, if necessary, using the **Data Format...** function. A **Custom Table Cell Format...** option is available for access to the **Table Cell Data Format** dialog box from the **Table Style** dialog box.

Ribbon
Table Cell
> Cell Format
> Cell Locking

Ribbon
Table Cell
> Cell Format
> Data Format

PROFESSIONAL TIP

The current table style controls most cell style and format properties. If you plan to make significant changes to cell properties, modify the table style or create a new style.

Right-click and select **Remove All Property Overrides** to restore selected cells to their original properties defined in the selected table style.

Inserting Fields and Blocks

In addition to text, table cells can contain fields, formulas, and blocks. To insert these items, use the options available from the **Insert** panel of the **Table Cell** ribbon tab, or from the **Insert** cascading menu of the shortcut menu. You will learn about formulas later in this chapter. Choose the **Field...** option to insert a field into a table cell using the same **Field** dialog box available for creating mtext and text objects. You can also insert fields into a cell using the text editor after double-clicking inside the cell to activate it.

Blocks are AutoCAD symbols that are useful in tables for applications such as creating a legend, displaying a view or flag note symbol in a parts list, and adding tags to a schedule. This chapter briefly describes options for inserting a block in a table. You will learn about blocks later in this textbook. Select the **Block...** option to insert a block into a table cell using the **Insert a Block in a Table Cell** dialog box. **Figure 21-23** briefly describes the options available in this dialog box. Double-click on a block to reopen the **Insert a Block in a Table Cell** dialog box to make changes. A cell can contain both text and blocks.

Adding and Resizing Columns and Rows

You can add, delete, and resize existing columns and rows as needed. Select a single cell or a group of cells to add, depending on the requirement. To delete and resize columns and rows, you do not need to select entire columns and rows. The following options are available from the **Columns** panel of the **Table Cell** ribbon tab or the **Columns** cascading menu of the shortcut menu:

- **Insert Left**. Add a new column to the left of the selection.
- **Insert Right**. Add a new column to the right of the selection.
- **Delete**. Eliminate entire columns.
- **Size Equally**. Size multiple columns to the width of the widest column.

The following options are available from the **Rows** panel of the **Table Cell** ribbon tab or the **Rows** cascading menu of the shortcut menu:

- **Insert Above**. Add a new row above the selection.
- **Insert Below**. Add a new row below the selection.
- **Delete**. Delete entire rows.
- **Size Equally**. Size multiple rows to the height of the tallest row.

Figure 21-23.
Options in the **Insert a Block in a Table Cell** dialog box.

Feature	Description
Name	Used to choose the block from a drop-down list of the blocks stored in the current drawing.
Browse	Displays the **Select Drawing File** dialog box, where a drawing file can be selected and inserted into the table cell as a block.
AutoFit	Scales the block automatically to fit inside the cell.
Scale	Sets the block insertion scale. For example, a value of 2 inserts the block at twice its original size. A value of .5 inserts the block at half its created size. The **Scale** option is not available if the **AutoFit** check box is checked.
Rotation angle	Rotates the block to the specified angle.
Overall cell alignment	Determines the justification of the block in the cell and overrides the current cell alignment setting.

To insert a new row at the bottom of a table that uses a **Down** direction, position the cursor in the lower-right cell and press [Tab]. To insert a new row at the top of a table that uses an **Up** direction, position the cursor in the upper-right cell and press [Tab].

PROFESSIONAL TIP

You can also adjust column and row size using grips. Grip boxes appear in the middle of cell border lines. To resize a column or row, select a grip, move the crosshairs, and pick.

Merging Cells

Ribbon

Table Cell
> Merge

Merge Cells

Merging allows you to combine adjacent cells. The default title cell style is an example of merged cells. Merge options are available from the **Merge** panel of the ribbon and from the **Merge** cascading menu in the shortcut menu. Select the cells to merge and select the appropriate **Merge cells** option. Select the **All** option to merge all cells into one cell. The **By Row** and **By Column** options allow you to merge cells in multiple rows or columns without removing the horizontal or vertical borders. Use the **Unmerge Cells** option to separate merged cells back to individual cells.

The **Delete All Contents** option deletes the contents in the selected cell. You can accomplish the same task by picking a cell and pressing [Delete].

Exercise 21-6

Complete the exercise on the companion website.
www.g-wlearning.com/CAD

Picking a Cell Edge to Edit Table Layout

Pick the edge, or border, of a cell to access additional methods for adjusting table layout. The display includes the table indicator grid, grips you can use to adjust row height and column width, and the table break function. Once you pick a cell border, right-click to display the shortcut menu shown in **Figure 21-24**.

Adjusting Table Style

Select a table style from the **Table Style** cascading submenu to apply a different table style to the selected table. Pick the **Set as Table in Current Table Style** option to create a starting table style based on the selected table and the current table style. This is a convenient technique for creating a starting table style without opening the **Table Style** dialog box. If the selected table was drawn using a starting table style, selecting the **Set as Table in Current Table Style** redefines the starting table. To save modifications made to the table as a new table style, pick the **Save as New Table Style...** option and enter a name for the style in the **Save as New Table Style** dialog box.

Figure 21-24.
Pick a cell border to access several additional options for modifying table layout.

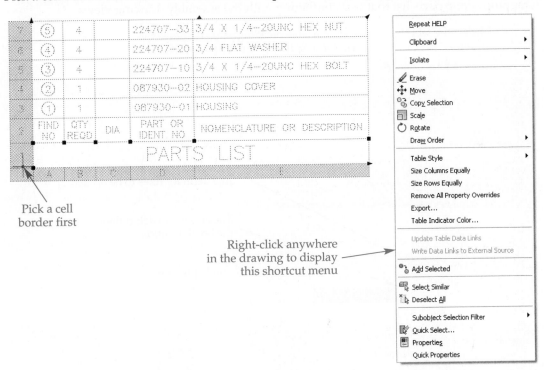

Pick a cell border first

Right-click anywhere in the drawing to display this shortcut menu

Resizing Columns and Rows

Use the grip boxes or arrowheads at the corners of columns and rows to adjust column and row size. To resize a column or row, select a grip box, move the crosshairs, and pick. Use the arrowhead grips to increase or decrease row height and/or column width uniformly.

The **Size Columns Equally** option sizes all columns to the same width. AutoCAD divides the total width of the table evenly among the columns. The **Size Rows Equally** option sizes all rows to match the height of the tallest row in the table.

> The default grip box stretches the column or row without changing the size of the table. Hold [Ctrl] while stretching to increase or decrease the size of the table with the column or row resize.

Table Breaks

Use the table break function to break a table into separate sections while maintaining a single table object. Breaking a table is common when it is necessary to fit a long table in a specific area or on a certain size sheet. The table breaking grip is located midway between the sides of the table at the top or the bottom of the table, depending on the table direction. See **Figure 21-25**.

To break a table, select the table breaking grip and move the crosshairs into the table to display a preview of the table sections and a vector line. The crosshairs determines the location of the break. The closer to the table title and headers you move the crosshairs, the more sections you create, as shown in the table preview. When the preview of the table looks correct, pick the location to form the table breaks.

Figure 21-25.
The procedure for breaking, or wrapping, a table into sections. This example shows wrapping a long parts list that was conflicting with the assembly drawing views.

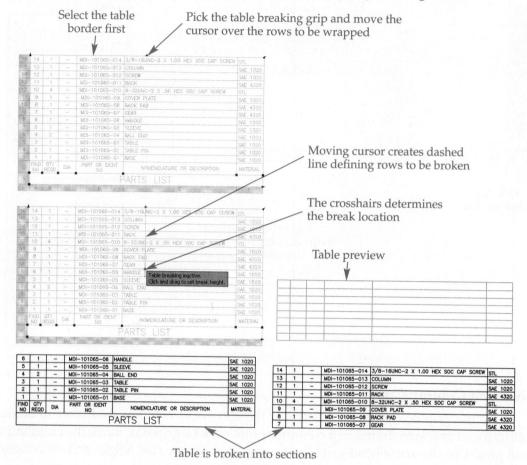

Select the table border first

Pick the table breaking grip and move the cursor over the rows to be wrapped

Moving cursor creates dashed line defining rows to be broken

The crosshairs determines the break location

Table preview

Table is broken into sections

After you add table breaks, several options become available from the **Properties** palette for adjusting the table sections, as explained later in this chapter.

Additional Table Layout Options

The following additional table options are available from the shortcut menu:

- **Remove All Property Overrides**. Restores the original properties to the table, defined according to the selected table style.
- **Export**. Exports the table as a CSV file.
- **Table Indicator Color...**. Allows you to change the color of the table indicator shown when you pick inside a cell.

Exercise 21-7

Complete the exercise on the companion website.
www.g-wlearning.com/CAD

Table Properties

The **Properties** palette displays certain table properties depending on whether you select inside a cell or pick a cell edge. See **Figure 21-26**. The **Cell** and **Content** categories appear when you select inside a cell, and allow you to adjust the properties of the selected cell. The **Table** and **Table Breaks** categories appear when you select a cell edge, and include common table settings and table break properties. The **Quick Properties** panel also lists certain table-specific properties.

PROFESSIONAL TIP

If the height of table rows is taller than desired, or if rows become unequal in height, enter a very small value in the **Table height** row of the **Table** category to return all rows to the smallest height possible based on the margin spacing between cell content and cell borders.

Several useful options are available for adjusting table breaks in the **Table Breaks** category of the **Properties** palette. The **Enabled** option toggles between the broken and unbroken display. The **Yes** value appears when you create table breaks and enable breaking. Pick **No** to return the table to an unbroken display. The **Direction** option defines direction of broken table flow, or wrap. The default **Right** option wraps the table to the right. Select **Left** to wrap the table to the left, or pick **Up** to wrap the table above.

The **Repeat top labels** option repeats cells that use a Label cell type at the beginning of each table section. Typically, the title cell and header cells use a Label cell type. Choose **Yes** to add the title and header cells to the wrapped table sections. The **Repeat bottom labels** option repeats cells that use a Label cell type at the end of each section. The **Manual**

Figure 21-26.
Table properties in the **Properties** palette. A—The properties displayed when you select inside a cell. B—The properties displayed when you pick a cell edge.

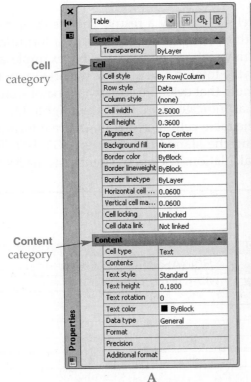

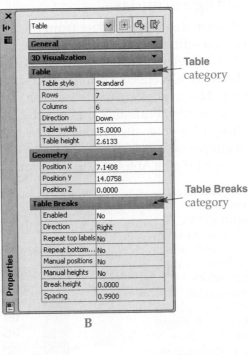

positions option allows you to move table sections independently while maintaining the table as a single object. When you select **No**, table sections move as a group.

The **Manual height** option adds a table-breaking grip to each section, which allows you to adjust the number of rows in each section independently and add additional breaks. When you select **No**, the table-breaking grip appears at the original section only and controls the number of breaks. The **Break height** text box allows you to define the height of each table section. The selected height determines the number of sections. The **Spacing** text box allows you to define the spacing between table sections. A value of 0 places the sections together.

Exercise 21-8

Complete the exercise on the companion website.
www.g-wlearning.com/CAD

Calculating Values in Tables

formulas:
Mathematical expressions that allow you to perform calculations within table cells.

Performing calculations on table data using *formulas* is a common requirement. Examples include calculating and showing the total number of parts in a parts list, the total cost of items in a bill of materials, or the total glazing area in a window schedule. Formulas calculate operations based on numeric data in table cells. AutoCAD allows you to write formulas for sums, averages, counts, and other mathematical functions.

The table indicator grid that appears when you edit a table cell provides an identification system for cells. Letters identify columns, and numbers identify rows. Use the combination of column letter and row number to describe cells in formulas. For example, C6 identifies the cell located in Column C, Row 6. See **Figure 21-27**.

Creating Formulas

A formula evaluates data from other cells to display a result. The result updates when you edit data in the table linked to the formula. You often write a formula to calculate all cells in a row or column, or in a range of continuous cells. For example, add the values of all cells in a column to display the total in a new cell at the bottom of the column. However, you can also write a formula that evaluates cells that do not share a common border. Formulas are field objects, as indicated by a light gray highlight.

You must enter the proper syntax in the table cell to create an accurate formula. An example of a complete expression using the standard syntax is =(C3+D4). The equal sign (=) tells AutoCAD to perform a calculation. The open parenthesis marks the

Figure 21-27.
Column letters and row numbers identify table cells. The table indicator grid provides a reference for identifying each cell. This example shows a room schedule, which often includes square footage calculations for various areas.

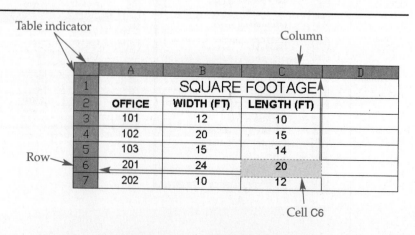

AutoCAD and Its Applications—Basics

beginning of the expression. Type C3+D4 to tell AutoCAD to add the value of cell C3 to the value of cell D4. The closing parenthesis marks the end of the expression.

When identifying a cell in an expression, you must enter the letter before the number. For example, you cannot enter 3C to designate the cell C3. Common operators used for mathematical functions include + for addition, – for subtraction, * for multiplication, / for division, and ^ for exponentiation. If you enter an incorrect expression or an expression evaluating cells without numeric data, AutoCAD displays the pound character (#) to indicate an error.

Parentheses are unnecessary in some expressions, but other expressions cannot be calculated without them. It is good practice to use parentheses in all expressions.

Input a formula using the text editor. Figure 21-28 shows an example of typing the multiplication formula =(B3*C3) in cell D3. The result appears when you close the text editor of the cell, such as when you move to another cell.

You can use grouped expressions in formulas by enclosing expression sets in parentheses. For example, the expression =(E1+F1)*E2 multiplies the sum of E1 and F1 by E2. Another example: =(E1+F1)*(E2+F2)/G6 multiplies the sum of E1 and F1 by the sum of E2 and F2 and divides the product by G6.

Sum, Average, and Count Formulas

An alternative to entering formulas manually is to select a formula from AutoCAD to calculate the sum, average, or count of a range of cells. Pick inside a cell where the calculation is to occur, and then choose a formula using the **Formula** drop-down list, which is also available by right-clicking and selecting from the **Formula** cascading menu on the **Insert** cascading menu.

Ribbon

Table Cell
> Insert
    > Formula

Figure 21-28.
Entering a multiplication formula to calculate the square footage of a room. A—Type the expression in the table cell using the correct syntax. B—The calculation occurs when you close the text editor of the cell.

Expression

	A	B	C	D
1	SQUARE FOOTAGE			
2	OFFICE	WIDTH (FT)	LENGTH (FT)	
3	101	12	10	=(B3*C3)
4	102	20	15	
5	103	15	14	
6	201	24	20	
7	202	10	12	

A

Result

SQUARE FOOTAGE			
OFFICE	WIDTH (FT)	LENGTH (FT)	
101	12	10	120
102	20	15	
103	15	14	
201	24	20	
202	10	12	

B

Select the **Sum** option and then use window selection to add the values of selected cells. See **Figure 21-29A**. The range, or window, can include cells from several columns and rows. Be sure to select all of the cells to be included in the calculation. When you select the second point, the expression appears in the cell. Notice in **Figure 21-29B** that the resulting expression is =Sum(D3:D7). This formula specifies that the selected cell is equal to the sum of cells D3 through D7. The colon (:) indicates the range of cells for the calculation. **Figure 21-29** shows an example of calculating the square footage for each office, and then calculating the total square footage.

The **Average** option creates a formula that calculates the average value of selected cells. The average is the sum of the selected cells divided by the number of cells selected. The **Count** option creates a formula that counts the number of selected cells. The count only includes cells that contain a value.

You can type sum, average, and count formulas directly into a cell without using the supplied formulas. If you calculate a value over a range of cells, use the colon symbol (:) to designate the range. You can also write an expression that evaluates individual cells instead of a range. The cells do not have to share a common border. To write an expression using nonadjacent cells, use a comma to separate the cell names. For example, to average cells D1, D3, and D6, type =Average(D1,D3,D6).

You can include a range of cells and individual cells in the same expression. For example, to count cells A1 through B10 in addition to cells C4 and C6, enter =Count(A1:B10,C4,C6). **Figure 21-30** shows examples of sum, average, and count formulas.

When using architectural units in a drawing, you can type the foot (') and inch (") symbols in table cells for use in values and formulas. When you use the foot symbol for a cell value, a formula in another cell automatically converts the resulting value to inches and feet.

Figure 21-29.
Creating a sum formula in a table cell. A—Pick a cell to hold the formula and select a range of cells for the formula by windowing around the cells. B—After you pick the second point of the window, the formula displays in the cell.

Pick first point

	A	B	C	D
1		SQUARE FOOTAGE		
2	OFFICE	WIDTH (FT)	LENGTH (FT)	
3	101	12	10	120
4	102	20	15	300
5	103	15	14	210
6	201	24	20	480
7	202	10	12	120
8				
9	OFFICE COUNT			TOTAL SQ FT
10				

Pick second point

Select cell first

A

	A	B	C	D
1		SQUARE FOOTAGE		
2	OFFICE	WIDTH (FT)	LENGTH (FT)	
3	101	12	10	120
4	102	20	15	300
5	103	15	14	210
6	201	24	20	480
7	202	10	12	120
8				
9	OFFICE COUNT			TOTAL SQ FT
10				=Sum(D3:D7)

Sum formula

B

Figure 21-30.
Examples of sum, average, and count formulas and their resulting values.

	A	B	C	D
1	SQUARE FOOTAGE			
2	OFFICE	WIDTH (FT)	LENGTH (FT)	SQ FT
3	101	12	10	120
4	102	20	15	300
5	103	15	14	210
6	201	24	20	480
7	202	10	12	120
8				
9	OFFICE COUNT	AVERAGE SQ FT PER ROOM		TOTAL SQ FT
10	5	246		1230

=Count(A3:A7)   =Average(D3:D7)   =Sum(D3:D7)

Other Formula Options

The **Formula** drop-down list and **Insert** cascading menu contain additional options for writing table formulas. The **Cell** option allows you to select a cell from a different table in the drawing to insert the contents in the current cell. You can then use the cell value in a formula. Select the **Equation** option to place an equal sign (=) in the current cell. You can then type the expression manually.

You can use the **Field** command to insert and edit table cell formulas. Select **Formula** from the **Field names** list in the **Field** dialog box to display buttons for creating sum, average, and count formulas. You can also select a cell value from a different table as a starting point. Select table cells in the drawing area to define the formula. You can use the **Formula** text box in the **Field** dialog box to add to or edit the formula. Unit format options are also available.

Exercise 21-9

Complete the exercise on the companion website.
www.g-wlearning.com/CAD

Supplemental Material *Linking a Table to Excel Data*
For information about using existing data entered in a Microsoft® Excel spreadsheet or a CSV file to create an AutoCAD table, go to the companion website (www.g-wlearning.com/CAD), select this chapter, and select **Linking a Table to Excel Data**.

Supplemental Material *Extracting Table Data*
For information about using existing AutoCAD text to create a table, go to the companion website (www.g-wlearning.com/CAD), select this chapter, and select **Extracting Table Data**.

Template Development
Chapter 21

Adding Table Styles

For detailed instructions on adding table styles to each drawing template, go to the companion website (www.g-wlearning.com/CAD), select this chapter, and select **Template Development**.

Chapter Review

Answer the following questions. Write your answers on a separate sheet of paper or complete the electronic chapter review on the companion website.
www.g-wlearning.com/CAD

1. What is the purpose of creating a table style?
2. Briefly describe the procedure for creating a table style based on an existing table style.
3. What is the purpose of the **Alignment** setting in the **New Table Style** dialog box?
4. Which setting would you adjust in the **New Table Style** dialog box to increase the spacing between the text and the top of the cell?
5. How can creating a new table using a starting table style save time?
6. How can you make a table style current without opening the **Table Style** dialog box?
7. List two ways to open the **Insert Table** dialog box.
8. Describe the two ways to insert an empty table and explain how the methods differ.
9. By default, what two types of rows are at the top of a table?
10. What ribbon tab opens when you insert a table?
11. If you finish typing in a cell and want to move to the next cell in the same row, what two keyboard keys can you use?
12. List two ways to make a cell active for editing.
13. Explain how to insert a field into a table cell.
14. How can you insert a new row at the bottom of a table?
15. How are table cells identified in formulas?
16. Write the table cell formula that adds the value of C3 and the value of D4.
17. What is the function of the colon (:) in the formula =Sum(D3:D7)?
18. What is the difference between a sum formula and a count formula?
19. Write the table cell formula that averages the values of cells D1, D3, and D6.
20. Explain how to write a formula that calculates a function for cells that do not share common borders.

Drawing Problems

Start AutoCAD if it is not already started. Start a new drawing using an appropriate template of your choice. The template should include layers, text styles, and table styles when necessary for drawing the given objects. Add layers, text styles, and table styles as needed. Draw all objects using appropriate layers, text styles, table styles, justification, and format. Follow the specific instructions for each problem. Use only drawing commands and techniques you have already learned. Use your own judgment and approximate dimensions when necessary.

Note: Some of the problems in this chapter are built on problems from previous chapters. If you have not yet completed those problems, complete them now.

▼ Basic

1. Create the electronic schematic reference designations list shown. Save the drawing as P21-1.

REFERENCE DESIGNATIONS	
LAST USED	DATE
R9	1/30/2010
C1	1/30/2010
CRS	1/30/2010
Q2	1/30/2010

2. Open P18-7 and save the file as P21-2. The P21-2 file should be active. Complete the drawing of the chain link by adding the table shown.

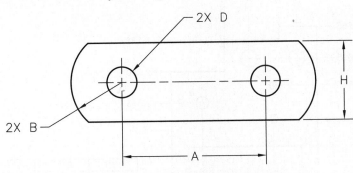

CHAIN NO.	A	B	D	H
SST1000	2.1	.75	.44	1.125
SST1001	3.0	.90	.48	1.525
SST1002	3.0	1.17	.95	2.125

3. Open P18-9 and save the file as P21-3. The P21-3 file should be active. Complete the drawing of the chassis spacer by adding the table shown.

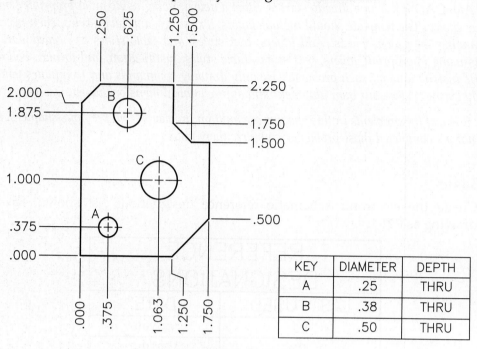

KEY	DIAMETER	DEPTH
A	.25	THRU
B	.38	THRU
C	.50	THRU

Title: Chassis Spacer
Material: .008 Aluminum

4. Open P18-10 and save the file as P21-4. The P21-4 file should be active. Complete the drawing of the chassis by adding the table shown.

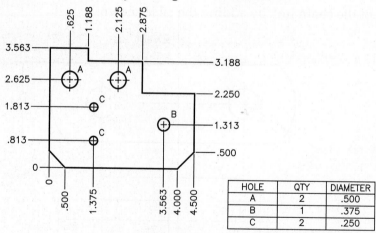

HOLE	QTY	DIAMETER
A	2	.500
B	1	.375
C	2	.250

Title: Chassis
Material: Aluminum .100 THK

5. Open P18-17 and save the file as P21-5. The P21-5 file should be active. Complete the drawing of the part view by adding the table shown.

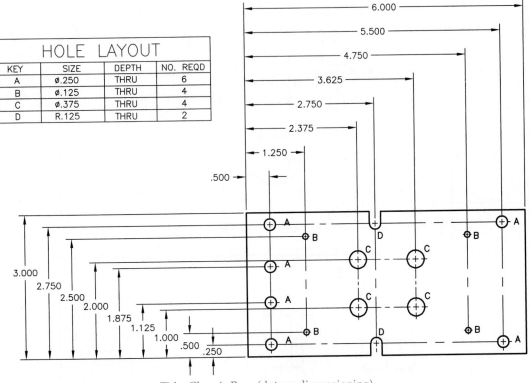

HOLE LAYOUT			
KEY	SIZE	DEPTH	NO. REQD
A	ø.250	THRU	6
B	ø.125	THRU	4
C	ø.375	THRU	4
D	R.125	THRU	2

Title: Chassis Base (datum dimensioning)
Material: 12 gage Aluminum

6. Open P18-18 and save the file as P21-6. The P21-6 file should be active. Complete the drawing of the part view by adding the table shown.

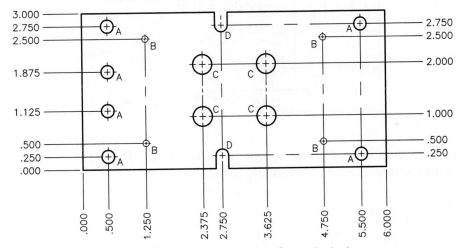

HOLE LAYOUT			
KEY	SIZE	DEPTH	NO. REQD
A	ø.250	THRU	6
B	ø.125	THRU	4
C	ø.375	THRU	4
D	R.125	THRU	2

Title: Chassis Base (arrowless dimensioning)
Material: 12 gage Aluminum

7. Open P18-20 and save the file as P21-7. The P21-7 file should be active. Complete the drawing of the part views by adding the table shown.

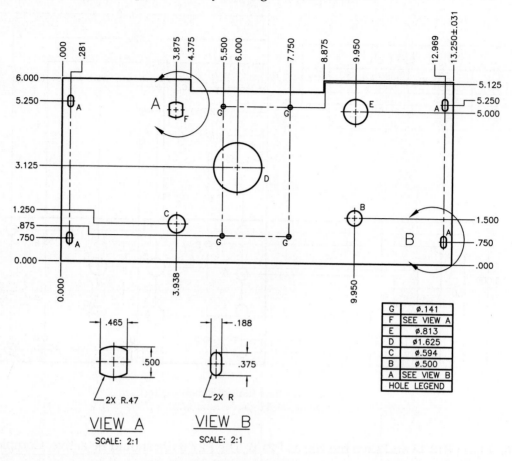

G	⌀.141
F	SEE VIEW A
E	⌀.813
D	⌀1.625
C	⌀.594
B	⌀.500
A	SEE VIEW B
HOLE LEGEND	

VIEW A
SCALE: 2:1

VIEW B
SCALE: 2:1

8. Create the parts list shown in Figure 21-1B. Do not draw the circles around the find numbers. Save the drawing as P21-8.

9. Create the parts list shown. Save the drawing as P21-9.

FIND NO	QTY REQD	DIA	PART OR IDENT NO	NOMENCLATURE OR DESCRIPTION
5	1	–	210014–29	1/2–12UNC HEX NUT
4	1	–	320014–33	1/2 FLAT WASHER
3	2	–	632043–43	7/16 EXTERNAL SNAP RING
2	2	–	255010–41	1/4–20UNC WING NUT
1	2	–	803010–11	3/4 X 1/4–20UNC BOLT

PARTS LIST

10. Create the window schedule shown in Figure 21-1A. Do not draw the polygons around the marks. Save the drawing as P21-10.

▼ Intermediate

11. Open **P18-19** and save the file as **P21-11**. The P21-11 file should be active. Complete the drawing of the part view by adding the table shown.

HOLE LAYOUT				
KEY	X	Y	SIZE	TOL
A1	.500	2.750	ø.250	±.002
A2	.500	1.875	ø.250	±.002
A3	.500	1.125	ø.250	±.002
A4	.500	.250	ø.250	±.002
A5	5.500	2.750	ø.250	±.002
A6	5.500	.250	ø.250	±.002
B1	1.250	2.500	ø.125	±.001
B2	1.250	.500	ø.125	±.001
B3	4.750	2.500	ø.125	±.001
B4	4.750	.500	ø.125	±.001
C1	2.375	2.000	ø.375	±.005
C2	2.375	1.000	ø.375	±.005
C3	3.625	2.000	ø.375	±.005
C4	3.625	1.000	ø.375	±.005
D1	2.750	2.750	R.125	±.002
D2	2.750	.250	R.125	±.002

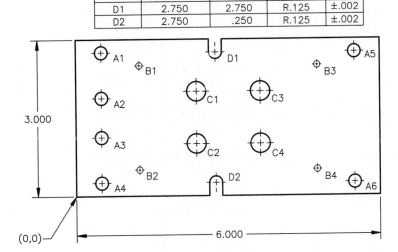

Title: Chassis Base (arrowless tabular dimensioning)
Material: 12 gage Aluminum

12. Convert the drawing shown to a drawing with the holes located using the **DIMORDINATE** command based on the X and Y coordinates given in the table. Create a table with columns for Hole (identification), Quantity, Description, and Depth (Z axis). Save the drawing as P21-12.

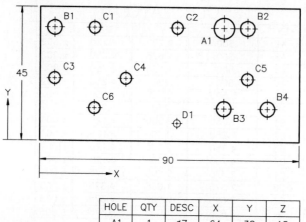

HOLE	QTY	DESC	X	Y	Z
A1	1	⌀7	64	38	18
B1	1	⌀5	5	38	THRU
B2	1	⌀5	72	38	THRU
B3	1	⌀5	64	11	THRU
B4	1	⌀5	79	11	THRU
C1	1	⌀4	19	38	THRU
C2	1	⌀4	48	38	THRU
C3	1	⌀4	5	21	THRU
C4	1	⌀4	30	21	THRU
C5	1	⌀4	72	21	THRU
C6	1	⌀4	19	11	THRU
D1	1	⌀2.5	48	6	THRU

Title: Base
Material: Bronze

13. Draw the finish schedule shown. Use the **Symbol** option of the text editor to locate and insert bullet symbols as shown. Save the drawing as P21-13.

INTERIOR FINISH SCHEDULE

ROOM	FLOOR				WALLS				CEIL		
	CARPET	VINYL	TILE	HARDWOOD	PAINT	PAPER	TEXTURE	SPRAY	SMOOTH	BROCADE	PAINT
FOYER			•		•		•			•	•
KITCHEN			•			•		•	•		•
DINING				•	•		•			•	•
FAMILY	•				•		•			•	•
LIVING	•				•		•			•	•
MASTER BED	•				•		•			•	•
MASTER BATH			•			•		•	•		•
BATH 2		•				•		•	•		•
BED 2	•				•		•			•	•
BED 3	•				•		•			•	•
UTILITY		•				•		•	•		

14. Create the door schedule from the architect's sketch exactly as shown. Use a text style assigned the CountryBlueprint font. Save the drawing as P21-14. Then make the following changes:
    - Edit the text style to use the SansSerif font.
    - Replace the abbreviation SYM. with the word MARK (no period).
    - Remove the period after the abbreviation QTY.
    - Remove the periods from the abbreviation S.C.
    - Replace the abbreviation RP. with the words RAISED PANEL.
    - Remove the period after the abbreviation SLDG.

    Resave the drawing.

DOOR SCHEDULE			
SYM.	SIZE	TYPE	QTY.
1	36x80	S.C. RP. METAL INSULATED	1
2	36x80	S.C. FLUSH METAL INSULATED	2
3	32x80	S.C. SELF CLOSING	2
4	32x80	HOLLOW CORE	5
5	30x80	HOLLOW CORE	5
6	30x80	POCKET SLDG.	2

15. Create the window schedule from the architect's sketch exactly as shown. Use a text style assigned the CityBlueprint font. Save the drawing as P21-15. Then make the following changes:
    - Edit the text style to use the SansSerif font.
    - Replace the abbreviation SYM. with the word MARK (no period).
    - Remove the period after the abbreviations QTY, SLDG, and AWN.
    - Replace the abbreviation CSM. with the abbreviation CSMT (no period).

    Resave the drawing.

WINDOW SCHEDULE				
SYM.	SIZE	MODEL	ROUGH OPEN	QTY.
A	12x60	JOB BUILT	VERIFY	2
B	96x60	W4N5 CSM.	8'-0 3/4" x 5'-0 7/8"	1
C	48x60	W2N5 CSM.	4'-0 3/4" x 5'-0 7/8"	2
D	48x36	W2N3 CSM.	4'-0 3/4" x 3'-6 1/2"	2
E	42x42	2N3 CSM.	3'-6 1/2" x 3'-6 1/2"	2
F	72x48	G6A SLDG.	6'-0 1/2" x 4'-0 1/2"	1
G	60x42	G536 SLDG.	5'-0 1/2" x 3'-6 1/2"	4
H	48x42	G436 SLDG.	4'-0 1/2" x 3'-6 1/2"	1
J	48x24	A41 AWN.	4'-0 1/2" x 2'-0 7/8"	3

16. Create the door schedule shown using a table break. Save the drawing as P21-16. Make the following changes:
    - Change the schedule so items 11, 12, 13, and 14 continue directly below SYMBOL items 1 through 10, with only one DOOR SCHEDULE heading at the top.
    - Replace the word SYMBOL with the word MARK (no period).
    - Replace the word QUANTITY with the abbreviation QTY (no period).
    - Replace the abbreviation S.C.R.P. with the words SOLID CORE RAISED PANEL (no period).
    - Replace the abbreviation S.C.- with the words SOLID CORE (no hyphen, but include a space).
    - Replace the abbreviation H.C. with the words HOLLOW CORE (no period).
    - Replace WOOD FRAME-TEMP. SLDG GL. with the abbreviations TMPD SGD (no period).

    Resave the drawing.

DOOR SCHEDULE				DOOR SCHEDULE			
SYMBOL	SIZE	MODEL	QUANTITY	SYMBOL	SIZE	MODEL	QUANTITY
1	3'-0" X 6'-8"	S.C. R.P. METAL INSULATED	1	11	4'-0" X 6'-8"	BI-FOLD	1
2	3'-0" X 6'-8"	S.C.-FLUSH-METAL INSULATED	2	12	2'-0" X 6'-0"	SHATTER PROOF	1
3	2'-8" X 6'-8"	S.C.-SELF CLOSING	2	13	6'-0" X 6'-8"	WOOD FRAME-TEMP. SLDG GL.	1
4	2'-8" X 6'-8"	H.C.	5	14	9'-0" X 7'-0"	OVERHEAD GARAGE	2
5	2'-6" X 6'-8"	H.C.	3				
6	2'-6" X 6'-8"	POCKET	2				
7	2'-4" X 6'-8"	POCKET	1				
9	5'-0" X 6'-0"	BI-PASS	2				
10	3'-0" X 6'-8"	BI-FOLD	1				

▼ Advanced

17. Open P21-9 and save the file as P21-17. The P21-17 file should be active. Make the following changes:
    - Change PARTS LIST to PURCHASE PARTS LIST.
    - Change Part Number 803010-11 as follows: Find No: 7, Name: HEX HD, Description: $\frac{1}{4}$-20UNC-2 X $\frac{3}{4}$ BOLT.
    - Change Part Number 255010-41 as follows: Find No: 11, Name: WING NUT, Description: $\frac{1}{4}$-20UNC.
    - Change Part Number 632043-43 as follows: Find No: 15, Name: SNAP RING, Description: $\emptyset \frac{7}{16}$ EXTERNAL.
    - Change Part Number 320014-33 as follows: Find No: 19, Name: WASHER, Description: $\emptyset \frac{1}{2}$ FLAT.
    - Change Part Number 210014-29 as follows: Find No: 21, Name: NUT, Description: $\frac{1}{2}$-12UNC-2 HEX.

    Resave the drawing.

18. Create a table of your own design and use at least six of the applications of calculating values in tables described in this chapter. Save the drawing as P21-18.

19. Access the companion website content for this chapter and review Supplement 21A, "Linking a Table to Excel Data." Use this information to create a table by linking to Microsoft® Excel data. To do this, create your own Excel spreadsheet or find an existing Excel spreadsheet containing suitable data. Link a table to the Excel data. Save the drawing as P21-19.

20. Access the companion website content for this chapter and review Supplement 21B, "Extracting Table Data." Use the description to create a table similar to the given examples and then extract the table data into an AutoCAD drawing. Save the drawing as P21-20.

21. Research the requirements for a set of working drawings for a mechanical assembly. Identify a mechanical assembly consisting of several parts and possibly subassemblies. Use the **TABLE** command to create a parts list identifying each component. Save the drawing as P21-21.

22. Research an example of a residential home design. Use the **TABLE** command to create a door and window schedule identifying all of the doors and windows in the home. Save the drawing as P21-22.

23. Research an example of an electronic wiring harness schematic. Use the **TABLE** command to create a wire list identifying all of the required wires. Save the drawing as P21-23.

24. Obtain a copy of leveling field notes prepared by a surveyor. Use the **TABLE** command to reproduce the notes and calculate unspecified values. Save the drawing as P21-24.

AutoCAD Certified Associate Exam Practice

Answer the following questions. Write your answers on a separate sheet of paper.

1. What is the meaning of the formula =((C3*D2)/4)? *Select the one item that best answers the question.*
   A. add the values of C3 and D2 and divide the result by 4
   B. multiply the values of both C3 and D2 by 4 and add the results
   C. multiply the value of C3 by the value of D2 and divide the result by 4
   D. divide the value of D2 by 4 and multiply the result by the value of C3

2. How can you remove existing breaks in a table? *Select all that apply.*
   A. select the table, right-click, and select **Remove All Property Overrides**
   B. move the table breaking grip down to the last row
   C. select the table breaking grip and press [Delete]
   D. use the **Properties** palette to set **Table Breaks Enabled** to **No**

3. In a table that uses the **Down** direction, what is the cell identification for a cell in the fourth row from the top, in the fifth column from the left? *Select the one item that best answers the question.*
   A. 4E
   B. 5D
   C. D4
   D. D5
   E. E4

AutoCAD Certified Professional Exam Practice

Follow the instructions for the following problem. Write your answer on a separate sheet of paper.

1. **Open CPE-21formula.dwg. This file is available on the companion website.**
   Perform the following tasks on the existing table:
   A. In the first data row of the AREA column, create a formula to calculate the area (length × width) of each room listed in the table.
   B. Use auto-fill to copy the formula to the remaining rows in the AREA column.
   C. Add a row at the bottom of the table.
   D. In the new bottom row, merge cells A11 through E11, set the alignment to **Top Right**, and enter the text TOTAL AREA.
   E. In the bottom-right cell, create a formula to calculate the total area of all the rooms.
   What is the total area of the house?

Chapter

Parametric Drafting 22

Learning Objectives

After completing this chapter, you will be able to:

✓ Explain parametric drafting processes and applications.
✓ Add and manage geometric constraints.
✓ Add and manage dimensional constraints.
✓ Adjust the form of dimensional constraints.

Parametric drafting tools allow you to assign *parameters*, or *constraints*, to objects. The parametric concept, also known as *intelligence*, provides a way to associate objects and limit design changes. You cannot change a constraint so that it conflicts with other parametric geometry. A database stores and allows you to manage all parameters. You typically use parametric tools with standard drafting practices to create a more interactive drawing.

Parametric Fundamentals

Parametric drafting can increase your ability to control every aspect of a drawing during and after the design and documentation process. Parametric tools can change the way you construct and edit geometry. However, in general, use parametric tools as a supplement to standard drafting practices and drawing aids. When used correctly, this technique allows you to produce accurate parametric drawings efficiently.

Understanding Constraints

Add parameters using *geometric constraints* and *dimensional constraints*. Well-defined constraints allow you to incorporate and preserve specific design intentions and increase revision efficiency. For example, if two holes through a part, drawn as circles, must always be the same size, use a geometric constraint to make the circles equal and add a dimensional constraint to size one of the circles. The size of both circles changes when you modify the dimensional constraint value. See **Figure 22-1**.

You must add constraints to make an object parametric. Dimensional constraints create parameters that direct object size and location. In contrast, an associative dimension is associated with an object, but it does not control object size or location. **Figure 22-2** shows an

parametric drafting: A form of drafting in which parameters and constraints drive object size and location to produce drawings with features that adapt to changes made to other features.

parameters (constraints): Geometric characteristics and dimensions that control the size, shape, and position of drawing geometry.

geometric constraints: Geometric characteristics applied to restrict the size or location of geometry.

dimensional constraints: Measurements that numerically control the size or location of geometry.

Figure 22-1.
An example of a parametric relationship. The dimensional constraint controls the size of both circles with the aid of an equal geometric constraint.

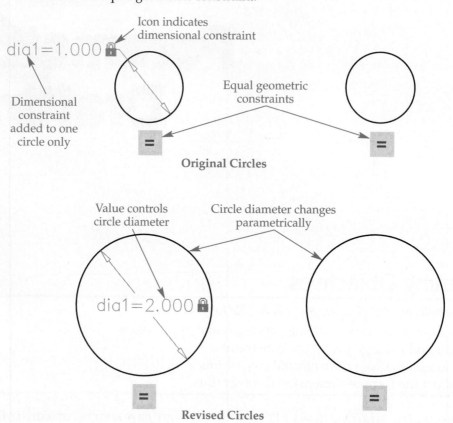

Original Circles

Revised Circles

example of a drawing that is *under-constrained*, *fully constrained*, and *over-constrained*. As you progress through the design process, you will often fully or almost fully constrain the drawing to ensure that the design is accurate. However, a message appears if you attempt to over-constrain the drawing. See **Figure 22-3.** AutoCAD does not allow you to over-constrain a drawing, as shown by the *reference dimension* in **Figure 22-2.**

Figure 22-4 shows an extreme example of constraining, for reference only. Study the figure to understand how constraints work, and how applying constraints differs from and compliments traditional drafting with AutoCAD. Typically, you should prepare initial objects as accurately as possible using standard commands and drawing aids. Add geometric constraints while you are drawing, or add them later to existing objects. Apply dimensional constraints after creating the geometry.

Parametric Applications

Parametric tools aid the design and revision process, place limits on geometry to preserve design intent, and help form geometric constructions. Consider using constraints to help maintain relationships between objects in a drawing, especially during the design process, when changes are often frequent. However, you must decide if the additional steps required to make a drawing parametric are appropriate and necessary for the application.

Product Design and Revision

Figure 22-5 shows an example of a front view of a spacer, well suited to parametric construction. First, as shown in **Figure 22-5A**, use standard commands to create the geometry accurately. Next, as shown in **Figure 22-5B**, use geometric and dimensional

Figure 22-2.
Levels of parametric constraint.

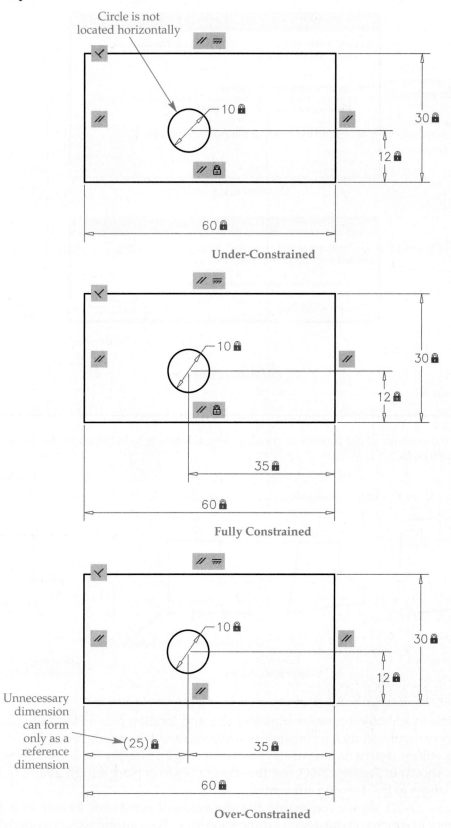

Circle is not
located horizontally

10

30

12

60

Under-Constrained

10

30

12

35

60

Fully Constrained

10

30

12

Unnecessary
dimension
can form
only as a
reference
dimension

(25)

35

60

Over-Constrained

Figure 22-3.
Error messages appear when you attempt to over-constrain objects.

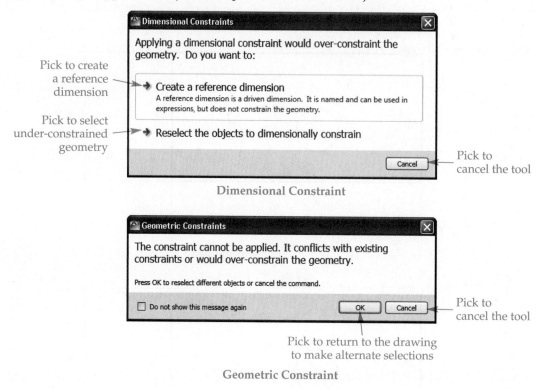

Pick to create a reference dimension

Pick to select under-constrained geometry

Pick to cancel the tool

Dimensional Constraint

Pick to cancel the tool

Pick to return to the drawing to make alternate selections

Geometric Constraint

Figure 22-4.
An extreme example of the process of constraining a drawing to help you understand how to apply constraints.

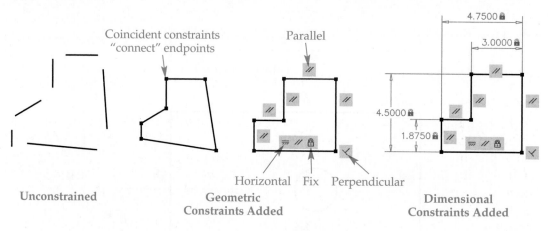

Coincident constraints "connect" endpoints

Parallel

Horizontal Fix Perpendicular

Unconstrained

Geometric Constraints Added

Dimensional Constraints Added

constraints to add object relationships and size and location parameters. You also have the option to apply geometric constraints while you are drawing the view. This example uses centerlines, drawn on a separate construction layer, to apply the correct constraints. Then, as shown in Figure 22-5C, use the constraints to explore design alternatives and make changes to the drawing efficiently.

Figure 22-5D shows converting the dimensional constraint format to a formal appearance to which you can assign a dimension style. You can still use converted dimensions to adjust geometry parametrically. This example shows converting all dimensions except the .250 radius and diameter dimensions. Using standard associative .250 radial and diameter dimensions allows you to add the 6X prefix and relocate the dimensions. You can then hide the constraint information to view the finished drawing.

AutoCAD and Its Applications—Basics

Figure 22-5.
The front view of this spacer is a good candidate for parametric drafting. A—Accurate view geometry constructed using standard AutoCAD practices. B—Adding geometric and dimensional constraints to constrain the drawing. C—Changing the values of a few dimensional constraints to update the entire drawing. D—Reusing dimensional constraints to help prepare a formal drawing.

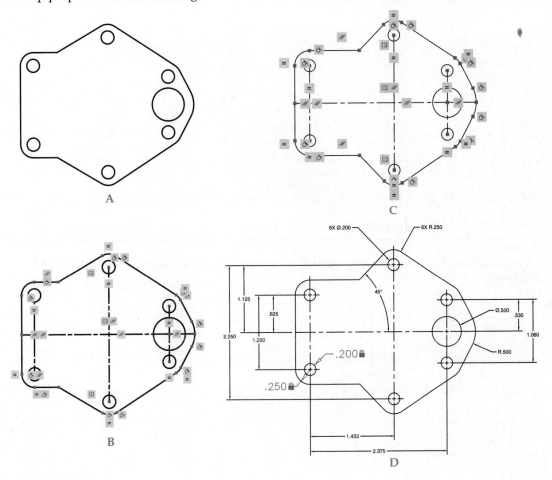

Geometric Construction

Use constraints to form geometric constructions when standard commands are inefficient or ineffective. For example, suppose you know that the angle of a line is 30°, and you know the line is tangent to a circle. However, you do not know the length of the line or the location of the line endpoints. One option is to position a 30° construction line, using the **Ang** option of the **XLINE** command, anywhere in the drawing. See **Figure 22-6A**. Use a **Tangent** geometric constraint to form a tangent relationship between the xline and circle. See **Figure 22-6B**. Then hide or delete the constraint if necessary. See **Figure 22-6C**.

Unsuitable Applications

You may find that parametric drafting is unsuitable or ineffective for some applications. For example, it may be unsuitable to add parameters to a drawing if the drawing is of a finalized product that will not require revision, or if you can easily modify drawing geometry without associating objects.

In addition, if your drawing includes a large number of objects, you may find it cumbersome to add the constraints required to form a fully intelligent drawing. For instance, you can use constraints to form all necessary relationships between objects in a floor plan. See **Figure 22-7**. In this example, constraints connect walls, specify walls as perpendicular or parallel, control wall thickness, position windows between

Figure 22-6.
A—A 30° xline placed near an associated circle. B—Using a **Tangent** geometric constraint to form a tangent construction. Notice the appropriate selection process. C—The final drawing.

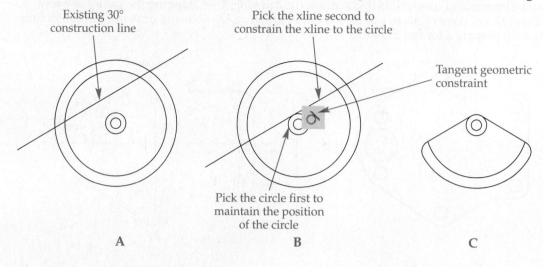

Existing 30° construction line

Pick the xline second to constrain the xline to the circle

Tangent geometric constraint

Pick the circle first to maintain the position of the circle

A

B

C

Figure 22-7.
An architectural floor plan usually includes too many objects to constrain effectively and efficiently.

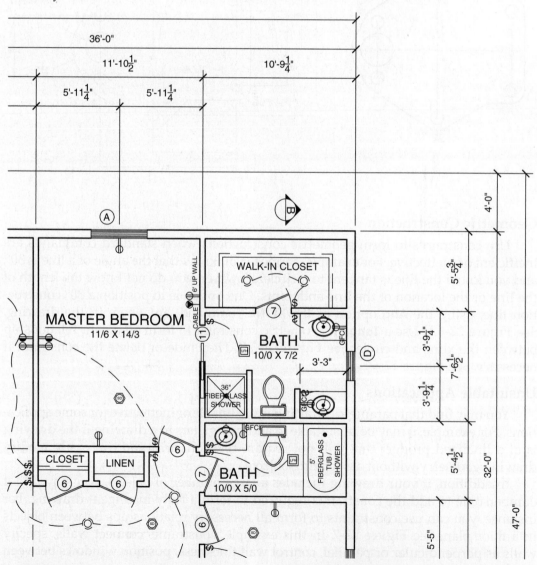

walls, locate sinks on vanities, and form many other parametric relationships. You can then adjust dimensional constraints as needed to update the drawing.

If you effectively constrain all objects shown in **Figure 22-7**, you have the option, for example, to change the 11'-10 1/2" dimensional constraint to increase the width of the master bedroom. The entire floor plan adjusts to the modified room size. Consider, however, what this process requires. You must constrain all wall endpoints; the points where doors and windows meet walls; the distance between walls and objects, such as cabinets, sinks, and fixtures; and form all other geometric and dimensional constraints.

NOTE

You can also use parametric tools when constructing blocks, as described later in this textbook.

PROFESSIONAL TIP

Constraints can also be used in multiview layout, to help maintain alignment between views.

Exercise 22-1

Complete the exercise on the companion website.
www.g-wlearning.com/CAD

Geometric Constraints

Geometric constraint tools allow you to add the geometric relationships required to build a parametric drawing. You typically add geometric constraints, or at least a portion of the necessary geometric constraints, before dimensional constraints to help preserve design intent. You can *infer* certain geometric constraints while drawing and editing, or manually apply geometric constraints to existing unconstrained geometry.

infer: Automatically detect and apply using logic.

Inferring constraints is the fastest way to add geometric relationships. However, often a combination of inferred and manually added geometric constraints is necessary. By default, a constraint-specific icon is visible when you infer constraints to indicate the presence of a geometric constraint. View, adjust, and remove geometric constraints as needed. **Figure 22-8** describes each geometric constraint.

PROFESSIONAL TIP

Placing too many geometric constraints can cause problems as you progress through the design process. Apply only the geometric constraints necessary to generate the required geometric constructions.

Figure 22-8.
Geometric constraints form geometric relationships between points and/or objects. Only some of the geometric constraints can be inferred.

Constraint	Icon	Inferable	Description
Horizontal	Object / Point-to-point	Yes	Horizontally aligns two points or a line, polyline, ellipse axis, mtext, or text object; positions geometry along the X axis on the default XY plane.
Vertical	Object / Point-to-point	Yes	Vertically aligns two points or a line, polyline, ellipse axis, mtext, or text object; positions geometry along the Y axis on the default XY plane.
Parallel	Object-to-object	Yes	Creates a parallel constraint between a line, polyline, ellipse axis, mtext, or text object with another line, polyline, ellipse axis, mtext, or text object.
Perpendicular	Object-to-object	Yes	Forms a perpendicular constraint between a line, polyline, ellipse axis, mtext, or text object with another line, polyline, ellipse axis, mtext, or text object.
Tangent	Object-to-object	Yes	Forms a tangent constraint between a circle, arc, or ellipse and a line, polyline, circle, arc, or ellipse.
Collinear	Object-to-object	No	Aligns a line, polyline, ellipse axis, mtext, or text object with another line, polyline, ellipse axis, mtext, or text object.
Concentric	Object-to-object	No	Constrains the center of a circle, arc, or ellipse to the center of another circle, arc, or ellipse.
Equal	Object-to-object	No	Sizes and locates an object in reference to another object.
Symmetric	Object-to-object / Point-to-point / Line of symmetry	No	Establishes symmetry between objects or points and a line, polyline, ellipse axis, mtext, or text object as the line of symmetry.
Fix	Object / Point	No	Secures an object to its current location in space.
Smooth	Object-to-object	No	Connects and creates a curvature-continuous situation, or G2 curve, between a spline and a line, polyline, spline, or arc.

Inferring Geometric Constraints

The **CONSTRAINTINFER** system variable controls whether AutoCAD infers constraints as you create new geometry. The **Infer Constraints** button on the status bar provides a quick way to toggle **Infer Constraints** on and off. When **Infer Constraints** is on, constraints are inferred when you draw a new object. Appropriate object snaps, other drawing aids, and **Infer Constraints** must be active in order to infer constraints, except when you use the **RECTANGLE** command. For example, use the **LINE** command and a horizontal (0° or 180°) or vertical (90° or 270°) polar tracking angle, or activate **Ortho** mode to constrain a horizontal or vertical line. See Figure 22-9A.

Constraints are also inferred when you edit an object. Some commands, such as **FILLET** and **CHAMFER**, infer constraints automatically. However, you must use grip editing or appropriate object snaps to infer constraints when stretching, moving, or copying. For example, use grip editing and the center point and midpoint grips or use the **MOVE** command and **Center** and **Midpoint** object snap modes to move the center point of a circle coincident to the midpoint of a line. See Figure 22-9B.

Figure 22-9.
A—Inferring a horizontal and vertical constraint while drawing a line. B—Editing a drawing to infer a coincident constraint.

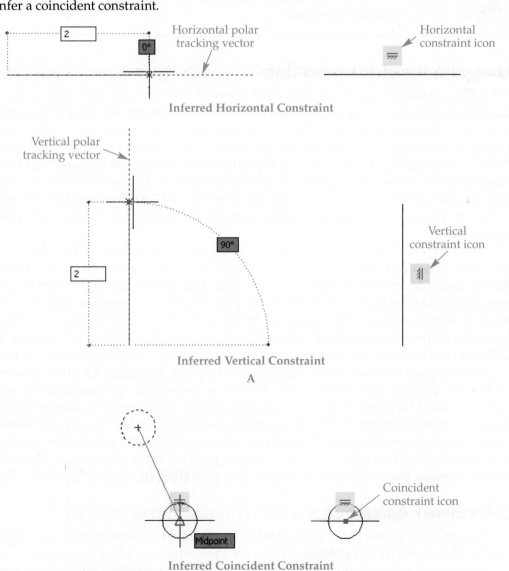

NOTE

The **Quadrant, Intersection, Extension,** and **Apparent extension** object snaps do not infer constraints. The **OFFSET, BREAK, TRIM, EXTEND, SCALE, MIRROR, ARRAY,** and **MATCHPROP** commands do not infer constraints. Exploding a polyline removes all inferred constraints.

CAUTION

The **Infer Constraints** function can save drafting time by placing geometric constraints while you draw or edit. However, use caution to ensure that appropriate geometric constructions occur. You may have to replace certain constraints manually.

Exercise 22-2

Complete the exercise on the companion website.
www.g-wlearning.com/CAD

Manual Geometric Constraints

Infer constraints when possible and appropriate. You can also add geometric constraints manually to apply geometric constraints that are not inferred automatically, as indicated in **Figure 22-8**, to include additional geometric constraints as needed, or to constrain a nonparametric drawing. The quickest way to place geometric constraints manually is to pick the appropriate button from the **Geometric** panel of the **Parametric** ribbon tab. You can also type GC followed by the name of the constraint, such as GCTANGENT, or select an option from the **GEOMCONSTRAINT** command. Follow the prompts to make the required selection(s), form the constraint, and exit the command.

Type
**GEOMCONSTRAINT
GCON**

Select objects, points, or an object and a point, depending on the objects and geometric constraint. **Figure 22-10** shows the point markers that appear as you move the crosshairs on an object to select a point. The object associated with a specific point highlights, allowing you to confirm that the point is on the appropriate object. Pick the marked location to apply the constraint to the object at that point. **Figure 22-10** also shows the ellipse axes and mtext and text constraint lines that appear for selection. Pick when you see the appropriate line to apply the constraint to the object using the line.

The **Fix** constraint requires a single object or point selection. All other geometric constraints require you to select two objects or points, or an object and a point. Generally, the first object or point you select remains the same. The second object or point you select changes in relation to the first selection, unless the second selection is fixed. For example, to create perpendicular lines using the **Perpendicular** constraint, first select the line that will remain in the same position at the same angle. Then select the line to make perpendicular to the first line, assuming the second line is not fixed.

Coincident Constraints

coincident:
A geometric construction that specifies two points sharing the same position.

Figure 22-11 shows *coincident* constraints applied to a basic multiview drawing. Refer to this figure as you explore options for assigning coincident constraints.

Infer coincident constraints by turning on **Infer Constraints** and using specific drawing and editing commands and object snaps. When you use the **LINE** command

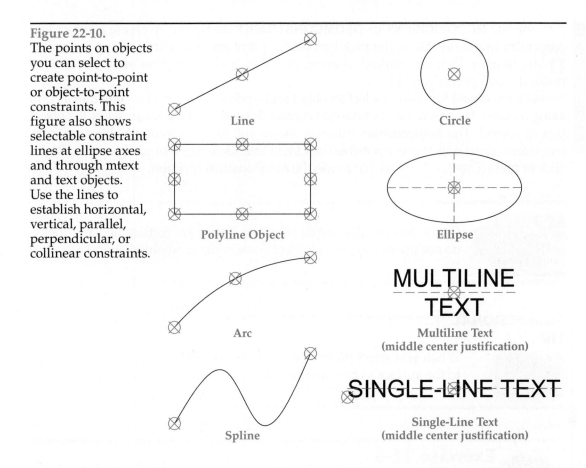

Figure 22-10.
The points on objects you can select to create point-to-point or object-to-point constraints. This figure also shows selectable constraint lines at ellipse axes and through mtext and text objects. Use the lines to establish horizontal, vertical, parallel, perpendicular, or collinear constraints.

Line

Circle

Polyline Object

Ellipse

Arc

MULTILINE
TEXT

Multiline Text
(middle center justification)

SINGLE-LINE TEXT

Single-Line Text
(middle center justification)

Spline

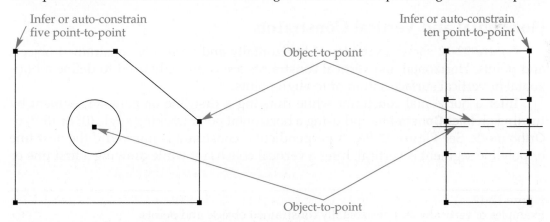

Figure 22-11.
Coincident constraints required for a basic multiview drawing. For clarity, labels do not indicate all required coincident constraints. This drawing is used as an example throughout this chapter.

Infer or auto-constrain
five point-to-point

Infer or auto-constrain
ten point-to-point

Object-to-point

Object-to-point

and draw a series of connected line segments, coincident constraints are inferred at adjoined endpoints. Use the **Close** option or an **Endpoint** object snap to infer a coincident constraint between the first and last points. When you use the **CHAMFER** command, coincident constraints are inferred at each endpoint.

The **Endpoint**, **Midpoint**, **Center**, **Insertion**, and **Node** object snaps infer a point-to-point coincident constraint. For example, snap to the midpoint of a line to constrain the center of a donut to the midpoint of the line. The **Nearest** object snap infers an object-to-point coincident constraint. For example, snap to a nearest location on a line to constrain the center of a circle to the line. The center of the circle is free to move anywhere aligned with the line, and does not have to touch the line.

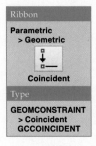

Ribbon

**Parametric
> Geometric**

Coincident

Type

**GEOMCONSTRAINT
> Coincident
GCCOINCIDENT**

Use the **GCCOINCIDENT** or **GEOMCONSTRAINT** command to create a coincident constraint manually. Move the pick box near a point on an existing object to display a point marker. Pick the marked location, and then pick a point on another object to make the two points coincide.

Use the **Object** function to select an object and a point, as required to constrain a point along a curve. The point does not have to contact the curve, and you can select the object first or second. The **Autoconstrain** function allows you to select multiple objects to form coincident constraints at every possible coincident intersection in a single operation. Right-click or press [Enter] or the space bar to use the **Autoconstrain** function.

 Connected polyline segments act as if they are constrained, but they do not use or accept coincident constraints at adjoined endpoints.

PROFESSIONAL TIP

 When you select points, be sure to select the points corresponding to the surface to be constrained.

 Exercise 22-3

Complete the exercise on the companion website.
www.g-wlearning.com/CAD

Horizontal and Vertical Constraints

Figure 22-12 shows examples of horizontally and vertically constrained objects and points. Horizontal and vertical constraints are commonly used to define a horizontal or vertical surface datum or to align points.

Infer a horizontal constraint while drawing a first line or polyline segment by turning on **Infer Constraints** and using a horizontal polar tracking angle (0° or 180°) or **Ortho** mode. See **Figure 22-9A**. A perpendicular constraint is inferred if the next line or polyline segment is vertical. Infer a vertical constraint while drawing a first line or

Figure 22-12.
Examples of vertically and horizontally constrained objects and points.

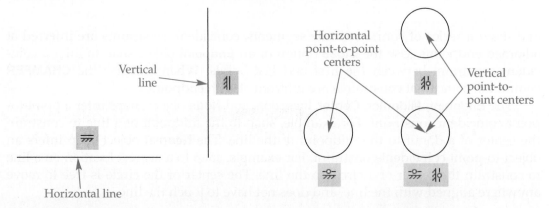

polyline segment using a vertical polar tracking angle (90° or 270°) or **Ortho** mode. A perpendicular constraint is inferred if the next line or polyline segment is horizontal.

Use the **GCHORIZONTAL**, **GCVERTICAL**, or **GEOMCONSTRAINT** command to create a horizontal or vertical constraint manually. Select a line, polyline, ellipse axis, mtext, or text object to constrain, or use the **2Points** function to pick two points to align horizontally or vertically.

Parallel and Perpendicular Constraints

Figure 22-13 shows *parallel* and *perpendicular* constraints applied to a basic multiview drawing. Parallelism and perpendicularity are common geometric characteristics that provide specific controls related to the orientation of features.

Infer parallel or perpendicular constraints by turning on **Infer Constraints** and using drawing and editing commands and object snaps. A perpendicular constraint is inferred if the next segment of a connected line or polyline is horizontal or vertical. When you use the **RECTANGLE** command to draw a polyline object, parallel and perpendicular constraints are inferred to create a true rectangle. Additional parallel and perpendicular constraints are inferred when you use the **Chamfer** option of the **RECTANGLE** command, depending on the specified distances. An additional perpendicular constraint is inferred when you use the **Fillet** option.

The **Parallel** object snap infers a parallel constraint between objects. The **Perpendicular** object snap infers a perpendicular and coincident constraint between objects.

Use the **GCPARALLEL**, **GCPERPENDICULAR**, or **GEOMCONSTRAINT** command to create a parallel or perpendicular constraint manually. Select a line, polyline, ellipse axis, mtext, or text object and a line, polyline, ellipse axis, mtext, or text object.

Ribbon

Parametric > Geometric

Horizontal

Type

GEOMCONSTRAINT > Horizontal GCHORIZONTAL

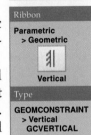

Ribbon

Parametric > Geometric

Vertical

Type

GEOMCONSTRAINT > Vertical GCVERTICAL

parallel: A geometric construction that specifies that objects such as lines will never intersect, no matter how long they become.

perpendicular: A geometric construction that defines a 90° angle between objects such as lines.

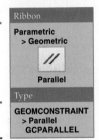

Ribbon

Parametric > Geometric

Parallel

Type

GEOMCONSTRAINT > Parallel GCPARALLEL

Ribbon

Parametric > Geometric

Perpendicular

Type

GEOMCONSTRAINT > Perpendicular GCPERPENDICULAR

Exercise 22-4

Complete the exercise on the companion website.
www.g-wlearning.com/CAD

Figure 22-13.
Parallel and perpendicular constraints required for the example multiview drawing.

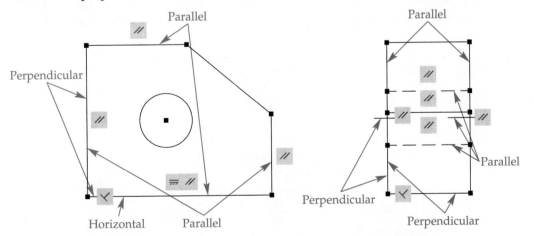

Tangent and Collinear Constraints

Figure 22-14 shows tangent and collinear constraints applied to a basic multiview drawing. Collinear constraints cannot be inferred.

Infer tangent constraints by turning on **Infer Constraints** and using the **FILLET** command or the **Fillet** option of the **RECTANGLE** command with a radius greater than 0. You can also infer tangent constraints on arc segments of polylines, and by using the **Tangent** object snap.

> The **Continue** option of the **ARC** command does not infer a coincident or tangent constraint between the existing object and the new arc.

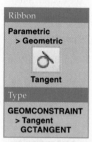

Ribbon

Parametric > Geometric

Tangent

Type

GEOMCONSTRAINT > Tangent GCTANGENT

Ribbon

Parametric > Geometric

Collinear

Type

GEOMCONSTRAINT > Collinear GCCOLLINEAR

Use the **GCTANGENT** or **GEOMCONSTRAINT** command to create a tangent constraint manually. Select a circle, arc, or ellipse and then a line, polyline, circle, arc, or ellipse. Use the **GCCOLLINEAR** or **GEOMCONSTRAINT** command to create a collinear constraint manually. A collinear constraint is commonly used for applications such as

Figure 22-14.
A—Collinear and tangent constraints required for the example multiview drawing. B—Common tangent constraint applications.

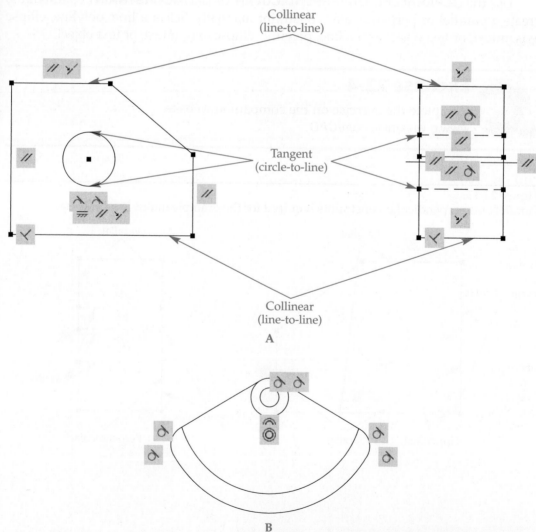

aligning multiview surfaces. Select a line, polyline, ellipse axis, mtext, or text object and then another line, polyline, ellipse axis, mtext, or text object. Activate the **Multiple** function to select multiple objects to align in a single operation. Right-click or press [Enter] or the space bar to complete a multiple constrain operation.

Exercises 22-5 and 22-6

Complete the exercises on the companion website.
www.g-wlearning.com/CAD

Concentric and Equal Constraints

Use the **GCCONCENTRIC** or **GEOMCONSTRAINT** command to assign a *concentric* constraint. Select a circle, arc, or ellipse and a second circle, arc, or ellipse. Use the **GCEQUAL** or **GEOMCONSTRAINT** command to size objects equally and, in some cases, to locate objects. Select two objects, or activate the **Multiple** function to select multiple objects to equalize in a single operation. Right-click or press [Enter] or the space bar to complete a multiple constrain operation. **Figure 22-15** shows examples of concentric and equal constraints.

Symmetric, Fix, and Smooth Constraints

Use the **GCSYMMETRIC** or **GEOMCONSTRAINT** command to establish symmetry between objects or points and a line, polyline, ellipse axis, mtext, or text object as the line of symmetry. Select one object, followed by another, and finally, a line of symmetry. Use the **2Points** option to pick points followed by the line of symmetry to constrain symmetrical points.

Use the **GCFIX** or **GEOMCONSTRAINT** command to secure a point or object to its current location in space to help preserve design intent. A single fix constraint is often required to fully constrain a drawing. Use the default method to fix a point, or activate the **Object** function to select an object to fix. **Figure 22-16** shows examples of symmetric and fix constraints. Use the **GCSMOOTH** or **GEOMCONSTRAINT** command to connect and create a curvature-continuous situation, or G2 curve, between a selected spline and a line, polyline, spline, or arc.

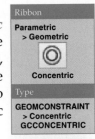

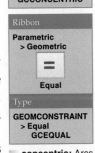

concentric: Arcs, circles, and/or ellipses sharing the same center point.

Figure 22-15.
Examples of concentric and equal constraints. All circles are concentric to an arc. All small arcs are equal, and all small circles are equal. Use the **Multiple** option to help make multiple objects equal in a single operation.

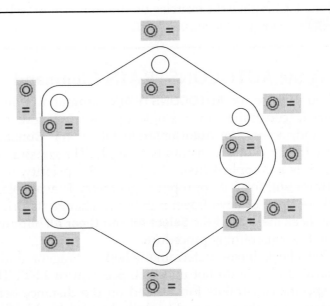

Figure 22-16.
A symmetrical parametric drawing created by adding symmetric constraints to circles and arcs. A fix constraint secures the drawing in space.

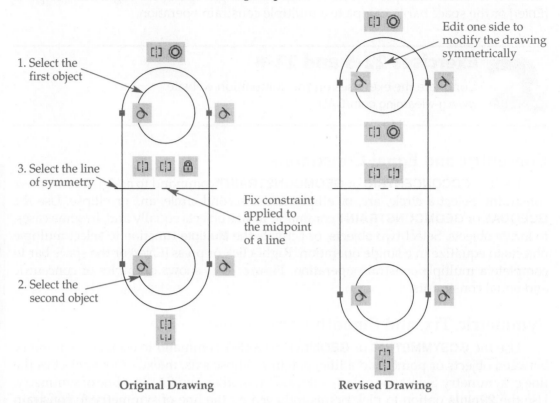

Original Drawing Revised Drawing

When you apply a **Coincident** or **Smooth** constraint to a fit points spline, the spline is converted to a control vertices spline.

Exercise 22-7

Complete the exercise on the companion website.
www.g-wlearning.com/CAD

Using the AUTOCONSTRAIN Command

Ribbon
Parametric
> Geometric

Auto Constrain

Type
AUTOCONSTRAIN

Ribbon
Parametric
> Geometric

Constraint Settings

Type
**CONSTRAINT-
SETTINGS**

You can use the **AUTOCONSTRAIN** command in an attempt to add all required geometric constraints in a single operation. Before using the **AUTOCONSTRAIN** command, access the **AutoConstrain** tab of the **Constraint Settings** dialog box to specify the geometric constraints to apply. The constraint priority determines which constraints are applied first. The higher the priority, the more likely and often the constraint will form if appropriate geometry is available. Select a constraint and use the **Move Up** and **Move Down** buttons to change its priority. Use the corresponding **Apply** check marks and the **Select All** and **Clear All** buttons to omit specific constraints during the constraining procedure.

The check boxes determine whether tangent and perpendicular constraints can form if objects do not intersect. See **Figure 22-17**. The **Tolerances** area controls how specific constraints form based on the distance between and angle of objects. A distance less than or equal to the value specified in the **Distance** text box receives

Figure 22-17.
Examples of constraints that will form when you select the **Tangent objects must share an intersection** or the **Perpendicular objects must share an intersection** check box.

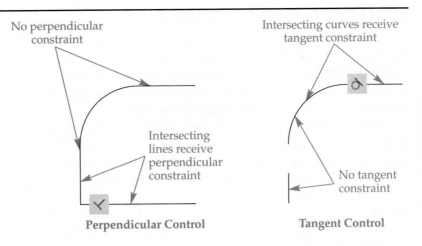

No perpendicular constraint

Intersecting curves receive tangent constraint

Intersecting lines receive perpendicular constraint

No tangent constraint

Perpendicular Control

Tangent Control

constraints. An angle less than or equal to the value specified in the **Angle** text box receives constraints. See **Figure 22-18**.

Once you specify the settings, access the **AUTOCONSTRAIN** command and select the objects to constrain. The **Settings** option is available before selection to access the **AutoConstrain** tab of the **Constraint Settings** dialog box. A fix constraint does not occur.

You can also access the **Constraint Settings** dialog box by right-clicking on the **Infer Constraints** button on the status bar and picking **Settings...**.

CAUTION

The **AUTOCONSTRAIN** command can save drafting time by placing geometric constrains in a single operation. However, use caution to ensure that appropriate geometric constructions occur and geometry does not shift. You may have to replace certain constraints.

Figure 22-18.
Examples of constraints that form based on **Distance** and **Angle** tolerance values.

	Distance between Objects > Distance Tolerance	Distance between Objects ≤ Distance Tolerance	Angle > Angle tolerance	Angle ≤ Angle tolerance
Before				
After				

Managing Geometric Constraints

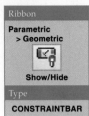

geometric constraint bars: Toolbars that allow you to view and remove geometric constraints.

Ribbon

Parametric > Geometric

Show/Hide

Type

CONSTRAINTBAR

Geometric constraint bars appear by default to indicate geometric constraints. See **Figure 22-19**. Refer to **Figure 22-8** for help recognizing constraint icons. The **CONSTRAINTBAR** command includes options to show or hide constraint bars. The quickest way to access these options is to pick the appropriate button from the **Geometric** panel of the **Parametric** ribbon tab.

Pick the **Show/Hide** button, pick objects to manage, and then right-click or press [Enter] or the space bar to access options. Choose the **Show** option to display hidden constraint bars, the **Hide** option to hide visible constraint bars, or the **Reset** option to display hidden constraint bars and move constraint bars back to the default positions. Select the **Show All** button to display all constraint bars. Pick the **Hide All** button to hide all constraint bars. Hiding constraint bars does not remove geometric constraints.

Access the **Geometric** tab of the **Constraint Settings** dialog box to specify the geometric bars that appear and other geometric bar characteristics. Select the check boxes for individual constraints to display the constraints in constraint bars. Use the **Select All** and **Clear All** buttons to select or deselect all constraint type check boxes. By default, all constraint types are displayed. Limiting constraint bar visibility to specific constraint types often helps to locate and adjust constraints.

Use the **Constraint bar transparency** slider or text box to adjust the constraint bar transparency. Check **Show constraint bars after applying constraints to selected objects** to display constraint bars after you have applied geometric constraints. Check **Show constraint bars when objects are selected** to display constraint bars when you select objects, even if constraint bars are hidden.

A coincident constraint initially appears as a dot. Hover over the dot to show the coincident constraint bar. All other constraints appear in constraint bars. Hover over an object to highlight constraint bars associated with constraints applied to the object. Hover over or select a constraint bar to highlight the corresponding constrained objects and display markers that identify constrained points. See **Figure 22-19**. The visual effects of hovering over an object or geometric constraint bar helps you recognize the objects and points associated with the constraint.

If a constraint bar blocks your view, drag it to a new location. To hide a specific constraint bar, pick the **Hide Constraint Bar** button, or right-click and choose **Hide**. The right-click shortcut menu also includes options for hiding all constraint bars and for accessing the **Geometric** tab of the **Constraint Settings** dialog box.

Design changes sometimes require deleting existing constraints. To delete geometric constraints, hover over an icon in the constraint bar and press [Delete], or right-click and select **Delete**.

Figure 22-19.
Use geometric constraint bars to view and delete geometric constraints. Horizontal, vertical, symmetric, and fix constraints use different icons depending on whether the constraint is a point-to-point or object-to-point constraint.

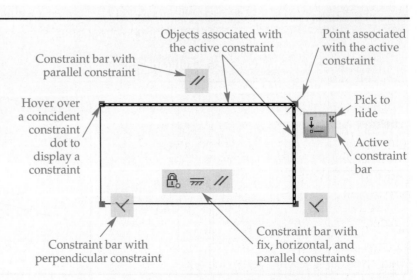

Objects associated with the active constraint

Point associated with the active constraint

Constraint bar with parallel constraint

Hover over a coincident constraint dot to display a constraint

Pick to hide

Active constraint bar

Constraint bar with perpendicular constraint

Constraint bar with fix, horizontal, and parallel constraints

NOTE
You can also right-click with no objects selected to access a **Parametric** cascading menu that provides options for displaying and hiding constraints and for accessing the **Constraint Settings** dialog box.

PROFESSIONAL TIP

To confirm that constraints, especially geometric constraints, are present and appropriate, select an object to display grips and attempt to stretch a grip. As an object becomes constrained, you should observe less freedom of movement. Stretching or attempting to stretch grips is one of the fastest ways to assess design options and to analyze where a constraint is still required. You know the drawing is constrained when you are no longer able to stretch the geometry.

Exercise 22-8
Complete the exercise on the companion website.
www.g-wlearning.com/CAD

Dimensional Constraints

Dimensional constraints establish size and location parameters. You must include dimensional constraints to create a truly parametric drawing. Dimensional constraints use a *dynamic format* by default, and their appearance is different from traditional associative dimensions. The easiest way to preset dimensional constraints to use the dynamic format is to pick the **Dynamic Constraint Mode** button from the expanded **Dimensional** panel of the **Parametric** ribbon tab.

You cannot modify how dynamic dimensional constraints appear, but you can change them to an *annotational format*. To preset new dimensional constraints to use an annotational format, pick the **Annotational Constraint Mode** button from the expanded **Dimensional** panel of the **Parametric** ribbon tab. The annotational format uses the current dimension style. This textbook focuses on using the **Dynamic Constraint Mode** function and then assigning annotational format later. View, adjust, and remove dimensional constraints as needed.

Adding Dimensional Constraints

The quickest way to create dimensional constraints is to pick the appropriate button on the **Dimensional** panel of the **Parametric** ribbon tab. You can also type DC followed by the name of the constraint, such as DCLINEAR, or select an option from the **DIMCONSTRAINT** command. To create a dimensional constraint, follow the prompts to make the required selections, pick a location for the dimension line, enter a value to form the constraint, and exit the command.

The process of selecting points or objects to locate dimensional constraint extension lines is the same as that for adding geometric constraints. When a dimensional constraint command requires you to pick two points or objects, the first point or object you select generally remains the same. In some cases, you can consider the first point

dynamic format: A dimensional constraint format specifically for controlling the size or location of geometry.

Ribbon
Parametric > Dimensional > Dynamic Constraint Mode

Type
CCONSTRAINT-FORM
DCFORM

annotational format: A dimensional constraint format in which the constraints look like traditional dimensions, using a dimension style. Annotational dimensional constraints can still control the size or location of geometry.

Ribbon
Parametric > Dimensional > Annotational Constraint Mode

Type
CCONSTRAINT-FORM
DCFORM

or object the datum. The second point or object you select changes in relation to the first selection, unless the second selection is fixed. When selecting points, be sure to select the points corresponding to the surface to constrain.

A text editor appears after you select the location for the dimension line, allowing you to specify the dimension value. See Figure 22-20A. Each dimensional constraint is a parameter with a specific name, expression, and value. By default, linear dimensions receive d names, angular dimensions receive ang names, diameter dimensions receive dia names, and radial dimensions receive rad names. Every parameter must have a unique name. The name of the first of each type of dimension includes a 1, such as d1. The next dimension includes a 2, such as d2, and so on. Use the text editor to enter a more descriptive name, such as Length, Width, or Diameter. Follow the name with the = symbol and then the dimension value. Changing the current dimension value modifies object size or location. Press [Enter] or pick outside of the text editor to form the constraint. See Figure 22-20B.

The most basic way to specify the value of a dimensional constraint is to type a value in the text editor. Dimensional constraint units reflect the current work environment and unit settings, including length, angle type, and precision. Accept the current value if the drawing is accurate. Enter a different value if the drawing is inaccurate, or to change object size.

Another option is to enter an expression in the text editor. You can use an expression if you do not know an exact value, much like using a calculator. Usually, however, expressions include parameters to associate dimensional constraints. This enables the drawing to adapt according to parameter changes. In the active text editor, move the text cursor to the location to add the parameter, and then pick another existing dimensional constraint to copy its name to the expression. An alternative is to type the parameter name in the expression. An fx: in front of the dimension text indicates an expression. Figure 22-21 shows an example of using an existing parameter in an expression to control the size of an object. In this example, the design requires that the height of the object always be half the value of the length.

If you add an existing parameter to the text of a different dimensional constraint, without making a calculation, such as d1 = d2, the dimensional constraint uses the same value as the reference parameter. This is necessary for many applications, but use an equal geometric constraint when possible.

Figure 22-20.
A—Use the text editor that appears after you establish the location of the dimension line to specify a parametric dimension. B—An example of modifying a parameter name and value. Notice that the dimension controls the object size.

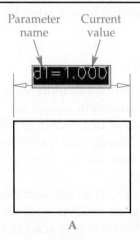

Parameter name   Current value

d1=1.000

A

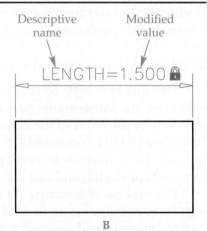

Descriptive name   Modified value

LENGTH=1.500

B

Figure 22-21.
A—A modified parameter name and expression that references another parameter.
B—The dimension that includes a parameter references the parameter when changes occur.

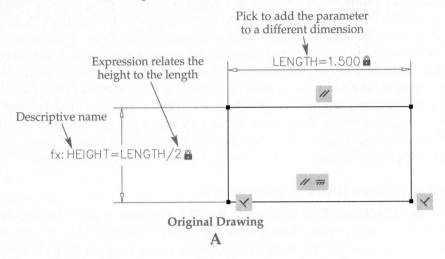

Pick to add the parameter
to a different dimension

Expression relates the
height to the length

LENGTH=1.500 🔒

Descriptive name

fx: HEIGHT=LENGTH/2 🔒

Original Drawing
A

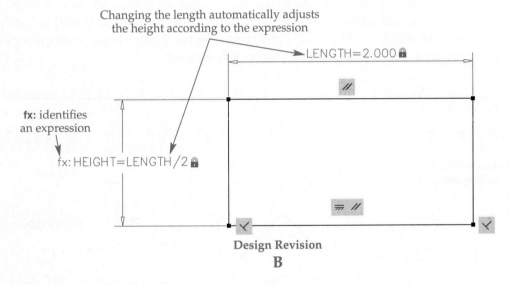

Changing the length automatically adjusts
the height according to the expression

LENGTH=2.000 🔒

fx: identifies
an expression

fx: HEIGHT=LENGTH/2 🔒

Design Revision
B

Linear Dimensional Constraints

Use the **DCLINEAR** or **DIMCONSTRAINT** command to place a horizontal or vertical linear dimensional constraint. The **Horizontal** option sets the command to constrain only a horizontal distance. The **Vertical** option sets the command to constrain only a vertical distance. The **Horizontal** and **Vertical** options are helpful when it is difficult to produce the appropriate linear dimensional constraint, such as when you are dimensioning the horizontal or vertical distance of an angled surface.

Pick two points to specify the origin of the dimensional constraint, or use the **Object** function to select a line, polyline, or arc to constrain. Move the dimension line to an appropriate location and pick. Specify the dimension value and adjust the parameter name if desired. Press [Enter] or pick outside of the text editor to form the constraint. **Figure 22-20** and **Figure 22-21** show examples of linear dimensions.

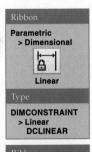

Ribbon
Parametric
> Dimensional

Linear

Type
DIMCONSTRAINT
> Linear
DCLINEAR

Ribbon
Parametric
> Dimensional

Horizontal

Type
DIMCONSTRAINT
> Horizontal
DCHORIZONTAL

Ribbon
Parametric
> Dimensional

Vertical

Type
DIMCONSTRAINT
> Vertical
DCVERTICAL

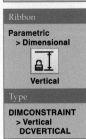

Exercise 22-9

Complete the exercise on the companion website.
www.g-wlearning.com/CAD

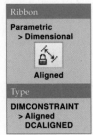

Ribbon

**Parametric
> Dimensional**

Aligned

Type

**DIMCONSTRAINT
> Aligned
DCALIGNED**

Aligned Dimensional Constraints

Use the **DCALIGNED** or **DIMCONSTRAINT** command to place a linear dimensional constraint with a dimension line that is aligned with an angled surface, with extension lines perpendicular to the surface. Pick two points to specify the origin of the dimensional constraint, or use the **Object** function to select a line, polyline, or arc to constrain.

Often when you apply aligned dimensions, such as when you dimension an auxiliary view, it is necessary to pick a point and an aligned surface or two aligned surfaces. Use the **Point & Line** function to select a point and an alignment line. Use the **2Lines** function to select two alignment lines. Move the dimension line to an appropriate location and pick. Specify the dimension value, and adjust the parameter name if desired. Press [Enter] or pick outside of the text editor to form the constraint. **Figure 22-22** shows examples of aligned dimensional constraints created using each method.

Angular Dimensional Constraints

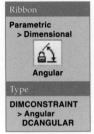

Ribbon

**Parametric
> Dimensional**

Angular

Type

**DIMCONSTRAINT
> Angular
DCANGULAR**

Use the **DCANGULAR** or **DIMCONSTRAINT** command to place an angular dimension between two objects or three points. Pick two lines, polylines, or arcs, or use the **3Point** function to select the angle vertex, followed by two points to locate each side of the angle. Move the dimension line to an appropriate location and pick. Specify the dimension value, and adjust the parameter name if desired. Press [Enter] or pick outside of the text editor to form the constraint. **Figure 22-23** shows examples of angular dimensional constraints created using each method.

Figure 22-22.
An auxiliary view with full dimensional constraints created using the **Aligned** option.

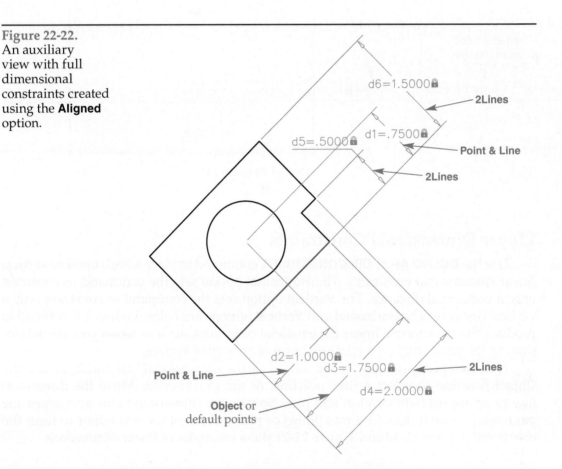

Figure 22-23.
Forming angular dimensional constraints.

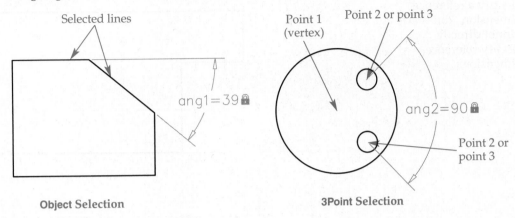

Object Selection

3Point Selection

Exercise 22-10

Complete the exercise on the companion website.
www.g-wlearning.com/CAD

Diameter and Radius Dimensional Constraints

Access the **DCDIAMETER** or **DIMCONSTRAINT** command to create a diameter dimensional constraint. Access the **DCRADIAL** or **DIMCONSTRAINT** command to form a radius dimensional constraint. You can select a circle or arc when using either command. In formal drafting, diameter constraints are applied to circles and radius constraints are applied to arcs. See **Figure 22-24**. In some parametric applications, however, most often when you are dimensioning construction geometry, it is appropriate to constrain arcs using a diameter and circles using a radius.

Reference Dimensional Constraints

When you try to constrain a fully constrained object, the drawing should become over-constrained. However, AutoCAD does not allow over-constraining to occur. You can either cancel the command without accepting the dimension, or accept the dimension and allow it to become a reference dimension. See **Figure 22-25**. Reference dimensions are sometimes required to form specific parameters or expressions. You cannot edit a reference dimension to change the size of an object, but a reference dimension changes when you modify related dimensions.

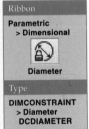

Ribbon

Parametric
> Dimensional

Diameter

Type

DIMCONSTRAINT
> Diameter
DCDIAMETER

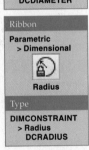

Ribbon

Parametric
> Dimensional

Radius

Type

DIMCONSTRAINT
> Radius
DCRADIUS

Figure 22-24.
Forming diameter and radius dimensional constraints. In this example, equal geometric constraints control the size of the undimensioned arcs.

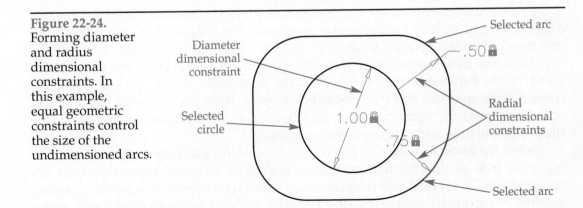

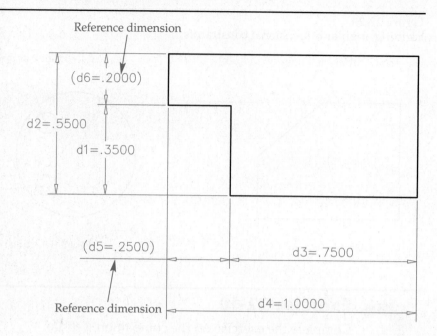

Figure 22-25.
Parentheses identify a reference dimension. You cannot directly modify reference dimensions.

Reference dimension

(d6=.2000)

d2=.5500

d1=.3500

(d5=.2500)

d3=.7500

d4=1.0000

Reference dimension

Dimensional constraints define and constrain drawing geometry. They use a unique AutoCAD style and do not comply with ASME standards. Do not be overly concerned about the placement or display characteristics of dimensional constraints, but if possible, apply dimensional constraints just as you would add dimensions to a drawing, using correct drafting practices. In addition, move and manipulate dimensional constraints so the drawing is as uncluttered as possible. Use grips to make basic adjustments to dimensional constraint position.

Converting Associative Dimensions

Ribbon

Parametric > Dimensional

Convert

Type

DIMCONSTRAINT

Use the **DCCONVERT** or **DIMCONSTRAINT** command to convert an associative dimension to a dimensional constraint. This allows you to prepare a parametric drawing using existing associative dimensions. Pick the associative dimensions to convert and press [Enter] or pick outside of the text editor. By default, new dimensional constraints and converted dimensions use the dynamic format. See **Figure 22-26**.

Managing Dimensional Constraints

Ribbon

Parametric > Dimensional

Show/Hide Dynamic Constraints

By default, all dimensional constraints are displayed in the dynamic format, include a lock icon, and include the parameter name and dimension value. The **DCDISPLAY** command provides methods to show or hide dimensional constraints. The quickest way to access these options is to pick the appropriate button from the **Dimensional** panel of the **Parametric** ribbon tab. Pick the **Show/Hide** button, pick dimensional constraints to manage, and then right-click or press [Enter] or the space bar to access options. Choose the **Show** option to display hidden dimensional constraints, or the **Hide** option to hide visible dimensional constraints.

Select the **Show All** button to display all dimensional constraints. Pick the **Hide All** button to hide all dimensional constraints. Hiding dimensional constraints does not remove them. You typically hide dimensional constraints to prepare a formal drawing, or when you no longer need to see dimensional constraints for the current design phase.

Figure 22-26.
Converting a dimension drawn using the **DIMDIAMETER** command to a dimensional constraint. The default dimensional constraint format is dynamic.

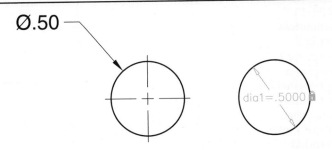

Use the **Dimensional** tab of the **Constraint Settings** dialog box to adjust additional dimensional constraint settings. The **Dimension name format** drop-down list allows you to display dimensional constraints with the parameter name and value, parameter name, or value. Use the **Show lock icon for annotational constraints** check box to toggle the lock icon on or off for new dimensional constraints. If you hide dimensional constraints and select the **Show hidden dynamic constraints of selected objects** check box, you can pick an object to show associated dimensional constraints temporarily.

Design changes sometimes require deleting existing constraints. Use the **ERASE** command to eliminate specific dimensional constraints.

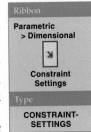

Ribbon

Parametric > Dimensional

Constraint Settings

Type

CONSTRAINT-SETTINGS

You can also right-click with no objects selected to access a **Parametric** cascading menu that provides options for displaying and hiding constraints, changing the dimension name format, and accessing the **Constraint Settings** dialog box. The options are also available when you pick a dimensional constraint and then right-click.

Working with Parameters

A *dimensional constraint parameter* automatically forms every time you add a dimensional constraint. You can adjust parameters by changing the dimensional constraint value or by using the options in the **Constraint** category of the **Properties** palette. The **Parameters Manager** allows you to manage all parameters in the drawing. See **Figure 22-27**.

The list view pane on the right side of the **Parameters Manager** lists parameters in a table and provides parameter controls. Pick a parameter in the **Parameters Manager** to highlight the corresponding dimensional constraint in the drawing. To change the name of a parameter, pick inside a text box in the **Name** column to activate it, type the new name, and press [Enter] or pick outside of the text box. Enter a new value or expression for the parameter in an **Expression** column text box. The value appears in the **Value** column display box for reference. Use the **Delete** button or right-click on a parameter and select **Delete** to remove the parameter and the corresponding dimensional constraint.

To specify *user parameters*, pick the **User Parameters** button to display a **User Variables** node. User parameters function like dimensional constraints in the **Parameters Manager**. Create user parameters in order to access specific parameters throughout the design process. For example, if you know the thickness of a part will always be twice a certain dimensional constraint, create a user parameter similar to the parameter shown in **Figure 22-27** to define the thickness. You can then use the custom parameter for reference and in expressions when you place additional dimensional constraints.

Pick the **Expand Parameters** filter tree button or right-click in the list pane and select **Show Filter Tree** to display the tree view pane on the left side of the **Parameters Manager**. *Parameter filters* are typically appropriate when it becomes difficult to manage a very large number of parameters. Filter a large list of parameters to make it easier to work with the parameters needed for a specific drawing task.

dimensional constraint parameters: Parameters that form when you insert a dimensional constraint.

Ribbon

Parametric > Manage

Parameters Manager

Type

PARAMETERS

user parameters: Additional parameters you define.

parameter filters: Settings that screen out, or filter, parameters you do not want to display in the list view pane of the **Parameters Manager**.

Figure 22-27.
The **Parameters Manager** is a good resource for reviewing and editing parameters and for creating user parameters. This figure shows the expanded parameters filter tree.

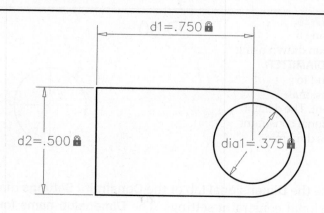

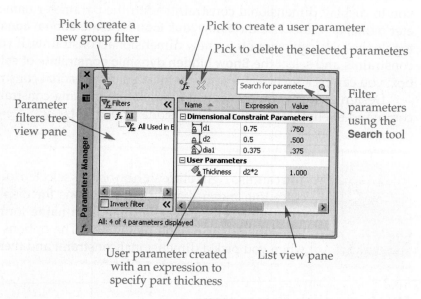

Pick to create a new group filter

Pick to create a user parameter

Pick to delete the selected parameters

Parameter filters tree view pane

Filter parameters using the **Search** tool

User parameter created with an expression to specify part thickness

List view pane

Parameter filters are listed in alphabetical order in the **All** node. Select the **All** node to display all parameters in the drawing. Pick the **All Used in Expressions** filter to display only parameters used in or referenced by expressions. Pick the **Invert filter** check box to invert, or reverse, a parameter filter. Select the **Filter** button or right-click on a filter name and select **New Group Filter** to create a custom parameter group filter. A new group filter appears in the filter tree view ready to accept a custom name, if desired. Select the **All** node at the top of the filter tree area to display all the parameters in the drawing. Then, to add a parameter to the group filter, drag a parameter from the list and drop it onto the group filter name. Right-click on a filter name in the tree view or a parameter name in the list view to access options for managing group filters.

Use the **Search** text box to search for specific parameters according to parameter name.

Exercise 22-11

Complete the exercise on the companion website.
www.g-wlearning.com/CAD

Dynamic and Annotational Form

This textbook focuses on using the default dynamic format for placing dimensional constraints. **Figure 22-28A** shows the dimensional constraints required to fully constrain the example multiview drawing in their dynamic format. The annotational format is more appropriate than the dynamic format for formal dimensioning practices. Annotational constraints still control object size and location.

The **CCONSTRAINTFORM** system variable sets the dimensional constraint form applied when you add new dimensional constraints. The easiest way to adjust the **CCONSTRAINTFORM** system variable is to pick the **Dynamic Constraint Mode** or **Annotational Constraint Mode** button from the expanded **Dimensional** panel of the **Parametric** ribbon tab. You can also preset the dimensional constraint format using the **DIMCONSTRAINT** command by selecting the **Form** option, or by using the **DCFORM** command before creating a dimensional constraint.

> The specified **Annotational** or **Dynamic** form is a system variable and applies to new dimensional constraints until you change it to the alternate setting.

Use the **Constraint Form** drop-down list in the **Constraint** category of the **Properties** palette to change the dimensional constraint form assigned to existing dimensions. **Figure 22-28B** shows the dimensional constraints in dynamic format that require changing to the annotational format for the example multiview drawing. Annotational dimensions, especially those converted from dynamic dimensions, often require that you make format and organizational changes to prepare the final drawing. You may also have to add non-parametric dimensions. Make the following changes to create the final drawing shown in **Figure 22-28C**.

- Hide all geometric constraints and dynamic dimensions.
- Disable the lock icon display.
- Change the dimension name format to **Value**.
- Make basic dimension style overrides and dimension location adjustments if necessary.

Exercise 22-12

Complete the exercise on the companion website.
www.g-wlearning.com/CAD

Parametric Editing

You can adjust existing parametric drawings in a variety of ways. Drawing additional unconstrained objects adds geometry that requires constraining to make the drawing fully parametric once again. You may need to delete or replace existing constraints to constrain new objects. Erasing objects and exploding polylines removes constraints. If you erase geometry associated with an expression, an alert appears asking if you want to convert the dimensional constraint to a user parameter, maintain the information, or remove the parameter with the dimensional constraint.

Figure 22-28.
The typical process for changing a parametric drawing to a final, formal appearance. A—A fully constrained drawing with dynamic dimensional constraints. Notice that dimensions apply to all items, including centerline extensions and the distance between views. B—Changing dynamic dimensions to annotational dimensions. Change only those dimensions required for formal dimensioning. C—The appearance of the final drawing.

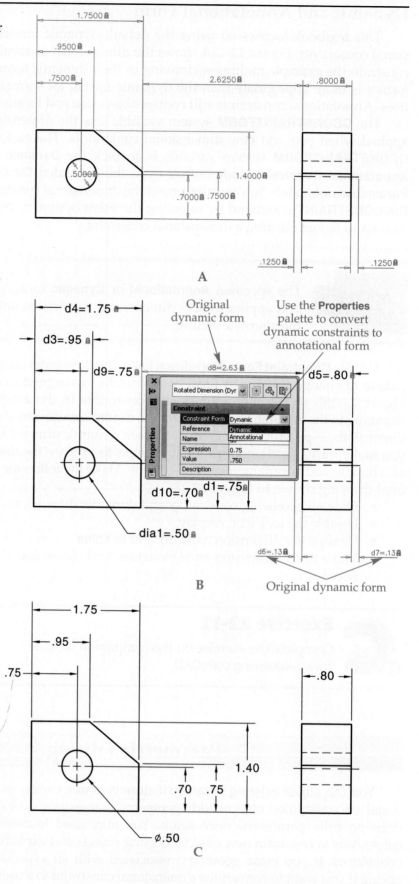

Modifying Dimensional Constraints

Adjust dimensional constraints to make design changes to a constrained drawing. To edit a dimensional constraint value, double-click on the value, or select the dimensional constraint, right-click, and pick **Edit Constraint**. The text editor appears, allowing you to make changes. Press [Enter] or pick outside of the text editor to complete the operation.

Another method is to pick a dimensional constraint and use parameter grips to change the value. See **Figure 22-29**. Parameter grip-editing is most appropriate for analyzing design options, when you do not know a specific value, or when using drawing aids such as object snaps to adjust the value. You can also modify a dimensional constraint value by entering a different expression in the **Expression** text box in the **Constraint** category of the **Properties** palette, or in the **Expression** text box of the **Parameters Manager**.

If a drawing includes enough geometric constraints, and the geometric constraints are accurate, changing dimensions should maintain all geometric relationships.

Figure 22-29.
Using a parameter grip to change the dimensional constraint value.

Selected dimensional constraint

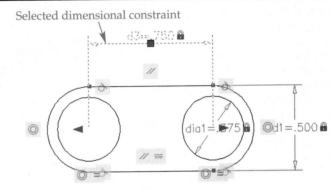

Original Drawing

Value changes according to the second stretch point

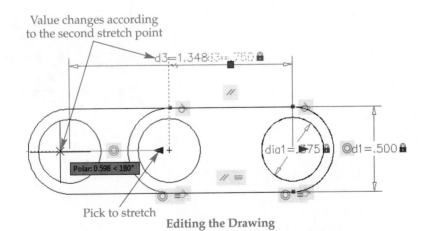

Pick to stretch

Editing the Drawing

Removing Constraints

Constraints limit your ability to make changes to a drawing. For example, you cannot rotate a horizontally or vertically constrained line. You may have to relax or delete constraints to make significant changes to a drawing. Relax constraints using standard editing practices. When you use a basic editing command such as **ROTATE**, a message appears asking if you want to relax constraints. To relax constraints while stretching, moving, or scaling with grips, you may need to press [Ctrl] to toggle relaxing on and off. See **Figure 22-30A**. Other edits automatically remove constraints. See **Figure 22-30B**. The only constraints, if any, that you remove are those required to achieve the edit.

A **DELCONSTRAINT** command is available in addition to the methods previously described for deleting individual constraints. Access the **DELCONSTRAINT** command and select the geometric and dimensional constraints to remove. The **DELCONSTRAINT** command is especially effective for removing a significant number of constraints. Use the **All** option to remove all constraints from the drawing.

Figure 22-30.
Examples of relaxing constraints. A—Access the **MOVE** command and then press [Ctrl] to remove the concentric constraint. The process is similar for stretching and scaling. B—Rotating removes a vertical constraint in this example.

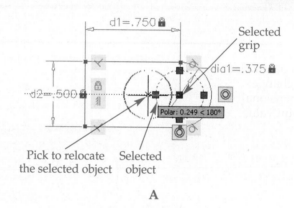

A

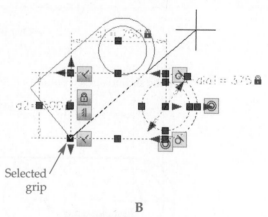

B

NOTE

Some editing techniques, such as mirroring or moving, may only require removal of fix constraints, especially when you are editing an entire drawing. Exploding a polyline removes all constraints.

PROFESSIONAL TIP

Occasionally, constraining geometry causes objects to twist out of shape, making it difficult to control the size and position of the drawing. Use the **UNDO** command to return to the previous design. Consider the following suggestions to help avoid this situation:

- Use standard and accurate drafting practices to construct objects at or close to their finished size.
- Add as many geometric constraints as appropriate before dimensional constraints.
- Dimensionally constrain the largest objects first.
- Move objects to a more appropriate location, if necessary, and change object size before constraining.

Exercise 22-13

Complete the exercise on the companion website.
www.g-wlearning.com/CAD

Chapter Review

Answer the following questions. Write your answers on a separate sheet of paper or complete the electronic chapter review on the companion website.
www.g-wlearning.com/CAD

1. Give an example demonstrating how to use constraints to form a geometric construction when standard AutoCAD commands are inefficient or ineffective.
2. Describe two applications in which parametric drafting may be unsuitable or ineffective.
3. Briefly describe the purpose of geometric constraint tools, and identify what you see on-screen that indicates the presence of a geometric constraint.
4. Name the function that forms geometric constraints while you draw or edit.
5. Name the tools that allow you to assign geometric constraints manually.
6. When you use geometric constraint commands that allow you to pick two objects or points, describe what generally happens to the first and second objects you select.
7. List the object snaps that infer coincident constraints.
8. Identify common uses for horizontal and vertical constraints.
9. Describe how to infer horizontal and vertical constraints.
10. List the object types that can form parallel or perpendicular constraints.
11. Name the types of objects you can constrain with the tangent constraint.
12. What does the collinear constraint allow you to do?
13. Explain the basic function of the equal constraint.
14. Describe the default function of the symmetric constraint.
15. Name the command you can use to attempt to add all required geometric constraints in a single operation.
16. Briefly describe how to specify the appearance and characteristics of geometric constraint bars.
17. Compare the appearance of a coincident constraint with the display of other constraints.
18. Explain how to determine which objects and points are associated with a constraint.
19. What should you do if constraint bars block your view or if you want to hide constraint bars?
20. Name the command that allows you to assign linear, diameter, radius, and angular dimensional constraints.
21. What is the most basic method to specify dimension values when you create a dimensional constraint?
22. Name the tools that allow you to place a linear dimensional constraint with a dimension line aligned with an angled surface with extension lines perpendicular to the surface.
23. Which commands allow you to place an angular dimension between two objects or three points?
24. Explain the options AutoCAD provides when you try to over-constrain a drawing.
25. Briefly explain how to convert an associative dimension to a dimensional constraint, and give the advantage of using this option.
26. What happens every time you add a dimensional constraint?
27. How do you adjust parameters?
28. Explain how to edit a dimensional constraint value.
29. Briefly describe how to relax constraints.
30. Which command provides an efficient method of removing a significant number of constraints in a single operation?

Drawing Problems

Start AutoCAD if it is not already started. Start a new drawing for each problem using an appropriate template of your choice. The template should include layers and text, dimension, multileader, and table styles, when necessary, for drawing the given objects. Add layers and text, dimension, multileader, and table styles as needed. Draw all objects using appropriate layers and text, dimension, multileader, and table styles, justification, and format. Follow the specific instructions for each problem. Use only drawing and editing commands and techniques you have already learned. Use your own judgment and approximate dimensions when necessary. Apply formal dimensions accurately using ASME or appropriate industry standards.

Note: Dimensional constraints shown for reference are created using AutoCAD and may not comply with ASME standards.

▼ Basic

1. Use the **POLYGON** command to draw the hexagon as shown. Fully constrain the hexagon as shown. All sides are equal. Fix the midpoint of the construction line. Edit the d1 parameter to change the distance across the flats to 4.000. Save the drawing as P22-1.

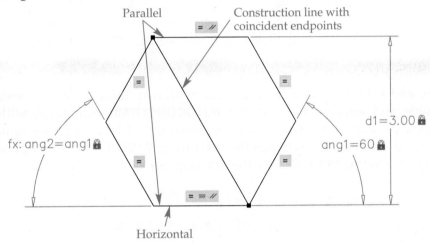

2. Use the **RECTANGLE** and **CIRCLE** commands to draw the view shown. Fully constrain the view as shown. The circle is tangent to the rectangle in two locations. Edit the d2 parameter to change the distance to 4.500. Edit the d3 parameter to change the distance to 2.000. Save the drawing as P22-2.

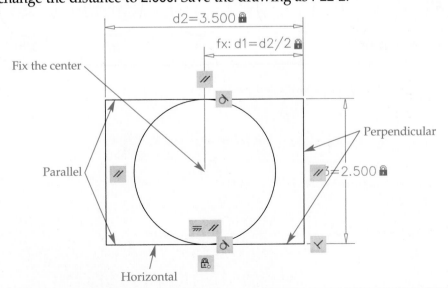

Chapter 22   Parametric Drafting

3. Draw and fully constrain the circles shown. Both of the smaller circles are tangent to the larger circle. Edit the dia1 parameter to change the diameter to 2.000. Save the drawing as P22-3.

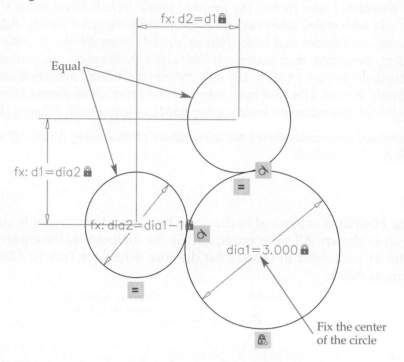

4. Draw and fully constrain the pipe spacer shown. Infer as many constraints as possible and appropriate. Use the **AUTOCONSTRAIN** command, with default settings, to apply additional geometric constraints. Make all circles equal in size. Edit the d1 parameter to change the distance to 2.000. Edit the rad1 parameter to change the radius to 1.250. Save the drawing as P22-4.

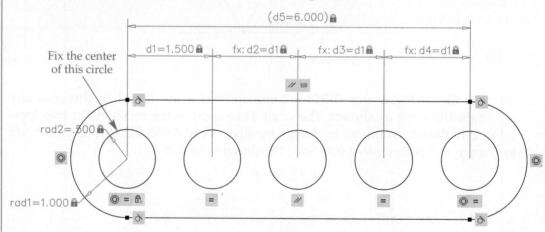

▼ Intermediate

5. Draw and fully constrain the view shown. Do not infer constraints. Apply geometric constraints in the following order: fix the center of the center circle; use the **Autoconstrain** function of the **Coincident** option to apply all coincident constraints; make all circles equal; make all small arcs equal; make all large arcs concentric to the appropriate circles; use the **AUTOCONSTRAIN** command with default settings to apply the remaining geometric constraints. Edit the d1 parameter to change the distance to 1.750. Save the drawing as P22-5.

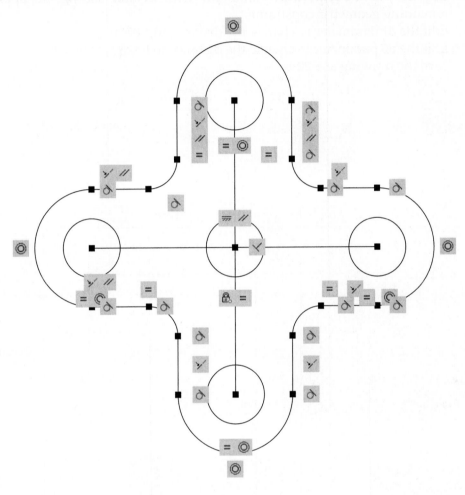

Chapter 22 Parametric Drafting

701

Drawing Problems - Chapter 22

6. Draw and fully constrain the view shown. Follow the guidelines below.
   - Infer constraints when possible and appropriate.
   - Use the **Fillet** option of the **RECTANGLE** command to create the 4.6250 by 6.3750 rectangle with .6250 rounded corners.
   - Use the **RECTANGLE** command to create the construction rectangle and then add the circles. Notice that the circles are not concentric to the arcs.
   - Apply geometric constraints in the following order: fix the center of the lower-left circle; make all circles equal; make all arcs equal.
   - Use the **AUTOCONSTRAIN** command with default settings to apply the remaining geometric constraints.
   - Edit the d1 parameter to change the distance to 1.0000.
   - Edit the d5 parameter to change the distance to 5.1250.
   - Save the drawing as P22-6.

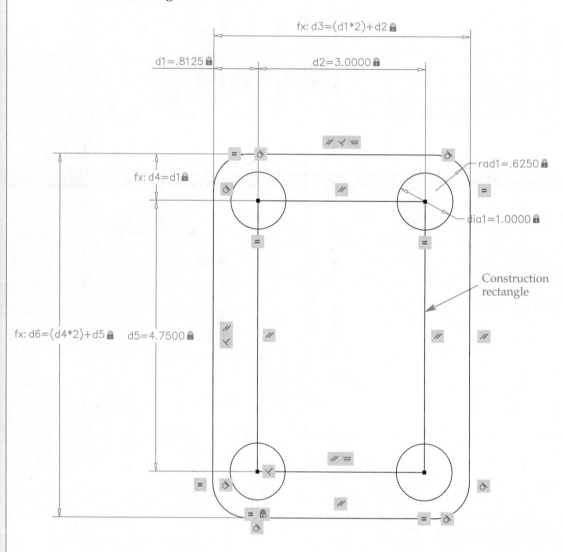

Drawing Problems – Chapter 22

▼ Advanced

7. Draw and fully constrain the view shown. Save the drawing as **P22-7**.

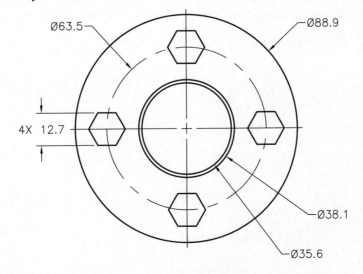

8. Use the information in B to draw the non-parametric view shown in A. Then fully constrain the view as shown in B. Edit the view as shown in C. Finish by converting dimensional constraints to create the formal drawing shown in D. Save the drawing as **P22-8**.

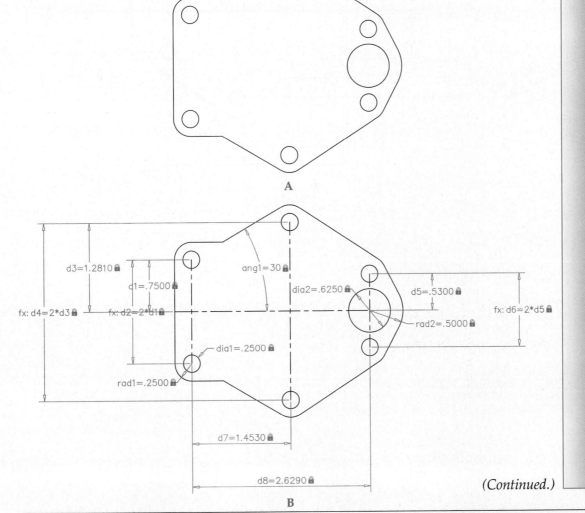

(Continued.)

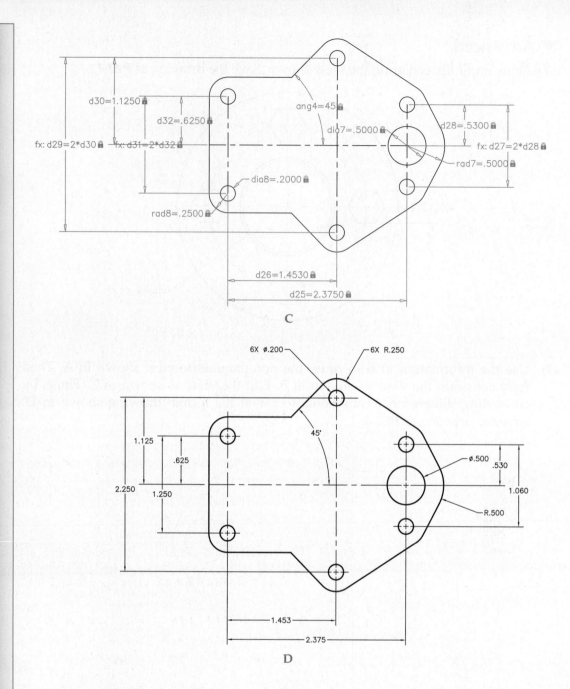

C

D

9. Draw and fully constrain the view shown. Convert dynamic dimensional constraints to annotational dimensional constraints. Adjust the drawing and add associative dimensions as needed to create the formal drawing shown. Save the drawing as P22-9.

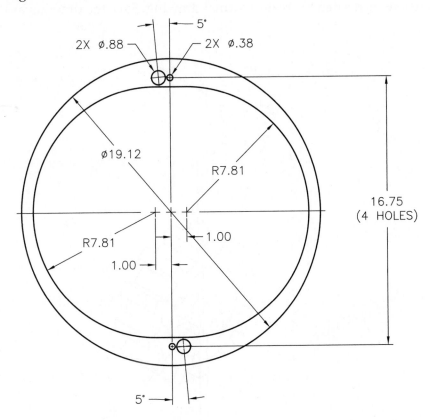

10. Draw and fully constrain the multiview drawing of the support shown. Convert dynamic dimensional constraints to annotational dimensional constraints. Adjust the drawing and add associative dimensions as needed to create the formal drawing shown. Save the drawing as P22-10.

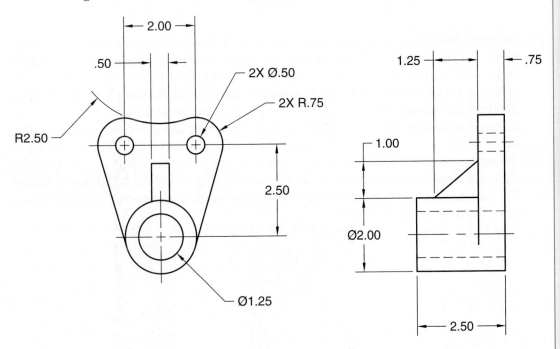

11. Use the isometric drawing provided to create a multiview orthographic drawing for the part. Include only the views necessary to fully describe the object. Draw and fully constrain the drawing. Convert dynamic dimensional constraints to annotational dimensional constraints. Adjust the drawing and add associative dimensions as needed to create a formal drawing. Save the drawing as P22-11.

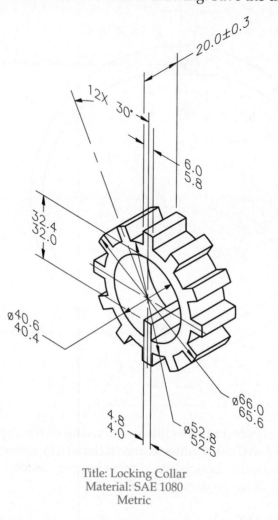

Title: Locking Collar
Material: SAE 1080
Metric

12. Research the design of an existing manifold with the following specifications: one-piece; nickel-plated 6061 aluminum, NPT female connections; two 1/2" inlets; eight 3/8" outlets on one side, 1" center-to-center; four mounting holes; for use with air, water, or hydraulic oil. Create a dimensioned 2D sketch of the existing design from manufacturer's specifications, or from measurements taken from an actual manifold. Start a new drawing from scratch or use a decimal-unit template of your choice. Draw and fully constrain a multiview orthographic drawing of the manifold. Include only the views necessary to fully describe the part. Convert dynamic dimensional constraints to annotational dimensional constraints. Adjust the drawing and add associative dimensions as needed to create a formal drawing. Save the drawing as P22-12.

AutoCAD Certified Associate Exam Practice

Answer the following questions. Write your answers on a separate sheet of paper.

1. Which of the following are forms of dimensional constraints? *Select all that apply.*
   A. annotational
   B. construction
   C. dynamic
   D. geometric
   E. static

2. What type of dimension can you place instead of a dimensional constraint if a dimensional constraint would over-constrain the drawing? *Select the one item that best answers the question.*
   A. annotational dimension
   B. construction dimension
   C. dynamic dimension
   D. inferred dimension
   E. reference dimension

3. Which of the following actions can you perform using the **DCLINEAR** command? *Select all that apply.*
   A. constrain a horizontal distance
   B. constrain a vertical distance
   C. constrain the horizontal distance of an angled surface
   D. place a horizontal geometric constraint
   E. place a vertical geometric constraint

AutoCAD Certified Professional Exam Practice

Follow the instructions in each problem. Write your answers on a separate sheet of paper.

1. **Open CPE-22parameter.dwg. This file is available on the companion website.** Constrain the lines as shown. Begin by applying a fix constraint to the lower-left intersection. Then apply the appropriate constraints to achieve the figure shown. Do not change the length of any of the lines. What are the coordinates of Point A?

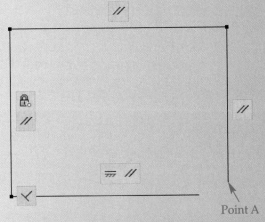

Point A

2. **Open CPE-22edit.dwg. This file is available on the companion website.**
   Show the dimensional constraints and edit the LENGTH dimension to 7.500. Do not change any other settings. What are the coordinates of Point A (the center of the right arc)?

Point A

Section Views and Graphic Patterns

23

Learning Objectives

After completing this chapter, you will be able to:

✓ Identify sectioning techniques.
✓ Add graphic patterns using the **HATCH** command.
✓ Insert hatch patterns using **DesignCenter**.
✓ Insert hatch patterns using tool palettes.
✓ Edit existing hatch patterns.

Drawings often require *graphic patterns* to describe specific information. For example, the front elevation of the house shown in **Figure 23-1** contains patterns and fills that graphically represent building materials and shading. One of the most common graphic patterns is a group of section lines added to a section view. This chapter focuses on using the **HATCH** command to draw graphic patterns. You will also learn other methods of adding and edit hatches.

graphic pattern: A patterned arrangement of objects or symbols.

Section Views

It is poor practice, and often not possible, to dimension internal, hidden features. *Section views*, also called *sectional views* or *sections*, clarify hidden features. Typically, you add section views to a multiview drawing to describe the exterior and interior features of a product. Often, as shown in **Figure 23-2**, a primary view is a section or includes a section.

section view (sectional view, section): A view that shows internal features as if a portion of the object is cut away.

When a drawing includes a section, one of the other views contains a *cutting-plane line* to show the location of the cut. The cutting-plane line is a thick dashed or phantom line in accordance with the ASME Y14.2 *Line Conventions and Lettering* standard. The standard cutting-plane line terminates with arrowheads that point toward the cutting plane, indicating the line of sight when you are looking at the section view.

cutting-plane line: The line that cuts through the object to expose internal features.

Each end of the cutting-plane line is labeled with a letter. The letters correlate to the section view title below the section view, such as SECTION A-A, to key the cutting plane with the section view. When you section more than one view, labels continue with B-B through Z-Z, if necessary. Do not use letters I, O, Q, S, X, and Z, because they can be confused with numbers. Labeling section views is necessary for drawings with multiple sections. You can often omit the label when only one section view is present and its location is obvious.

Figure 23-1.
Graphic patterns describe repetitive drawing information, such as the siding, brick, roofing, concrete, and shading added to the front elevation of a house. In this illustration, all of the items shown in color are graphic patterns.

Figure 23-2.
A two-view mechanical part drawing with a full section view (front) and a broken-out section (left-side).

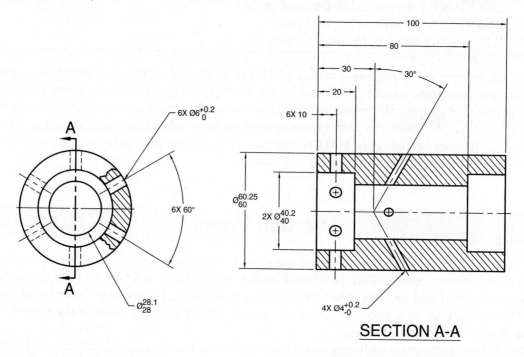

SECTION A-A

Other drafting fields, including architectural, structural, and civil drafting, also use sectioning. Cross sections through buildings show construction methods and materials. See **Figure 23-3**. Profiles and cross sections on a civil engineering drawing show the contour and construction of land for utility, transportation, and other civil engineering projects. Cutting-plane lines used in nonmechanical fields are often composed of letter and number symbols to help coordinate the large number of sections often found in a set of drawings.

AutoCAD and Its Applications—Basics

Figure 23-3.
An architectural section view showing the construction of a proposed bathroom for a residential renovation project.

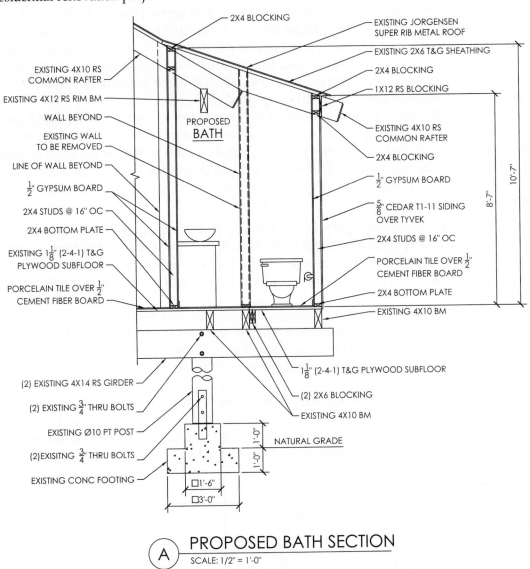

2X4 BLOCKING

EXISTING JORGENSEN SUPER RIB METAL ROOF

EXISTING 4X10 RS COMMON RAFTER

EXISTING 2X6 T&G SHEATHING

EXISTING 4X12 RS RIM BM

2X4 BLOCKING

WALL BEYOND

1X12 RS BLOCKING

EXISTING WALL TO BE REMOVED

PROPOSED BATH

EXISTING 4X10 RS COMMON RAFTER

LINE OF WALL BEYOND

2X4 BLOCKING

$\frac{1}{2}$" GYPSUM BOARD

$\frac{1}{2}$" GYPSUM BOARD

$\frac{5}{8}$" CEDAR T1-11 SIDING OVER TYVEK

2X4 STUDS @ 16" OC

2X4 STUDS @ 16" OC

2X4 BOTTOM PLATE

EXISTING 1$\frac{1}{8}$" (2-4-1) T&G PLYWOOD SUBFLOOR

PORCELAIN TILE OVER $\frac{1}{2}$" CEMENT FIBER BOARD

PORCELAIN TILE OVER $\frac{1}{2}$" CEMENT FIBER BOARD

2X4 BOTTOM PLATE

EXISTING 4X10 BM

10'-7"

8'-7"

(2) EXISTING 4X14 RS GIRDER

1$\frac{1}{8}$" (2-4-1) T&G PLYWOOD SUBFLOOR

(2) EXISTING $\frac{3}{4}$" THRU BOLTS

(2) 2X6 BLOCKING

EXISTING Ø10 PT POST

EXISTING 4X10 BM

(2)EXISITNG $\frac{3}{4}$" THRU BOLTS

NATURAL GRADE

EXISTING CONC FOOTING

1'-0"

1'-0"

1'-0"

□1'-6"

□3'-0"

Ⓐ **PROPOSED BATH SECTION**
SCALE: 1/2" = 1'-0"

Section Lines

Section lines are the graphic pattern used in a section view. Section lines distinguish hidden features from exterior objects. Section views include section line symbols to show where material is cut to reveal hidden features. The standard on mechanical drawings is 45° section lines, unless another angle is required to satisfy other section line rules. Avoid drawing section lines at angles greater than 75° or less than 15° from horizontal. Section lines should never be parallel or perpendicular to adjacent lines on the drawing. In addition, section lines should not cross object lines.

Equally spaced section lines, with a minimum .063" (1.5 mm) spacing, are standard on mechanical drawings, and are adequate for basic applications such as the front view in **Figure 23-2.** Depending on the drawing and discipline, you may use different patterns to clarify the drawing or to represent the specific material cut by the section. A specific material pattern is not necessary on a part drawing if the title block or a note clearly indicates the material. However, use different or coded section lines on an assembly drawing to represent each component or different material. Section line symbols on a nonmechanical drawing can be lines or can consist of graphic patterns

section lines: Lines that show where material is cut away.

that represent specific materials, such as insulation, earth, and concrete on an architectural or structural section. Section lines may also be omitted for clarity, as is common on civil engineering profiles.

AutoCAD provides standard graphic patterns and section line symbols known as *hatches*, or *hatch patterns*. The acad.pat file stores standard hatch patterns. The ANSI31 pattern is a general section line symbol and is the default pattern in some templates. The ANSI31 pattern also represents cast iron in a section. The ANSI32 symbol identifies steel in a section. When you change to a different hatch pattern, the new pattern becomes the default in the current drawing until changed.

Use the Solid hatch pattern whenever it is necessary to create a solid filled area. The ASME Y14.3 *Multiview and Sectional View Drawings* standard recommends omitting section lines on section views showing thin features, such as ribs and lugs. The Solid hatch pattern is appropriate for thin sections if you do not follow the ASME standard.

Types of Sections

Choose the appropriate type of section based on the application and features to be sectioned. For example, an object that includes a significant number of hidden features may require a section that cuts completely through the object. In contrast, a drawing may require that you remove a small portion to expose and dimension a single, minor interior feature.

The front view in **Figure 23-2** shows an example of a *full section*. The cutting plane applied to a full section passes completely through the view, typically along the center plane, as shown by the cutting-plane line. *Offset sections* are similar to full sections, except the cutting plane staggers to cut through features that are not in a straight line. See **Figure 23-4**.

Figure 23-5 provides an example of an *aligned section*. The cutting plane cuts through the feature and then rotates to align with the center plane before projecting onto the section view. An aligned section shows the true size and shape of the aligned features. If you use direct projection, such as with a full or offset section, the section will appear foreshortened.

hatches (hatch patterns): AutoCAD section line symbols and graphic patterns.

full sections: Sections that show half the object removed.

offset sections: Sections that have a staggered cutting plane.

aligned sections: Sections used when a feature is out of alignment with the center plane.

Figure 23-4.
A two-view mechanical part drawing with an offset section view.

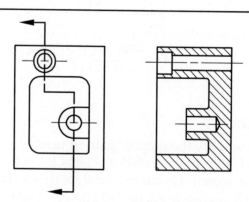

Figure 23-5.
A two-view
mechanical part
drawing with an
aligned section view.

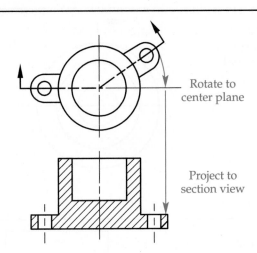

Rotate to
center plane

Project to
section view

Figure 23-6.
The same revolved section drawn in place and with broken object lines. Breaking the view is appropriate for this part to clarify the profile. Revolving in place is most common when the section shows a cylindrical or similar feature.

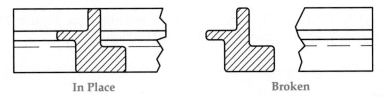

In Place   Broken

Figure 23-6 shows an example of a *revolved section*. A revolved section may appear in place within the object, or a portion of the view may be broken away to make dimensioning easier. Omit cutting-plane lines with revolved sections. Use a *removed section* when it is not possible to show a standard section in direct projection from the cutting plane, or when a revolved section is inappropriate. See **Figure 23-7**. A cutting-plane line identifies the location of the section. Multiple removed sections require labeled cutting-plane lines and related views. Drawing only the ends of the cutting-plane lines simplifies the views.

Figure 23-8 shows an example of a *half section*. The term *half* describes how half of the view appears in section, while the other half remains as an exterior view. Half sections are commonly used to draw symmetrical objects. A centerline separates the sectioned portion of the view from the unsectioned portion. You normally omit hidden lines from the unsectioned half.

revolved sections:
Sections that clarify the contour of objects that have the same shape throughout their length.

removed sections:
Standard section views, but removed from direct projection from the cutting plane.

half sections:
Sections that show one-quarter of the object removed.

Figure 23-7.
A drawing with two removed sections (Section A-A and Section C-C) and a full section (Section B-B).

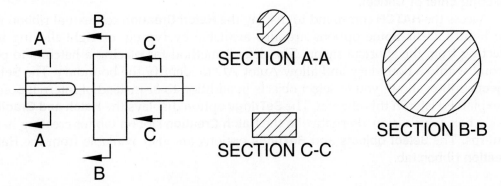

SECTION A-A

SECTION C-C

SECTION B-B

Figure 23-8.
A two-view
mechanical part
drawing with a half
section.

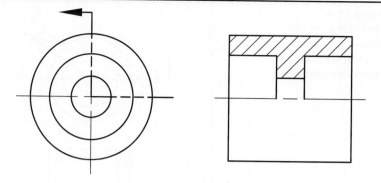

Figure 23-9.
Use a broken-out
section to display
specific hidden
features.

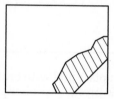

**broken-out
sections:** Sections
that show a small
portion of the object
removed.

Broken-out sections clarify specific hidden features. See **Figure 23-9**. The left-side view in **Figure 23-2** also shows a broken-out section.

Section lines are a basic application for hatch patterns. Many other drawing requirements also use hatching, such as material and shading representation on an architectural elevation, shading on a technical illustration, and artistic patterns for graphic layouts.

boundary: The area
filled by a hatch.

HATCH

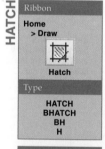

Ribbon
**Home
> Draw**

Hatch

Type
**HATCH
BHATCH
BH
H**

Ribbon
**Hatch Creation
> Close**

**Close Hatch
Creation**

Using the HATCH Command

The **HATCH** command simplifies the process of creating section lines and graphic patterns by forming a single hatch object that fills an existing *boundary*. The boundary is typically a closed object or group of connected objects, but a small opening is possible with an appropriate gap tolerance, as explained later in this chapter. Review the previous figures to identify the boundaries associated with each view. The typical approach to hatching is to specify boundaries, adjust hatch properties and other settings, and then create the hatch and exit the **HATCH** command by picking the **Close Hatch Creation** button, pressing [Esc], [Enter], or the space bar, or right-clicking and selecting **Enter** or **Cancel**.

Access the **HATCH** command to display the **Hatch Creation** contextual ribbon tab. See **Figure 23-10**. Some options are also available by typing, or right-clicking and selecting from the shortcut menu. The default method for placing a hatch is to pick a point within a boundary and allow AutoCAD to identify the boundary. The **Select objects** option allows you to select objects in addition to or instead of internal points, as explained later in this chapter. The **SeTtings** option displays the **Hatch and Gradient** dialog box, which is an alternative to the **Hatch Creation** ribbon tab for creating hatch patterns. The **Select objects** and **SeTtings** options are also available from the **Hatch Creation** ribbon tab.

Hatch Creation tab

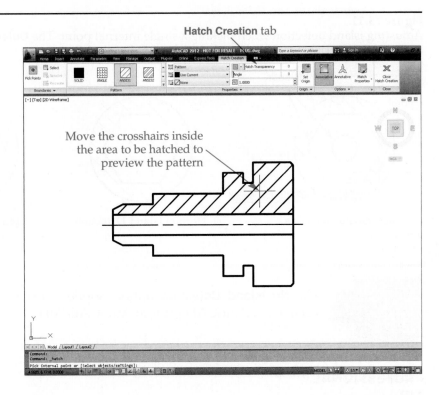

Move the crosshairs inside the area to be hatched to preview the pattern

The **Hatch Creation** ribbon tab is contextual. As a result, pressing [Esc], [Enter], or the space bar or right-clicking and selecting **Enter** or **Cancel** exits the **HATCH** command and creates a hatch if you have specified boundaries.

Specifying Boundaries

The default and typically the easiest method for placing a hatch is to pick a point within a boundary and allow AutoCAD to detect the boundary. Move the crosshairs inside a boundary to preview the hatch, as shown in **Figure 23-10**. Use the preview to help identify the boundary. The initial preview may not use the correct hatch pattern or boundary properties, but should allow you to determine if selecting the point will fill the acceptable boundary. You can adjust the pattern and boundary properties interactively after you specify boundaries.

If the boundary looks correct, pick the point to apply the hatch and highlight the boundary. Continue picking points to specify additional boundaries to include with the hatch object. If a preview does not appear, AutoCAD cannot detect the boundary or the current hatch scale is too large for the size of the boundary. The most common reason AutoCAD cannot detect a boundary is because the boundary contains gaps. Exit the **HATCH** command and edit the drawing to create a closed boundary.

Island Detection

When picking points to specify hatch boundaries, you may need to adjust how AutoCAD treats *islands*, as shown in **Figure 23-11**. Use the flyout in the expanded **Options** panel to specify island detection. Pick the **Normal Island Detection** option to hatch every other boundary, stepping inward from the outer boundary. Select the **Outer Island Detection** option to hatch only the outermost boundary. Choose the **Ignore Island Detection** option to ignore all islands and hatch everything within the outer boundary.

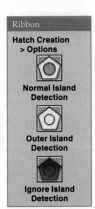

Ribbon

Hatch Creation > Options

Normal Island Detection

Outer Island Detection

Ignore Island Detection

islands: Boundaries inside another boundary.

Figure 23-11.
Adjusting island detection when picking a single internal point. The **Outer Island Detection** option is appropriate for this example and for most applications.

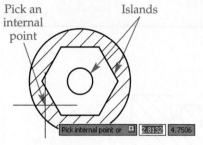
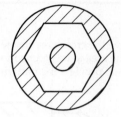
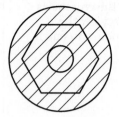

Outer Island Detection   Normal Island Detection   Ignore Island Detection

The **No Island Detection** option applies to hatches created with versions of AutoCAD prior to AutoCAD 2011.

PROFESSIONAL TIP

Pick the **Outer** island display style to ensure that inner islands are not hatched unintentionally.

Exercise 23-1

Complete the exercise on the companion website.
www.g-wlearning.com/CAD

Selecting Objects

Ribbon
Hatch Creation
> Boundaries

Select Boundary
Objects

An alternative method to specify a boundary is to use the **Select Boundary Objects** option to select objects that form a boundary. A common example is selecting a closed object such as a polyline, circle, or group of connected objects to hatch an area that would be difficult or time-consuming to hatch by picking points because of numerous internal boundaries. See **Figure 23-12**. When necessary, select objects within the hatch boundary to exclude from the hatch pattern. See **Figure 23-13**. The preview automatically updates according to the selections. Use the **Pick Points** option to return to internal point selection mode.

Ribbon
Hatch Creation
> Boundaries

Pick Points

If you specify the wrong boundary, use the **U** command as needed to undo the selection without exiting the **HATCH** command.

Removing Boundaries

Ribbon
Hatch Creation
> Boundaries

Remove

The **Remove boundaries** option is available after you specify a boundary. Use the **Remove boundaries** option to select unwanted boundaries or objects to remove from the selection set.

Figure 23-12.
An example of an architectural elevation on which it is easier to use the **Select Boundary Objects** option to specify a hatch boundary. The closed polyline around the area to be hatched is for construction purposes only.

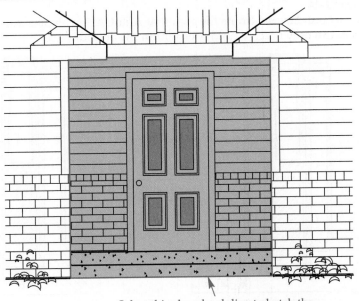

Select this closed polyline to hatch the area instead of picking multiple points

Figure 23-13.
Using the **Select Boundary Objects** option to exclude a polyline hexagon and text object from the hatch pattern.

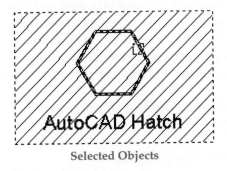

Selected Objects

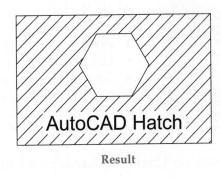

Result

Exercise 23-2

Complete the exercise on the companion website.
www.g-wlearning.com/CAD

Boundary Set

By default, the **HATCH** command evaluates the entire current viewport to detect boundaries. To limit the evaluation area, possibly increasing hatch performance, pick the **Select new boundary set** option and use a window to define the area to evaluate. See **Figure 23-14A**. Right-click or press [Enter] or the space bar to create the boundary set. Toggle between the **Use Boundary Set** and **Use Current Viewport** option from the **Specify Boundary Set** drop-down list in the expanded **Boundaries** panel. Creating a new boundary set overrides the previous boundary set. **Figure 23-14B** shows a new hatch pattern applied to boundaries within the boundary set.

Ribbon

Hatch Creation
> Boundaries

Select new boundary set

Figure 23-14.
A—The boundary set limits the area that AutoCAD evaluates during a hatching operation.
B—You can only hatch boundaries within the boundary set.

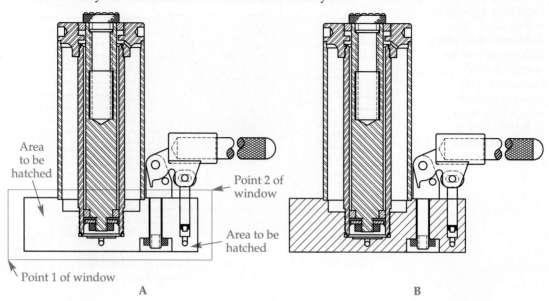

Area to be hatched

Point 2 of window

Area to be hatched

Point 1 of window

A

B

PROFESSIONAL TIP

Apply the following techniques to specify a boundary area and save time, especially when you are hatching large, complex drawings:
- Zoom in on the boundary area to be hatched to aid selection.
- Use previews to confirm correct boundaries before picking.
- Turn off layers assigned to objects that might interfere with boundary definition.
- Create boundary sets of small areas within a complex drawing.

Hatching Unclosed Areas and Correcting Boundary Errors

The **HATCH** command works well unless there is a gap in a boundary, or you pick a point or select objects outside a likely boundary. When you select a point or objects where no boundary can form, an error message states that a valid boundary cannot be determined. Close the message and try again to specify the boundary. When you try to hatch an area that does not close because of a small gap, you will see the error message and circles shown in **Figure 23-15**. Close the message and eliminate the gap to create the hatch. Use the **REGEN** or **REDRAW** command to hide the circles at the probable gap.

For most applications, it is best to identify and close a gap. However, you can hatch an unclosed boundary by setting a *gap tolerance* using the **Gap Tolerance** option. Use the slider or enter a value in the text box up to 5000. AutoCAD ignores any gaps in the boundary less than or equal to the tolerance value.

Retaining Boundaries

By default, a hatch pattern boundary forms according to objects in the drawing. The boundary is temporary, which is appropriate for most applications. Select the **Retain Boundary - Polyline** option to form a separate polyline object overlapping the boundary. Choose the **Retain Boundary - Region** option to form a separate *region* object overlapping the boundary. Retaining a boundary forms additional boundary objects that you can use even if you remove or edit the original objects.

gap tolerance:
The amount of gap allowed between segments of a boundary to be hatched.

Ribbon
**Hatch Creation
> Options
> Gap Tolerance**

Ribbon
**Hatch Creation
> Boundaries**

Retain Boundary - Polyline
Retain Boundary - Region

region: A closed two-dimensional area.

Figure 23-15.
Close a boundary to create a hatch pattern.

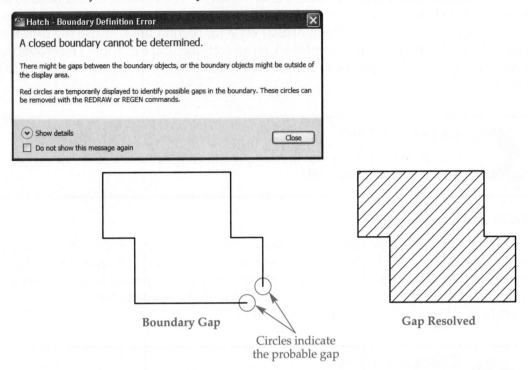

Boundary Gap **Gap Resolved**

Circles indicate
the probable gap

Selecting a Hatch Pattern

The **Pattern** panel includes all of the hatch patterns and fills supplied with AutoCAD. Use the scroll buttons to the right of the patterns to locate a pattern, or pick the expansion arrow to display patterns in a temporary window. See **Figure 23-16**. The **Hatch Type** drop-down list in the **Properties** panel includes Solid, Gradient, Pattern, and User defined categories to help you locate specific hatches in the **Pattern** panel. Select a category to display related hatches in the **Pattern** panel. Select a hatch pattern from the **Pattern** panel to apply to the specified boundaries.

Ribbon

Hatch Creation
> Properties

Pattern

Figure 23-16.
Select a hatch pattern from the **Pattern** panel of the **Hatch Creation** ribbon tab.

Choose a category to view related
hatches in the **Pattern** panel

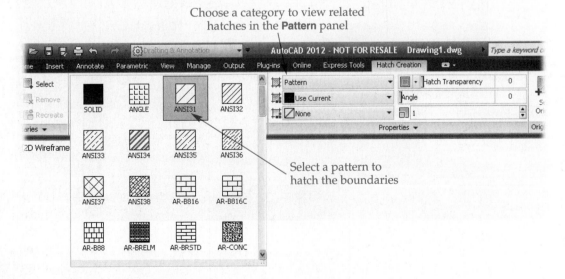

Select a pattern to
hatch the boundaries

Patterns

The **Pattern** hatch type provides patterns stored in the acad.pat and acadiso.pat files. AutoCAD includes many different patterns to accommodate most hatch requirements. For example, use the default ANSI31 style for section lines, or use the BRICK or one of the AR-B patterns to represent brick on an architectural elevation. The **Properties** panel provides settings specific to the selected hatch pattern, as explained later in this chapter.

Exercise 23-3

Complete the exercise on the companion website.
www.g-wlearning.com/CAD

Solid and Gradient Fills

The **Solid** hatch type and corresponding SOLID hatch provide an effective way to fill a boundary with a solid. **Figure 23-17** shows an example of several boundaries filled with the SOLID hatch. This example uses specific transparent layers assigned to different hatch objects. Use an option associated with the **Gradient** hatch type to create a gradient fill, as explained later in this chapter.

Exercise 23-4

Complete the exercise on the companion website.
www.g-wlearning.com/CAD

Figure 23-17.
A portion of a storm water pollution control plan with a transparent SOLID hatch. The solid fills represent different impervious and non-impervious surfaces.

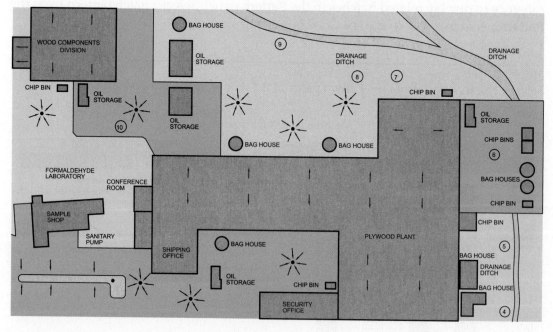

The **SOLID** command allows you to draw basic solid shapes without creating a boundary. Access the **SOLID** command and pick points in a specific sequence to form different shapes. The **SOLID** hatch applied using the **HATCH** command is typically a better method of creating solid fills.

User Defined Hatch

The **User defined** hatch type and corresponding USER hatch create a pattern of equally spaced lines for basic hatching applications. The lines use the linetype assigned to the current layer. Use options in the **Properties** panel of the **Hatch Creation** ribbon tab to adjust the USER hatch appearance, as explained later in this chapter.

Create and save custom hatch patterns in PAT files. Add the files to the AutoCAD search path to have access to custom hatch from the **Pattern** panel. *AutoCAD and Its Applications—Advanced* provides more information about customizing hatch-related dialog boxes.

Pattern Size

The **Properties** panel provides hatch pattern appearance control. Scale, or spacing, is a primary hatch property. Use the **Hatch Pattern Scale** option to adjust hatches of the **Pattern** type. The default scale is 1. If the pattern appears too small or large, specify a different scale. See **Figure 23-18**. For example, by default, the ANSI31 hatch is a pattern of lines spaced .125″ (3 mm) apart. If you change the scale to 2, the pattern of lines is spaced .25″ (6 mm) apart.

The **Hatch Spacing** option replaces the **Hatch Pattern Scale** option when you select the USER hatch. Specify the exact distance between lines. Changes to scale or spacing update automatically in the preview of the specified boundaries.

Ribbon

Hatch Creation
> Properties

Hatch Pattern
Scale

PROFESSIONAL TIP

Use a smaller hatch size for small objects and a larger hatch size for larger objects. This makes section lines look appropriate for the drawing scale. Often you must use your best judgment when selecting a hatch size.

Use an appropriate hatch scale or spacing to display the hatch pattern on-screen and plot correctly according to the drawing scale. To understand the concept of hatch size, look at the section view shown in **Figure 23-19**. In this example, which uses the ANSI31 hatch pattern, the section line spacing should be the same distance apart regardless of drawing scale. The section lines on the full-scale (1:1) drawing display

Figure 23-18.
Hatch pattern scale applies to hatches of the **Pattern** type and to custom hatch patterns.

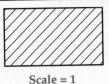

Scale = 1

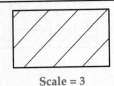

Scale = 2

Scale = 3

correctly. However, the section lines are too close on the half-scale (1:2) drawing, and they are too far apart on the double-scale (2:1) drawing. To obtain the correct results, you must adjust the hatch size according to the drawing scale. You can calculate scale factor manually and apply it to hatch scale or spacing, or you can allow AutoCAD to calculate the scale factor using annotative hatch patterns.

Scaling Hatch Patterns Manually

To adjust hatch size manually according to a specific drawing scale, you must first calculate the drawing scale factor. Then multiply the scale factor by the plotted hatch scale or spacing to get the model space hatch scale or spacing. Specify the scale of predefined or custom hatch patterns using the **Hatch Pattern Scale** option. Specify the spacing of the USER hatch using **Hatch Spacing** option. **Figure 23-20** shows examples of adjusting hatch scale according to drawing scale. Refer to Chapter 9 for information on determining the drawing scale factor.

Annotative Hatch Patterns

Ribbon

Hatch Creation > Options

Annotative

Pick the **Annotative** button in the **Options** panel to make the hatch pattern annotative. AutoCAD scales annotative hatches according to the annotation scale you select, which eliminates the need for you to calculate the scale factor. Once you choose an annotation scale, AutoCAD applies the corresponding scale factor to annotative hatches and all other annotative objects.

The hatch pattern is displayed at the proper size regardless of the drawing scale, much like the example shown in **Figure 23-20**, but without requiring you to change the hatch scale or spacing. For example, if you specify a value using the **Hatch Pattern Scale** or **Hatch Spacing** option appropriate for an annotation scale of 1/4″ = 1′-0″, and then change the annotation scale to 1″ = 1′-0″, the appearance of the hatch pattern relative to the drawing scale does not change. It looks the same on the 1/4″ = 1′-0″ scale drawing as it does on the 1″ = 1′-0″ scale drawing. Refer to Chapter 9 for information on setting annotation scale.

Relative to Paper Space Option

Ribbon

Hatch Creation > Properties

Relative to Paper Space

The **Relative to Paper Space** option in the expanded **Properties** panel allows you to scale the hatch pattern relative to the scale of the active layout viewport. You must enter a floating layout viewport in order to select the **Relative to Paper Space** option. The hatch scale automatically adjusts according to the viewport scale. For example, a floating viewport scale set to 4:1 uses a scale factor of .25 (1 ÷ 4 = .25). If you enter a hatch scale of 1, the hatch automatically appears at a scale of .25 (1 × .25 = .25).

Figure 23-19.
The hatch pattern may appear incorrect if the drawing scale changes.

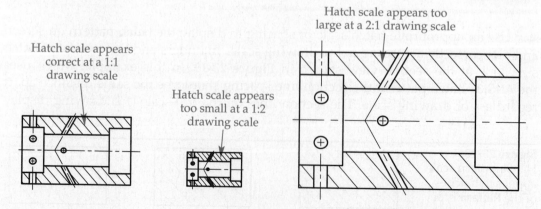

You can control the ISO pen width for predefined ISO patterns using the **ISO Pen Width:** drop-down list in the expanded **Properties** panel.

Ribbon
Hatch Creation
> Properties
> ISO Pen
Width

Additional Pattern Properties

The **Properties** panel provides other settings that you can preview interactively as you make changes. The **Hatch Angle** option controls pattern rotation. Use the slider or enter a value up to 359 in the text box to rotate the pattern relative to the X axis. For example, the ANSI31 hatch is a pattern of 45° lines. Change the angle to 15° to form a pattern of 60° (45 + 15 = 60) lines. The **Double** option is available with the USER hatch and allows you to create a pattern of double lines. **Figure 23-21** shows examples of user-defined hatch patterns.

The **Background Color** option allows you to fill the specified boundaries with the hatch pattern and a background color similar to a solid fill. See **Figure 23-22**. Hatching with a solid fill and pattern is not a common drafting practice, but is appropriate in some applications. Choose a color from the drop-down list or pick the **Select Colors...**

Ribbon
Hatch Creation
> Properties
> Hatch Angle

Ribbon
Hatch Creation
> Properties

Double

Ribbon
Hatch Creation
> Properties

Background color

Figure 23-20.
The hatch pattern scale may require adjusting, depending on the drawing scale.

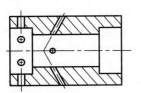

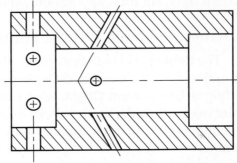

Drawing scale: 1:1
Drawing scale factor: 1
Hatch scale: 1

Drawing scale: 1:2
Drawing scale factor: 2
Hatch scale: 2

Drawing scale: 2:1
Drawing scale factor: .5
Hatch scale: .5

Figure 23-21.
Examples of user-defined hatch patterns with different hatch angles and spacing.

Angle	0°	45°	0°	45°
Spacing	.125	.125	.250	.250
Single Hatch				
Double Hatch				

Figure 23-22.
Use the **Background Color** option on the **Properties** panel to fill boundaries with a color and a hatch pattern. This example shows an architectural window elevation with background Color 9 and the ANSI34 pattern.

option to pick a color from the **Select Color** dialog box. The color must be different from the color assigned to the pattern layer.

Ribbon

Hatch Creation > Properties

Hatch Layer Override

Use the **Hatch Layer Override** option to apply a different layer to the hatch without exiting the **HATCH** command and making the different layer current. The **Hatch Color** and **Hatch Transparency** options override the color and transparency assigned to the current layer. For most applications, you should not override color or transparency.

Setting the Hatch Origin Point

Ribbon

Hatch Creation > Origin

Set Origin

The **Origin** panel includes options that control the position of hatch patterns. Select an option from the expanded **Origin** panel, or pick the **Set Origin** button. The default setting is **Use Current Origin**, which refers to the current UCS origin, to define the point from which the hatch pattern forms and how the pattern repeats. In some cases, it is important that a hatch pattern align with, or originate from, a specific point. A common example is hatching a representation of bricks. To specify a different origin point, pick the **Set Origin** button and select an origin point. See **Figure 23-23**.

The expanded **Origin** panel includes **Bottom Left**, **Bottom Right**, **Top Right**, **Top Left**, and **Center** options. Select an option to align the hatch origin point with the corresponding point on the hatch boundary. For example, use the **Bottom Right** option to create the pattern shown in **Figure 23-23B**. Select the **Store as Default Origin** option to save the custom origin point.

Figure 23-23.
A—The default **Use current origin** setting. B—Pick the **Set Origin** button and select the lower-left corner (endpoint) of the rectangle. Notice that the pattern, or in this example the first brick, starts exactly at the corner of the hatched area.

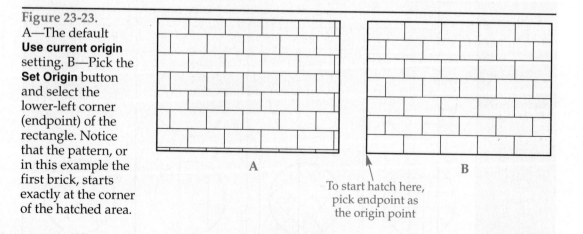

A

B

To start hatch here, pick endpoint as the origin point

Gradient Fill

Select the **Gradient** hatch type and choose a gradient style from the **Pattern** panel to create a *gradient fill*. See **Figure 23-24**. Gradient fills are commonly used to simulate color-shaded objects and to create the appearance of a lit surface with a gradual transition from an area of highlight to a darker area. Use two colors to simulate a transition from light to dark between the colors. Several different gradient fill patterns are available to create linear sweep, spherical, radial, or curved shading.

The **Gradient Colors** button is active by default to specify a fill using a smooth transition between two colors. Use the **Gradient Color 1** and **Gradient Color 2** drop-down lists to specify gradient colors. Deselect the **Gradient Colors** button to create a fill that has a smooth transition between the darker *shades* and lighter *tints* of one color. The **Gradient Tint and Shade** option becomes enabled when you use the single-color option. Use the slider or enter a value up to 100 in the text box to specify the tint or shade of a color used for a one-color gradient fill.

The **Centered origin** button is active by default and applies a symmetrical configuration. If you deselect the **Centered origin** button, the gradient fill shifts to simulate the projection of a light source from the left of the boundary. Use the **Angle** option to specify the gradient fill angle relative to the current UCS. The default angle is 0°.

gradient fill: A shading transition between the tones of one color or two separate colors.

shade: A specific color mixed with black.

tint: A specific color mixed with white.

Ribbon
Hatch Creation > Properties
Gradient Colors

Ribbon
Hatch Creation > Properties
Gradient Tint and Shade

Exercise 23-5

Complete the exercise on the companion website.
www.g-wlearning.com/CAD

Hatch Composition Options

The **HATCH** command creates an *associative hatch pattern* by default. To create a *non-associative hatch pattern*, deselect the **Associative** button in the **Options** panel. An associative hatch is appropriate for most applications. If you stretch, scale, or otherwise edit the objects that define the boundary of an associative hatch, the pattern automatically adjusts to fill the modified boundary. A non-associative hatch pattern does

associative hatch pattern: A hatch pattern that updates automatically when you edit associated objects.

non-associative hatch pattern: A hatch that is independent of objects and updates when the boundary changes, but not when you make changes to objects.

Figure 23-24.
Select a gradient style from the **Pattern** panel to access options for creating gradient fills, such as this two-color representation of a tropical lagoon during sunset.

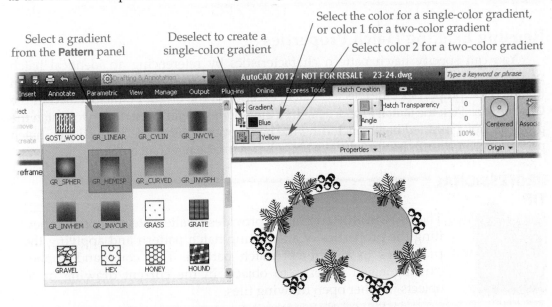

Select a gradient from the **Pattern** panel

Deselect to create a single-color gradient

Select the color for a single-color gradient, or color 1 for a two-color gradient

Select color 2 for a two-color gradient

not respond to changes made to the original boundary. Instead, non-associative hatch boundary grips are available for changing the extents of the hatch, separate from the original boundary objects.

You can select multiple points and objects during a single hatch operation. By default, multiple boundaries form a single hatch object. Selecting and editing one of the hatch patterns selects and edits all patterns created during the same operation. If this is not the preferred result, select the **Create Separate Hatches** option before applying the hatch. Individual hatch patterns form for each boundary.

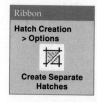
PROFESSIONAL TIP

When you create an associative hatch, it is often best to specify a single internal point per hatch. If you specify more than one internal point in the same operation, AutoCAD creates one hatch object from all points you pick. This can cause unexpected results when you try to edit what appears to be a separate hatch object.

Controlling the Draw Order

The draw order drop-down list in the **Options** panel provides options for controlling the order of display when a hatch pattern overlaps other objects. The **Send behind boundary** option is the default and makes the hatch pattern appear behind the boundary. Select the **Bring in front of boundary** option to make the hatch pattern appear on top of the boundary. Select the **Do not assign** option to have no automatic drawing order setting assigned to the hatch.

Use the **Send to back** option to send the hatch pattern behind all other objects in the drawing. Any objects that are in the hatched area appear as if they are on top of the hatch pattern. Use the **Bring to front** option to bring the hatch pattern in front of, or on top of, all other objects in the drawing. Any objects that are in the hatched area appear as if they are behind the hatch pattern.

Use the **DRAWORDER** command to change the draw order setting after creating a hatch pattern.

Reusing Existing Hatch Properties

You can specify hatch pattern characteristics by referencing an identical hatch pattern from the drawing. Pick the **Match Properties - Use the Current Origin Point** option to match the properties of a selected hatch, except use the origin point specified in the **Origin** panel. Pick the **Match Properties - Use the Source Origin Point** option to match the properties of a selected hatch, including the hatch origin point. Select an existing hatch pattern to match, and then specify the boundaries for the new hatch object.

PROFESSIONAL TIP

The **MATCHPROP** command provides an alternate method of inheriting the properties of an existing hatch pattern and applying the properties to a different hatch pattern. This command applies existing hatch patterns to objects in the current drawing or to objects in other open drawing files.

AutoCAD and Its Applications—Basics

Exercise 23-6

Complete the exercise on the companion website.
www.g-wlearning.com/CAD

Hatching Using DesignCenter

To pattern a boundary using **DesignCenter**, use the **Tree View** pane to locate and select a PAT file to display the patterns in the **Content** pane. See **Figure 23-25A**. The most effective technique to transfer a hatch pattern from **DesignCenter** to the active drawing is to use a drag-and-drop operation. Press and hold down the pick button on the pattern to import, and then drag the cursor to the drawing window. A hatch pattern symbol appears with the cursor. See **Figure 23-25B**. Release the pick button in a boundary to apply the hatch pattern. See **Figure 23-25C**.

An alternative to the drag-and-drop method is copy and paste. Right-click on a hatch pattern in **DesignCenter** and pick **Copy**. Move the cursor into the active drawing, right-click, and select **Paste from the Clipboard** from the cascading menu. A hatch pattern symbol appears with the crosshairs. Pick in a boundary to apply the hatch pattern. You can also use **DesignCenter** with the **HATCH** command. Right-click on a hatch pattern in **DesignCenter** and select **BHATCH...** to access the **Hatch Creation** ribbon tab with the selected hatch pattern active.

Ribbon
View
> Palettes
DesignCenter
Type
ADCENTER
ADC

ADCENTER

Figure 23-25.
A—Pick a PAT file in **DesignCenter** to display the available hatch patterns in the **Content** pane.
B—The hatch pattern symbol appears under the cursor during the drag-and-drop and paste operations. C—Pick a point to apply the hatch pattern.

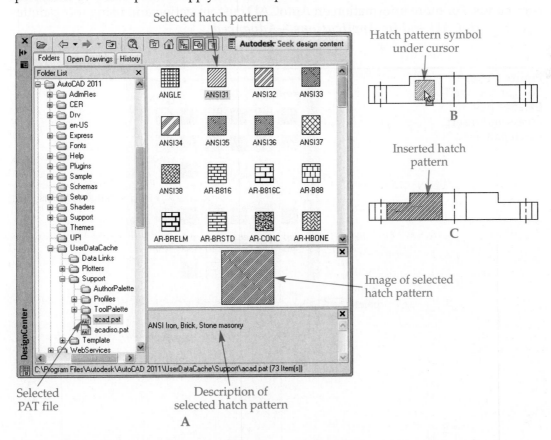

Selected hatch pattern

Hatch pattern symbol under cursor

Inserted hatch pattern

Image of selected hatch pattern

Selected PAT file

Description of selected hatch pattern

A

The same rules that apply to the **HATCH** command also apply to dragging and dropping or copying and pasting hatches. When you insert hatch patterns using **DesignCenter**, the angle, scale, and island detection settings match the settings of the previous hatch pattern. Edit the hatch pattern, as described later in this chapter, to change the settings after you insert the hatch pattern.

 AutoCAD includes two PAT files: acad.pat and acadiso.pat. To verify the location of AutoCAD support files, access the **Files** tab in the **Options** dialog box and check the path listed under the **Support File Search Path**.

 Exercise 23-7

Complete the exercise on the companion website.
www.g-wlearning.com/CAD

Hatching Using Tool Palettes

Ribbon

View
> Palettes

Tool Palettes

Type

TOOLPALETTES
TP

tool palette: A palette that contains tabs to help organize commands and other features.

The **Tool Palettes** palette, shown in **Figure 23-26**, provides an alternative means of storing and inserting hatch patterns. *Tool palettes* can also store and activate other drawing content and tools, such as blocks, images, tables, external reference files, drawing and editing commands, user-defined macros, script files, and Visual Lisp expressions. The **Command Tools Samples** tool palette contains examples of custom commands. For more information on AutoCAD customization and using tool palettes, refer to *AutoCAD and Its Applications—Advanced*.

Figure 23-26.
You can use the **Tool Palettes** palette to access and insert hatch patterns.

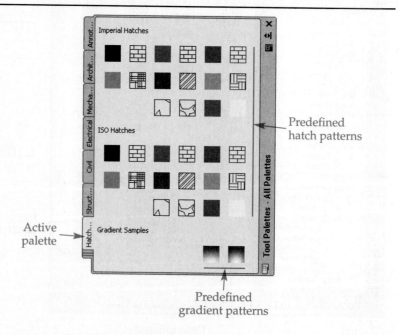

Predefined hatch patterns

Active palette

Predefined gradient patterns

Locating and Viewing Content

Tabs along the side of the **Tool Palettes** palette divide each tool palette. Pick a tab to view related content in a tool palette. If the **Tool Palettes** palette contains more palettes than can be displayed on-screen, pick on the edge of the lowest tab to display a menu listing the palette tabs. Select the name of the tab to access the related tool palette. Use the scroll bar or scroll hand to view large lists of tool palette content. The scroll hand appears when you place the cursor in an empty area in the tool palette. Picking and dragging scrolls the tool palette up and down.

 By default, icons represent the tools in each tool palette. Several tool palette view options are available, including displaying a tool as an image of your choice. For more information on adjusting tool palette display, refer to *AutoCAD and Its Applications—Advanced*.

Inserting Hatch Patterns

To insert a hatch pattern from the **Tool Palettes** palette, access a tool palette containing hatches and fills. To drag and drop a pattern, press and hold down the pick button on the hatch to be inserted, and then drag the cursor into the drawing. A hatch pattern symbol appears with the cursor. Release the pick button in a boundary to apply the hatch pattern. An alternative to the drag-and-drop method is to pick once on the hatch image to attach the hatch to the crosshairs, and then pick a boundary in the drawing to apply the hatch pattern. Edit the hatch pattern, as described later in this chapter, to change the settings after you insert the hatch pattern.

 You can add tool palettes to the **Tool Palettes** palette and add tools to tool palettes. For more information on creating and modifying tool palettes, refer to *AutoCAD and Its Applications—Advanced*.

 ### Exercise 23-8

Complete the exercise on the companion website.
www.g-wlearning.com/CAD

Editing Hatch Patterns

A hatch pattern is a single object that you can edit using the **Properties** palette or standard editing commands such as **ERASE**, **COPY**, and **MOVE**. AutoCAD also provides specific tools to edit hatches. Edit a hatch object to apply a different pattern or pattern properties, or when the drawing changes, such as when you add or remove objects that change existing hatch boundaries.

Click on a hatch object to display the **Hatch Editor** contextual ribbon tab, or double-click to display both the **Quick Properties** panel and the **Hatch Editor** ribbon tab. The **Hatch Editor** ribbon tab is similar to the **Hatch Creation** ribbon tab, but focuses on options for adjusting the existing hatch.

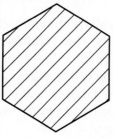

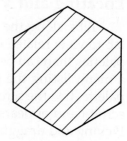

Original Hatched
Polygon

Polygon Erased

Recreated Polygon
Boundary

NEW

Ribbon
Hatch Editor
> Boundaries

Recreate

Ribbon
Hatch Editor
> Boundaries

Display Boundary
Objects

Ribbon
Home
> Modify

Edit Hatch

Type
HATCHEDIT
HE

HATCHEDIT

The **Recreate Boundary** and **Display Boundary Objects** options are specific to the **Hatch Editor** ribbon tab. The most practical application for the **Recreate Boundary** option is to recreate erased boundary geometry. See Figure 23-27. Pick the **Recreate Boundary** button and follow the prompts to recreate the boundary as a region or polyline. You can also specify whether to associate the hatch with the objects. The **Display Boundary Objects** option, available for associative hatches, selects the associative boundary. Use grips to change the size and shape of the boundary.

> The **HATCHEDIT** command also allows you to change hatch characteristics, but uses the **Hatch Edit** dialog box instead of the **Hatch Editor** ribbon tab.

Exercise 23-9

Complete the exercise on the companion website.
www.g-wlearning.com/CAD

Adding and Removing Boundaries

Use the **Pick Points**, **Select Objects**, and **Remove Boundaries** options of the **Hatch Editor** ribbon tab to add boundaries to and remove boundaries from existing hatch objects. For example, Figure 23-28 shows drawing a rectangle to create a window on a portion of an architectural elevation. To add the window as an island in the boundary, double-click the hatch pattern to display the **Hatch Editor** ribbon tab, and then use the **Select Objects** option to pick the rectangle. Close the **Hatch Editor** to complete the operation.

Editing Associative Hatch Patterns

When you edit an object associated with a hatch pattern, the hatch pattern changes to adapt to the edit. Figure 23-29 shows examples of stretching an associated object and removing an island from an associative boundary. As long as you edit the objects associated with the boundary, the hatch pattern updates.

Exercise 23-10

Complete the exercise on the companion website.
www.g-wlearning.com/CAD

Figure 23-28.
Adding an object to an existing hatch pattern boundary.

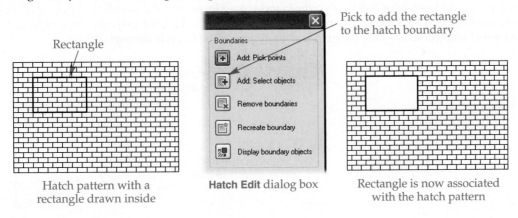

Rectangle

Pick to add the rectangle
to the hatch boundary

Boundaries

Add: Pick points

Add: Select objects

Remove boundaries

Recreate boundary

Display boundary objects

Hatch pattern with a
rectangle drawn inside

Hatch Edit dialog box

Rectangle is now associated
with the hatch pattern

Figure 23-29.
Editing objects with associative hatch patterns. A—The hatch pattern stretches with the object. B—The hatch pattern revises to fill the area of an erased island.

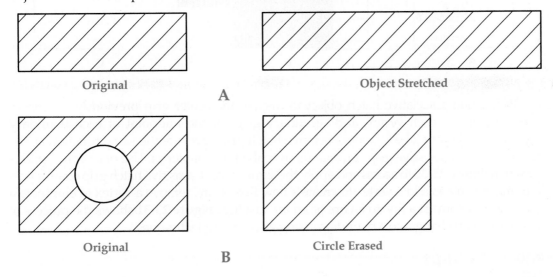

Original

Object Stretched

A

Original

B

Circle Erased

Center Grip Editing

The circle center grip that appears when you select a hatch object with the crosshairs provides convenient access to hatch editing options. There are three different ways to access and apply the same context-sensitive hatch grip commands. **Figure 23-30** briefly explains and illustrates each technique. The **Stretch** option functions as a primary grip, providing access to the standard **STRETCH, MOVE, ROTATE, SCALE,** and **MIRROR** commands. Use the **Origin Point** option to specify a new origin point for the hatch pattern. Use the **Hatch Angle** option to rotate the hatch, and use the **Hatch Scale** option to edit the pattern size.

Editing Non-Associative Hatch Patterns

Create a non-associative hatch pattern by deselecting the **Associative Boundaries** option in the **Hatch Creation** or **Hatch Editor** ribbon tab. A non-associative hatch also forms when you move an associative hatch pattern away from or erase the associated boundary objects. The objects you reference to create a non-associative hatch pattern do not control the size and shape of the hatch. However, you can edit non-associative hatch patterns using standard and grip editing commands.

Figure 23-30.
Use the center hatch object grip to edit a hatch pattern. Apply one of the following methods to access context-sensitive hatch grip commands. A—Hover over the unselected grip to display a menu of options, and then pick an option from the list. B—Pick the grip and then right-click and select an option. C—Pick the grip and then press [Ctrl] to cycle through options.

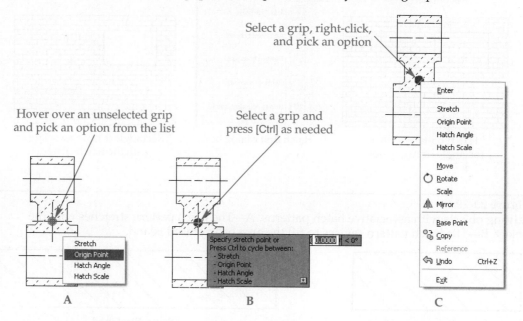

Hover over an unselected grip and pick an option from the list

Select a grip and press [Ctrl] as needed

Select a grip, right-click, and pick an option

A   B   C

Pick a non-associative hatch object to display the center grip previously described, primary grips at each vertex, and secondary grips at the midpoint of each boundary segment. See **Figure 23-31**. Non-associative hatch grips provide the same function as the polyline context-sensitive grip commands described in Chapter 14. Use one of the options shown in **Figure 23-30** to access and apply the same context-sensitive hatch grip commands. You may be able to add a vertex to create a new line or arc, remove a vertex to eliminate a line or arc, or convert a line to an arc or an arc to a line. **Figure 23-31** shows the process of making several changes to a boundary using grip editing techniques.

PROFESSIONAL TIP

When working with associative and non-associative hatch patterns, remember that associative hatch patterns are associated with objects. The objects define the hatch boundary. Non-associative hatch patterns are not associated with objects, but they do show association with the hatch boundary.

The **MIRRHATCH** system variable is set to 0 by default. This setting prevents a hatch from reversing during a mirror operation. Change the **MIRRHATCH** value to 1 to mirror a hatch in relation to the original object. See **Figure 23-32**.

Exercise 23-11

Complete the exercise on the companion website.
www.g-wlearning.com/CAD

Figure 23-31.
Using the grips that appear when you select a non-associative hatch pattern. A—The original non-associative hatch. B—Moving an island out of the boundary and adjusting edge grips. C—Adjusting edge and point grips. D—Moving an island back into the boundary and adjusting edge and point grips.

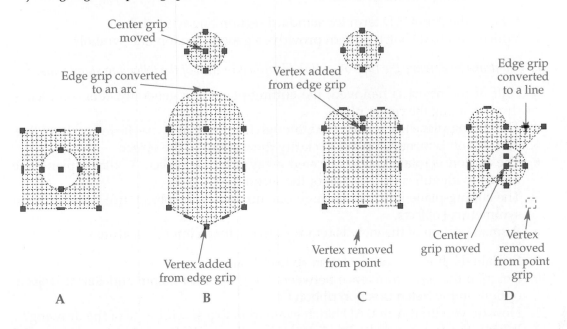

Center grip moved

Edge grip converted to an arc

Vertex added from edge grip

Edge grip converted to a line

Vertex added from edge grip

Vertex removed from point

Center grip moved

Vertex removed from point grip

A   B   C   D

Figure 23-32.
The **MIRRHATCH** system variable options.

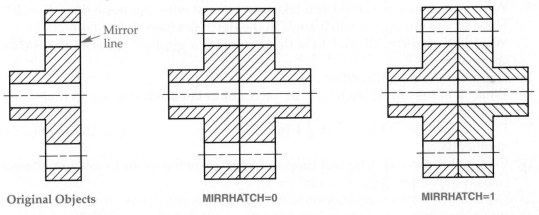

Mirror line

Original Objects   MIRRHATCH=0   MIRRHATCH=1

Express Tools
Chapter 23

The SUPERHATCH Command
The **Draw** panel of the **Express Tools** ribbon tab includes a **SUPERHATCH** command. For information about the **SUPERHATCH** command, go to the companion website (www.g-wlearning.com/CAD), select this chapter, and select **The SUPERHATCH Command**.

Chapter Review

Answer the following questions. Write your answers on a separate sheet of paper or complete the electronic chapter review on the companion website.
www.g-wlearning.com/CAD

1. What is the AutoCAD term for standard section line symbols?
2. Which AutoCAD hatch pattern provides a general section line symbol?

For Questions 3–8, name the type of section identified in each of the following statements:

3. Half of the object is removed; the cutting-plane line generally cuts completely through along the center plane.
4. The cutting-plane line is staggered through features that do not lie in a straight line.
5. The section is turned in place to clarify the contour of the object.
6. The section is rotated and is located away from the object. A cutting-plane line normally identifies the location of the section.
7. The cutting-plane line cuts through one-quarter of the object; used primarily on symmetrical objects.
8. A small portion of the view is removed to clarify an internal feature.

9. Explain the three island detection style options.
10. Describe the basic difference between using the **Pick Points** and **Select Objects** options in the **Hatch Creation** ribbon tab.
11. How do you limit AutoCAD hatch evaluation to a specific area of the drawing?
12. What is the purpose of the **Gap Tolerance** setting in the **Hatch Creation** ribbon tab?
13. Explain how to select a hatch pattern or fill using the **Hatch Creation** ribbon tab.
14. Name the two files supplied with AutoCAD that contain hatch patterns.
15. What considerations should you take into account when choosing a hatch scale?
16. How do you change the hatch angle in the **Hatch Creation** ribbon tab?
17. What is a gradient fill, and how do you create a gradient fill with the **HATCH** command?
18. Define *associative hatch pattern*.
19. What is the result of stretching an object associated with an associative hatch pattern?
20. Explain how to use an existing hatch pattern on a drawing as the pattern for another hatch.
21. Explain how to use drag and drop to insert a hatch pattern from **DesignCenter** into an active drawing.
22. Explain two ways to use drag and drop to insert a hatch pattern from a tool palette into a drawing.
23. What is typically the easiest way to access the **Hatch Editor** ribbon tab?
24. How does the **Hatch Editor** ribbon tab compare to the **Hatch Creation** ribbon tab?
25. What happens if you erase an island inside an associative hatch pattern?

Drawing Problems

Start AutoCAD if it is not already started. Start a new drawing for each problem using an appropriate template of your choice. The template should include layers and text, dimension, multileader, and table styles, when necessary, for drawing the given objects. Add layers and text, dimension, multileader, and table styles as needed. Draw all objects using appropriate layers and text, dimension, multileader, and table styles, justification, and format. Follow the specific instructions for each problem. Use only drawing and editing commands and techniques you have already learned. Use your own judgment and approximate dimensions when necessary. Apply formal dimensions accurately using ASME or appropriate industry standards.

▼ Basic

1. Draw the game board shown. Save the drawing as P23-1.

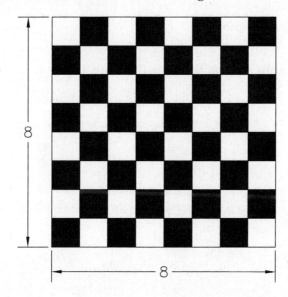

2. Draw the bar graph shown. Save the drawing as P23-2.

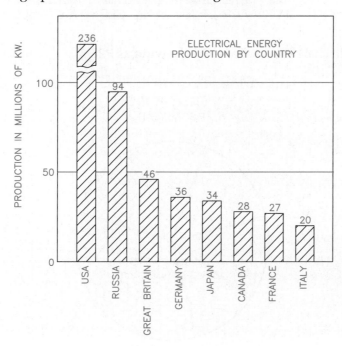

3. Draw the component layout shown. Save the drawing as P23-3.

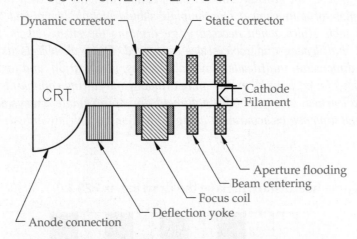

4. Draw the bar graphs shown. Save the drawing as P23-4.

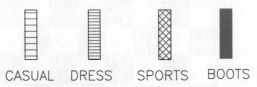

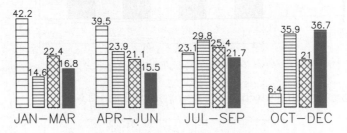

5. Draw the pie chart shown. Save the drawing as P23-5.

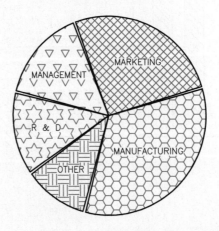

Drawing Problems - Chapter 23

6. Draw the bar graph shown. Save the drawing as P23-6.

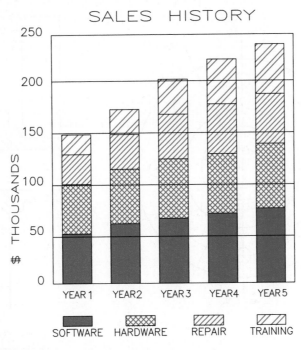

▼ Intermediate

7. Draw and dimension the views shown, which include aligned and broken-out sections. Save the drawing as P23-7.

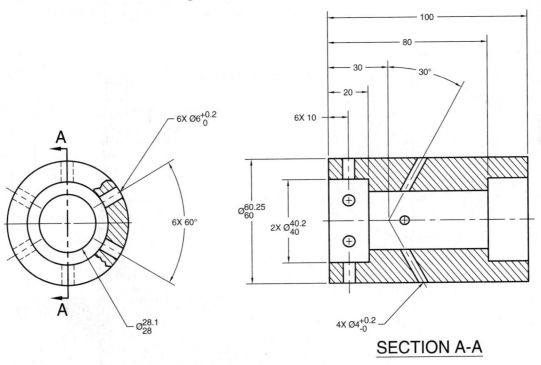

SECTION A-A

8. Draw and dimension the views shown. Save the drawing as P23-8.

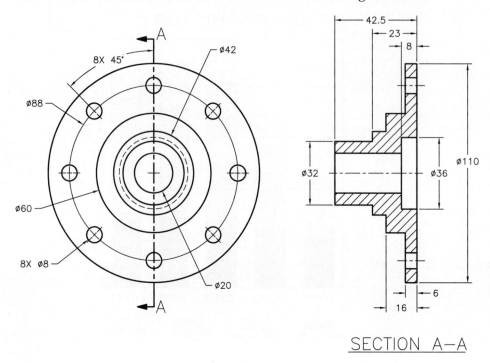

SECTION A—A

9. Draw and dimension the views shown. Save the drawing as P23-9.

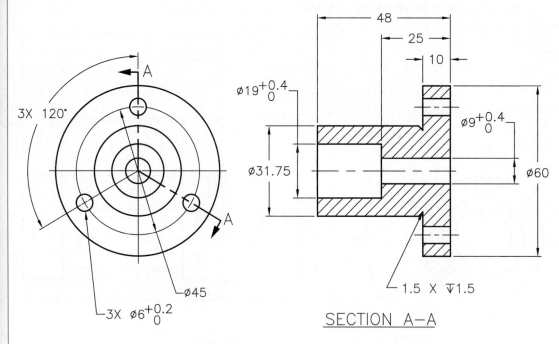

SECTION A—A

10. Draw and dimension the views of the chain guide as shown. Save the drawing as P23-10.

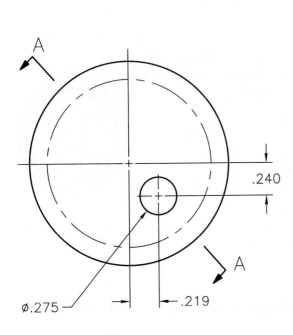

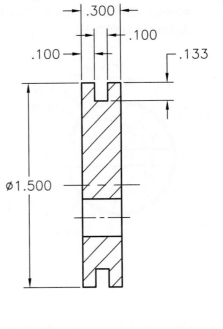

.300
.100
.100
.133
Ø1.500

SECTION A–A

.240
Ø.275
.219

11. Draw and dimension the views of the sleeve and add the notes as shown. Save the drawing as P23-11.

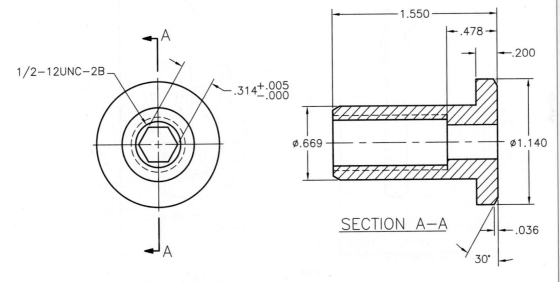

1/2–12UNC–2B
$.314^{+.005}_{-.000}$

1.550
.478
.200
Ø.669
Ø1.140
.036
30°

SECTION A–A

1. DIMENSIONS AND TOLERANCES PER ASME Y14.5–2009.
2. REMOVE ALL BURRS AND SHARP EDGES.
3. CASE HARDEN 45–50 ROCKWELL.
4. PAINT ACE GLOSS BLACK ALL OVER.

▼ **Advanced**

12. Draw and dimension the views shown. Add the following notes: OIL QUENCH 40-45C, CASE HARDEN .020 DEEP, and 59-60 ROCKWELL C SCALE. Save the drawing as P23-12.

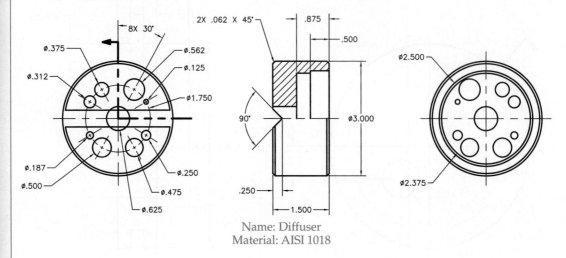

Name: Diffuser
Material: AISI 1018

13. Draw and dimension the views of the tow hook as shown. Save the drawing as P23-13.

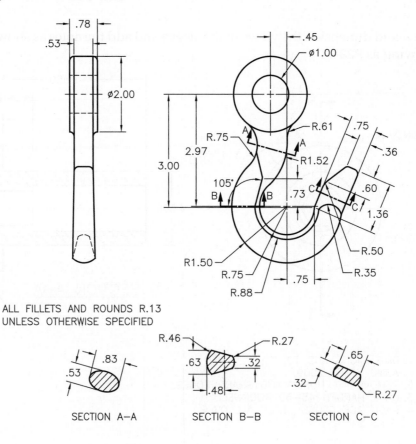

ALL FILLETS AND ROUNDS R.13
UNLESS OTHERWISE SPECIFIED

SECTION A–A   SECTION B–B   SECTION C–C

14. Draw the stair detail as shown. Save the drawing as P23-14.

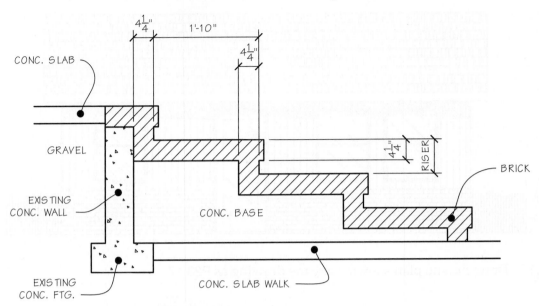

15. Draw the foundation detail shown, but make the following changes:
   A. Use the SansSerif font
   B. Use .125″ leader shoulders
   C. When text is on the left side of a leader, position text at the middle of the bottom line
   D. When text is on the right side of a leader, position text at the middle of the top line

   Save the drawing as P23-15.

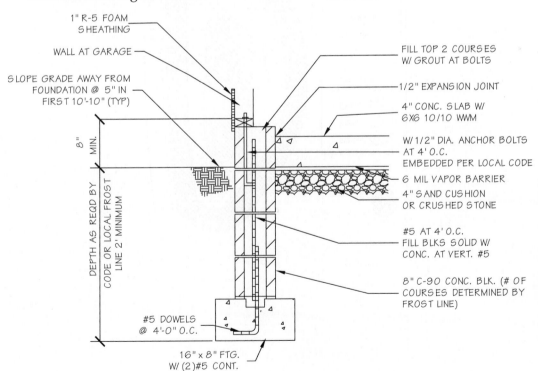

16. Draw the front elevation shown. Save the drawing as P23-16.

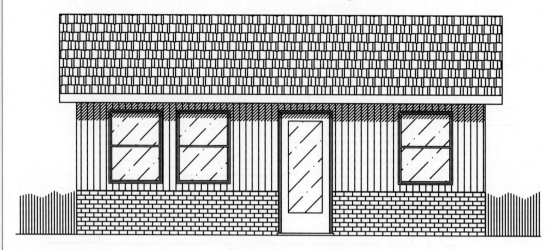

17. Draw the site plan shown. Save the drawing as P23-17.

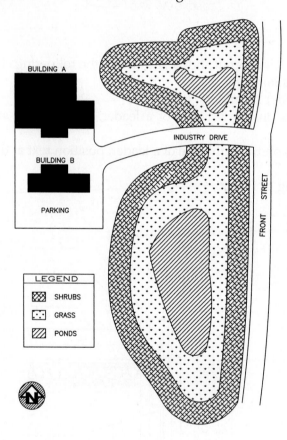

18. Draw the site plan shown. Save the drawing as P23-18.

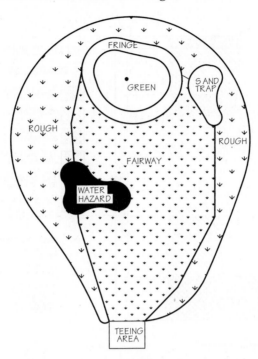

19. Draw the site plan shown. Save the drawing as P23-19.

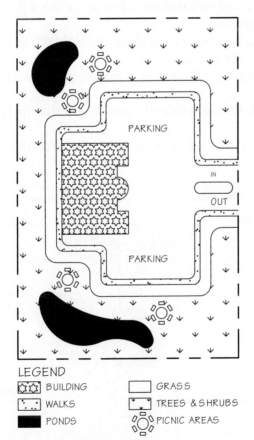

20. Draw the profile shown. Save the drawing as P23-20.

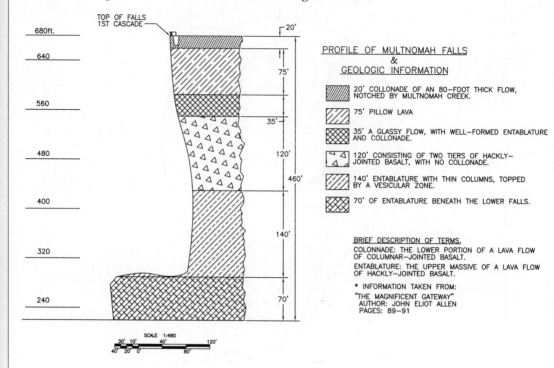

PROFILE OF MULTNOMAH FALLS
&
GEOLOGIC INFORMATION

20' COLLONADE OF AN 80—FOOT THICK FLOW, NOTCHED BY MULTNOMAH CREEK.

75' PILLOW LAVA

35' A GLASSY FLOW, WITH WELL—FORMED ENTABLATURE AND COLLONADE.

120' CONSISTING OF TWO TIERS OF HACKLY—JOINTED BASALT, WITH NO COLLONADE.

140' ENTABLATURE WITH THIN COLUMNS, TOPPED BY A VESICULAR ZONE.

70' OF ENTABLATURE BENEATH THE LOWER FALLS.

BRIEF DESCRIPTION OF TERMS.
COLLONADE: THE LOWER PORTION OF A LAVA FLOW OF COLUMNAR—JOINTED BASALT.
ENTABLATURE: THE UPPER MASSIVE OF A LAVA FLOW OF HACKLY—JOINTED BASALT.

* INFORMATION TAKEN FROM:
"THE MAGNIFICENT GATEWAY"
  AUTHOR: JOHN ELIOT ALLEN
  PAGES: 89—91

21. Draw the drop cleanout detail shown, but make the following changes:
    A. Use the SansSerif font
    B. Use diagonal leader lines at a maximum angle of 60° and a minimum angle of 30°
    C. Use .125" leader shoulders
    D. When text is on the left side of a leader, position text at the middle of the bottom line
    E. When text is on the right side of a leader, position text at the middle of the top line
    F. Use the **PLINE** command to create the FLOW arrowheads.
    Save the drawing as P23-21.

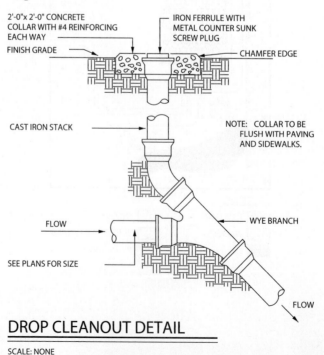

DROP CLEANOUT DETAIL

SCALE: NONE

22. Draw the map shown, using the **SPLINE** command to create the curved shapes. Draw the map at full scale, following the map shown as closely as possible. Save the drawing as P23-22.

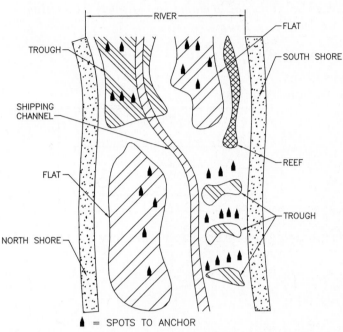

CATFISHING ANCHOR LOCATIONS

▲ = SPOTS TO ANCHOR

AutoCAD Certified Associate Exam Practice

Answer the following questions. Write your answers on a separate sheet of paper.

1. Which of the following sections would show all of the internal features of the part most efficiently? *Select the one item that best answers the question.*
   A. broken-out section
   B. full section
   C. offset section
   D. removed section
   E. revolved section

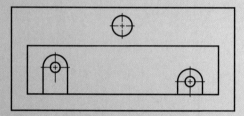

2. Which of the following can be used as a boundary for a hatch? *Select all that apply.*

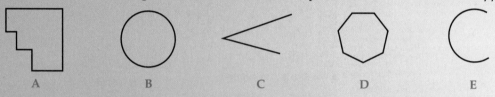

3. Which of the following options would you choose to create the hatch shown? *Select the one item that best answers the question.*
   A. **Ignore Island Detection**
   B. **No Island Detection**
   C. **Normal Island Detection**
   D. **Outer Island Detection**

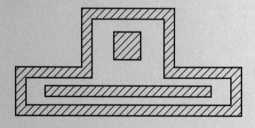

AutoCAD Certified Professional Exam Practice

Follow the instructions in each problem. Write your answers on a separate sheet of paper.

1. **Open CPE-23scale.dwg. This file is available on the companion website.** What is the scale of the hatch in this drawing?

2. **Open CPE-23boundary.dwg. This file is available on the companion website.**

   Recreate the boundary for the hatch. Use the boundary and the appropriate measurement command to answer the following question: What is the total area of the hatched object?

Standard Blocks 24

Learning Objectives

After completing this chapter, you will be able to:

✓ Create and save blocks.
✓ Insert blocks into a drawing.
✓ Edit a block and update the block in a drawing.
✓ Create blocks as drawing files.
✓ Construct and use a symbol library.
✓ Purge unused content from a drawing.

The ability to create and use *blocks* is a major benefit of drawing with AutoCAD. The **BLOCK** command stores a block within a drawing as a *block definition*. The **WBLOCK** command saves a *wblock* as a separate drawing file. You can insert blocks as often as needed and share blocks between drawings. You also have the option to scale, rotate, and adjust blocks to meet specific drawing requirements.

block: Objects, such as a symbol, saved and stored in a drawing for future use.

block definition: Information about a block stored within the drawing file.

wblock: A block definition saved as a separate drawing file.

Constructing Blocks

A block can consist of any object or group of objects, including annotation, or can be an entire drawing. Review each drawing and project to identify items you can use more than once. Screws, punches, subassemblies, plumbing fixtures, and appliances are examples of items to consider converting to blocks. Draw the objects once and then save them as a block for multiple use.

Selecting a Layer

Identify the appropriate layer on which to create block elements before drawing the objects. To do this, you must understand how layers and object properties apply to blocks. The 0 layer is the preferred layer on which to draw block objects. If you originally create block objects on the 0 layer, the block inherits the properties of any layer you assign to the block. Draw the objects for all blocks on the 0 layer and then assign the appropriate layer to each block when you insert the block. If you draw block objects on a layer other than layer 0, place all the objects on layer 0 before creating the block.

A second method is to create block objects using one or more layers other than layer 0. If you originally create block objects on a layer other than layer 0, the block belongs to the layer you assign to the block, but the objects retain the properties of the layers you use to create the objects. The difference is only noticeable if you place the block on a layer other than the layer you use to draw the block objects.

A third technique is to create block objects using the ByBlock color, linetype, lineweight, and transparency. If you originally create block objects using ByBlock properties, the block belongs to the layer you assign to the block, but the objects take on the color, linetype, lineweight, and transparency you assign to the block, regardless of the layer on which you place the block. Using the ByBlock setting is only noticeable if you assign absolute values to the block using the properties in the **Properties** panel of the **Home** ribbon tab, the **Quick Properties** panel, or the **Properties** palette.

Another option is to create block objects using an absolute color, linetype, lineweight, and transparency. If you originally create block objects using absolute values, such as a Blue color, a Continuous linetype, a 0.05 mm lineweight, and a transparency value of 50, the block belongs to the layer you assign to the block, but the objects display the specified absolute values regardless of the properties assigned to the drawing or the layer on which you place the block.

CAUTION

Drawing block objects on a layer other than layer 0, or using ByBlock or absolute properties, can cause significant confusion. The result is often a situation in which a block belongs to a layer, but the block objects display properties of a different layer, or absolute values. In most cases, you should draw block objects on layer 0, and then assign a specific layer to each block.

Drawing Block Elements

Draw the elements of a block as you would any other geometry. If you plan to define a block as annotative, to scale the block according to the drawing scale, you can include annotative or non-annotative objects, such as text, with the block. As long as you specify the block as annotative, all objects act annotative, even if some objects are non-annotative. However, you must use non-annotative objects when preparing a non-annotative block.

insertion base point: The point on a block that defines where the block is positioned during insertion.

When you finish drawing the objects, determine the best location for the *insertion base point*. When you insert the block into a drawing, the insertion base point positions the block. **Figure 24-1** shows examples of common blocks and a possible insertion base point for each.

PROFESSIONAL TIP

A single block allows you to create multiple features that are identical except for scale. In these cases, draw the base block to fit inside a one-unit square. This makes it easy to scale the block when you insert it into a drawing to create variations of the block.

Creating Blocks

Once you draw objects and identify an appropriate insertion base point, you are ready to save the objects as a block. Use the **BLOCK** command and the corresponding **Block Definition** dialog box to create a block. See **Figure 24-2**.

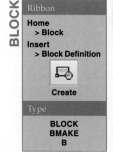

BLOCK

Ribbon
Home
> Block
Insert
> Block Definition

Create

Type
BLOCK
BMAKE
B

Figure 24-1.
Common drafting symbols and their insertion points for placement on drawings. Colored grips indicate the insertion points.

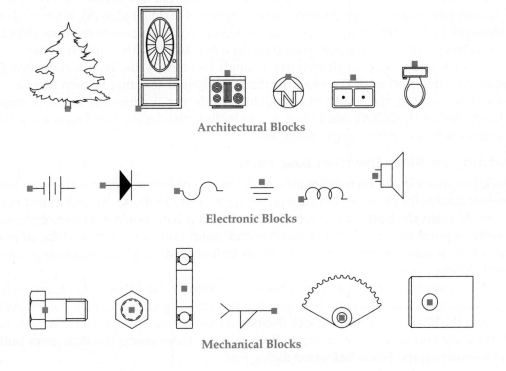

Architectural Blocks

Electronic Blocks

Mechanical Blocks

Figure 24-2.
Use the **Block Definition** dialog box to create a block.

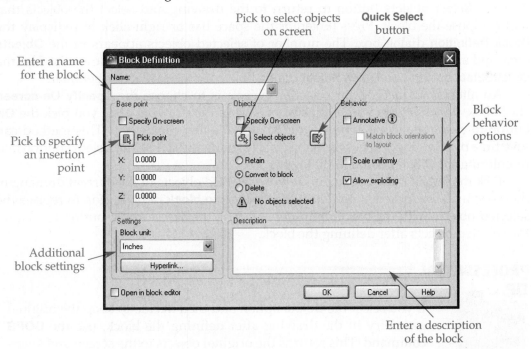

Pick to select objects on screen

Quick Select button

Enter a name for the block

Pick to specify an insertion point

Block behavior options

Additional block settings

Enter a description of the block

Naming and Describing the Block

Enter a descriptive name for the block in the **Name:** text box. For example, name a vacuum pump PUMP or a 3′ × 6′-8″ door DOOR_3068. The block name cannot exceed 255 characters. It can include numbers, letters, spaces, the dollar sign ($), hyphen (-), and underscore (_). Use the drop-down list to access an existing name to recreate a block or to use a block name as reference when naming a new block with a similar name.

A block name is often descriptive enough to identify the block. However, you can enter a description of the block in the **Description:** text box to help identify the block. For example, the PUMP block might include the description This is a vacuum pump symbol, or the DOOR_3068 block might include the description This is a plan view, 3′ wide by 6′-8″ tall, interior, single-swing door.

Defining the Block Insertion Base Point

Use options in the **Base point** area to define the insertion base point. If you know the coordinates for the insertion base point, type values in the **X:**, **Y:**, and **Z:** text boxes. However, often the best way to specify the insertion base point is to use object snap to select a point on an object. Choose the **Pick point** button to return to the drawing and select an insertion base point. The **Block Definition** dialog box reappears after you select the insertion base point.

An alternative technique is to choose the **Specify On-screen** check box, which allows you to pick an insertion base point in the drawing after you pick the **OK** button to create the block and exit the **Block Definition** dialog box. This method can save time by allowing you to pick the insertion base point without using the **Pick point** button and re-entering the **Block Definition** dialog box.

Selecting Block Objects

The **Objects** area includes options for selecting objects for the block definition. Pick the **Select objects** button to return to the drawing and select the objects that will compose the block. Press [Enter] or the space bar or right-click to redisplay the **Block Definition** dialog box. The number of selected objects appears in the **Objects** area, and an image of the selection appears next to the **Name:** drop-down list. Use the **QuickSelect** button and **Quick Select** dialog box to define a selection set filter.

An alternative method for selecting objects is to choose the **Specify On-screen** check box, which allows you to pick objects from the drawing after you pick the **OK** button to create the block and exit the **Block Definition** dialog box. This method can save time by allowing you to select objects without using the **Select objects** button and re-entering the **Block Definition** dialog box.

Pick the **Retain** radio button to keep the selected objects in the current drawing in their original, unblocked state. Select the **Convert to block** radio button to replace the selected objects with the block definition. Choose the **Delete** radio button to remove the selected objects after defining the block.

PROFESSIONAL TIP

If you select the **Delete** option and then decide to keep the original geometry in the drawing after defining the block, use the **OOPS** command. This returns the original objects to the screen and keeps the block definition. Using the **UNDO** command removes the block definition from the drawing.

Block Scale Settings

Pick the **Annotative** check box in the **Behavior** area to make the block annotative. AutoCAD scales annotative blocks according to the annotation scale you select, which eliminates the need for you to calculate the scale factor. Pick the **Match block orientation to layout** check box, which becomes enabled when you select the **Annotative** check box, to keep annotative blocks planar to the layout in a floating viewport, even if the drawing view rotates, such as if you rotate the UCS. Selecting the **Match block orientation to layout** option also prohibits you from using the **ROTATE** command to rotate a block.

If you check **Scale uniformly** in the **Behavior** area, you do not have the option of specifying different X and Y scale factors when you insert the block. You will learn options for scaling blocks later in this chapter.

Additional Block Definition Settings

If you check **Allow exploding** in the **Behavior** area, you have the option of exploding the block. If you do not check **Allow exploding**, you cannot explode the block after inserting it. Select a unit type from the **Block unit** drop-down list in the **Settings** area to specify the insertion units of the block. Pick the **Hyperlink...** button to access the **Insert Hyperlink** dialog box to insert a hyperlink in the block. If you check **Open in block editor**, the new block immediately opens in the **Block Editor** when you create the block and exit the **Block Definition** dialog box. The **Block Editor** is described later in this chapter.

To verify that a block has been saved properly, reopen the **Block Definition** dialog box. Pick the **Name:** drop-down list to display a list of blocks in the current drawing.

PROFESSIONAL TIP

You can use blocks to create other blocks. Insert existing blocks into a view and then save all of the objects as a block. This is a process known as *nesting*. You must give the top-level block a name that is different from any nested block. Proper planning and knowledge of all existing blocks can speed up the drawing process and the creation of complex views.

nesting: Creating a block that includes other blocks.

Exercise 24-1

Complete the exercise on the companion website.
www.g-wlearning.com/CAD

block reference: A specific instance of a block inserted into a drawing.

dependent symbols: Named objects in a drawing that have been inserted or referenced into another drawing.

named objects: Blocks, dimension styles, groups, layers, linetypes, materials, multileader styles, plot styles, shapes, table styles, text styles, and visual styles that have specific names.

Inserting Blocks

AutoCAD provides several options for inserting a block into a drawing. Remember to make the layer you want to assign to the block current before inserting the block. You should also determine the proper size and rotation angle for the block before insertion. The term *block reference* describes an inserted block. *Dependent symbols* are any *named objects*, such as blocks and layers. AutoCAD automatically updates dependent symbols in a drawing the next time you open the drawing.

Using the INSERT Command

The **INSERT** command provides a common method for inserting a block into a drawing. Access the **INSERT** command to display the **Insert** dialog box. See **Figure 24-3**.

Selecting the Block to Insert

Use the **Name:** drop-down list to show the blocks defined in the current drawing and select the name of the block you want to insert. You can also type the name of the block in the **Name:** text box. Another option is to pick the **Browse...** button to display the **Select Drawing File** dialog box. This allows you to locate and select a drawing or DXF file (wblock) to insert as a block. Inserting a file as a block is described later in this chapter.

Specifying the Block Insertion Point

The **Insertion point** area contains options for specifying where to insert the block. Select the **Specify On-screen** check box to specify a location in the drawing when you pick the **OK** button. To insert the block using absolute coordinates, deselect the **Specify On-screen** check box and enter coordinates in the **X:**, **Y:**, and **Z:** text boxes.

Scaling Blocks

The **Scale** area allows you to specify scale values for the block in relation to the X, Y, and Z axes. Deselect the **Specify On-screen** check box to enter scale values in the **X:**, **Y:**, and **Z:** text boxes. Activate the **Uniform Scale** check box to specify a scale value for the X axis that also applies to the scale of the Y and Z axes. The X value is the only active axis value if you created the block with **Scale uniformly** checked in the **Block Definition** dialog box. Select the **Specify On-screen** check box to receive prompts for scaling the block during insertion.

It is possible to create a mirror image of a block by entering negative scale factor values. For example, enter –1 for the X and Y scale factor to mirror the block to the opposite quadrant of the original orientation, but retain the original size. **Figure 24-4** shows examples of scale and mirroring techniques.

Blocks are classified as *real blocks*, *schematic blocks*, or *unit blocks*, depending on how you scale the block during insertion. Examples of real blocks include a bolt, a bathtub, a pipe fitting, and the car shown in **Figure 24-5A**. Examples of schematic blocks include notes, detail bubbles, tags, and section symbols. See **Figure 24-5B**. Schematic blocks typically include annotative blocks. When you insert an annotative schematic block, AutoCAD automatically determines the block scale based on the annotation scale. When you insert a non-annotative schematic block, you must specify the scale factor.

real block: A block originally drawn at a 1:1 scale and then inserted using 1 for both the X and Y scale factors.

schematic block: A block originally drawn at a 1:1 scale and then inserted using the drawing scale factor for both the X and Y scale values.

unit block: A 1D, 2D, or 3D block drawn to fit in a 1-unit, 1-unit-square, or 1-unit-cubed area so that it can be scaled easily.

Figure 24-3.
The **Insert** dialog box allows you to select and prepare a block for insertion. Select a block from the drop-down list or enter the block name in the **Name:** text box.

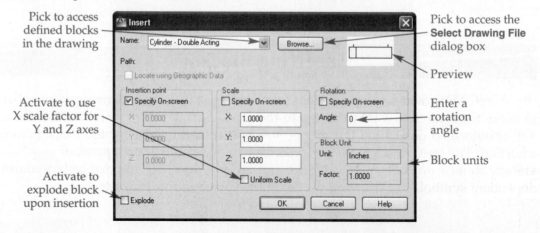

Pick to access defined blocks in the drawing

Pick to access the **Select Drawing File** dialog box

Preview

Activate to use X scale factor for Y and Z axes

Enter a rotation angle

Block units

Activate to explode block upon insertion

Figure 24-4.
Negative and positive scale factors have different effects when used to insert a block. Colored grips indicate the insertion points.

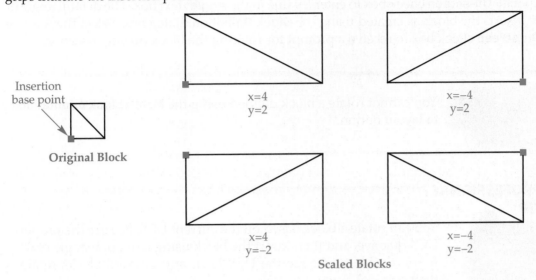

Insertion base point

Original Block

x=4
y=2

x=−4
y=2

x=4
y=−2

x=−4
y=−2

Scaled Blocks

Figure 24-5.
A—Real blocks, such as this car, are drawn at full scale and inserted using a scale factor of 1 for both the X and Y axes. B—A schematic block is inserted using the scale factor of the drawing for the X and Y axes. C—A 2D unit block is often inserted at different scales for the X and Y axes.

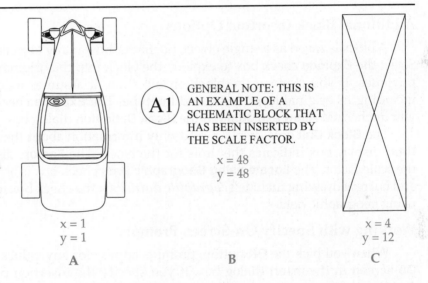

A1

GENERAL NOTE: THIS IS AN EXAMPLE OF A SCHEMATIC BLOCK THAT HAS BEEN INSERTED BY THE SCALE FACTOR.

x = 48
y = 48

x = 1
y = 1

x = 4
y = 12

A   B   C

For most applications, insert annotative blocks at a scale of 1 to apply the annotation scale correctly. Entering a scale other than 1 adjusts the scale of the block by multiplying the scale value by the annotative scale factor.

1D unit block:
A 1-unit, one-dimensional object, such as a straight line segment, saved as a block.

There are three general types of unit blocks. An example of a *1D unit block* is a 1″ blocked line object. An example of a *2D unit block* is a 1″ × 1″ square. A 3D unit block is any blocked object that can fit inside a 1-unit × 1-unit × 1-unit (1″ × 1″ × 1″, for example) cube. To use a unit block, insert the block and determine the individual scale factors for each axis. For example, insert a 1″ 1D unit block line at a scale of 4 to create a 4″ line. When inserting a 2D unit block, assign different scale factors for the X and Y axes to change the block dimensions. For example, specify 4 for the X axis and 12 for the Y axis to create the 4″ × 12″ beam shown in **Figure 24-5C**. A *3D unit block* allows you to adjust the scale of the X, Y, and Z axes.

2D unit block: A 2D object that fits into a 1-unit × 1-unit square, saved as a block.

3D unit block: A 3D object that fits into a 1-unit × 1-unit × 1-unit cube, saved as a block.

Rotating Blocks

The **Rotation** area allows you to insert the block at a specific angle. Deselect the **Specify On-screen** check box to enter a value in the **Angle:** text box. The default angle of 0° inserts the block as created using the **Block Definition** dialog box. Select the **Specify On-screen** check box to receive a prompt for rotating the block during insertion.

You cannot rotate a block defined using the **Match block orientation to layout** option.

PROFESSIONAL TIP

You can rotate a block based on the current UCS. Be sure the proper UCS is active, and then insert the block using a rotation angle of 0°. If you decide to change the UCS later, any inserted blocks retain their original angle.

Additional Block Insertion Options

A block is saved as a single object, no matter how many objects the block includes. Select the **Explode** check box to explode the block into the original objects for editing purposes. If you explode the block on insertion, it assumes its original properties, including its original layer, color, and linetype. The **Explode** check box is disabled if you unchecked **Allow exploding** in the **Block Definition** dialog box.

The **Block Unit** area displays read-only information about the selected block. The **Unit:** display box indicates the units for the block. The **Factor:** display box indicates the scale factor. The **Locate using Geographic Data** check box is active when the block and current drawing include *geographic data*. Pick the check box to position the block using geographic data.

geographic data: Information added to a drawing to describe specific locations and directions on Earth.

Working with Specify On-Screen Prompts

When you pick the **OK** button, prompts appear for any values defined as **Specify On-screen** in the **Insert** dialog box. If you specify the insertion point on-screen, the Specify insertion point or [Basepoint/Scale/X/Y/Z/Rotate/PScale/PX/PY/PZ/PRotate]: prompt appears. Enter or select a point to insert the block. The options allow you to specify a different base point; enter a value for the overall scale; enter independent scale factors for the X, Y, and Z axes; enter a rotation angle; and preview the scale of the X, Y, and Z axes or the rotation angle before entering actual values. If you use one of these options, the new value overrides the related setting in the **Insert** dialog box.

If you specify the X scale factor on-screen, the Enter X scale factor, specify opposite corner, or [Corner/XYZ] <1>: prompt appears. Pick a point or enter a value for the scale. You can also use the **Corner** option to scale the block. The Enter Y scale factor <use X scale factor>: prompt appears if you enter an X scale factor. Specify a value different from the X scale factor, or press [Enter] or the space bar, or right-click to accept the same scale specified for the X axis.

The X and Y scale factors allow you to stretch or compress the block to create modified versions of the block. See **Figure 24-6**. This is why it is a good idea to draw blocks to fit inside a one-unit square when appropriate. It makes the block easy to scale because you can enter the exact number of units for the X and Y dimensions. For example, if you want the block to be three units long and two units high, enter 3 when prompted to enter the X scale factor, and enter 2 when prompted to enter the Y scale factor.

AutoCAD and Its Applications—Basics

Figure 24-6.
A comparison of different X and Y scale factors used for inserting a 2D unit block. This example shows a block of a plan view window symbol inserted into a 6″ wall.

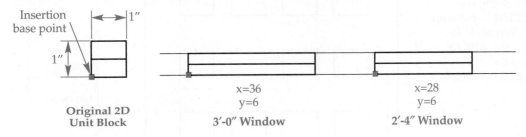

Insertion base point

1″

1″

Original 2D
Unit Block

x=36
y=6
3′-0″ Window

x=28
y=6
2′-4″ Window

The insertion base point specified when the block was created may not always be the best point when you actually insert the block. Instead of inserting and then moving the block, use the **Basepoint** option to specify a different base point before locating the block. Select the **Basepoint** option when prompted to specify the insertion point. The block temporarily appears on-screen, allowing you to choose an alternate insertion base point. The block reattaches to the crosshairs at the new point and a message appears indicating that the command is resuming, allowing you to pick the insertion point in the drawing.

Exercise 24-2

Complete the exercise on the companion website.
www.g-wlearning.com/CAD

Inserting Multiple Arranged Copies of a Block

The **MINSERT** command combines the functions of the **INSERT** and **ARRAY** commands. **Figure 24-7** shows an example of an **MINSERT** command application. To follow this example, set architectural units, draw a 4′ × 3′ rectangle, and save the rectangle as a block named DESK. Then access the **MINSERT** command and enter DESK. Pick a point as the insertion point and then accept the X scale factor of 1, the Y scale factor of use X scale factor, and the rotation angle of 0. The arrangement is to be three rows and four columns. In order to make the horizontal spacing between desks 2′ and the vertical spacing 4′, you must consider the size of the desk when entering the distance between rows and columns. Enter 7′ (3′ desk depth + 4′ space between desks) at the Enter distance between rows or specify unit cell: prompt. Enter 6′ (4′ desk width + 2′ space between desks) at the Specify distance between columns: prompt.

The complete pattern takes on the characteristics of a block, except that you cannot explode the pattern. Therefore, you must use the **Properties** palette to modify the number of rows and columns, change the spacing between objects, or change other properties. If you rotate the initial block, all objects in the pattern rotate about their insertion points. If you rotate the patterned objects about the insertion point while using the **MINSERT** command, all objects align on that point.

Type

MINSERT

Exercise 24-3

Complete the exercise on the companion website.
www.g-wlearning.com/CAD

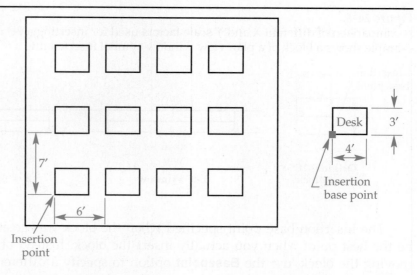

Figure 24-7.
Creating an arrangement of desks using the **MINSERT** command. An alternative is to use the **ARRAY** command after you position one block.

Desk | 3'
4'
Insertion base point

7'

6'

Insertion point

Inserting Entire Drawings

The **INSERT** command also allows you to insert an entire drawing into the current drawing as a block. Access the **INSERT** command and pick the **Browse...** button in the **Insert** dialog box. Use the **Select Drawing File** dialog box to select a drawing or DXF file to insert.

When you insert a drawing into another drawing, the inserted drawing becomes a block reference and functions as a single object. The drawing is inserted on the current layer, but only objects drawn on the 0 layer inherit the color, linetype, lineweight, and transparency properties of the current layer. Explode the inserted drawing if necessary. When the block is exploded, the objects revert to their original layers. Inserting a drawing can bring any existing block definitions and other drawing content, such as layers and dimension styles, into the current drawing.

By default, every drawing has an insertion base point of 0,0,0 when you insert it into another drawing. To change the insertion base point of the drawing, access the **BASE** command and select a new insertion base point. Save the drawing before inserting it into another drawing.

When inserting a drawing as a block, you have the option of using the existing drawing to create a block with a different name. For example, to define a block named BOLT from an existing drawing named Fastener.dwg, access the **INSERT** command and use the **Browse...** button to select the Fastener.dwg file. Use the **Name:** text box to change the name from Fastener to Bolt, and pick the **OK** button. You can then insert the file into the drawing or press [Esc] to exit the command. A BOLT block definition is now available for use.

BASE

Ribbon

Home
> Block

Set Base Point

Type

BASE

Exercise 24-4

Complete the exercise on the companion website.
www.g-wlearning.com/CAD

Inserting Blocks Using DesignCenter

DesignCenter provides an effective way to insert blocks or entire drawings as blocks in the current drawing. To insert a block using **DesignCenter**, use the folder list to locate and select a file containing the block to be inserted. Select the **Blocks** branch in the folder list or double-click on the **Blocks** icon in the content pane to display blocks defined in the file. See **Figure 24-8**. The most effective technique to transfer a block from **DesignCenter** to the active drawing is to use a drag-and-drop operation.

Figure 24-8.
Use **DesignCenter** to insert blocks from files or drawings from folders. Several example blocks are available in the AutoCAD Sample folder shown.

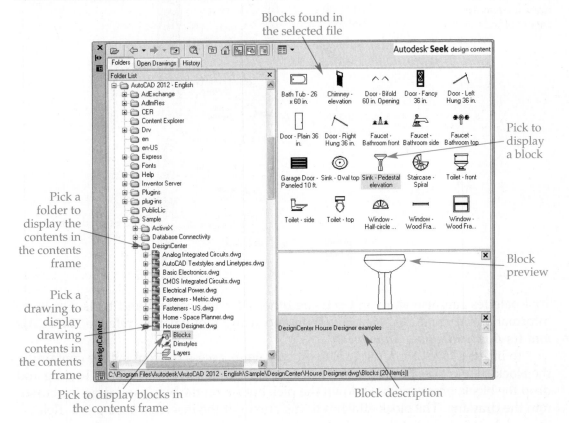

Blocks found in the selected file

Pick to display a block

Block preview

Block description

Pick a folder to display the contents in the contents frame

Pick a drawing to display drawing contents in the contents frame

Pick to display blocks in the contents frame

Press and hold down the pick button on the block and drag the cursor to the drawing window. The block attaches to the cursor at the insertion base point. Release the pick button to insert the block at the location of the cursor.

An alternative to the drag-and-drop method is copy and paste. Right-click on a block in **DesignCenter** and pick **Copy**. Move the cursor into the active drawing, right-click, and select **Paste** from the Clipboard cascading menu. The block attaches to the cursor at the insertion base point. Specify a point to insert the block.

You can also use **DesignCenter** with the **Insert** dialog box. Right-click on a block in **DesignCenter** and select **Insert Block...** to access the **Insert** dialog box with the selected block active. This technique allows you to scale, rotate, or explode the block during insertion.

To insert a drawing or DXF file using **DesignCenter**, use the folder list to locate and select a folder to display the contents of the folder in the content pane. Drag and drop or copy and paste the file from the content pane into the current drawing. You can also right-click a file icon in the **Content** pane and select **Insert as Block...**.

Blocks are inserted from **DesignCenter** based on the type of block units you specify when you created the block. For example, if the original block was a 1 × 1 square and you specified the block units as feet when you created the block, then the block inserts as a 12″ × 12″ square.

Inserting Blocks Using Tool Palettes

The **Tool Palettes** palette, shown in **Figure 24-9**, provides another means of storing and inserting blocks. AutoCAD refers to blocks in a tool palette as *block insertion tools*.

block insertion tools: Blocks located on a tool palette.

Figure 24-9.
The **Tool Palettes** palette provides another way to insert blocks.

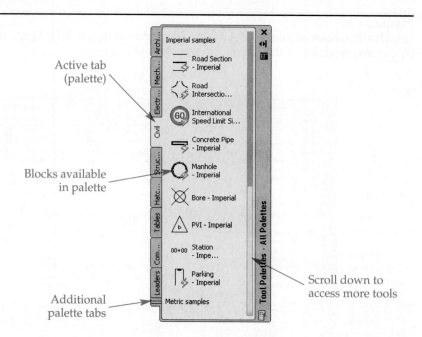

Active tab (palette)

Blocks available in palette

Additional palette tabs

Scroll down to access more tools

Tool palettes can also store and activate other drawing content and tools. For more information on AutoCAD customization and using tool palettes, refer to *AutoCAD and Its Applications—Advanced.*

To insert a block from the **Tool Palettes** palette, access the tool palette containing the block. Hover over the block icon to display the name and description. To drag and drop the block, press and hold down the pick button on the block and drag the cursor into the drawing. The block attaches to the cursor at the insertion base point. Release the pick button to insert the block at the location of the cursor.

An alternative to the drag-and-drop method is to pick once on the block image to attach the block to the crosshairs, and then specify a point in the drawing to insert the block. This method offers an advantage over the drag-and-drop method by presenting options for adjusting the insertion base point, scale, and rotation.

Exercise 24-5

Complete the exercise on the companion website.
www.g-wlearning.com/CAD

Editing Blocks

A block reference, or inserted block, is a single object that you can edit using grip editing, the **Properties** palette, or standard editing commands such as **ERASE**, **COPY**, and **ROTATE**. The grip box for a block appears at the insertion base point of the block.

A different form of block editing involves redefining the block by modifying the block definition or changing the objects that compose the block. You can redefine a block using the **Block Editor** or by exploding and then recreating the block.

Changing Block Properties to ByLayer

If you originally set block element properties such as color, linetype, lineweight, and transparency to absolute values, and you want to change the properties to ByLayer, you can either edit the block definition or use the **SETBYLAYER** command to accomplish the same task without editing the block definition. Access the **SETBYLAYER** command and use the **Settings** option to display the **SetByLayer Settings** dialog box. Select the check boxes that correspond to the object properties you want to convert to ByLayer. Pick the **OK** button to exit the **SetByLayer Settings** dialog box.

Next, select the blocks with the properties you want to set to ByLayer and press [Enter] or the space bar, or right-click to display the Change ByBlock to ByLayer? prompt. Select the **Yes** option to change all object properties currently set to ByBlock to ByLayer. Pick the **No** option to change all object properties set to values other than ByBlock to ByLayer. The next prompt asks if you want to include blocks in the conversion. If the selection is a block, choose **Yes** to convert the properties of all references of the same block in the drawing to ByLayer. If you pick **No**, only the properties of the selected block are converted. All other references of the same block remain unchanged.

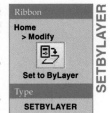
SETBYLAYER

Ribbon
Home
> Modify
Set to ByLayer
Type
SETBYLAYER

PROFESSIONAL TIP

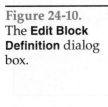

To change the properties of several blocks, use the **Select Similar** or **Quick Select** command to create a selection set of block references.

Using the Block Editor

Use the **BEDIT** command to make changes to a block definition using the **Block Editor**. Access the **BEDIT** command to display the **Edit Block Definition** dialog box. See **Figure 24-10**. Select the name of an existing block to edit from the list box. Pick the <Current Drawing> option to edit a block saved as the current drawing, such as a wblock. A preview and description of the selected block appear. You can create a new block by typing a unique name in the **Block to create or edit** field. Pick the **OK** button to open the selected block in the **Block Editor**. See **Figure 24-11**. If you typed a new block name, the drawing area is empty, allowing you to create a new block.

Ribbon
Home
> Block
Insert
> Block Definition
Edit
Type
BEDIT

Modifying a Block

Use drawing and editing commands to modify or create the block definition. Specify the block insertion base point at the UCS origin, or 0,0,0 point. The tools in the panels of the **Block Editor** ribbon tab are specifically for modifying and creating block

Figure 24-10.
The **Edit Block Definition** dialog box.

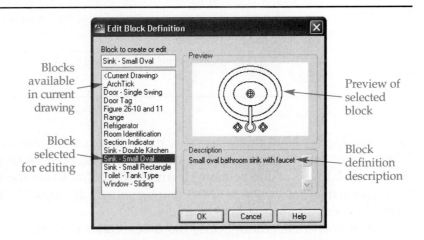

Blocks available in current drawing

Block selected for editing

Preview of selected block

Block definition description

Figure 24-11.
The context-sensitive **Block Editor** ribbon tab and the **Block Authoring Palettes** palette are available in block editing mode. Only the block geometry appears in the **Block Editor**.

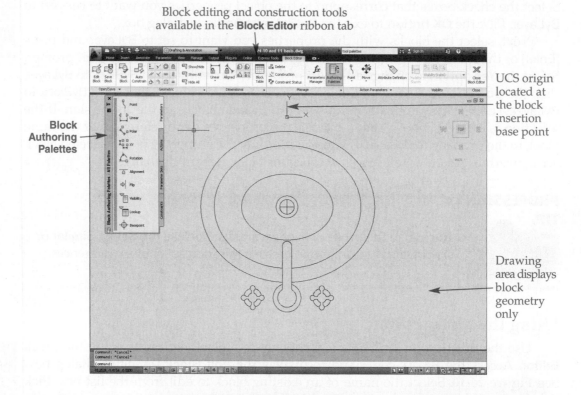

Block editing and construction tools available in the **Block Editor** ribbon tab

Block Authoring Palettes

UCS origin located at the block insertion base point

Drawing area displays block geometry only

Figure 24-12.
The **Block Editor** ribbon tab contains several tools and options specifically for editing and constructing blocks. This table describes the most basic functions.

Button	Description
	Saves changes to the block and updates the block definition.
	Opens the **Save Block As** dialog box, allowing you to save the block as a new block, using a different name.
	Opens the **Edit Block Definition** dialog box, which is the same dialog box displayed when you enter block editing mode. You can select a different block to edit or specify the name of a new block to create from scratch.
	Toggles the **Block Authoring Palettes** palette off and on.
	Closes the **Block Editor**.

geometry. **Figure 24-12** describes some of the basic commands available in the **Block Editor** ribbon tab. Parametric tools allow you to constrain block geometry and form block tables. Many of the commands and options found on the **Block Editor** ribbon tab relate to dynamic blocks. This textbook explains dynamic blocks, block tables, and other block editing tools in later chapters.

When you finish editing, close the **Block Editor** to return to the drawing. If you have not saved your changes, a dialog box appears asking if you want to save changes. Pick the appropriate option to save or discard changes, or pick the **Cancel** button to return to the **Block Editor**.

Double-click on a block to display the **Edit Block Definition** dialog box with the block selected. Open a block directly in the **Block Editor** by selecting the block and then right-clicking and choosing **Block Editor**. Another option is to open a block directly in the **Block Editor** when you create the block by selecting the **Open in block editor** check box in the **Block Definition** dialog box.

Adding a Block Description

To change the description assigned to the original block definition, open the block in the **Block Editor** and display the **Properties** palette with no objects selected. Make changes to the description using the **Description** property in the **Block** category. Pick the **Save Block Definition** button and the **Close Block Editor** button to return to the drawing.

You can also edit blocks in place using the **REFEDIT** command. Chapter 31 describes in-place editing using the **REFEDIT** command as it applies to external references. Use the same techniques to edit blocks.

 ### Exercise 24-6

Complete the exercise on the companion website.
www.g-wlearning.com/CAD

Exploding and Redefining a Block

You can explode a block during insertion by checking **Explode** in the **Insert** dialog box. This is useful when you want to edit the individual objects of the block. You can also use the **EXPLODE** command after inserting the block to break it into the original objects, but only if you check **Allow exploding** in the **Block Definition** dialog box. Access the **EXPLODE** command, select the objects to be exploded, and press [Enter] or the space bar or right-click to complete the operation.

Follow this procedure to redefine an existing block using the **EXPLODE** and **BLOCK** commands:

1. Insert the block to redefine.
2. Make sure you know the exact location of the insertion base point, which is lost during the explosion.
3. Use the **EXPLODE** command to explode the block.
4. Edit the elements of the block as needed.
5. Recreate the block definition using the **BLOCK** command.
6. Assign the block the same original name and, if appropriate, the same insertion point.
7. Select the objects to include in the block.
8. Pick the **OK** button in the **Block Definition** dialog box to save the block. When a message appears asking if you want to redefine the block, pick **Yes**.

A common mistake is to forget to use the **EXPLODE** command before redefining the block. When you try to recreate the block using the same name, an alert box indicates that the block references itself. This means you are trying to create a block that already exists. Press the **Cancel** button, explode the block, and try again to redefine the block.

Once a block is modified, whether from changes made using the **BEDIT** command or from redefinition using the **EXPLODE** and **BLOCK** commands, all instances of that block in the drawing update according to the changes.

Exercise 24-7

Complete the exercise on the companion website.
www.g-wlearning.com/CAD

Understanding the Circular Reference Error

circular reference error: An error that occurs when a block definition references itself.

A *circular reference error* occurs when you try to redefine a block that already exists using the same name. AutoCAD informs you that the block references itself or has not been modified. A block can be composed of many objects, including other blocks. When you use the **BLOCK** command to incorporate an existing block into a new block, AutoCAD detects all objects that compose the new block, including existing block definitions. A problem occurs if you select an instance of the redefined block as an element of the new definition. The new block refers to a block of the same name, or references itself. **Figure 24-13A** illustrates the process of correctly redefining a block named BOX to avoid a circular reference error. **Figure 24-13B** shows an incorrect redefinition resulting in a circular reference error.

Renaming Blocks

Type
RENAME
REN

Access the **RENAME** command to rename a block using the **Rename** dialog box, without editing the block definition. See **Figure 24-14**. Select **Blocks** from the **Named Objects** list, and then pick the block to be renamed in the **Items** list. The current name appears in the **Old Name:** text box. Type the new block name in the **Rename To:** text box. Pick the **Rename To:** button to display the new name in the **Items** list. Pick the **OK** button to exit the **Rename** dialog box.

Updating Block Icons

Type
BLOCKICON

A block icon forms when you define a block. The icon appears when you insert and edit blocks to help you recognize the block. Block icons require updating when an icon does not appear, as is often the case when you store a block in a drawing created with an older version of AutoCAD, or when the icon does not reflect changes made to the block. To create or update a block icon, open the drawing that contains the block, access the **BLOCKICON** command, enter the name of the block, and right-click or press [Enter] or the space bar.

Figure 24-13.
A—The correct procedure for redefining a block. B—Redefining a block without first exploding it creates an invalid circular reference.

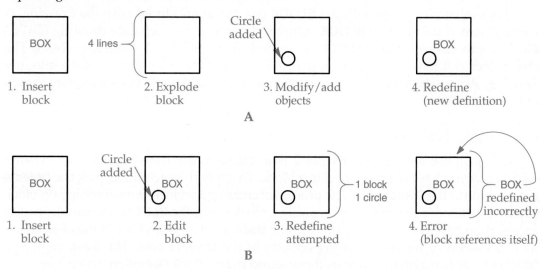

Figure 24-14.
The **Rename** dialog box allows you to change the name of blocks and other named objects.

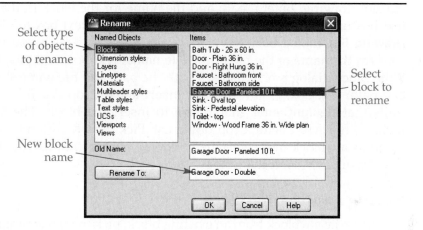

Copying Nested Objects

Some modify commands, such as **TRIM** and **EXTEND**, allow you to pick individual objects nested within a block reference or external reference (xref). However, most other commands recognize a block reference as a single block reference object and an xref as a single external reference object. For example, use the **COPY** command to create a copy of a block reference or xref. In contrast, the **NCOPY** command allows you to copy individual objects nested within a block reference or xref, which is more efficient than exploding a block and copying objects, for example.

Access the **NCOPY** command and use the **Settings** option to specify the method for copying nested objects. The difference between the default **Insert** option and the **Blind** option is most apparent when you are copying named objects nested within an xref. Chapter 31 explains xrefs. Use the **Insert** option to copy objects nested within a block. Individually select objects nested within a block reference or xref to copy, specify a base point, pick a location to place the copy, and exit the command. The **Multiple** option is available before you specify a base point and allows you to create several copies of the same object using a single **NCOPY** operation. The **NCOPY** command provides the same options as the standard **COPY** command, allowing you to specify a base point and a second point, choose a displacement using the **Displacement** option, or define the first point as the displacement.

Creating Blocks as Drawing Files

Blocks that you create with the **BLOCK** command are stored with the drawing. A wblock created using the **WBLOCK** command is saved as a separate drawing (DWG) file. You can also use the **WBLOCK** command to create a block from any object. The object does not have to be a block definition. Insert the wblock as a block into any drawing. Access the **WBLOCK** command to display the **Write Block** dialog box shown in **Figure 24-15**.

Creating a New Wblock

One purpose for creating a wblock is to create a new drawing file from existing objects that you have not converted to a block. To apply this technique, pick the **Objects** radio button in the **Source** area. The process of creating a wblock from existing non-block objects is similar to the process of creating a block using the **BLOCK** command. Specify an insertion base point using options in the **Base point** area, and select the objects and the disposition of the objects using options in the **Objects** area. The **Base point** and **Objects** areas function the same as those found in the **Block Definition** dialog box.

In contrast to a block, a wblock is saved as a drawing file, not as a block in the current drawing. Enter a path and file name for the block in the **File name and path:** text box or pick the ellipsis (**...**) button next to the text box to display the **Browse for Drawing File** dialog box. Navigate to the folder in which you want to save the file, confirm the name of the file in the **File name:** text box, and pick the **Save** button. The **Write Block** dialog box redisplays with the path and file name shown in the **File name and path:** text box. Finally, use the **Insert units:** drop-down list to select the type of units that **DesignCenter** should use to insert the block. The **Destination** area also includes the **Insert units:** drop-down list. Pick the **OK** button to finish. The objects are saved as a wblock in the specified folder. Now you can use the **INSERT** command in any drawing to insert the wblock.

Saving an Existing Block As a Wblock

To create a wblock from an existing block, pick the **Block** radio button in the **Source** area. See **Figure 24-16**. Select the block to save as a wblock from the drop-down list. Use the options in the **Destination** area to locate the wblock, and pick the **OK** button to finish.

Figure 24-15.
Using the **Write Block** dialog box to create a wblock from selected objects without first defining a block.

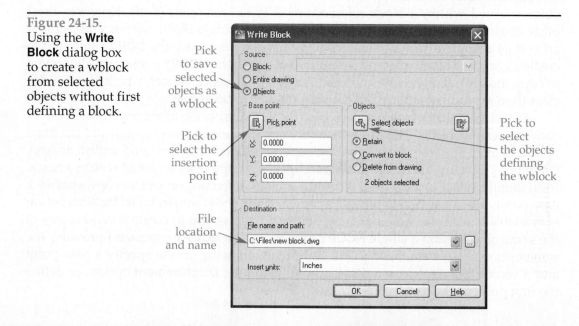

Pick to save selected objects as a wblock

Pick to select the insertion point

File location and name

Pick to select the objects defining the wblock

Figure 24-16.
Using the **Write Block** dialog box to create a wblock from an existing block definition.

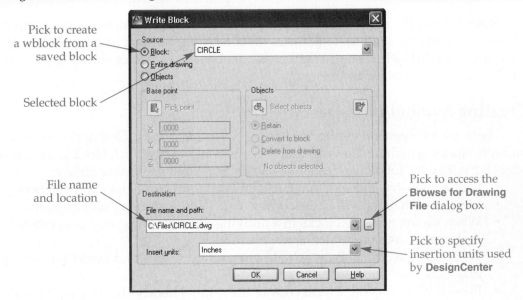

Pick to create a wblock from a saved block

Selected block

File name and location

Pick to access the **Browse for Drawing File** dialog box

Pick to specify insertion units used by **DesignCenter**

Storing a Drawing As a Wblock

To store an entire drawing as a wblock, pick the **Entire drawing** radio button in the **Source** area. Use the options in the **Destination** area to locate the wblock. In this case, the whole drawing is saved as if you are using the **SAVE** command. However, all uninserted, or unused, blocks in the drawing are deleted. If the drawing contains any unused blocks, the **Entire drawing** method may reduce the size of a drawing considerably. Pick the **OK** button to finish.

Exercise 24-8

Complete the exercise on the companion website.
www.g-wlearning.com/CAD

Revising an Inserted Drawing

If you insert a wblock into multiple drawings and then need to make changes to the wblock, use the **INSERT** command to access the original drawing file with the **Select Drawing File** dialog box. Then activate the **Specify On-screen** check box in the **Insertion point** area and pick the **OK** button. When a message asks to redefine the block, pick the **Yes** option. All of the wblock references update. Press [Esc] to cancel the command so that you do not insert a new block.

PROFESSIONAL TIP

If you work on projects that use inserted drawings that require revision, use reference drawings (xrefs) instead of inserting drawing files. Chapter 31 explains reference drawings placed using the **XREF** command. All xrefs automatically update when you open a drawing file that contains the xref content.

Symbol Libraries

Store frequently used symbols in *symbol libraries* to increase productivity. Continue to add new symbols to libraries to increase the number of symbols available for reuse. Establish whether you will store symbols as blocks or drawing files, and identify a storage location and system.

Creating Symbol Libraries

There are two general methods to create a symbol library. One option is to save multiple blocks in a single drawing. The other option is to save each block to a separate wblock file. Use the following guidelines when developing a symbol library:

- Follow industry and company or school standards for blocks and symbols.
- Identify each block with a descriptive name and insertion point location.
- When saving multiple blocks in a drawing file, save one group of symbols per drawing file and use folders to organize files.
- When using wblocks, give each file a meaningful name and use folders to organize files.
- Provide all users with a hard copy of the symbol library showing each symbol, insertion points, storage locations, and any other necessary information. See Figure 24-17.

Figure 24-17.

A printed copy of a typical symbol library distributed to architectural drafters. The "X" symbols indicate insertion points and are not part of the blocks. *(Courtesy of Ron Palma, 3D-DZYN)*

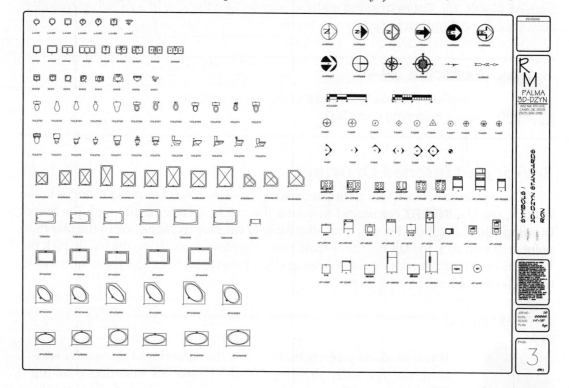

AutoCAD and Its Applications—Basics

- If a network is not in use, place symbol library files on each workstation in the classroom or office.
- Keep backup copies of all files in a secure location.
- When you revise symbols, update all files containing the edited symbols as appropriate.
- Inform all users of any changes to saved symbols.

Storing Symbol Libraries

Store symbol libraries on the local or network hard drive. This location is easy to access, quick, and more convenient to use than portable media. Removable media, such as a removable hard drive, USB flash drive, or CD, are appropriate for backup purposes if a network drive with an automatic backup function is not available. In the absence of a network, use removable media to transport files from one workstation to another.

Use a system of folders and files to store and organize symbol libraries. Store content outside of the AutoCAD system folders to keep the system folders uncluttered, and to differentiate your folders and files from AutoCAD system folders and files. A good method is to create a \Blocks folder for storing blocks, as shown in **Figure 24-18**.

If you save multiple symbols within a single drawing, use **DesignCenter** or the **Tool Palettes** palette to insert the symbols as needed. This system often works best when you use several drawing files to group similar symbols. For example, create different symbol libraries based on fastener, electronic, electrical, piping, mechanical, structural, architectural, landscaping, and mapping symbols. Limit the symbols in a drawing to a reasonable number so you can easily find and load the symbols.

Figure 24-18.
An efficient way to store blocks saved as drawing files is to set up a \Blocks folder containing folders for each type of block on the hard drive.

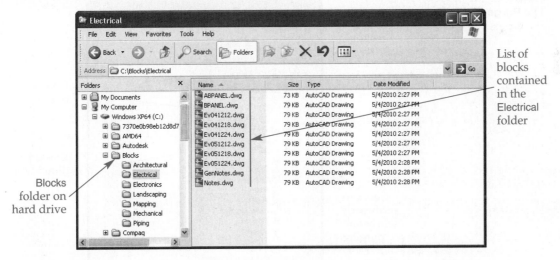

Arrange drawing files saved on the hard drive in a logical manner. All workstations in a non-networked classroom or office should have folders with the same names. Assign one person to update and copy symbol libraries to all workstations. Copy drawing files to each workstation from a master CD. Keep the master and backup versions of the symbol libraries in separate locations.

Purging Named Objects

A drawing typically accumulates unused named objects that may be unnecessary. Unused named objects increase drawing file size and may make it more difficult to locate and use content that is common and necessary. Use the **PURGE** command to *purge* unused objects from the drawing if necessary. Access the **PURGE** command to display the **Purge** dialog box. See **Figure 24-19.**

Select the appropriate radio button at the top of the dialog box to view content that can or cannot be purged. Before purging, select the **Confirm each item to be purged** check box to have an opportunity to review each item before deleting. Check **Purge nested items** to purge nested items. Selecting the **Purge zero-length geometry and empty text objects** check box is an effective way to erase all zero-length objects, such as a line or arc drawn as a dot or text that only includes spaces. These objects are often mistakes or unintended results of the drawing and editing processes.

To purge specific unused items, use the tree view to locate and highlight the items to purge, and then pick the **Purge** button. To purge all unused items, pick the **Purge All** button. Purging may cause other named objects to become unreferenced. As a result, you may need to purge more than once to purge the drawing of all unused named objects. Messages appear to guide you through the purge operation.

Figure 24-19.
The **Purge** dialog box.

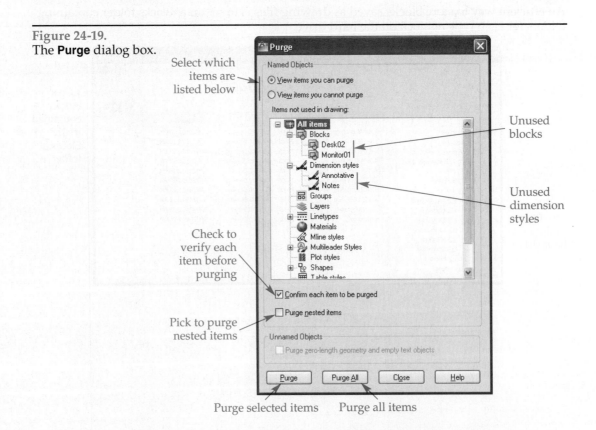

Select which items are listed below

Unused blocks

Unused dimension styles

Check to verify each item before purging

Pick to purge nested items

Purge selected items   Purge all items

Chapter Review

Answer the following questions. Write your answers on a separate sheet of paper or complete the electronic chapter review on the companion website.
www.g-wlearning.com/CAD

1. Why would you draw blocks on the 0 layer?
2. What properties do blocks drawn on a layer other than layer 0 assume when they are inserted?
3. What characters can be used in a block name?
4. Define the term *nesting* in relation to blocks.
5. What is a block reference?
6. How can you access a listing of all blocks in the current drawing?
7. Describe the effect of entering negative scale factors when inserting a block.
8. What type of block is a one-unit line object?
9. How do you preset block insertion variables using the **Insert** dialog box?
10. Name a limitation of an array pattern created with the **MINSERT** command.
11. What is the purpose of the **BASE** command?
12. Briefly explain how to insert a block into a drawing from **DesignCenter**.
13. What command allows you to change a block's layer without editing the block definition?
14. Identify the command that allows you to break an inserted block into its individual objects for editing purposes.
15. How can you edit all instances of an inserted block quickly?
16. What is the primary difference between blocks created with the **BLOCK** and **WBLOCK** commands?
17. Explain the advantage of storing a drawing as a wblock if you anticipate the need to insert the drawing into other drawings.
18. Define *symbol library*.
19. What is the purpose of the **PURGE** command?
20. Explain how to remove all unused blocks from a drawing.

Drawing Problems

Start AutoCAD if it is not already started. Start a new drawing for each problem using an appropriate template of your choice. The template should include layers and text, dimension, multileader, and table styles, when necessary, for drawing the given objects. Add layers and text, dimension, multileader, and table styles as needed. Draw all objects using appropriate layers and text, dimension, multileader, and table styles, justification, and format. Follow the specific instructions for each problem. Use only drawing and editing commands and techniques you have already learned. Use your own judgment and approximate dimensions when necessary. Apply formal dimensions accurately using ASME or appropriate industry standards.

Note: *Some of the problems in this chapter are built on problems from previous chapters. If you have not yet completed those problems, complete them now.*

▼ Basic

1. Open P12-16 and save as P24-1. The P24-1 file should be active. The sketch for this drawing is shown. Erase all copies of the symbols, leaving the original objects intact. These include the steel column symbols and the bay and column line tags. Then do the following:
   A. Make blocks of the steel column symbol and the tag symbols.
   B. Use the **MINSERT** command or an **ARRAY** command to place the symbols in the drawing.
   C. Dimension the drawing as shown.
   D. Resave the drawing.

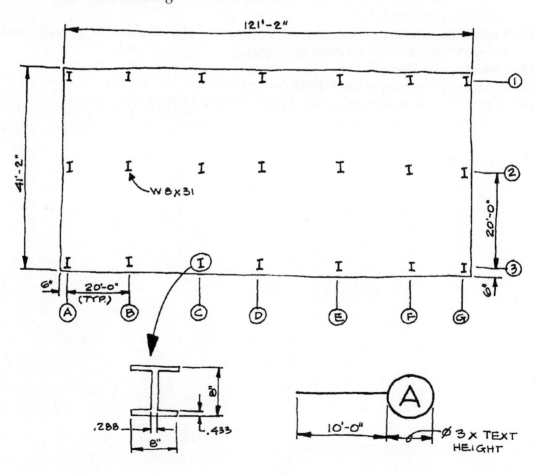

2. Open P12-17 and save as P24-2. The P24-2 file should be active. The sketch for this drawing is shown. Erase all of the desk workstations except one. Then do the following:
   A. Create a block of the workstation.
   B. Insert the block into the drawing using the **MINSERT** command.
   C. Dimension one of the workstations as shown.
   D. Resave the drawing.

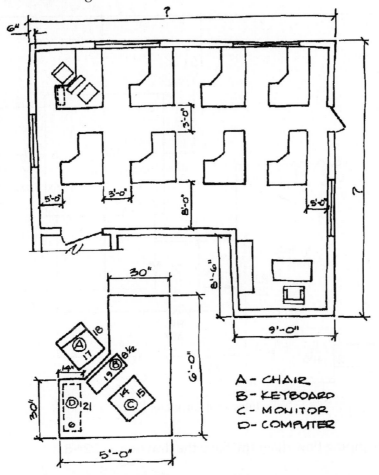

A - CHAIR
B - KEYBOARD
C - MONITOR
D - COMPUTER

3. Complete this problem after completing Problem 24-7. Open P24-7 and save as P24-3. The P24-3 file should be active. Modify the NAND gates to become XNOR gates, as shown, by modifying the block definition. Resave the drawing.

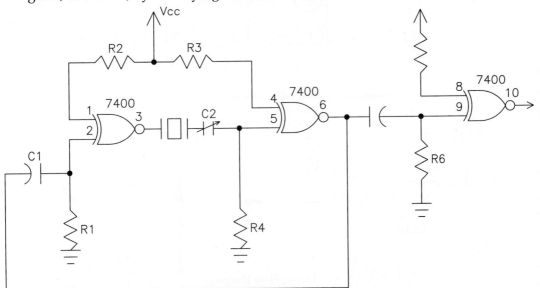

▼ **Intermediate**

Problems 4–7 represent a variety of diagrams created using symbols as blocks. Create each drawing as shown. The drawings are not to scale. Create the symbols first as blocks or wblocks and then save the symbols in a symbol library using one of the methods described in this chapter.

4. Draw the integrated circuit schematic for a clock. Save the drawing as **P24-4**.

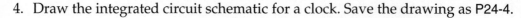
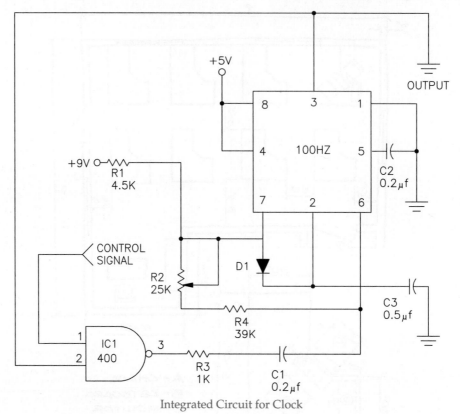

Integrated Circuit for Clock

5. Draw the piping flow diagram. Save the drawing as **P24-5**.

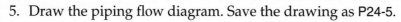
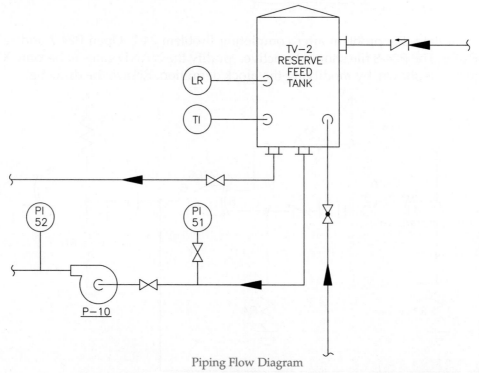

Piping Flow Diagram

6. Draw the logic diagram of a marking system. Save the drawing as P24-6.

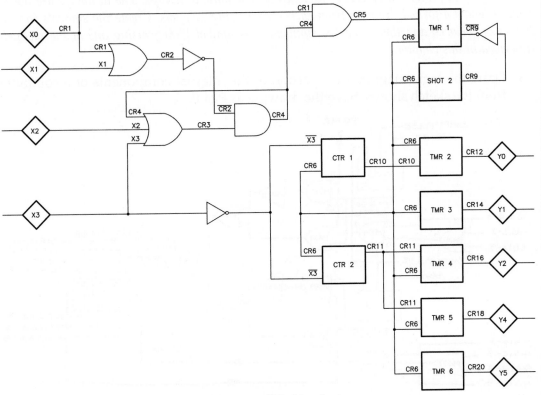

Logic Diagram of Marking System

7. Draw the digital logic circuit. Create each type of component in the circuit as a block. Save the drawing as P24-7.

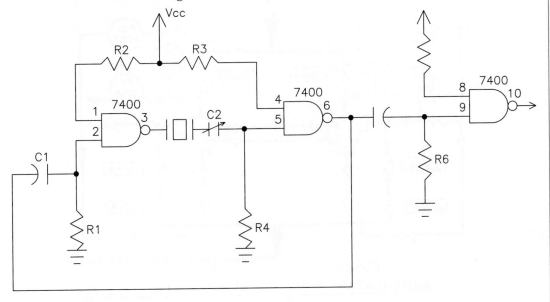

▼ Advanced

Problems 8–12 present engineering sketches of schematic drawings. The drawings are not to scale. Prepare a formal drawing from each sketch using symbols. Create the symbols first as blocks or wblocks and then save the symbols in a symbol library using one of the methods described in this chapter.

8. Draw the logic diagram of a portion of the internal components of a computer from the sketch shown. Save the drawing as P24-8.

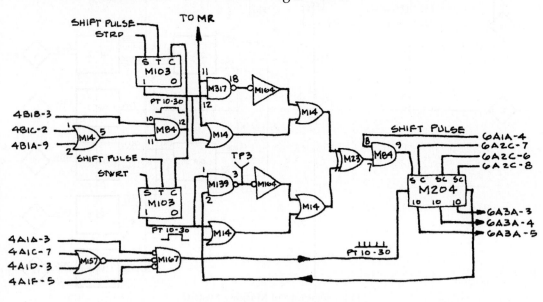

9. Draw the piping flow diagram of a cooling water system from the sketch shown. Look closely at this drawing before you begin. Draw the thick flow lines with polylines. Save the drawing as P24-9.

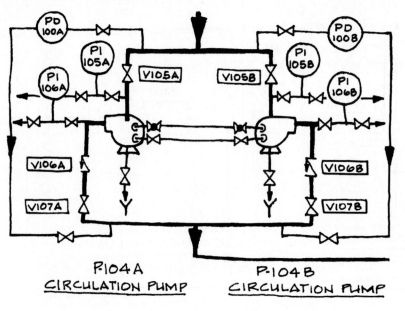

10. Draw the general arrangement of a basement floor plan for a new building from the sketch shown. The engineer has shown one example of each type of equipment. Use the following instructions to complete the drawing.
   A. All text should be 1/8" high, except the text for the bay and column line tags, which should be 3/16" high. The diameter of the line balloons for the bay and column lines should be twice the diameter of the text height.
   B. The column and bay steel symbols represent wide-flange structural shapes and should be 8" wide × 12" high.
   C. The PUMP and CHILLER installations (except PUMP #4 and PUMP #5) should be drawn per the dimensions given for PUMP #1 and CHILLER #1. Use the dimensions shown for the other PUMP units.
   D. TANK #2 and PUMP #5 (P-5) should be drawn per the dimensions given for TANK #1 and PUMP #4.
   E. Tanks T-3, T-4, T-5, and T-6 are all the same size and are aligned 12' from column line A.
   F. Plan this drawing carefully and create as many blocks as necessary to increase your productivity. Dimension the drawing as shown, and provide location dimensions for all equipment not shown in the engineer's sketch.
   G. Save the drawing as P24-10.

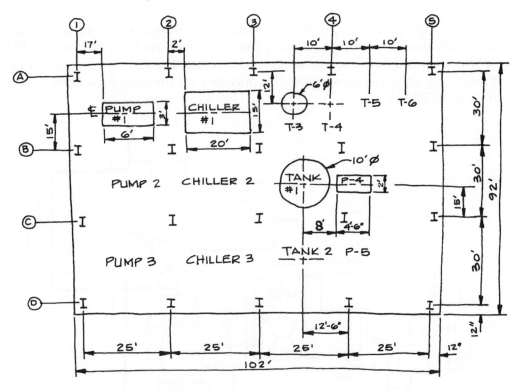

11. Open P24-10 and save as P24-11. The P24-11 file should be active. The engineer has provided you with a sketch of the necessary revisions to the drawing. It is up to you to alter the drawing as quickly and efficiently as possible. Do not add the dimensions shown on the sketch; the dimensions are provided for construction purposes only. Revise the drawing so all chillers and the four tanks reflect the changes. Save the drawing as P24-11.

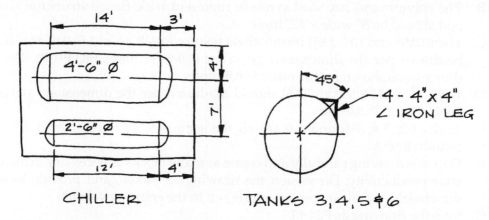

12. Draw the piping flow diagram of an industrial effluent treatment system from the sketch shown. Eliminate as many bends in the flow lines as possible. Place arrowheads at all flow line intersections and bends. The flow lines should not run through any valves or equipment. Use polylines for the thick flow lines. Save the drawing as P24-12.

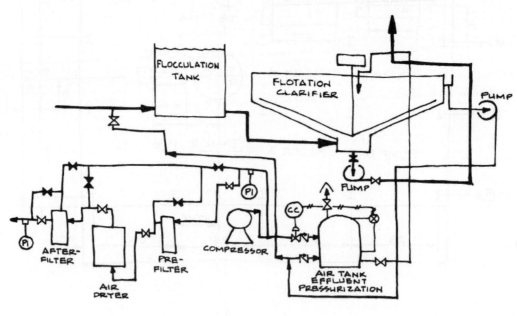

13. Create computer, plotter, and printer/copier blocks and then draw the network diagram shown. Save the drawing as P24-13.

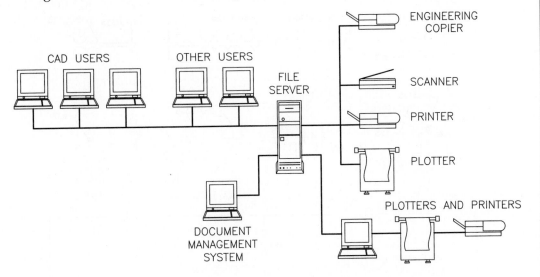

CAD USERS   OTHER USERS   FILE SERVER   ENGINEERING COPIER   SCANNER   PRINTER   PLOTTER   PLOTTERS AND PRINTERS   DOCUMENT MANAGEMENT SYSTEM

14. Draw the piping diagram, creating blocks for each type of fitting. Save the drawing as P24-14.

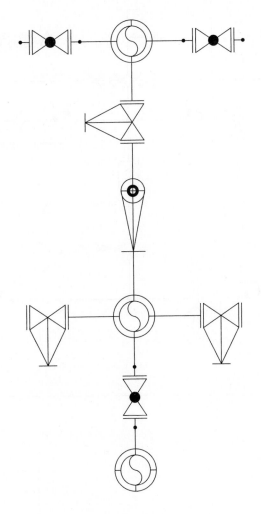

15. Create component blocks based on the dimensions shown. Then use the blocks to draw the schematic below. Save the drawing as P24-15.

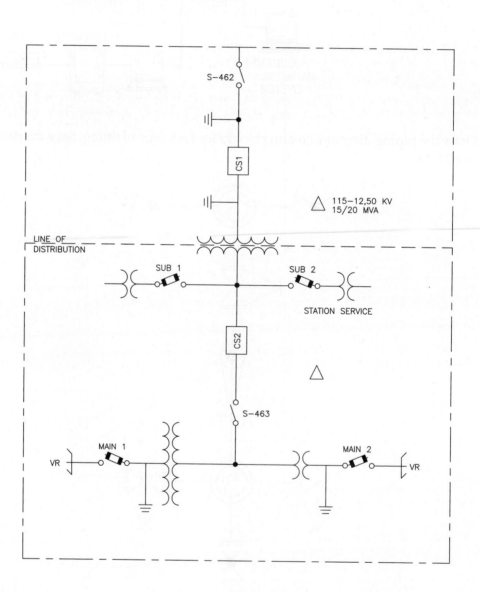

16. Create a symbol library for one of the drafting disciplines listed and save the symbol library as a template or drawing file. Then, after checking with your supervisor or instructor, draw a problem using the library. If you save the symbol library as a template, start the problem with the template. If you save the symbol library as a drawing file, start a new drawing and insert the symbol library into the new file. Specialty areas you might create symbols for include:

- Mechanical (machine features, fasteners, tolerance symbols)
- Architectural (doors, windows, fixtures)
- Structural (steel shapes, bolts, standard footings)
- Civil (mapping symbols, survey markers, utilities)
- Industrial piping (fittings, valves)
- Piping flow diagrams (tanks, valves, pumps)
- Electrical schematics (resistors, capacitors, switches)
- Electrical one-line (transformers, switches)
- Electronics (IC chips, test points, components)
- Logic diagrams (AND gates, NAND gates, buffers)
- GD&T (GD&T symbols)

Save the drawing as P24-16 or choose an appropriate file name, such as ARCH-PRO or ELEC-PRO. Display the symbol library created in this problem and print a hard copy. Put the printed copy in your notebook for reference.

AutoCAD Certified Associate Exam Practice

Answer the following questions. Write your answers on a separate sheet of paper.

1. A drawing has five layers: 0 (white), Objects (blue), Dimensions (red), Center (yellow), and Hidden (green). If you create objects on the 0 layer, assign the objects an absolute color of green, create a block of the objects, and then insert the block on the Objects layer, what color will the block be? *Select the one item that best answers the question.*
   A. blue
   B. green
   C. red
   D. white
   E. yellow

2. You have a 1″ × 1″ unit block of a window. To use the block to represent a 3′ window in a 4″ wall, which of the following scale factors would you use? *Select the one item that best answers the question.*
   A. x = 1, y = 1
   B. x = 3, y = 4
   C. x = 4, y = 30
   D. x = 30, y = 4
   E. x = 36, y = 4

3. Which of the following can you use to insert a block into a drawing? *Select all that apply.*
   A. **BLOCK** command
   B. **DesignCenter**
   C. **INSERT** command
   D. **Tool Palettes** palette
   E. **WBLOCK** command

AutoCAD Certified Professional Exam Practice

Follow the instructions in each problem. Write your answers on a separate sheet of paper.

1. **Open CPE-24block.dwg. This file is available on the companion website.**
   This drawing contains a block named Hole. Insert the block using the default settings. Select the lower-left corner of the existing object as the insertion point. What are the coordinates of the center of the hole?

2. **Open CPE-23insert.dwg. This file is available on the companion website.**
   This drawing contains a 2D unit block named Door. Use the appropriate scale factors and rotation to place a 2′-4″ door 6″ above the intersection of the two walls. Edit the wall lines to finish the doorway as shown. What are the coordinates of Point A?

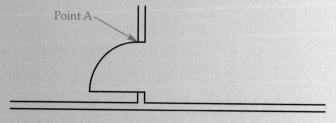

Chapter 25

Block Attributes

Learning Objectives

After completing this chapter, you will be able to:

✓ Define attributes.
✓ Create blocks that contain attributes.
✓ Insert blocks with attributes into a drawing.
✓ Edit attribute values in existing blocks.
✓ Edit single and multiple attribute references.
✓ Create title blocks, revision history blocks, and parts lists with attributes.
✓ Display attribute values in fields.

Attributes enhance blocks that require text or numerical information. For example, a door tag block contains a letter or number that links the door to a door schedule. Adding an attribute to the door tag block allows you to include a unique letter or number with the symbol, without adding block definitions for each door tag. You can also extract attribute data to automate drawing requirements, such as preparing schedules, parts lists, and bills of materials.

attributes: Text-based data assigned to a specific object. Attributes turn a drawing into a graphical database.

Defining Attributes

Attributes and geometry are often used together to create a block. See **Figure 25-1**. However, you can prepare blocks that include only attributes. Create attributes with other objects during the initial phase of block development. Add as many attributes as needed to describe the symbol or product, such as the name, number, manufacturer, type, size, price, and weight of an item. Access the **ATTDEF** command to assign attributes using the **Attribute Definition** dialog box. See **Figure 25-2**.

Ribbon

Home
> Block
Insert
> Attributes
Block Editor
> Action
  Parameters

Define Attributes

Type

ATTDEF
ATT

ATTDEF

Figure 25-1.
Examples of blocks with defined attributes.

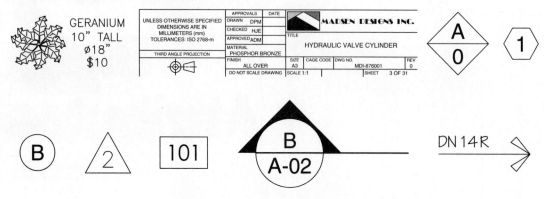

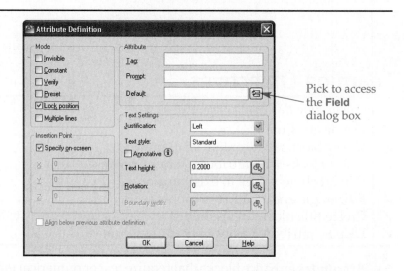

Figure 25-2.
Use the **Attribute Definition** dialog box to assign attributes to blocks.

Pick to access the **Field** dialog box

Attribute Modes

Use the **Mode** area to set attribute modes. Symbols often require attributes to appear with the block. An alternative is to check **Invisible** to hide attributes, but still include attribute data in the drawing that you can reference and *extract*. The geranium symbol in **Figure 25-1** is an example of a block with attributes that might be invisible, depending on the application. The other symbols in **Figure 25-1** are examples of blocks with visible attributes. Blocks often include both visible and invisible attributes.

Pick the **Constant** check box if the value of the attribute should always be the same. All insertions of the block display the same value for the attribute, without prompting for a new value when you insert the block. Deselect the **Constant** check box to use different attribute values for multiple insertions of the block. Pick the **Verify** check box to display a prompt that asks if the attribute value is correct when you insert the block. Choose the **Preset** check box to have the attribute assume preset values during block insertion. The **Preset** option disables the attribute prompt when you insert the block. Uncheck **Preset** to display the normal prompt.

Deselect the **Lock position** check box to have the ability to move the attribute independently of the block after insertion. In addition, you must deselect the **Lock position** check box to include the attribute with the action selection set when you assign an action to a dynamic block. If **Lock position** is checked, the attribute is filtered out when you assign the action to the dynamic block. You will learn about dynamic blocks later in this textbook.

extract: Gather content from the drawing file database to display in the drawing or in an external document.

You can create single-line or multiple-line attributes. Pick the **Multiple lines** check box to activate options for creating a multiple-line attribute. Deselect the **Multiple lines** check box to create a single-line attribute.

Tag, Prompt, and Value

The **Attribute** area provides text boxes for assigning a tag, prompt, and default value to the attribute. Use the **Tag** text box to enter the attribute name, or tag. For example, the tag for a size attribute for a valve block could be SIZE. You must enter a tag in order to create an attribute. The tag cannot include spaces. The attribute definition applies uppercase characters to the tag, even if you type lowercase characters in the text box.

Type a statement in the **Prompt** text box that will display when you insert or edit the block. For example, if you specify SIZE as the attribute tag, you might specify What is the valve size? or Enter valve size: as the prompt. You can also leave the prompt blank. The **Prompt** text box is disabled when you select the **Constant** attribute mode.

Use the **Default** text box to specify a default attribute value or a description of an acceptable value for reference. For example, you might type the most common size for the SIZE attribute, or a message regarding the type of information needed, such as 10 SPACES MAX or NUMBERS ONLY. If you deselect the **Multiple** lines attribute mode, enter the default value directly in the text box.

Attribute values can include up to 255 characters. If the first character in an entry is a space, start the string with a backslash (\). If the first character is a backslash, begin the entry with two backslashes (\\). If you select the **Multiple lines** attribute mode, pick the ellipsis (...) button to enter the drawing area and place multiline text using the **Text Formatting** toolbar and text editor. See **Figure 25-3**. Enter the default text, and then pick the **OK** button on the toolbar to return to the **Attribute Definition** dialog box. Use the Insert field button to include a field in the default value. You have the option to leave the default value blank.

The abbreviated **Text Formatting** toolbar shown in **Figure 25-3** appears by default. Set the **ATTIPE** system variable to 1 to display the complete **Text Formatting** toolbar. The **ATTIPE** system variable is set to 0 by default.

Text Settings

Use options in the **Text Settings** area to specify attribute text format. Use the **Justification** drop-down list to select a justification for the attribute text. The default

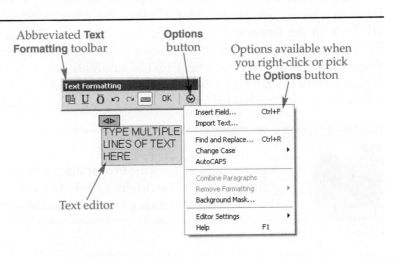

Figure 25-3.
Define multiple-line attributes directly on-screen. The abbreviated **Text Formatting** toolbar appears instead of the **Text Editor** ribbon tab.

Abbreviated **Text Formatting** toolbar

Options button

Options available when you right-click or pick the **Options** button

Text editor

justification is **Left**. In single-line attributes, the text itself is justified. In the **Multiple lines** attribute mode, the text boundary is justified.

Use the **Text Style** drop-down list to select a text style available in the current drawing to assign to the attribute. Pick the **Annotative** check box to make the attribute text height annotative. AutoCAD scales annotative attributes according to the specified annotation scale, which reduces the need for you to calculate the scale factor. If you plan to define a block as annotative, you can include annotative or non-annotative objects, such as attributes, with the block. However, you must use non-annotative objects when preparing a non-annotative block.

Specify the height of the attribute text in the **Height** text box, or pick the **Text Height** button next to the text box to specify two points in the drawing to set the text height. Identify the rotation angle for the attribute text in the **Rotation** text box, or pick the **Rotation** button next to the text box to pick two points in the drawing to set the text rotation. The **Boundary width** option is available in the **Multiple lines** attribute mode. Type a width in the **Boundary width** text box, or pick the **Boundary width** button next to the text box to specify two points in the drawing to set a text boundary width.

Defining the Attribute Insertion Point

Use options in the **Insertion Point** area to define how and where to position the attribute during insertion. Choose the **Specify On-screen** check box to pick an insertion point in the drawing after you pick the **OK** button to create the attribute and exit the **Attribute Definition** dialog box. An alternative is to type values in the **X:**, **Y:**, and **Z:** text boxes if you know the coordinates for the insertion point.

The **Align below previous attribute definition** check box is enabled if the drawing already contains at least one attribute. Check the box to place the new attribute directly below the most recently created attribute using the justification of that attribute. This is an effective technique for placing a group of different attributes in the same block. The **Text Options** and **Insertion Point** areas are deactivated.

Placing the Attribute

After defining all elements of the attribute, pick the **OK** button to close the **Attribute Definition** dialog box. The attribute tag appears on-screen if you specified coordinates for the insertion point, or if you used the **Align below previous attribute definition** option. Otherwise, AutoCAD prompts you to specify an insertion point. If the attribute mode is set to **Invisible**, do not be concerned that the tag is visible. The tag disappears when you include the attribute with a block definition.

Editing Attribute Properties

The **Properties** palette provides options for editing attributes before you include them in a block. See **Figure 25-4**. Change the color, linetype, or layer of the selected attribute in the **General** category. Use options in the **Text** category to adjust the **Tag**, **Prompt**, or **Value**. If the value contains a field, the value appears as normal text in the **Properties** palette. Modified field text is automatically converted to text and is disassociated from the field. The **Text** category also contains options to change the attribute text settings. Additional text and attribute options are available in the **Misc** section.

PROFESSIONAL TIP

A powerful feature of the **Properties** palette for editing attributes is the ability to change the original attribute modes. The **Invisible**, **Constant**, **Verify**, and **Preset** mode settings are available in the **Misc** category.

Figure 25-4.
The **Properties** palette allows you to modify attributes.

Selected object to edit

Pick to change the attribute tag

Pick to change an attribute mode setting

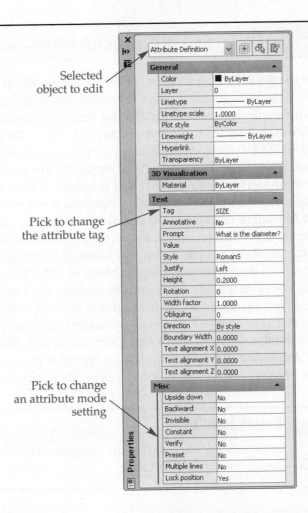

Attribute Definition		
General		▲
Color	■ ByLayer	
Layer	0	
Linetype	—— ByLayer	
Linetype scale	1.0000	
Plot style	ByColor	
Lineweight	—— ByLayer	
Hyperlink		
Transparency	ByLayer	
3D Visualization		▲
Material	ByLayer	
Text		▲
Tag	SIZE	
Annotative	No	
Prompt	What is the diameter?	
Value		
Style	RomanS	
Justify	Left	
Height	0.2000	
Rotation	0	
Width factor	1.0000	
Obliquing	0	
Direction	By style	
Boundary Width	0.0000	
Text alignment X	0.0000	
Text alignment Y	0.0000	
Text alignment Z	0.0000	
Misc		▲
Upside down	No	
Backward	No	
Invisible	No	
Constant	No	
Verify	No	
Preset	No	
Multiple lines	No	
Lock position	Yes	

Creating Blocks with Attributes

Attributes are useful only when they are included with a block definition. Use the **BLOCK** or **WBLOCK** command to define a block with attributes. Select all objects and attributes to be included with the block. The order in which you select attribute definitions is the order of prompts, or the order in which the attributes appear in the **Edit Attributes** dialog box. If you select the **Convert to Block** radio button in the **Block Definition** dialog box, the **Edit Attributes** dialog box appears when you create the block. See **Figure 25-5**. The **Edit Attributes** dialog box allows you to adjust attribute values when you insert or edit the block.

PROFESSIONAL TIP

If you create attributes in the order in which you want to receive prompts and then use window or crossing selection to select the attributes, the attribute prompts display in the reverse order of the desired prompting. To change the order, insert, explode, and then redefine the block using window or crossing selection again to pick the attributes. The attribute prompt order reverses again, placing the prompts in the desired order.

Inserting Blocks with Attributes

Use the **INSERT** command or another block insertion method, such as **DesignCenter** or a tool palette, to insert a block that contains attributes. The process of inserting a block with attributes is similar to that for inserting a block without attributes. The only difference is that additional prompts request values for each attribute.

The **ATTDIA** system variable is set to 0 by default, which displays single-line attribute prompts at the command line or dynamic input and displays multiple-line attribute prompts using the AutoCAD text window. A better method of entering attribute values is to set the **ATTDIA** system variable to 1 before inserting blocks. This enables the **Edit Attributes** dialog box. The dialog box appears after you specify the insertion point, scale, and rotation angle, allowing you to answer each attribute prompt. Type single-line attribute values in the text boxes. To define multiple-line attributes, select the ellipsis (...) button next to the text boxes to enter values on-screen as multiline text. If a value includes a field, you can right-click on the field to edit it or convert it to text.

Press [Tab] to move quickly through the attributes and buttons in the **Edit Attributes** dialog box. Press [Shift]+[Tab] to cycle through attributes and buttons in reverse order. If the block includes more than eight attributes, pick the **Next** button at the bottom of the **Edit Attributes** dialog box to display the next page of attributes. When you finish entering values, pick the **OK** button to close the dialog box and create the block.

Exercise 25-1

Complete the exercise on the companion website.
www.g-wlearning.com/CAD

Attribute Prompt Suppression

Some blocks may include attributes that always retain default values. In this case, there is no need to receive prompts for attribute values when you insert the block. Assign the **Constant** mode to a specific attribute during attribute definition if you are confident the attribute value will not change, or turn off prompts for all attributes by setting the **ATTREQ** system variable to 0. To display attribute prompts again, change the setting back to 1. The **ATTREQ** system variable setting is saved with the drawing.

Figure 25-5.
The **Edit Attributes** dialog box allows you to enter attribute definitions when you insert or edit a block.

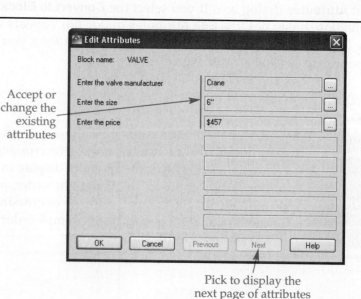

Accept or change the existing attributes

Pick to display the next page of attributes

Project and drawing planning should involve setting system variables such as **ATTREQ**. Setting **ATTREQ** to 0 before using blocks can save time in the drawing process. Always remember to set **ATTREQ** back to 1 to receive prompts instead of accepting defaults. If anticipated attribute prompts do not appear, check the current **ATTREQ** setting and adjust it if necessary.

Controlling Attribute Display

Some attributes only provide content to generate parts lists or bills of materials and to speed accounting. These types of attributes usually do not display on-screen or plot. Use the **ATTDISP** command to control the display of attributes on-screen. The easiest way to activate the **ATTDISP** command option is to pick the corresponding button from the ribbon.

Use the default **Retain display (Normal)** option to display attributes exactly as created. Use the **Display all (ON)** option to display all attributes, both visible and invisible. Apply the **Hide all (OFF)** option to suppress the display of all attributes, including visible attributes.

Use the **ATTDISP** command to display invisible attributes when necessary or to hide all attributes if they interfere with the current drawing session. In a drawing in which attributes should be visible but are not, check the current setting of **ATTDISP** and adjust it if necessary.

Editing Attribute References

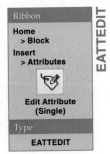

Once you insert a block with attributes, tools are available for editing attribute values and settings. One option is modify the attributes of a single block using the **EATTEDIT** command. Access the **EATTEDIT** command and pick the block to display the **Enhanced Attribute Editor**. See **Figure 25-6**. If you want to edit attributes in other blocks, pick the **Select block** button to return to the drawing to select a different block to modify.

The **Attribute** tab, shown in **Figure 25-6**, displays all of the attributes assigned to the selected block. Pick the attribute you want to modify and enter a new value in the **Value:** text box. If the attribute is a multiple-line attribute, pick the ellipsis (...) button to modify the text on-screen. Pick the **Apply** button after adjusting the value.

A quick way to access the **EATTEDIT** command is to double-click on a block containing attributes. You can also edit multiple-line attribute values without accessing the **Enhanced Attribute Editor** using the **ATTIPEDIT** command.

Figure 25-6.
Select the attribute to modify and change the value in the **Attribute** tab of the **Enhanced Attribute Editor.**

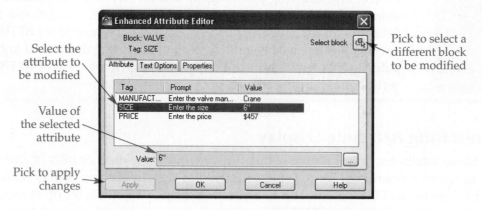

Select the attribute to be modified

Pick to select a different block to be modified

Value of the selected attribute

Pick to apply changes

Use the **Text Options** tab, shown in **Figure 25-7A**, to modify the text options of an attribute. The **Properties** tab, shown in **Figure 25-7B**, provides general property adjustments for an attribute. Each attribute in a block is a separate item. The settings you apply in the **Text Options** and **Properties** tabs affect the active attribute in the **Attribute** tab. Pick the **Apply** button to view changes made to attributes. Pick the **OK** button to close the dialog box.

Figure 25-7.
A—The **Text Options** tab provides options in addition to those set in the **Attribute Definition** dialog box. B—The **Properties** tab allows you to modify the properties of an attribute.

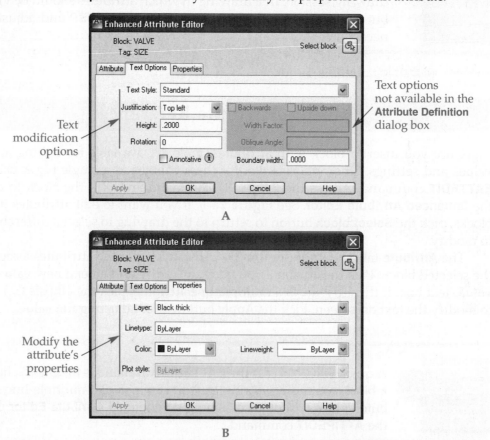

Text modification options

Text options not available in the **Attribute Definition** dialog box

A

Modify the attribute's properties

B

Using the FIND Command to Edit Attributes

The **FIND** command, which displays the **Find and Replace** dialog box described in Chapter 10, provides one of the quickest ways to edit attributes. You can also access the **FIND** command when no command is active by right-clicking in the drawing area and selecting **Find...**, or by entering the text in the **Find text** text box in the **Text** panel of the **Annotation** ribbon tab and pressing [Enter]. Specify the portion of the drawing to search, the attribute value to find, and the replacement value.

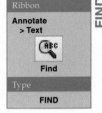

Ribbon
Annotate
> Text

Find

Type
FIND

FIND

Editing Multiple Attribute References

The **Enhanced Attribute Editor** allows you to edit attribute values and settings by selecting blocks one at a time. The **-ATTEDIT** command edits the attributes of several blocks. When you access the **-ATTEDIT** command, a prompt asks if you want to edit attributes individually. Use the default **Yes** option to select specific blocks with attributes to edit. Use the **No** option to apply *global attribute editing*.

If you choose the **Yes** option, prompts appear to specify the block name, attribute tag, and attribute value. To edit attribute values selectively, respond to each prompt with the correct name or value, and then select one or more attributes. If you see the message 0 found after selecting attributes, you picked an incorrectly specified attribute. It is often quicker to press [Enter] at each of the three specification prompts and then pick the attribute to edit. Select an option and follow the prompts to edit the attribute.

If you choose the **No** option, the Edit only attributes visible on screen? prompt appears. Select the **Yes** option to edit all visible attributes, or choose the **No** option to edit all attributes, including invisible attributes. The same three prompts previously described for individual block editing appear.

Figure 25-8A shows a VALVE block inserted three times with the manufacturer specified as CRANE. In this example, the manufacturer was supposed to be POWELL. To change the attribute for each insertion, access the **-ATTEDIT** command and specify global editing. Press [Enter] at each of the three specification prompts. When the Select attributes: prompt appears, pick CRANE on each of the VALVE blocks and press [Enter]. At the Enter string to change: prompt, enter CRANE, and at the Enter new string: prompt, enter POWELL. See the result in **Figure 25-8B**.

Ribbon
Home
> Block
Insert
> Attributes

Edit Attribute
(Multiple)

Type
-ATTEDIT
-ATE

-ATTEDIT

global attribute editing: Editing or changing all insertions, or instances, of the same block in a single operation.

PROFESSIONAL TIP

Use care when assigning the **Constant** mode to attribute definitions. The **-ATTEDIT** command displays 0 found if you attempt to edit a block attribute that has a **Constant** mode setting. Assign the **Constant** mode only to attributes you know will not change.

You can also use the **-ATTEDIT** command to edit individual attribute values and properties. However, it is more efficient to use the **Enhanced Attribute Editor** to change individual attributes.

Figure 25-8.
Using the global
editing technique
with the **-ATTEDIT**
command allows
you to change the
same attribute
on several block
insertions.

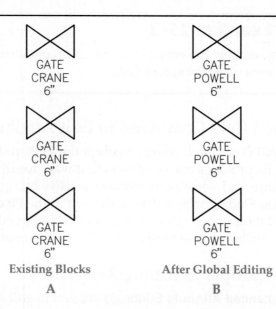

GATE CRANE 6"	GATE POWELL 6"
GATE CRANE 6"	GATE POWELL 6"
GATE CRANE 6"	GATE POWELL 6"
Existing Blocks	**After Global Editing**
A	**B**

Exercise 25-3

Complete the exercise on the companion website.
www.g-wlearning.com/CAD

Editing Attribute Definitions

Ribbon
Home
> **Block**
Insert
> **Attributes**

Manage Attributes

Type
BATTMAN

Once you create a block with attributes, tools are available for modifying attribute definitions. One option is to edit attribute definitions using the **BATTMAN** command, which displays the **Block Attribute Manager**. See **Figure 25-9**. To manage the attributes in a block, choose the block name from the **Block:** drop-down list or pick the **Select block** button to return to the drawing and pick a block.

The tag, prompt, default value, and modes for each attribute are listed by default. To select the attribute properties listed in the **Block Attribute Manager**, pick the **Settings...** button to open the **Block Attribute Settings** dialog box. See **Figure 25-10**. Check the

Figure 25-9.
Use the **Block Attribute Manager** to change attribute definitions, delete attributes, and change the order of attribute prompts.

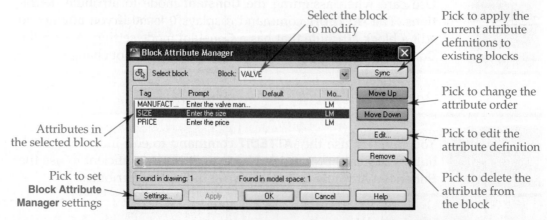

Figure 25-10.
The **Block Attribute Settings** dialog box controls the types of attributes displayed in the **Block Attribute Manager**.

Select the attribute properties to list in the **Block Attribute Manager**

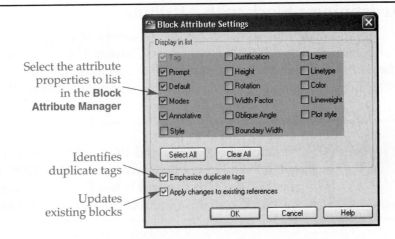

Identifies duplicate tags

Updates existing blocks

properties to list in the **Display in list** area. Select the **Emphasize duplicate tags** check box to highlight attributes with identical tags in red. Check **Apply changes to existing references** to apply changes made in the **Block Attribute Manager** to existing blocks. Pick the **OK** button to return to the **Block Attribute Manager**.

The attribute list in the **Block Attribute Manager** reflects the order in which prompts appear when you insert a block. Use the **Move Up** and **Move Down** buttons to change the order of the selected attribute within the list, modifying the prompt order. To delete an attribute, pick the **Remove** button. To modify an attribute, select the attribute and pick the **Edit...** button to display the **Edit Attribute** dialog box. See Figure 25-11. Use the **Attribute** tab to modify the modes, tag, prompt, and default value. The **Text Options** and **Properties** tabs of the **Edit Attribute** dialog box are identical to the tabs found in the **Enhanced Attribute Editor**. Check **Auto preview changes** at the bottom of the dialog box to display changes to attributes immediately in the drawing area.

After modifying the attribute definition in the **Edit Attribute** dialog box, pick the **OK** button to return to the **Block Attribute Manager**. Then pick the **OK** button to return to the drawing. When you modify attributes within a block, future insertions of the block reflect the changes. Existing blocks update only if you select the **Apply changes to existing references** check box in the **Settings** dialog box.

The **Block Attribute Manager** modifies attribute definitions, not attribute reference values. Modify attribute values using the **Enhanced Attribute Editor**.

Figure 25-11.
Use the **Edit Attribute** dialog box to modify attribute definitions and properties.

Use these tabs to modify attribute properties

Select modes

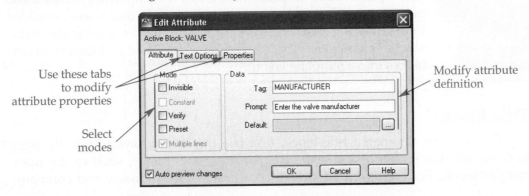

Modify attribute definition

Chapter 25   Block Attributes

Redefining a Block with Attributes

To add attributes to or revise the geometry of a block, edit the block definition using the **BEDIT** or **REFEDIT** commands. Chapter 31 explains the **REFEDIT** command. The **BEDIT** and **REFEDIT** commands allow you to make changes to a block definition, including attributes assigned to the block, without exploding the block.

You can also explode and then redefine the block using the same name. Another option is to use the **ATTREDEF** command. However, the **ATTREDEF** command is text-based, and you must first explode the block. Use **BEDIT** or **REFEDIT** to edit the block.

Synchronizing Attributes

Redefining a block automatically updates the properties of all of the same blocks in the drawing, but does not apply changes made to attributes. For example, if you add an object to a block, all existing blocks of the same name update to display the new object. However, if you add an attribute to a block, all existing blocks of the same name continue to display the original attributes, without the new attribute. Synchronize the blocks to update the attribute redefinition.

Synchronize blocks in the **Block Attribute Manager** by picking the **Sync** button. This method is convenient because it allows you to make changes to and remove attributes using the **Block Attribute Manager**. To synchronize attributes without using the **Block Attribute Manager**, use the **ATTSYNC** command. Access the **ATTSYNC** command and use the default **Select** option to pick a block that contains the attributes you want to synchronize. An alternative is to use the **Name** option to type the block name, or use the **?** option to list the names of all blocks in the drawing. Then choose the **Yes** option to synchronize attributes, or the **No** option to select a different block.

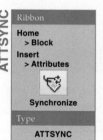

ATTSYNC

Ribbon

Home
> Block

Insert
> Attributes

Synchronize

Type

ATTSYNC

Automating Drafting Documentation

Attributes automate the process of placing symbols that require textual information. Attributes are especially useful for automating common detailing or documentation tasks such as preparing title block information, revision history block data, schedules, or a parts list or bill of materials. Filling out these items is usually one of the more time-consuming tasks associated with drafting documentation.

You typically draw or place title blocks, revision history blocks, and parts lists in a layout, because they are content that is usually added to the drawing sheet. However, it is common to develop the initial blocks or wblocks of these items in model space, and then insert the blocks into a template layout.

Title Blocks

To create an automated title block, first use the correct layer, typically layer 0, to draw title block objects and add text that does not change, such as the titles of compartments. Format the title block in accordance with industry and company or school standards. Include your company or school logo if appropriate. If you work in an industry that produces items for the federal government, also include the applicable *Commercial and Government Entity Code (CAGE Code).* **Figure 25-12** shows a title block, dimensioning and tolerancing block, and angle of projection block drawn in accordance with the ASME Y14.1 *Decimal Inch Drawing Sheet Size and Format* standard.

Next, define attributes for each area of the title block. As you create attributes, determine the appropriate text height and justification for each definition. Common title block attributes include drawing title, drawing number, drafter, checker, dates, drawing scale, sheet size, material, finish, revision letter, and tolerance information. See **Figure 25-13**. Create approval attributes with a prompt such as ENTER INITIALS OR SEEK SIGNATURE, providing the flexibility to type initials or leave the cell blank for written initials. Apply the same practice to date attributes. Include any other information that may be specific to the organization or drawing application. Assign default values to the attributes wherever possible. For example, if you consistently specify the same tolerances for dimensions, assign default values to the tolerance attributes.

PROFESSIONAL TIP

The size of each area within the title block limits the number of characters displayed in a line of text. Include a reminder about the maximum number of characters in the attribute prompt, such as Enter drawing name (15 characters max). Each time you insert a block or drawing containing the attribute, the prompt displays the reminder.

Figure 25-12.
Sheet blocks must comply with applicable standards. This title block, dimensioning and tolerancing block, and angle of projection block comply with the ASME Y14.1 *Decimal Inch Drawing Sheet Size and Format* standard.

			ENGINEERING DRAFTING & DESIGN, INC.			
	APPROVALS	DATE	*Drafting, design, and training for all disciplines.*			
	DRAWN		*Integrity - Quality - Style*			
	CHECKED		TITLE			
	APPROVED					
THIRD ANGLE PROJECTION	MATERIAL					
	FINISH		SIZE	CAGE CODE	DWG NO.	REV
	DO NOT SCALE DRAWING		SCALE		SHEET OF	

Figure 25-13.
Define attributes for each area of the title block. Attributes should define all information that might possibly change, including general tolerances. The dimensioning and tolerance block in this example uses a multiple line attribute. This figure shows the attributes in color for illustrative purposes only.

			ENGINEERING DRAFTING & DESIGN, INC.			
TOLERANCE	APPROVALS	DATE	*Drafting, design, and training for all disciplines.*			
	DRAWN DRAWN	DRAWN	*Integrity - Quality - Style*			
	CHECKED CHECKED	CHECKED	TITLE			
	APPROVED APPROVED	APPROVED	TITLE			
THIRD ANGLE PROJECTION	MATERIAL MATERIAL					
	FINISH FINISH		SIZE SIZE	CAGE CODE CAGE CODE	DWG NO. DRAWING	REV REVISION
	DO NOT SCALE DRAWING		SCALE SCALE		SHEET SHEET OF TOTAL	

Block insertion base point

After you define each attribute in the title block, you are ready to create the block. One option is to use the **BLOCK** command to create a block of the title block within the current file. When specifying the insertion base point, pick a corner of the title block that is convenient to use each time you insert the block. **Figure 25-13** shows the insertion base point that is appropriate for that particular title block. Use the **Delete** option in the **Block Definition** dialog box to remove the selected objects from the drawing.

Another option is to use the **WBLOCK** command to save the drawing as a file. Give the file a descriptive name, such as TITLE_B or FORMAT_B for a B-size title block. **Figure 25-14** shows the attribute block created in **Figure 25-13**, inserted and completely filled out using attributes.

If you are creating a template, insert the block at the appropriate location and save the file as a drawing template. Edit the values in an existing title block using the **Enhanced Attribute Editor**.

Revision History Blocks

It is almost certain that a detail drawing will require revision. Typical changes include design improvements and the correction of drafting errors. The first revision is usually assigned the revision letter *A*. If necessary, revision letters continue with *B* through *Y*, but the letters *I, O, Q, S, X,* and *Z* are not used because they might be confused with numbers.

Drawing layout formats include an area with columns specifically designated to record drawing changes. This area, commonly called the *revision history block*, is normally located at the upper-right corner of the drawing sheet. A column for *zones* is included only if applicable.

The **TABLE** command is an excellent tool for preparing a revision history block. An alternative is to use blocks and attributes to document revisions. The process is similar to creating a title block, but a revision history block requires two separate blocks. The first block consists of only lines and text and forms the title and heading rows. See **Figure 25-15A**. Insert the second block, which includes attributes, whenever a revision is required. See **Figure 25-15B**.

Format the revision history block according to industry and company or school standards, and use the correct layer, typically layer 0. As you create attributes, determine the appropriate text height and justification for each definition. Define attributes for the zone (if necessary), revision letter, description, date, and approval. Assign the APPROVED attribute a prompt such as ENTER INITIALS OR SEEK SIGNATURE, providing

revision history block: A block that provides space for the revision letter, a description of the change, the date, and approvals.

zones: A system of letters and numbers used on large drawings to help direct the attention of the person reading the print to a location on the drawing.

Figure 25-14.
The title block after insertion of the attributes.

UNLESS OTHERWISE SPECIFIED DIMENSIONS ARE IN INCHES (IN) TOLERANCES: 1 PLACE ±.1 2 PLACE ±.01 3 PLACE ±.005 4 PLACE ±.0050 ANGLES 30' FINISH 62 u IN	APPROVALS		DATE	**ENGINEERING DRAFTING & DESIGN, INC.** *Drafting, design, and training for all disciplines.* *Integrity - Quality - Style*				
	DRAWN	DPM	08-30-11					
	CHECKED	HJE	09-02-11	TITLE				
	APPROVED	DAM	09-03-11	LENS HOUSING				
THIRD ANGLE PROJECTION	MATERIAL AL ALY 6061-T6							
	FINISH PASSIVATE			SIZE B	CAGE CODE	DWG NO. 290-0013		REV 0
	DO NOT SCALE DRAWING			SCALE 1:2			SHEET 4 OF 29	

Figure 25-15.
Creating a revision history block using two separate blocks. A—The first block forms the title and heading rows. B—The second block includes attributes and is added each time an engineering change is employed. The revision history block shown complies with the ASME Y14.1 *Decimal Inch Drawing Sheet Size and Format* standard. This figure shows the attributes in color for illustrative purposes only.

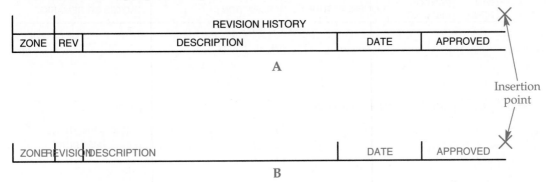

the flexibility to type initials or leave the cell blank for written initials. Apply the same practice to the date attribute.

Use the **BLOCK** or **WBLOCK** command to create the blocks. If you create wblocks, use descriptive file names such as REVBLK or REV. Figure 25-16 shows an example of revision information added by inserting the two blocks created in Figure 25-15 in the upper-left inside corner of the border.

Parts Lists

Assembly drawings require a parts list, or bill of materials, that provides information about each component of the assembly or subassembly. Columns can identify a variety of information depending on the organization and application. Common elements include the list type (parts list, index list, application list, data list, wire list), design activity, contract number, find or item number, quantity required, CAGE Code (when necessary), part or identification number, and nomenclature or description. Companies often include the parts list on the face of the assembly drawing. A parts list on an assembly drawing usually appears directly above the title block, depending on industry and company standards. Other organizations create the parts list as a separate document, often in an 8-1/2" × 11" format.

The **TABLE** command is an excellent tool for preparing a parts list. An alternative is to use blocks and attributes. The process is very similar to creating a revision history block. The first block consists of only lines and text and forms the title (if used) and heading rows. See Figure 25-17A. The second block includes attributes and is inserted as many times as necessary to document each assembly component. See Figure 25-17B.

Format the parts list according to industry and company or school standards, and use the correct layer, typically layer 0. As you create attributes, select the appropriate

Figure 25-16.
The completed revision history block after inserting one title and heading block and two change blocks.

		REVISION HISTORY		
ZONE	REV	DESCRIPTION	DATE	APPROVED
C3	A	ADDED .125 CHAMFER	11-16-11	ADM
B3	B	REDUCED SHAFT LENGTH FROM 4.580	03-18-11	BAE

Figure 25-17.
Creating a parts list using two separate blocks. A—The first block forms the title (if used) and heading rows. B—The second block includes attributes and is inserted as many times as necessary to define each assembly component. The revision block shown complies with the ASME Y14.1 *Decimal Inch Drawing Sheet Size and Format* standard.

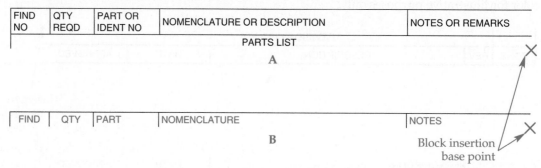

FIND NO	QTY REQD	PART OR IDENT NO	NOMENCLATURE OR DESCRIPTION	NOTES OR REMARKS
			PARTS LIST	

A

FIND	QTY	PART	NOMENCLATURE	NOTES

B

Block insertion base point

text height and justification for each definition. Define attributes for the column or cell in the parts list.

Use the **BLOCK** or **WBLOCK** command to create the blocks. If you save wblocks, use descriptive file names, such as PL for parts list or BOM for bill of materials. **Figure 25-18** shows an example of the beginning of a parts list developed by inserting the blocks created in **Figure 25-17**.

Using Fields to Reference Attributes

Use fields to link text to attribute values. To display a field as an attribute value, access the **Field** dialog box from within the **MTEXT** or **TEXT** command, or use the **FIELD** command. In the **Field** dialog box, pick **Objects** from the **Field category:** drop-down list, and pick **Object** in the **Field names:** list box. Then pick the **Select object** button to return to the drawing window and select the block containing the attribute.

When you select the block, the **Field** dialog box reappears with the available properties (attributes) listed. Pick an attribute tag to display the corresponding value in the **Preview:** box. Select the format and pick the **OK** button to insert the attribute value as a field in the text object.

Figure 25-18.
A parts list after inserting blocks and editing attribute values.

FIND NO	QTY REQD	PART OR IDENT NO	NOMENCLATURE OR DESCRIPTION	NOTES OR REMARKS
4	4	30-004579-04	ULTRA THIN RETAINER SLEEVE	SAE 1020
3	12	85741K4R	8-32UNC-2 X .500 HEX SOCKET HEAD CAP SCREW	SAE 4320
2	2	30-004579-02	HIGH PRESSURE RACK PAD	BLACK UHMW
1	1	30-004579-01	MAIN MOUNTING PLATE	AL ALY 6061-T6
FIND NO	QTY REQD	PART OR IDENT NO	NOMENCLATURE OR DESCRIPTION	NOTES OR REMARKS
			PARTS LIST	

UNLESS OTHERWISE SPECIFIED
DIMENSIONS ARE IN INCHES (IN)
TOLERANCES: 1 PLACE ±.1   2 PLACE ±.01
3 PLACE ±.005   4 PLACE ±.0050
ANGLES 30'   FINISH 62 u IN

THIRD ANGLE PROJECTION

APPROVALS		DATE
DRAWN	DPM	07-20-11
CHECKED	ADM	07-23-11
APPROVED	DAM	07-25-11

MATERIAL   NOTED

FINISH   ALL OVER

ENGINEERING DRAFTING & DESIGN, INC.
Drafting, design, and training for all disciplines.
Integrity - Quality - Style

TITLE
CRYOGENIC PLATE SUBASSEMBLY

SIZE C	CAGE CODE	DWG NO. 30-004579	REV 0

DO NOT SCALE DRAWING   SCALE 1:1   SHEET   1 OF 5

Supplemental Material *Extracting Attribute Data*
For information about using attributes to create a table and exporting attribute data to an external file, go to the companion website (www.g-wlearning.com/CAD), select this chapter, and select **Extracting Attribute Data**.

Chapter Review

Answer the following questions. Write your answers on a separate sheet of paper or complete the electronic chapter review on the companion website.
www.g-wlearning.com/CAD

1. What is an attribute?
2. Explain the purpose of the **ATTDEF** command.
3. Describe the function of the following attribute modes:
   A. **Invisible**
   B. **Constant**
   C. **Verify**
   D. **Preset**
4. What is the purpose of the **Default** text box in the **Attribute Definition** dialog box?
5. How can you edit attributes before including the attributes with a block?
6. How can you change an existing attribute from visible to invisible?
7. If you select attributes using the **Window** or **Crossing** selection method to define a block, in what order will attribute prompts appear?
8. What purpose does the **ATTREQ** system variable serve?
9. List the three options for attribute display.
10. Explain how to change the value of an inserted attribute.
11. What does *global attribute editing* mean?
12. After you save a block with attributes, what method can you use to change the order of prompts when you insert the block?
13. List examples of detailing or documentation tasks that attributes can help automate.
14. What element of an assembly drawing provides information about each component of the assembly or subassembly?
15. How can you link text to an attribute value?

Drawing Problems

Start AutoCAD if it is not already started. Start a new drawing for each problem using an appropriate template of your choice. The template should include layers and text, dimension, multileader, and table styles, when necessary, for drawing the given objects. Add layers and text, dimension, multileader, and table styles as needed. Draw all objects using appropriate layers and text, dimension, multileader, and table styles, justification, and format. Follow the specific instructions for each problem. Use only drawing and editing commands and techniques you have already learned. Use your own judgment and approximate dimensions when necessary. Apply dimensions accurately using ASME or appropriate industry standards.

Note: *Some of the problems in this chapter are built on problems from previous chapters. If you have not yet completed those problems, complete them now.*

▼ Basic

1. Use a word processor to list each attribute mode. Provide a brief description of each.

2. Draw the structural steel wide flange shape shown. Do not dimension the drawing. Create attributes for the drawing using the information given. Make a block of the drawing and name it W12 X 40. Insert the block once to test the attributes. Save the drawing as P25-2.

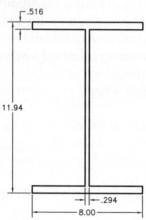

Attributes			
	Steel	W12 × 40	Visible
	Mfr.	Ryerson	Invisible
	Price	$.30/lb	Invisible
	Weight	40 lbs/ft	Invisible
	Length	10′	Invisible
	Code	03116WF	Invisible

▼ Intermediate

3. Open P25-2 and save it as P25-3. The P25-3 file should be active. Construct the floor plan shown. Insert the block W12 X 40 six times as shown. The chart below the drawing provides the required attribute data. Enter the appropriate information for the attributes as prompted. Note that the steel columns labeled 3 and 6 require slightly different attribute data. You can speed the drawing process by using **ARRAY** or **COPY**. Dimension the drawing. Resave the drawing.

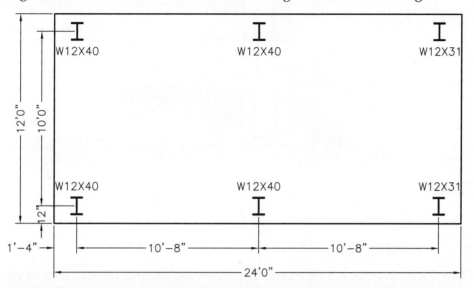

	Steel	Mfr.	Price	Weight	Length	Code
Blocks ①, ②, ④, & ⑤	W12 × 40	Ryerson	$.30/lb	40 lbs/ft	10'	03116WF
Blocks ③ & ⑥	W12 × 31	Ryerson	$.30/lb	31 lbs/ft	8.5'	03125WF

4. Open P25-2 and save it as P25-4. The P25-4 file should be active. Edit the W12 X 40 block in the newly saved drawing according to the following information. Resave the drawing.

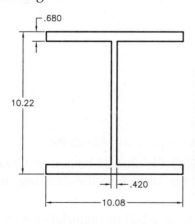

	Steel	W10 × 60	Visible
Attributes	Mfr.	Ryerson	Invisible
	Price	$.25/lb	Invisible
	Weight	60 lbs/ft	Invisible
	Length	10'	Invisible
	Code	02457WF	Invisible

▼ Advanced

5. Draw the structural detail shown. Save the drawing as P25-5.

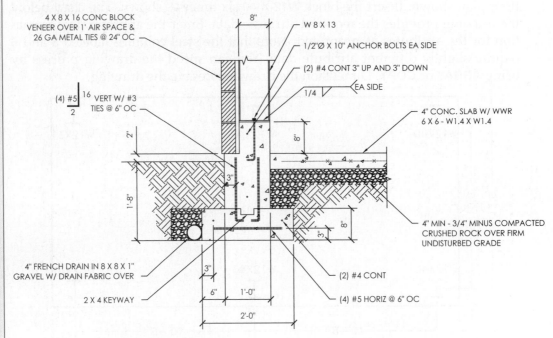

4 X 8 X 16 CONC BLOCK
VENEER OVER 1' AIR SPACE &
26 GA METAL TIES @ 24" OC

W 8 X 13

1/2"Ø X 10" ANCHOR BOLTS EA SIDE

(2) #4 CONT 3" UP AND 3" DOWN

1/4   EA SIDE

(4) #5 — 16 — VERT W/ #3
       2   TIES @ 6" OC

4" CONC. SLAB W/ WWR
6 X 6 - W1.4 X W1.4

4" MIN - 3/4" MINUS COMPACTED
CRUSHED ROCK OVER FIRM
UNDISTURBED GRADE

4" FRENCH DRAIN IN 8 X 8 X 1"
GRAVEL W/ DRAIN FABRIC OVER

2 X 4 KEYWAY

(2) #4 CONT

(4) #5 HORIZ @ 6" OC

2"   1'-8"   3"   3"   6"   1'-0"   2'-0"   8"   8"   3"   8"

6. Open P8-21 and save it as P25-6. The P25-6 file should be active. Add constant, invisible attributes to each view of the glulam (glued laminated) beam hanger to identify all necessary specifications. Create a block of each view with the attributes. Insert the block of each view. Resave the drawing.

7. Open P11-17 and save it as P25-7. The P25-7 file should be active. Add constant, invisible attributes to each view of the mudsill anchor to identify all necessary specifications. Create a block of each view with the attributes. Insert the block of each view. Resave the drawing.

8. Open P25-2 and save it as P25-8. The P25-8 file should be active. Create a tab-separated extraction file for the blocks in the drawing. Extract the following information for each block:
   - Block name
   - Steel
   - Manufacturer
   - Price
   - Weight
   - Length
   - Code

   Resave the drawing. Save the tab-separated extraction file as P25-8.

9. Open P25-2 and save it as P25-9. The P25-9 file should be active. Create a table from the block attribute data and insert the table into the drawing. Resave the drawing.

10. Select a drawing from Chapter 24 and create a bill of materials for the drawing using the **Data Extraction** wizard. Use the comma-separated format to display the file. Display the file in Windows Notepad. Save the drawing and the comma-separated extraction file as P25-10.

Drawing Problems – Chapter 25

11. Create a drawing of the computer workstation layout in the classroom or office in which you are working. Provide attribute definitions for all of the items listed here.
    - Workstation ID number
    - Computer brand name
    - Model number
    - Processor chip
    - Amount of RAM
    - Hard disk capacity
    - Video graphics card brand and model
    - CD-ROM/DVD-ROM speed
    - Date purchased
    - Price
    - Vendor phone number
    - Other data as you see fit

Generate and extract a file for all of the computers in the drawing. Save the drawing and extracted file as P25-11.

AutoCAD Certified Associate Exam Practice

Answer the following questions. Write your answers on a separate sheet of paper.

1. Which group of **ATTDEF** settings is most efficient for an attribute that is not expected to change and that will be used only to tabulate data? *Select the one item that best answers the question.*
   A. **Invisible, Constant, Lock position**
   B. **Invisible, Verify, Preset**
   C. **Visible, Constant, Preset**
   D. **Visible, Constant, Verify**
   E. **Visible, Verify, Lock position**

2. Which of the following tools can you use to change attribute modes after an attribute has been created? *Select all that apply.*
   A. **-ATTEDIT**
   B. **ATTREDEF**
   C. **BATTMAN**
   D. **EATTEDIT**
   E. **Properties** palette

3. Which of the following commands allow you to add attributes to a block without exploding the block? *Select all that apply.*
   A. **ATTREDEF**
   B. **BEDIT**
   C. **EATTEDIT**
   D. **FIELD**
   E. **REFEDIT**

AutoCAD Certified Professional Exam Practice

Follow the instructions in the problem. Write your answers on a separate sheet of paper.

1. **Open CPE-25attribute.dwg. This file is available on the companion website.** Insert the existing BUSH block into the drawing. Without exploding the block, change the attribute mode from **Invisible** to **Visible**, and change the text height to 12. What type of plant is this, and how much does it cost?

Introduction to Dynamic Blocks 26

Learning Objectives

After completing this chapter, you will be able to:

✓ Explain the function of dynamic blocks.
✓ Assign action parameters and actions to blocks.
✓ Modify parameters and actions.
✓ Assign value sets to parameters.
✓ Create and use chain actions.

A standard block typically represents a very specific item, such as a hex head bolt that is 1″ long. In this example, if the same hex head bolt is available in three other lengths, you must create three additional standard blocks. An alternative is to create a single *dynamic block* that can be adjusted to show each different bolt length. Dynamic blocks increase productivity and reduce the size of symbol libraries, making symbol libraries more manageable.

dynamic block:
An adjustable block to which you can assign parameters, actions, and geometric constraints and constraint parameters.

Dynamic Block Fundamentals

A dynamic block is a parametric symbol that you can adjust to change size, shape, and geometry without drawing additional blocks, and without affecting other instances of the block reference. **Figure 26-1** shows an example of a dynamic block of a plan view single-swing door symbol. In this example, the dynamic properties of the block allow you to create many different single-swing door symbols according to specific parameters, such as door size, wall thickness, swing location, swing angle representation, wall angle, and exterior or interior usage.

The process of constructing and using dynamic blocks is identical to the process of creating standard blocks, except for the addition of *action parameters* and (usually) *actions* that control block geometry. AutoCAD refers to action parameters as *parameters* in the context of dynamic blocks. A dynamic block can contain multiple parameters, and a single parameter can include multiple actions. You can use geometric constraints and *constraint parameters* as an alternative or in addition to parameters and actions. Many different commands and options exist for constructing dynamic blocks, depending on the purpose of the block.

action parameter (parameter): A specification for block construction that controls block characteristics such as positions, distances, and angles of dynamic block geometry.

action: A definition that controls how dynamic block parameters behave.

constraint parameters: Dimensional constraints that control the size or location of block geometry numerically.

Figure 26-1.
A—A dynamic block of a single-swing door symbol. B—The dynamic block allows you to create many different door symbols without creating new blocks or affecting other instances of the same block reference.

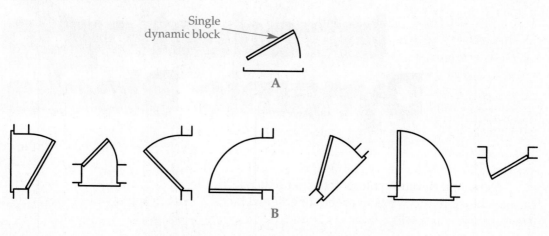

Single dynamic block

A

B

Figure **26-2A** shows an example of a bolt symbol created as a dynamic block and selected for grip editing. The bolt shaft objects include a linear parameter with a stretch action, as indicated by the *parameter grips*. The length of the bolt increases when you stretch the right-hand linear parameter grip to the right. See **Figure 26-2B**.

parameter grips:
Special grips that allow you to change the parameters of a dynamic block.

As you learn to create and use dynamic blocks, you will notice that many actions function like editing commands with which you are already familiar, allowing operations such as stretch, move, scale, array, and rotate.

Figure 26-2.
A linear parameter with a stretch action assigned to the shaft objects in the block of a bolt. A—Selecting the block displays the linear grips. B—Selecting and moving a linear grip stretches the bolt shaft.

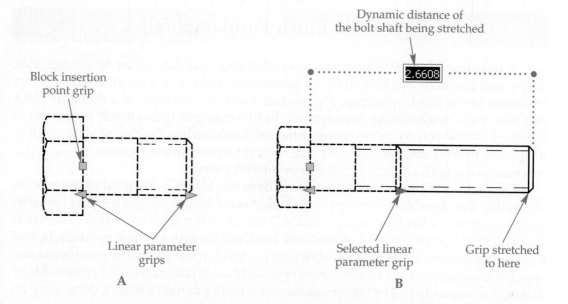

Dynamic distance of the bolt shaft being stretched

2.6608

Block insertion point grip

Linear parameter grips

Selected linear parameter grip

Grip stretched to here

A

B

Assigning Dynamic Properties

Edit an existing block in the **Block Editor** to assign dynamic properties. Access the **BEDIT** command to display the **Edit Block Definition** dialog box shown in Figure 26-3 and select the block from the list box. A preview and the description of the selected block appear. Pick the **OK** button to open the selection in the **Block Editor**. See Figure 26-4.

To create a dynamic block from scratch within the **Block Editor**, type a name for the new block in the **Block to create or edit** text box. To edit a block saved as the current drawing, such as a wblock, pick the **<Current Drawing>** option.

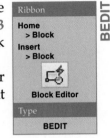

Ribbon
Home
> Block
Insert
> Block

Block Editor

Type
BEDIT

NOTE

Double-click on a block to display the **Edit Block Definition** dialog box with the block selected. You can open a block directly in the **Block Editor** by selecting the block, right-clicking, and choosing **Block Editor**. Another option is to open a block directly in the **Block Editor** during block creation by selecting the **Open in block editor** check box in the **Block Definition** dialog box.

The **Block Editor** ribbon tab and **Block Authoring Palettes** palette provide easy access to commands and options for assigning dynamic block properties and creating attributes. The **Block Authoring Palettes** palette contains parameter, action, and constraint commands. Although you can type **BPARAMETER** or **BACTION** to activate the **BPARAMETER** or **BACTION** command and then select a parameter or action as an option, it is easier to use the **Block Editor** ribbon tab or **Block Authoring Palettes**.

You can also assign actions to certain parameters, such as point parameters, by double-clicking on the parameter and selecting an action option. You can assign only specific actions to a given parameter. The process of assigning an action is slightly different, depending on the method used to access the action. If you type **BACTION**, you must first select the parameter and then specify the action type. If you pick the

Figure 26-3.
The **Edit Block Definition** dialog box.

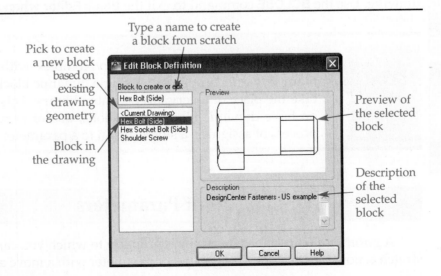

Type a name to create a block from scratch

Pick to create a new block based on existing drawing geometry

Block in the drawing

Preview of the selected block

Description of the selected block

Figure 26-4.
The **Block Editor** ribbon tab and the **Block Authoring Palettes** are available in block editing mode. Notice the drawing window background, which has changed color to indicate block editing mode, and the location of the block relative to the origin.

Block editing tools are located in the **Block Editor** tab

Position the insertion base point for the block at the origin (0,0,0), as indicated by the UCS icon

Block Authoring Palettes palette

Drawing area displays block geometry

action from the **Action Parameters** panel in the **Block Editor** ribbon tab or the **Block Authoring Palettes**, the specific action is active and a prompt asks you to pick the parameter. If you double-click on the parameter, the parameter becomes selected, but you must choose the action type.

Saving a Block with Dynamic Properties

Once you add one or more parameters to a block and assign actions to the parameters, you are ready to save and use the dynamic block. Use the **BSAVE** command to save the block, or use the **BSAVEAS** command to save the block using a different name. Remember that saving changes to a block updates all blocks of the same name in the drawing. Use the **BCLOSE** command to exit the **Block Editor** when you are finished.

Dynamic blocks can become very complex with the addition of many dynamic properties. A single dynamic block can potentially take the place of a very large symbol library. This chapter focuses on basic dynamic block applications, use of parameters, and the process of assigning a single action to a parameter.

BSAVE

Ribbon
Block Editor > Open/Save
Save Block
Type
BSAVE

BSAVEAS

Ribbon
Block Editor > Open/Save
Save Block As
Type
BSAVEAS

Point Parameters

point parameter: A parameter that defines an XY coordinate location in the drawing.

A *point parameter* creates a position property to which you can assign move and stretch actions. For example, assign a point parameter with a move action to a door tag that is part of a door block so you can move the tag independently of the door. Point parameters also provide multiple insertion point options. For example, you can add point parameters to the ends of a weld symbol reference line to create two insertion point options.

Figure 26-5 provides an example of adding a point parameter. Access the **Point parameter** option and specify a location for the parameter. The parameter location determines the base point from which dynamic actions occur. **Figure 26-5** shows picking the center of the door tag circle to identify the base point of a move action. Adjust the parameter location after initial placement if necessary. The yellow alert icon indicates that no action is assigned to the parameter.

After you specify the parameter location, pick a location for the *parameter label*. All parameters include a parameter label. The label appears only in block editing mode. By default, the label for the first point parameter is Position. Move the label as needed after initial placement.

Next, enter the number of grips to associate with the parameter by right-clicking on the parameter and picking **Grip Display**. The default **1** option creates a single grip at the parameter location that allows you to use grip editing to carry out the assigned action. If you choose the **0** option, you can only use the **Properties** palette to adjust the block.

Parameter options are available before you specify the parameter location. Most of the options are also available from the **Properties** palette if you have already created the parameter. Use the **Label** option to enter a more descriptive label name. The **Name** option allows you to specify a name for the parameter that displays as the **Parameter type** in the **Properties** palette. The **Chain** option specifies whether a chain action can affect the parameter. Chain actions are described later in this chapter. The **Description** option allows you to type a description, such as the purpose of or application for the parameter. The description displays in the drawing area as a tooltip. The **Palette** option determines whether the label appears in the **Properties** palette when you select the block.

Ribbon
Block Editor
> Action
Parameters
Point
Type
BPARAMETER
> Point

parameter label: A label that indicates the purpose of a parameter.

PROFESSIONAL TIP

Change the parameter label name to something more descriptive, and add a description. Naming labels and describing parameters helps organize parameters and helps you identify each parameter when you are controlling a block dynamically. This is especially important when you add multiple parameters to a block. Consider keeping the default parameter type as part of the name. For example, change the name of the door symbol point parameter from Point to Point – Tag Center.

Figure 26-5.
A point parameter consists of the grip location and a label. This example shows adding a point parameter to the tag of a plan view door symbol to allow the tag to be positioned independently of the door.

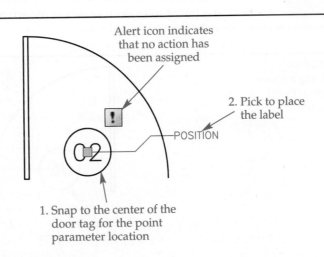

Alert icon indicates that no action has been assigned

2. Pick to place the label

POSITION

1. Snap to the center of the door tag for the point parameter location

Exercise 26-1

Complete the exercise on the companion website.
www.g-wlearning.com/CAD

Assigning a Move Action

Figure 26-6 illustrates the process of adding a *move action* to the door block example. First, access the **Move action** option and pick the point parameter. Then select the objects that make up the door tag and the associated parameter. Press [Enter] or the space bar or right-click to assign the action. Test the block, as explained later in this chapter. Save the block and exit the **Block Editor**. The dynamic block is now ready to use.

When you select objects to include with an action, you should also select the associated parameter. Otherwise, the parameter grip will be left behind when you apply the action.

PROFESSIONAL TIP

After you create an action, use the **Properties** palette to change the action name to something more descriptive, but keep the default action type with the name. For example, change the name of the door symbol move action from Move to Move – Tag.

Using a Move Action Dynamically

Figure 26-7A shows the door block reference, selected for editing. The point parameter grip appears as a light blue square in the center of the door tag. The insertion base point specified when the block was created appears as a standard unselected grip. Select the point parameter grip and move the door tag as shown in **Figure 26-7B**. Pick a point to specify a new location for the door tag. See **Figure 26-7C**.

During block insertion, press [Ctrl] to cycle through the positions of any parameters added to the dynamic block. This is one method of selecting a different insertion base point, corresponding to the position of a parameter, to use when inserting the block.

Figure 26-6.
Assigning a move action to a point parameter.

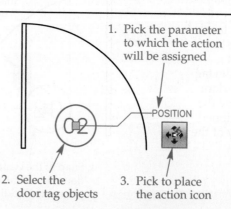

1. Pick the parameter to which the action will be assigned

—POSITION

2. Select the door tag objects

3. Pick to place the action icon

Figure 26-7.
Dynamically moving an action assigned to a point parameter. A—Select the block to display grips. The point parameter grip is shown as a light blue square. B—Select and move the point parameter grip. C—The door tag is at a new location, but it is still part of the block.

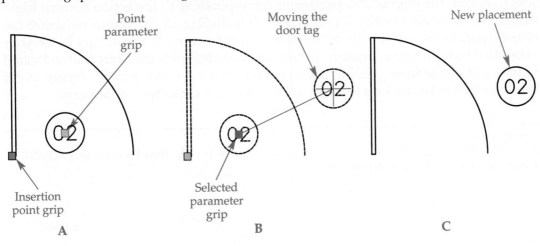

PROFESSIONAL
TIP

After block insertion, use the **Properties** palette to adjust dynamic properties. If you do not include grips with a parameter, the options in the **Custom** category are the only way to adjust the block.

Testing and Adjusting Dynamic Properties

After you exit the **Block Editor**, you can insert a block and use grip editing or the **Properties** palette to confirm appropriate dynamic function. If the block does not respond as desired, however, you must re-enter the **Block Editor** to make changes. A more convenient option is to access the **BTESTBLOCK** command from inside the **Block Editor** to enter the **Test Block Window**. The **Test Block Window** provides standard AutoCAD commands and options, allowing you to test dynamic function without exiting block editing mode. Pick the **Close Test Block Window** button to re-enter the **Block Editor**. AutoCAD discards changes made while testing so that you can adjust the original block as needed. Block testing is especially important when a block includes multiple dynamic properties.

Adjusting Parameters

Use standard editing commands such as **MOVE** to make changes to existing parameter labels or grips. You can move parameter grips independently of the parameter label, which is often required if multiple grips are stacked or are near the same location. Use the **ERASE** command to remove a parameter or parameter grips.

Grip editing is especially effective for adjusting parameters. When you select a parameter, grips appear at the parameter location and label. Use the **Properties** palette to adjust the properties of the selected parameter. The settings in the **Properties** palette change depending on the type of parameter. Limited property options are also available by selecting a parameter and right-clicking. Use the **Grip Display** cascading menu to redefine the number of grips or to relocate grips with the parameter location. Use the **Rename** option to change the name of the label.

Adjusting Actions

action bars:
Toolbars that allow you to view, remove, and adjust actions.

Action bars appear by default when you assign actions. See **Figure 26-8A**. Each action displays an icon to identify and control the action. When you hover over or select an icon, the objects and parameter corresponding to the action become highlighted and markers identify action points. This display allows you to recognize the objects, parameter, and points associated with the action. If action bars block your view, drag them to a new location. To hide an action bar, pick the **Close** button located to the right of the icons. Hiding action bars does not remove actions. **Figure 26-8B** briefly describes the options available when you right-click on an action icon.

Figure 26-8.
A—Action bars appear by default when you add actions. B—Options for adjusting actions when you right-click on an action icon.

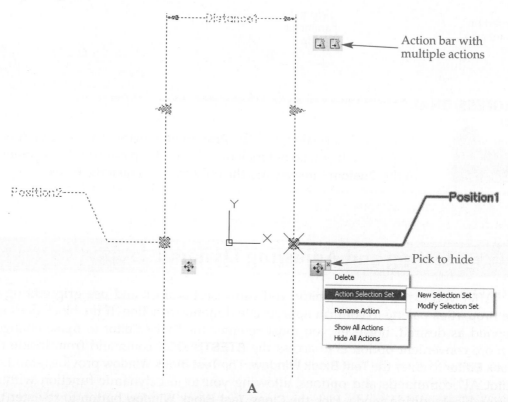

A

Option	Description
Delete	Deletes the action.
New Selection Set	Allows you to select objects to associate with the action; eliminates the original selection set.
Modify Selection Set	Adds objects to associate with the action.
Rename Action	Provides a text box for renaming the action.
Show All Actions	Displays all actions.
Hide All Actions	Hides all actions.

B

AutoCAD and Its Applications—Basics

To hide or show all actions, pick the **Hide All Actions** or **Show All Actions** button from the **Manage Parameters** panel of the **Block Editor** ribbon tab. You can also right-click with no objects selected to access an **Action Bars** cascading menu that provides options for displaying and hiding action bars.

Exercise 26-2

Complete the exercise on the companion website.
www.g-wlearning.com/CAD

Linear Parameters

A *linear parameter* creates a distance property to which you can assign move, scale, stretch, and array actions. For example, assign a linear parameter with a stretch action to a block of a bolt symbol to make the bolt shaft longer or shorter. Assign a second linear parameter and stretch action to the bolt head to control the bolt head diameter.

Figure 26-9 provides an example of adding a linear parameter. For this example, activate the **Linear parameter** option and use the **Label** function to name the linear parameter Shaft Length. Next, pick the start and endpoints of the linear parameter to determine the locations from which dynamic actions occur. If you plan to assign a single action to the parameter, select the point associated with the action second. **Figure 26-9** shows picking the endpoint of the lower edge of the shaft and then using polar tracking or the extension object snap to pick the point where the edge of the shaft would meet the end if extended. You must select points that are horizontal or vertical to each other to create a horizontal or vertical linear parameter.

Once you select the start and endpoints, pick a location for the parameter label. Next, enter the number of grips to associate with the parameter. The default **2** option creates grips at the start and endpoints, allowing you to use grip editing to carry out the action assigned to either point. Select the **1** option to assign a grip at the endpoint only, as shown in **Figure 26-9**. You will be able to grip-edit the block only if an action is

> **linear parameter:**
> A parameter that creates a measurement reference between two points.

Ribbon

Block Editor
> Action
Parameters

Linear

Type

BPARAMETER
> Linear

Figure 26-9.
Defining a linear parameter.

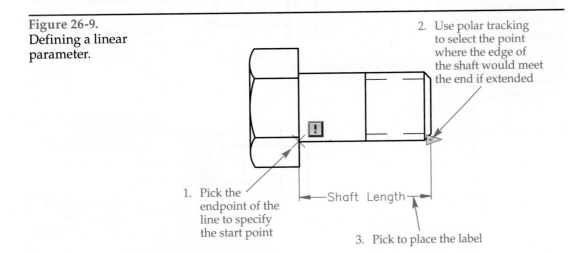

1. Pick the endpoint of the line to specify the start point

2. Use polar tracking to select the point where the edge of the shaft would meet the end if extended

Shaft Length

3. Pick to place the label

associated with the endpoint. If you choose the **0** option, you can adjust the block only by using the **Properties** palette.

Name, Label, Chain, Description, Base, Palette, and Value set options are available before you specify points. Most of these options are also available from the **Properties** palette if you have already created the parameter. The **Base** option allows you to assign the start point or midpoint of the linear parameter as the action base point. The **Value set** option allows you to specify values for the action. The **Base** and **Value set** options are described later in this chapter.

Assigning a Stretch Action

Figure 26-10 illustrates the process of adding a *stretch action* to the bolt symbol example. Access the **Stretch action** option and pick the Shaft Length parameter. Then specify a parameter point to associate with the action. Move the crosshairs near the appropriate parameter point to display the red snap marker, and pick to select. Alternatives include choosing the **sTart point** option to pick the start point of the linear parameter or the **Second point** option to select the endpoint. If you plan to use grip editing to control the block, and added a single grip, specify the point with the grip.

Next, create a window to define the stretch frame. This is the same technique you apply when using the **STRETCH** command. See **Figure 26-10A**. Then pick the objects to stretch, including the associated parameter. You do not need to use a crossing window, because the previous operation defines the stretch. However, crossing selection is often quicker. See **Figure 26-10B**. Press [Enter] or the space bar or right-click to place the action icon. Test and save the block, and exit the **Block Editor**. The dynamic block is now ready to use.

Using a Stretch Action Dynamically

Figure 26-11 shows the bolt block reference selected for editing. The linear parameter grip is a light blue arrow at the far end of the bolt shaft. The insertion base point specified when the block was created appears as a standard unselected grip. Select the parameter grip and stretch the shaft to the new length. Use dynamic input to view the stretch dimension, and enter an exact length value in the distance field. You can also use the **Properties** palette to define the distance.

Figure 26-10.
Assigning a stretch action to a linear parameter.
A—Specify the parameter and parameter grip, and create a window for the stretch frame.
B—Select the objects to be included in the stretch action.

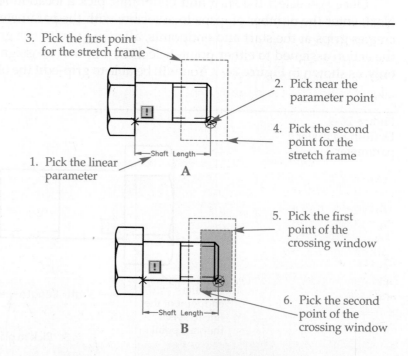

3. Pick the first point for the stretch frame

2. Pick near the parameter point

4. Pick the second point for the stretch frame

1. Pick the linear parameter

Shaft Length

A

5. Pick the first point of the crossing window

6. Pick the second point of the crossing window

Shaft Length

B

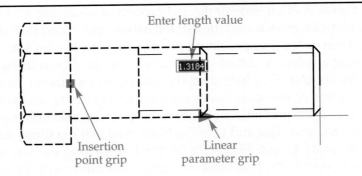

Figure 26-11.
Selecting the inserted block in the drawing displays parameter grips.

Enter length value

1.3184

Insertion point grip

Linear parameter grip

PROFESSIONAL TIP

The dynamic input distance field is a property of the linear parameter, allowing you to enter an exact distance. To get the best results when using a linear parameter, it is important that you locate the first and second points correctly.

Exercise 26-3

Complete the exercise on the companion website.
www.g-wlearning.com/CAD

Stretching Objects Symmetrically

The **Linear parameter** option includes a **Base** function that allows you to assign the start point or midpoint of the linear parameter as the action base point. Use the **Midpoint** setting to specify the midpoint as the action base point. The midpoint base point maintains symmetry when you adjust the block. You can set the base preference before picking the first point or later using the **Properties** palette.

Figure 26-12 shows an example of a linear parameter with a stretch action assigned to the objects composing the bolt head. For this example, activate the **Linear parameter** option and use the **Base** option to choose the **Midpoint** setting. Next, use the **Label** option to change the label name to Head Diameter. Select the start and endpoints of the linear parameter to define the parameter and automatically calculate the midpoint. This example uses the upper-right and lower-left corners of the bolt head. After you select the start and endpoints, pick a location for the parameter label. Enter the number of grips to associate with the parameter. The **Figure 26-12** example uses the default **2** option to create grips at the start and endpoints.

Figure 26-12.
The base point of a linear parameter appears as an X. Use the **Midpoint** option to locate the base point halfway between the start and endpoints.

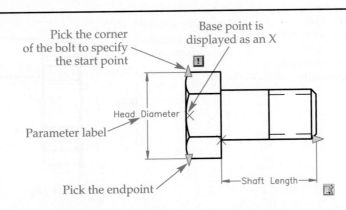

Pick the corner of the bolt to specify the start point

Base point is displayed as an X

Head_Diameter

Parameter label

Pick the endpoint

Shaft Length

Figure 26-13 demonstrates the process of assigning a stretch action to one side of the bolt head. First, access the **Stretch action** option and pick the Head Diameter parameter. Then pick the upper linear parameter point to associate with the action. Create a crossing window to define the stretch frame, as shown in **Figure 26-13A**. Then select the objects to stretch, including the associated parameter, as shown in **Figure 26-13B**. Press [Enter] or the space bar or right-click to place the action icon.

Repeat the previous sequence to assign a second stretch action to the opposite side of the bolt head. Test and save the block, and exit the **Block Editor**. The dynamic block is now ready to use. **Figure 26-14** illustrates using the lower grip point or dynamic input to stretch the bolt block reference. Use either grip to stretch the bolt head. You can also use the **Properties** palette to define the distance.

Assigning a Scale Action

Figure 26-15 shows a block of a vanity symbol that includes a sink. In this example, a *scale action* is assigned to a linear parameter to adjust the size of the sink while maintaining the dimensions of the vanity. Activate the **Linear parameter** option and use the **Base** option to choose the **Midpoint** setting. Next, use the **Label** option to change the label name to SINK LENGTH. Select the start and endpoints of the linear parameter to define the parameter and automatically calculate the midpoint. This example uses two quadrants of the sink. After you select the start and endpoints, pick a location for the parameter label. Next, choose the number of grips to associate with

Figure 26-13.

Assigning a stretch action to one side of the bolt head. A—Create a crossing window around the top of the bolt head. B—Select the objects to be included in the stretch action.

Figure 26-14.
Dynamically stretching the bolt head. Notice that the head stretches symmetrically.

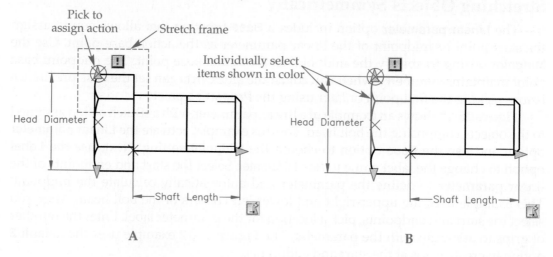

Figure 26-15.
A linear parameter assigned to the sink objects in a block of a vanity with a sink. Locate the parameter base point and independent base type at the center of the sink to scale the sink symmetrically.

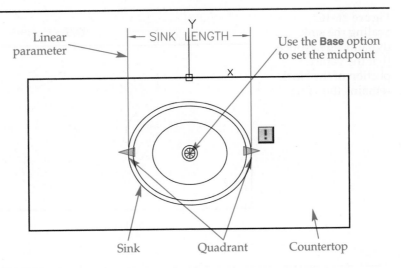

Linear parameter

SINK LENGTH

Use the **Base** option to set the midpoint

Sink   Quadrant   Countertop

the parameter. The Figure 26-15 example uses the default **2** option to create grips at the start and endpoints.

Now assign a scale action to the parameter. First, access the **Scale action** option and pick the SINK LENGTH linear parameter. Then select the objects to include in the scale action, including the associated parameter. Press [Enter] or the space bar or right-click to assign the action.

When using a scale action, it is critical to scale objects relative to the correct base point. Access the **Properties** palette and display the properties of the scale action. The **Overrides** category includes options for adjusting the base point. The default **Base type** option is **Dependent**, which scales the objects relative to the base point of the associated parameter. Choose the **Independent** option to specify a different location. The **Base X** and **Base Y** values default to the parameter start point. Enter the coordinates relative to the block insertion base point, or use the pick button that appears when you select the **Base X** and **Base Y** values to choose points on-screen. For the sink example, it is important that the objects be scaled relative to the exact center of the sink so that the sink will be centered within the vanity as the scale changes. Test and save the block, and exit the **Block Editor**. The dynamic block is now ready to use.

PROFESSIONAL TIP

In the previous example, the **Independent** option allows you to set the center of the sink as the base point for the scale action. This is necessary because the base point of the linear parameter is not the specified midpoint. The parameter uses a midpoint base to scale the parameter and parameter grips from the parameter midpoint. The **Independent** option of the scale action controls the point from which the geometry, not the parameter, is scaled.

Using a Scale Action Dynamically

Figure 26-16 shows using the right grip or dynamic input to scale the sink in the vanity block reference. You can use either grip to scale the sink. You can also use the **Properties** palette to define the distance.

Figure 26-16.
Scaling the sink
dynamically. Notice
that the vanity
portion of the block
remains the same.

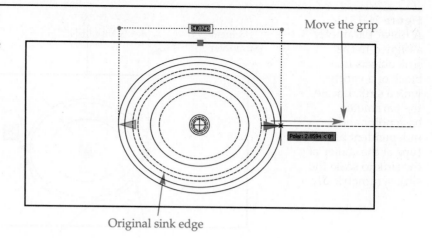

Move the grip

Original sink edge

Exercise 26-4

Complete the exercise on the companion website.
www.g-wlearning.com/CAD

Polar Parameters

polar parameter:
A parameter that
includes a distance
property and an
angle property.

A *polar parameter* provides the same function as a linear parameter, but creates an angle, or rotation, property in addition to a distance property. You can assign move, scale, stretch, polar stretch, and array actions to a polar parameter. For example, assign a polar parameter with a polar stretch action to the steel channel shape shown in **Figure 26-17** to adjust the depth to create different size channels, and rotate the block at the same time if necessary. See **Figure 26-18**.

To insert the polar parameter, access the **Polar parameter** option. Use the **Label** option to change the label name to Depth, but keep the default angle name. Then specify the base point, such as the left endpoint shown. Pick an endpoint aligned with the base point, such as the opposite endpoint shown. After you select the start and endpoints, pick a location for the parameter label. Next, enter the number of grips to associate with the parameter. Select the **1** option to assign a grip at the endpoint only, as shown in **Figure 26-17**. You will be able to grip-edit the block only if an action is associated with the endpoint.

Ribbon

**Block Editor
> Action
Parameters**

Polar

Type

**BPARAMETER
> Polar**

Figure 26-17.
Adding a polar
parameter.

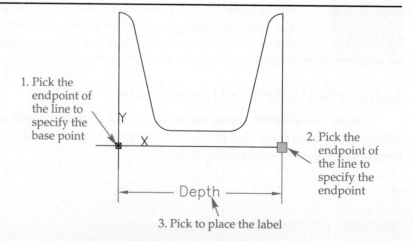

1. Pick the
endpoint of
the line to
specify the
base point

2. Pick the
endpoint of
the line to
specify the
endpoint

Depth

3. Pick to place the label

Figure 26-18.
An example of a portion of a motor mount with side views of different steel channel shapes created and rotated using a single block with a polar parameter, assigned a polar stretch action.

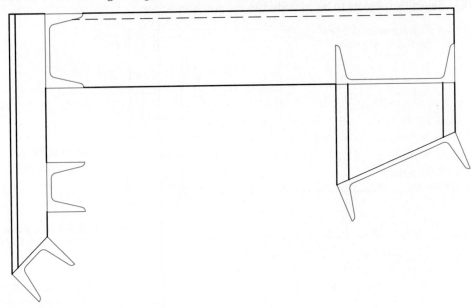

NOTE

Name, **Label**, **Chain**, **Description**, **Palette**, and **Value set** options are available before you specify the parameter. Most of these options are also available from the **Properties** palette if you have already created the parameter.

Assigning a Polar Stretch Action

Figure 26-19 illustrates the process of adding a *polar stretch action* to the steel channel shape example. Access the **Polar Stretch** action option and pick the **Depth** parameter. Then specify a parameter point to associate with the action. Move the cross-hairs near the appropriate parameter point to display the red snap marker, and pick to select. Alternatives include choosing the **sTart point** option to pick the start point of the polar parameter, or the **Second point** option to select the endpoint. If you plan to use grip editing to control the block, and added a single grip, specify the point with the grip.

Next, create a window to define the stretch frame. This is the same technique you apply when using the **STRETCH** command. See **Figure 26-19A**. Pick the objects to stretch, including the associated parameter. You do not need to use a crossing window, because the previous operation defines the stretch. However, crossing selection is often quicker. See **Figure 26-19B**. Then select the objects to rotate, which are often the objects of a block. See **Figure 26-19C**. Press [Enter] or the space bar or right-click to place the action icon. Test and save the block, and exit the **Block Editor**. The dynamic block is now ready to use.

polar stretch action: An action used to change the size, shape, and rotation of block objects with a stretch operation.

Ribbon

Block Editor
> Action
   Parameters

Polar Stretch

Type

BACTIONTOOL

Using a Polar Stretch Action Dynamically

Figure 26-20 shows the steel channel shape block reference selected for editing. The polar parameter grip appears as a light blue arrow at the far end of the channel depth. The insertion base point specified when the block was created appears as a standard unselected grip. Select the parameter grip and stretch the channel depth to the new length and angle. Use dynamic input to view the stretch dimension and rotation angle, and enter an exact length value in the distance field and rotation angle in the angle field. You can also use the **Properties** palette to define the distance.

Figure 26-19.
Assigning a polar stretch action to a polar parameter. A—Specify the parameter and parameter grip, and create a window for the stretch frame. B—Select the objects to be included in the stretch action. C—Select the objects to be included in the rotation action.

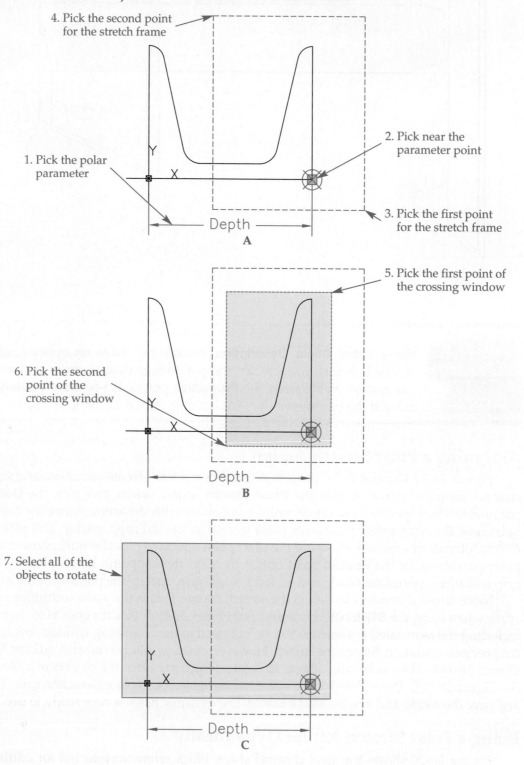

4. Pick the second point for the stretch frame

1. Pick the polar parameter

2. Pick near the parameter point

3. Pick the first point for the stretch frame

Depth

A

5. Pick the first point of the crossing window

6. Pick the second point of the crossing window

Depth

B

7. Select all of the objects to rotate

Depth

C

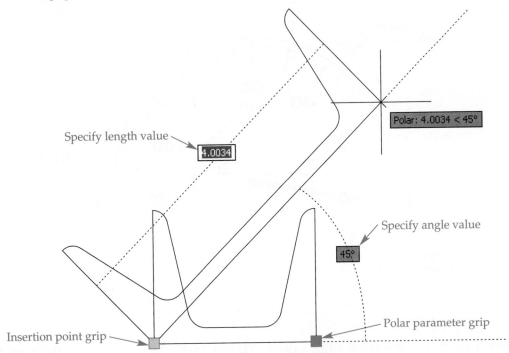

Specify length value — 4.0034

Polar: 4.0034 < 45°

Specify angle value

45°

Polar parameter grip

Insertion point grip —

The **Properties** palette includes **Multiplier** and **Angle Offset** options for
move, stretch, and polar stretch actions. Enter a value in the **Multiplier**
text box to multiply by the parameter value when adjusting the block.
For example, if you assign a distance multiplier of 2 to a move action
and move an object 4 units, the object actually moves 8 units. Enter an
angle in the **Angle Offset** text box to change the parameter grip angle.
For example, if you assign an offset angle of 45 to a move action and
move an object 10°, the object actually moves 55°.

Exercise 26-5

Complete the exercise on the companion website.
www.g-wlearning.com/CAD

Rotation Parameters

A *rotation parameter* creates a rotation property to which you can assign a *rotate
action*. For example, assign a rotation parameter with a rotate action to the needle in
the speedometer block shown in **Figure 26-21** to rotate the needle around the circumference of the dial. To insert the rotation parameter, access the **Rotation parameter**
option and pick the center of the circular base of the needle as the rotation base point.
Then pick a point, such as the needle endpoint shown, to specify the parameter radius.
Set the default rotation angle from 0° east, or if rotation should originate from an angle

rotation parameter:
A parameter that
allows objects in
a block to rotate
independently of the
block.

rotate action:
An action used to
rotate objects within
a block without
affecting the other
objects in the block.

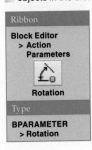

Ribbon

**Block Editor
> Action
Parameters**

Rotation

Type

**BPARAMETER
> Rotation**

Figure 26-21.
A rotation parameter with a rotate action allows you to rotate the needle in a speedometer block to indicate different speeds. Use the **Base angle** option to set a base angle other than 0°.

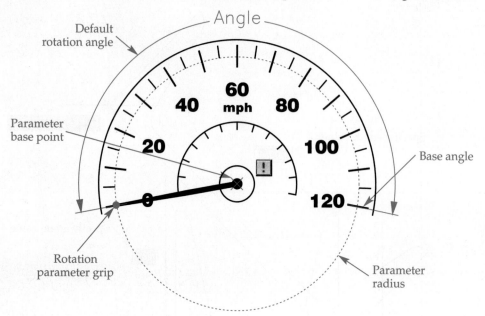

other than 0°, use the **Base angle** option. **Figure 26-21** shows using the **Base angle** option to base the rotation at –10° and specifying a default rotation angle of 200° to align the rotation with the 120 and 0 mph marks.

After you define the rotation parameter, pick a location for the parameter label. Next, enter the number of grips to associate with the parameter. The default **1** option creates a single grip at the parameter radius that allows you to use grip editing to carry out the rotate action.

Name, **Label**, **Chain**, **Description**, **Palette**, and **Value set** options are available before you specify the parameter. Most of these options and the **Base angle** setting are also available from the **Properties** palette if you have already created the parameter.

Assigning a Rotate Action

Ribbon

Block Editor > Action Parameters

Rotate

Type

BACTIONTOOL > Rotate

To assign a rotate action to the speedometer example, access the **Rotate Action** option and pick the rotation parameter. Then select the objects that make up the needle and the rotation parameter. Press [Enter] or the space bar, or right-click to place the action. If necessary, access the **Properties** palette and adjust the **Base type** option. The default **Dependent** option sets the rotation point as the base point of the rotation parameter, which is appropriate for the speedometer example. Test and save the block, and exit the **Block Editor**. The dynamic block is now ready to use.

Using a Rotate Action Dynamically

Figure 26-22 shows using the rotation parameter grip or dynamic input to rotate the needle inside a reference of the speedometer block. An endpoint object snap is most appropriate to select a specific speed for this example. You can also use the **Properties** palette to define the angle.

AutoCAD and Its Applications—Basics

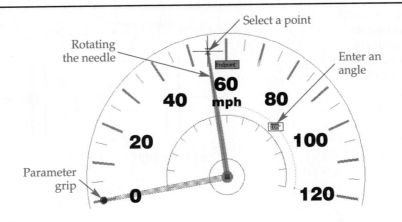

Figure 26-22.
Dynamically rotating the needle in a speedometer block using a rotate action assigned to a rotation parameter.

Exercise 26-6

Complete the exercise on the companion website.
www.g-wlearning.com/CAD

Alignment Parameters

An *alignment parameter* creates an alignment property. When you move a block with an alignment parameter near another object, the block rotates to align with the object based on the angle and alignment line defined in the block. An alignment parameter saves time by eliminating the need to rotate a block or assign a rotation parameter. An alignment parameter affects the entire block, and therefore requires no action.

alignment parameter: A parameter that aligns a block with another object in the drawing.

Creating an Alignment Parameter

Figure 26-23 provides an example of adding an alignment parameter to the block of a gate valve symbol to align the gate valve with pipes. Access the **Alignment parameter** option and pick the point in the center of the valve to locate the parameter grip and define the first point of the alignment line. Next, specify the alignment direction, or use the **Type** option to specify the alignment type. Alignment type does not affect how the block aligns; it determines the direction of the alignment grip. Select the **Perpendicular** option to point the grip perpendicular to the alignment line, or choose the **Tangent** option to point the grip tangent to the alignment line. Set the **Tangent** option for the gate valve example.

After specifying the base point and alignment type, pick a second point to set the alignment direction. The angle between the first point and the second point defines the alignment line. The alignment line determines the default rotation angle. **Figure 26-23** shows selecting the endpoint of the valve symbol. The alignment parameter grip is an arrow that points in the direction of alignment, perpendicular or tangent to the object with which the block will align. Test the block by drawing a line in the **Test Block Window** and attempting to align the block with the line. When you are finished, save the block and exit the **Block Editor**. The dynamic block is now ready to use.

Ribbon

Block Editor > Action Parameters

Alignment

Type

BPARAMETER > Alignment

Use the **Name** option before you specify the parameter to rename the parameter. Alignment parameters do not include labels. You can also adjust the alignment type from the **Properties** palette if you have already created the parameter.

Figure 26-23.
Adding an alignment parameter to a gate valve block.

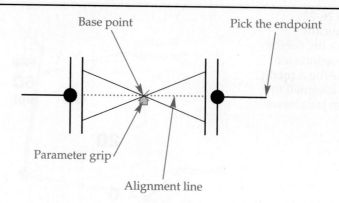

Base point

Pick the endpoint

Parameter grip

Alignment line

Using an Alignment Parameter Dynamically

Figure 26-24 shows using the alignment parameter grip to align the gate valve block with a pipeline. Select the block to display grips and then pick the parameter grip. Move the block near another object to align the block with the object. The rotation depends on the alignment path and type and the angle of the other object.

When you manipulate a block with an alignment parameter, the **Nearest** object snap temporarily turns on, if it is not already on.

Exercise 26-7
Complete the exercise on the companion website.
www.g-wlearning.com/CAD

Figure 26-24.
Move the gate valve block near a line to align the block with the line.

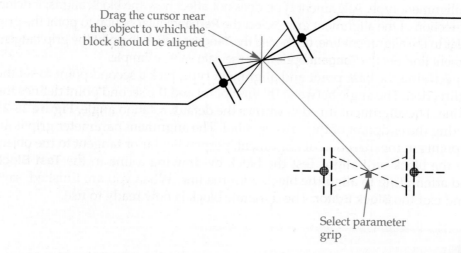

Drag the cursor near the object to which the block should be aligned

Select parameter grip

Flip Parameters

A *flip parameter* creates a flip property to which you can assign a *flip action*. For example, assign a flip parameter with a flip action to a door symbol to provide the option to place the door on either side of a wall. Another example is using a flip parameter to control the side of a reference line on which a weld symbol appears for arrow side or other side applications.

Figure 26-25 shows an example of adding a flip parameter. Access the **Flip Parameter** option and pick the base point, followed by the endpoint of the reflection line. See **Figure 26-25A**. Pick a location for the parameter label, and then enter the number of grips to associate with the parameter. The default **1** option creates a single flip grip that allows you to use grip editing to carry out the flip action.

Flipping a block mirrors the block over the reflection line. However, for the door symbol, with the line in the position shown in **Figure 26-25A**, an incorrect flip will result when you flip the block to the other side of a wall. To mirror the block properly, you must locate the reflection line to account for wall thickness. To place the door on a 4″ wall, for example, use the **MOVE** command to move the reflection line 2″ lower than the door. The label and parameter grip also move. In addition, move the parameter grip horizontally to the middle of the door opening to help place and flip the block properly. See **Figure 26-25B**.

Ribbon

**Block Editor
> Action
Parameters**

Flip

Type

**BPARAMETER
> Flip**

flip parameter: A parameter that mirrors selected objects within a block.

flip action: An action used to flip the entire block.

Name, **Label**, **Description**, and **Palette** options are available before you specify the parameter. Most of these options are also available from the **Properties** palette if you have already created the parameter.

PROFESSIONAL TIP

A block reference with a flip parameter mirrors about the reflection line. You must place the reflection line in the correct location so the flip creates a symmetrical, or mirrored, copy. This typically requires the reflection line to be coincident with the block insertion point.

Assigning a Flip Action

To assign a flip action to the door example, access the **Flip action** option and pick the flip parameter. Then select the objects that make up the door and the flip parameter.

Ribbon

**Block Editor
> Action
Parameters**

Flip

Type

**BACTIONTOOL
> Flip**

Figure 26-25.
A—Inserting a flip parameter. B—Moving the parameter so the block will flip correctly about the centerline of a wall.

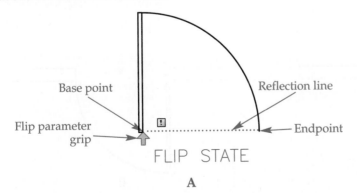

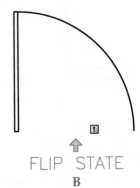

Press [Enter] or the space bar or right-click to place the action. Test and save the block, and exit the **Block Editor**. The dynamic block is now ready to use.

Using a Flip Action Dynamically

Figure 26-26A shows a reference of the door block selected for editing. Pick the flip parameter grip to flip the block to the other side of the reflection line, as shown in **Figure 26-26B**. Unlike other parameters and actions that require stretching, moving, or rotating, a single pick initiates a flip action.

PROFESSIONAL TIP

Add another flip parameter with a flip action to a door symbol to flip the door from side to side. This one block takes the place of four blocks to accommodate different door positions.

Exercise 26-8

Complete the exercise on the companion website.
www.g-wlearning.com/CAD

XY Parameters

An XY parameter creates horizontal and vertical distance properties. You can assign move, scale, stretch, and array actions to XY parameters. The XY parameter can include up to four parameter grips—one at each corner of a rectangle defined by the parameter. You can use the XY parameter for a variety of applications, depending on the assigned actions.

Figure 26-27 provides an example of inserting an *XY parameter*. Access the **XY parameter** option and pick the base point, which is the origin of the X and Y distances. Next, pick a point to specify the XY point, which is the corner opposite the base point. Finally, enter the number of grips to associate with the parameter. The default **2** option

XY parameter:
A parameter that specifies distance properties in the X and Y directions.

Ribbon

Block Editor > Action Parameters

XY

Type

BPARAMETER > XY

Figure 26-26.
A—Select the block to display the flip parameter grip. B—Pick the flip parameter grip to flip the block about the reflection line. The entire block flips because all of the objects within the block are included in the selection set for the action.

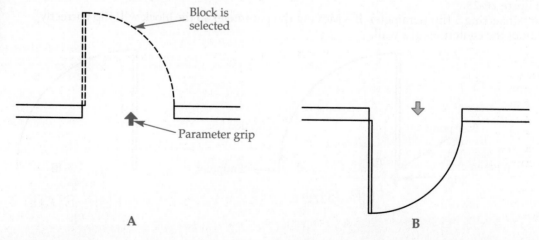

Block is selected

Parameter grip

A

B

Figure 26-27.
Adding an XY parameter to a block of an architectural glass block. The XY parameter consists of X and Y distance properties and four grips.

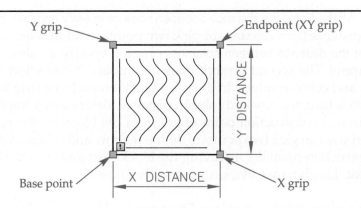

creates grips at the start and endpoints, allowing you to use grip editing to carry out the action assigned to either point. Select the **4** option, as shown in **Figure 26-27**, to assign a grip at each XY corner to maximize flexibility, or choose a smaller number to limit dynamic options. If you choose the **0** option, you can only adjust the block using the **Properties** palette.

> **Name**, **Label**, **Chain**, **Description**, **Palette**, and **Value set** options are available before you specify the parameter. Most of these options and the **Base angle** setting are also available from the **Properties** palette if you have already created the parameter.

Assigning an Array Action

Figure 26-28 illustrates using an *array action* assigned to an XY parameter. This example shows dynamically arraying the block of an architectural glass block to create an architectural feature of glass blocks, such as a wall, without using a separate array operation. Access the **Array action** option and pick the XY parameter. Then select the objects to be included in the array, and press [Enter] or the space bar or right-click to accept the selection.

array action:
An action used to array objects within the block based on preset specifications.

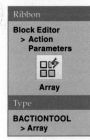

Figure 26-28.
Dynamically creating an array of architectural glass blocks using a block with an XY parameter and an array action. The pattern of rows and columns forms as you move the XY parameter. Notice the mortar joints that form between the glass blocks because of proper action definition.

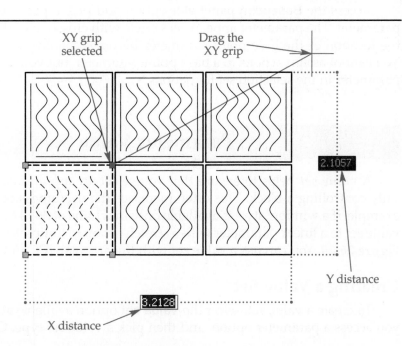

At the Enter the distance between rows or specify unit cell: prompt, enter a value for the distance between rows or pick two points to set the row and column values. At the Enter the distance between columns: prompt, specify a value for the distance between columns. The second prompt does not appear if you select two points to define the row and column values. In the glass block example, be sure to allow for a mortar joint when setting the row and column distance. Before assigning the action, you may want to draw a construction point offset from the block by the width of the mortar joint. Then you can pick two points to define the row and column values. Be sure to erase the construction point before saving the block. Test and save the block, and exit the **Block Editor**. The dynamic block is now ready to use.

Using an Array Action Dynamically

Figure 26-28 illustrates using the upper-right grip or dynamic input to array a reference of the architectural glass block. You can use any available grip to apply an array, depending on where you want the array to occur. You can also use the **Properties** palette to define the array. Notice that proper action definition produces mortar joints. The resulting array remains a single block.

Exercise 26-9

Complete the exercise on the companion website.
www.g-wlearning.com/CAD

Base Point Parameters

The **Block Editor** origin (0,0,0 point) determines the default location of the block insertion base point. Typically, you construct blocks in the **Block Editor** in reference to the origin, using the origin as the location of the insertion base point. The base point you choose when creating a block using the **BLOCK** command attaches to the origin when you open the block in the **Block Editor**. Add a *base point parameter* to override the default origin base point.

Access the **Basepoint parameter** option and pick a point to place the base point parameter. The parameter appears as a circle with crosshairs. After you save the block, the location of the base point parameter becomes the new base point for the block. You cannot assign actions to a base point parameter, but you can include a base point parameter in the selection set for actions.

base point parameter: A parameter that defines an alternate base point for a block.

Ribbon
**Block Editor
> Action
Parameters**

Basepoint

Type
**BPARAMETER
> Basepoint**

value set: A set of allowed values for a parameter.

Parameter Value Sets

A *value set* helps to ensure that you select an appropriate value when dynamically controlling a block. This often increases the usefulness of a dynamic block. For example, if a window style is available only in widths of 36″, 42″, 48″, 54″, and 60″, add a value set to a linear parameter with a stretch action to limit selection to these sizes. See **Figure 26-29**. You can use a value set with linear, polar, XY, and rotation parameters.

Creating a Value Set

To create a value set, select the **Value set** option available at the first prompt after you access a parameter option, and then pick a value set type. Choose the **List** option

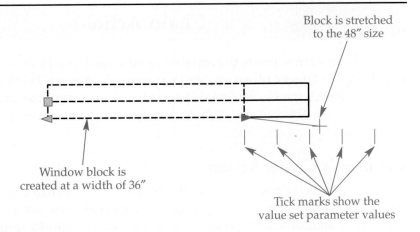

Figure 26-29.
When you adjust a block that includes a value set, tick marks appear at locations corresponding to the values in the value set. You can only adjust the block to one of the tick marks.

Block is stretched to the 48" size

Window block is created at a width of 36"

Tick marks show the value set parameter values

to create a list of possible sizes. Type all of the valid values for the parameter separated by commas. For the window block example, enter 36,42,48,54,60. Then press [Enter] or the space bar or right-click to return to the initial parameter prompt, and add the parameter. After you insert the parameter, the valid values appear as tick marks.

Select the **Increment** option to specify an incremental value. With this option, you can set minimum and maximum values to provide a limit for the increments. For the window block example, use the **Value set** option again to set 6" width increments. This time, choose the **Increment** option and type 6 for the distance increment, 36 for the minimum distance, and 60 for the maximum distance. The initial parameter prompt returns after you enter the maximum distance.

After you add a parameter with a value set, assign an action to the parameter. For the window block in **Figure 26-29**, assign a stretch action to the linear parameter. This allows the window to stretch to the valid widths specified in the value set. Test and save the block, and exit the **Block Editor**. The dynamic block is now ready to use.

You can also use the options in the **Value Set** category of the **Properties** palette to specify value sets during block definition.

Using a Value Set with a Parameter

Figure 26-29 shows using a linear parameter grip with a stretch action to specify the width of a window block reference. Tick marks appear at the positions of valid values. As you stretch the grip, the modified block snaps to the nearest tick mark. When using dynamic input, you can also enter a value in the input field. If you type a value that is not included in the value set, AutoCAD applies the nearest valid value. You can also use the **Value Set** category of the **Properties** palette to select a value.

Exercise 26-10

Complete the exercise on the companion website.
www.g-wlearning.com/CAD

Chain Actions

A *chain action* limits the number of edits that you have to perform by allowing one action to trigger other actions. For example, **Figure 26-30** shows using a chain action to stretch the block of a table and chairs, and array the chairs along the table at the same time. You can use a chain action with point, linear, polar, XY, and rotation parameters.

Creating a Chain Action

To create a chain action, select the **Chain** option available at the first prompt after you access a parameter option to display the Evaluate associated actions when parameter is edited by another action? [Yes/No]: prompt. The default **No** setting does not create a chain action. Select the **Yes** option to create a chain action.

Figure 26-31 shows the default arrangement of the table and chairs block example. For this example, access the **Linear parameter** option and use the **Label** option to change the label name to CHAIR ARRAY. Next, choose the **Chain** option and select **Yes**. To complete the parameter, select the start and endpoints shown in **Figure 26-31A**, and assign a single grip to the endpoint. Then assign an array action to the parameter, selecting the chairs on the top and bottom of the table as the objects to array. At the Enter the distance between rows or specify unit cell: prompt, use object snaps to snap to the endpoint of one of the chairs and then snap to the equivalent endpoint on the chair next to the first chair.

Add another single-grip linear parameter, labeled TABLE STRETCH, as shown in **Figure 26-31B**. Assign a stretch action to the parameter associated with the TABLE STRETCH parameter grip. Use a crossing window around the right end of the table and the CHAIR ARRAY parameter grip. See **Figure 26-31C**. Select the table, the chair at the right end of the table, and the CHAIR ARRAY parameter as the objects to stretch. Test and save the block, and exit the **Block Editor**. The dynamic block is now ready to use.

Figure 26-30.
A—A block of a table with six chairs. B—Use a chain action with a linear parameter to array the chairs automatically when you stretch the table.

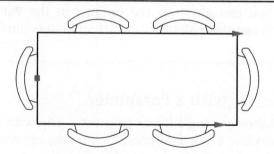

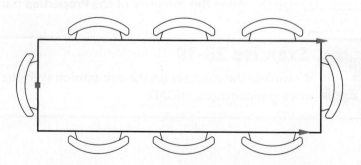

Figure 26-31.
A—Inserting a linear parameter to use with an array action for the chairs. B—Inserting a linear parameter to stretch the table. C—Assigning a stretch action to the linear parameter. When you specify the crossing window, be sure to include the **CHAIR ARRAY** parameter grip in the frame.

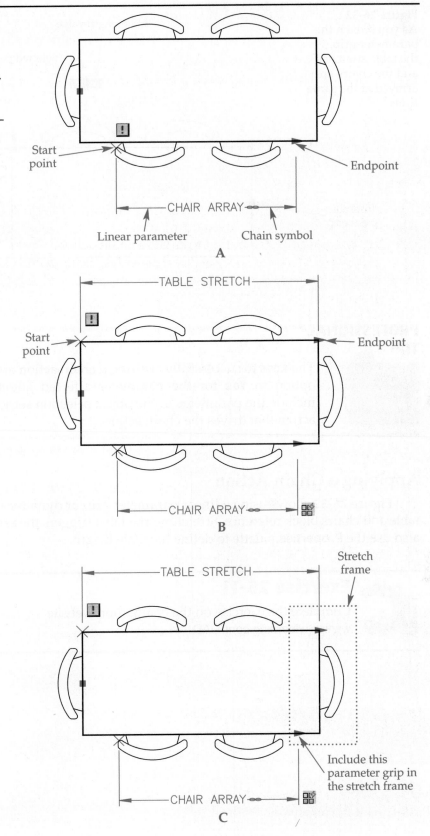

Figure 26-32.
As you stretch the parameter grip, the table stretches and the chairs are arrayed at the same time.

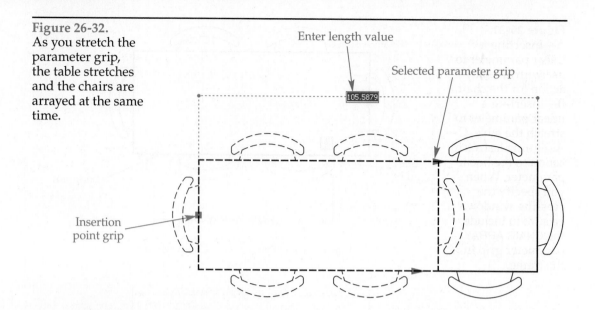

Enter length value

Selected parameter grip

105.5879

Insertion point grip

Applying a Chain Action

Figure 26-32 shows using a linear parameter grip or dynamic input to stretch the table and chairs block reference. Stretching the table triggers the array action. You can also use the **Properties** palette to define the table length.

Exercise 26-11

Complete the exercise on the companion website.
www.g-wlearning.com/CAD

Chapter Review

Answer the following questions. Write your answers on a separate sheet of paper or complete the electronic chapter review on the companion website.
www.g-wlearning.com/CAD

1. Define *dynamic block*.
2. What is the function of a dynamic block?
3. How are standard blocks and dynamic blocks the same? How are they different?
4. Identify the property that forms when you create a point parameter and list the actions you can assign.
5. What is the purpose of a move action?
6. When do action bars appear?
7. What information can you get by hovering over or selecting an action bar?
8. What is the basic function of a linear parameter?
9. Explain the function of a stretch action.
10. Briefly describe how to use a stretch action symmetrically.
11. What is the basic function of a scale action?
12. Describe the polar parameter type and list the actions that you can assign to the parameter.
13. Identify the property that forms when you create a rotation parameter and list the actions you can assign.
14. Describe what happens when you move a block with an alignment parameter near another object in the drawing. How does this save drawing time?
15. Give at least one practical example of using a flip parameter.
16. Briefly describe the properties an XY parameter creates and list the actions that you can assign to the parameter.
17. Give an example of using an array action assigned to an XY parameter.
18. When would you add a base point parameter?
19. Describe the basic use of a value set.
20. Explain the function of a chain action.

Drawing Problems

Start AutoCAD if it is not already started. Start a new drawing for each problem using an appropriate template of your choice. The template should include layers and text, dimension, multileader, and table styles, when necessary, for drawing the given objects. Add layers and text, dimension, multileader, and table styles as needed. Draw all objects using appropriate layers and text, dimension, multileader, and table styles, justification, and format. Follow the specific instructions for each problem. Use only drawing and editing commands and techniques you have already learned. Use your own judgment and approximate dimensions when necessary. Apply dimensions accurately using ASME or appropriate industry standards.

Note: *Some of the problems in this chapter are built on problems from previous chapters. If you have not yet completed those problems, complete them now.*

▼ Basic

1. Open P24-1 and save it as P26-1. The P26-1 file should be active. Erase all copies of the steel column symbols except for the one in the lower-left corner. Insert an XY parameter into the steel column block and associate an array action with the parameter. Use the proper values for the array action to array the block dynamically to match the drawing. Use the dynamic block to create the rest of the steel columns in the drawing. Resave the drawing.

2. Create a block named WIRE ROLL as shown. Do not include the dimensions. Insert a linear parameter on the entire length of the roll. Use a value set with the following values: 36″, 42″, 48″, and 54″. Assign a stretch action to the parameter and associate the action with either parameter grip. Create a crossing window that will allow the length of the roll to stretch. Select all of the objects on one end and the length lines as the objects to stretch. Insert the WIRE ROLL block four times into a drawing and stretch each block to use a different value set length. Save the drawing as P26-2.

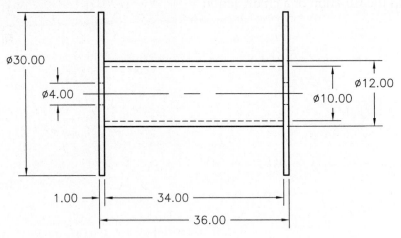

3. Create a block named 90D ELBOW as shown on the left. Do not include the dimensions. Insert two flip parameters and two flip actions. The purpose of one of the flip parameter/action combinations is to flip the elbow horizontally. The second flip parameter/action combination flips the elbow vertically. Use the dynamic block to create the drawing shown on the right. Save the drawing as P26-3.

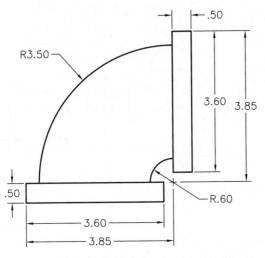

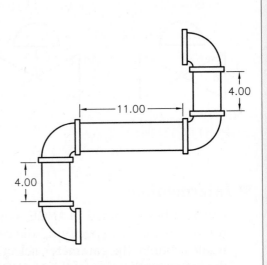

4. Create a block of the 48″ window shown on the left. Do not include the dimensions. Insert an alignment parameter so the length of the window can align with a wall. Then draw the walls shown on the right. Insert the window block as needed. Use the alignment parameter to align the window to the walls. Center the windows on wall segments unless dimensioned. Save the drawing as P26-4.

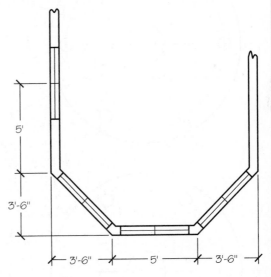

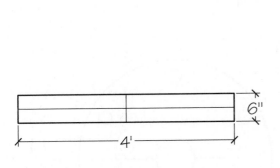

5. Create a block named **CONTROL VALVE** as shown on the left. Include the label in the block. Insert a point parameter and assign a move action to the parameter. Select the two lines of text as the objects to which the action applies. Insert the **CONTROL VALVE** block into the drawing three times. Use the point parameter to move the text to match the three positions shown. Save the drawing as **P26-5**.

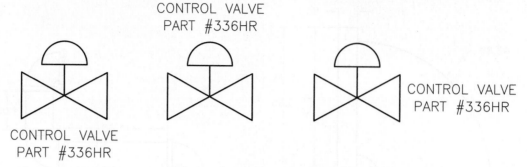

▼ **Intermediate**

6. Create a block named **FLANGE** as shown. Do not include dimensions. Insert a rotation parameter specifying the center of the flange as the base point. Assign a rotate action to the parameter, selecting the six ∅.20 circles as the objects to which the action applies. Insert the **FLANGE** block into the drawing twice. Use the rotation parameter to create the two configurations shown. Save the drawing as **P26-6**.

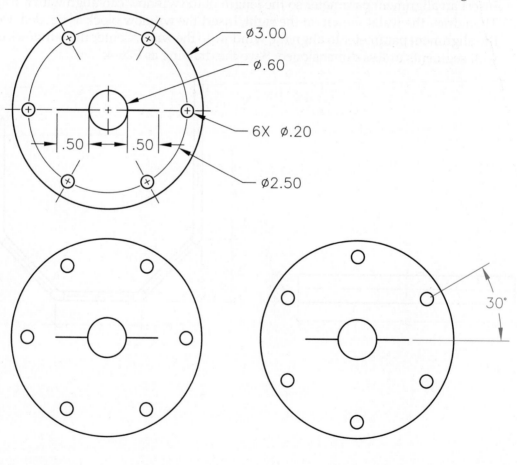

7. Open P26-6 and save it as P26-7. The P26-7 file should be active. Open the FLANGE block in the **Block Editor** and use the **Properties** palette to apply the following settings to the rotation parameter:
   A. **Angle label**—BOLT HOLES
   B. **Angle description**—ROTATION OF BOLT HOLE PATTERN
   C. **Ang type**—INCREMENT
   D. **Ang increment**—30
   E. Save the changes and exit the **Block Editor**. Save the drawing.

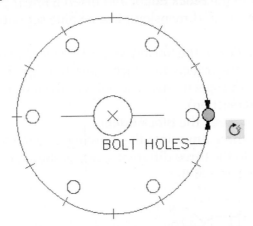

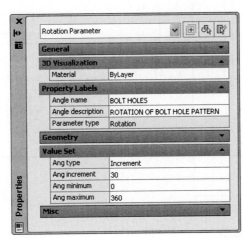

▼ Advanced

8. The drawing shown is a fan with an enlarged view of the motor. This fan can have one of three motors of different sizes. Create the fan as a dynamic block.
   A. Draw all the objects. Do not dimension the drawing or draw the enlarged view.
   B. Create a block named FAN consisting of the objects shown in the enlarged view.
   C. Open the block in the **Block Editor** and insert a linear parameter along the top of the motor (the 1.50″ dimension). Use a value set with the following values: 1.5, 1.75, and 2.
   D. Assign a scale action to the linear parameter. Select all of the objects that make up the motor as the objects to which the action applies. Use an independent base point type and specify the base point as the lower-left corner of the motor (the implied intersection).
   E. Save the block and exit the **Block Editor**.
   F. Insert the block three times into the drawing. Use the linear parameter grip to scale the motor to the three different sizes, as shown below on the right.
   G. Save the drawing as P26-8.

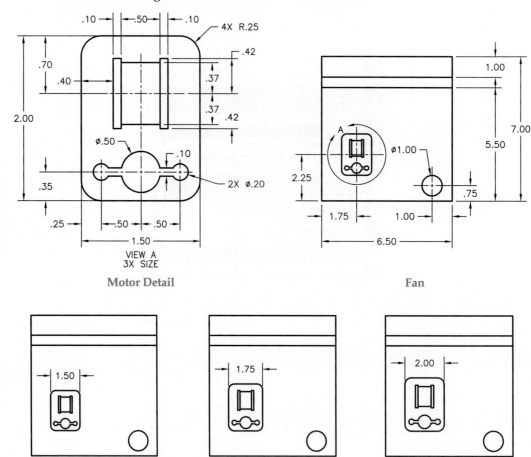

Motor Detail   Fan

9. Construct a block of an architectural glass block similar to the block shown. Use an XY parameter with an array action to create a glass block wall of 10 rows and 15 columns. Construct the block to include an appropriate mortar joint. Save the drawing as P26-9.

10. Construct a block of a steel C4×5.4 shape. Reference the American Institute of Steel Construction (AISC) manual for dimensions. Add a polar parameter with a polar stretch action to adjust the depth and angle of block references. Insert and adjust the block as needed to create a portion of a motor mount similar to the drawing shown. Save the drawing as P26-10.

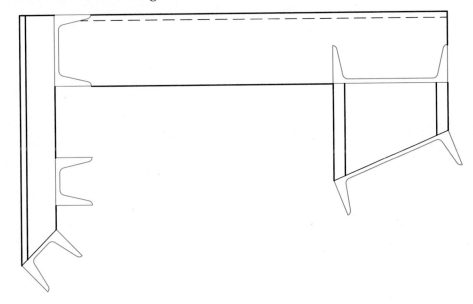

11. Construct a block of a speedometer similar to the speedometer shown. Add a rotation parameter with a rotate action to adjust the needle reading for block references. Insert the block four times and use the dynamic needle to create a speedometer that reads 5 mph, 25 mph, 55 mph, and 110 mph. Save the drawing as P26-11.

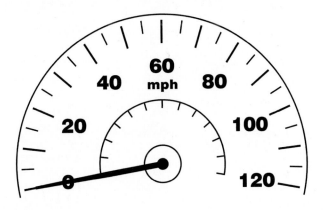

12. Construct a block of a table with six chairs similar to the block shown. Use linear parameters with array, stretch, and chain actions as needed to stretch and add chairs to block references. Insert the block twice and use the dynamic stretch and array to create a table with eight chairs and a table with ten chairs. Save the drawing as P26-12.

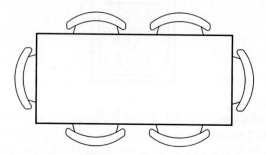

AutoCAD Certified Associate Exam Practice

Answer the following questions. Write your answers on a separate sheet of paper.

1. Which of the following commands can be used to save a dynamic block? *Select all that apply.*
   A. **BACTION**
   B. **BLOCK**
   C. **BPARAMETER**
   D. **BSAVE**
   E. **BSAVEAS**

2. Which of the following actions can be assigned to a point parameter? *Select all that apply.*
   A. flip
   B. move
   C. rotate
   D. scale
   E. stretch

3. Which of the following block properties can be changed using a polar stretch action? *Select all that apply.*
   A. color
   B. layer
   C. shape
   D. size
   E. rotation

AutoCAD Certified Professional Exam Practice

Follow the instructions in the problem. Write your answers on a separate sheet of paper.

1. **Open CPE-26align.dwg. This file is available on the companion website.**
   Edit the existing SOFA block to include an alignment parameter that will align the middle of the back of the sofa with a wall. Insert the block into the drawing. Use the alignment parameter and other appropriate tools to center the sofa on the window as shown, exactly 4″ from the wall. What are the coordinates of the middle of the sofa back? (Use the coordinates of the quadrant point on the sofa back.)

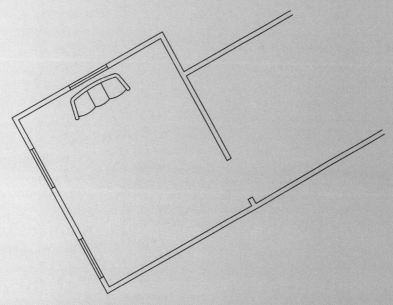

The American Design Drafting Association (ADDA) provides many services for professionals and drafting educators. ADDA offers professional certification in several areas, including mechanical drafting and architectural drafting. See the ADDA website at www.adda.org for additional information.

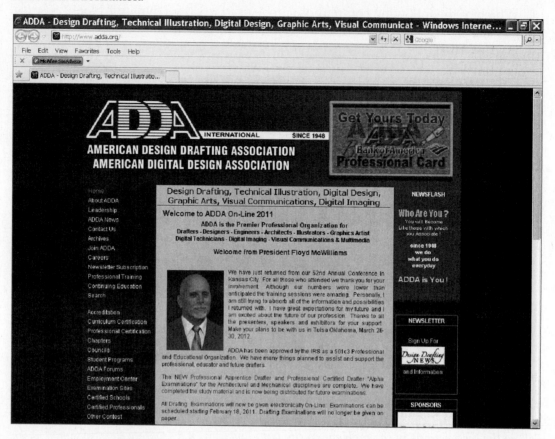

Additional Dynamic Block Tools

Learning Objectives

After completing this chapter, you will be able to:

✓ Apply visibility parameters.
✓ Create and use lookup parameters.
✓ Use parameter sets.
✓ Constrain block geometry.
✓ Use a block properties table.

This chapter describes adding visibility and lookup parameters to enhance the usefulness of blocks. It also explains how to apply geometric constraints and constraint parameters to blocks as an alternative or in addition to using parameters and actions. Finally, this chapter explores the process of using a block properties table.

Visibility Parameters

A *visibility parameter* allows you to assign *visibility states* to objects within a block. Selecting a visibility state displays the only objects in the block associated with the visibility state. Visibility states expand the capacity of blocks in a symbol library by allowing you to hide or make visible specific objects and even completely different symbols. A block can include only one visibility parameter. Visibility parameters do not require an action.

Figure 27-1 provides an example of using a visibility parameter to create four different valve symbols from a single block. To create the block, draw all of the objects representing the different variations, as shown in **Figure 27-1A**. Then assign a visibility parameter and add visibility states that identify the objects that are visible in each variation. Insert the block and select a visibility state to display the corresponding objects. See **Figure 27-1B**.

To add a visibility parameter, access the **Visibility Parameter** option and pick a location for the parameter label. The parameter automatically includes a single grip. When you insert the block and select the grip, a shortcut menu appears listing visibility states. There is no prompt to select objects because the visibility parameter is associated with the entire block.

visibility parameter: A parameter that allows you to assign multiple views to objects within a block.

visibility states: Views created by selecting block objects to display or hide.

Ribbon

Block Editor > Action Parameters

Visibility

Type

BPARAMETER > Visibility

Figure 27-1.
A—All of the objects composing each unique valve symbol shown together. B—Create each different valve from a single block using a visibility parameter with different visibility states.

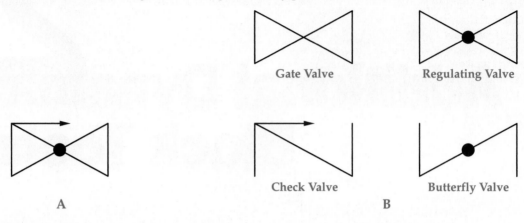

A   B

Name, **Label**, **Description**, and **Palette** options are available before you specify the parameter. Most of the options are also available from the **Properties** palette if you have already created the parameter.

Creating Visibility States

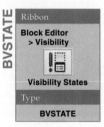

The tools in the **Visibility** panel of the **Block Editor** ribbon tab become enabled when you add a visibility parameter. See **Figure 27-2**. To create a visibility state, access the **BVSTATE** command to display the **Visibility States** dialog box. See **Figure 27-3A**. Pick the **New...** button to open the **New Visibility State** dialog box shown in **Figure 27-3B**. Type the name of the new visibility state in the **Visibility state name:** text box. For the valve block example shown in **Figure 27-1**, an appropriate name could be GATE VALVE, REGULATING VALVE, CHECK VALVE, or BUTTERFLY VALVE, depending on which valve the visibility state represents.

Pick the **Hide all existing objects in new state** radio button to make all of the objects in the block invisible when you create the new visibility state. This allows you to choose only the objects that should be visible for the visibility state. Pick the **Show all existing objects in new state** radio button to make all of the objects in the block visible when you create the new visibility state. This allows you to hide objects that should be invisible for the visibility state. Select the **Leave visibility of existing objects unchanged in new state** radio button to display the objects that are currently visible when you create the new visibility state.

Figure 27-2.
The visibility tools in the **Visibility** panel of the **Block Editor** ribbon tab.

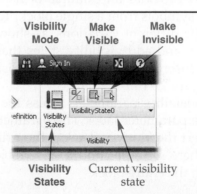

Figure 27-3.
A—Manage visibility states using the **Visibility States** dialog box.
B—Create new visibility states using the **New Visibility State** dialog box.

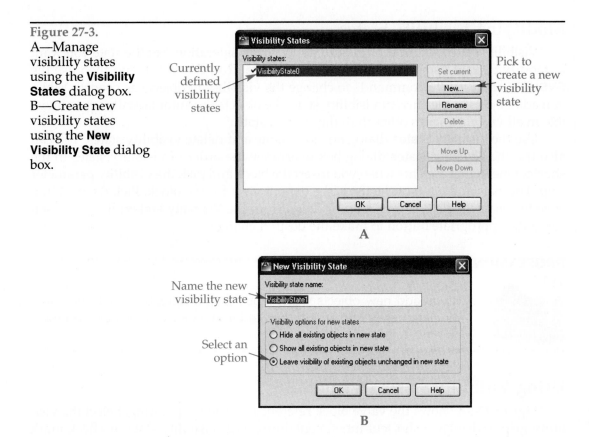

A

B

Pick the **OK** button to create the new visibility state. The new state is added to the list in the **Visibility States** dialog box and becomes the current state, as indicated by the check mark next to the name. Pick the **OK** button to return to block editing mode.

Next, use the **BVSHOW** and **BVHIDE** commands to display only the objects that should be visible in the current state. Pick the **Make Visible** button to select objects to make visible. Invisible objects are temporarily displayed semi-transparently for selection. Pick the **Make Invisible** button to select objects to make invisible. For example, to make a visibility state to depict the gate valve shown in **Figure 27-4B** from the valve block shown in **Figure 27-4A**, use the **Make Invisible** command to turn off the filled circle and the arrow. The changes are saved to the visibility state automatically. Use the **BVMODE** command to toggle the visibility mode on and off. Turn on visibility mode to display invisible objects as semi-transparent. Turn off visibility mode to display only visible objects.

Repeat the process to create additional visibility states for the block. The valve block example requires four visibility states. The **Current visibility state** drop-down list displays the current visibility state. Select a state from the drop-down list to make the state current. After you create all visibility states, test and save the block and exit the **Block Editor**. The dynamic block is now ready to use.

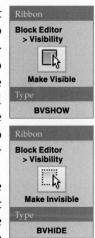

Figure 27-4.
A—The VALVE block with all objects visible.
B—The VALVE block after making the arrow and filled circle invisible to display the GATE VALVE visibility state.

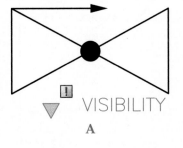

A

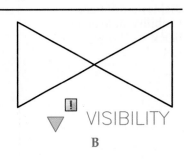

B

Modifying Visibility States

Visibility state modification requires special consideration. Set the state you want to modify current using the **Current visibility state** drop-down list, and then use the **BVSHOW** and **BVHIDE** commands to change the visibility of objects as needed. When you add objects to the current visibility state, the objects are automatically set as invisible in all visibility states other than the current state.

Use the **Visibility States** dialog box to rename and delete visibility states. You can also use the **Visibility States** dialog box to arrange the order of visibility states in the shortcut menu that appears when you insert the block and pick the visibility parameter grip. The state at the top of the list is the default view for the block. Pick the visibility state to rename, delete, or move up or down from the **Visibility states:** list box. Then select the appropriate button to make the desired change.

PROFESSIONAL TIP

If you add new objects when modifying a state, be sure to update the parameters and actions applied to the block to include the new objects, if needed.

Using Visibility States Dynamically

Figure 27-5A shows the valve block reference selected for editing. Select the visibility grip to display a shortcut menu containing each visibility state. A check mark indicates the current visibility state. To switch to a different view of the block, select the name of the visibility state from the list. See **Figure 27-5B**. You can also use the **Properties** palette to select a visibility state.

Exercise 27-1

Complete the exercise on the companion website.
www.g-wlearning.com/CAD

Figure 27-5.
A—Pick the visibility parameter grip to display a shortcut menu with the available visibility states. The current state is checked.
B—Select a different visibility state from the shortcut menu to display a different visibility state of the block.

Parameter grip

Visibility states

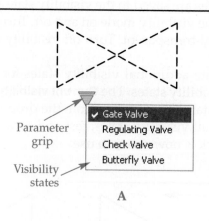

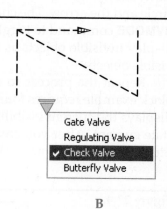

A

B

Lookup Parameters

A *lookup parameter* creates a lookup property to which you can assign a *lookup action*. For example, **Figure 27-6** shows three valve symbols created from a single block by adjusting the rotation parameter of the middle line. The lookup action allows the middle line rotation to control the length of the start and end lines.

To create the valve block shown in **Figure 27-7**, first draw the geometry of the 0° symbol. Then add a linear parameter and label it Start Line. Select the start point as the bottom of the start line, and the endpoint as the top of the start line. Assign a stretch action to the parameter, associated with the top parameter grip. Draw the crossing window around the top of the start line and select the start line as the object to stretch.

Add another linear parameter, labeled End Line. Select the bottom of the end line as the start point and the top of the end line as the endpoint. Assign a stretch action to the parameter, associated with the top parameter grip. Draw the crossing window around the top of the end line and select the end line as the object to stretch.

Next, add a rotation parameter labeled Middle Line. Specify the center of the circle as the base point. Select the right endpoint of the middle line to set the radius, and specify the default rotation angle as 0. Assign a rotation action to the parameter. Pick the center of the circle as the rotation base point, and select the middle line as the object to rotate.

To add a lookup parameter, access the **Lookup Parameter** option and pick a location for the parameter label. Then enter the number of grips to associate with the parameter. The default **1** option creates a single lookup grip that allows you to use grip editing to carry out the lookup action. When you insert the block and select the grip, a shortcut menu appears listing rotation options. There is no prompt to select objects because a lookup parameter is associated with the entire block.

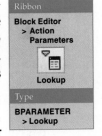

Ribbon

Block Editor
 > Action
  Parameters

Lookup

Type

BPARAMETER
 > Lookup

Figure 27-6.
A lookup parameter allows you to create these three valve symbols using the same block. Notice how the geometry changes.

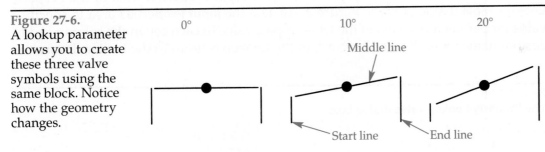

Figure 27-7.
The block of the valve symbol example with linear parameters and stretch actions assigned to the start and end lines and a rotation parameter and rotate action assigned to the middle line.

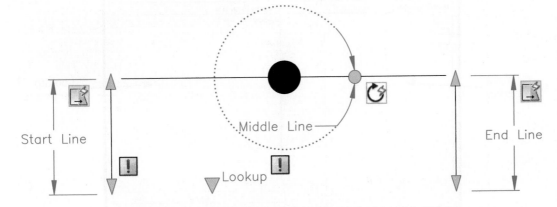

Name, Label, Description, and **Palette** options are available before you specify the parameter. Most of the options are also available from the **Properties** palette if you have already created the parameter.

Assigning a Lookup Action

To assign a lookup action, access the **Lookup Action** option and select a lookup parameter. The **Property Lookup Table** dialog box appears, allowing you to create a lookup table. See **Figure 27-8**.

Creating a Lookup Table

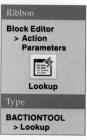

Ribbon

Block Editor > Action Parameters

Lookup

Type

BACTIONTOOL > Lookup

A *lookup table* groups parameter properties into custom-named lookup records. The **Action name:** display box indicates the name of the lookup action associated with the table. The table is initially blank. To add a parameter property, pick the **Add Properties...** button to open the **Add Parameter Properties** dialog box. See **Figure 27-9**.

All parameters in the block that contain property values appear in the **Parameter properties:** list. Lookup, alignment, and base point parameters do not contain property values. Notice that the property name is the parameter label. The **Property type** area determines the type of property parameters shown in the list. By default, the **Add input properties** radio button is active, which displays available input property parameters. To display available lookup property parameters, select the **Add lookup properties** radio button.

To add parameter properties to the lookup table, select the properties in the **Parameter properties:** list and pick the **OK** button. A new column, named as the parameter property, forms for each parameter in the **Input Properties** area of the **Property Lookup Table** dialog box. See **Figure 27-10**. Use the **Input Properties** area to specify a value for parameters added to the table. Type a value in each cell in the column. Add a custom name for each row, or record, in the **Lookup** column in the **Lookup Properties**

Figure 27-8.
The **Property Lookup Table** dialog box.

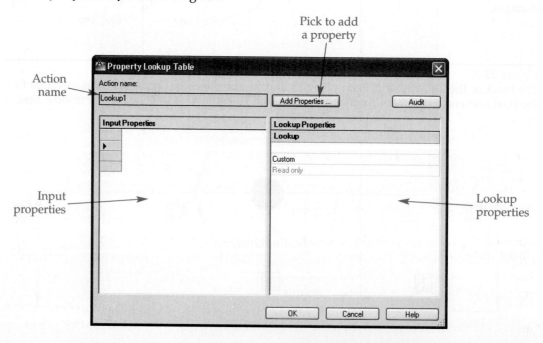

AutoCAD and Its Applications—Basics

Figure 27-9.
Parameter properties are listed in the **Add Parameter Properties** dialog box.

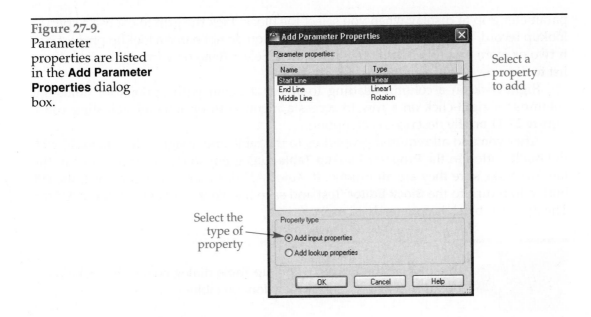

Select a property to add

Select the type of property

For the valve symbol example, add the Middle Line, Start Line, and End Line parameter properties to the table.

Figure 27-10.
A lookup table with multiple parameters and values added.

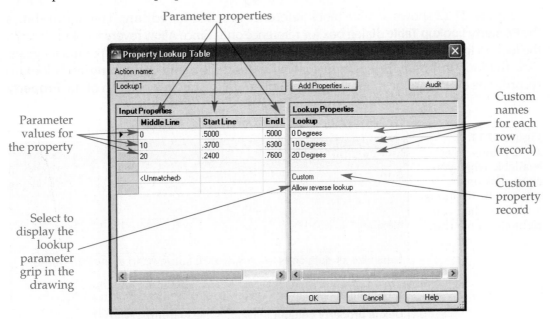

Parameter properties

Parameter values for the property

Select to display the lookup parameter grip in the drawing

Custom names for each row (record)

Custom property record

area. This area displays the name that appears in the shortcut menu when you insert the block and select the lookup parameter grip.

For the valve symbol example, add the Middle Line, Start Line, and End Line parameter properties to the table. Then complete the lookup table as shown in **Figure 27-10**. Start with the Middle Line values. Press [Enter] after typing the value to add a new blank row and then type the remaining values in each cell. Use the [Enter], [Tab], or arrow keys, or pick in a different cell to navigate through the table.

The row, or record, that contains the <Unmatched> value, named Custom in the **Lookup** column, applies when the current parameter values of the block do not match a record in the table. This allows you to adjust the block using parameter values other than those specified in the lookup table. You cannot add any values to the row, but you can change the name of Custom.

The **Allow reverse lookup** setting at the bottom of the **Lookup** column is available only if all of the names in the lookup table are unique. This option allows the lookup parameter grip to display when you select the block. Pick the grip to choose a specific lookup record. The **Read only** setting appears if you do not name a lookup property, or if two or more properties have the same name. Select **Read only** from the drop-down list to disallow selecting a lookup record.

Right-click on a column heading to access a menu with options for adjusting columns, or right-click on a row to access a menu with options for adjusting rows. **Figure 27-11** briefly describes each option.

After you add all required properties to the table and assign values to each, pick the **Audit** button in the **Property Lookup Table** dialog box to check each record in the table to make sure they are all unique. If AutoCAD does not find errors, pick the **OK** button to return to the **Block Editor**. Test and save the block, and exit the **Block Editor**. The dynamic block is now ready to use.

To redisplay the **Property Lookup Table** dialog box, right-click on a lookup action and pick **Display lookup table**.

Using a Lookup Action Dynamically

Figure 27-12 shows a valve block reference selected for editing. The figure shows the **Property Lookup Table** dialog box for reference only. Since **Allow reverse** lookup is set in the lookup table, the lookup parameter grip appears along with the other parameter grips. Pick the lookup parameter grip to display a shortcut menu containing each lookup record. The entries in the menu match the entries in the **Lookup** column of the **Property**

Figure 27-11.
A—Options available when you right-click on a column. B—Options available when you right-click on a row.

Menu Option	Function
Sort	Sorts the records (rows) in ascending or descending order. Pick again to reverse the sort order.
Maximize all headings	Adjusts all columns to the width of the column headings.
Maximize all data cells	Adjusts all columns to the width of the values in the cells.
Size columns equally	Makes all columns equal in width.
Delete property column	Deletes the column.
Clear contents	Deletes the cell values.

A

Menu Option	Function
Insert row	Inserts a new row above the selected row.
Delete row	Deletes the record (row).
Clear contents	Deletes the cell values.
Move up	Moves the row up by one row.
Move down	Moves the row down by one row.
Range syntax examples	Displays the online documentation examples of how to enter values into a lookup table.

B

Figure 27-12.
The lookup parameter grip appears when you select the block. The list of available lookup records is displayed when you pick the lookup parameter grip. Notice the correlation between the available options and the lookup property names in the **Property Lookup Table** dialog box.

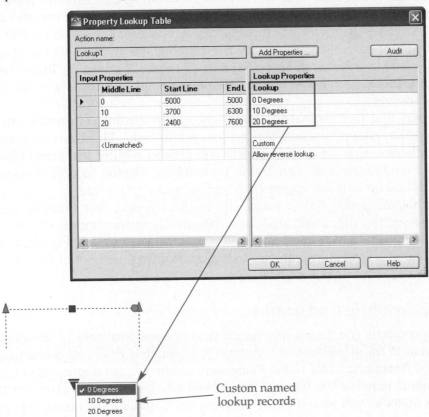

Custom named lookup records

Lookup Table dialog box. A check mark indicates the current record. To switch to a different view of the block, select the name of the record from the list.

You can change other parameters assigned to the block, such as the linear and rotation parameters of the example block, independently of the named records. When you change any of the parameters, the lookup parameter becomes Custom, because the current parameter values do not match one of the records in the lookup table.

Exercise 27-2

Complete the exercise on the companion website.
www.g-wlearning.com/CAD

Parameter Sets

The **Parameter Sets** tab of the **Block Authoring Palettes** window contains common parameters and actions grouped to enhance productivity. Follow the prompts to create a parameter and automatically associate an action with the parameter. The action forms without any selected objects, as is indicated by the yellow alert icon. If the parameter set contains an action that must include associated objects, as most do, double-click on the action icon and select objects. The prompts may differ depending on the type of action.

Constraining Block Geometry

constraint parameters: Dimensional constraints available for block construction to control the size or location of block geometry numerically.

Geometric constraints and *constraint parameters* can directly replace action parameters and actions. For example, the block of the cut framing member shown in **Figure 27-13A** uses geometric constraints to maintain geometric relationships and two linear constraint parameters to specify the member size. When you insert and select the block to edit, use the constraint parameter grips or options in the **Properties** palette to adjust the block. See **Figure 27-13B**. An alternative is to create the block using two linear parameters.

You may find that geometric constraints and constraint parameters are easier to use than action parameters and actions for certain tasks. However, for some blocks, you will discover that action parameters and actions require less effort than adding geometric constraints and constraint parameters. Decide which dynamic block commands and options are appropriate for the blocks you create.

A combination of dynamic properties is also effective. For example, parameters and actions such as alignment, array, and flip offer dynamic controls that are often not possible using geometric constraints and constraint parameters. **Figure 27-14** shows how adding an alignment parameter to the cut framing member block allows you to size and align instances of the block.

Using Geometric Constraints

The geometric constraint commands and options available in the **Block Editor** are identical to those you use to constrain a parametric drawing geometrically. The tools in the **Geometric** panel of the **Parametric** ribbon tab are duplicates of the tools in the **Geometric** panel of the **Block Editor** ribbon tab. Use the geometric constraints in the **Block Editor** as you would in the drawing environment, including the options for

Figure 27-13.
A—A cut framing member block made dynamic using geometric constraints and constraint parameters. B—Using the default 2x4 block to create a 4x4 symbol.

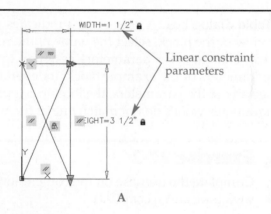

A

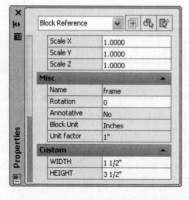
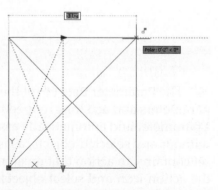

B

Figure 27-14.
Using geometric constraints and constraint parameters to adjust the size of a cut framing member symbol. An alignment parameter aligns each member for specific applications.

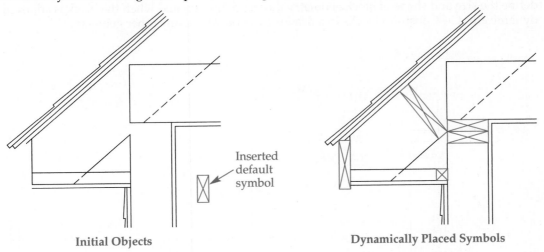

Inserted default symbol

Initial Objects

Dynamically Placed Symbols

relaxing and deleting constraints. The same shortcut menu, **Constraint Settings** dialog box, and **Properties** palette functions apply. Review Chapter 22 for information on adding geometric constraints.

Assign constraints to block objects before you define the block or during block editing to create a dynamic block. See **Figure 27-15A**. Once you define and insert the block, only constraint parameters, action parameters, or actions influence geometric constraints. This allows you to use blocks as objects in parametric drawings. For example, you can insert and rotate the block, as shown in **Figure 27-15B**, even though the block definition includes a horizontal constraint. Use constraints in the drawing to locate blocks and establish geometric relationships between blocks and other objects. See **Figure 27-15C**.

Use geometric constraints in the block environment to form geometric constructions in specific situations when standard AutoCAD commands are inefficient or ineffective.

Exercise 27-3

Complete the exercise on the companion website.
www.g-wlearning.com/CAD

Using Constraint Parameters

Constraint parameters replace dimensional constraints in the **Block Editor**. To help avoid confusion, remember that dimensional constraints constrain a parametric drawing, including block references, as shown in **Figure 27-15C**. Constraint parameters constrain the size and location of block components. By default, dimensional constraints are gray and constraint parameters are blue. You also have the option of converting dimensional constraints to constraint parameters.

You can often use constraint parameters instead of action parameters and actions. If you do not use action parameters, you must include constraint parameters to create

Figure 27-15.
A—A wide flange block made dynamic using geometric constraints and constraint parameters. B—You can rotate the block reference in the drawing, because the constraints define the size and shape of block geometry during definition and when the block is adjusted dynamically. C—Constrain blocks in a drawing as you would any other geometry.

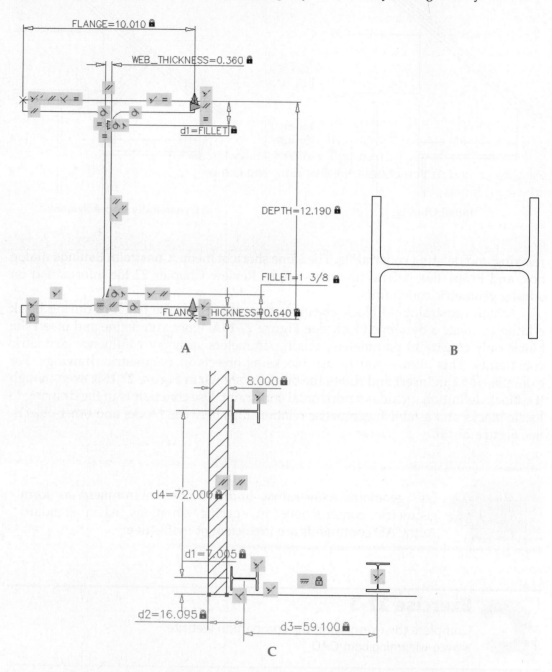

A dynamic block. The constraint parameter commands and options available in the **Block Editor** function much like those you use to constrain a parametric drawing dimensionally. Review Chapter 22 for information on adding dimensional constraints.

The **BCPARAMETER** command replaces the **DIMCONSTRAINT** command in the **Block Editor** and provides **Linear**, **Horizontal**, **Vertical**, **Aligned**, **Diameter**, and **Radius** options. The **Linear** option is the default in the **Block Editor** ribbon tab. You can also use the **BCPARAMETER** command to convert dimensional constraints to constraint parameters. Each constraint parameter is a separate **DIMCONSTRAINT** command option. The quickest way to add or convert constraint parameters using the **DIMCONSTRAINT**

Ribbon
**Block Editor
> Dimensional**

Linear

Type
**BCPARAMETER
> (option)**

command is to pick the appropriate button from the **Dimensional** panel of the **Block Editor** ribbon tab.

The process of adding constraint parameters is identical to that for adding dimensional constraints, except that constraint parameters can include grips. Constraint parameters are essentially a combination of dimensional constraints and action parameters. The constraint parameters given custom names in **Figure 27-16** are those that can be adjusted for specific block references. As when creating a parametric drawing, the other constraint parameters are required to define the block and define specific geometric relationships. Notice the expressions applied to these values.

To create a constraint parameter, follow the prompts to make the required selections, pick a location for the dimension line, and enter a value to form the constraint. When prompted, specify the number of grips. The radius constraint parameter allows you to add 0 or 1 grip. All other constraint parameters can include 0, 1, or 2 grips. If you plan to assign a single grip to a constraint parameter, select the point associated with the grip second. If you choose the **0** option, you can only use the **Properties** palette to adjust the block.

NOTE

If you attempt to over-constrain a block, a message appears indicating that adding the geometric constraint or constraint parameter is not allowed. You cannot create reference constraint parameters.

PROFESSIONAL TIP

As when adding dimensional constraints or action parameters, change the constraint parameter name to a custom, more descriptive name. Naming labels helps organize parameters and identify each parameter when you control the block dynamically. Custom parameters also appear in the **Custom** category of the **Properties** palette.

Figure 27-16.
Using constraint parameters to form a dynamic block of a spacer. A single grip is all that is required for each constraint parameter. Do not assign or rename constraint parameters that do not control geometry.

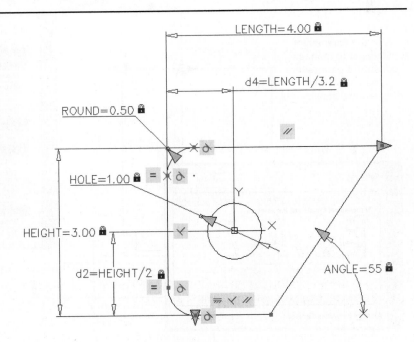

Ribbon

Block Editor
> Dimensional

Convert

Type

BCPARAMETER
> Convert

Use the **Convert** option of the **BCPARAMETER** command to convert a dimensional constraint to a constraint parameter. This allows you to prepare a dynamic block using existing dimensional constraints. Access the **Convert** option and pick the dimensional constraint to convert. The dimensional constraint becomes the corresponding constraint parameter and includes the default number of grips.

Controlling Constraint Parameters

Control and adjust constraint parameters using a combination of the same techniques you use to manage dimensional constraints and action parameters. Many of the options from shortcut menus, the **Constraint Settings** dialog box, and the **Properties** palette apply. Right-click with no objects selected to access options for displaying and hiding parametric constraints and for accessing the **Constraint Settings** dialog box. Select a constraint parameter and then right-click to display a shortcut menu with options for editing the constraint, changing the name format, and redefining the grips.

As with dimensional constraints and the action parameters, the **Properties** palette provides an effective way to control and enhance constraint parameters. You can also use the **Parameters Manager**. Figure 27-17 shows a foundation detail block with linear constraint parameters. Notice the multiple options available in the **Properties** palette for adjusting the selected constraint parameter.

Use the options in the **Value set** category of the **Properties** palette to assign value sets to a constraint parameter. Each constraint parameter in the Figure 27-17 example uses an incremental value to help ensure that you select an appropriate value when adjusting a block reference. You can also create a list of possible sizes. The processes of creating a value set in the **Properties** palette and using value sets are identical for constraint parameters and action parameters.

Figure 27-17.

Adjust constraint parameters as you would dimensional constraints and action parameters. Use the **Properties** palette to add value sets.

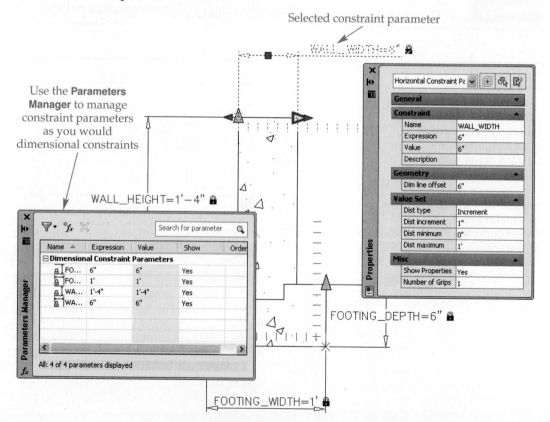

Exercise 27-4

Complete the exercise on the companion website.
www.g-wlearning.com/CAD

Additional Parametric Tools

The **Block Editor** offers additional options for adding constraints to blocks. Many of the tools, such as the **DELCONSTRAINT** command, function the same in block editing mode as in drawing mode. However, the **Block Editor** does offer some unique parametric construction commands.

The **BCONSTRUCTION** command allows you to create construction geometry to aid geometric construction and constraining. Construction geometry appears only in the block definition. See **Figure 27-18**. Access the **BCONSTRUCTION** command and select the objects to convert to or revert from construction geometry. Press [Enter] or the space bar, or right-click and pick **Enter**. Next, choose the **Convert** option to convert non-construction objects to the construction format, or choose **Revert** to return construction geometry to the standard format. You can also use the **Hide all** option to hide all existing construction geometry before selecting objects, or use the **Show all** option to display all construction geometry.

Use the **BCONSTATUSMODE** command to toggle constraint status identification on and off. When you turn constraint status mode on, objects with no constraints appear white (black) by default, objects assigned some form of constraints are blue, and fully constrained geometry is magenta. If the block contains a constraint error, objects associated with the error are red. Using constraint status is helpful, especially if you want to constrain objects in a certain order or confirm that geometry has been fully constrained.

Ribbon
Block Editor
>Manage
Construction
Geometry

Type
BCONSTRUCTION

BCONSTRUCTION

Ribbon
Block Editor
> Manage
Constraint Display
Status

Type
BCONSTATUSMODE

CONSTATUSMODE

> **NOTE**
>
> Use the **BESETTINGS** command to access the **Block Editor Settings** dialog box. There you can adjust parameter and parameter grip color and appearance, constraint status colors, and other **Block Editor** settings.

Figure 27-18.
A weld nut block in which construction geometry aids geometric construction and constraining.

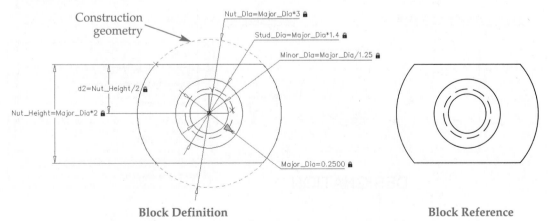

Block Definition   Block Reference

Chapter 27   Additional Dynamic Block Tools

855

block properties table: A table of action parameters and/or constraint parameters that allows you to create multiple block properties and then select them to create block references.

A *block properties table* allows you to assign specific values to multiple block properties, and then select a specific group, or row, of properties to create block references. The concept is similar to using a lookup action parameter. A block properties table can include action parameters, constraint parameters, or both. You can also add attributes to the table, which is often appropriate for naming each record, or row.

Figure 27-19 shows the block of the front view of a heavy hex nut in the **Block Editor**. The block includes an appropriate level of constraints and includes constraint parameters to direct dynamic changes. The block also includes an invisible and preset attribute for defining the designation of each different nut and, as shown in the **Parameters Manager**, a user-defined parameter for the nut thickness.

PROFESSIONAL TIP

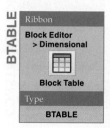

It is critical that you assign the **Preset** mode to attributes that you include in a block properties table. This allows the attribute value to adjust to the selected block record. The **Preset** mode requires no default value, and you will not receive a prompt to adjust the value.

BTABLE

Ribbon
Block Editor > Dimensional
Block Table
Type
BTABLE

After you create parameters and attributes, access the **BTABLE** command and select the parameter location. Then enter the number of grips to associate with the parameter. The default **1** option creates a single grip that allows you to select a table record from the grip shortcut menu. If you choose the **0** option, you can only use the **Properties** palette to select a record. The **Palette** option, available before you specify the parameter location or from the **Properties** palette, determines whether the label appears in the **Properties** palette when you select the block reference. The **Block Properties Table** dialog box appears, allowing you to create a block properties table. See **Figure 27-20**.

Creating a Block Properties Table

A block properties table groups the properties of parameters into custom records, or rows. To add parameter properties, pick the **Add Properties...** button to open the **Add Parameter Properties** dialog box. See **Figure 27-21**. All parameters in the block

Figure 27-19.
A heavy hex nut block definition ready to use to create a block properties table.

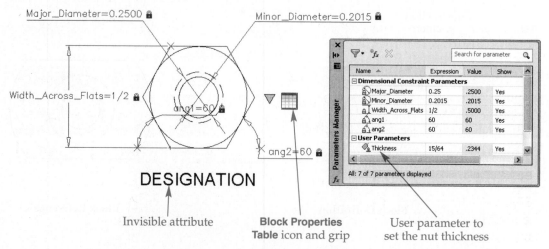

Invisible attribute

Block Properties Table icon and grip

User parameter to set the nut thickness

Figure 27-20.
The **Block Properties Table** dialog box.

Pick to add
block properties

Pick to specify
a user property

Block
properties
appear here

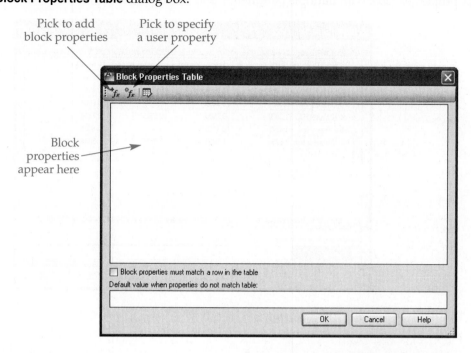

Figure 27-21.
Parameter
properties are listed
in the **Add Parameter
Properties** dialog
box.

Select the
properties
to include
in the table

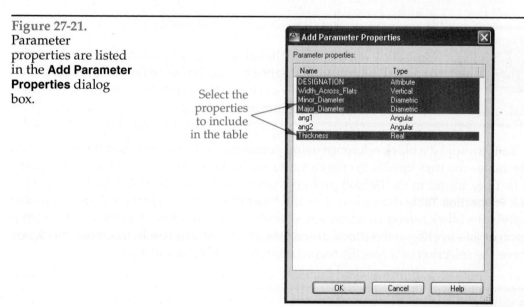

that contain property values appear in the **Parameter properties:** list. Lookup, alignment, and base point parameters do not contain property values. Notice that the property name is the parameter label.

To add parameter properties to the table, select the properties in the **Parameter properties:** list and pick the **OK** button. A column appears in the table for each parameter property. Type a value in each cell in the column. A new row forms automatically when you enter a value in a cell. See **Figure 27-22.** Press [Enter], [Tab], [Shift]+[Enter], or the arrow keys, or pick in a different cell to navigate through the table.

For the nut block example, complete the table as shown in **Figure 27-22.** The **DESIGNATION** column references the attribute property. The value you enter in the **DESIGNATION** text box in each row specifies the record name. This value appears in the shortcut menu when you insert the block and select the block properties table parameter grip.

Figure 27-22.
A block properties table with multiple parameters and values added.

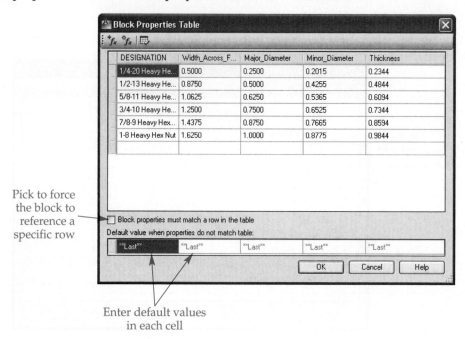

DESIGNATION	Width_Across_F...	Major_Diameter	Minor_Diameter	Thickness
1/4-20 Heavy He...	0.5000	0.2500	0.2015	0.2344
1/2-13 Heavy He...	0.8750	0.5000	0.4255	0.4844
5/8-11 Heavy He...	1.0625	0.6250	0.5365	0.6094
3/4-10 Heavy He...	1.2500	0.7500	0.6525	0.7344
7/8-9 Heavy Hex...	1.4375	0.8750	0.7665	0.8594
1-8 Heavy Hex Nut	1.6250	1.0000	0.8775	0.9844

Pick to force
the block to
reference a
specific row

☐ Block properties must match a row in the table
Default value when properties do not match table:

Last	**Last**	**Last**	**Last**	**Last**

OK   Cancel   Help

Enter default values
in each cell

Right-click on a column heading to access a menu with options for adjusting columns. Right-click on a row to access a menu with options for adjusting rows. The options are the same as those for adjusting lookup table columns and rows.

You can adjust a block reference using parameter values other than those specified in the table. You may be able to enter a value, such as the value of an attribute property, in a text box found in the **Default** property when values do not match table area of the **Block Properties Table** dialog box. Use the ****Last**** option to use the value assigned to the previous block reference when you specify a value not found in the table. Often it is appropriate to choose the **Block properties must match a row in the table** check box to force the selection of a specific record, matching all values in a row.

It is critical that all block definition values match the values specified in the default block row in the block properties table, especially if you force the selection of a specific record.

After you add all required properties to the table and assign values to each, pick the **Audit** button in the **Block Properties Table** dialog box to check each record in the table. Make sure the records are unique and that there are no discrepancies between the block definition and the table values. If AutoCAD does not find errors, pick the **OK** button to return to the **Block Editor**. Test and save the block, and exit the **Block Editor**. The dynamic block is now ready to use.

To redisplay the **Block Properties Table** dialog box, double-click on the parameter, or access the **BTABLE** command.

Using a Block Properties Table Dynamically

Figure 27-23 shows the inserted nut block example selected for editing. Since the block table parameter includes a grip, a grip appears that you can select to choose a specific block style. The entries in the grip menu match the rows in the **Block Properties Table** dialog box. A check mark indicates the current record. To switch to a different view of the block, select the name of the record from the list. You can also pick the **Properties Table...** option to display the **Block Properties Table** in drawing mode. Double-click a row to activate it. In this example, no other grips were assigned to blocks. This makes the table and the **Properties** palette the only two methods to select a block reference format.

PROFESSIONAL TIP

The options for developing dynamic blocks and creating parametric drawings can become confusing. Keep the following concepts in mind as you proceed:

- Use constraints as an alternative or in addition to action parameters and actions.
- Constraints allow you to create a parametric drawing or a dynamic block.
- Assign constraints to create a dynamic block during block definition or while editing the block.
- Treat inserted blocks like any other object when preparing a parametric drawing.

Exercise 27-5

Complete the exercise on the companion website.
www.g-wlearning.com/CAD

Figure 27-23.
The block table parameter grip displays when you select the block. The list of available records appears when you pick the parameter grip. Notice the correlation between the available options and the names in the **Block Properties Table** dialog box.

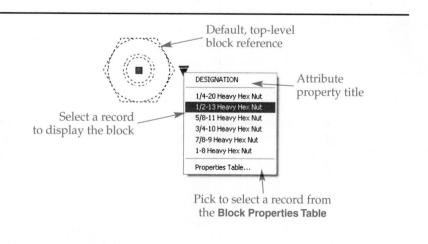

Express Tools
Chapter 27

Block Express Tools
The **Block** and **Draw** panels of the **Express Tools** ribbon tab include additional commands related to blocks. For information about the most useful block express tools, go to the companion website (www.g-wlearning.com/CAD), select this chapter, and select **Block Express Tools.**

Chapter Review

Answer the following questions. Write your answers on a separate sheet of paper or complete the electronic chapter review on the companion website.
www.g-wlearning.com/CAD

1. Define *visibility parameter.*
2. What are visibility states?
3. How do you display the shortcut menu that allows you to select from the existing visibility states of a block?
4. When you select a block and display the visibility parameter shortcut menu, what indicates the current visibility state?
5. Briefly describe a lookup parameter.
6. Explain the basic function of a lookup action.
7. Identify the basic function of a lookup table.
8. What is a parameter set?
9. What ribbon tab, in addition to the **Parametric** ribbon tab, contains geometric constraint tools?
10. When can you assign constraints to block objects to create a dynamic block?
11. What takes the place of dimensional constraints in the **Block Editor**?
12. What command and option allow you to convert a dimensional constraint to a constraint parameter?
13. How can you create construction geometry to aid geometric construction in the **Block Editor**?
14. Explain how to toggle constraint status identification on and off in the **Block Editor.**
15. What does a block properties table allow you to do?

Drawing Problems

Start AutoCAD if it is not already started. Start a new drawing for each problem using an appropriate template of your choice. The template should include layers and text, dimension, multileader, and table styles, when necessary, for drawing the given objects. Add layers and text, dimension, multileader, and table styles as needed. Draw all objects using appropriate layers and text, dimension, multileader, and table styles, justification, and format. Follow the specific instructions for each problem. Use only drawing and editing commands and techniques you have already learned. Use your own judgment and approximate dimensions when necessary. Apply dimensions accurately using ASME or appropriate industry standards.

Note: *Constraint parameters shown for reference are created using AutoCAD and may not comply with ASME standards.*

Note: *Some of the problems in this chapter are built on problems from previous chapters. If you have not yet completed those problems, complete them now.*

▼ Basic

1. Use **DesignCenter** to insert the HEAVY HEX NUT block you created in Exercise 27-5. Insert or copy the block to create six total symbols. Use the block table parameter to display each size nut as shown. Save the drawing as P27-1.

2. Open P22-2 and save as P27-2. The P27-2 file should be active. Create a block named FIXTURE, select all objects, and pick the center of the circle as the insertion base point. Open the block in the **Block Editor** and convert the dimensional constraints to constraint parameters. Insert the FIXTURE block into the drawing three times to create the 4x4, 6x6, and 4x5 symbols as shown. Resave the drawing.

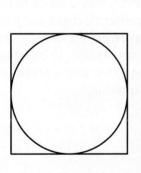 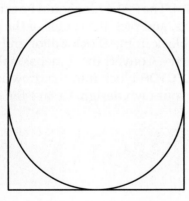 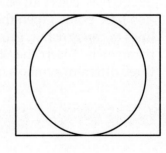

3. Open P22-6 and save as P27-3. The P27-3 file should be active. Create a block named PLATE, select all objects and pick the center of the plate as the insertion base point. Open the block in the **Block Editor** and convert the construction rectangle to construction geometry. Convert the dimensional constraints to constraint parameters. Insert the PLATE block into the drawing and create the drawing shown. Do not add dimensions. Resave the drawing.

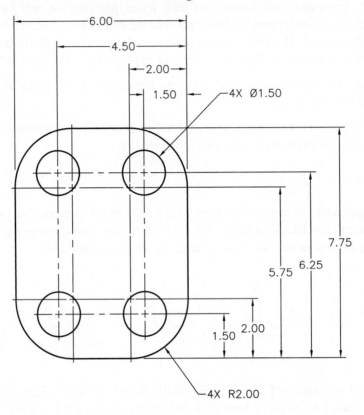

▼ Intermediate

4. Open P22-5 and save as P27-4. The P27-4 file should be active. Create a block named SELECTOR, select all objects, and pick the center of the center circle as the insertion base point. Open the block in the **Block Editor** and convert the construction lines to construction geometry. Convert the dimensional constraints to constraint parameters. Insert the SELECTOR block into the drawing three times and create three different symbols of your own design. Resave the drawing.

5. Create a single block that can be used to represent each of the three door blocks shown below. Name the block 30 INCH DOOR. Do not include labels. Create an appropriately named visibility state for each view: 90 OPEN, 60 OPEN, and 30 OPEN. Insert the 30 INCH DOOR block into the drawing three times. Set each block to a different visibility state. Save the drawing as P27-5.

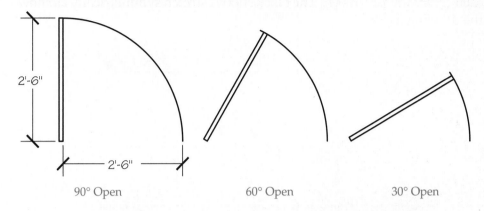

90° Open   60° Open   30° Open

6. Create a cut framing member block that can be used to represent each of the symbols shown below. Name the block FRAME. Add geometric constraints and constraint parameters as needed, and assign an alignment parameter. Use the block to create the portion of the detail shown. Save the drawing as P27-6.

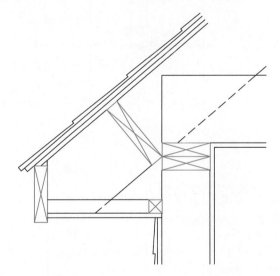

7. Open P25-6 and save it as P27-7. The P27-7 file should be active. Explode the glulam beam hanger blocks. Use the geometry and attributes as needed to create a single block using a visibility parameter. Stack the views on top of each other using an appropriate insertion base point location in the block definition so that each view occurs at the correct location when you select a visibility state. Insert the block multiple times and adjust the visibility to display each view. Resave the drawing.

8. Open P25-7 and save it as P27-8. The P27-8 file should be active. Explode the mudsill anchor blocks. Use the geometry and attributes as needed to create a single block using a visibility parameter. Stack the views on top of each other using an appropriate insertion base point location in the block definition so that each view occurs at the correct location when you select a visibility state. Insert the block multiple times and adjust the visibility to display each view. Resave the drawing.

Drawing Problems - Chapter 27

▼ Advanced

9. Create a foundation footing and wall block that can be used to represent various construction requirements. Name the block Foundation. Add the geometric constraints and constraint parameters shown. Include 1″ increment value sets for each constraint parameter. The block should stretch symmetrically as shown. Save the drawing as P27-9.

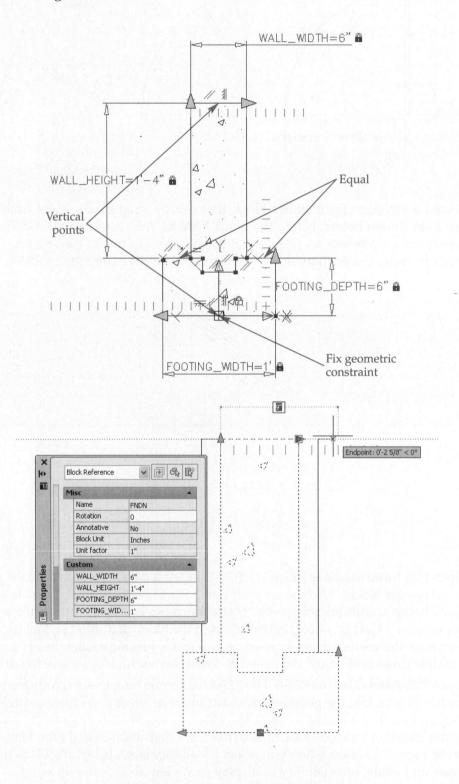

10. The bolt shown in the drawing below is available in four different lengths. As the length increases, the size of the bolt head increases for added strength. Create a dynamic block that will allow the length of the shaft and the size of the bolt head to be changed in a single operation.
    A. Draw the objects that make up the bolt and create a block named BOLT. Do not include dimensions.
    B. Insert a linear parameter along the length of the shaft from the bottom of the bolt head to the end of the shaft. Label it SHAFT LENGTH.
    C. Assign a stretch action to the SHAFT LENGTH parameter. Associate the action with the parameter grip at the end of the shaft. Create a crossing window around the end of the shaft that includes the threads. Select the end of the shaft, threads, and edges of the shaft.
    D. Insert a linear parameter along the depth of the bolt head (the .3″ dimension). Label it HEAD THICKNESS.
    E. Assign a scale action to the HEAD THICKNESS parameter and select the objects that compose the bolt head. Use an independent base point type and specify the midpoint of the vertical line where the shaft meets the bolt head.
    F. Insert a lookup parameter and assign a lookup action to it.
    G. Add the SHAFT LENGTH and the HEAD THICKNESS parameters to the lookup table. Complete the table with the following properties:

Shaft Length	Head Thickness	Lookup
1	0.3	1″ Length
1.5	0.333	1.5″ Length
2	0.366	2″ Length
2.5	0.4	2.5″ Length

    H. Set the table to allow reverse lookup, save the block, and exit the **Block Editor**.
    I. Insert the block four times into the drawing. Specify a different lookup property for each block.
    J. Save the drawing as P27-10.

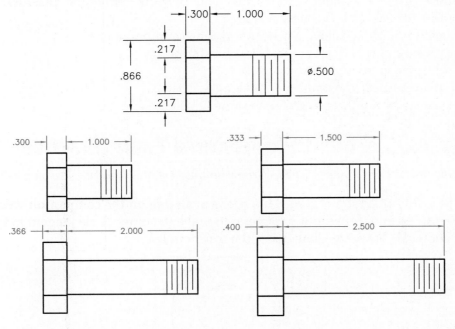

11. Repeat Problem 27-10, but this time use geometric constraints, constraint parameters, and a block properties table instead of action parameters. Save the drawing as P27-11.

AutoCAD Certified Associate Exam Practice

Answer the following questions. Write your answers on a separate sheet of paper.

1. How many visibility parameters can you assign to a block? *Select the one item that best answers the question.*
   A. 1
   B. 2
   C. 4
   D. 16
   E. unlimited number

2. Study the property lookup table for a pipeline valve. Which option would you choose from the lookup parameter grip to display the valve in the 45° position? *Select the one item that best answers the question.*
   A. Full Flow
   B. No Flow
   C. Normal Flow
   D. Reduced Flow

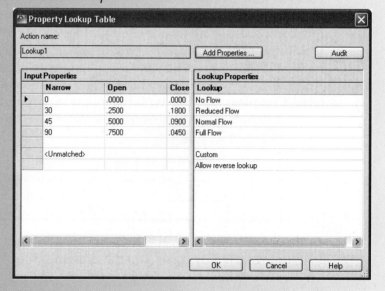

3. Which of the following actions can you perform using the **BCPARAMETER** command? *Select all that apply.*
   A. convert a dimensional constraint to constraint parameter
   B. create a perpendicular geometric constraint
   C. create a radius constraint parameter
   D. place a linear constraint parameter
   E. place a parallel geometric constraint

AutoCAD Certified Professional Exam Practice

Follow the instructions in the problem. Write your answers on a separate sheet of paper.

1. **Open CPE-27constraint.dwg. This file is available on the companion website.** Use the existing constraint parameter to make the outer circle tangent to the line, as shown. What is the diameter of the inner circle?

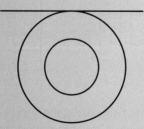

Learning Objectives

After completing this chapter, you will be able to:

✓ Describe the purpose for and proper use of layouts.
✓ Begin to prepare layouts for plotting.
✓ Manage layouts.
✓ Use the **Page Setup Manager** to define plot settings.
✓ Use plot styles and plot style tables.

You may often print or plot a drawing in model space to make a quick hard copy for check or reference purposes. Ordinarily, however, you create a drawing in model space and then lay out the drawing for plotting in paper space. Hard-copy plots are required for a variety of reasons. For example, it is typically easier for workers in a machine shop or a construction crew in the field to refer to a print than to use a computer to view the drawing file. Most of the same steps you follow to prepare a drawing for plotting apply to exporting a drawing to an electronic format.

Introduction to Layouts

The first step to make an AutoCAD drawing is to create a *model* in *model space*. See **Figure 28-1A**. Model space is usually active by default. You have been using model space throughout this textbook to create objects and dimensioned drawing views. Once you complete a model, use a *layout* in *paper space* to prepare the final drawing for plotting. See **Figure 28-1B**. A layout represents the sheet of paper used to lay out and plot a drawing. A layout often includes the following items, depending on the drawing:

- Floating viewports to display content from model space
- Border
- Sheet blocks such as a title block and revision history block
- General notes
- Bill of materials, parts list, schedules, legend, and other associated lists
- Sheet annotation and symbols such as titles, north arrow, and graphic scale
- Page setup information

model: A 2D or 3D drawing or model composed of various objects, such as lines, circles, and text, usually created at full size.

model space: The environment in AutoCAD in which you create drawings and designs.

layout: A specific arrangement of views or drawings for plotting or printing on paper.

paper space: The environment in AutoCAD in which you create layouts.

Figure 28-1.
A—Design and draft objects in model space. B—Use paper space to finalize and lay out drawings and designs on paper for plotting or export.

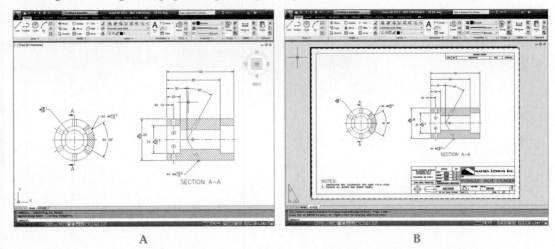

A   B

floating viewport:
A viewport added to a layout in paper space to display objects drawn in model space.

A major element of the layout system is the *floating viewport*. Consider a layout to be a virtual sheet of paper and a floating viewport as a window cut into the paper to show objects drawn in model space. In **Figure 28-1B**, a single viewport exposes objects drawn in model space. Draw the floating viewport on a layer that you can turn off or freeze so the viewport does not plot and is not displayed on-screen. Chapter 29 explains using floating viewports.

Layouts with floating viewports offer the ability to construct properly scaled drawings. A single drawing can have multiple layouts, each representing a different paper space, or plot, definition. Each layout can include multiple floating viewports to provide additional or alternate drawing views, prepared at different scales if necessary. You can use a single drawing file to prepare several different final drawings and drawing views. For example, an architectural drawing file might include several details that are too large to place on a single sheet of paper. You can use multiple layouts, and if necessary differently scaled floating viewports, to prepare as many sheets as needed to plot all of the details in the drawing.

Working with Layouts

Learn to use tools and options for displaying and managing layouts before you prepare a layout for plotting. The layout and model tabs and model space and paper space tools in the status bar are available by default. These are the most effective tools for navigating between model space and layouts, and for managing layouts. You can also type or enter MODEL to return to model space from a layout.

Model and Layout Tabs

The model and layout tabs appear directly below the drawing window. See **Figure 28-2A**. The model space tab is to the left of the layout tabs. Layout tabs are arranged in the order they were created, from left to right. If the drawing includes so many layouts that the tabs spread past the screen, use the forward and reverse buttons to the left of the tabs to access model space or the appropriate layout. Hover over an inactive tab to display a preview of the contents. Pick a layout tab to enter paper space with the selected layout current, or pick the **Model** tab to re-enter model space.

Figure 28-2.
A—Using the layout and model tabs to activate model space and paper space. B—Options available when you right-click on a tab.

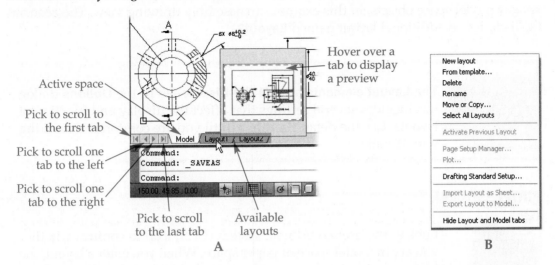

A

B

Right-click on a tab to access a shortcut menu with options for controlling layouts and the layout and model tabs. See **Figure 28-2B**. Pick **Activate Model Tab** to enter model space. Select **Activate Previous Layout** to make the previously current layout current. Pick **Select All Layouts** to select all layouts in the drawing for purposes such as publishing or deleting. The shortcut menu is also the primary resource for adding layouts and moving, renaming, and deleting existing layouts.

Figure 28-3 shows an example of using the supplied acad.dwt drawing template and picking the **Layout1** tab to display the default **Layout1** layout. The layout uses

Figure 28-3.
Pick the **Layout1** tab to display **Layout1** provided in the default acad.dwt template.

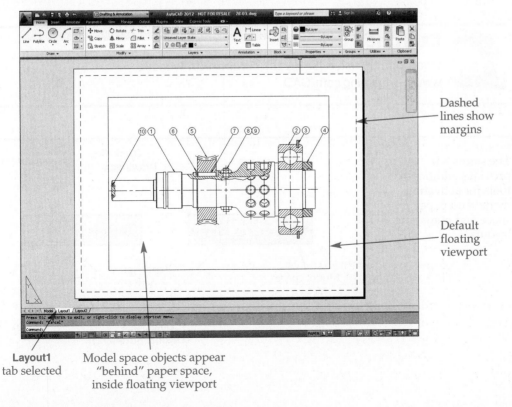

Dashed lines show margins

Default floating viewport

Layout1 tab selected

Model space objects appear "behind" paper space, inside floating viewport

default settings based on an 8.5″ × 11″ sheet of paper in a landscape (horizontal) orientation. The white rectangle you see on the gray background is a representation of the sheet. Dashed lines mark the sheet *margin*. A large, rectangular floating viewport reveals model space objects, in this example an assembly drawing view. The acad.dwt file includes an additional layout named **Layout2**.

margin: The extent of the printable area; objects drawn past the margin (dashed lines) do not print.

The **Layout elements** area of the **Display** tab in the **Options** dialog box includes several settings that affect the display and function of layouts. Use the default settings until you are comfortable working with layouts.

PROFESSIONAL TIP

Look at the user coordinate system (UCS) icon to confirm whether you are in model space or paper space. When you enter a layout, the UCS icon changes from two lines to a triangle that indicates the X and Y coordinate directions.

Model and Paper Buttons

The status bar provides other convenient tools for managing layouts. See Figure 28-4. The **Quick View Layouts** and **Quick View Drawings** tools are described later in this chapter. Pick the **MODEL** button to exit model space and enter paper space. If the file contains multiple layouts, the top-level layout is displayed unless you previously accessed a different layout. While a layout is active, pick the **PAPER** button to use the **MSPACE** command, which activates a floating viewport. The **MSPACE** command does not return you to model space. Select the **PAPER** button to deactivate a floating viewport.

Type
MSPACE

Exercise 28-1

Complete the exercise on the companion website.
www.g-wlearning.com/CAD

Figure 28-4.
The status bar provides additional tools for activating model and paper space and managing layouts.

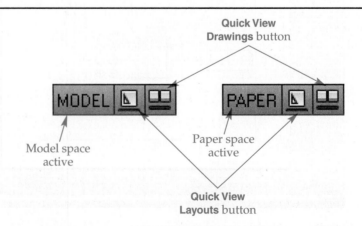

The Quick View Layouts Tool

The **Quick View Layouts** tool is similar to the layout and model tabs, but provides additional options and a visual format for displaying and adjusting layouts in the current file. The quickest way to access the **Quick View Layouts** tool is to pick the **Quick View Layouts** button on the status bar. The **Quick View Layouts** tool appears in the lower center of the AutoCAD window. See **Figure 28-5**.

The **Model** thumbnail image is on the left, followed by layout thumbnails in the order they were created, from left to right. If the drawing includes so many layouts that thumbnails spread past the screen, hover the cursor over the furthest right and left thumbnails to scroll though the options. Hover over a thumbnail to highlight the image and show additional options. See **Figure 28-6**. To adjust the size of the thumbnails, press and hold [Ctrl] then roll the mouse wheel forward to enlarge, or back to reduce. Pick a layout thumbnail to enter paper space with the selected layout current, or pick the **Model** thumbnail to re-enter model space.

Type
QVLAYOUT

> Icons represent model and layout thumbnails until you enter a layout for the first time (initialize the layout). The icon then changes to a thumbnail image of the layout.

The **Quick View Layouts** tool includes a small toolbar below the thumbnail images, as shown in **Figure 28-6**. By default, the **Quick View Layouts** tool disappears when you pick a thumbnail to switch layouts or enter model space. To keep the tool on-screen

Figure 28-5.
The **Quick View Layouts** tool offers an effective visual method for changing between model space and paper space and provides options for managing layouts.

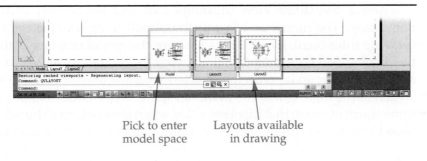

Pick to enter model space

Layouts available in drawing

Figure 28-6.
Hover over a thumbnail image to display **Plot...** and **Publish...** buttons.

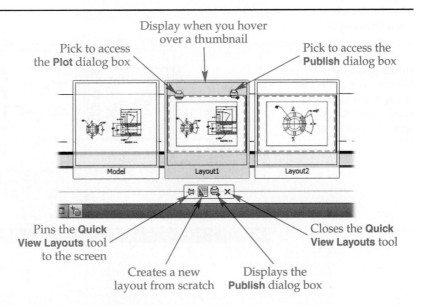

Display when you hover over a thumbnail

Pick to access the **Plot** dialog box

Pick to access the **Publish** dialog box

Pins the **Quick View Layouts** tool to the screen

Closes the **Quick View Layouts** tool

Creates a new layout from scratch

Displays the **Publish** dialog box

after you select a thumbnail, pick the **Pin Quick View Layouts** button. Pick the **New Layout** button to create a new layout from scratch, as described later in this chapter. Pick the **Publish…** button to access the **Publish** dialog box, explained later in this textbook. Select the **Close** button to exit the **Quick View Layouts** tool.

If you pin the **Quick View Layouts** tool to the screen, close the tool and then access the tool again, the **Quick View Layouts** tool will still be in the pinned state.

Right-click on a thumbnail image to access the same shortcut menu available when you right-click on the model or a layout tab, excluding the **Hide Layout and Model tabs** option. See **Figure 28-2B**. You can also access some menu options from a shortcut menu that displays when you right-click directly on the **Quick View Layouts** button on the status bar. An option selected from this menu applies to the current file and the current layout.

The Quick View Drawings Tool

Type

QVDRAWING

The **Quick View Drawings** tool provides the same features for working with layouts as the **Quick View Layouts** tool, but allows you to manage the layouts in all open files without changing drawing windows. Use **Quick View Drawings** tool to increase productivity when you are working between existing drawings. The quickest way to access the **Quick View Drawings** tool is to pick the **Quick View Drawings** button on the status bar. Refer to Chapter 2 for information on basic **Quick View Drawings** tool features, such as using the tool to work with multiple open documents.

Access the **Quick View Drawings** tool and hover over a drawing file thumbnail image. The model and layout thumbnail images appear above the highlighted drawing thumbnail. Move the cursor over the model or a layout thumbnail to enlarge the display. See **Figure 28-7**. Pick a layout thumbnail to switch to the highlighted file and enter paper space with the selected layout active, or pick the **Model** thumbnail to switch to the highlighted file in model space. Right-click on a thumbnail to access the same shortcut menu that displays when you right-click on a thumbnail using the **Quick View Layouts** tool. Right-click on a thumbnail to make the associated file current.

Figure 28-7.
Use the **Quick View Drawings** tool to manage layouts located in other open files. Hover over a file thumbnail to display model and layout thumbnails for the file.

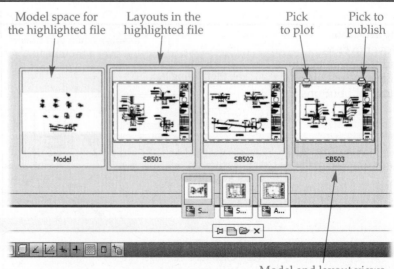

Model space for the highlighted file

Layouts in the highlighted file

Pick to plot

Pick to publish

Model and layout views enlarge when hovered over

The **LAYOUT** command allows you to accomplish the same tasks as the **Quick View Layouts** and **Quick View Drawings** tools. For most applications, the **Quick View Layouts** and **Quick View Drawings** tools are much easier and more convenient to use than the **LAYOUT** command.

Adding Layouts

To add a new layout to a drawing, you can create a new layout from scratch, use the **Create Layout** wizard, or reference an existing layout. Referencing an existing, preset layout is often the most effective approach. You can also insert a layout from a different DWG, DWT, or DXF file into the current file or create a copy of a layout in the current file.

Starting from Scratch

To create a new layout from scratch, right-click on the model or a layout tab, a **Quick View Layouts** or **Quick View Drawings** thumbnail image, or the **Quick View Layouts** button on the status bar, and pick **New Layout**. A new layout appears on the far right of the layout list. The settings applied to the new layout depend on the template used to create the original file. The name of the layout is set according to the names of other existing layouts. For example, when you add a new layout to a default drawing started from the acad.dwt template, a new layout named **Layout3** appears and includes an 8.5″ × 11″ sheet of paper, a landscape (horizontal) orientation, and a large floating viewport.

Using the Create Layout Wizard

Use the **Create Layout** wizard to build a layout from scratch using values and options you enter in the wizard. The wizard provides options for naming the new layout and selecting a printer, paper size, drawing units, paper orientation, title block, and viewport configuration. The pages of the wizard guide you through the process of developing the layout.

Type
LAYOUTWIZARD

Using a Template

To create a new layout from a layout stored in an existing DWG, DWT, or DXF file, right-click on the model or a layout tab, a **Quick View Layouts** or **Quick View Drawings** thumbnail image, or on the **Quick View Layouts** button on the status bar, and pick **From Template...**. The **Select Template From File** dialog box appears. See Figure 28-8A. The Template folder in the path set by the AutoCAD Drawing Template File Location appears by default. Select the file containing the layout you want to add to the current drawing and pick the **Open** button. The **Insert Layout(s)** dialog box appears, listing all layouts in the selected file. See Figure 28-8B. Highlight the layout(s) to copy and pick the **OK** button.

Using DesignCenter

DesignCenter provides an effective way to add existing layouts to the current drawing. Use the folder list to locate and select a drawing or template file. Pick the Layouts branch in the folder list, or double-click on the **Layouts** icon in the content pane. See Figure 28-9. Select one or more layouts to copy from the content pane and then drag and drop, or use the **Add Layout(s)** or **Copy** option from the shortcut menu to insert the layouts into the current drawing.

Copying and Moving Layouts

To create a copy of a layout, right-click on a layout tab, a **Quick View Layouts** or **Quick View Drawings** thumbnail image, or make the layout to copy current and right-click on the **Quick View Layouts** button on the status bar. Then pick **Move or Copy...** to display the **Move or Copy** dialog box. See Figure 28-10. To create a copy, select the

Figure 28-8.
Adding a layout using an existing layout stored in a different drawing, drawing template, or DXF file. A—Select the file containing the layout. B—Highlight the layout to add to the current drawing.

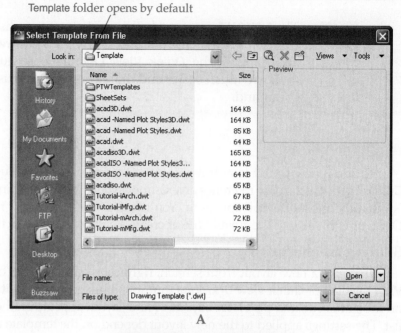

Template folder opens by default

A

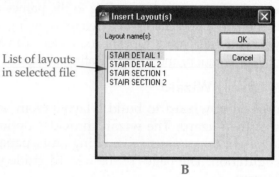

List of layouts in selected file

B

Figure 28-9.
Using **DesignCenter** to share layouts between drawings.

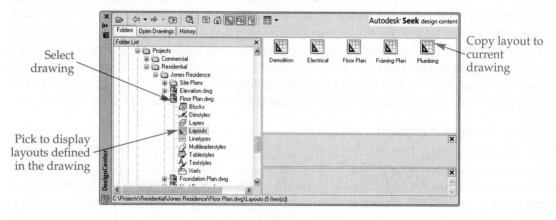

Select drawing

Pick to display layouts defined in the drawing

Copy layout to current drawing

Create a copy check box and pick the layout that will appear to the right of the new layout, or pick **(move to end)** to place the copy to the right of all other layouts. The default name of the new layout is the name of the current or selected layout plus a number in parentheses.

To move a layout using the **Move or Copy...** dialog box, uncheck the **Create a copy** check box. When you add and rename layouts, the layouts do not automatically rearrange

AutoCAD and Its Applications—Basics

Figure 28-10.
The **Move or Copy**
dialog box allows
you to reorganize
layouts and copy
layouts within a
drawing.

Select location
of new layout tab

Check to create
a copy of the
current layout

Click to move a new or existing
layout to the end of the list

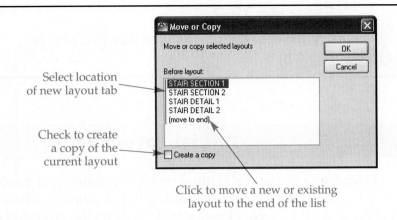

into a predetermined order. Organize layouts in an appropriate order to reduce confusion
and aid in the publishing process, described in Chapter 33.

Renaming Layouts

Layouts are easier to recognize and use when they have descriptive names. To
rename a layout, right-click on a layout tab or a **Quick View Layouts** or **Quick View
Drawings** thumbnail image and pick **Rename**. You can also slowly double-click on the
current name to activate it for editing. When the layout name highlights, type a new
name and press [Enter].

Deleting Layouts

To delete an unused layout from the drawing, right-click on the layout tab or the
Quick View Layouts or **Quick View Drawings** thumbnail image and pick **Delete**. An
alert message warns that the layout will be deleted permanently. Pick the **OK** button
to remove the layout.

Exporting a Layout to Model Space

The **EXPORTLAYOUT** command allows you to save the entire layout display as
model space objects in a separate DWG file. The **EXPORTLAYOUT** command produces
a "snapshot" of the current layout display that you can use for applications in which it
is necessary to combine model space and paper space objects, such as when exporting
a file as an image. Model space and paper space are not exported together as an image.

A quick way to access the **EXPORTLAYOUT** command is to right-click on a layout
tab or a **Quick View Layouts** or **Quick View Drawings** thumbnail image and pick **Export
Layout to Model....** The **Export Layout to Model Space** dialog box appears and functions much like the **Save As** dialog box. Pick a location for the file, use the default file
name or enter a different name, and pick the **SAVE** button. Everything shown in the
layout, including objects drawn in model space, are converted to model space and are
saved as a new file.

Application	
Print Menu	
Save Layout as a Drawing	
Ribbon	
Output >Export	
Type	
EXPORTLAYOUT	

CAUTION

The **EXPORTLAYOUT** command eliminates the relationship between
model space and paper space. Export a layout only when it is necessary to export model space and paper space together as a single
unit. Use a descriptive name to differentiate between the exported
layout file and the original file.

Exercise 28-2

Complete the exercise on the companion website.
www.g-wlearning.com/CAD

Initial Layout Setup

Preparing a layout for plotting involves creating and modifying floating viewports, adjusting plot settings, and adding layout content such as annotation, symbols, a border, and a title block. This chapter focuses on the process of preparing layouts for plotting using the **Page Setup Manager**. When you complete this initial phase, you will be better prepared to add content to layouts and create and manage floating viewports, as described in Chapter 29.

page setup: A saved collection of settings required to create a finished plot of a drawing.

A *page setup* establishes most of the settings that determine how a drawing plots. Plot settings include plot device selection, paper size and orientation, plot area and offset, plot scale, and plot style. The **Page Setup Manager** and related **Page Setup** dialog box allow you to create and modify saved page setups that control how layouts appear on-screen and plot. This is where initial layout setup occurs. You then use the **Plot** dialog box to create the actual plot using the saved page setup. The **Page Setup** and **Plot** dialog boxes include most of the same settings.

PROFESSIONAL TIP

Layout setup usually involves several steps. A well-defined page setup decreases the amount of time required to prepare a drawing for plotting. Once a layout is set up, only a few steps are required to produce a plot. Add fully defined layouts to drawing templates for convenient future use.

Page Setups

PAGESETUP

Application Menu
Print

Page Setup
Output
> Plot

Page Setup Manager

Type
PAGESETUP

Access the **PAGESETUP** command to create page setups using the **Page Setup Manager**. See **Figure 28-11**. The **Page setups** area of the **Page Setup Manager** contains a list box that lists available page setups, and includes buttons to add and modify page setups. The **Selected page setup details** area provides information about the highlighted page setup.

NOTE

You can also access the **Page Setup Manager** by right-clicking on the model or a layout tab, a **Quick View Layouts** or **Quick View Drawings** thumbnail image, or the **Quick View Layouts** button on the status bar, and selecting **Page Setup Manager...**.

When you access the **Page Setup Manager** in model space, *Model* appears in the **Page Setups** list box. When you access the **Page Setup Manager** in paper space, the name of the current layout appears in the **Page Setups** list box. When preparing a layout for plotting, check to be sure that you are in paper space and that the appropriate layout

Figure 28-11.
Use the **Page Setup Manager** to modify existing page setups and to create and import page setups.

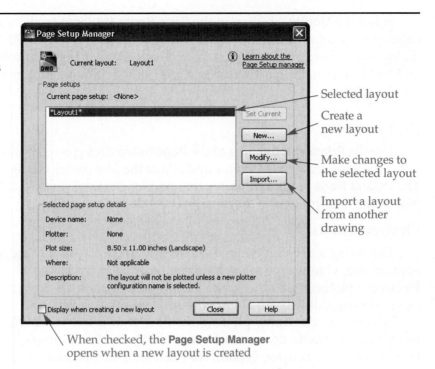

Selected layout

Create a new layout

Make changes to the selected layout

Import a layout from another drawing

When checked, the **Page Setup Manager** opens when a new layout is created

is current. Each layout can have a unique page setup. Asterisks (*) before and after the layout name indicate the page setup assigned to the current layout. You have the option to create or use other page setups instead of the page setup associated with the layout.

Create a new page setup to use different plot characteristics without overriding plot settings or spending time making page setup changes. For example, create two page setups to plot to two different printers or plotters. Pick the **New...** button to create a new page setup using the **New Page Setup** dialog box. See **Figure 28-12**. Type a name for the new page setup and choose an option from the **Start with:** list box. Pick the **OK** button to create the page setup and display the **Page Setup** dialog box. Pick the **Import...** button in the **Page Setup Manager** to use existing page setups from a DWG, DWT, or DXF file.

To attach a different page setup to the current layout, select a page setup from the list in the **Page Setup Manager** and pick the **Set Current** button, or right-click on a page setup and choose **Set Current**. The layout will now plot according to the selected page setup. When you make a different page setup current, the selected page setup overrides the layout page setup. The page setup name appears in parentheses next to the layout name. To rename or delete an existing page setup, right-click on the page setup and pick **Rename** or **Delete**.

Figure 28-12.
The **New Page Setup** dialog box appears when you create a new page setup. Selecting **None** in the **Start with:** list box does not select a printer. Default selects the default printer assigned to the computer.

New page setup name

Select a layout to copy its settings

Select the **Modify...** button in the **Page Setup Manager** to change the settings of an existing page setup using the **Page Setup** dialog box. See **Figure 28-13**. The **Page Setup** dialog box defines page setup characteristics. The settings control layout appearance and plot function. Each area, as described in the following sections, controls a specific plot setting.

Plot Device

plot device: The printer, plotter, or alternative plotting system to which the drawing is sent.

Use the **Printer/plotter** area of the **Page Setup** dialog box, shown in **Figure 28-14**, to select the appropriate *plot device* and adjust the plot device configuration if necessary. The default **None** setting indicates that no plot device is specified. Select a *configured* plot device to print or plot to from the **Name:** drop-down list.

configured: Installed and ready to use.

exporting: Transferring electronic data from a database, such as a drawing file, to a different format used by another program.

Electronic Plots

Exporting a drawing is an effective way to display and share a drawing for some applications. One way to export a drawing is to plot a layout to a different file type. Electronic plotting uses the same general process as hard-copy plotting, but the plot exists electronically instead of on an actual sheet of paper.

A common electronic plotting method is to select the **DWG To PDF.pc3** option to plot to a portable document format (PDF) file. For example, send a PDF file of a layout to a manufacturer, vendor, contractor, agency, or plotting service. The recipient uses the free Adobe® Reader software to view the plot electronically, and to plot the drawing to scale without having AutoCAD software. This method also helps avoid inconsistencies that sometimes occur when sharing AutoCAD files. Select the **DWF6 ePlot.pc3** or **DWFX ePlot (XPS compatible).pc4** option to plot to the appropriate design web format (DWF) file. The recipient of a DWF file uses a viewer such as the Autodesk® Design Review software to view and mark up the plot.

Figure 28-13.
The **Page Setup** dialog box allows you to adjust the plot settings for the selected page setup. This is where the initial phase of layout setup begins.

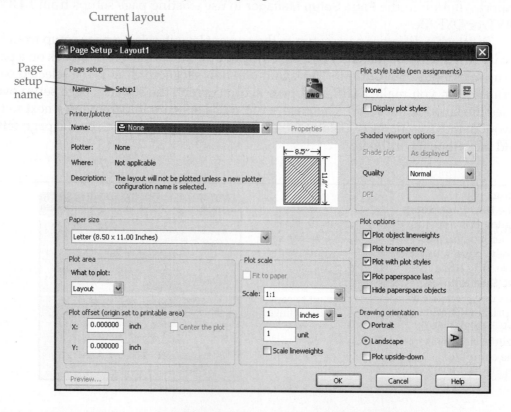

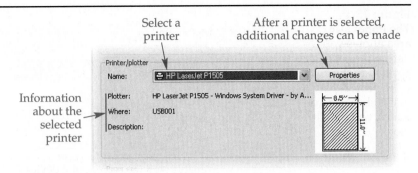

Figure 28-14.
The **Printer/plotter** area of the **Page Setup** dialog box.

Select a printer

After a printer is selected, additional changes can be made

Information about the selected printer

Printer/plotter
Name: HP LaserJet P1505 — Properties
Plotter: HP LaserJet P1505 - Windows System Driver - by A...
Where: USB001
Description:

8.5"

11.0"

You will explore additional information on creating electronic plots later in this textbook and in *AutoCAD and Its Applications—Advanced*.

PROFESSIONAL TIP

If you use a plotting service or blueprint shop to plot large sheets, ask if the printer accepts PDF files. Plotting a layout to a PDF file and then using the PDF file to create a hard copy is often the most convenient way to share a drawing and help ensure that the hard copy plots exactly as expected.

CAUTION

You can use the **PLOTTERMANAGER** tool to add and configure printers or plotters to the **Printer/plotter** list. However, avoid editing and saving modified plot configuration (PC3) files unless you know doing so is appropriate. These files are critical to the proper functioning of your plotter.

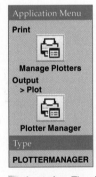

Application Menu
Print

Manage Plotters
Output
> Plot

Plotter Manager

Type
PLOTTERMANAGER

sheet size: The size of the paper used to lay out and plot drawings.

Sheet Size

The **Paper size** area, shown in **Figure 28-15**, controls the *sheet size*. The sheet size determines the size of the virtual sheet of paper displayed in a layout and corresponds to the actual sheet size you plan to use when plotting. To determine sheet size, take into account the size of the drawing and additional space for dimensions, notes, border, clear space between the drawing and border, sheet blocks, such as the title block and revision history block, zoning, and an area for general notes and other annotations and symbols. Select the appropriate sheet size from the drop-down list in the **Paper size** area.

Standard Sheet Sizes

The ASME Y14.1, *Decimal Inch Drawing Sheet Size and Format*, and ASME Y14.1M, *Metric Drawing Sheet Size and Format*, standards specify the American Society of Mechanical Engineers (ASME) standard sheet sizes and formats. **Figure 28-16** shows the ASME Y14.1 sheet size specifications in inches. **Figure 28-17** shows the ASME Y14.1M sheet size specifications in millimeters.

Figure 28-15.
The **Paper size** area allows you to define the sheet size applied to the layout, which corresponds to the sheet size on which you plan to plot.

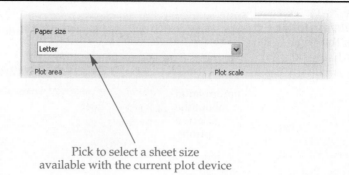

Pick to select a sheet size
available with the current plot device

Figure 28-16.
ASME Y14.1 standard sheet sizes for U.S. customary (inch) drawings.

Size Designation	Size (in inches)	
	Vertical	Horizontal
A	8 1/2 11	11 (horizontal format) 8 1/2 (vertical format)
B	11	17
C	17	22
D	22	34
E	34	44
F	28	40
Sizes G, H, J, and K are roll sizes.		

Figure 28-17.
ASME Y14.1M standard sheet for metric drawings.

Size Designation	Size (in millimeters)	
	Vertical	Horizontal
A0	841	1189
A1	594	841
A2	420	594
A3	297	420
A4	210	297

Figure 28-18 shows standard ASME sheet sizes and layout. To describe sheet size values verbally, state the vertical measurement and then the horizontal measurement. For example, describe a C-size sheet as 22 (horizontal) × 17 (vertical). Longer lengths are known as *elongated* and *extra-elongated* drawing sizes. These are available in multiples of the short side of the sheet size.

Figure 28-18.
A—Standard decimal inch drawing sheet sizes and layout based on ASME Y14.1. B—
Standard metric drawing sheet sizes and layout based on ASME Y14.1M.

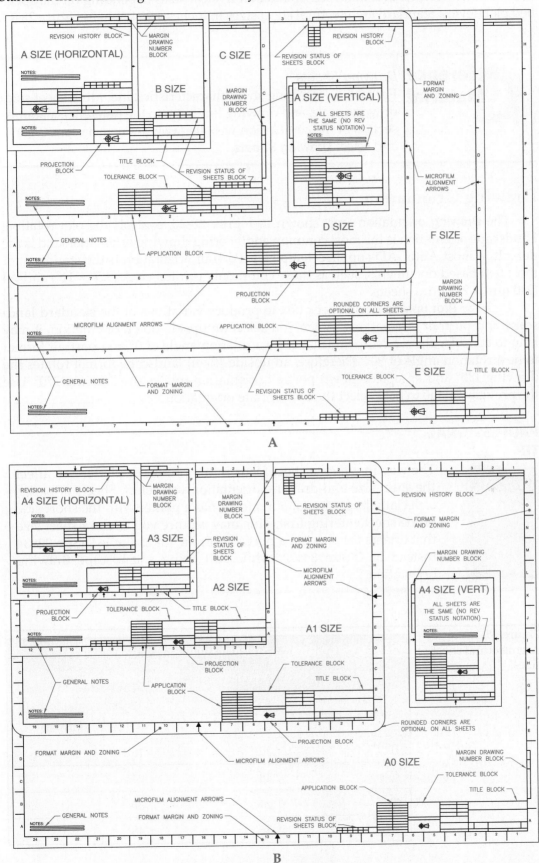

Chapter 28 Layout Setup

881

Some companies that prepare architectural or related drawings prefer inch-unit architectural sheet sizes, which vary slightly from the ASME standard. AutoCAD includes architectural and modified architectural sheet sizes depending on the specified plot device. Standard architectural sheet sizes in inches are shown in **Figure 28-19**.

Reference Material

Drawing Sheets

For tables describing sheet characteristics, including sheet size, drawing scale, and drawing limits, go to the **Reference Material** section of the companion website (www.g-wlearning.com/CAD) and select **Drawing Sheets**.

Drawing Orientation

The **Drawing orientation** area, shown in **Figure 28-20**, controls the plot rotation. Landscape orientation is the most common engineering drawing orientation and is the default in most AutoCAD-supplied templates. Portrait orientation is the standard for most text-based documents, such as those commonly printed on A (8.5″ × 11″) and A4 (210 mm × 297 mm) sheets.

Pick the **Plot upside-down** check box to produce variations of the standard landscape and portrait orientations. When you select an upside-down orientation, it may help to consider the landscape format to be a rotation angle of 0° and portrait format to be a rotation angle of 90°. Therefore, an upside-down landscape format rotates the drawing 180° and an upside-down portrait orientation rotates the drawing 270°. Use the preview image to help select the appropriate orientation.

PROFESSIONAL TIP

The way in which a sheet feeds into a printer or plotter can affect the sheet size and drawing orientation you select. Sheets of paper, especially large sheets, often feed into a plotter with the short side of the sheet entering first. This may require you to use a sheet size that orients the sheet in a portrait format, for example D-Size 22x34 instead of D-Size 34x22, while still using a landscape drawing orientation.

Figure 28-19.
Architectural sheet sizes.

Size Designation	Size (in inches)	
	Vertical	Horizontal
A	9	12 (horizontal format)
	12	9 (vertical format)
B	12	18
C	18	24
D	24	36
E	36	48

AutoCAD and Its Applications—Basics

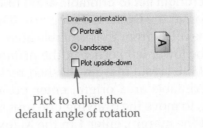

Pick to adjust the
default angle of rotation

Exercise 28-3

Complete the exercise on the companion website.
www.g-wlearning.com/CAD

Plot Area

The **Plot area** section, shown in **Figure 28-21**, allows you to choose the portion of the drawing to plot. Select an option from the **What to plot:** drop-down list. The **Layout** option is available when you plot a layout. When you select this option, everything inside the margins of the layout plots. The **Layout** option is the default and is a common setting for plotting a layout. Use other plot area options primarily for plotting in model space, or to adjust the area to plot in paper space.

The **View** option is available when named views exist in the drawing, and if the model space or paper space environment associated with the current page setup includes a named view. When you pick the **View** option, an additional drop-down list appears in the **Plot area** section, allowing you to select a specific view to define as the plot area. Refer to Chapter 6 for information about other **Plot area** options.

Plot Offset

The **Plot offset** area, shown in **Figure 28-22**, controls the distance the drawing is offset from the plot origin. You can specify the plot origin as the lower-left corner of the printable area or the lower-left corner of the sheet by selecting the appropriate radio button in the **Specify plot offset relative to** area on the **Plot and Publish** tab of the **Options** dialog box.

Figure 28-21.
Use the **Plot area** section to define the portion of the drawing to plot.

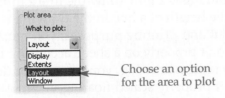

Choose an option
for the area to plot

Figure 28-22.
The **Plot offset** area controls how far the drawing is offset from the lower-left corner of the printable area or layout border.

Enter offset from
lower-left corner

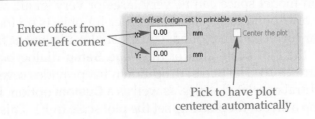

Pick to have plot
centered automatically

The **Plot offset (origin set to printable area)** title appears when you use the default **Printable area** option in the **Options** dialog box. The values you enter in the **X:** and **Y:** text boxes define the offset from the printable area. Use the default values of 0 to locate the plot origin at the lower-left corner of the printable area, which corresponds to the lower-left corner of the layout margin (dashed rectangle). To move the drawing away from the default printable area origin, enter positive or negative values in the text boxes. For example, to move the drawing one unit to the right and two units above the lower-left corner of the margin, enter 1 in the **X:** text box and 2 in the **Y:** text box.

Select the **Edge of paper** radio button in the **Options** dialog box to display the **Plot offset (origin set to layout border)** title. The values you enter in the **X:** and **Y:** text boxes define the offset from the edge of the sheet. Use the default values of 0 to locate the plot origin at the lower-left corner of the sheet. Change the values in the text boxes to move the drawing away from the sheet origin.

When you select any option other than the **Layout** option from the **What to plot:** drop-down list in the **Plot area** section, the **Center the plot** check box becomes active. Check **Center the plot** to shift the plot origin automatically as needed to center the selected plot area in the printable area.

PROFESSIONAL TIP

If a drawing does not center on the sheet when you plot using the **Layout** option, open the **Options** dialog box and select the **Plot and Publish** tab. Pick the **Edge of paper** radio button in the **Specify plot offset relative to** area. Then use values of 0 in the X and Y offset text boxes in the **Page Setup** dialog box to locate the plot origin at the lower-left corner of the sheet. Depending on the specific plot configuration, you may also need to pick the **Plot upside-down** check box in the **Drawing orientation** area of the **Page Setup** dialog box to locate the origin exactly at the lower-left corner of the sheet. Alternatively, resolve the issue by changing the plot offset to the X and Y values of the lower-left corner of the printable area.

Plot Scale

Always draw objects at their actual size, or full scale, in model space, regardless of the size of the objects. For example, if you draw a small machine part and the length of a line in the drawing is 2 mm, draw the line 2 mm long in model space. If you draw a building and the length of a line in the drawing is 80′, draw the line 80′ long in model space. For layout and printing purposes, you must then scale drawings with features of these sizes to fit properly on a sheet, according to a specific *drawing scale*.

drawing scale: The ratio between the actual size of objects in the drawing and the size at which the objects plot on a sheet of paper.

When you scale a drawing, you increase or decrease the displayed size of model space objects. A properly scaled floating viewport in a layout allows for this process, as described in Chapter 29. When setting up a layout for plotting, remember that the layout is also at full scale. One difference between model space and paper space is that objects in model space can be very large or very small, while objects in paper space always correspond to sheet size. In order for objects on the layout and in model space to appear correct when plotted, you must plot a layout at full scale, or 1:1.

Use the **Plot scale** area of the **Page Setup** dialog box to specify the plot scale. See **Figure 28-23**. The **Scale:** drop-down list provides several predefined decimal and architectural or related scales, as well as a **Custom** option. For most applications, when setting up a layout for plotting, set the plot scale to 1:1. This ensures that the layout and scaled floating viewports plot correctly. If you choose a scale other than 1:1, the layout does not plot to scale.

Figure 28-23.
Use the **Plot scale** area to adjust the scale at which the drawing plots. The plot scale is typically set to 1:1 to plot a layout even though the drawing scale may not be 1:1.

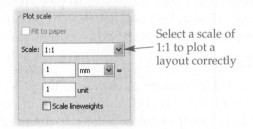

Select a scale of 1:1 to plot a layout correctly

When preparing to plot in model space, you may choose to select a scale other than 1:1. For example, to plot an architectural floor plan, you might set the plot scale to 1/4″ = 1′-0″. If the desired scale is not available, enter values in the text boxes below the drop-down list of predefined scales and select the correct unit of measure from the drop-down list. The **Custom** option automatically displays when you enter values. For example, 1 inch = 600 units is a custom scale entry used to plot at a scale of 1″ = 50′ (50′ × 12″ = 600). Refer to Chapter 9 for more information about drawing scale and scale factors.

Pick the **Fit to paper** check box to adjust the plot scale automatically to fit on the selected sheet. Fitting to paper is useful if you are not concerned about plotting to scale, such as when you are creating a check copy on a sheet that is too small to plot at the appropriate scale. Select the **Scale lineweights** check box to scale (increase or decrease the weight of) lines when the plot scale changes.

PROFESSIONAL TIP

If you need to plot an inch drawing on a metric sheet, use a custom scale of 1 inch = 25.4 units. Conversely, use a custom scale of 25.4 mm = 1 unit to plot a metric drawing on an inch sheet.

Exercise 28-4
Complete the exercise on the companion website.
www.g-wlearning.com/CAD

Plot Styles

Object properties control the appearance of objects on-screen. By default, what you see on-screen is what plots. For example, if you draw objects on a layer that uses a Red color, Continuous linetype, and 0.60 mm lineweight, the objects display and plot red, continuous, and thick, assuming you show lineweights on-screen and use a color plotter. If this is the result you want, you are ready to continue with the page setup and plotting process.

To define exactly how objects plot regardless of what displays on-screen, you must assign *plot styles* to objects. Use plot styles to maintain object properties in the drawing, but plot objects according to specific plotting properties. For example, to plot all objects in a drawing as dark as possible, you should plot them using the color black.

plot styles: Properties, including color, linetype, lineweight, line end treatment, and fill style, that are applied to objects for plotting purposes only.

In this example, plot styles allow you to plot all objects black without making them black on-screen. Plot styles also allow you to plot objects using shades of gray instead of color, or to plot objects lighter or darker than they display on-screen.

Plot Style Tables

Plot style tables contain plot styles. Choose to use either a *color-dependent plot style table* or a *named plot style table*. A color-dependent plot style table forces objects to plot according to object color. Color-dependent plot style tables contain 255 preset plot styles—one for each AutoCAD Color Index (ACI) color. Each color-dependent plot style is linked to an index color. Plot style properties control how to treat objects of a certain color when the objects plot. For example, the plot style Color 1, which is Red, defines how to plot all objects that are red on-screen. If you assign the Black plot color to plot style Color 1, all red objects are plotted black, even though the objects are red on-screen.

A named plot style table forces objects to plot according to named plot style values, which you can assign to a layer or object. Any layer or object assigned a named plot style plots using the settings specified for that plot style. For example, create a layer named OBJECT that uses a Red color and a plot style named BLACK that uses a Black color. Then assign the BLACK plot style to the OBJECT layer. Objects drawn on the OBJECT layer plot using the BLACK plot style and plot black in color, even though the objects are red on-screen.

Ideally, decide which plot style table is appropriate before you begin drawing. The templates created in the Template Development feature of this textbook, for example, assume that drawings plot so that all objects appear dark, or black, with different object linetypes and lineweights. You can use a color-dependent plot style table or a named plot style table to create this effect. Default AutoCAD plot style behavior uses color-dependent plot style tables. For most applications, it is usually best to use color-dependent plot style tables, because they are the default and do not require you to assign named plot styles to layers or objects.

Configuring Plot Style Table Type

To configure the plot style type used by default when you create new drawings, pick the **Plot Style Table Settings...** button on the **Plot and Publish** tab of the **Options** dialog box. This opens the **Plot Style Table Settings** dialog box shown in **Figure 28-24**.

To use a named plot style table, pick the **Use named plot styles** radio button from the **Default plot style behavior for new drawings** area before you start a new drawing file. Select a specific plot style table to use as the default for new drawings from the

Figure 28-24.
Use the **Plot and Publish** tab of the **Options** dialog box to set up the default plot style types and tables for new drawings.

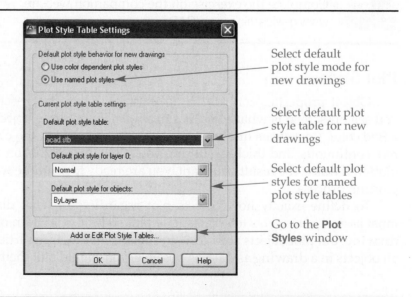

Select default plot style mode for new drawings

Select default plot style table for new drawings

Select default plot styles for named plot style tables

Go to the **Plot Styles** window

Default plot style table: drop-down list in the **Current plot style table settings** area. When you select the **Use named plot styles** radio button, the **Default plot style for layer 0:** and **Default plot style for objects:** options become active.

When you use a template to create a new drawing, the plot style settings defined in the template override the settings you specify in the **Plot Style Table Settings** dialog box. For example, if you configure the template to use named plot style tables, you can only select a named plot style table to apply to the plot, even if you select the **Use color dependent plot styles** radio button in the **Plot Style Table Settings** dialog box.

Pick the **Add or Edit Plot Style Tables...** button in the **Plot Style Table Settings** dialog box to open the **Plot Styles** window. See **Figure 28-25**. The **Plot Styles** window lists available color-dependent and named plot style tables saved in the Plot Style Table Search Path, as defined in the expanded **Printer Support File Path** option in the **Files** tab of the **Options** dialog box. Color-dependent plot style table files (CTB) use the .ctb extension. Named plot style table files (STB) include an .stb extension. Double-click on Add-A-Plot Style Table Wizard to create a new plot style table, or double-click on an existing plot style table file to edit the file.

Application Menu
Print

Manage Plot Styles

Type
STYLESMANAGER

Creating and Editing Plot Style Tables
For detailed information about creating and editing plot style tables, go to the companion website (www.g-wlearning.com/CAD), select this chapter, and select **Creating and Editing Plot Style Tables**.

Figure 28-25.
The **Plot Styles** window lists available plot style files and allows you to create and edit plot style tables.

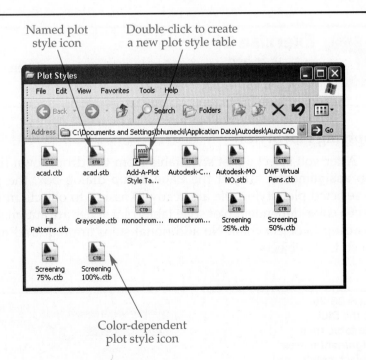

Selecting a Plot Style Table

Use the **Plot style table (pen assignments)** area of the **Page Setup** dialog box, shown in Figure 28-26, to activate and manage plot style tables. Select a plot style table from the drop-down list. Use the default **None** option instead of selecting a specific color-dependent or named plot style table. The **None** option plots exactly what appears on-screen without using plot styles, assuming you use a color plotter to plot objects with color.

Only color-dependent or named plot style tables appear, depending on the type of plot style table assigned to the current drawing. The most often used color-dependent plot style tables are:

- Monochrome.ctb—plots the drawing in monochrome (black and white).
- Grayscale.ctb—plots the drawing using shades of gray.
- Screening files—plot the drawing using faded, or screened, colors.

When a drawing uses named plot style tables, it is common to select one of the following:

- Autodesk-Color.stb—provides access to named plot styles for plotting the drawing using solid and faded colors.
- Monochrome.stb—references a named plot style for plotting the drawing in monochrome.
- Autodesk-MONO.stb—provides access to named plot styles for plotting objects monochrome, in color, and in faded monochrome.

PROFESSIONAL TIP

You cannot select named plot style tables if you start the drawing with color-dependent plot style tables. Conversely, you cannot select color-dependent plot style tables if you start the drawing with named plot style tables. You can type CONVERTPSTYLES to access the **CONVERTPSTYLES** command, which allows you to switch between table modes in a drawing. However, you should avoid converting plot style tables when possible. Instead, use the appropriate plot style type when beginning a new drawing.

Exercise 28-5

Complete the exercise on the companion website.
www.g-wlearning.com/CAD

Applying Plot Styles

After you select a plot style table from the drop-down list in the **Plot style table (pen assignments)** area of the **Page Setup** dialog box, the plot styles contained in the selected plot style table are ready to assign to objects in the drawing. When you select a color-dependent plot style table, plot styles are automatically applied to objects according to object color. No additional steps are required to apply color-dependent plot styles to objects.

Figure 28-26.
Use the **Plot style table (pen assignments)** area to select, create, and edit plot style tables.

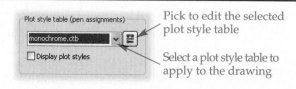

Pick to edit the selected plot style table

Select a plot style table to apply to the drawing

When you select a named plot style table, the named plot styles contained in the table are ready to assign to layers or individual objects. Any layer or object assigned a named plot style plots using the settings specified for that style. Named plot style tables contain as many plot styles as have been created. For example, the monochrome.stb plot style table contains the default Normal plot style and a style named Style 1. When you apply the Normal plot style to layers or objects, objects plot exactly as they appear on-screen. Style 1 assigns the color Black to all layers or objects that use Style 1 for plotting. In order to plot all objects in the drawing black, even if the objects have different colors on-screen, you must apply Style 1 to all layers.

Assign named plot styles to layers in the **Layer Properties Manager** palette. See Figure 28-27. Select a layer and then pick the current plot style, such as Normal, from the **Plot Style** column. The **Select Plot Style** dialog box appears. See Figure 28-28. Select a named style to assign to the highlighted layer. Any object drawn on a layer assigned to a named plot style plots using the settings in the named plot style.

You can also assign named plot styles to individual objects. If you assign a plot style to an object drawn on a layer that has already been assigned a named plot style, the plot style assigned to the object overrides the plot style settings for the layer. Assign plot styles to objects using the **Properties** palette or the **Plot Style Control** drop-down list in the **Properties** panel on the **Home** ribbon tab. See Figure 28-29.

If the current drawing is set to use a named plot style table, the default plot style for all layers is Normal. The default plot style for all objects is ByLayer. Objects plotted with these settings keep their original properties.

Figure 28-27.
Assign named plot styles to layers using the **Layer Properties Manager**. This example uses the monochrome.stb plot style table and the Style 1 plot style assigned to all layers to plot all objects black.

Pick plot style name for layer
to select a different plot style

S..	Name	O..	Fre...	L...	Color	Linetype	Lineweight	Trans...	Plot Style	Plot	N..	Description
✓	0				wh...	Continuo...	Default	0	Normal			
	Break-Long				blue	Continuo...	0.30 mm	0	Style 1			
	Break-Short				blue	Continuo...	0.60 mm	0	Style 1			
	Center				red	CENTER	0.30 mm	0	Style 1			
	Chain				m...	CENTER	0.60 mm	0	Style 1			
	Construction				wh...	Continuo...	0.30 mm	0	Style 1			
	Defpoints				cyan	Continuo...	Default	0	Style 1			
	Dimension				red	Continuo...	0.30 mm	0	Style 1			
	Hidden				gr...	HIDDEN	0.30 mm	0	Style 1			
	Object				blue	Continuo...	0.60 mm	0	Style 1			
	Phantom				wh...	PHANTOM	0.30 mm	0	Style 1			
	Plane				m...	PHANTOM	0.60 mm	0	Style 1			
	Section				m...	Continuo...	0.30 mm	0	Style 1			
	Sheet				gr...	Continuo...	0.60 mm	0	Style 1			
	Stitch				ye...	DOT	0.30 mm	0	Style 1			
	Title Block				red	Continuo...	0.30 mm	0	Style 1			
	Viewports				ye...	Continuo...	0.30 mm	0	Style 1			
	Xref				wh...	Continuo...	0.30 mm	0	Style 1			

Current layer: 0

Search for layer

Filters — All — All Used Layers

Invert filter

All: 18 layers displayed of 18 total layers

Figure 28-28.
Use the **Select Plot Style** dialog box to assign a plot style to a layer.

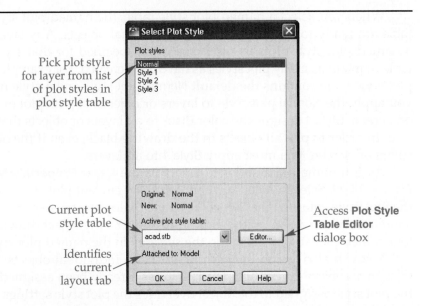

Pick plot style for layer from list of plot styles in plot style table

Current plot style table

Identifies current layout tab

Access **Plot Style Table Editor** dialog box

Figure 28-29.
You can assign a plot style to individual objects using
A— the **Properties** palette or
B—the **Plot Style Control** drop-down list in the ribbon.

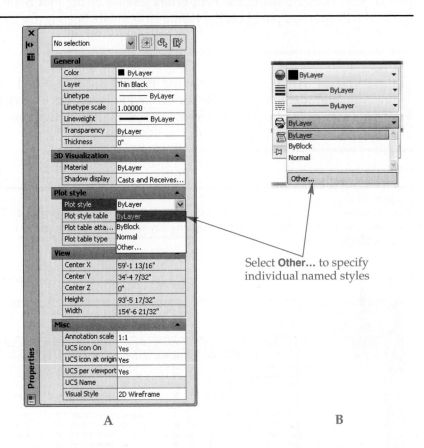

Select **Other...** to specify individual named styles

A

B

Exercise 28-6

Complete the exercise on the companion website.
www.g-wlearning.com/CAD

Viewing Plot Style Effects On-Screen

To display plot style effects on-screen, pick the **Display plot styles** check box in the **Plot style table** area in the **Page Setup** dialog box. Objects appear on-screen as they will plot. Typically, it is not appropriate to work with objects displayed as they will plot, especially if you print in monochrome. In this example, the color assigned to a layer appears black, which defeats the purpose of assigning colors to layers. A better practice is to use the preview feature of the **Page Setup** or **Plot** dialog box, as described later in this chapter, to preview the effects of plot styles before plotting.

Other Plotting Options

The **Plot options** area of the **Page Setup** dialog box contains additional options that affect how specific items plot. If you plan to plot objects using a plot style table, be sure to select the **Plot with plot styles** check box. This check box toggles the use of plot styles on and off, allowing you to create a plot quickly with or without using plot styles.

When you deselect the **Plot with plot styles** check box, the **Plot object lineweights** check box becomes enabled and selected. When this box is checked, all objects with a lineweight greater than 0 plot using the assigned lineweight. Deselect the check box to plot all objects using a 0, or thin, lineweight.

The **Plot transparency** check box provides the flexibility to plot transparent objects transparent or opaque. Check **Plot transparency** for objects that are assigned to a transparent layer and objects in which the style has been overridden to appear transparent. For many drawings, transparency should be reproduced on the plot, as shown in Figure 28-30, especially if you are using transparent fills. However, sometimes transparency is only for on-screen display proposes. For this application, deselect the **Plot transparency** check box.

Pick the **Plot paperspace last** check box to plot paper space objects after model space objects. This option ensures that objects in paper space that overlap objects in model space plot over the model space objects. The **Plot paperspace last** check box is disabled when you plot in model space, because model space does not include paper space objects. Select the **Hide paperspace objects** check box to remove hidden lines from 3D objects created in paper space. This option is only available when you plot from a layout tab and affects only objects drawn in paper space. It does not affect any 3D objects in a viewport.

The **Shaded viewport options** area of the **Page Setup** dialog box provides settings that control viewport shading. These options set the type and quality of shading for plotting 3D models from a shaded or rendered viewport. *AutoCAD and Its Applications—Advanced* explains 3D models.

Completing Page Setup

After you select all appropriate page setup options, pick the **Preview** button in the lower-left corner of the **Page Setup** dialog box to preview the effects of the page setup. What you see on-screen is the exact plot appearance, assuming you use a color plotter to make color prints. The **Realtime Zoom** command is automatically activated in the preview window. Additional view commands are available from the toolbar near the top of the window or from a shortcut menu. Use view commands to help confirm that the plot settings are correct. When you finish previewing the plot, pick the **Close** button on the toolbar, press [Esc] or [Enter], or right-click and select **Exit** to return to the **Page Setup** dialog box. Pick the **OK** button to exit the **Page Setup** dialog box, and pick **Close** button to exit the **Page Setup Manager**.

Figure 28-30.
An example of an architectural elevation with transparent layers plotted to show the existing structure. Opaque layers identify the proposed structure.

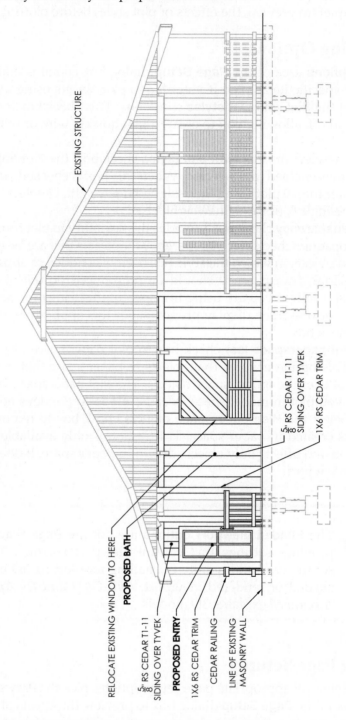

Exercise 28-7

Complete the exercise on the companion website.
www.g-wlearning.com/CAD

Chapter Review

Answer the following questions. Write your answers on a separate sheet of paper or complete the electronic chapter review on the companion website.
www.g-wlearning.com/CAD

1. What is a model?
2. Define *model space*.
3. What is the purpose of a layout?
4. What is paper space?
5. What is the purpose of a floating viewport?
6. What is the purpose of the dashed rectangle that appears on the default layout?
7. If you pick the **MODEL** button to enter paper space and the file contains multiple layouts, none of which you previously accessed, which layout is displayed by default?
8. Briefly describe the function of the **Quick View Layouts** tool.
9. What does the **Quick View Drawings** tool allow you to do?
10. What is a page setup?
11. Briefly describe the basic function of the **Page Setup Manager** and the related **Page Setup** dialog box.
12. Briefly describe the function of the **Plot** dialog box and explain when it is used in relation to the **Page Setup Manager** and **Page Setup** dialog box.
13. Explain the importance of a well-defined page setup.
14. How can you identify which page setup is tied to the current layout?
15. What is a plot device?
16. Define *sheet size*.
17. What factors should you consider when you select a sheet size for a drawing?
18. Which ASME standards specify sheet sizes?
19. What is the plot offset of a layout?
20. Define *drawing scale*.
21. What are plot styles?
22. Briefly explain the purpose of a plot style table.
23. How do you assign plot styles in a color-dependent plot style?
24. Briefly explain the function of a named plot style table.
25. How can you be certain that your page setup options will produce the desired plot?

Drawing Problems

Start AutoCAD if it is not already started. Follow the specific instructions for each problem.

Note: Some of the problems in this chapter are built on problems from previous chapters. If you have not yet completed those problems, complete them now.

▼ Basic

1. Use a word processor to list five items commonly found in a layout. Provide a brief description of each item.

2. Start a new drawing using the acad.dwt template. Create a plot style table named Black35mm.cbt that will plot all colors in the AutoCAD drawing in black ink on the paper, with a lineweight of 0.35 mm. (*Hint:* To make the same change to a property of all the plot styles, select the first plot style in the list, in this case Color 1, then scroll to the end of the list, hold down the [Shift] key, and select the last plot style in the list, in this case Color 255.) Save the drawing as P28-2.

3. Start a new drawing using the acad -Named Plot Styles.dwt template. Create a plot style table named BlackShades.stb. Create the following plot styles:
   * Black100% with color set to black, all other properties set to their default values.
   * Black50% with color set to black, screening set to 50, all other properties set to their default values.
   * Black25% with color set to black, screening set to 25, all other properties set to their default values.
   Save the drawing as P28-3.

▼ Intermediate

4. Open P11-15 and save as P28-4. The P28-4 file should be active. Delete the default **Layout2**. Create a new B-size sheet layout by following these steps:
   A. Rename the default **Layout1** to **B-SIZE**.
   B. Select the **B-SIZE** layout and access the **Page Setup Manager**.
   C. Modify the **B-SIZE** page setup according to the following settings:
      * **Printer/Plotter:** Select a printer or plotter that can plot a B-size sheet
      * **Paper size:** Select the appropriate B-size sheet (varies with printer or plotter)
      * **Plot area:** Layout
      * **Plot offset:** 0,0
      * **Plot scale:** 1:1 (1 inch = 1 unit)
      * **Plot style table:** monochrome.ctb
      * **Plot with plot styles**
      * **Plot paperspace last**
      * Do not check **Hide paperspace objects**
      * **Drawing orientation:** Select the appropriate orientation (varies with printer or plotter)
      Resave P28-4.

5. Open P11-16 and save as **P28-5**. The P28-5 file should be active. Delete the default **Layout2**. Create a new A2-size sheet layout according to the following steps:
   A. Rename the default **Layout1** to **A2-SIZE**.
   B. Select the **A2-SIZE** layout and access the **Page Setup Manager**.
   C. Modify the **A2-SIZE** page setup according to the following settings:
      - **Printer/Plotter:** Select a printer or plotter that can plot an A2-size sheet
      - **Paper size:** Select the appropriate A2-size sheet (varies with printer or plotter)
      - **Plot area:** Layout
      - **Plot offset:** 0,0
      - **Plot scale:** 1:1 (1 mm = 1 unit)
      - **Plot style table:** monochrome.ctb
      - **Plot with plot styles**
      - **Plot paperspace last**
      - Do not check **Hide paperspace objects**
      - **Drawing orientation:** Select the appropriate orientation (varies with printer or plotter)

      Resave P28-5.

6. Open P8-18 and save as **P28-6**. The P28-6 file should be active. Delete the default **Layout2**. Create a new B-size sheet layout according to the following steps:
   A. Rename the default **Layout1** to **B-SIZE**.
   B. Select the **B-SIZE** layout and access the **Page Setup Manager**.
   C. Modify the **B-SIZE** page setup according to the following settings:
      - **Printer/Plotter:** Select a printer or plotter that can plot a B-size sheet
      - **Paper size:** Select the appropriate B-size sheet (varies with printer or plotter)
      - **Plot area:** Layout
      - **Plot offset:** 0,0
      - **Plot scale:** 1:1 (1 inch = 1 unit)
      - **Plot style table:** monochrome.ctb
      - **Plot with plot styles**
      - **Plot paperspace last**
      - Do not check **Hide paperspace objects**
      - **Drawing orientation:** Select the appropriate orientation (varies with printer or plotter)

      Resave P28-6.

▼ Advanced

7. Use a word processor to write a report of approximately 250 words explaining the difference between model space and paper space and describing the importance of using layouts. Cite at least three examples from actual industry applications of using layouts to prepare a multi-sheet drawing. Use at least four drawings to illustrate your report.

For Problems 8 through 11: Start a new drawing for each problem using an appropriate template of your choice. The template should include layers and text, dimension, multileader, and table styles, when necessary, for drawing the given objects. Add layers and text, dimension, multileader, and table styles as needed. Draw all objects using appropriate layers and text, dimension, multileader, and table styles, justification, and format. Follow the specific instructions for each problem. Use only drawing and editing commands and techniques you have already learned. Use your own judgment and approximate dimensions when necessary. Apply dimensions accurately using ASME or appropriate industry standards.

8. Draw and dimension the wood beam details shown. Establish the missing information using your own specifications, or determine the correct size of items not dimensioned. Prepare a single layout for plotting on a C-size sheet using the monochrome.ctb plot style. Delete all other layouts. Save the drawing as P28-8.

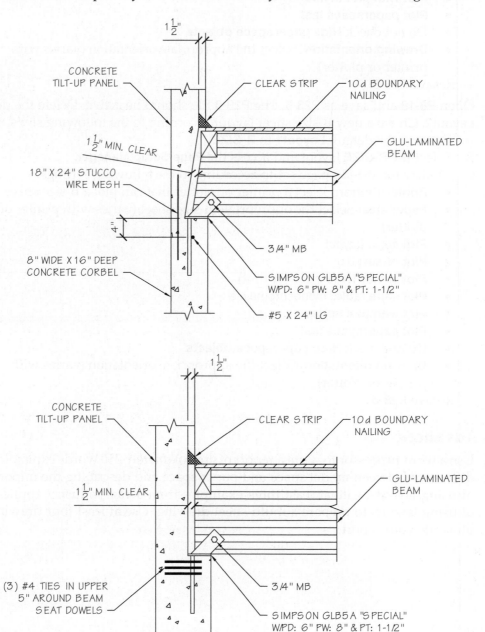

9. Draw and dimension the views of the support shown. Prepare a single layout for plotting on an A3-size sheet using the monochrome.ctb plot style. Delete all other layouts. Save the drawing as P28-9.

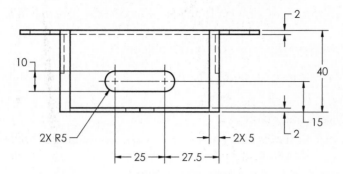

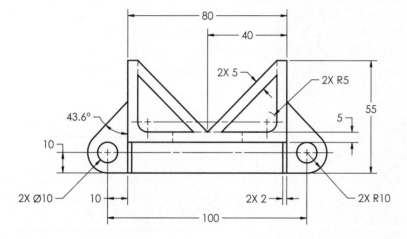

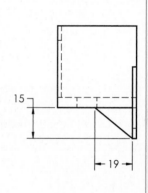

10. Research the design of an existing drive sprocket with the following specifications: single strand with hub, accepts ANSI 50 chain, 28 teeth, 5/8" pitch, finished 1" bore with 1/4" × 1/8" ANSI keyway and two set screws, steel material. Create a dimensioned 2D sketch of the existing design from manufacturer's specifications or from measurements taken from an actual sprocket. Start a new drawing from scratch or use a decimal-unit template of your choice. Draw and dimension the views required to document the design. Prepare a single layout for plotting on an appropriate sheet using the monochrome.ctb plot style. Delete all other layouts. Save the drawing as P28-10.

11. Create a dimensioned 2D sketch of the floor plan of a precast concrete restroom for a public park. Design the restroom to fit a 16'-0" × 16'-0" concrete slab foundation. Divide the restroom to provide separate stalls for men and women. Use dimensions based on your experience, research, and measurements. Start a new drawing from scratch or use an architectural template of your choice. Draw and dimension the floor plan from your sketch. Prepare a single layout for plotting on an appropriate sheet using the monochrome.ctb plot style. Delete all other layouts. Save the drawing as P28-11.

AutoCAD Certified Associate Exam Practice

Answer the following questions. Write your answers on a separate sheet of paper.

1. At what scale should you draw objects in model space? *Select the one item that best answers the question.*
   A. choose a scale based on the true size of the objects
   B. full scale (1:1)
   C. half scale (1:2)
   D. quarter scale (1:4)
   E. no particular scale is necessary

2. Which of the following methods can you use to create a new layout? *Select all that apply.*
   A. pick **New** and **Layout** from the **Application Menu**
   B. pick the **New...** button in the **Page Setup Manager**
   C. pick the **New Layout** button in the **Quick View Drawings** tool
   D. pick the **New Layout** button in the **Quick View Layouts** tool
   E. right-click on a current layout and select **New layout**

3. Which of the following plot styles can you use in a drawing that uses named plot styles? *Select all that apply.*
   A. acad.stb
   B. Fill Patterns.ctb
   C. grayscale.ctb
   D. monochrome.stb
   E. My Plot Style Table.stb

AutoCAD Certified Professional Exam Practice

Follow the instructions in the problem. Write your answers on a separate sheet of paper.

1. **Open CPE-28pagesetup.dwg. This file is available on the companion website.** Assuming that you print this drawing using a color printer, in what color will objects drawn on the **Objects** layer plot?

2. **Open CPE-28sheetsize.dwg. This file is available on the companion website.** Use appropriate inquiry commands to determine the answer to this question. What sheet size is **Layout1** most likely designed to plot?

Chapter 29

Plotting Layouts

Learning Objectives

After completing this chapter, you will be able to:

✓ Add layout content.
✓ Use floating viewports to create properly scaled final drawings.
✓ Preview and plot layouts.

This chapter explores the additional steps needed to complete layout setup. You will learn to add content to a layout and to place and use floating viewports. You will also use the **PLOT** command to preview the plot and send the layout to a printer, plotter, or alternate electronic format.

Layout Content

Model space provides an environment to create drawing views and add dimensions and annotations directly to views. Layouts provide an effective method to display model space content using floating viewports and to add items such as:

• Border
• Sheet blocks, such as title blocks and revision history blocks
• General notes
• Bill of materials, parts list, schedules, legends, and other associated lists
• Sheet annotation and symbols such as titles, north arrow, and graphic scale

Layouts provide flexibility to lay out, scale, and prepare a final drawing. Consider the objects placed in model space to be model content and the objects you add to a layout to be sheet content. A complete drawing forms when you bring model and sheet content together. See **Figure 29-1**.

Drawing in Paper Space

A layout is a representation of a flat piece of paper. As a result, paper space is a 2D drawing environment. Most 2D drawing and editing commands and options described throughout this textbook function the same in paper space as in model space. However, some tools are specific to or most commonly used in either model space or paper space. For example, floating viewports are specific to paper space.

Figure 29-1.
Dimensioned drawing views created in model space, combined with a border, title block,
revision history block, and general notes created in paper space, form the final drawing.

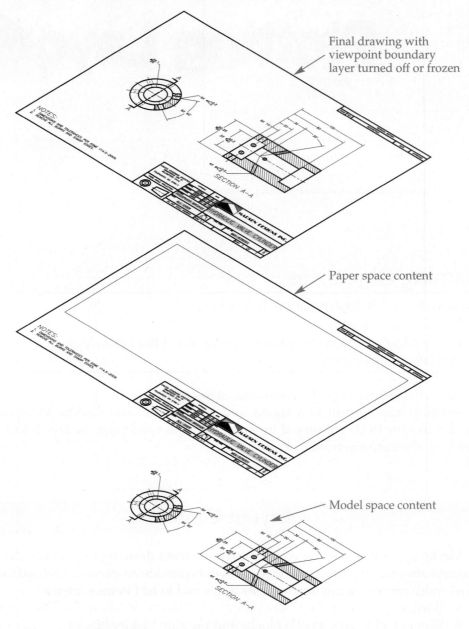

Final drawing with
viewpoint boundary
layer turned off or frozen

Paper space content

Model space content

 Although paper space is a 2D environment, you can display 3D
models created in model space in paper space floating viewports.

As in model space, you typically draw geometry on a layout at full scale. One
difference between model space and paper space is that objects in model space may
be very large or very small, while paper space objects always correspond to the sheet
size. Draw all layout content using the actual size you want the objects to appear on the
plotted sheet. Use multiline or single-line text or blocks with attributes to add layout
content such as general notes, view titles, and similar annotations to a layout. Place a
bill of materials, parts list, schedules, or similar tabular information using the **TABLE**
command or blocks with attributes. Typically, you will create items such as a border,

title block, and revision history block as blocks, often with attributes, and then insert these items into the layout.

Use layers appropriate for the layout and the objects added to the layout. Consider using a single layer named SHEET, for example, on which you draw all layout content. Another option is to use layers specific to layout items, such as a BORDER layer for the border and a TITLE or A-ANNO-TTBL layer for the title block. Assign a layer named Viewport, VPORT, or A-ANNO-NPLT, for example, to floating viewports so you can turn off, freeze, or set the viewport boundary to "no plot" before plotting.

 The default floating viewport boundary assumes the layer that is current when you first access a layout. If you use the default viewport, it may be necessary to change the layer on which it is drawn.

The Layout Origin

Most drawing and editing in paper space occurs on the sheet, which is the white rectangle you see on the gray background. However, it is possible, and necessary in some applications, to create objects off the sheet. When drawing and editing on a layout, remember that the origin (0,0) is controlled by the X and Y values you enter in the **Plot offset** area of the **Page Setup** dialog box. For example, if you draw a line with a start point of 1,1, the line begins 1 unit to the left and 1 unit up from the plot origin. The default origin position is at the lower-left corner of the printable area. See **Figure 29-2**. Refer to Chapter 28 for more information about plot origin and the **Plot offset** area of the **Page Setup** dialog box.

Layout content often references the edge of the sheet. For example, you might position a border 1/2" inside of the sheet edge. In this situation, it is usually best to define the layout origin as the lower-left corner of the sheet. The best option is to select the **Edge of paper** radio button in the **Specify plot offset relative to** area in the **Publish and Plot** tab of the **Options** dialog box. In the **Plot offset** area of the **Page Setup** dialog box, use values of 0.000 in the X and Y offset text boxes to locate the plot origin at the lower-left corner of the sheet. This is an excellent way to center the drawing on the sheet.

 Depending on the specific plot configuration, you may need to pick the **Plot upside-down** check box in the **Drawing orientation** area of the **Page Setup** dialog box in addition to using the **Edge of paper** offset option. Draw an object to see if 0,0 is actually located exactly at the lower-left corner of the sheet. If not, pick the **Plot upside-down** check box to solve the problem.

Figure 29-2.
The default location of the plot origin is often not appropriate, because all point entry is in reference to the lower-left corner of the printable area. Change the location of the plot origin to define the lower-left corner of the sheet at coordinates 0,0.

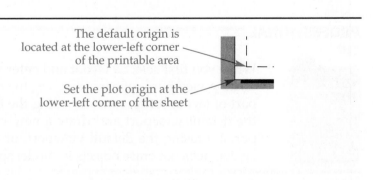

The default origin is located at the lower-left corner of the printable area

Set the plot origin at the lower-left corner of the sheet

Defining the plot offset relative to the edge of the paper maintains the plot offset location from the sheet edge even when you use a different plotter or plot configuration, as long as you use the same sheet size.

A second option to relate content to the lower-left corner of the sheet is to select the **Printable area** radio button in the **Specify plot offset relative to** area in the **Publish and Plot** tab of the **Options** dialog box. Then identify the values of the lower-right corner of the printable area. Information about the printable area is available in the **Device and Document Settings** tab of the plotter **Configuration Editor**. Next, in the **Plot offset** area of the **Page Setup** dialog box, change the X and Y plot offset values to the values of the lower-left corner of the printable area. The values you enter must be negative. The origin offsets from the printable area, so if you use a different plotter or plot configuration, the printable area may change, causing the offset to shift.

Exercise 29-1

Complete the exercise on the companion website.
www.g-wlearning.com/CAD

Floating Viewports

The primary advantage of using floating viewports in a paper space layout is the ability to prepare scaled drawings without increasing or decreasing the actual size of drawing views or sheet content. You can also create multiple viewports on a single layout to show differently scaled or alternate drawing views. For example, a single sheet might contain a floor plan drawn at a 1/4″ = 1′-0″ scale, an eave detail drawn at a 3/4″ = 1′-0″ scale, and a foundation detail drawn at a 3/4″ = 1′-0″ scale. See **Figure 29-3**.

A floating viewport boundary is the portion of the viewport that you see. Everything inside the viewport shows through from model space. Use commands such as **MOVE**, **ERASE**, **STRETCH**, and **COPY** to modify the viewport boundary in paper space. Use display commands such as **VIEW**, **PAN**, and **ZOOM** to modify the display of model space objects in the floating viewport. Additional options are available for adjusting how objects appear, according to the layers you assign to objects. Layer control allows you to define how the drawing appears within the viewport. Be sure a layout tab is current as you work through the following sections describing floating viewports.

When you first select a layout and enter paper space, a rectangular floating viewport appears, showing the objects in model space. As part of layout setup, consider using the **ERASE** command to erase the default viewport and draw a new viewport (or several viewports). Erasing the default viewport, or erasing everything in the layout, does not erase objects in model space.

Figure 29-3.
An example of an architectural layout, ready to plot, that includes three floating viewports used to display drawing views at different scales. The layer on which the viewports are drawn is off or frozen for plotting.

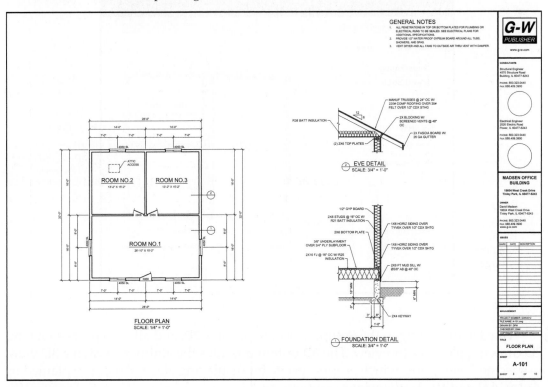

Specifying a Single Rectangular Viewport

The easiest way to place a single rectangular floating viewport is pick the **Rectangular** button from the **Viewports** panel of the **View** ribbon tab. Specify a first and second corner to create the floating viewport and exit the command. **Figure 29-1** shows an example of a single rectangular floating viewport used to display multiviews drawn in model space.

Using the Viewports Dialog Box

The **Rectangular** button previously described offers a shortcut for creating a single rectangular viewport using the **VIEWPORTS** command. Access the **VIEWPORTS** command to display the **Viewports** dialog box with multiple viewport configuration options. See **Figure 29-4**. The **Viewports** dialog box is similar in paper space and model space.

The **Standard viewports:** list contains preset viewport configurations. Select the ***Active Model Configuration*** option to convert a saved model space viewport configuration into individual floating viewports. For example, if model space displays two tiled viewports, use the ***Active Model Configuration*** option to create two floating viewports.

The remaining items in the **Standard viewports:** list are preset viewport configurations. The configuration name identifies the number of viewports and the arrangement or location of the largest viewport. Select a configuration to see a preview of the floating viewports in the **Preview** area. The **Viewport spacing:** text box is available when you select a standard viewport configuration that includes two or more viewports, as shown in **Figure 29-4**. Enter a value to define the space between multiple viewports.

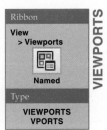

Figure 29-4.
The **Viewports** dialog box with the **New Viewports** tab selected. The **Viewport Spacing:** setting is available when paper space is active.

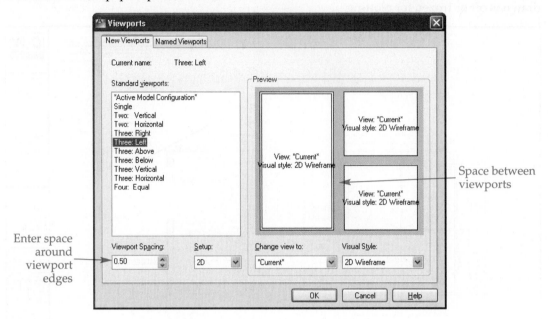

Enter space around viewport edges

Space between viewports

The default setting in the **Setup:** drop-down list is **2D**, and all viewports show the 2D drawing plane, or top view. The **3D** option provides the ability to assign a 3D view to specified viewports, which is appropriate for displaying 3D models, as explained in *AutoCAD and Its Applications—Advanced*.

The viewport configuration is displayed in the **Preview** image. To change a view in a viewport, pick the viewport in the **Preview** image and select from the **Change view to:** and **Visual Style** drop-down lists. The **Change view to:** drop-down list includes any named model views saved with the file. Pick a named view from the list to apply to the selected viewport. Use the **2D Wireframe** option of the **Visual Style** drop-down list for standard 2D drafting applications. *AutoCAD and Its Applications—Advanced* explains the visual style settings appropriate for 3D modeling.

Pick the **OK** button to specify a first and second corner to define the area occupied by the viewport configuration. See **Figure 29-5**. An alternative is to choose the **Fit** option before specifying the first point to fill the printable area with the viewport configuration.

The text-based **–VPORT** and **MVIEW** commands offer the same floating viewport options that you can more easily access from the **Viewports** panel of the **View** ribbon tab, the **Viewports** dialog box, or other methods such as a shortcut menu and the **Properties** palette. These techniques are described thoroughout this chapter.

Exercise 29-2

Complete the exercise on the companion website.
www.g-wlearning.com/CAD

Figure 29-5.
Select two points on the layout to specify the area to be filled by the viewport configuration. Notice that the viewport spacing forms as specified in the **Viewports** dialog box. This example shows the initial placement of three structural details using a three-viewport configuration. Each viewport will eventually display a single detail.

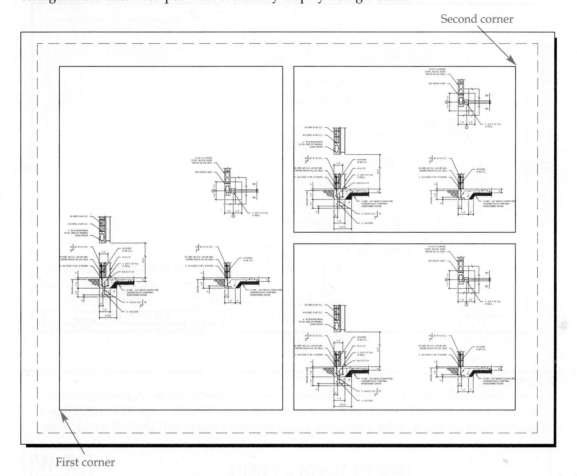

Second corner

First corner

Forming Polygonal Floating Viewports

A rectangular floating viewport is common and suitable for many applications. An alternative is to form a polygonal floating viewport boundary. The easiest way to construct a polygonal floating viewport is pick the **Polygonal** button from the **Viewports** panel of the **View** ribbon tab. Construct a polygonal viewport using the same techniques you use to draw a closed polyline object. The viewport can be any closed shape composed of lines and arcs. **Figure 29-6** shows a polygonal floating viewport used to define the maximum drawing view area 1/2″ in from the border and title block.

Ribbon

View
> Viewports

Polygonal

Converting Objects to Floating Viewports

AutoCAD also allows you to convert a closed object drawn in paper space to a floating viewport. The easiest way to convert an object to a floating viewport is pick the **From Object** button from the **Viewports** panel of the **View** ribbon tab. Select a closed shape, such as a circle, ellipse, or closed polyline shape, to convert the object to a viewport. **Figure 29-7** shows an example of a circle and rectangle converted to floating viewports.

Ribbon

View
> Viewports

From Object

Figure 29-6.
A structural detail displayed in a polygonal floating viewport. The layer on which the viewport is drawn is turned off or frozen for plotting.

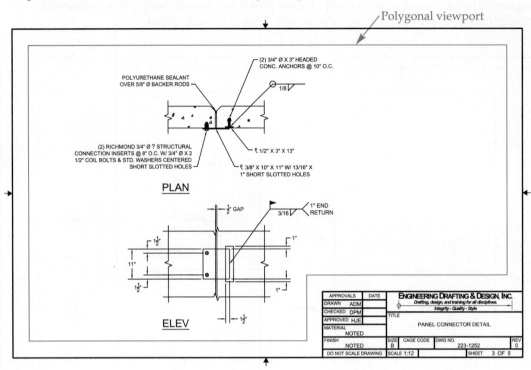

Polygonal viewport

Figure 29-7.
You can convert any closed object to a floating viewport. This example shows a cover sheet with a plot plan viewport converted from a rectangle and a vicinity map converted from a circle. The layer on which these viewports are drawn remains on and thawed for plotting.

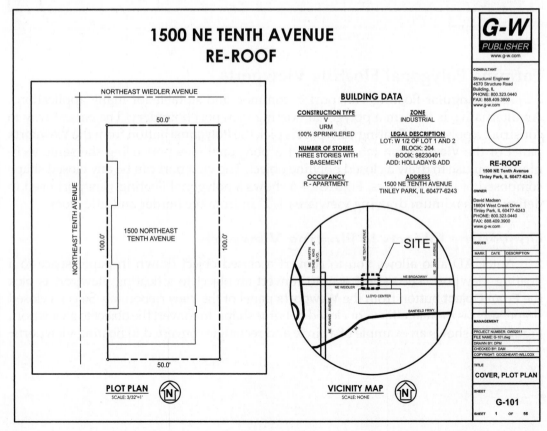

Exercise 29-3

Complete the exercise on the companion website.
www.g-wlearning.com/CAD

Adjusting the Floating Viewport Boundary

For purposes of adjusting a floating viewport boundary, you should consider the boundary a closed object. For example, treat rectangular viewports like rectangles, polygonal viewports like closed polyline objects, circular viewports like circles, and elliptical viewports like ellipses. Use grip editing and editing commands such as **MOVE**, **ERASE**, **STRETCH**, and **COPY** as needed to modify the size, shape, and location of floating viewports. When you adjust a floating viewport, the "hole" cut through the sheet changes.

Exercise 29-4

Complete the exercise on the companion website.
www.g-wlearning.com/CAD

Clipping Viewports

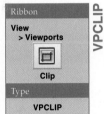

The **VPCLIP** command allows you to redefine the boundary of an existing viewport. To access the command from a shortcut menu, select a viewport and then right-click and pick **Viewport Clip**. You can clip a floating viewport to an existing closed object that you draw before accessing the **VPCLIP** command, or you can clip the viewport to a polygonal shape that you create while using the command.

After you access the **VPCLIP** command, select the viewport to clip. Then select an existing closed shape, such as a circle, ellipse, or closed polyline, to recreate the viewport in the shape of the selected object. See Figure 29-8. An alternative, after you select the existing closed shape, is to use the **Polygonal** option to redefine the viewport to a polygonal shape. This option works the same as the **Polygonal** option previously described, except the existing viewport transforms into the new shape.

AutoCAD recognizes a clipped floating viewport as clipped. The **VPCLIP** command offers a **Delete** option when you select a clipped viewport. Use the **Delete** option to remove the clipped definition and convert the shape into a viewport sized to fit the extents of the original clipping object or polygonal shape.

A clipping object or polygonal shape does not need to be on or overlap the viewport to be clipped.

PROFESSIONAL TIP

Create floating viewports and adjust the size and shape of viewport boundaries after you insert the border, title block, and other layout content. This will allow you to position viewports so they do not interfere with layout information.

Figure 29-8.
Clipping a viewport showing a wall section to a closed polyline to create an eave detail.
AutoCAD removes the original viewport and converts the rectangle to a viewport.

Polyline created using the **RECTANGLE** tool

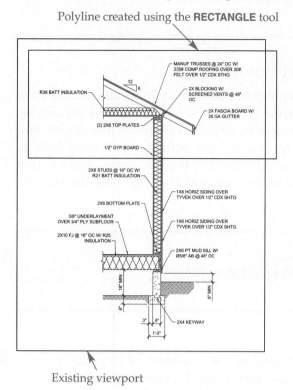

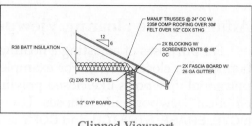

Clipped Viewport

Existing viewport

Rotating Model Space Content

The **VPROTATEASSOC** system variable setting determines what happens to model space content when you rotate a floating viewport. The variable has a value of 1 by default. As a result, when you rotate a viewport, the display of objects in model space rotates to align with the viewport. See **Figure 29-9A**. The orientation of objects in model space does not change. In order to maintain the original alignment of model space content in a floating viewport, as shown in **Figure 29-9B**, access the **VPROTATEASSOC** system variable before rotating the viewport and enter a value of 0.

Other commands, such as **UCS** and **MVSETUP**, include options for rotating items shown through a viewport. The **VPROTATEASSOC** system variable automates the process. The entire display of model space rotates with the rotation of the viewport. Drawing characteristics such as unidirectional dimensions do not update according to the rotation.

Activating and Deactivating Floating Viewports

Activate a floating viewport to work with model space objects while in paper space. This allows you to adjust the display of the model space drawing shown in the viewport. Repeat the process of activating and adjusting a viewport for every floating viewport in the layout to achieve the final drawing.

To activate a floating viewport, double-click inside the viewport boundary, press the **PAPER** button on the status bar, or type **MSPACE** or **MS**. If the layout contains a single viewport, the viewport appears highlighted, indicating that it is current. On

Figure 29-9.
A highway cut and fill plan shown through a viewport. A—The default **VPROTATEASSOC** setting of 1 rotates model space content on a floating viewport to align with viewport rotation. B—Change the value to 0 to maintain the original model space orientation when you rotate a viewport.

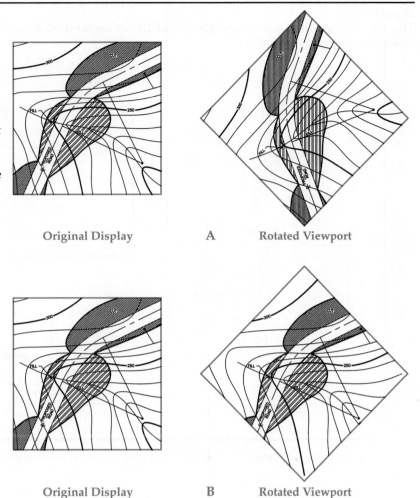

Original Display A Rotated Viewport

Original Display B Rotated Viewport

the layout, the UCS icon disappears, and the model space UCS icon is displayed in the corner of each layout viewport. You are now working directly in model space, through the paper space viewport. The active and highlighted viewport is the viewport you double-click on or the newest viewport, depending on how you access the **MSPACE** command. By default, each floating viewport includes in-canvas viewport controls, the navigation bar, and the ViewCube. See **Figure 29-10**. To make a different viewport active, pick once inside the viewport.

> The in-canvas viewport controls are initially difficult to see, but they are located in the upper-left corner of the active floating viewport.

After you adjust the display of all floating viewports, you must re-enter paper space to plot and continue working with the layout. To activate paper space, double-click outside a viewport, press the **MODEL** button on the status bar, or type **PSPACE** or **PS**. The layout space UCS icon reappears, and the model space UCS icon disappears from the corners of the viewports.

Scaling a Floating Viewport

The scale you assign to a floating viewport is the same as the drawing scale. The quickest way to set viewport scale is to activate a viewport, or pick a viewport

Figure 29-10.
The active viewport appears highlighted and allows you to work in model space while AutoCAD displays paper space.

Highlighted viewport is active

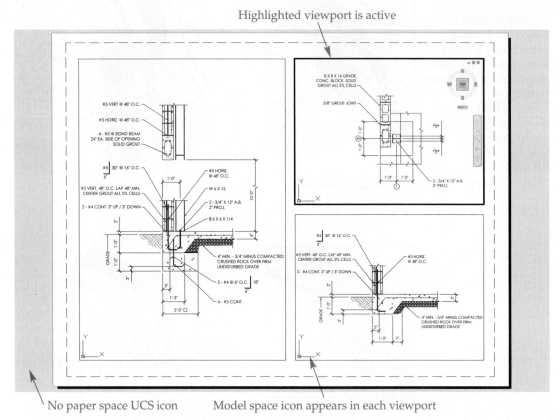

No paper space UCS icon   Model space icon appears in each viewport

boundary in paper space without activating the viewport. Then select the appropriate scale from the **Viewport Scale** flyout on the status bar. See **Figure 29-11**. An alternative is to pick the viewport to be scaled in paper space and access the **Properties** palette. Then choose a viewport scale from the **Standard scale** drop-down list.

If a scale is unavailable, or to change an existing scale, pick the **Annotation Scale** flyout on the status bar and choose **Custom...** to access the **Edit Scale List** dialog box. Move the highlighted scale up or down in the list using the **Move Up** or **Move Down** button. To remove the highlighted scale from the list, pick the **Delete** button.

Select the **Edit...** button to open the **Edit Scale** dialog box, where you can change the name of the scale and adjust the scale by entering the paper and drawing units. For example, a scale of 1/4" = 1'-0" uses a paper units value of .25 or 1 and a drawing units value of 12 or 48. To create a new annotation scale, pick the **Add...** button to display the **Add Scale** dialog box, which provides the same options as the **Edit Scale** dialog box. Pick the **Reset** button to restore the list to display the default scales.

Changes you make in the **Edit Scale List** dialog box are stored with the drawing and are specific to the drawing. To make changes to the default scale list saved to the system registry, pick the **Default Scale List...** button on the **User Preferences** tab of the **Options** dialog box to access the **Default Scale List** dialog box. The options are the same as those in the **Edit Scale List** dialog box, but changes are saved as the default for new drawings, resetting the **Edit Scale List** dialog box.

Figure 29-11.
Using the **Viewport Scale** flyout button on the status bar to set the drawing scale. This example
shows scaling dimensioned multiviews in a single viewport to finalize a part drawing.

Active or selected viewport   Select a drawing scale from the list

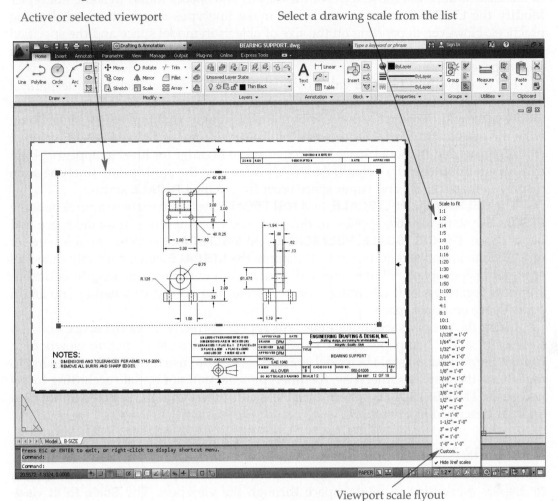

Viewport scale flyout

CAUTION

Setting viewport scale is a zoom function that increases or decreases the displayed size of the drawing in the viewport. You can also use the **XP** option of the **ZOOM** command to specify the scale of the active viewport. If you use an option of the **ZOOM** command other than a specific XP value to adjust the drawing inside an active floating viewport, the drawing loses the correct scale. Once the viewport scale is set, do not zoom in or out. Lock the viewport, as described later in this chapter, to help ensure that the drawing remains properly scaled.

Scaling Annotations

You should always draw objects at their actual size, or full scale, in model space, regardless of the size of the objects. However, this method requires special consideration for annotations, hatches, and similar items added to objects in model space. You can adjust the appearance of annotations manually, but it is often best to use annotative objects to automate the process. Scaled viewports and annotative objects function together to scale drawings properly and increase multiview drawing flexibility. Chapter 30 explains annotative objects.

Controlling Linetype Scale

Adjust the **LTSCALE** system variable to make a global change to the linetype scale to increase or decrease the lengths of the dashes and spaces found in some linetypes. Modify the **LTSCALE** value as needed to make linetypes match standard drafting practices. However, depending on the size of objects in model space and the specified floating viewport scale, an **LTSCALE** value in model space may not be appropriate for paper space.

For example, an **LTSCALE** value of .5 is appropriate for an inch unit mechanical part drawing plotted at full scale. In this example, apply an **LTSCALE** value of .5 to model space and paper space because both environments function at full scale. If you scale the drawing to 2:1, a linetype scale of .25 (scale factor of $1/2 \times$ **LTSCALE** value of .5 = .25) is needed in model space and paper space in order for lines to appear correct in both environments. By default, AutoCAD calculates the appropriate linetype scale display in model space and paper space according to the **LTSCALE** setting.

The **CELTSCALE**, **PSLTSCALE**, and **MSLTSCALE** system variables control how the **LTSCALE** system variable applies, or does not apply, to linetypes in model space and paper space. The **CELTSCALE**, **PSLTSCALE**, and **MSLTSCALE** system variables are set to **1** by default, and should be set to **1** to apply the **LTSCALE** value correctly in model space and paper space. All linetypes will then appear with the same lengths of dashes and dots regardless of the floating viewport scale, and no matter whether you are in paper space or model space.

Using the previous example, lines will appear correctly in model space and at a scale of 1:1 and 2:1 in paper space. However, when you scale a floating viewport or change the annotation scale in model space, remember to use the **REGEN** command to regenerate the display. Otherwise, the linetype scale will not update according to the new scale. The **MSLTSCALE** system variable is associated with the selected annotation scale, as described in Chapter 30.

Adjusting a View

When you first create a floating viewport, AutoCAD performs a **ZOOM Extents** to display everything in model space through the viewport. The **Scale to fit viewport scale** option accomplishes the same task. When you scale a viewport, AutoCAD adjusts the view from the center of the viewport, which often results in the appropriate display. However, you must adjust the view when you change the size or shape of the viewport, when a centered view is not appropriate, or to display a specific portion of the drawing. Use the **PAN** command in an active viewport to redefine the displayed location of the view.

Boundary Adjustment

The viewport boundary can "cut off" a scaled model space drawing. This may be acceptable to display a portion of a view. However, to display the entire view, you can either increase the size of the viewport boundary or select a different scale to reduce the displayed size of the view to fit the viewport. If it is not appropriate to increase the size of the viewport or decrease the scale, use a larger sheet size. **Figure 29-12** shows a drawing with two viewports. The rectangular viewport shows everything in model space at full scale. The circular viewport cuts off model space objects and displays objects at a 2:1 scale to create a detail.

Precision Adjustment and Alignment

The **PAN** command is effective for adjusting model space content in a floating viewport, but it offers limited precision, especially when you are attempting to align views in different viewports. One way to adjust and align views precisely is to use the **MOVE** command to move viewports, because objects shown in a viewport move with

Figure 29-12.
An example of a multiview part drawing in which it is appropriate to show all model space objects, but also necessary to display only a portion of model space to create a view enlargement.

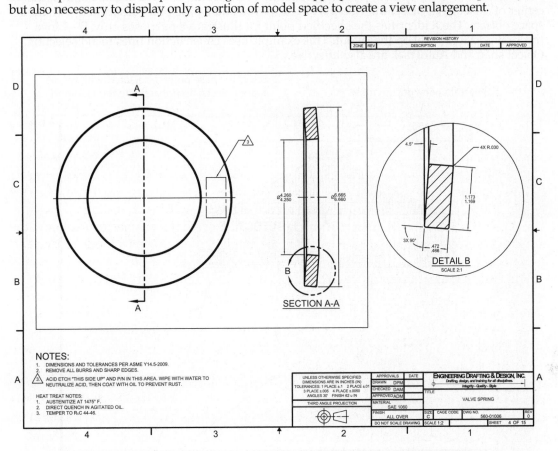

the viewport. Draw construction geometry, such as a line, on the layout or use object snaps and AutoTrack to reference specific points between views. Access the **MOVE** command and select the viewport as the object to move, but specify the base point and second point on objects in model space or on construction geometry on the layout. Use object snaps to aid point selection. See **Figure 29-13**.

MVSETUP

The **MVSETUP** command provides another way to adjust and align views precisely. Choose the **Align** option and then select the **Horizontal** option to align views horizontally or the **Vertical** alignment option to align views vertically. A viewport becomes activated when you select an alignment option. Pick inside the viewport with the model space view to pan. Then specify a base point in model space from which model space pans. Use object snaps, AutoTrack, or construction geometry to aid point selection. Activate the viewport that contains the model space view with which to align, and specify a stationary point in model space to which the base point will pan horizontally or vertically. See **Figure 29-14**. Continue using the **MVSETUP** command or cancel to exit.

The **MVSETUP** command includes several additional options that are outdated, perform operations that you can accomplish more easily using other commands, or have limited application.

Figure 29-13.
Moving a floating viewport to position the center of the front elevation of this home in the center of a sheet. A—Select the viewport to be moved and a base point associated with model space objects. The X identifies the specified point for illustrative purposes only. B—Move the viewport to a point on the layout. This example uses construction lines to aid selection. Object snap and AutoTrack are also effective.

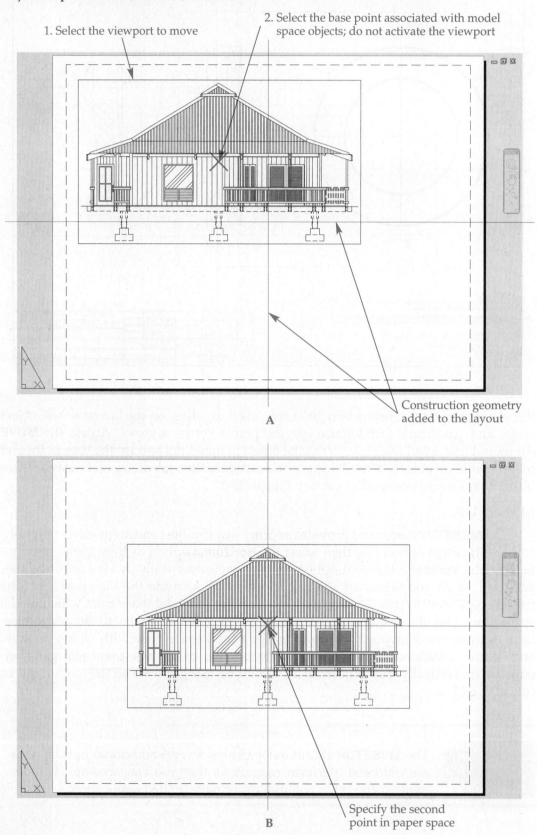

1. Select the viewport to move

2. Select the base point associated with model space objects; do not activate the viewport

Construction geometry added to the layout

A

Specify the second point in paper space

B

Figure 29-14.
Using the **MVSETUP** command, **Align** option, and **Vertical** alignment function to align views vertically in different floating viewports. The same process applies to using the **Horizontal** function of the **Align** option to align views horizontally.

Selected base point

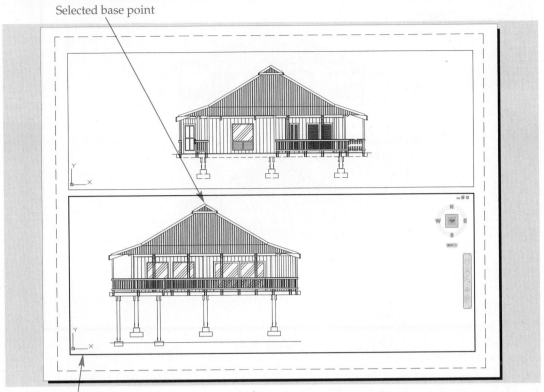

Active viewport

1. Specify the Base Point

Selected alignment point   Active viewpoint

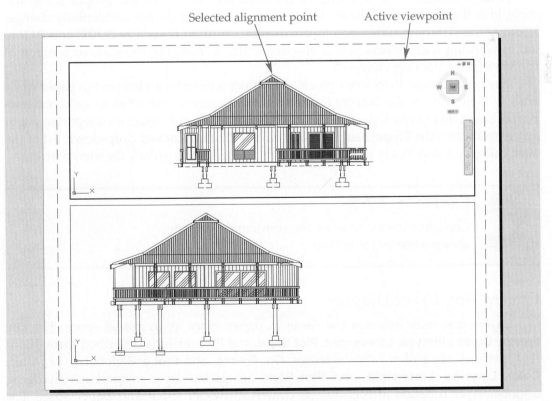

2. Specify the Alignment Point

(Continued).

Figure 29-14.
Continued.

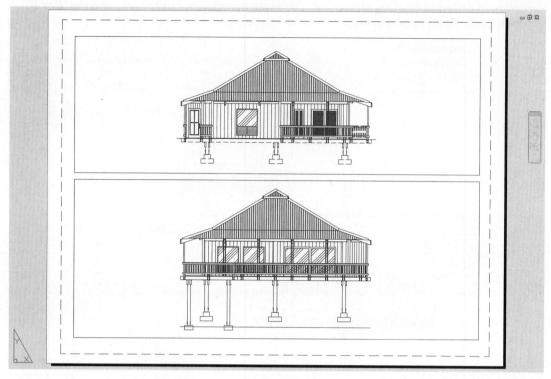

Aligned View

Locking and Unlocking Floating Viewports

After you adjust the drawing in the viewport to reflect the proper scale and view, lock the viewport so the scale and view orientation do not accidentally change. Locking the viewport allows you to use display commands such as **ZOOM** and **PAN** to aid in working with objects in model space without changing the scale or position of the view in the floating viewport.

The quickest way to lock or unlock a viewport is to select a viewport in paper space and right-click. From the **Display Locked** cascading menu, select **Yes** to lock the viewport or select **No** to unlock the viewport. A second option is to select a viewport in paper space and access the **Properties** palette. From the **Display Locked** drop-down list of the **Misc** category, select **Yes** to lock the viewport or select **No** to unlock the viewport.

Exercise 29-5

Complete the exercise on the companion website.
www.g-wlearning.com/CAD

Controlling Layer Display

Layers generally function the same in paper space as in model space. The **On**, **Freeze**, **Color**, **Linetype**, **Lineweight**, **Plot Style**, and **Plot** settings described throughout this textbook are *global layer settings*. **On**, **Freeze**, and **Plot** are global layer states. **Color**, **Linetype**, **Lineweight**, and **Plot Style** are global layer properties. Changing a global layer function affects objects drawn in model space and paper space. For example, if you change the color of a layer in model space and lock the layer, all objects drawn on that layer in paper space also change color and become locked.

global layer settings: Layer settings applied to both model space and paper space.

AutoCAD provides the option to freeze layers in a floating viewport and apply *layer property overrides*. These features expand the function of the layer system and improve your ability to reuse drawing content.

Use the **LAYER** command and the corresponding **Layer Properties Manager** to control layer display in floating viewports. See **Figure 29-15**. This is the same **Layer Properties Manager** used to manage layers throughout this textbook. The **NEW VP Freeze, VP Freeze, VP Color, VP Linetype, VP Lineweight, VP Transparency,** and **VP Plot Style** columns control layer display options for floating viewports. Except for the **NEW VP Freeze** column, these columns appear only in layout mode. You probably need to use the scroll bar at the bottom of the palette to see the columns. The options can apply to layout content, such as the viewport boundary. However, layer settings typically apply to an active floating viewport. Be sure the floating viewport to which you want to apply layer control settings is active as you work through the following sections.

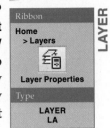

The text-based **VPLAYER** command also controls layer display in floating viewports. The **Layer Properties Manager** is faster and easier to use than the **VPLAYER** command.

Freezing and Thawing

Freeze layers in the active viewport to create different views using a single drawing. For example, **Figure 29-16A** shows the model space display of a floor plan with electrical plan content added directly to the floor plan using electrical plan layers. **Figure 29-16B** shows two layouts from the same drawing file. One layout displays a floor plan with no electrical information, and the other layout displays an electrical plan without specific floor plan content.

In the **Figure 29-16** example, you draw many objects, such as doors, walls, and windows, on layers that maintain the global **Thaw** setting. As a result, these objects appear in model space and in both floating viewports. Freeze layers in specific viewports (**VP Freeze**) to create two different drawings. This example shows viewport layer

Figure 29-15.
Use the **Layer Properties Manager** to control the display of layers in floating viewports.

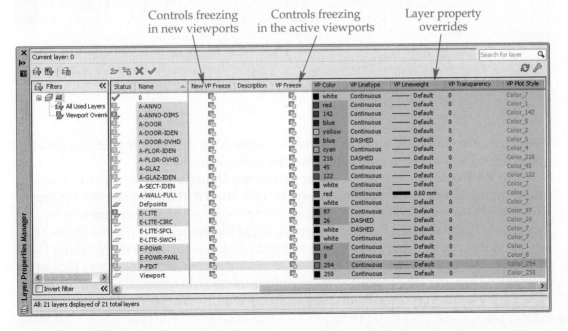

Creating a floor plan and separate electrical plan using the same drawing file. A—An example of "overlapping" layers in model space. B—Layers frozen in separate layouts to create two different drawing views.

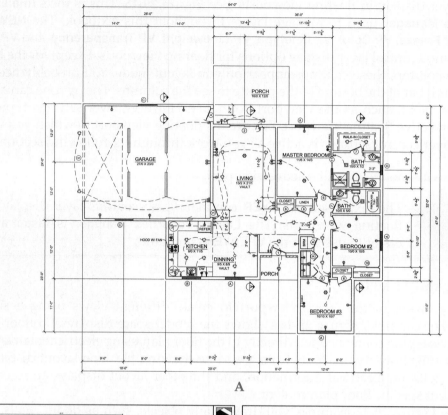

A

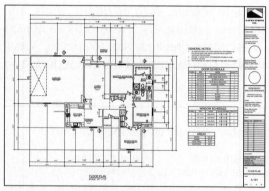

Floor Plan Layout
Created by freezing electrical layers
in the floating viewport

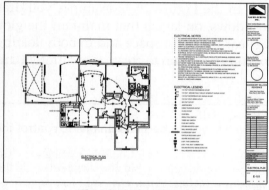

Electrical Plan Layout
Created by freezing floor plan
layers in the floating viewport

B

freezing in two different viewports, each viewport in a different layout, but you can apply the same concept to multiple viewports in the same layout.

The **VP Freeze** column of the **Layer Properties Manager** palette controls freezing and thawing layers in the current viewport. Pick the **VP Thaw** icon or the **VP Freeze** icon to toggle freezing and thawing in the current viewport. The **VP Freeze** icon freezes layers only in the selected floating viewport, while the **Freeze** icon freezes layers globally in all floating viewports. To freeze a layer in all layout viewports, including those created before picking the **VP Freeze** icon or **VP Thaw** icon, right-click and pick **In All Viewports** from the **VP Freeze Layer** cascading menu. To freeze a layer in all layout viewports except the current viewport, right-click and select **In All Viewports Except**

VP Freeze

VP Thaw

Current from the **VP Freeze Layer** cascading menu. To thaw a layer in all layout viewports, right-click and choose **VP Thaw Layer In All Viewports**.

PROFESSIONAL TIP

The **VP Freeze** function is also available in the **Layer Control** drop-down list in the **Layers** panel on the **Home** ribbon tab. This provides a quick way to freeze and thaw layers in a viewport without accessing the **Layer Properties Manager**.

New VP Freeze

The **New VP Freeze** column of the **Layer Properties Manager** controls freezing and thawing of layers in newly created floating viewports. Pick the **New VP Thaw** icon or the **New VP Freeze** icon to toggle freezing and thawing in any new floating viewport. This feature has no effect on the active viewport.

New VP Thaw

Right-click on a layer in the **Layer Properties Manager** and select **New Layer VP Frozen in All Viewports** to create a new layer preset with the **VP Freeze** and **New VP Freeze** icons selected.

Exercise 29-6

Complete the exercise on the companion website.
www.g-wlearning.com/CAD

Layer Property Overrides

You can use layer property overrides to create different views without changing individual object properties, creating separate drawing files, or readjusting global layer properties. For example, **Figure 29-17A** shows the model space display of a hopper and conveyer system with unique layers assigned to the hopper and conveyer. **Figure 29-17B** shows a layout with two floating viewports. The viewport on the left shows the hopper and conveyer with global layer settings applied, as in model space. The viewport on the right shows the hopper with a layer color override and the conveyer with a layer color and linetype override. In this example, layer property overrides create a view that clearly shows the two separate components. Phantom lines highlight the conveyer as the mechanism.

The **VP Color**, **VP Linetype**, **VP Lineweight**, **VP Transparency**, and **VP Plot Style** columns in the **Layer Properties Manager** control the property overrides assigned to layers. The **VP Plot Style** column appears only when a named plot style is in use. Layer property overrides apply only to floating viewports in paper space. AutoCAD does not identify layers that contain layer property overrides in model space.

The general process of overriding a layer property is just like that of changing a global value. For example, to override the color assigned to a layer, pick the color swatch and choose a color from the **Select Color** dialog box. The difference is that layer property overrides apply only to specific layers in an active floating viewport. Object properties do not change from Bylayer, and the model space display does not change.

When viewed in paper space, the **Properties** palette, **Layer Properties Manager**, and **Layer Control** drop-down list of the **Layers** panel of the **Home** ribbon tab indicate which layers include layer property overrides. See **Figure 29-18**. The **Layer Properties Manager** identifies layers that contain layer property overrides with a sheet and viewport icon

Figure 29-17.
A—A hopper and conveyer drawn in model space. B—Using property overrides to create different layout views.

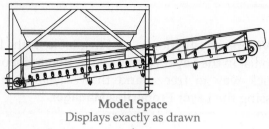

Model Space
Displays exactly as drawn

A

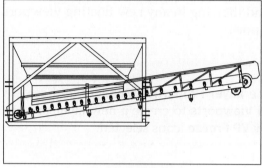

Layer color override applied to hopper

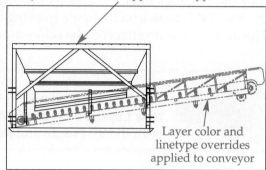

Layer color and linetype overrides applied to conveyor

Left Viewport
Displays model space objects exactly as drawn

Right Viewport
Layer property overrides applied

B

in the status column. The layer names, global properties affected by the overrides, and the property overrides are highlighted. Use the **Viewport Overrides** filter to display and manage only those layers that include layer property overrides. You can also save layer property overrides in a layer state.

The **Properties** palette highlights layer names that contain layer property overrides. Properties affected by the override are also highlighted and are identified as Bylayer (VP). The **Layer Control** drop-down list in the **Layers** panel of the **Home** ribbon tab also highlights layers that include layer property overrides.

The **Viewport Overrides** icon appears in the status bar when you activate a floating viewport or assign layer property overrides to the active viewport.

If layer property overrides are no longer necessary, you should remove the overrides from the layer. Changing a property back to the original, or global, value does not remove the override. Right-click on a layer that contains layer property overrides in the **Layer Properties Manager** and pick **Remove Viewport Layer Overrides** to access a cascading menu of options for removing layer property overrides. Pick **Selected Layers** and then **In Current Viewport Only** or **In All Viewports** to remove layer property overrides from the current viewport or from all viewports that include overrides.

You can also use the **Layer** option of the **MVIEW** command or the **Reset** option of the **VPLAYER** command to remove layer property overrides.

AutoCAD and Its Applications—Basics

Figure 29-18.
Layers with property overrides are highlighted in the **Properties** palette, the **Layer Properties Manager**, and the **Layer Control** drop-down list on the ribbon.

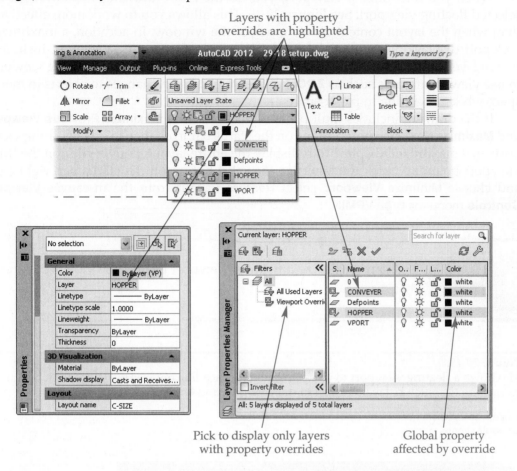

Layers with property overrides are highlighted

Pick to display only layers with property overrides

Global property affected by override

Exercise 29-7

Complete the exercise on the companion website.
www.g-wlearning.com/CAD

Turning Off Floating Viewport Objects

By default, objects appear in floating viewports, allowing you to view model space through the viewports. You can hide objects in the floating viewport without removing the viewport, which is convenient, for example, to plot a certain view, but still have access to the viewport. One option to toggle the display of objects in the viewport on and off is to select a viewport in paper space and right-click. From the **Display Viewport Objects** cascading menu, select **No** to hide objects or **Yes** to display objects. You can also select a viewport in paper space, access the **Properties** palette, and from the **On** drop-down list, select **No** to hide objects or **Yes** to show objects.

Maximizing Floating Viewports

When you activate a floating viewport, you are working in model space from within the paper space display. The primary function of activating a floating viewport is to adjust the display of model space to prepare a final drawing. Avoid working inside an active viewport to make changes to model space objects. One alternative to activating a floating viewport is to maximize the viewport. The quickest way to maximize a viewport is to double-click on the viewport or pick the **Maximize Viewport** button on the status bar.

When you maximize a viewport, you fill the entire drawing window with the selected floating viewport. See **Figure 29-19**. This allows you to work more effectively than when the layout content covers much of the window. In addition, a maximized viewport displays objects exactly as they appear in the floating viewport, including frozen layers and layer overrides. Typically, you should maximize a floating viewport to use view commands such as **ZOOM** and **PAN** and make changes to objects in model space while remaining in paper space.

If the drawing includes multiple viewports, use the **Maximize Previous Viewport** and **Maximize Next Viewport** buttons on the status bar to change to other floating viewports in a maximized display. To redisplay the entire layout, double-click on the thick viewport boundary, pick the **Minimize Viewport** button on the status bar, right-click and choose **Minimize Viewport**, select **Restore Layout** from the in-canvas **Viewport Controls** menu, or type **VPMIN**.

You can maximize a floating viewport even if the viewport is not active.

Figure 29-19.
Maximize a floating viewport to work in a model-space-like environment, but with layout characteristics, such as layers frozen in the viewport and layer property overrides.

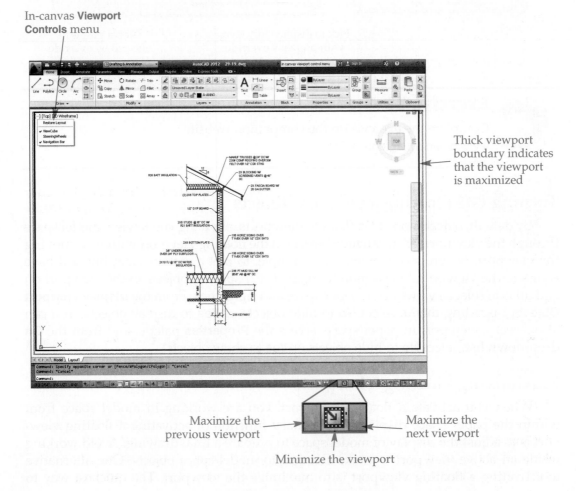

In-canvas **Viewport Controls** menu

Thick viewport boundary indicates that the viewport is maximized

Maximize the previous viewport

Maximize the next viewport

Minimize the viewport

PROFESSIONAL TIP

If you do not want to see floating viewport boundaries on the plot, remember to freeze or turn off the layer assigned to the viewport before plotting.

Exercise 29-8

Complete the exercise on the companion website.
www.g-wlearning.com/CAD

Using the CHSPACE Command

The **CHSPACE** command provides a convenient way to move objects between model space and paper space without having to copy and paste between the different environments. The **CHSPACE** command is only available in paper space. Activate a floating viewport to move objects from model space to paper space, or deactivate floating viewports to move objects from paper space to model space. Then, access the **CHSPACE** command, select the objects to transfer, and right-click or press [Enter] or the space bar. To move objects to model space, select a viewport to activate and then right-click or press [Enter] or the space bar to complete the operation. To move objects to paper space, right-click or press [Enter] or the space bar twice to complete the operation.

Type
CHSPACE

Plotting

After you prepare a layout for plotting, you are ready to plot. If you develop an appropriate page setup and layout, the process of creating the actual print should be almost automatic. Select the layout to plot and access the **PLOT** command. The **Plot** dialog box appears with the name of the layout displayed on the title bar. See **Figure 29-20**.

You can also access the **Plot** dialog box by selecting from the shortcut menu available from the model or layout tab, or by picking the **Plot** button in a **Model** or a layout thumbnail image in the **Quick View Layouts** or **Quick View Drawings** tool display.

The **Page Setup** and **Plot** dialog boxes are very similar, except the **Plot** dialog box provides additional options specific to creating a plot. All the settings in the **Plot** dialog box correspond to those in the **Page Setup** dialog box. Pick the **>**, or **More Options**, button in the lower-right corner of the **Plot** dialog box to toggle the display of additional dialog box areas, as shown in **Figure 29-20**. Specify a number in **Number of copies** text box to indicate how many copies of the layout to plot. The **Plot options** area provides additional plot settings. Pick the **Plot in background** check box to continue working while the plot processes.

Changing plot settings in the **Plot** dialog box overrides the page setup for a specific plotting requirement. This is a convenient way to make a plot using slightly modified plot settings without creating a new page setup. For example, you can make a "check print" by selecting a printer, using an A- or B-size sheet and scaling the plot to fit the

Figure 29-20.
Use the **Plot** dialog box to finalize the layout and send the drawing to a printer, plotter, or file.

Select a different page
setup from the list

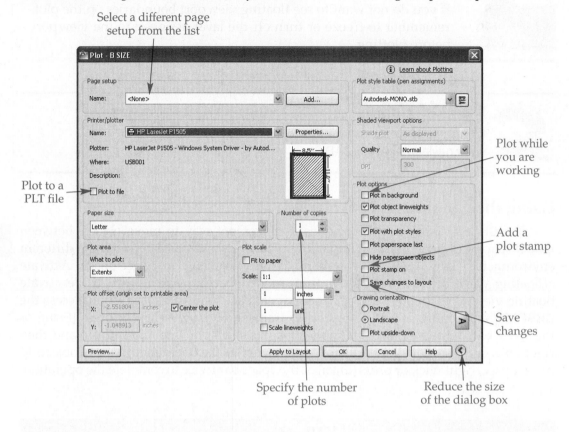

Plot to a
PLT file

Plot while
you are
working

Add a
plot stamp

Save
changes

Specify the number
of plots

Reduce the size
of the dialog box

paper. After the drawing prints, the settings return to those originally assigned in the page setup, allowing you to plot the final drawing using the appropriate printer, sheet size, and 1:1 scale.

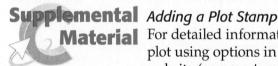

Supplemental Material

Adding a Plot Stamp

For detailed information about adding a *plot stamp* to the plot using options in the plot dialog box, go to the companion website (www.g-wlearning.com/CAD), select this chapter, and select **Adding a Plot Stamp**.

plot stamp: Text added only to the hard copy that includes information such as the drawing name or the date and time the drawing was printed.

Saving Changes to the Layout

If you make changes in the **Plot** dialog box and want to save changes to the layout page setup for future plots, pick the **Save changes to layout** check box in the **Plot options** area. You can also save changes by picking the **Apply to Layout** button. If you do not save the changes or apply them to the layout, changes made in the **Plot** dialog box are discarded, and the original page setup appears the next time you open the **Plot** dialog box.

Page Setup Options

The **Plot** dialog box provides an alternate means of creating a page setup. To apply this technique, access the **Plot** dialog box and make changes to plot settings, just as you would in the **Page Setup** dialog box. Then select the **Add...** button in the **Page setup** area to display the **Add Page Setup** dialog box. Enter a name for the page setup in the

New page setup name: text box. All current settings in the **Plot** dialog box are saved with the new page setup. Select a page setup from the **Name:** drop-down list to restore the settings in the **Plot** dialog box. Pick the **<Previous plot>** option to reference the settings used to create the last plot, or pick the **Import...** button to import a page setup from a DWG, DWT, or DXF file.

When using the **Plot** dialog box to define settings for a page setup, name the page setup after you make changes to settings. If you name the page setup and want to make changes later, such as changes to a plot style, use the **Page Setup Manager** dialog box instead.

Previewing the Plot

The final step before plotting is to preview the plot. The plot preview shows exactly what the plot should look like based on plot and layout settings. Always preview the plot to check the drawing for errors and view the effects of plot settings before sending the information to the plot device. This will help you eliminate unnecessary plots. To preview the plot, pick the **Preview** button in the lower-left corner of the **Plot** dialog box to enter preview mode. See **Figure 29-21**. What you see on-screen is exactly what will plot, assuming you use a color plotter to make color prints and load the correct sheet size in the plot device.

The **Realtime ZOOM** command is automatically active in the preview window. Additional view commands are available from the toolbar near the top of the window or from a shortcut menu. Use view commands to help confirm that the plot settings are correct. When you finish previewing the plot and are ready to plot, pick the **Plot** button on the toolbar, or right-click and select **Plot**. When you finish previewing the plot, pick the **Close** button on the toolbar, press [Esc] or [Enter], or right-click and select **Exit** to return to the **Plot** dialog box.

Figure 29-21.
Previewing a plot is an excellent way to help confirm that the plot will be correct before sending the information to the plot device.

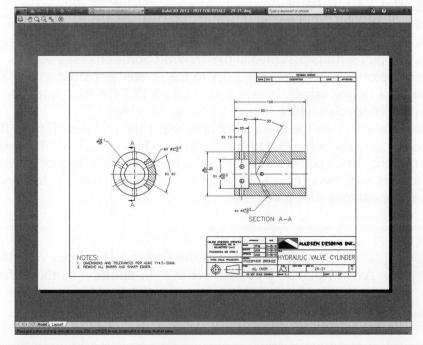

 You must select a plot device other than **None** in order to activate the **Preview...** button.

Output

When plotting to a plotter or printer, pick the **OK** button to send the plot to the plot device and close the **Plot** dialog box. When plotting to a PDF file, pick the **OK** button to display the **Browse for Plot File** dialog box. Specify a location and name for the file and then pick the **Save** button to close the **Plot** dialog box and open the PDF file automatically using installed Adobe® software. When plotting to a DWF or DWFX file, pick the **OK** button to display the **Browse for Plot File** dialog box. Specify a location and name for the file and then pick the **Save** button to close the **Plot** dialog box. View the DWF or DWFX file using Autodesk® Design Review software.

AutoCAD notifies you of the success or failure of a hard copy or electronic plot with a message from the status bar tray. To view additional details about the plot, pick **Click to view plot and publish details...** to display the **Plot and Publish Details** dialog box. You can also access the **Plot and Publish Details** dialog box using the **VIEWPLOTDETAILS** command.

 AutoCAD provides additional tools for exporting drawings, automating the process of transmitting drawings electronically (eTransmit), and *publishing*. Publishing a set of drawing sheets is described later in this textbook.

publishing:
Preparing a sequential set of multiple drawings for hard copy or electronic plotting of the set.

 Exercise 29-9

Complete the exercise on the companion website.
www.g-wlearning.com/CAD

Plotting to a PLT File

If a plot device is not available, but you are ready to plot, an alternative is to plot to a file. A plot file saves with a .plt extension. The file stores all the drawing geometry, plot styles, and plot settings assigned to the drawing. Some offices or schools with only one printer or plotter attach a *plot spooler* to the printer or plotter to plot a PLT file. The plot spooler device usually allows you to take a PLT file from a storage disk and copy it to the plot spooler, which in turn plots the drawing.

plot spooler: A disk drive with memory that allows you to plot files.

To plot to a file, open the **Plot** dialog box, select the plot device from the **Name:** drop-down list, and check the **Plot to file** check box. The setting in the **Plot and Publish** tab of the **Options** dialog box determines the location in which the plot file is saved. To specify the path, pick the ellipsis (...) button to display the **Select default location for all plot-to-file operations** dialog box.

 The **Plot and Publish** tab of the **Options** dialog box contains other general plot and publish settings in addition to those described in this textbook. Most plot and publish settings seldom require adjustment.

Express Tools
Chapter 29

Layout Express Tools
    The **Layout** panel of the **Express Tools** ribbon tab includes additional layout commands. For information about the most useful layout express tools, go to the companion website (www.g-wlearning.com/CAD), select this chapter, and select **Layout Express Tools**.

Template Development
Chapter 29

Adding Layouts
    For detailed instructions on adding layouts to each drawing template, go to the companion website (www.g-wlearning.com/CAD), select this chapter, and select **Template Development**.

Chapter Review

Answer the following questions. Write your answers on a separate sheet of paper or complete the electronic chapter review on the companion website.
www.g-wlearning.com/CAD

1. Name the two types of content that are brought together to create a complete drawing.
2. What commands can you use to modify the boundary of a floating viewport?
3. Briefly explain how to create a polygonal viewport.
4. How can you convert an object created in paper space into a floating viewport?
5. How do you activate a floating viewport?
6. How can you tell that a viewport is active in paper space?
7. How do you reactivate paper space after activating a floating viewport for editing?
8. How does the scale you assign to a floating viewport compare with the drawing scale?
9. To what value should the **CELTSCALE**, **PSLTSCALE**, and **MSLTSCALE** system variables be set so that the **LTSCALE** value will be applied correctly in model space and paper space?
10. Viewport edges may cut off the drawing when the viewport is correctly scaled. List three options to display the entire view.
11. Why should you lock a viewport after you adjust the drawing in the viewport to reflect the proper scale and view?
12. Give an example of why you would hide objects in a floating viewport without removing the viewport.
13. What is a plot stamp?
14. If you make changes to the page setup using the **Plot** dialog box, how can you save these changes to the page setup so that the changes apply to future plots?
15. Give at least two reasons why you should always preview a plot before sending the information to the plot device.

Drawing Problems

Start AutoCAD if it is not already started. Start a new drawing for each problem using an appropriate template of your choice. The template should include layers and text, dimension, multileader, and table styles, when necessary, for drawing the given objects. Add layers and text, dimension, multileader, and table styles as needed. Draw all objects using appropriate layers and text, dimension, multileader, and table styles, justification, and format. Follow the specific instructions for each problem. Use only drawing and editing commands and techniques you have already learned. Use your own judgment and approximate dimensions when necessary. Apply dimensions accurately using ASME or appropriate industry standards.

Note: *Some of the problems in this chapter are built on problems from previous chapters. If you have not yet completed those problems, complete them now.*

▼ Basic

1. Follow the instructions in the Template Development portion of the companion website for this chapter to add and set up layouts for the Mechanical-Inch template file.

2. Follow the instructions in the Template Development portion of the companion website for this chapter to add and set up layouts for the Mechanical-Metric template file.

3. Follow the instructions in the Template Development portion of the companion website for this chapter to add and set up layouts for the Architectural-US template file.

4. Follow the instructions in the Template Development portion of the companion website for this chapter to add and set up layouts for the Architectural-METRIC template file.

5. Follow the instructions in the Template Development portion of the companion website for this chapter to add and set up layouts for the Civil-US template file.

6. Follow the instructions in the Template Development portion of the companion website for this chapter to add and set up layouts for the Civil-METRIC template file.

7. Open P28-4 and save as P29-7. The P29-7 file should be active. Make the **B-SIZE** layout current. Create a new layer named VPORT. Delete the default floating viewport and create a single floating viewport .5" in from the edges of the sheet on the VPORT layer. Scale model space in the viewport to 1:1. Plot the layout, leaving the VPORT layer on and thawed. Resave the file.

8. Open P28-5 and save as P29-8. The P29-8 file should be active. Activate the **A2-SIZE** layout. Create a new layer named VPORT. Delete the default floating viewport and create a single floating viewport 10 mm from the edges of the sheet on the VPORT layer. Scale model space in the viewport to 1:1. Plot the layout, leaving the VPORT layer on and thawed. Resave the file.

9. Open P28-6 and save as P29-9. The P29-9 file should be active. Activate the **B-SIZE** layout. Create a new layer named VPORT. Delete the default floating viewport and create a single floating viewport .5" from the edges of the sheet on the VPORT layer. Scale model space in the viewport to 1:1. Plot the layout, leaving the VPORT layer on and thawed. Resave the file.

10. Open P28-8 and save as P29-10. The P29-10 file should be active. Create a floating viewport and scale model space in the viewport using an appropriate scale. Plot the layout, leaving the VPORT layer on and thawed. Resave the file.

11. Open P8-1 and save as P29-11. The P29-11 file should be active. Delete the default **Layout2**. Create a new A-size sheet layout according to the following steps:
    A. Rename the default **Layout1** to **A-SIZE**.
    B. Select the **A-SIZE** layout and access the **Page Setup Manager**.
    C. Modify the **A-SIZE** page setup according to the following settings:
       • **Printer/Plotter:** Select a printer or plotter that can plot an A-size sheet
       • **Paper size:** Select the appropriate A-size sheet (varies with printer or plotter)
       • **Plot area:** Layout
       • **Plot offset:** 0,0
       • **Plot scale:** 1:1 (1 in. = 1 unit)
       • **Plot style table:** monochrome.ctb
       • **Plot with plot styles**
       • **Plot paper space last**
       • Do not check **Hide paper space objects**
       • **Drawing orientation:** Select the appropriate orientation (varies with printer or plotter)
    D. Create a new layer named VPORT.
    E. Delete the default floating viewport and create a single floating viewport on the VPORT layer .5" from the edges of the sheet.
    F. Scale model space in the viewport to 1:2. Plot the layout, leaving the VPORT layer on and thawed.
    G. Resave the file.

12. Open P8-11 and save as P29-12. The P29-12 file should be active. Create layouts and floating viewports as needed to plot the drawing at an appropriate scale.

▼ **Advanced**

13. Open P29-13 from the companion website. Create a layout, plot style, and page setup so the layout can be plotted as follows: Using color-dependent plot styles, have the equipment (shown in color in the diagram) plot with a lineweight of 0.8 mm and 80% screening on an A-size sheet oriented horizontally. Plotted text height should be 1/8". Plot in paper space at 1:1. Save the drawing as P29-13.

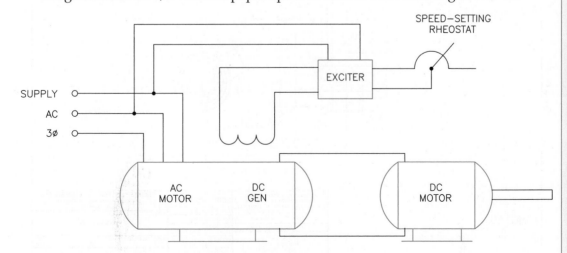

14. Open P29-14 from the companion website. Create four layouts with names and displays as follows:
    - The **Entire Schematic** layout plots the entire schematic on a B-size sheet.
    - The **3 Wire Control** layout plots only the 3 Wire Control diagram on an A-size sheet, horizontally oriented.
    - The **Motor** layout plots the motor symbol and connections in the lower center of the schematic on an A-size sheet, oriented vertically.
    - The **Schematic** layout plots schematic without the 3 Wire Control and motor components on an A-size sheet, oriented horizontally.

    Set up the layouts so they will plot with a text height of 1/8". Plot in paper space at a scale of 1:1. Save the drawing as P29-14.

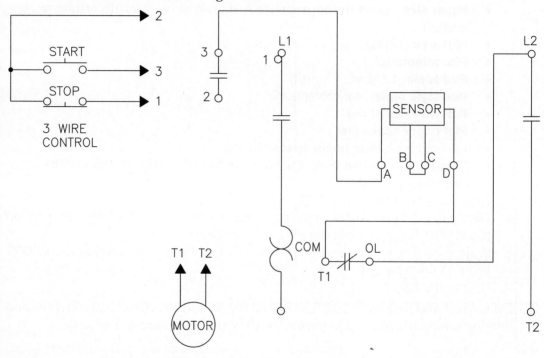

15. Draw and dimension the female insert shown. Use a layout to plot the drawing. Save the drawing as P29-15.

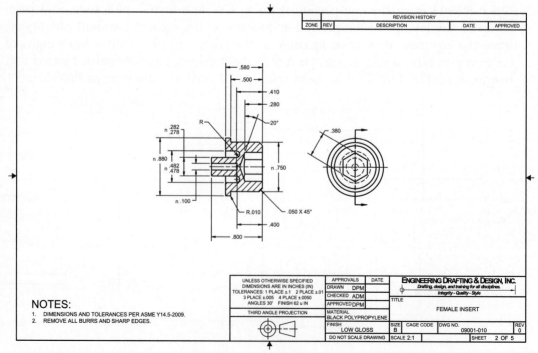

NOTES:
1. DIMENSIONS AND TOLERANCES PER ASME Y14.5-2009.
2. REMOVE ALL BURRS AND SHARP EDGES.

AutoCAD Certified Associate Exam Practice

Answer the following questions. Write your answers on a separate sheet of paper.

1. Which of the following drawing elements should you typically create in paper space? *Select all that apply.*
   A. border
   B. size and location dimensions
   C. general drawing notes
   D. thread notes
   E. title block

2. Which of the following methods can you use to create a floating viewport? *Select all that apply.*
   A. **EXPORTLAYOUT** command
   B. **MVIEW** command
   C. **Viewports** dialog box
   D. **Viewports** panel on the **View** ribbon tab
   E. **VPCLIP** command

3. Which of the following features of the **Layer Properties Manager** apply layer property overrides? *Select all that apply.*
   A. **Freeze**
   B. **New VP Freeze**
   C. **New VP Thaw**
   D. **Thaw**
   E. **VP Plot Style**

AutoCAD Certified Professional Exam Practice

Follow the instructions in the problems. Write your answers on a separate sheet of paper.

1. **Open CPE-29adjust.dwg. This file is available on the companion website.**
   Move the polygonal viewport up so its top edge aligns with the top edge of the rectangular viewport and the left edge of the upper portion of the polygonal viewport is .25" from the rectangular viewport, as shown. What are the 2D coordinates of point A?

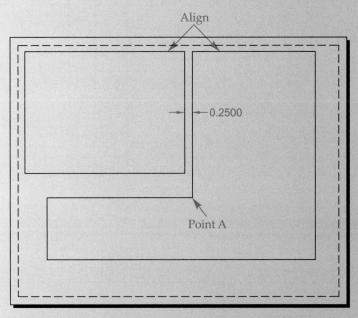

2. **Navigate to this chapter on the companion website and open CPE-29rotate.dwg.** Make the Stair Detail layout current. Set the contents of the existing viewport to display at a scale of 1/4″ = 1′-0″. Do not make any other changes. What are the 2D coordinates of point B in the layout?

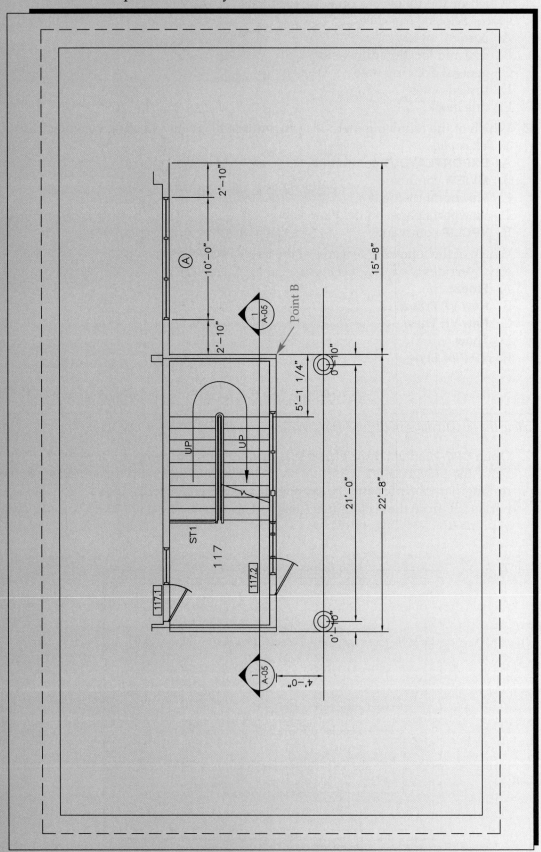

Chapter 30

Annotative Objects

Learning Objectives

After completing this chapter, you will be able to:

✓ Explain the differences between manual and annotative object scaling.
✓ Specify objects as annotative.
✓ Create and use annotative objects in model space.
✓ Display annotative objects in scaled layout viewports.
✓ Adjust the scale of annotations according to a new drawing scale.
✓ Use annotative objects to help prepare multiview drawings.

Annotations and related items, such as dimension symbols and hatch patterns, must be scaled so that information appears on-screen and plots correctly relative to scaled objects. AutoCAD provides annotative tools to automate the process of scaling *annotative objects*. Annotative tools also provide flexibility for working with layouts to create multiview drawings.

> **annotations:**
> Letters, numbers, words, and notes used to describe information on a drawing.

> **annotative objects:**
> AutoCAD objects that can adapt automatically to the current drawing scale.

Introduction to Annotative Objects

Always draw objects at their actual size, or full scale, in model space, regardless of the size of the objects. For example, if you draw a small machine part and the length of a line in the drawing is 2 mm, draw the line 2 mm long in model space. If you draw a building and the length of a line in the drawing is 80′, draw the line 80′ long in model space. These examples describe features on a drawing that are too small or too large for layout and printing purposes. To fit these objects properly on a sheet, you *scale* them to a specific drawing scale.

When you scale a drawing, you increase or decrease the displayed size of model space objects. A properly scaled floating viewport in a layout allows for this process. Scaling a drawing greatly affects the display of items added to objects in model space, such as annotations, because these items should be the same size on a plotted sheet, regardless of the displayed size, or scale, of the rest of the drawing. See **Figure 30-1**.

Traditional manual scaling of annotations, hatches, and other objects requires determining the drawing scale factor and then multiplying the scale factor by the plotted size of the objects. In contrast, annotative objects are scaled automatically

> **scale:** (verb) The process of enlarging or reducing objects to fit properly on a sheet of paper. (noun) The ratio between the actual size of drawing objects and the size at which objects plot on a sheet of paper.

Figure 30-1.
The large drawing features in this example of a residential site plan require scaling in order to fit on a standard size sheet. Annotations are scaled according to the plotted size of the drawing; otherwise, they would be too small to see.

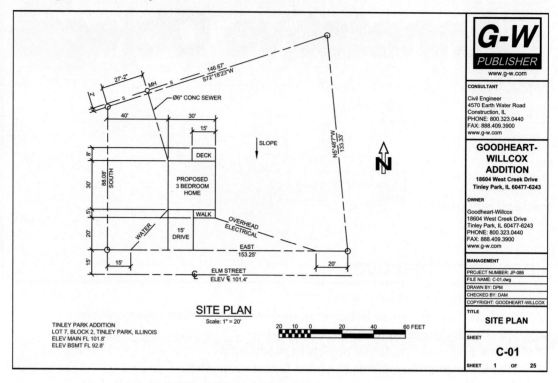

according to the selected annotation scale, which is the same as the drawing scale. This reduces the need to focus on the scale factor and manually adjust the size of objects according to the drawing scale.

PROFESSIONAL TIP

Use annotative objects instead of traditional manual scaling even if you do not anticipate using a drawing scale other than 1:1.

Defining Annotative Objects

Annotative objects include single-line and multiline text, dimensions, leaders and multileaders, GD&T symbols created using the **TOLERANCE** command, hatch patterns, blocks, and attributes. The method you use to define objects as annotative varies depending on the object type. You can make objects annotative when you first draw them or convert non-annotative objects to annotative status as needed.

Creating New Annotative Objects

Single-line and multiline text is annotative when it is drawn using an annotative text style. To make a text style annotative, pick the **Annotative** check box in the **Size** area of the **Text Style** dialog box. See **Figure 30-2**. A drawing may include a combination of annotative and non-annotative text, dimension, and multileader styles. An example of text that is typically not annotative is text added directly to a layout, which is usually printed at a scale of 1:1.

Dimensions, standard leaders, and GD&T symbols created using the **TOLERANCE** command are annotative when they are drawn using an annotative dimension style. To make a dimension style annotative, pick the **Annotative** check box in the **Fit** tab of the **New** (or **Modify**) **Dimension Style** dialog box. See **Figure 30-3**.

Multileaders are annotative when they are drawn using an annotative multileader style. To make a multileader style annotative, pick the **Annotative** check box in the **Leader Structure** tab of the **Modify Multileader Style** dialog box. See **Figure 30-4**.

Figure 30-2.
Single-line and multiline text objects are annotative when you draw them using an annotative text style.

Pick to make the text style annotative

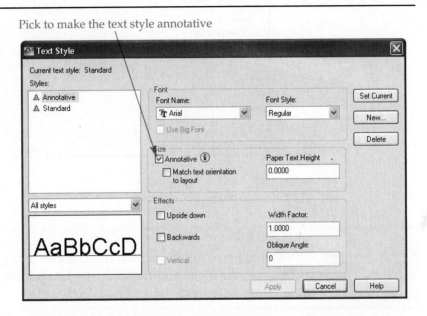

Figure 30-3.
Dimensions, leaders, and GD&T symbols created using the **TOLERANCE** command are annotative when you draw them using an annotative dimension style.

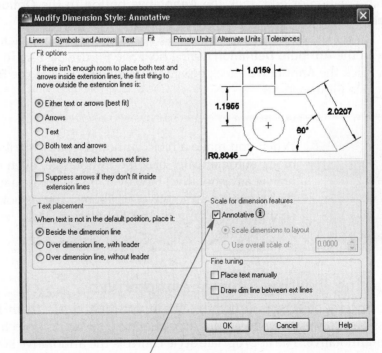

Pick to make the dimension style annotative

Figure 30-4.
Multileaders are
annotative when
you draw them
using an annotative
multileader style.

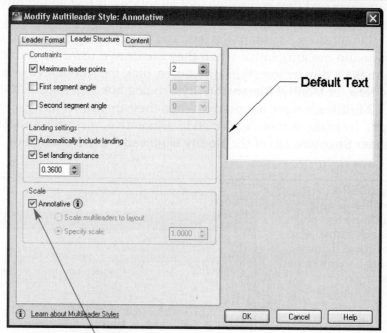

Pick to make the multileader
style annotative

When you create an annotative multileader using the block multil-eader type, the block automatically becomes annotative, even if the block is not set as annotative.

Hatch patterns are annotative when you set the hatch scale as annotative during hatch creation or edit. Pick the **Annotative** button in the **Options** panel of the **Hatch Creation** or **Hatch Editor** ribbon tab. See **Figure 30-5**.

To make attribute text height and spacing annotative, pick the **Annotative** check box in the **Attribute Definition** dialog box. See **Figure 30-6A**. To make a block annota-tive, pick the **Annotative** check box in the **Behavior** area of the **Block Definition** dialog box. See **Figure 30-6B**.

When you make a block annotative, any attributes included in the block automatically become annotative, even if the attributes are not set as annotative. However, if you create a non-annotative block that contains annotative attributes, the annotative attribute scale changes according to the annotation scale, while the size of the block remains fixed.

Making Existing Objects Annotative

Specify objects as annotative when you first create them in model space when possible. However, you can assign annotative status to objects originally drawn as non-annotative. The appropriate style controls the annotative status of single-line and multiline text, dimensions, standard leaders and multileaders, and GD&T symbols created using the **TOLERANCE** command. Change the style assigned to the object to an annotative style to make the object annotative. You must edit or recreate existing hatch patterns, blocks, and attributes in order to make the objects annotative.

Figure 30-5.
Set the hatch pattern scale to **Annotative** when you create or edit the hatch pattern.

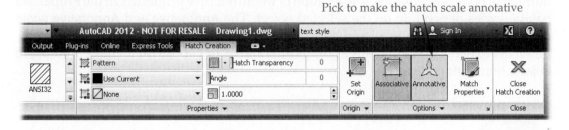

Pick to make the hatch scale annotative

Figure 30-6.
Set attributes (A) and blocks (B) as annotative during definition.

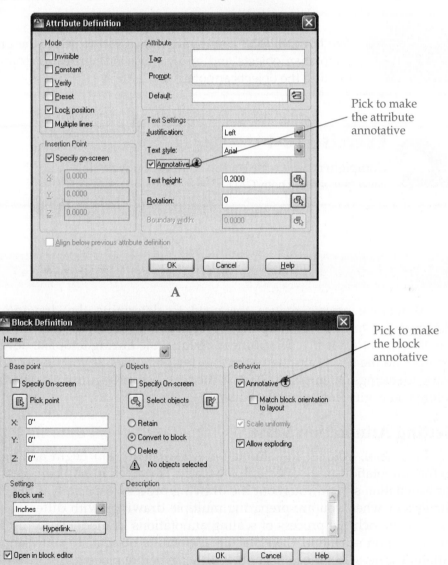

Pick to make the attribute annotative

Pick to make the block annotative

A

B

One way to make existing objects annotative is to override the non-annotative status using the **Properties** palette. This technique is most effective to make a limited number of objects annotative. The location of the annotative properties in the **Properties** palette varies depending on the selected object. The **Annotative** and **Annotative scale** properties are common to all annotative objects. Select **Yes** from the **Annotative** drop-down list to make non-annotative objects annotative, or choose **No** to make annotative objects non-annotative.

CAUTION

Use caution when overriding an object to annotative status. When possible, assign an appropriate annotative style or status to annotative objects instead of overriding specific objects.

The **MATCHPROP** command allows you to select the properties of annotative objects and apply those properties to existing objects, making the objects annotative.

Exercise 30-1

Complete the exercise on the companion website.
www.g-wlearning.com/CAD

Drawing Annotative Objects

Annotative objects reduce the need to determine the drawing scale factor. However, you must still identify the appropriate drawing scale, which is the same as the *annotation scale*. Ideally, determine drawing scale during template development and incorporate the scale into the settings in template files. If you do not apply drawing scales to settings in templates, identify the scale before beginning a drawing, or at least before you begin placing annotations.

annotation scale: The scale AutoCAD uses to calculate the scale factor applied to annotative objects.

Setting Annotation Scale

In general, you set the annotation scale before you begin adding annotations so that annotations are scaled automatically. However, it may be necessary to adjust the annotation scale throughout the drawing process, especially if the drawing scale changes or when you are preparing multiple drawings with different scales on one sheet. Approach the process of scaling annotations in model space by first selecting an annotation scale and then placing annotative objects. To draw annotations at a different scale, select the new annotation scale before placing annotative objects.

The **Select Annotation Scale** dialog box may appear when you add an annotative object. This dialog box provides a convenient way to set annotation scale before creating the object. The other primary means of specifying the annotation scale is to choose a scale from the **Annotation Scale** flyout on the status bar. See **Figure 30-7**. The annotation scale is typically the same as the drawing scale. You can also set the annotation scale in the **Properties** palette when no objects are selected by choosing the annotation scale from the **Annotation Scale** option in the **Misc** category.

AutoCAD and Its Applications—Basics

Figure 30-7.
The status bar
includes several
annotation scale
options. If you
display the drawing
status bar, the
Annotation Scale
button moves from
the application
status bar to the
drawing status bar.

1:1
1:2
1:4
1:5
1:8
1:10
1:16
1:20
1:30
1:40
1:50
1:100
2:1
4:1
8:1
10:1
100:1
1/128" = 1'-0"
1/64" = 1'-0"
1/32" = 1'-0"
1/16" = 1'-0"
3/32" = 1'-0"
1/8" = 1'-0"
3/16" = 1'-0"
1/4" = 1'-0"
3/8" = 1'-0"
1/2" = 1'-0"
3/4" = 1'-0"
1" = 1'-0"
1-1/2" = 1'-0"
3" = 1'-0"
6" = 1'-0"
1'-0" = 1'-0"
Custom...

Hide Xref scales

Current
annotation
scale

Pick the **Custom...**
option to add an
annotation scale
to the list

Pick this flyout to select
an annotation scale

If the scale you need is unavailable, or to change an existing scale, pick the **Annotation Scale** flyout on the status bar and choose **Custom...** to access the **Edit Scale List** dialog box. The **Edit Scale List** dialog box is also available by picking the **Edit Scale List...** button in the **User Preferences** tab of the **Options** dialog box. The **Edit Scale List** dialog box is the same dialog box used to edit floating viewport scales, as explained in Chapter 29.

Annotation scale sets the drawing scale in model space for controlling annotative objects. Viewport scale sets the drawing scale in a layout floating viewport to define the drawing scale. Both scales should be the same and should match the drawing scale.

Controlling Model Space Linetype Scale

The **CELTSCALE**, **PSLTSCALE**, and **MSLTSCALE** system variables control how the **LTSCALE** system variable applies to linetypes in model space and paper space. Leave the **CELTSCALE**, **PSLTSCALE**, and **MSLTSCALE** system variables at their default setting of 1 to apply the **LTSCALE** value correctly according to the current annotation scale. However, when you change the annotation scale, remember to use the **REGEN** command to regenerate the display. Otherwise, the linetype scale will not update according to the new scale.

Annotative Text

Draw annotative text using the same commands you use to draw non-annotative text. The difference is the value you enter for text height. To create annotative multiline text, use an annotative text style or pick the **Annotative** button. Then enter the paper text height, such as 1/4″, in the **Size** text box. See **Figure 30-8**. The text scale, which includes spacing, width, and paragraph settings, automatically adjusts according to the current annotation scale.

To create annotative single-line text, use an annotative text style. After you pick the start point, specify the paper text height. The text scale automatically adjusts according to the current annotation scale.

The **Properties** palette contains specific annotative text properties in addition to those displayed for all annotative objects. For example, use the **Paper text height** property to specify a paper text height. The **Model text height** property is a reference value that identifies the height of the text after the scale factor is applied.

Annotative Dimensions and Multileaders

Draw annotative dimensions, leaders, GD&T symbols created using the **TOLERANCE** command, and multileaders using the same commands you use to draw non-annotative dimensions and multileaders. Once you activate an annotative dimension or multileader style and select the appropriate annotation scale, the process of placing correctly scaled dimensions and multileaders is automatic.

Figure 30-8.
Create annotative multiline text using an annotative text style, or pick the **Annotative** option.

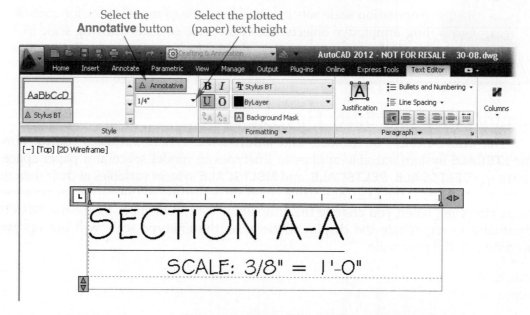

However, you must still determine the correct dimension and text location and spacing from objects when you add dimensions and text to scaled drawings. This involves multiplying the scale factor by the plotted spacing. For example, if the first dimension line should be 3/4″ from an object when plotted at a 1/4″ = 1′-0″ scale (scale factor = 48), the correct spacing in model space is 36″ (48 × 3/4″ = 36″) from the object.

Annotative Hatch Patterns

The difference between annotative and non-annotative hatch patterns is the way in which the drawing scale affects the hatch scale. When you create annotative hatch patterns, the scale you enter in the **Scale:** text box produces the same results regardless of the specified annotation scale. For example, if you enter a value in the **Scale:** text box that is appropriate for an annotation scale of 1:1, and then change the annotation scale to 4:1, the hatch pattern scale does not change relative to the drawing display. It looks the same on the 1:1 scaled drawing as on the 4:1 scaled drawing.

In contrast, when you create non-annotative hatch patterns, if you enter a value in the **Scale:** text box that is appropriate for a drawing scaled to 1:1 and then change the drawing scale to 4:1, the displayed scale of the hatch pattern increases. The hatch looks four times as large on the 4:1 drawing as on the 1:1 drawing.

Annotative Blocks and Attributes

Annotative blocks, often classified as *schematic blocks*, are commonly used for annotation purposes. When you insert an annotative schematic block, AutoCAD determines the block scale based on the current annotation scale, eliminating the need to enter a scale factor. For most applications, insert annotative blocks at a scale of 1 to apply the annotation scale correctly. Entering a scale other than 1 adjusts the scale of the block by multiplying the block scale by the annotation scale.

schematic block:
A block originally drawn at a 1:1 scale.

PROFESSIONAL TIP

When you create unit and schematic blocks that contain text and attributes, you should usually not make the text and attributes annotative. The text height you specify is set according to the full-scale size of the block, not necessarily the paper height. Any non-annotative text and attributes you select when you make a block annotative also automatically become annotative.

Exercise 30-2

Complete the exercise on the companion website.
www.g-wlearning.com/CAD

Displaying Annotative Objects in Layouts

Once you create drawing features and symbols and add annotative objects according to the appropriate annotation scale, you are ready to display and plot the drawing using a paper space layout. Refer to Chapter 29 to review the process of using and scaling floating viewports. **Figure 30-9** shows a drawing scaled to 3/8″ = 1′-0″. In this example, drawing features are drawn at full scale in model space. The annotation scale in model space was set to 3/8″ = 1′-0″, and annotative text, dimensions, multileaders, hatch patterns,

Figure 30-9.
Scaling a drawing in a floating paper space viewport. The **Viewport Scale** flyout provides
one of the easiest ways to set the viewport scale. A button is also available to synchronize the
viewport and annotation scale if they do not match.

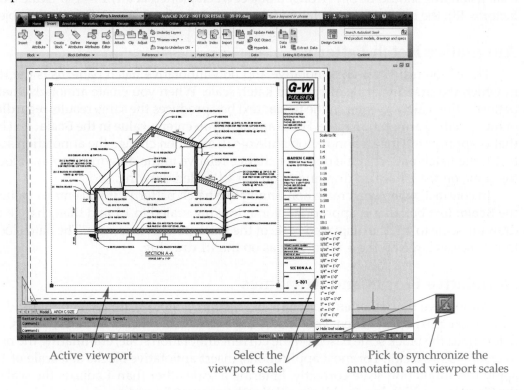

Active viewport   Select the   Pick to synchronize the
                       viewport scale   annotation and viewport scales

and blocks were added. The annotative objects in paper space are automatically scaled
according to the 3/8″ = 1′-0″ annotation scale.

In the **Figure 30-9** example, the viewport scale and the annotation scale are the
same, which is typical when scaling annotative objects. If you select a different viewport scale from the **Viewport Scale** flyout, the annotation scale automatically adjusts
according to the viewport scale. However, if you adjust the viewport scale using a
view tool such as zooming, the annotation scale does not change. The viewport scale
and the annotation scale must match in order for the drawing and annotative objects
to be scaled correctly. Pick the button to the right of the **Viewport Scale** flyout, identified in **Figure 30-9**, to synchronize the viewport and annotation scales.

The **Properties** palette also provides viewport and annotation scale controls. You
must be in paper space and pick a floating viewport to access viewport properties.
Choose a viewport scale from the **Standard scale** drop-down list. Adjust the annotation scale using the **Annotation scale** option. See **Figure 30-10**.

PROFESSIONAL TIP

Lock the viewport display to avoid zooming and disassociating the
viewport scale from the annotation scale. Refer to Chapter 29 for
more information on locking and unlocking floating viewports.

Exercise 30-3

Complete the exercise on the companion website.
www.g-wlearning.com/CAD

Figure 30-10.
The **Properties** palette also allows you to set the viewport and annotation scale.

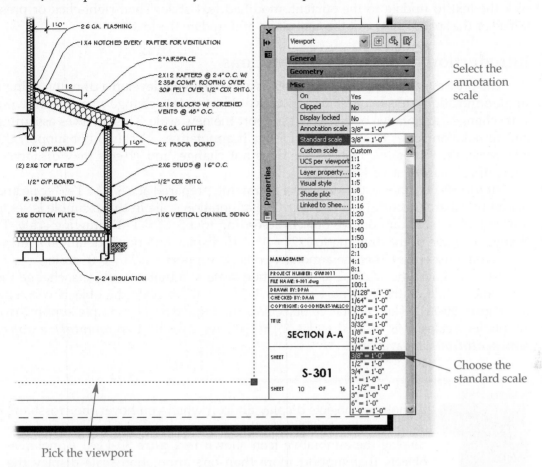

Select the annotation scale

Choose the standard scale

Pick the viewport

Changing Drawing Scale

No matter how much you plan a drawing, drawing scale can change throughout the drawing process. Reduce the drawing scale if it is necessary to use a smaller sheet. Increase the drawing scale if drawing features are redesigned and become larger, or if additional drawing detail is required.

Changing the drawing scale affects the size and position of annotations. If you change the drawing scale, remember that the annotation scale is the same as the drawing scale.

To change the annotation scale in model space, select a new annotation scale from the **Annotation Scale** flyout. To change the annotation scale in an active viewport in a layout, adjust the viewport scale by selecting the drawing scale from the **Viewport Scale** flyout. Again, the viewport and annotation scales should be the same for most applications.

Using the ANNOUPDATE Command

When you create single-line text using a non-annotative text style and then change the style to annotative, text drawn using the style becomes annotative. However, the properties of the annotative text remain set according to the non-annotative text style. When you create annotative text using an annotative text style and then change the style to non-annotative, text drawn in the style becomes non-annotative. However, the properties of the non-annotative text remain set according to the annotative style.

Type

ANNOUPDATE

Use the **ANNOUPDATE** command to update text properties to reflect the current properties of the text style in which the text is drawn. When prompted to select objects, pick the text to update to the current, modified text style. Then right-click or press [Enter] or the space bar to exit the command and update the text.

Introduction to Scale Representations

The previous sections in this chapter assume that you develop a drawing using a single annotation scale. In order for annotative object scale to change when the drawing scale changes, annotative objects must support the new scale. This involves assigning new annotation scales to annotative objects. If annotative objects do not support the new scale, the annotative object scale does not change, and objects may disappear, depending on annotative settings.

Figure 30-11A shows an example of a drawing prepared at a 3/8″ = 1′-0″ scale and placed on an architectural C-size sheet. The annotation scale in this example is set to 3/8″ = 1′-0″, to scale annotative objects according to a 3/8″ = 1′-0″ drawing scale. To change the scale of the drawing to 1/2″ = 1′-0″ to display additional detail on a larger sheet, you must ensure that the annotative objects support a 1/2″ = 1′-0″ scale.

After you add the 1/2″ = 1′-0″ annotation scale to annotative objects, change the annotation scale or the viewport scale to 1/2″ = 1′-0″ to scale the objects correctly. See **Figure 30-11B**. The annotative objects in the **Figure 30-11** example support two annotation scales: 3/8″ = 1′-0″ and 1/2″ = 1′-0″. As a result, two *annotative object representations* are available.

annotative object representation: Display of an annotative object at an annotation scale that the object supports.

Annotative objects display an icon when you hover the crosshairs over the objects. Objects that support a single annotation scale display the annotative icon shown in **Figure 30-12A**. Annotative objects that support more than one annotation scale display the annotative icon shown in **Figure 30-12B**. The icons appear by default according to selection preview settings in the **Selection** tab of the **Options** dialog box.

Understanding Annotation Visibility

Before changing the current annotation scale, you should understand how the annotation scale affects annotative object visibility. The annotative object scale does not change if annotative objects do not support the selected annotation scale. In addition, annotative objects disappear when an annotation scale that the objects do not support is current. For example, if annotative objects only support an annotation scale of 3/8″ = 1′-0″, and you set an annotation scale of 1/2″ = 1′-0″, the annotative object scale remains set at 3/8″ = 1′-0″, and the objects disappear.

The easiest way to turn annotative object visibility on and off according to the current annotation scale is to pick the **Annotation Visibility** button on the status bar. See **Figure 30-13**. Turning on annotation visibility is most effective when you are adding annotation scales to or deleting them from annotative objects. If you add multiple annotation scales to annotative objects, the annotative object representation is based on the current scale.

Deselect the **Annotation Visibility** button to display only the annotative objects that support the current annotation scale. Any annotative objects unsupported by the current annotation scale disappear. See **Figure 30-14**. Turning annotation visibility off is most effective when you are annotating a drawing, or a portion of a drawing, using a different annotation scale without showing annotative object representations

Figure 30-11.
A—A drawing created using an annotation scale of 3/8″ = 1′-0″ on an architectural C-size sheet. Annotative objects automatically appear at the correct scale. B—The same drawing shown in A, modified to an annotation scale of 1/2″ = 1′-0″ and placed on an architectural D-size sheet. An annotation scale of 1/2″ = 1′-0″ is added to all of the annotative objects, allowing the objects to adapt to the new scale automatically.

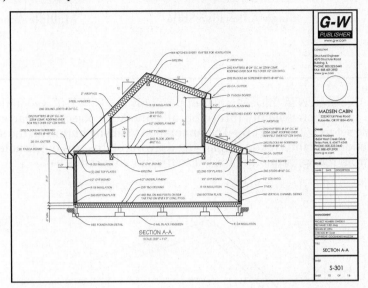

A

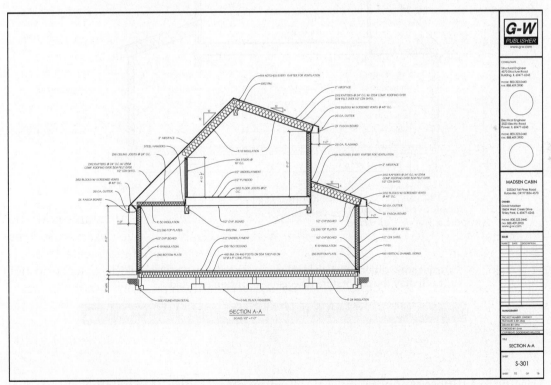

B

Figure 30-12.
The icons displayed when annotative objects support single or multiple annotation scales.

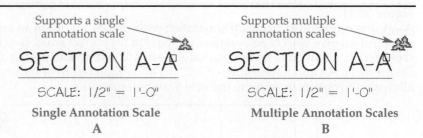

Supports a single annotation scale

Supports multiple annotation scales

SECTION A-A

SECTION A-A

SCALE: 1/2" = 1'-0"

SCALE: 1/2" = 1'-0"

Single Annotation Scale
A

Multiple Annotation Scales
B

Figure 30-13.
A—The annotative objects in this example support only a 3/8" = 1'-0" annotation scale. However, with annotation visibility turned on, all annotative objects appear, even with the annotation scale set to 1/2" = 1'-0". B—The **Annotation Visibility** button on the status bar controls annotation visibility. If you display the drawing status bar, the **Annotation Visibility** button moves from the application status bar to the drawing status bar.

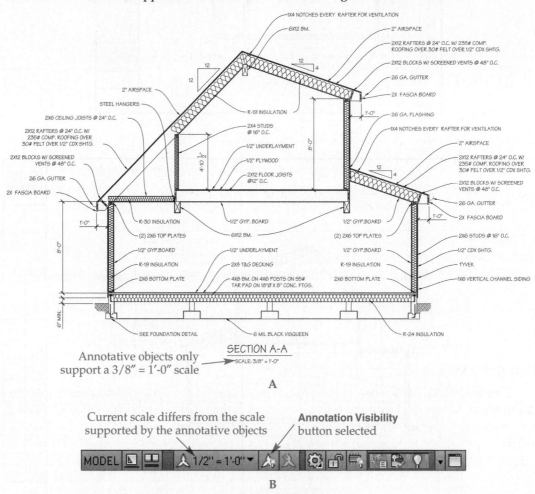

Annotative objects only support a 3/8" = 1'-0" scale

SECTION A-A
SCALE: 3/8" = 1'-0"

A

Current scale differs from the scale supported by the annotative objects

Annotation Visibility button selected

B

specific to a different annotation scale. Turning off visibility of annotative objects that do not support the current annotation scale is also effective for preparing multiview drawings because it eliminates the need to create separate layers for objects displayed at different scales. This practice is described later in this chapter.

Adding and Deleting Annotation Scales

One method for assigning additional annotation scales to annotative objects is to add the scales to selected objects. This method is appropriate whenever the drawing

Figure 30-14.
Deselect the **Annotation Visibility** button to display only the annotative objects that support the current annotation scale. The annotative objects in this example do not appear because they support only a 3/8″ = 1′-0″ annotation scale, and the current annotation scale is 1/2″ = 1′-0″.

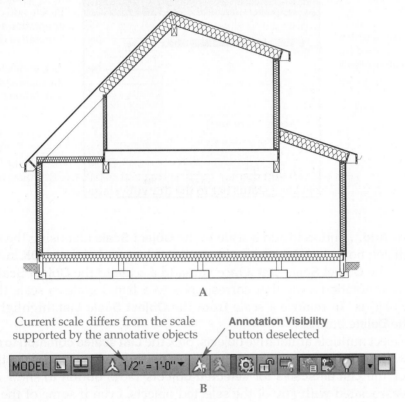

A

Current scale differs from the scale
supported by the annotative objects

Annotation Visibility
button deselected

B

scale changes, but it is especially effective for adding annotation scales only to specific objects, such as when you are creating multiview drawings.

Delete an annotation scale from annotative objects if the annotation scale is no longer in use, should not be displayed in a specific view, or makes it difficult to work with annotative objects. When you delete an annotation scale from annotative objects, the scale can no longer be applied to the objects. Add or delete annotation scales from selected objects using annotation scaling tools or the **Properties** palette.

Using the OBJECTSCALE Command

The **OBJECTSCALE** command provides one method of adding and deleting annotation scales supported by annotative objects. A quick way to access the **OBJECTSCALE** command is to select an annotative object and then right-click and pick **Add/Delete Scale…** from the **Annotative Objects Scales** cascading menu. If you activate the **OBJECTSCALE** command by right-clicking on objects, the **Annotation Object Scale** dialog box appears, allowing you to add or remove annotation scales from the selected objects. See **Figure 30-15**. If you access the **OBJECTSCALE** command before selecting objects, all annotative objects are displayed, even those objects that do not support the current annotation scale. Select the annotative objects to modify and right-click or press [Enter] or the space bar to display the **Annotation Object Scale** dialog box.

The **Object Scale List** shows the annotation scales associated with the selected annotative objects. A scale must appear in the list in order to apply to the annotative objects. If you select a different annotation scale, and that scale is not displayed in the **Object Scale List**, annotative objects do not adapt to the new annotation scale, and you have the option to make the objects invisible. In the example shown in **Figure 30-13** and **Figure 30-14**, 1/2″ = 1′-0″ must appear in the **Object Scale List** in order for the annotative objects to adapt to the new annotation scale of 1/2″ = 1′-0″.

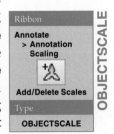

Ribbon

Annotate
> Annotation
Scaling

Add/Delete Scales

Type

OBJECTSCALE

OBJECTSCALE

Figure 30-15.
Use the **Annotation Object Scale** dialog box to add annotation scales to and delete them from annotative objects.

Annotation scales currently supported by the annotative object

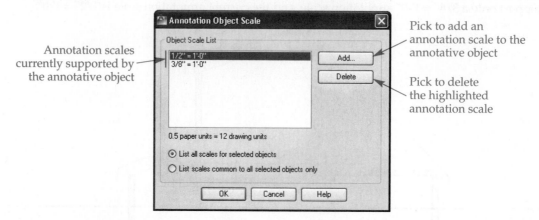

Pick to add an annotation scale to the annotative object

Pick to delete the highlighted annotation scale

Pick the **Add...** button to add a scale to the **Object Scale List** using the **Add Scales to Object** dialog box. Highlight scales in the **Scale List** and pick the **OK** button to add the scales to the **Object Scale List**. Once you add a scale to the **Object Scale List**, you can pick an annotation scale that corresponds to a listed scale to scale the selected annotative objects. To remove a scale from the **Object Scale List**, highlight the scale and pick the **Delete** button.

If you select multiple annotative objects, pick the **List scales common to all selected objects only** radio button to display only the annotative scales common to the selected objects. Pick the **List all scales for selected objects** radio button to show all annotation scales associated with any of the selected objects, even if some of the objects do not support the listed scales. Listing all scales for selected objects is helpful when you want to delete a scale that applies only to certain objects.

If a desired scale is not available in the **Add Scales to Object** dialog box, close the **Annotation Object Scale** dialog box and access the **Edit Scale List** dialog box to add a new scale to the list of available scales.

Using the Properties Palette

The **Properties** palette also allows you to add annotation scales to selected annotative objects. See **Figure 30-16**. The location of the annotative properties in the **Properties** palette varies depending on the selected object. The **Annotative scale** property displays the annotation scale currently applied to the selected annotative objects and contains an ellipsis button (...) that you can pick to open the **Annotation Object Scale** dialog box.

Automatically Adding Annotation Scales

Another technique for assigning additional annotation scales to annotative objects is to add a selected annotation scale automatically to all annotative objects in the drawing. This eliminates the need to add annotation scales to individual annotative objects and quickly produces newly scaled drawings.

The **ANNOAUTOSCALE** system variable controls the ability to add an annotation scale to all existing annotative objects. Enter 1, –1, 2, –2, 3, –3, 4, or –4, depending on the desired effect. **Figure 30-17A** describes each option. After you enter the initial value, the easiest way to toggle the **ANNOAUTOSCALE** system variable on and off is to pick the button on the status bar. See **Figure 30-17B**.

Figure 30-16.
The **Annotative scale** property in the **Properties** palette provides another way to access the **Annotation Object Scale** dialog box.

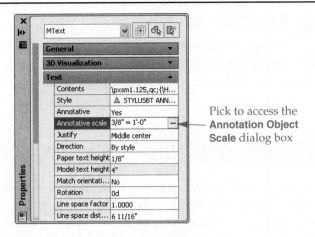

Pick to access the **Annotation Object Scale** dialog box

Figure 30-17.
A—**ANNOAUTOSCALE** system variable options. B—After you enter the initial **ANNOAUTOSCALE** system variable setting, use the button on the status bar to toggle **ANNOAUTOSCALE** on and off. If you display the drawing status bar, the **ANNOAUTOSCALE** button moves from the application status bar to the drawing status bar.

Value	Mode	Description
1	On	Adds the selected annotation scale to annotative objects, not including those drawn on a layer that is turned off, frozen, locked, or frozen in a viewport.
–1	Off	1 behavior is used when **ANNOAUTOSCALE** is turned back on.
2	On	Adds the selected annotation scale to annotative objects, not including those drawn on a layer that is turned off, frozen, or frozen in a viewport.
–2	Off	2 behavior is used when **ANNOAUTOSCALE** is turned back on.
3	On	Adds the selected annotation scale to annotative objects, not including those drawn on a layer that is locked.
–3	Off	3 behavior is used when **ANNOAUTOSCALE** is turned back on.
4	On	Adds the selected annotation scale to all annotative objects regardless of the status of the layer on which the annotative object is drawn. 4 is the AutoCAD default setting when toggled on.
–4	Off	4 behavior is used when **ANNOAUTOSCALE** is turned back on. –4 is the AutoCAD default setting when toggled off.

A

Pick to toggle the **ANNOAUTOSCALE** system variable on or off

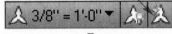

B

CAUTION

Use caution when adding annotation scales automatically. Due to the effectiveness and transparency of the tool, annotation scales are often added to annotative objects unintentionally. Although you can later delete scales, this causes additional work and confusion.

Exercise 30-4

Complete the exercise on the companion website.
www.g-wlearning.com/CAD

Preparing Multiview Drawings

Drawings for many different engineering fields often contain views, sections, and details drawn at different scales. Annotative objects offer several advantages for these drawings, especially when views in model space appear at different scales in layouts. Use scaled viewports to display multiple views using a single file. You can assign a different annotation scale to each drawing view that contains annotative objects, reducing the need to calculate multiple drawing scale factors, while maintaining the appropriate scale of previously drawn annotative objects. Additionally, by adjusting annotative scale representation visibility and position, you can prepare differently scaled multiview drawings, while eliminating the need to use separate, scale-specific layers and annotations.

Creating Differently Scaled Drawings

Figure 30-18A shows an example of two different drawing views, a full section (SECTION A-A) and a stair section, both drawn at full scale in model space. The full section in Figure 30-18A uses a 3/8″ = 1′-0″ scale. To prepare the full section, set the annotation scale in model space to 3/8″ = 1′-0″, and then add annotative objects. The annotative objects are automatically scaled according to the 3/8″ = 1′-0″ annotation scale. The stair section in Figure 30-18A uses a 1/2″ = 1′-0″ scale. To prepare the stair section, change the annotation scale in model space from 3/8″ = 1′-0″ to 1/2″ = 1′-0″, and then add annotative objects. These annotative objects are automatically scaled according to the 1/2″ = 1′-0″ annotation scale. If you look closely, you can see the different scales applied to the drawing views.

Figure 30-18A shows annotation visibility on, allowing you to see all annotative objects and observe the effects of using different scales. Figure 30-18B shows annotation visibility off to show only annotative objects that support the current annotation scale, which is 1/2″ = 1′-0″ in this example.

The next step is to display and plot the drawing using multiple paper space viewports. Figure 30-19 shows an architectural D-size sheet layout with two floating viewports. One viewport displays the full section at a viewport scale of 3/8″ = 1′-0″. The other viewport displays the stair section at a viewport scale of 1/2″ = 1′-0″. Notice that the annotative objects are the same size in both views.

Exercise 30-5

Complete the exercise on the companion website.
www.g-wlearning.com/CAD

Reusing Annotative Objects

Often the same drawing features appear in different views at different scales. For example, you may plot a drawing on a large sheet using a large scale, and plot the same drawing on a smaller sheet using a smaller scale. Another example is preparing a view enlargement or detail.

Figure 30-18.
Two different drawing views drawn at full scale in model space. The full section (SECTION A-A) uses an annotation scale of 3/8″ = 1′-0″, and the stair section uses an annotation scale of 1/2″ = 1′-0″. A—Annotation visibility is on. B—Annotation visibility is off with the current annotation scale set to 1/2″ = 1′-0″ (stair section view scale).

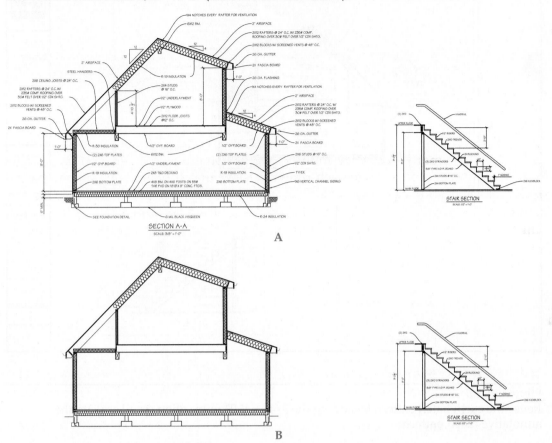

Annotative objects significantly improve the ability to reuse existing drawing features. Use annotation visibility to hide annotative objects not supported by the current annotation scale. You can also adjust the position of scale representations according to the appropriate annotation scale. These options allow you to include differently scaled annotative objects on the same sheet without creating copies of the objects and without using scale-specific layers.

Using Invisible Scale Representations

If annotative objects do not support an annotation scale, the annotative objects disappear when the annotation scale that the objects do not support is current. This is a valuable technique for displaying certain items at a specific scale. Pick the **Annotation Visibility** button on the status bar to turn on and off annotative object visibility.

The following example shows how adjusting the visibility of annotative objects that only support the current annotation scale allows you to create an additional view from existing drawing features. This example uses an annotation scale of 3/4″ = 1′-0″ to create a foundation detail. To begin constructing the foundation detail, add the 3/4″ = 1′-0″ annotation scale to the existing earth hatch pattern so it will appear on the full section and the foundation detail. See **Figure 30-20**. Next, with the current annotation scale set to 3/4″ = 1′-0″, add annotative objects specific to the foundation detail. See **Figure 30-21**. These objects support only the 3/4″ = 1′-0″ annotation scale, hiding the objects on the full section, which uses a 3/8″ = 1′-0″ scale.

Figure 30-19.
Using viewports with different scales to create a multiview drawing. Notice that the annotative objects are the same size in both views.

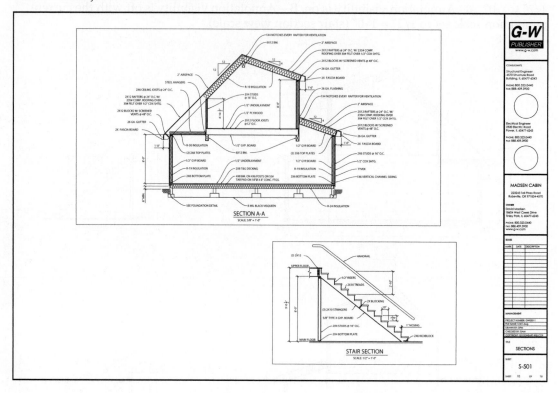

Figure 30-20.
Reuse the earth hatch pattern by adding the 3/4″ = 1′-0″ foundation detail scale to the annotative hatch pattern.

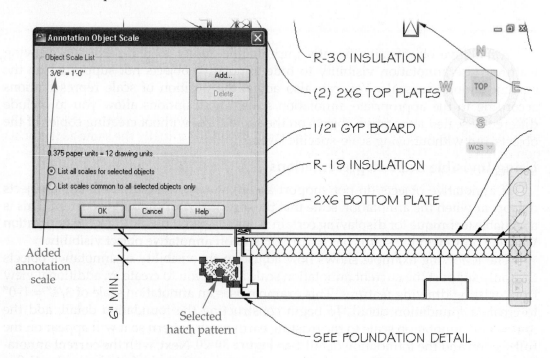

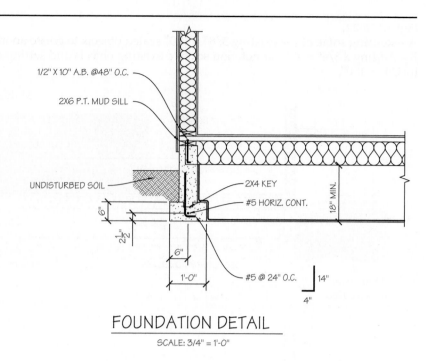

Figure 30-21.
Adding annotative text, dimensions, multileaders, and hatch patterns specific to the foundation detail using a 3/4″ = 1′-0″ annotation scale.

1/2″ X 10″ A.B. @48″ O.C.

2X6 P.T. MUD SILL

UNDISTURBED SOIL

2X4 KEY

#5 HORIZ. CONT.

18″ MIN.

6″

2 1/2″

6″

1′-0″

#5 @ 24″ O.C.

14″

4″

FOUNDATION DETAIL

SCALE: 3/4″ = 1′-0″

PROFESSIONAL TIP

If objects already support an annotation scale, but you do not want to display those annotations at the current scale, delete the annotation scale from the objects.

Adjusting Scale Representation Position

When you reuse annotative objects, the location and spacing of annotative objects on one scale are often not appropriate for another scale. Reposition each scale representation to overcome this issue.

In the foundation detail example in **Figure 30-21**, some of the existing 3/8″ = 1′-0″ scaled dimensions and multileaders from the full section are reused in the foundation detail. See **Figure 30-22A**. The first step is to add a 3/4″ = 1′-0″ annotation scale to the objects. Next, with **Annotation Visibility** turned off, as shown in **Figure 30-22B**, you can see the resulting position of the selected objects, which is initially the same as the position of the 3/8″ = 1′-0″ objects. The only difference is that now the 3/8″ = 1′-0″ objects also support a 3/4″ = 1′-0″ scale.

Use grip editing to adjust the position of annotation scale representations. When you select annotative objects that support more than one annotation scale, all scale representations appear by default. See **Figure 30-23**. An annotative object is a single object, but it can contain several scale representations. Grips are displayed on the scale representation that corresponds to the current annotation scale.

Using grips to edit scale representations is similar to editing the object used to create the scale representation. The difference when editing a scale representation is that you adjust a scaled copy of the object. **Figure 30-24** shows the effects of editing the position of dimension and multileader scale representations on the foundation detail. The figure shows selecting the representations to help demonstrate the effects of editing scale representation position. Notice that you can edit all elements of the scale representation to produce the desired annotations at the appropriate locations.

Figure 30-22.
A—Reusing some of the existing 3/8″ = 1′-0″ scaled objects to create another drawing view.
B—Adding a 3/4″ = 1′-0″ annotation scale to existing objects and setting the annotation scale
to 3/4″ = 1′-0″.

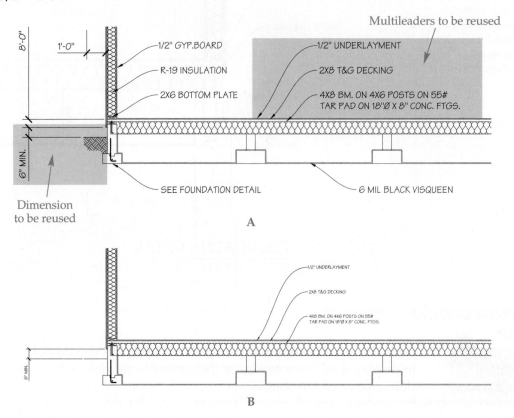

A

B

Figure 30-23.
Adjust the position
of annotation scale
representations
using grip editing.
When you select
annotative objects
that support more
than one annotation
scale, all scale
representations
appear by default.

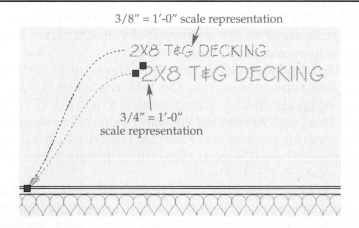

3/8″ = 1′-0″ scale representation

2X8 T&G DECKING
2X8 T&G DECKING

3/4″ = 1′-0″
scale representation

**PROFESSIONAL
TIP**

Use the **DIMSPACE** and **MLEADERALIGN** commands to adjust
dimension spacing and multileader alignment after changing the
drawing scale.

The **SELECTIONANNODISPLAY** system variable controls the display of selected
scale representations and is set to 1 by default. As a result, all scale representations
display and appear dimmed when you pick an annotative object that supports multiple
annotation scales. See Figure 30-24. The display can be confusing if the selected object

Figure 30-24.

Editing the position of scale representations is much like creating scaled copies of existing annotations.

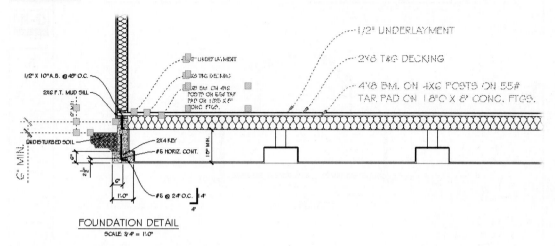

FOUNDATION DETAIL
SCALE 3/4" = 1'-0"

supports several annotation scales. Set the **SELECTIONANNODISPLAY** system variable to 0 to display only the scale representation that corresponds to the current annotation scale.

 You can edit scale representations individually only by using grip editing. When you use modify commands to edit an annotative object, all of the scale representations change at once.

Resetting Scale Representation Position

The **ANNORESET** command removes multiple scale representation positions, allowing you to change the position of all selected scale representations to the position of the scale representation that is set for the current annotation scale. A quick way to access the **ANNORESET** command is to select annotative objects, right-click and pick the option from the **Annotative Object Scale** cascading menu. If you activate the **ANNORESET** command by right-clicking on objects, the position of the selected objects resets. If you access the command before selecting objects, pick the annotative objects. Then right-click or press [Enter] or the space bar to exit the command and reset the scale representation positions.

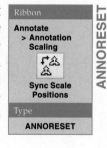

Ribbon
Annotate
> Annotation
Scaling
Sync Scale Positions
Type
ANNORESET

ANNORESET

Completing a Multiview Drawing

The last step in creating a multiview drawing is to display and plot the drawing using multiple paper space viewports. Figure 30-25 shows an architectural D-size sheet layout with three floating viewports. One viewport displays the full section at a 3/8" = 1'-0" viewport scale. A second viewport displays the stair section at a 1/2" = 1'-0" viewport scale. A third viewport displays the foundation detail at a 3/4" = 1'-0" viewport scale.

PROFESSIONAL TIP

 When you save drawings using annotative objects to earlier versions of AutoCAD that do not support annotative objects, scale representations may convert to non-annotative objects, but automatically become assigned to unique layers. To use this function, select the **Maintain visual fidelity for annotative objects** check box in the **Open and Save** tab of the **Options** dialog box.

Figure 30-25.
A complete multiview drawing created using annotative objects.

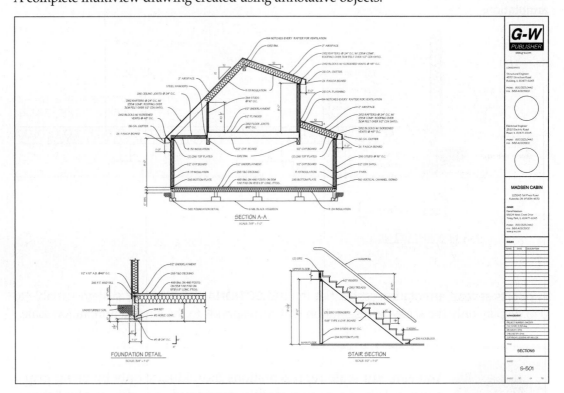

  Exercise 30-6

Complete the exercise on the companion website.
www.g-wlearning.com/CAD

Chapter Review

Answer the following questions. Write your answers on a separate sheet of paper or complete the electronic chapter review on the companion website.
www.g-wlearning.com/CAD

1. What are annotative objects?
2. Explain the practical differences between manual and annotative object scaling.
3. Identify at least four types of objects that you can make annotative.
4. How do you set text scale, including spacing, width, and paragraph settings, to adjust automatically according to the current annotation scale?
5. Identify an important relationship between the viewport scale and the annotation scale.
6. Which **MSLTSCALE** system variable setting should you use so you do not have to calculate the drawing scale factor when entering an **LTSCALE** value?
7. Name the command used to update text properties according to the current properties of the text style on which the text is drawn.
8. What is an annotative object representation?
9. Briefly describe the result of setting the **ANNOAUTOSCALE** system variable to a value of 4.
10. Briefly explain the effect of turning annotation visibility on and off.

Drawing Problems

Start AutoCAD if it is not already started. Start a new drawing for each problem using an appropriate template of your choice. The template should include layers and text, dimension, multileader, and table styles, when necessary, for drawing the given objects. Add layers and text, dimension, multileader, and table styles as needed. Draw all objects using appropriate layers and text, dimension, multileader, and table styles, justification, and format. Follow the specific instructions for each problem. Use only drawing and editing commands and techniques you have already learned. Use your own judgment and approximate dimensions when necessary. Apply dimensions accurately using ASME or appropriate industry standards.

Note: Some of the problems in this chapter are built on problems from previous chapters. If you have not yet completed those problems, complete them now.

▼ Basic

1. Open P23-9 and save as P30-1. The P30-1 file should be active. Convert all the non-annotative objects to annotative objects. Resave the drawing.

2. Open P24-10 and save as P30-2. The P30-2 file should be active. Convert all of the non-annotative objects to annotative objects. Resave the drawing.

▼ Intermediate

3. Draw the section view and side view shown. Use annotative objects to prepare a full-scale drawing of the part. Change the annotation scale to 2:1 and adjust the scale representations as needed according to the new scale. Save the drawing as P30-3.

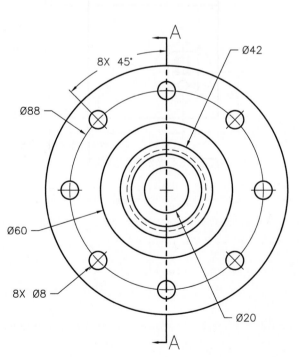

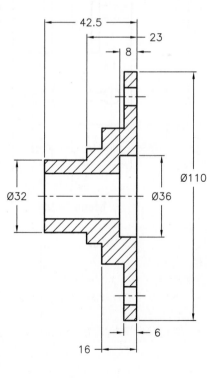

SECTION A—A

4. Draw the section view and side views shown. Use annotative objects to prepare a full-scale drawing of the part. Change the annotation scale to 2:1 and adjust the scale representations as needed according to the new scale. Save the drawing as P30-4.

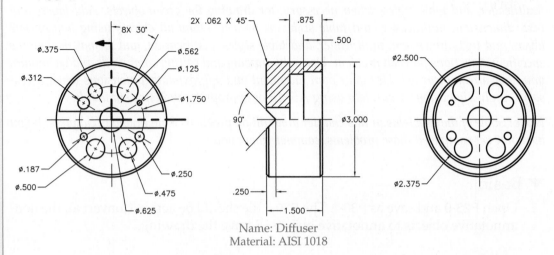

Name: Diffuser
Material: AISI 1018

5. Draw the fan shown at full scale in model space. Use annotative objects to prepare a full-scale view of the fan as shown and a view enlargement of the motor. You should not have to create a copy of the motor or develop scale-specific layers. Save the drawing as P30-5.

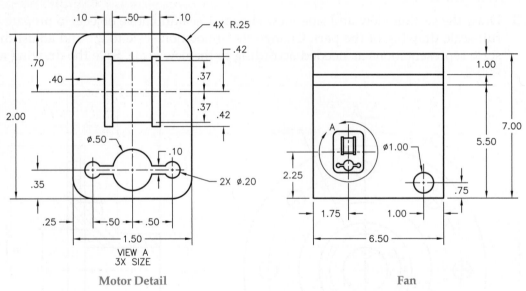

Motor Detail

Fan

6. Draw the floor plan shown at full scale in model space. Use annotative objects to prepare a 1/4″ = 1′-0″ view. Change the annotation scale to 1/8″ = 1′-0″ and adjust the scale representations as needed according to the new scale. Save the drawing as P30-6.

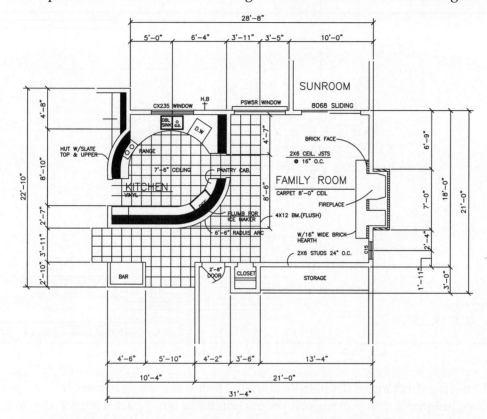

7. Draw the part shown at full scale in model space. Use annotative objects to prepare the full-scale view and the view enlargement shown. You should not have to create a copy of the part or develop scale-specific layers. Plot the layout. Save the drawing as P30-7.

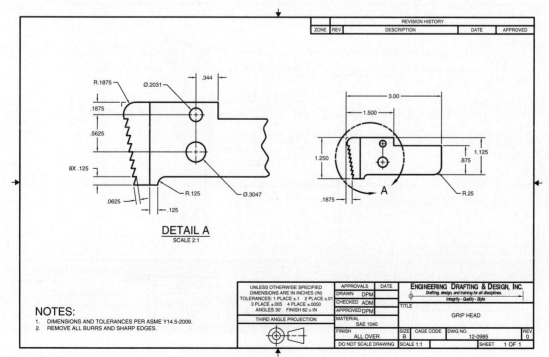

8. Draw, dimension, and plot using a layout the male insert shown. Use annotative objects. Use a larger sheet and annotative object scaling to plot a 4:1 scale drawing of the insert. You should not have to create a copy of the part or develop scale-specific layers. Save the drawing as P30-8.

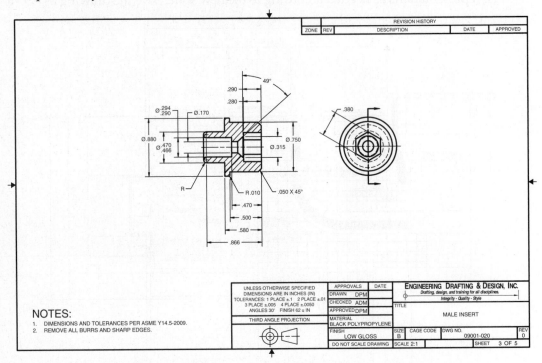

9. Draw the sheet metal flat pattern shown. Design the shape of punch A using your own judgment and dimensions proportionate to other part features. Use annotative objects to prepare the full-scale view of the flat pattern and the view enlargement of the punch. Dimension the punch. You should not have to create a copy of the part or develop scale-specific layers. Plot the layout. Save the drawing as P30-9.

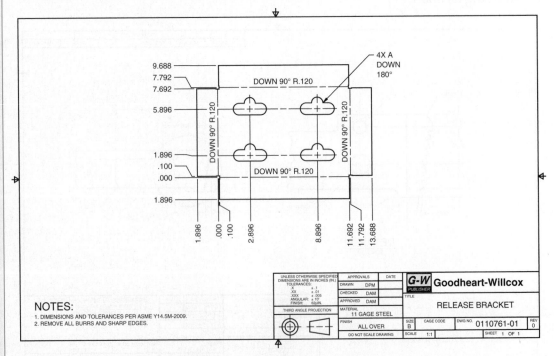

10. Research the design of an existing paper-cutting scissors consisting of at least three separate parts. Create dimensioned 2D sketches of the existing design from manufacturer's specifications, or from measurements taken from actual scissors. Start a new drawing from scratch or use a decimal template of your choice. Prepare detail drawings of each part and an assembly drawing with balloons and a parts list. Use annotative objects and separate layouts for each drawing to prepare a set of drawings in one file. Plot each drawing. Save the drawing as P30-10.

11. Create a dimensioned 2D sketch of a complete floor plan for a home with three bedrooms and two bathrooms. Start a new drawing from scratch or use an architectural template of your choice. Draw, dimension, and plot the floor plan from your sketch using a layout. Use annotative objects. Save the drawing as P30-11.

12. Obtain a hard copy of a plot plan of a small residential subdivision, or a portion of a subdivision. Draw, dimension, and plot the subdivision using a layout. Use annotative objects. Save the drawing as P30-12.

AutoCAD Certified Associate Exam Practice

Answer the following questions. Write your answers on a separate sheet of paper.

1. A drawing of a mechanical part is set up to be printed at a scale of 2:1 in a layout in paper space. The length of one feature on the part is 35.125″. How long should that feature be drawn in model space? *Select the one item that best answers the question.*
    A. 3.513″
    B. 17.563″
    C. 35.125″
    D. 70.250″

2. Which of the following statements are true about blocks? *Select all that apply.*
    A. Attributes included in an annotative block automatically become annotative even if they are not set to be annotative.
    B. Attributes included in an annotative block change according to the annotation scale, but the block remains fixed.
    C. You can make a block annotative when you originally create the block.
    D. You can make a block annotative after its creation by using the **Properties** palette.

3. How can you change the overall annotation scale of a layout? *Select all that apply.*
    A. right-click and select an annotation scale from the cascading list
    B. select **Scale List** in the **Annotation Scaling** panel of the **Annotate** ribbon tab
    C. use the **Annotation Scale** flyout
    D. use the **Properties** palette
    E. use the **Viewport Scale** flyout

AutoCAD Certified Professional Exam Practice

Follow the instructions in the problem. Write your answers on a separate sheet of paper.

1. **Open CPE-29annoscale.dwg. This file is available on the companion website.** Change the annotation scale to 1/2″=1′-0″ to display annotative hatch patterns created at that scale. What hatch pattern is used for the undisturbed soil in this drawing?

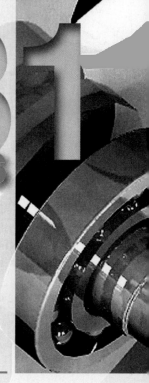

Chapter

31

External References

Learning Objectives

After completing this chapter, you will be able to:

✓ Explain the function of external references.
✓ Prepare drawings for xref insertion.
✓ Attach existing drawings to the current drawing.
✓ Work with xrefs and dependent objects.
✓ Manage multiple xrefs in a drawing.
✓ Bind external references and selected dependent objects to a drawing.
✓ Edit external references in the current drawing.

External references (xrefs) expand on the concept of reusing existing content in AutoCAD. Xrefs provide an effective way to relate existing base drawings, complex symbols, images, and details to other drawings. Xrefs also help multiple users share content. This chapter focuses on using xref drawings and provides common xref applications.

external reference (xref): A DWG, DWF, DWFx, raster image, DNG, or PDF file incorporated into a drawing for reference only.

Introduction to Xrefs

An xref is a drawing (DWG), design web format (DWF and DWFx), raster image, digital negative (DNG), or portable document format (PDF) file that you reference into a *host drawing*. Inserting an xref is similar to inserting an entire drawing as a block. However, unlike a block, which is actually stored in the file in which you insert the block, the file geometry in a *reference file* is not added to the host drawing. File data appears on-screen for reference only. The result is usable information, but the host file remains much smaller than if you insert a block or copy and paste objects. Xrefs are also easier to manage in a host drawing than blocks or pasted objects.

host drawing: The drawing into which xrefs are incorporated.

reference file: An xref; a file referenced by the host drawing.

Another major benefit of using xrefs is the link between reference and host files. Any changes you make to reference files are reflected in host drawings, so the host drawings display the most recent reference content. AutoCAD reloads each xref whenever the host drawing loads. This allows you or a design drafting team to work on a multi-file project, with the assurance that any revisions to reference files are displayed in host drawings.

You can add as many xrefs as needed for a drawing at any time during the drawing process. You can also include xrefs within referenced files. The host drawing updates to recognize each new xref.

Xref Files

Files that you can reference into a current drawing include existing DWG, DWF, DWFx, raster image, DNG, and PDF files. DWF and DWFx files are drawings compressed for publication, viewing, and mark-up using a viewer, such as the Autodesk® Design Review software. DWF and DWFx files are commonly used to share drawings with members of a design drafting team who do not use AutoCAD. The high compression also makes DWF and DWFx files easy to transmit electrically.

Raster image and DNG file reference is appropriate for adding an image to a drawing, such as for a company logo in a title block. PDF file reference allows you to reuse PDF file content. Externally referencing an image or PDF file into a drawing is an excellent technique, because the large file sizes often associated with images and PDF files do not affect the host drawing.

AutoCAD and Its Applications—Advanced explains using external reference DWF, image, DNG, and PDF files, and underlays.

Xref Applications

DWG files are the most common xref files and are the focus of this chapter. The term *xref* often applies specifically to referenced DWG files. In general, use xrefs to reuse existing drawing information and help develop other drawings. There are countless applications for xref drawings in every drafting field. The following sections provide typical xref drawing applications. As you work with AutoCAD, you will discover a variety of uses for xref drawings.

Reference Existing Geometry

One of the most common applications for xref drawings is to reference existing geometry to use as a pattern or source of needed information in the host drawing. For example, a floor plan includes size and shape information required to prepare

---

Figure 31-1.
Using a floor plan xref drawing as a pattern, or outline, to draw a roof plan.

Floor Plan Xref Drawing

Roof Plan Added

Final Roof Plan Geometry
with floor plan xref layers turned off

AutoCAD and Its Applications—Basics

additional plans, elevations, sections, and details. **Figure 31-1** shows an example of referencing a floor plan file into a new drawing to use as an outline for creating a roof plan file.

Figure 31-2 shows an example of the roof plan file created in **Figure 31-1** attached to a new drawing as an xref and then used to project an elevation. The roof plan xref includes a *nested xref* of the floor plan. In this example, the elevation file references the roof plan. The roof plan in turn references the floor plan.

nested xrefs: Xrefs contained within other xrefs.

Create a Multiview Drawing

Another xref application is to create commonly used drawings, such as sections and details, as separate drawing files and then attach each drawing as an xref to a host drawing, known as the *master drawing*. Use floating viewports and layer viewport freezing to create a multiview layout. You can prepare a multiview drawing entirely from existing xref drawings or from a combination of objects created "in place" in the master drawing and attached xrefs.

master drawing: A host drawing created by attaching several frequently used xrefs.

Figure 31-3A shows an example of five stock details referenced into the model space environment of a new drawing. Floating viewports arrange the details in a layout, as shown in **Figure 31-3B**. When you make changes to details in the referenced detail files, the files are updated in the host file.

Figure 31-2.
Using a roof plan xref drawing that contains a nested floor plan xref drawing as a pattern for projecting geometry needed to create an elevation. The projection lines, drawn using the **XLINE** command, are for reference.

Roof plan xref includes the nested floor plan xref

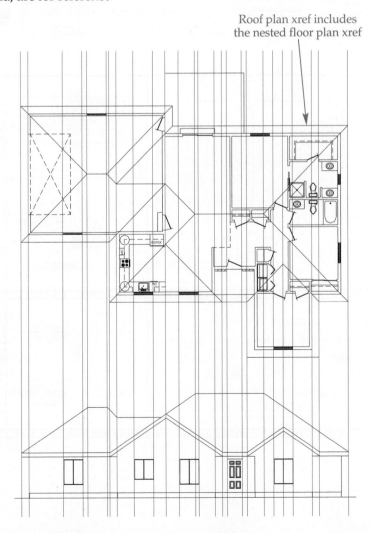

Add Layout Content

Layout content, such as a title block or general notes, typically has a standard format. If the format requires modification, such as adding a new note to a list of general notes, you can make changes to the xref drawing and easily update each host file that references the xref. This is the same concept as using xref drawings to build a multiview drawing. Figure 31-3B shows general notes added to the layout as an xref.

Arrange Sheet Views

You can use external references to arrange sheet views in layouts when you are working with sheet sets. Chapters 32 and 33 describe sheet sets.

Figure 31-3.
A—Xref frequently used drawing views into model space, reducing the size of the file and providing the ability to change instances of the view used in multiple host drawings.
B—Arrange referenced views in floating viewports like other model space objects.

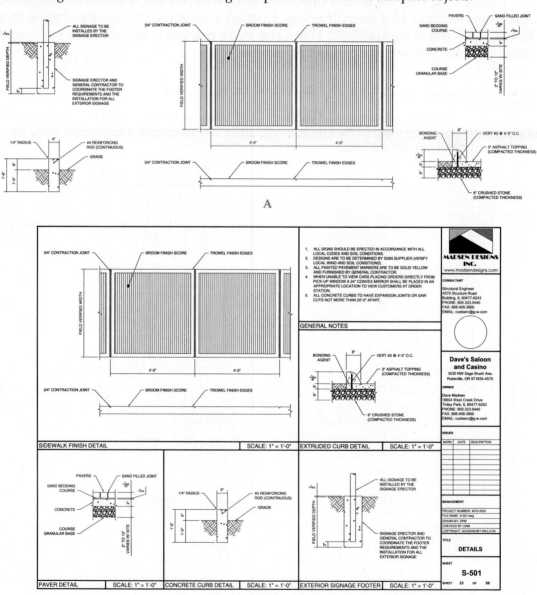

Preparing Xref and Host Drawings

Before you begin placing xref drawings, you should prepare the xref and host drawing files for xref insertion. When you place an xref drawing, everything you see in model space is inserted into the host file as a single item. Layout content is not included. The default insertion base point for an xref file is the model space origin, or 0,0,0. The insertion base point attaches to the crosshairs or appears at the specified insertion point when you insert the xref into the host drawing. If it is critical that xref objects coincide with the 0,0,0 point for insertion, move all objects in model space as needed in the xref file.

An alternative to moving objects to the origin is to use the **BASE** command to change the insertion base point of the drawing. Access the **BASE** command and select a new insertion base point. Save the drawing before using it as an xref.

If you use an appropriate template, little effort is necessary to prepare the host file to accept an xref. The host file should include a unique layer (named XREF or A-ANNO-REFR, for example) assigned to xrefs. As you will learn, layers in a referenced drawing file remain intact when you add the xref to a host drawing. Therefore, properties and states that you assign to the XREF layer have no effect on xref objects. Set the XREF layer current and proceed to place the xref drawing.

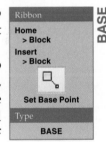

Ribbon	
Home	
> Block	
Insert	
> Block	
Set Base Point	
Type	
BASE	

BASE

Placing Xref Drawings

To place an xref, access the **ATTACH** command to display the **Select Reference File** dialog box. The **Drawing (*.dwg)** option in the **Files of type:** drop-down list limits the files displayed in the dialog box to drawings. Use the **Select Reference File** dialog box to locate the drawing file to add to the host file as an xref. Then pick the **Open** button to display the **Attach External Reference** dialog box. See **Figure 31-4.**

The **Attach External Reference** dialog box includes options for specifying how and where to place the selected file in the host drawing as an xref. If an external reference already exists in the current drawing, place another copy by choosing the file from the **Name:** drop-down list. To place a different xref drawing, pick the **Browse...** button and select the new file in the **Select Reference File** dialog box.

You can also place an xref using the **External References** palette shown in **Figure 31-5**. The **External References** palette is a complete external reference management tool. To place an xref drawing using the **External References** palette, pick the **Attach DWG** button from the **Attach** flyout, or right-click on the **File References** pane and select **Attach DWG...**. The **Select Reference File** dialog box appears, displaying only drawing files. Locate and select a file to add as an xref and pick the **Open** button to display the **Attach External Reference** dialog box.

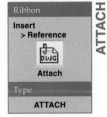

Ribbon	
Insert	
> Reference	
Attach	
Type	
ATTACH	

ATTACH

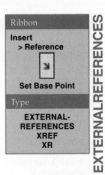

Ribbon	
Insert	
> Reference	
Set Base Point	
Type	
EXTERNAL-REFERENCES	
XREF	
XR	

EXTERNALREFERENCES

Figure 31-4.
Use the **Attach External Reference** dialog box to specify how to place an xref in the host drawing. Pick the **Show Details** button to display additional file details, as shown.

Pick to access existing xrefs   Pick to select a new file to attach

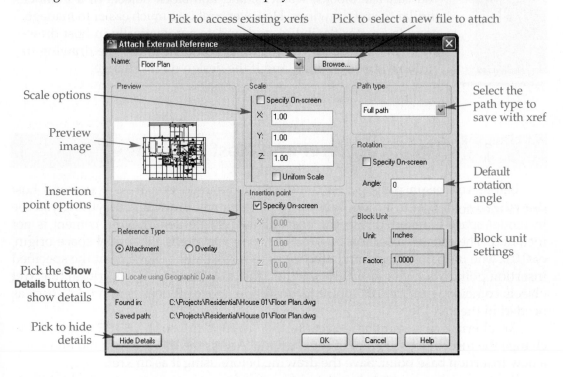

Scale options

Preview image

Insertion point options

Pick the **Show Details** button to show details

Pick to hide details

Select the path type to save with xref

Default rotation angle

Block unit settings

Figure 31-5.
The **External References** palette provides access to all options for externally referenced files.

Pick to add an xref drawing to the host file

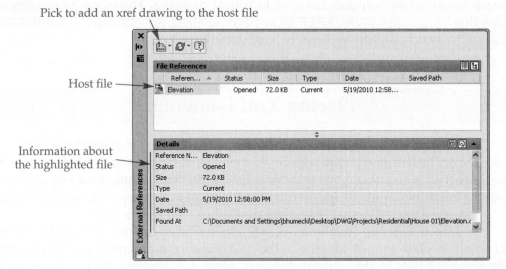

Host file

Information about the highlighted file

Type
XATTACH
XA

attachment: An xref linked with or referenced into the current drawing.

overlay: An xref displayed in the host drawing, but not attached to it.

The **XATTACH** command is identical to the **ATTACH** command, but it initially displays only drawing files in the **Select Reference File** dialog box.

Attachment vs. Overlay

You can choose to insert an xref drawing as an *attachment* or an *overlay* by selecting the **Attachment** or **Overlay** radio button in the **Reference Type** area. Attach xrefs for most applications. An xref overlay allows you to share content with others in a design drafting team, typically while working in a networked environment. You can overlay drawings without referencing nested xrefs.

Nesting occurs when an xref file references another xref file. An attached xref that has nested xrefs is the *parent xref*. When you attach an xref, the host drawing receives any nested xrefs that the xref contains. This does not happen when you overlay an xref. Furthermore, if you overlay an xref in a host drawing and then attach the host drawing to the current drawing, the overlaid xref does not appear in the current drawing.

parent xref: An xref that contains one or more other xrefs.

For example, suppose you attach a floor plan xref to a host file to create a foundation plan, and then attach the foundation plan xref to a host file to draw a section. Attaching the foundation plan brings the foundation and floor plan geometry into the section file for reference. If a member of your design drafting team uses your section, or is working on a drawing that already has the floor and/or foundation plan attached, she or he can overlay the section xref into a drawing without bringing in the floor plan and foundation plan.

You can change an overlay to an attachment or an attachment to an overlay after insertion. Managing xrefs is described later in this chapter.

Selecting the Path Type

Use the **Path type** drop-down list in the **Path Type** area to set how AutoCAD stores the path to the xref file. The path locates the xref file when you open the host file. The path appears in the **Attach External Reference** dialog box when you pick the **Show Details** button, and later appears in the **External References** palette. See **Figure 31-6**.

The default **Full path** option saves an *absolute path*. When using the **Full path** option, you must locate xref drawings in the drive and folder specified in the saved path. You can move the host drawing to any location, but the xref drawings must remain in the saved path. This option is acceptable if it is unlikely that you will move or copy the host and xref drawings to another computer, drive, or folder.

absolute path: A path to a file defined by the location of the file on the computer system.

The **Relative path** option is often more appropriate if you share drawings with a client or eventually archive drawings. The **Relative path** option saves a *relative path*.

relative path: A path to a file defined according to the location of the file relative to the host drawing.

Figure 31-6.
You can reference a file using a full path, a relative path, or no path. The path type is displayed in the **Save Path** column in the **External References** palette.

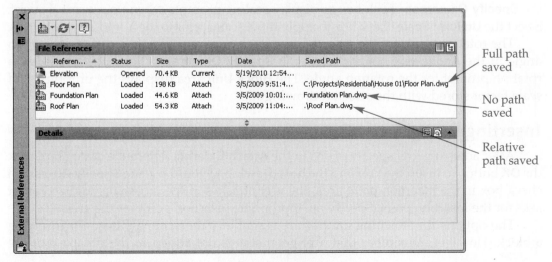

If the host drawing and xref files are located in a single folder and subfolders, you can copy the folder to any location without losing the connection between files. For example, copy the folder from the C: drive of one computer to the D: drive of another computer, to a folder on a CD, or to an archive server. If you perform these types of transfers with the **Full path** option, you need to open the host drawing after copying and redefine the saved paths for all xref files. You cannot use a relative path if the xref file is on a drive other than the drive on which the host file is stored.

Select the **No Path** option if you do not want to save the path to the xref file. If you choose the **No Path** option, the xref file loads only if you include the path to the file in one of the Support File Search Path locations or if the xref file is in the same folder as the host file. Specify the Support File Search Path locations in the **Files** tab of the **Options** dialog box.

AutoCAD searches for xref files in all paths of the current project name. Project search paths are listed under Project Files Search Path in the **Files** tab of the **Options** dialog box. Create a new project as follows:
1. Pick Project Files Search Path to highlight it, and then pick the **Add...** button.
2. Enter a project name.
3. Pick the plus sign icon (+), and then pick the word **Empty**.
4. Pick the **Browse...** button and locate the folder that is to become part of the project search path. Then pick **OK**.
5. Complete the project search path definition by entering the **PROJECTNAME** system variable and specifying the same name you specified in the **Options** dialog box.

Additional Xref Placement Options

The remaining items in the **Attach External Reference** dialog box allow you to control or identify xref insertion location, scaling, rotation angle, and block unit settings. Deselect the **Specify On-screen** check box in the **Insertion point** area to enter 2D or 3D coordinates in the text boxes for insertion of the xref. Activate the **Specify On-screen** check box to specify the insertion location on-screen. The **Locate using Geographic Data** check box is active if the xref and host drawings include *geographic data*. Pick the check box to position the xref using geographic data.

geographic data:
Information added to a drawing to describe specific locations and directions on Earth.

Use the **Scale** area to set the xref scale factor. AutoCAD sets the X, Y, and Z scale factors to 1 by default. Enter different values in the corresponding text boxes or activate the **Specify On-screen** check box to display scaling prompts when you insert the xref. Select the **Uniform Scale** check box to apply the X scale factor to the Y and Z scale factors.

The rotation angle for the inserted xref is 0 by default. Specify a different rotation angle in the **Angle:** text box, or select the **Specify On-screen** check box to display a rotation prompt for the rotation angle. The **Block Unit** area displays the unit type and scale factor stored with the selected drawing file.

Inserting the Xref

After adjusting xref specifications in the **Attach External Reference** dialog box, pick the **OK** button to insert the xref into the host drawing. If you chose the **Specify On-screen** check box in the **Insertion point** area, the xref attaches to the crosshairs and a prompt asks for the insertion point. Specify an appropriate insertion point for the xref.

The options for attaching an xref are essentially the same as those for inserting a block. However, remember that xref geometry is not added to the database of the host file, as are inserted blocks. Therefore, using external references helps keep your drawing file size to a minimum.

Exercise 31-1
Complete the exercise on the companion website.
www.g-wlearning.com/CAD

Placing Xrefs with DesignCenter and Tool Palettes

To place an xref into the current drawing using **DesignCenter**, first use the **Tree View** pane to locate the folder containing the drawing you want to attach. Then display the drawing files located in the selected folder in the **Content** pane. Right-click on the drawing file in the **Content** pane and select **Attach as Xref...**. Another method is to drag and drop the drawing into the current drawing area using the right mouse button. When you release the button, select the **Attach as Xref...** option to display the **Attach External Reference** dialog box. Enter the appropriate values and pick the **OK** button to place the xref.

You must add an xref or drawing file to a tool palette in order to use the **Tool Palettes** palette to place the file as an xref. To add an xref to a tool palette, drag an existing xref from the current drawing or an xref from the **Content** pane of **DesignCenter** into the **Tool Palettes** palette. Use drag and drop to attach the xref to the current drawing from the palette. Xref files in tool palettes display an external reference icon.

A drawing file (not an xref) added to a tool palette from the current drawing or using **DesignCenter** is a block tool. To convert the block tool to an xref tool, right-click on the image in the **Tool Palettes** palette and select **Properties...** to display the **Tool Properties** dialog box. Then change the **Insert as** field status from **Block** to **Xref** using the **Insert as** drop-down list. The **Reference type** row controls whether the xref is inserted as an attachment or an overlay.

Working with Xref Objects

An xref is inserted as a single object. Xref drawings appear faded by default to help differentiate the xref from the host drawing. Xref fading is an on-screen display function only and does not apply to plots. The **XDWGFADECTL** system variable controls fading of xref drawings on-screen. The easiest way to adjust fading is to use the options in the expanded **Reference** panel of the **Insert** ribbon tab. See Figure 31-7A. Pick the **Xref Fading** button to activate or deactivate xref fading, and use the slider or text box to increase or decrease fading. The default value of 70% creates significant fading. See Figure 31-7B.

Select an xref in the **External References** palette to highlight all visible instances of the xref in the drawing. Select an xref in the drawing to highlight the name in the **External References** palette. Use editing commands such as **MOVE** and **COPY** to modify the xref as needed. However, there are some significant differences between xrefs and other objects. For example, if you erase an xref, the xref definition remains in the file, similar to an erased block. You must detach an xref to remove it from the file completely.

To select an xref, pick an object displayed on-screen that is part of the xref.

Figure 31-7.
A—Use options in the expanded **Reference** panel of the **Insert** ribbon tab to control xref fading.
B—Default fading applied to xref objects.

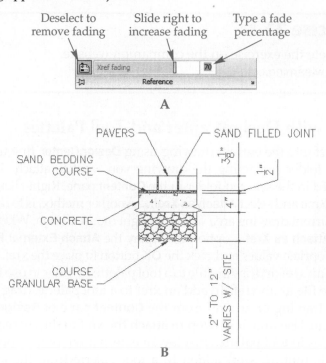

B

Dependent Objects

When you place an xref in a drawing, the host file receives all named objects in the xref file, such as layers and blocks, as *dependent objects*, even if the xref file does not use the objects. Dependent objects are displayed in the host drawing for reference only. The xref drawing stores the actual object definitions.

When you attach an xref, dependent objects are assigned unique names that consist of the xref file name followed by the actual object name, separated by a vertical bar symbol (|). For example, a layer named A-DOOR in a reference drawing named Floor Plan comes into the host drawing as Floor plan|A-DOOR. See **Figure 31-8**. This name distinguishes xref-dependent layers from layers that may have the same name in the host drawing. The names also make it easier to manage layers when several xrefs are attached to the host drawing, because the layers from each reference file are preceded by their file names. You cannot rename xref-dependent objects.

When you attach an xref, dependent objects such as layers are added to the host drawing only in order to support the display of the objects in the reference file. You cannot set xref layers current, and as a result, you cannot draw on xref layers. However, you can turn xref layers on and off, thaw and freeze them, and lock or unlock them as needed. You can also change the colors and linetypes of xref layers.

dependent objects: Objects displayed in the host drawing, but defined in the xref drawing.

Use the Xref filter in the **Layer Properties Manager** to display and manage dependent layers. You can also save dependent layers in a layer state.

Figure 31-8.
The xref drawing name and a vertical bar symbol (|) precede xref-dependent layer names in the host drawing.

Filters available by default when you place an xref

Pick to display only xref layers

Xref-dependent layers

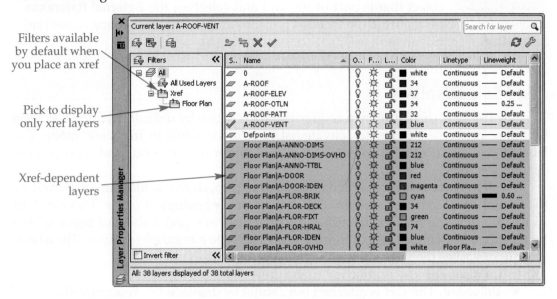

PROFESSIONAL TIP

When you attach a drawing as an xref, the reference file comes into the host drawing with the same layer colors and linetypes used in the original file. If you reference a drawing to check the relationship of objects between two drawings, consider changing the xref layer colors to make it easier to differentiate between the content of the host drawing and the xref drawing. Changing xref layer colors affects only the display in the current drawing and does not alter the original reference file.

Exercise 31-2

Complete the exercise on the companion website.
www.g-wlearning.com/CAD

Managing Xrefs

The **External References** palette is the primary tool for managing and accessing current information about xrefs found in a host drawing. You can also access the **External References** palette by picking an xref and selecting the **External Reference** button from the **Options** panel of the **External References** contextual ribbon tab. The **External References** palette displays an upper **File References** pane and a lower **Details** pane. See **Figure 31-9**. Display the **File References** pane in list view or tree view and with details or a preview.

 You can also access the **External References** palette by picking an object that is part of the xref and selecting the **External Reference** button from the **Options** panel of the **External Reference** contextual ribbon tab.

List View Display

The list view display shown in **Figure 31-9** is active by default. Pick the **List View** button or press the [F3] key to activate list view mode while in tree view mode. The labeled columns displayed in list view provide information about and management options for xrefs.

The **Reference Name** column displays the current drawing file name followed by the names of all existing xrefs in alphabetical or chronological order. The standard AutoCAD drawing file icon identifies the host drawing, and a sheet of paper with a paper clip icon identifies xref drawings. Each xref type has a different icon. The **Status** column describes the status of each xref, which can be:

- **Loaded.** The xref is attached to the drawing.
- **Unloaded.** The xref is attached but cannot be displayed or regenerated.
- **Unreferenced.** The xref has nested xrefs that are not found or are unresolved. An unreferenced xref is not displayed.
- **Not Found.** The xref file is not found in the specified search paths.
- **Unresolved.** The xref file is missing or cannot be found.
- **Orphaned.** The parent of the nested xref cannot be found.

The **Size** column lists the file size for each xref. The **Type** column indicates whether the xref is attached or referenced as an overlay. The **Date** column indicates the date the xref was last modified.

The **Saved Path** column lists the path name saved with the xref. If only a file name appears, the path was not saved. Prefixes describe the relative paths to xref files. In

Figure 31-9.
The **External References** palette allows you to view and manage referenced files. The **File References** pane appears in **List View** mode and the **Details** pane appears in **Details** mode.

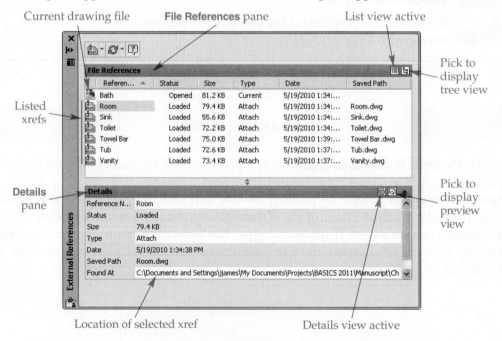

AutoCAD and Its Applications—Basics

Figure 31-6, the characters .\ precede the Roof Plan reference file. The period (.) represents the folder containing the host drawing. From that folder, AutoCAD looks in the House 01 folder that contains the Roof Plan drawing. The Elevation reference file in **Figure 31-10** uses a similar specification. In this example, the same folder contains the Elevation xref and the host drawing. The characters ..\ precede the specification for the Wall xref. The double period instructs AutoCAD to move up one folder level from the current location. The double period repeats to move up multiple folder levels. For example, AutoCAD locates the Panel xref in **Figure 31-10** by moving up two folder levels from the folder of the host drawing and opening the Symbols folder.

The path saved to the xref is one of several locations AutoCAD searches when you open a host drawing and an xref requires loading. AutoCAD searches path locations to load xref files in the following order:
1. The full or relative path associated with the xref
2. The current folder of the host drawing
3. The project paths specified in the Project Files Search Path
4. The support paths specified in the Support File Search Path
5. The Start in: folder path specified for the AutoCAD application shortcut associated using the **Properties** option in the desktop icon shortcut menu

Figure 31-10.
Relationship between the symbols in the **Saved Path** list and file locations within the folder structure.

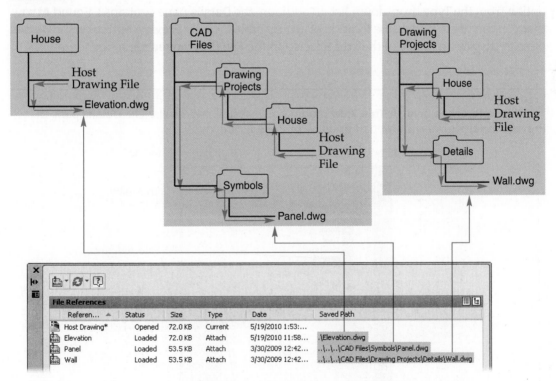

Adjust the column widths in list view mode as necessary to view complete information. To adjust the width of a column, move the cursor to the edge of the button at the top of the column until the cursor changes to a horizontal resizing cursor. Press and hold the left mouse button and drag the column to the desired width. If columns extend beyond the width of the dialog box, a horizontal scroll bar appears at the bottom of the list.

Tree View Display

Pick the **Tree View** button or press the [F4] key to see a list of xrefs in the **File References** pane, and show nesting levels. See **Figure 31-11**. Nesting levels are displayed in a format similar to the arrangement of folders. The status of the xref determines the appearance of the xref icon. An xref with an unloaded or not found status has a grayed-out icon. An upward arrow shown with the icon means the xref was reloaded, and a downward arrow means the xref was unloaded.

Viewing Details or a Preview

The **Details** mode, shown in **Figure 31-9**, is active by default. Pick the **Details** button to display details while in **Preview** mode. The information listed in the **Details** pane corresponds to the host file or xref selected in the **File References** pane. The rows displayed in **Detail** mode are the same as the columns found in **List View** mode of the **File References** pane. However, in the **Details** pane, you can modify the reference name by entering a new name in the **Reference Name** text box. You can also change the reference type from an attachment to an overlay or from an overlay to an attachment by picking the appropriate option from the **Type** drop-down list. In addition, the **Details** pane contains a **Found At** row that you can use to update the location of an xref path. Pick the **Preview** button on the **Details** pane to display an image of the xref selected in the **File References** pane. See **Figure 31-11**.

Figure 31-11.
The **File References** pane in **Tree View** mode shows nested xref levels. The **Details** pane in **Preview** mode shows a thumbnail preview of the selected xref.

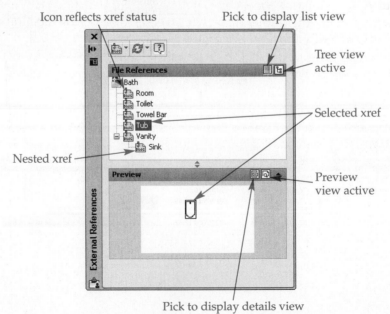

Icon reflects xref status

Pick to display list view

Tree view active

Selected xref

Nested xref

Preview view active

External References

Pick to display details view

You can also use the **External References** palette to manage data extractions.

Detaching, Reloading, and Unloading Xrefs

Each time you open a host drawing containing an attached xref, the xref loads and appears on-screen. This association remains permanent until you *detach* the xref. Erasing an xref does not remove the xref from the host drawing. To detach an xref, right-click on the reference name in the **File References** pane of the **External References** palette and pick **Detach**. All instances of the xref and all of its nested xrefs are detached from the current drawing, along with all referenced data.

In some situations, you may need to update, or *reload*, an xref file in the host drawing. For example, if you edit an xref while the host drawing is open, the updated version may be different from the version you see. To update the xref, right-click the reference name in the **File References** pane of the **External References** palette and pick **Reload**, or pick the **Reload All References** button from the flyout to reload all unloaded xrefs. Reloading xrefs forces AutoCAD to read and display the most recently saved version of each xref.

To *unload* an xref, right-click on the reference name in the **File References** pane of the **External References** palette and pick **Unload**. An unloaded xref is not displayed or regenerated, so performance increases. Reload the xref to redisplay it.

detach: Remove an xref from a host drawing.

reload: Update an xref in the host drawing.

unload: Suppress the display of an xref without removing the xref from the host drawing.

If AutoCAD cannot find an xref, an alert appears when you open the host drawing. Choose the appropriate option to ignore the problem or fix the problem using the **External References** palette.

Updating the Xref Path

A file path saved with an xref is displayed in the **Saved Path** column of the **File References** pane and the **Saved Path** row of the **Details** pane in the **External References** palette. If the **Saved Path** location does not include an xref file, when you open the host drawing, AutoCAD searches the *library path*. A link to the xref forms if AutoCAD finds a file with a matching name. In such a case, the **Saved Path** location differs from where AutoCAD actually found the file.

Check for matching paths in the **External References** palette by comparing the path listed in the **Saved Path** column of the **File References** pane and **Saved Path** row of the **Details** pane with the listing in the **Found At** row of the **Details** pane. When you move an xref and the new location is not in the library path, the xref status is **Not Found**. To update or find the **Saved Path** location, select the path in the **Found At** edit box and pick the **Browse...** button to the right of the edit box to access the **Select new path** dialog box. Use the **Select new path** dialog box to locate the new folder and select the desired file. Then pick the **Open** button to update the path.

library path: The path AutoCAD searches by default to find an xref file, including the current folder and locations set in the **Options** dialog box.

The Manage Xrefs Icon

By default, when you edit, save, and close an xref, and then open the host drawing, changes made to the xref automatically appear without any notification. If you make changes to an xref while the host drawing is open, a notification appears in the status bar tray. Changes are indicated by the appearance of the **Manage Xrefs** icon, a balloon message, or both.

The **Tray Settings** dialog box controls notifications in the status bar tray for xref changes and other system updates. Select **Tray Settings...** from the status bar shortcut menu to access the **Tray Settings** dialog box. Select the **Display** icons from services check box to display the **Manage Xrefs** icon in the status bar tray when you attach an xref to the current drawing. If you modified an xref in the current file since opening the file, the **Manage Xrefs** icon appears with an exclamation sign. Pick the **Manage Xrefs** icon or right-click on the **Manage Xrefs** icon and select **External References...** to open the **External References** palette to reload the xref.

Select the **Display notifications from services** check box in the **Tray Settings** dialog box to display a balloon message notification with the name of the modified xref file. See **Figure 31-12A**. You can then pick the xref name in the balloon message to reload the file. The example in **Figure 31-12** shows adding a Towel Bar xref to the Room parent xref drawing. The xref reloads in the host drawing named Bath. See **Figure 31-12B**. You can also reload xrefs by right-clicking on the **Manage Xrefs** icon and selecting **Reload DWG Xrefs**.

Exercise 31-3

Complete the exercise on the companion website.
www.g-wlearning.com/CAD

Clipping Xrefs

subregion: The displayed portion of a clipped xref.

Clip, or crop, an xref to display only a specific portion, or an xref *subregion*. All geometry that falls outside the clipping boundary is invisible, and objects that are partially within the subregion appear trimmed at the boundary. Although clipped objects appear trimmed, the xref file does not change. Clipping applies to a selected instance of an xref, not to the actual xref definition.

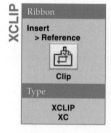

Use the **XCLIP** command to create and modify clipping boundaries. Quick ways to access the **XCLIP** command include picking an object that is part of the xref and selecting the **Create Clipping Boundary** button from the **Clipping** panel of the **External Reference** contextual ribbon tab, or right-clicking and selecting **Clip Xref**. If you access the **XCLIP** command before selecting an xref, pick an object associated with the xref to clip. Then press [Enter] to accept the default **New boundary** option and select the clipping boundary.

When you select the **New boundary** option, a prompt asks you to specify the clipping boundary. Use the default **Rectangular** option to create a rectangular boundary. Then pick opposite corners of the rectangular boundary. See **Figure 31-13**. Note that the geometry outside the clipping boundary no longer appears after clipping. The **New boundary** option includes additional methods for specifying the clip boundary and area to clip, as briefly described in **Figure 31-14**.

Edit a clipped xref as you would an unclipped xref. The clipping boundary moves with the xref. Note that nested xrefs are clipped according to the clipping boundary for the parent xref. Use the **XCLIPFRAME** system variable to toggle the display and plotting behavior of the clipping boundary frame. The default value of 2 displays the frame but does not plot it. Set the value to 0 to hide and not plot the frame. Set the value to 1 to display and plot the frame. The **FRAMESELECTION** system variable controls the ability to select a hidden frame (**XCLIPFRAME** system variable set to 0). The default value of 1 allows you to select hidden frames. The frame appears for reference when you select the xref. Change the value to 0 to disable selection of hidden frames. You can still pick an object that is part of the xref to select the xref.

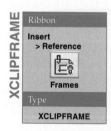

The other options of the **XCLIP** command apply after you define a clip boundary. The **ON** and **OFF** options turn the clipping feature on or off. The **Clipdepth** option allows you to define front and back clipping planes to control the portion of a 3D drawing that displays. Clipping 3D models is described in *AutoCAD and Its Applications—Advanced*.

Figure 31-12.
The **Manage Xrefs** icon in the status bar tray provides a notification when you modify and save an xref file. A—A balloon message and an exclamation point appear at the icon. B—Reloading the xref file updates the current drawing and changes the appearance of the icon.

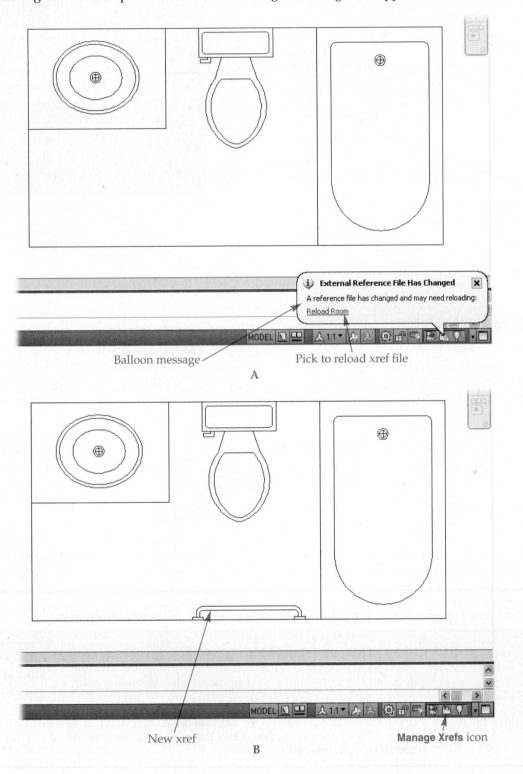

Balloon message · Pick to reload xref file

A

New xref

Manage Xrefs icon

B

Use the **Delete** option to remove an existing clipping boundary, returning the xref to its unclipped display. A quick way to remove clipping is to pick an object that is part of the xref and select the **Remove Clipping** button from the **Clipping** panel of the **External Reference** contextual ribbon tab. Use the **generate Polyline** option to create and display a polyline object at the clip boundary to frame the clipped portion.

Figure 31-13.
Clipping a large site plan xref to display a specific area. A—Using the **Rectangular** boundary selection option. B—The clipped xref.

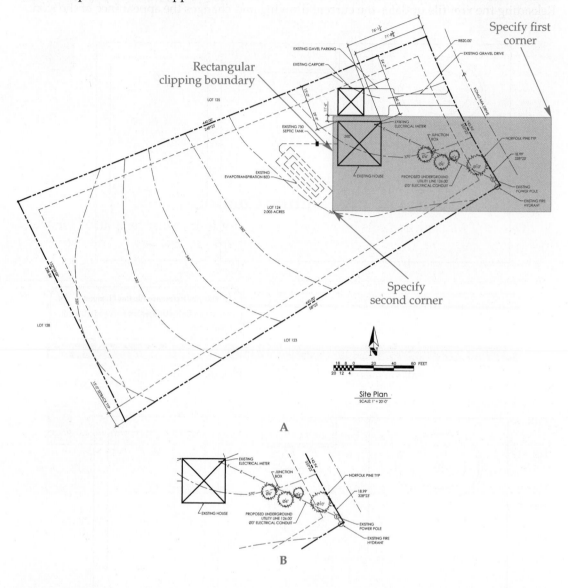

A

B

Figure 31-14.
Additional options available for the **New boundary** function of the **XCLIP** command.

Option	Description
Select Polyline	Select an existing polyline object as the clip boundary. If the polyline does not close, the start and endpoints of the boundary connect.
Polygonal	Draw an irregular polygon as a boundary.
Invert clip	Inverts the selection so that the portion of the xref that lies outside of the clipping boundary is clipped. Only the portion of the xref outside of the boundary is displayed.

You can also remove clipping by selecting an object that is part of the xref or a clipping frame, and picking the **Remove Clipping** button from the **Clipping** panel of the **External Reference** contextual ribbon tab.

Exercise 31-4

Complete the exercise on the companion website.
www.g-wlearning.com/CAD

Demand Loading and Xref Editing Controls

Demand loading controls how much of an xref loads when you attach the xref to the host drawing. It improves performance and saves disk space because only a portion of the xref file loads into the host drawing. For example, data on frozen layers and data outside of clipping regions does not load.

Demand loading occurs by default. Use the **Open and Save** tab of the **Options** dialog box to check or change the setting. The **Demand load Xrefs:** drop-down list in the **External References (Xrefs)** area contains each demand loading option. Select the **Enabled with copy** option to turn on demand loading. Other users can edit the original drawing because AutoCAD uses a copy of the referenced drawing. Alternatively, pick the **Enabled** option to turn on demand loading. If you use this option, the xref file is considered "in use" while you are referencing the drawing, preventing other users from editing the file. Select the **Disabled** option to turn off demand loading.

Two additional settings in the **Open and Save** tab of the **Options** dialog box control the effects of changes made to xref-dependent layers and in-place reference editing. The **Retain changes to Xref layers** check box allows you to keep all changes made to the properties and states of xref-dependent layers. Any changes to layers take precedence over layer settings in the xref file. Edited properties remain even after you reload the xref. The **Allow other users to Refedit current drawing** check box controls whether the current drawing can be edited in place by others while it is open and when it is referenced by another file. Both check boxes are selected by default.

> **demand loading:** Loading only the portion of an xref file necessary to regenerate the host drawing.

PROFESSIONAL TIP

If you plan to use a drawing as an external reference, save the file with *spatial indexes* and *layer indexes*. These lists help improve performance when you reference drawings with frozen layers and clipping boundaries. Use the following procedure to create spatial and layer indexes:

1. Access the **Save Drawing As** dialog box.
2. Pick **Options...** from the **Tools** flyout button and select the **DWG Options** tab of the **Saveas Options** dialog box.
3. Select the type of index required from the **Index type:** drop-down list.
4. Pick the **OK** button and save the drawing.

> **spatial index:** A list of objects ordered according to their location in 3D space.

> **layer index:** A list of objects ordered according to the layers to which they are assigned.

Bind an xref to make the xref a permanent part of the host drawing, as if you were inserting the file using the **INSERT** command. Binding is useful when you need to send the full drawing file to another location or user, such as a plotting service or client. To bind an xref using the **External References** palette, right-click on the reference name and pick **Bind…**. The **Bind Xrefs** dialog box that appears contains **Insert** and **Bind** radio buttons.

Using the Insert and Bind Options

The **Insert** option converts the xref into a normal block, as if you had used the **INSERT** command to place the file. In addition, the drawing is added to the block definition table, and all named objects, such as layers, blocks, and styles, are incorporated into the host drawing as named in the xref. For example, if you bind an xref file named PLATE that contains a layer named OBJECT, the xref-dependent layer PLATE|OBJECT becomes the locally defined layer OBJECT. All other xref-dependent objects lose the xref name and assume the properties of the locally defined objects with the same name. The **Insert** binding option provides the best results for most purposes.

The **Bind** option also converts the xref into a normal block. However, the xref name remains with all dependent objects. Two dollar signs with a number between them replace the vertical line in each name. For example, an xref layer named Title|Notes becomes Title0Notes. The number inside the dollar signs is automatically incremented if a local object definition with the same name exists. For example, if Title0Notes already exists in the drawing, the newly bound layer becomes Title1Notes. In this manner, all xref-dependent object definitions that are bound receive unique names. Use the **RENAME** command or other appropriate method to rename bound objects.

Binding Specific Dependent Objects

Binding an xref allows you to make all dependent objects in the xref file a permanent part of the host drawing. Dependent objects include named items such as blocks, dimension styles, layers, linetypes, and text styles. Before binding, you cannot directly use any dependent objects from a referenced drawing in the host drawing. For example, you cannot make an xref layer or text style current in the host drawing.

In some cases, you may only need to incorporate one or more specific named objects, such as a layer or block, from an xref into the host drawing, instead of binding the entire xref. If you only need selected items, it can be counterproductive to bind an entire drawing. Instead, use the **XBIND** command and corresponding **Xbind** dialog box, shown in **Figure 31-15**, to select specific named objects to bind.

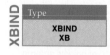

Type
    XBIND
    XB

Figure 31-15.
Use the **Xbind** dialog box to bind xref-dependent objects individually to the host drawing.

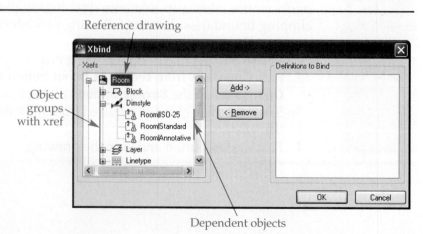

Xrefs have AutoCAD drawing file icons. Expand a group to select an individually named object. To select an object for binding, highlight the object and pick the **Add** button. The names of all objects selected and added are displayed in the **Definitions to Bind** list. Pick the **OK** button to complete the operation. A message displayed on the command line indicates how many objects of each type are bound.

Individual objects bound using the **XBIND** command are renamed in the same manner as objects bound using the **Bind** option in the **Bind Xrefs** dialog box. An automatic linetype bind performs so that a layer that includes a linetype not loaded in the host drawing can reference the required linetype definition. The linetype includes a new linetype name, such as xref1$0$hidden. In a similar manner, a previously undefined block may automatically bind to the host drawing because of binding nested blocks. Use the **RENAME** command or other appropriate method to rename objects.

Exercise 31-5

Complete the exercise on the companion website.
www.g-wlearning.com/CAD

Editing Xref Drawings

One option for editing an xref drawing is to use in-place editing, or *reference editing*, within the host drawing. You can save any changes made to the xref to the original xref drawing from within the host drawing. Alternatively, you can edit the xref in a separate drawing window as you would any other drawing file.

reference editing:
Editing reference drawings from within the host file.

Reference Editing

The **REFEDIT** command allows you to edit xref drawings in place. Quick ways to initiate reference editing include double-clicking on an xref, picking an object that is part of the xref and selecting the **Edit Reference In-Place** button from the **Edit** panel of the **External Reference** contextual ribbon tab, or selecting an xref and then right-clicking and selecting **Edit Xref In-place**. If you access the **REFEDIT** command without first selecting an xref, you must then pick the xref to edit. The **Reference Edit** dialog box opens with the **Identify Reference** tab active. See **Figure 31-16**. The example shows

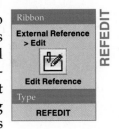

Ribbon
External Reference > Edit

Edit Reference

Type
REFEDIT

REFEDIT

Figure 31-16.
The **Reference Edit** dialog box lists the name of the selected reference drawing and displays an image preview.

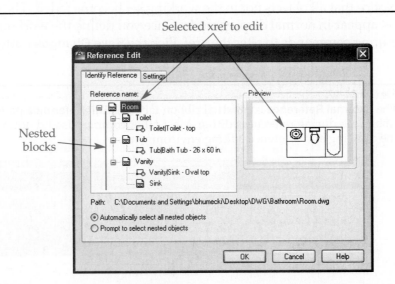

the Room reference drawing selected for editing. Notice that nested blocks, such as the Bath Tub 26 x 60 in. block found in the Tub reference, are listed under their parent xrefs.

The **Automatically select all nested objects** radio button in the **Path:** area is active by default. Use this option to make all xref objects available for editing. To edit specific xref objects, pick the **Prompt to select nested objects** radio button. The **Select nested objects:** prompt displays after you pick the **OK** button, allowing you to pick objects that belong to the selected xref. Pick all the geometry you want to edit and press [Enter]. The nested objects you select make up the *working set*. If multiple instances of the same xref appear, be sure to pick objects from the original xref you select.

Additional options for reference editing are available in the **Settings** tab of the **Reference Edit** dialog box. The **Create unique layer, style, and block names** option controls the naming of selected layers and *extracted* objects. Check the box to assign the prefix n, with n representing an incremental number, to object names. This is similar to the renaming method used when you bind an xref.

The **Display attribute definitions for editing** option is available if you select a block object in the **Identify Reference** tab of the **Reference Edit** dialog box. Check the box to edit any attribute definitions included in the reference. To prevent accidental changes to objects that do not belong to the working set, check the **Lock objects not in working set** option. This makes all objects outside of the working set unavailable for selection in reference editing mode.

If the selected xref file contains other references, the **Reference name:** area lists all nested xrefs and blocks in tree view. In the example given, Toilet, Tub, Vanity, and Towel Bar are nested xrefs in the Room xref. If you pick the drawing file icon next to Vanity in the tree view, for example, an image preview appears and the selected xref is highlighted in the drawing window.

When you finish adjusting settings, pick the **OK** button to begin editing the xref. The primary difference between the drawing and reference editing environments is the **Edit Reference** panel that appears in each ribbon tab, including the expanded **External Reference** contextual ribbon tab that appears when you pick an xref. See **Figure 31-17.** Use the tools in the **Edit Reference** panel to add objects to the working set, remove objects from the working set, and save or discard changes to the original xref file.

Any object you draw during the in-place edit is automatically added to the working set. Use the **Add to Working Set** button to add existing objects to the working set. When you add an object to the working set, the object is extracted, or removed, from the host drawing. The **Remove from Working Set** button allows you to remove selected objects from the working set. Removing a previously extracted object adds the object back to the host drawing.

Figure 31-18A shows reference-editing the Vanity xref nested in the Room xref. Notice that all objects not in the working set become faded. The objects in the working set appear in normal display mode. Once you define the working set, use drawing and editing commands to alter the xref. Pick the **Save Changes** button to save the changes.

Figure 31-17.
The **External Reference** contextual ribbon tab. The **Edit Reference** panel appears in each ribbon tab during reference editing. Most other commands and options function the same in the drawing and reference editing environments.

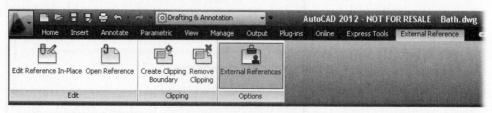

Pick the **OK** button when AutoCAD asks if you want to continue with the save and redefine the xref. All instances of the xref are updated. **Figure 31-18B** shows the xref after editing to redesign the sink and add a faucet. Pick the **Discard Changes** button to exit reference editing without saving changes.

In-place reference editing is best suited for minor revisions. Conduct major xref revisions in the reference drawing file.

CAUTION

All edits made using reference editing are saved back to the reference drawing file and affect any host drawing that references the file. For this reason, it is critically important that you edit external references only with the permission of your supervisor or instructor.

Figure 31-18.
Reference editing.
A—Objects in the drawing that are not a part of the working set appear faded during the reference-editing session. B—All instances of the xref update immediately after reference editing.

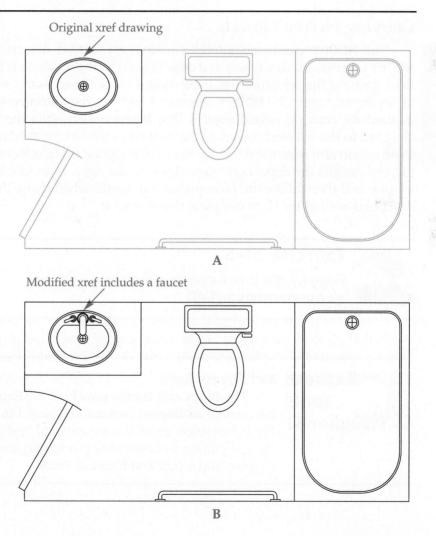

Original xref drawing

A

Modified xref includes a faucet

B

Opening an Xref File

Type

XOPEN

The **XOPEN** command allows you to select and open an xref in a separate drawing window from within the host drawing. This is essentially the same procedure as using the **OPEN** command, but faster. Quick ways to access the **XOPEN** command include picking an object that is part of the xref and selecting the **Open Reference** button from the **Edit** panel of the **External Reference** contextual ribbon tab, or right-clicking and picking **Open Xref**. If you access the **XOPEN** command without first selecting an xref, you must then pick the xref to open.

The xref drawing file opens in a separate drawing window. After you make changes and save the xref file, use the **Manage Xrefs** icon in the status bar or the **External References** palette to reload the modified xref file. Reloading ensures that the host file is up-to-date.

You can also open an xref in the **External References** palette by right-clicking the xref name and selecting **Open**.

Copying Nested Objects

NCOPY

Ribbon

Express Tools > Blocks

Copy Nested Objects

Type

NCOPY

The **NCOPY** command allows you to copy nested objects xref drawing. The **NCOPY** command extracts selected objects from the nest, which is much more efficient than opening the reference file, exploding a block if necessary, and copying objects, for example. Access the **NCOPY** command and use the **Settings** option to specify the method for copying nested objects. The **Insert** option adds the reference file layer assigned to the selected objects to the host file, and assigns the layer to the copy. The **Blind** option also adds the reference file layer assigned to the selected objects to the host file, but assigns the dependent object layer to the copy. Individually select the objects to copy and then follow the prompts as you would when using the **COPY** command, described in Chapter 12, to complete the operation.

Exercise 31-6

Complete the exercise on the companion website.
www.g-wlearning.com/CAD

Express Tools

Chapter 31

Xref Express Tools

The **Block** and **Modify** panels of the **Express Tools** ribbon tab include additional commands related to xrefs and blocks. For information about the most useful xref express tools, go to the companion website (www.g-wlearning.com/CAD), select this chapter, and select **Xref Express Tools**.

Chapter Review

Answer the following questions. Write your answers on a separate sheet of paper or complete the electronic chapter review on the companion website.
www.g-wlearning.com/CAD

1. What types of files can you reference into an AutoCAD drawing?
2. What effect does the use of referenced drawings have on drawing file size?
3. What is a nested xref?
4. List at least three common applications for xrefs.
5. On what layer should you consider inserting xrefs into a host drawing?
6. Which command allows you to attach an xref drawing to the current file?
7. What is the difference between an overlaid xref and an attached xref?
8. What is the difference between an absolute path and a relative path?
9. Describe the process of placing an xref using **DesignCenter**.
10. What must you do before you can use a tool palette to place an xref?
11. If you attach an xref file named FPLAN to the current drawing, and FPLAN contains a layer called ELECTRICAL, what name will appear for this layer in the **Layer Properties Manager**?
12. What is the purpose of the **Detach** option in the **External References** palette?
13. When are xrefs updated in the host drawing?
14. What could you do to suppress an xref temporarily without detaching it from the master drawing?
15. Which command allows you to display only a specific portion of an externally referenced drawing?
16. What are spatial and layer indexes, and what function do they perform?
17. Why would you want to bind a dependent object to a master drawing?
18. What does the layer name WALL0NOTES mean?
19. What command allows you to edit external references in place?
20. What command allows you to open a parent xref drawing into a new AutoCAD drawing window by selecting the xref in the host drawing?

Drawing Problems

Start AutoCAD if it is not already started. Start a new drawing for each problem using an appropriate template of your choice. The template should include layers and text, dimension, multileader, and table styles, when necessary, for drawing the given objects. Add layers and text, dimension, multileader, and table styles as needed. Draw all objects using appropriate layers and text, dimension, multileader, and table styles, justification, and format. Follow the specific instructions for each problem. Use only drawing and editing commands and techniques you have already learned. Use your own judgment and approximate dimensions when necessary. Apply dimensions accurately using ASME or appropriate industry standards.

Note: *Some of the problems in this chapter are built on problems from previous chapters. If you have not yet completed those problems, complete them now.*

▼ Basic

1. Attach a dimensioned problem from Chapter 17 into a new drawing as an xref. Save the drawing as P31-1.

2. Attach a dimensioned problem from Chapter 18 into a new drawing as an xref. Save the drawing as P31-2.

3. Attach a dimensioned problem from Chapter 19 into a new drawing as an xref. Save the drawing as P31-3.

▼ Intermediate

4. Attach the EX29-9.dwg file used in Exercise 29-9 into a new drawing as an xref. Copy the xref three times. Use the **XCLIP** command to create a clipping boundary on each view. Apply an inverted rectangular clip to the original xref, a polyline boundary on the first copy, and a polygonal boundary on the second copy. Save the drawing as P31-4.

5. Attach the EX29-9.dwg file used in Exercise 29-9 into a new drawing as an xref. Bind the xref to the new drawing. Rename the layers to the names assigned to the original EX29-9 (xref) file. Explode the block created by binding the xref. Save the drawing as P31-5.

6. Create the multi-detail drawing shown according to the following information:
   - Use the Mechanical-Inch.dwt drawing template file available on the companion website.
   - Set drawing units to fractional.
   - Xref the following files into model space: Detail-Item 1.dwg, Detail-Item 2.dwg, Detail-Item 3.dwg, Detail-Item 4.dwg, Detail-Item 5.dwg, and Detail-Item 6.dwg. These files are available on the companion website.
   - Use six floating viewports on the **C-SIZE** layout to arrange and scale the details. Use a 1:2 scale.
   - Adjust the title block information using the drawing property fields and attributes.
   - Plot the drawing.
   Save the drawing as P31-6.

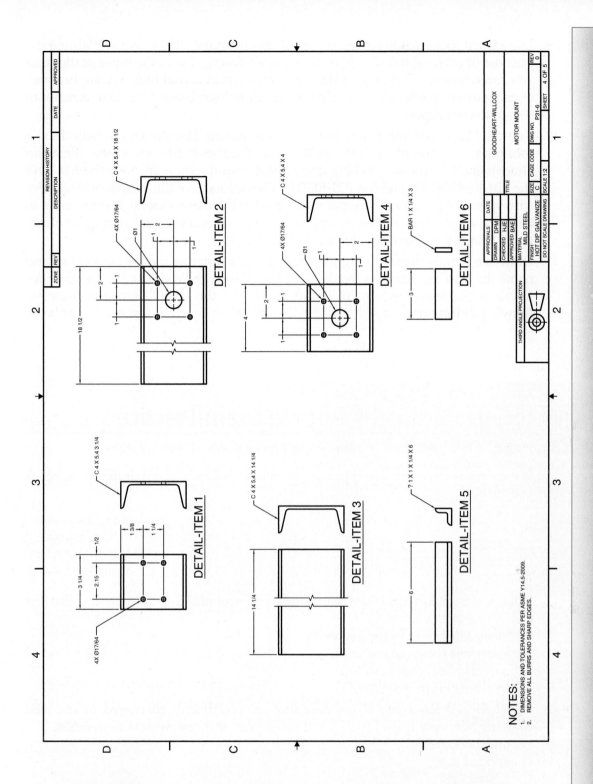

DETAIL-ITEM 2

DETAIL-ITEM 4

DETAIL-ITEM 6

DETAIL-ITEM 1

DETAIL-ITEM 3

DETAIL-ITEM 5

NOTES:
1. DIMENSIONS AND TOLERANCES PER ASME Y14.5-2009.
2. REMOVE ALL BURRS AND SHARP EDGES.

GOODHEART-WILLCOX	
TITLE	MOTOR MOUNT
	DWG NO. P31-6
CAGE CODE	SIZE C
	SCALE 1:2 SHEET 4 OF 5

▼ **Advanced**

7. Design and draw a basic residential floor plan using an appropriate template. Save the file as P31-7FLOOR. Xref the P31-7FLOOR file into a new file as an attachment. Use the xref to help draw a roof plan. Save the roof plan file as P31-7ROOF.

8. Xref the P31-7ROOF file into a new file as an attachment. Use the xref to help draw front and rear elevations. Save the elevation file as P31-8.

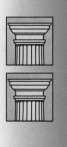

Drawing Problems - Chapter 31

9. Use a word processor to write a report of approximately 250 words explaining the purpose of external references. Include a brief description of the types of files that you can reference. Cite at least three examples from actual industry applications of using external references to help prepare drawings. Use at least four sketches to illustrate your report.

10. Xref the P28-11 file into a new file as an attachment. Use the xref to help draw a roof plan. Save the file as P31-10ROOF. Xref the P28-11 file into a new file as an attachment. Use the xref to help draw a slab foundation plan. Save the file. Xref the P31-10ROOF file and the P31-10FDTN files into a new file as attachments. Use corresponding insertion base points to overlap the plans exactly. Use the xrefs to help draw front, rear, right-side, and rear exterior elevations and an interior elevation for each wall. Save the elevation file as P31-10.

11. Draw the details shown in Figure 31-3A using a separate file for each detail and name the files according to the detail names shown in Figure 31-3B. Xref the files to create the layout shown in Figure 31-3B. Plot the layout. Save the complete drawing as P31-11.

AutoCAD Certified Associate Exam Practice

Answer the following questions. Write your answers on a separate sheet of paper.

1. Which of the following file types can be referenced into a host drawing? *Select all that apply.*
   A. DNG
   B. DOC
   C. DWFx
   D. PDF
   E. XLS

2. What is the default insertion base point for an xref file? *Select the one item that best answers the question.*
   A. lower-left corner of the geometry
   B. lower-left corner of the layout
   C. model space origin
   D. upper-right drawing limit

3. Which of the following terms describes an xref that is displayed in the host drawing but is not linked to it? *Select the one item that best answers the question.*
   A. attached xref
   B. clipped xref
   C. nested xref
   D. overlaid xref
   E. parent xref

AutoCAD Certified Professional Exam Practice

Follow the instructions in the problem. Write your answers on a separate sheet of paper.

1. **Open CPE-31backyard.dwg. This file is available on the companion website.**
   What layers in this drawing are former xrefs that have been bound to the drawing?

Chapter
Introduction to Sheet Sets
32

Learning Objectives

After completing this chapter, you will be able to:

✓ Describe the functions of an AutoCAD sheet set.
✓ Create sheet sets from examples and existing drawings.
✓ Manage sheet sets.
✓ Create and manage subsets.
✓ Add sheets to a sheet set.

A design project typically requires a set of drawings and documents that completely specify the design. Preparing accurate drawings and making revisions in a timely manner involves significant organization, especially when a project includes multiple related *sheets* and *views*. *Sheet sets* help organize a set of drawings and simplify project management. This chapter explains how to organize and manage sheet sets, subsets, and sheets. In order to work effectively with sheet sets, you must understand the elements of a sheet set such as drawing templates, blocks, and fields, as explained throughout this textbook.

sheet: A printed drawing or electronic layout that displays project design requirements.

view: 2D representation of an object.

sheet set: A collection of drawing sheets for a project; the AutoCAD tool that aids project organization.

Sheet Set Fundamentals

A sheet set is an electronic database of information about a project and the set of drawings required to document the design. A sheet set provides a way to organize files related to a set of drawings, similar to using Windows Explorer and a folder with subfolders to contain files. A basic application of a sheet set is to group drawings in the proper order for easy access and quick opening. A sheet set also provides functions that automate managing and creating a set of drawings, which improves speed and accuracy. **Figure 32-1** shows an example of using a sheet set to manage the drawings for a small architectural project of a Cappuccino Express drive-through coffee stand. The sheet set organizes the four required layouts in the appropriate order for viewing, editing, *publishing*, and *archiving*.

Every sheet set has a sheet set data (DST) file specific to the project. A DST file is essentially an electronic version of a design project. It is similar to a folder containing subfolders and each sheet in a set of drawings. The purpose of a DST file is to manage

publishing: Preparing a sequential set of multiple drawings for hard copy or electronic plotting of the set.

archiving: Gathering and storing all drawings and associated files related to a project.

Figure 32-1.
Using a sheet set to manage a set of architectural drawings. The **Sheet Set Manager** is a palette for creating and working with sheet sets.

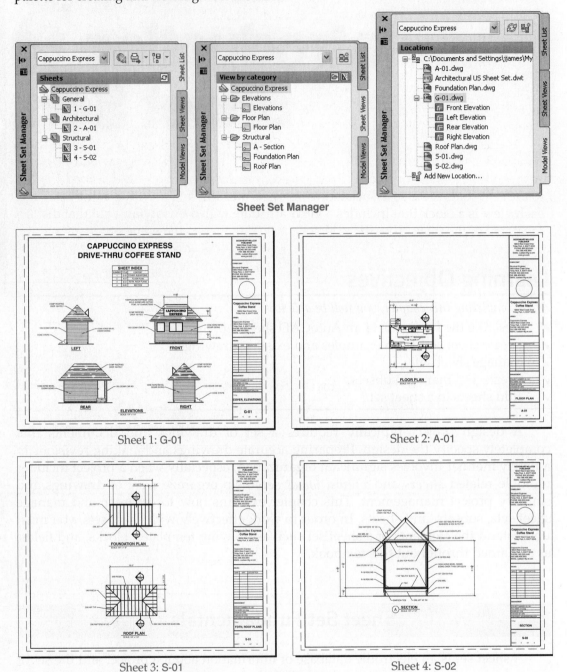

Sheet Set Manager

Sheet 1: G-01

Sheet 2: A-01

Sheet 3: S-01

Sheet 4: S-02

Sheet Set

the paths to files related to a set of drawings and the procedures required for the sheet set to function properly. A sheet set stores and displays all information about the project, including:

- Paths to required drawing layouts, or sheets
- The structure and organization of sheets in the sheet set
- Project properties common to all sheets and usually found in sheet blocks, such as the name and number of the project
- A path to a template file and storage location for creating new sheets
- Paths to drawing and model space views

- Paths to blocks with attributes containing fields that link information from one drawing or view to another drawing or view on different sheets of the sheet set
- A path to a page setup used to plot all sheets using the same settings, if appropriate

Introduction to Sheet Set Fields

One way to enhance the usefulness of a sheet set is to use text and blocks with attributes that include *fields* linked to elements of the sheet set. **Figure 32-2** highlights some of the fields used in the Cappuccino Express sheet set shown in **Figure 32-1**. Common applications for sheet set fields include titles, title blocks, callout blocks, view label blocks, and sheet list tables. You will learn about these items in Chapter 33.

Fields display properties and allow values to change during the course of the project. For example, the SHEET NUMBER attribute in the title block on each sheet uses a **CurrentSheetNumber** field that increments the sheet number when you add a sheet to the set. Fields are also associated with sheets and views. For example, the title under each view is a block that includes a VIEW attribute with a **SheetView** field that displays the name of the view. The title updates when you change the name of the view.

field: A text object that can display a specific property value, setting, or characteristic.

Figure 32-2.
Examples of fields added to text and attributes linked to sheet set properties. Use the same tools and options to add fields that reference sheet set data as you would for fields associated with other items.

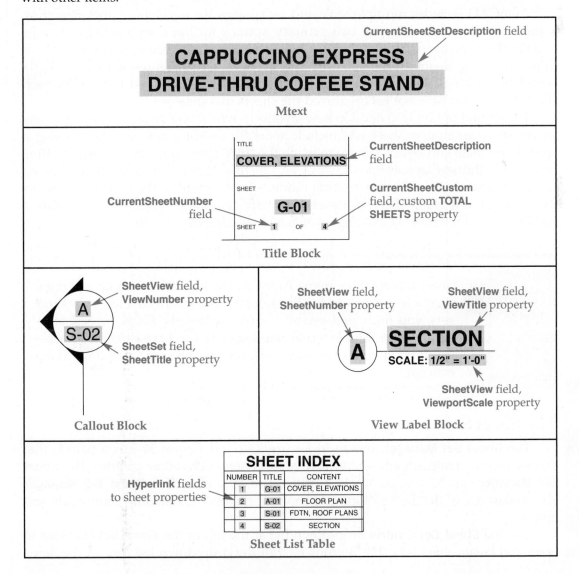

Preparing for a Sheet Set

Preparing to create a sheet set involves several processes. You can adjust every aspect of a sheet set whenever necessary, but for best efficiency, prepare for a sheet set in advance by identifying and creating all of the required elements. Preparing for a sheet set includes considering basic properties such as the project name and number, and identifying where you will store the DST file. Preparation also includes developing a drawing template (DWT) file for creating new sheets, drawing blocks with attributes containing sheet set fields, and defining a page setup to plot all sheets using the same settings.

A sheet in a sheet set is a layout in paper space. Although you can add multiple layouts to a drawing file, you can only open and use one layout in a drawing assigned to a sheet set. Therefore, prepare a single layout per drawing if you plan to use a sheet set. Delete all other layouts found in existing drawings and in the template you assign to the sheet set. Use an external reference to reuse model space content in other layouts, and in different drawing files when necessary.

If you plan to organize existing drawings in a sheet set, store all files related to the project in a designated folder when possible. Use a limited number of subfolders to group files, such as an Architectural folder to group architectural drawings and a Structural folder to group structural drawings for a building project. AutoCAD allows you to reference existing subfolders to structure a sheet set.

Developing a Sheet Set

AutoCAD includes multiple tools and techniques for developing and managing a sheet set. However, there are two primary options for building a set of drawings in a sheet set during or after you create a new sheet set file. One option is to link, or import, existing layouts to the sheet set. The layouts typically represent final or nearly complete drawings. Use this method if you have already prepared layouts for each sheet in a set, but have not yet organized the sheets in a sheet set.

The second option is to develop new layouts in which you display existing or new model space geometry. Sheet sets include tools for creating new drawings using a specific template and layout, similar to using the **NEW** command, and at the same time incorporate the new layouts into the sheet set. Use this technique to add a new drawing to the set. Externally reference existing model space content to the new sheet, or use the drawing as a blank sheet for developing new geometry. Projects often require a combination of referencing existing layouts and creating new sheets.

Sheet sets combine many AutoCAD features to automate and organize a set of drawings. To understand and effectively apply sheet sets, you must understand templates, layouts, fields, blocks, attributes, views, and external references. If you have difficulty understanding an aspect of sheet sets, review the associated underlying concept.

The Sheet Set Manager

The **Sheet Set Manager**, shown in Figure 32-1 and Figure 32-3, is a palette that allows you to create, organize, and access sheet sets. Like other palettes, the **Sheet Set Manager** can be resized, docked, and set to auto-hide. The **Sheet Set Manager** also makes use of detailed tooltips and shortcut menus for accessing commands and options.

Use the **Sheet Set Control** drop-down list at the top of the **Sheet Set Manager** to open and create sheet sets. The buttons next to the drop-down list control the items

SHEETSET

Quick Access

Sheet Set Manager

Ribbon

View
> Palettes

Sheet Set Manager

Type

SHEETSET
SSM

listed in the **Sheet Set Manager** and vary depending on the current tab. Right-click on the **Sheet list** area and select **Preview/Details Pane** to display the **Preview** or **Details** pane. Pick the **Details** or **Preview** button to toggle the corresponding display. The **Details** pane, shown in **Figure 32-3**, lists properties associated with the item selected in the **Sheet Set Manager**. The **Preview** pane displays an image of a selected sheet or view.

The **Sheet List** tab is the primary resource for managing sheets within a sheet set. Some projects require using only the **Sheet List** tab. The **Sheet Views** tab allows you to organize and place named views on layouts. Sheet views also provide a method to insert blocks that relate information on one sheet to a view on another sheet, such as the section bubbles shown in **Figure 32-1**, and sheet blocks associated with views, such as the view titles shown in **Figure 32-1**. The **Model Views** tab allows you to open files related to the current sheet set, and includes an option to create new layout views using existing model space content.

> When no drawing is open, the **Sheet Set Manager** button appears in the **Quick Access** toolbar. Pick the button to access the **Sheet Set Manager** without opening a drawing.

Creating Sheet Sets

To create a new sheet set, first begin a new file or open an existing file, even if the file does not relate to the sheet set. Then access the **NEWSHEETSET** command to display the **Create Sheet Set** wizard. See **Figure 32-4**. The **NEWSHEETSET** command is also available from the **Sheet Set Manager** by picking the **New Sheet Set...** option from the **Sheet Set Control** drop-down list. Use the **Create Sheet Set** wizard to create a sheet set from an *example sheet set*, from scratch without selecting existing layouts to include as sheets, or from scratch but with the option to select existing layouts to include as sheets.

Type
NEWSHEETSET

example sheet set:
An existing sheet set used as a template for developing a new sheet set.

An Example Sheet Set Option

Pick the **An example sheet set** radio button on the **Begin** page of the **Create Sheet Set** wizard to generate a new sheet set from an existing sheet set that has properties and settings similar to those you want to apply to the new sheet set. This concept is

Figure 32-3.
The **Sheet Set Manager** contains **Sheet List, Sheet Views,** and **Model Views** tabs. Pick the **Sheet Set Control** drop-down list to access options to create or open a sheet set.

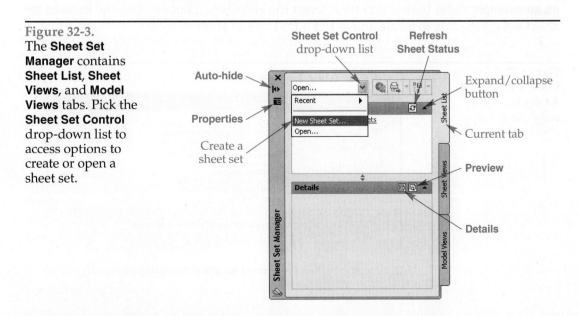

Figure 32-4.
On the **Begin** page, choose whether to start from an example sheet set or an existing drawing.

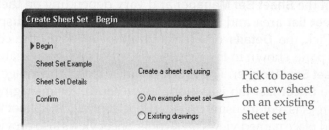

Pick to base the new sheet on an existing sheet set

much like using a drawing template to begin a new drawing. Later, you can modify the characteristics from the example sheet set in the new sheet set as needed. Beginning with an example sheet set is useful if a sheet set is available that closely matches the sheet set you intend to create. For example, use the Cappuccino Express sheet set shown in **Figure 32-1** as an example sheet set for projects with similar characteristics. All properties and many settings that you assign to the example sheet set, such as the General, Architectural, and Structural *subsets* shown in **Figure 32-1**, are reproduced in the new sheet set.

subsets: Groups of similar layouts, such as those in the same discipline, sometimes based on folder hierarchy.

PROFESSIONAL TIP

Avoid using an example sheet set if the new sheet set is significantly different from an available example sheet set. It may take more time to modify the sheet set than to create a new sheet set that is specific to the new project.

Sheet Set Example Page

Pick the **Next** button to display the **Sheet Set Example** page. See **Figure 32-5**. When you create a sheet set from an example sheet set, you start from an existing DST file. The **Select a sheet set to use as an example** radio button is active by default, and a list box displays all DST files in the default Template folder. Select an example sheet set from the list, or pick the **New Sheet Set** option in the **Select a sheet set to use as an example** list box to begin a new sheet set from scratch.

You can use any existing sheet set as an example sheet set. To use an example sheet set not saved in the Template folder, pick the **Browse to another sheet set to use as an example** radio button and then select the ellipsis (...) button. Use the **Browse for sheet set** dialog box to locate and select a DST file in another folder.

Figure 32-5.
Use the **Sheet Set Example** page to select an example sheet set.

List of sheet sets in **Template** folder

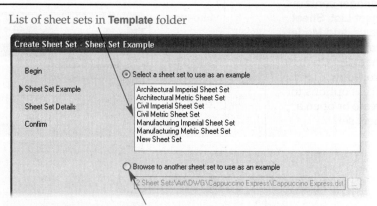

Select a template from the **Browse for Sheet Set** dialog box

Sheet Set Details Page

Select the **Next** button to display the **Sheet Set Details** page. See **Figure 32-6**. Use the **Sheet Set Details** page to modify the existing sheet set data and create settings for the new project. Most sheet set details are specific to a project. Enter the name, or title, of the sheet set in the **Name of new sheet set** text box. The name is typically the project number or a short description of the project. Type a description for the sheet set in the **Description (optional)** area. The **Store sheet set data file (.dst) here** text box determines where the sheet set file is saved. Pick the ellipsis (**...**) button and select a folder in the **Browse for sheet set folder** dialog box to change the default DST file location. Pick the **Create a folder hierarchy based on subsets** check box to allow AutoCAD to create subfolders that match the subsets in the example sheet set. Layouts in each folder are listed under each subset.

Pick the **Sheet Set Properties** button to adjust and create additional sheet set properties using the **Sheet Set Properties** dialog box. See **Figure 32-7**. Use the **Sheet Set Properties** dialog box to edit all properties except the Sheet set data file property. Methods of adjusting properties depend on the property, but include typing in a text box, selecting from a flyout, or picking the ellipsis (**...**) button to navigate to a specific location.

The **Sheet Set** category includes **Name**, **Sheet set data file**, and **Description** properties that correspond to the values you define at the **Sheet Set Details** page. The **Model view** property specifies the folder(s) containing drawing files with model space content that you intend to use for constructing new views on layouts in the sheet set. The **Label block for views** property specifies the block used to label views. The **Callout blocks** property specifies blocks available for use as callout blocks. The **Page setup overrides file** property determines the location of a DWT file containing a page setup that applies to all sheets in the sheet set. The **Project Control** category lists properties common to most projects, including **Project number**, **Project name**, **Project phase**, and **Project milestone**.

Figure 32-6.
On the **Sheet Set Details** page, type a name and description for the sheet set and specify the location of the DST file. You can also access additional sheet set properties.

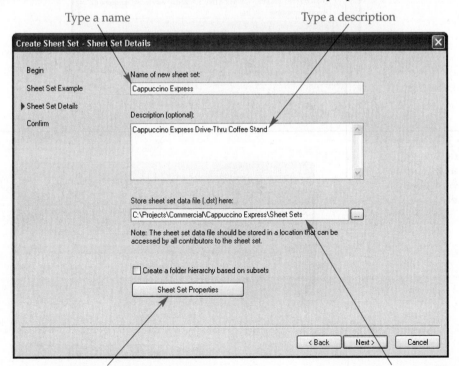

Type a name   Type a description

Pick to open the **Sheet Set Properties** dialog box   Specify a location for the DST file

The **Sheet Set Properties** dialog box includes an option to create a new sheet and drawing file without referencing an existing layout. Use the **Sheet storage location** property in the **Sheet Creation** category to specify where to store new sheets. Remember the specified location so that you can locate the new drawings. Use the **Sheet creation template** property to define the drawing template and layout for creating new sheets. The **Select Layout as Sheet Template** dialog box appears when you select the ellipsis (**...**) button. See **Figure 32-8**. All layouts in the selected template appear in the list box. Select the appropriate layout and pick the **OK** button. Set the **Prompt for template** property to **No** to use the specified **Sheet creation template** property to create all new sheets. Select **Yes** to have the option to choose a different template to create a new sheet.

The **Sheet Set Custom Properties** category lists custom properties that you create by picking the **Edit Custom Properties** button to access the **Custom Properties** dialog box. Custom properties are necessary to specify additional information on sheets, such as the properties shown in **Figure 32-7**, which are linked to fields in attributes of the

Figure 32-7.
The **Sheet Set Properties** dialog box stores properties related to the sheet set, which are typically properties and settings that apply to the entire project.

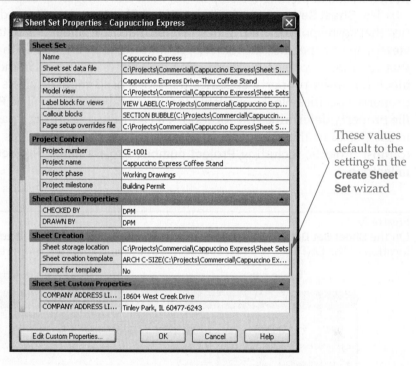

These values default to the settings in the **Create Sheet Set** wizard

Figure 32-8.
Selecting an existing layout as a template for new sheets in a sheet set.

Pick to select a different template file

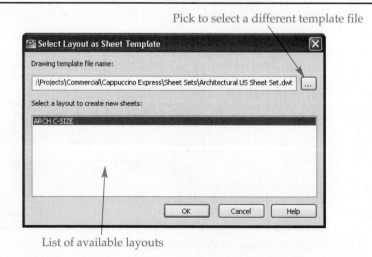

List of available layouts

title block. Use the **Add** and **Delete** buttons in the **Custom Properties** dialog box to add and delete custom properties. After you set all values in the **Sheet Set Properties** dialog box, pick the **OK** button to return to the **Sheet Set Details** page.

You are required to specify all relevant sheet set details if you choose the **New Sheet Set** option of the **Sheet Set Example** page.

Confirm Page

Pick the **Next** button to display the **Confirm** page. See **Figure 32-9**. The **Sheet Set Preview** area displays all information associated with the sheet set. Use the **Back** button to return to previous pages to make changes. Pick the **Finish** button to create the sheet set.

PROFESSIONAL TIP

Copy and paste the information in the **Sheet Set Preview** area of the **Confirm** page into a word processing program to save and print for reference.

Exercise 32-1

Complete the exercise on the companion website.
www.g-wlearning.com/CAD

Figure 32-9.
Use the **Confirm** page to preview settings before creating the sheet set.

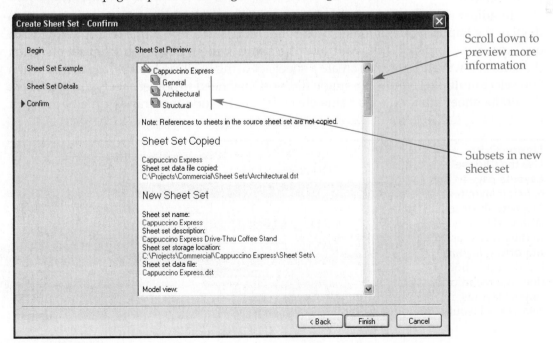

Existing Drawings Option

To add existing layouts to a new sheet set during the sheet set creation process, select the **Existing drawings** radio button on the **Begin** page of the **Create Sheet Set** wizard. See **Figure 32-4**. The **Existing drawings** option generates a new sheet set from scratch, without the option to use an example sheet set, but with the option to add existing layouts as sheets.

The **Existing drawings** option is useful if an example sheet set is unavailable or inappropriate, and if you have already developed multiple final or nearly final layouts. If you choose to apply the **Existing drawings** option, it is especially important that all drawings only include a single layout. In addition, organize files in a structured hierarchy of folders before creating the sheet set. This saves time because you can reproduce the subfolders as subsets in the new sheet set.

Sheet Set Details Page

Select the **Next** button to display the **Sheet Set Details** page. This is nearly the same page that appears when you create a new sheet set from an example sheet set, as shown in **Figure 32-6**. Use the **Sheet Set Details** page to define sheet set data and create properties for the new project. Because the **Existing drawings** option does not use an example sheet set, you must specify all relevant details, including custom properties.

Choose Layouts Page

Pick the **Next** button to display the **Choose Layouts** page. See **Figure 32-10**. The **Choose Layouts** page allows you to specify layouts in existing drawings to add to the sheet set as sheets. Pick the **Browse...** button to display the **Browse for Folder** dialog box, and select the folder containing the drawing files with the desired layouts. The list box displays the selected folder, subfolders, drawing files, and all layouts within the drawings. Repeat the process to add other folders to the list box as needed.

Check the boxes corresponding to the layouts you want to add to the new sheet set. Deselect layouts that you do not want to add to the sheet set, such as layouts in unneeded reference files or other drawings not directly associated with the set of drawings. Subfolders, files, and layouts are selected as one item. Uncheck a folder to uncheck all related drawing files and layouts. Uncheck a drawing file to uncheck all related layouts.

To adjust sheet naming and subset options, pick the **Import Options...** button to display the **Import Options** dialog box. See **Figure 32-11**. The name of a drawing with one layout, the name of the layout, and the name of the sheet in the sheet set are often the same. However, when you create a sheet set using existing layouts, especially when you select multiple layouts in a single file, you may need to adjust sheet naming. Select the **Prefix sheet titles with file name** check box to include the drawing file name with

Figure 32-10.
Use the **Choose Layouts** page to link existing layouts to a new sheet set. AutoCAD refers to the process as importing layouts, but the operation does not technically import layouts; it references layouts.

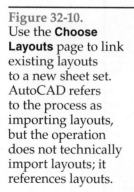

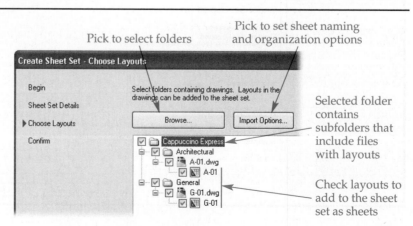

Pick to select folders

Pick to set sheet naming and organization options

Selected folder contains subfolders that include files with layouts

Check layouts to add to the sheet set as sheets

Figure 32-11.
The **Import Options** dialog box allows you to specify sheet naming conventions and folder structuring options.

Check to include drawing file name with layout name for new sheets

Check to create subsets from folders

Check to omit top folder name from subset structure

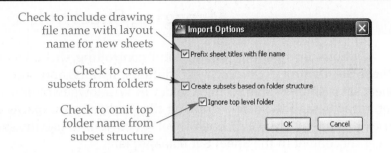

the name of the layouts that become sheets. For example, the sheet name Electrical Plan – First Floor Electrical forms from a layout named First Floor Electrical imported from the drawing file Electrical Plan.dwg.

Select the **Create subsets based on folder structure** check box to organize a sheet set so that subfolders are grouped into subsets. Layouts in each folder are listed in each subset. The **Ignore top level folder** option determines whether the top level folder creates a subset. **Figure 32-12A** shows the **Choose Layouts** page with layouts imported from the Cappuccino Express folder for the Cappuccino Express sheet set. This sheet set uses the **Create subsets based on folder structure** and **Ignore top level** folder options, but not the **Prefix sheet titles with file name** option. **Figure 32-12B** shows the result of the configuration in the **Sheet Set Manager**. As shown in **Figure 32-12B**, each sheet can include a number preceding the sheet name, separated by a dash.

Confirm Page

Pick the **Next** button to display the **Confirm** page. This is the same page that appears when you create a new sheet set from an example sheet set. The **Sheet Set Preview** area displays all information associated with the sheet set. Use the **Back** button to return to previous pages to make changes. Pick the **Finish** button to create the sheet set.

Exercise 32-2

Complete the exercise on the companion website.
www.g-wlearning.com/CAD

Figure 32-12.
Creating a sheet set named Cappuccino Express with subsets. A—Layouts imported from the Architectural and General subfolders in the Cappuccino Express folder. Subfolders are designated as subsets for the new sheet set. B—Subsets shown in the **Sheet Set Manager**.

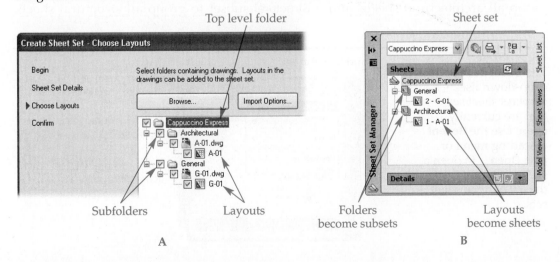

Managing Sheet Sets

Options for opening, closing, and controlling sheet sets are available from the **Sheet Set Control** drop-down list, buttons, flyouts, and shortcut menus. Opening a sheet set is similar to accessing a folder in Windows Explorer. You can open any sheet set, regardless of the active drawing file. A sheet set remains open until you close it or exit AutoCAD. Make an open sheet set current to use the sheet set and display the sheet set content in the **Sheet Set Manager** tabs.

The top portion of the **Sheet Set Control** drop-down list displays sheet sets that are open in the current AutoCAD session. See **Figure 32-13**. Pick a sheet set from the list to make it current. The list clears when you exit AutoCAD. Use the **Recent** cascading menu to open a recently opened sheet set. Select **Open...** to display the **Open Sheet Set** dialog box. Then navigate to a DST file and open the file in the **Sheet Set Manager**.

To close a sheet set, right-click on an open sheet set in the **Sheet Set Control** drop-down list or on the sheet set title in the **Sheet List** tab, and select **Close Sheet Set**. Sheet sets automatically close when you exit AutoCAD. Closing a sheet set removes the sheet set from the **Sheet Set Manager** and allows you to delete the DST file if necessary. To save all of the drawing files with layouts associated with the current sheet set at the same time, right-click on a sheet set title in the **Sheet List** tab and select **Resave All Sheets**.

Right-click on the sheet set and pick **Properties...** to display the **Sheet Properties** dialog box. This is the same dialog box that is available from the **Sheet Set Details** page when you create a new sheet set, as shown in **Figure 32-7**.

 Close drawing files with layouts associated with the current sheet set to update the sheet set according to changes made to the drawings. To update changes to the sheet list manually, pick the **Refresh Sheet Status** button. See **Figure 32-13**.

Subsets

Subsets provide a way to organize sheets in a sheet set, similar to using subfolders to organize files with Windows Explorer. The example in **Figure 32-10** uses a General subset to group all general sheets for a building project, an Architectural subset to group all architectural sheets, and a Structural subset to group all structural sheets.

Figure 32-13.
The **Sheet Set Control** drop-down list displays sheet sets that are currently open. Use the **Recent** cascading menu or pick **Open...** to open a sheet set that is not in the list of open sheet sets.

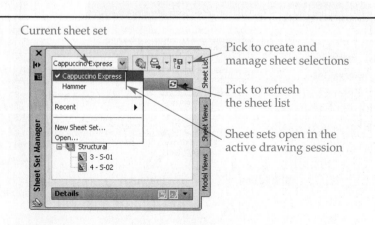

Current sheet set

Pick to create and manage sheet selections

Pick to refresh the sheet list

Sheet sets open in the active drawing session

Another example is using subsets to organize assemblies, subassemblies, and parts for a mechanical design project. Although subsets are not required, they can provide significant help in managing a sheet set, especially when the sheet set includes multiple sheets. For example, you can set a subset not to publish instead of taking the time to set each individual sheet not to publish.

Creating Subsets

The **Existing drawings** option for creating a new sheet set includes the **Create subsets based on folder structure** function that allows you to form subsets based on an existing folder hierarchy during the process of creating a sheet set. For existing sheet sets, use the **Sheet List** tab of the **Sheet Set Manager** to create new subsets. Add a subset to a sheet set or to an existing subset to create the desired subset hierarchy. Right-click on the sheet set or subset name in the **Sheet List** tab and select **New Subset...** to open the **Subset Properties** dialog box. See **Figure 32-14**. Type the name of the subset in the **Subset name** text box. For example, name a subset Structural to contain all structural sheets in a sheet set for a building project. Select **Yes** from the **Create Folder Hierarchy** drop-down list to create a new folder that corresponds to the subset.

The **Publish Sheets in Subset** property determines whether sheets in the subset are published. The **Do Not Publish Sheets** setting prevents sheets in the subset from being published. An icon identifies subsets set not to publish. The **New Sheet Location** property determines the path to which new sheets are saved when you add a new sheet from the subset. The default location is the folder associated with the sheet set or subset in which you create the new subset, or the folder you specified when you created the sheet set.

Use the **Sheet Creation Template** property to specify a drawing template and layout to use when creating new sheets in the subset. For example, assign a specific drawing template and layout to an Architectural subset to create new architectural discipline sheets, and assign a different drawing template and layout to a Structural subset to create new structural discipline sheets. The process for specifying a sheet creation template and layout for a subset is identical to the process for selecting the sheet creation template and layout for a sheet set. Use the **Prompt for Template** drop-down list to indicate whether a prompt should ask for a sheet template instead of using the specified sheet creation template.

Managing Subsets

Drag and drop subsets as needed to restructure the sheet set. Right-click on a subset in the **Sheet List** tab and pick **Collapse** to collapse the nodes of a subset. To make changes to an existing subset, right-click on the subset and select **Properties...** to redisplay the **Subset Properties** dialog box. The **Rename Subset...** shortcut menu option also opens the **Subset Properties** dialog box. To delete a subset, right-click on the subset and select **Remove Subset**. The **Remove Subset** option is not available if the subset contains sheets.

Figure 32-14.
The **Subset Properties** dialog box allows you to define a new subset.

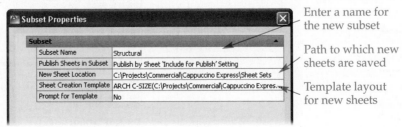

Enter a name for the new subset

Path to which new sheets are saved

Template layout for new sheets

Exercise 32-3

Complete the exercise on the companion website.
www.g-wlearning.com/CAD

Sheets

A sheet set or subset contains sheets, just as a folder or subfolder contains drawing files in Windows Explorer. A sheet is a path to a single layout in a drawing file. See **Figure 32-15.** To open a sheet, double-click on the sheet or right-click on the sheet and select **Open.** The drawing file that contains the referenced layout tab opens and displays the layout corresponding to the sheet. You can also open a sheet as *read-only.*

read-only:
Describes a drawing file opened for viewing only. You can make changes to the drawing, but you cannot save changes without using the **SAVEAS** command.

NOTE

When you use any technique to open a file, including opening files from the **Sheet Set Manager,** the files are added to the open files list. Use the **Quick View Drawings** tool or window control commands on the ribbon to view open files. Save and close any files that you do not need open.

Adding a Layout as a Sheet

The **Existing drawings** option for creating a new sheet set includes the **Choose Layouts** page that allows you to add existing layouts to a sheet set as sheets. You can also add a layout to an existing sheet set using the **Sheet Set Manager** or directly from an open drawing. To add an existing layout to a sheet set using the **Sheet Set Manager,** right-click on the sheet set or the subset that will contain the sheet, and select **Import Layout as Sheet....** The **Import Layouts as Sheets** dialog box appears. See **Figure 32-16.**

Figure 32-15.
Sheets in the set of drawings for the example Cappuccino Express building project. Hover over a sheet to display a detailed tooltip.

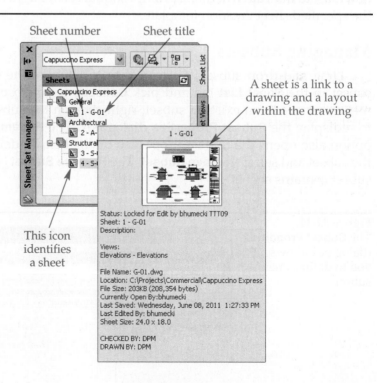

Pick the **Browse for Drawings** button to select a drawing file. All layouts in the drawing file appear in the list box and are checked for import by default. The **Status** column indicates whether the layout can be imported into the sheet set. You can only use a layout in one sheet set. In other words, you cannot import a layout that is already part of a sheet set. Each checked layout is imported as a separate sheet. Importing layouts only links layouts to the sheet set. The original file remains unchanged, and no new drawing files or layouts form. Uncheck a box to exclude a layout from importing. Select the **Prefix sheet titles with file name** check box to include the name of the file in the sheet title. Pick the **Import Checked** button to import the selected layouts as sheets.

A layout tab, the **Quick View Layouts** tool, and the **Quick View Drawings** tool provide a way to add existing layouts to a sheet set directly from an open drawing. Right-click on the layout tab or thumbnail image you want to import and select **Import Layout as Sheet...**. The **Import Layouts as Sheets** dialog box appears with the selected layout listed. You must save the drawing and set up the layout to make the **Import Layout as Sheet...** menu option available.

Right-click on a subset in the **Sheet Set Manager** and select **Import Layout as Sheet...** to import layouts into a subset. Importing a layout from a layout tab, the **Quick View Layouts** tool, or the **Quick View Drawings** tool does not provide an initial option to add layouts as sheets in a subset. You must drag and drop the sheets into the appropriate subset after importing.

PROFESSIONAL TIP

To use the same layout as a sheet in a different sheet set, create a new file and attach the existing drawing file as an xref. Create a layout of the model space drawing in the new file and access the layout through the **Import Layouts as Sheets** dialog box.

Figure 32-16.
Use the **Import Layouts as Sheets** dialog box to link existing layouts to a sheet set as sheets. AutoCAD refers to the process as importing layouts, but the operation does not technically import layouts; it references layouts.

Pick to locate and select a drawing file

Uncheck to exclude a layout from importing

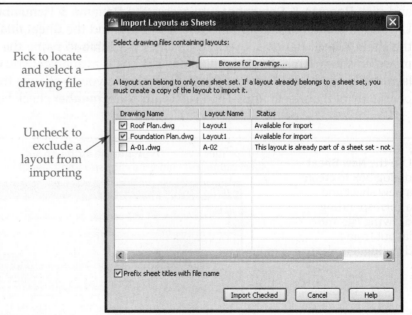

Adding a Sheet Using a Template

Another option for adding a sheet to a sheet set is to create a new drawing using a specific template and layout, and at the same time incorporate the new layout into the sheet set as a sheet. Externally reference existing model space content to the new sheet, or use the drawing as a blank sheet for developing new geometry.

Before adding a sheet from a template, consider assigning a specific drawing template to use for new sheet creation in the **Sheet creation template** setting in the **Sheet Set Properties** and **Subset Properties** dialog boxes. Then set the **Prompt for template** property to **No** to force AutoCAD to use only the specified template, or choose **Yes** to have the option to select a different template when you create a new sheet.

Right-click on the sheet set or subset that will contain the new sheet, and select **New Sheet…**. If you do not assign a drawing template to use for new sheet creation, or if you selected the **Prompt for template** option, AutoCAD displays the **Select Layout as Sheet Template** dialog box. The **New Sheet** dialog box appears if a specified template layout exists, or after you select the template layout. See Figure 32-17.

Type the sheet number in the **Number** text box and the sheet name in the **Sheet title** text box. A new sheet creates a new drawing file, and the sheet title becomes the name of the layout in the drawing file. Enter the file name in the **File name** text box. The file name is the sheet number and title by default. Edit the drawing file name, if necessary. For example, you might remove the sheet number from the name. The **Folder path** display box shows where the drawing file will be saved, as specified in the **Sheet Set Properties** or **Subset Properties** dialog box. Check the **Open in drawing editor** to open the file and display the layout.

Exercise 32-4

Complete the exercise on the companion website.
www.g-wlearning.com/CAD

Managing Sheets

Drag and drop sheets as needed to restructure the sheet set and move sheets from the sheet set to subsets or between subsets. Right-click on a sheet in the **Sheet Set Manager** and pick **Rename & Renumber…** to access the **Rename & Renumber Sheet** dialog box. Use the **Number:** text box to renumber the sheet and the **Sheet title:** text box to rename the sheet. Check **Rename layout to match: Sheet title** to name the layout to match the modified sheet name. The **Layout name:** text box is available if you deselect the **Rename layout to match: Sheet title** check box, and allows you to change the layout name independently of the sheet name. The **Prefix with sheet number** check box is also available.

Figure 32-17.
Use the **New Sheet** dialog box to create a new sheet and a corresponding drawing file and layout from a drawing template.

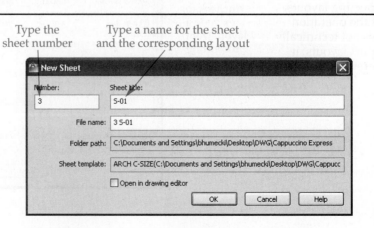

Options for renaming the file are available only if the drawing file is closed. Check **Rename drawing file to match: Sheet title** to name the file to match the modified sheet name. The **File name:** text box is available if you deselect the **Rename drawing file to match: Sheet title** check box, and allows you to change the file name independently of the sheet name. The **Prefix with sheet number** check box is also available. If the sheet is one of several in a subset, pick the **Next** and **Previous** buttons to access different sheets in the subset.

Right-click on a sheet and pick **Properties...** to display the **Sheet Properties** dialog box. See **Figure 32-18**. The **Sheet Properties** dialog box offers another way to change the sheet name and number, and provides additional properties. Use the text boxes to type the appropriate information. It is important that you specify all relevant information to define the project completely, so that text and blocks containing attributes with fields display the correct content. The **Include for publish** option determines whether the sheet is published or plotted with the sheet set. The default value is **Yes**.

The **Expected layout** and **Found layout** text boxes display the path to the file where you initially saved the sheet and the path to the file where AutoCAD found the sheet. If the paths are different, pick the ellipsis (...) button to update the **Expected layout** setting using the **Import Layout as Sheet** dialog box. The **Sheet Properties** dialog box includes the same **Rename** options area found in the **Rename & Renumber Sheet** dialog box.

Right-click on a sheet and select **Remove Sheet** to remove the sheet from the sheet set. Removing a sheet does not delete the drawing file.

 When you change the location of a drawing file that has layouts associated with a sheet set, the association with the sheet set is broken. Update the specified path to the drawing file in the **Sheet Properties** dialog box or re-import the layouts into the sheet set.

Figure 32-18.
The **Sheet Properties** dialog box allows you to modify sheet properties. Add all relevant sheet details that will appear in title blocks and similar content.

Add a sheet number if you imported the sheet from a layout

These locations match unless you move the original layout

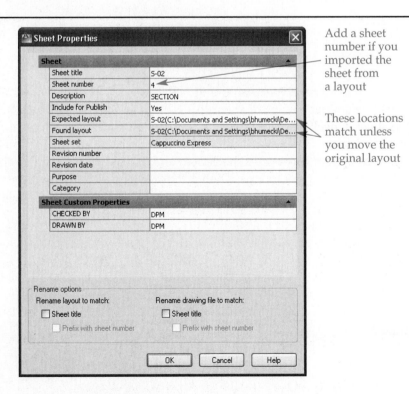

Exercise 32-5

Complete the exercise on the companion website.
www.g-wlearning.com/CAD

Chapter Review

Answer the following questions. Write your answers on a separate sheet of paper or complete the electronic chapter review on the companion website.
www.g-wlearning.com/CAD

1. What does the term *sheet* refer to in relation to a sheet set and a drawing file?
2. What is a sheet set?
3. What is the purpose of a sheet set data file, and what extension does the file have?
4. What is the purpose of fields in a sheet set?
5. Explain when you would add existing layouts to a sheet set and when you should add a new layout.
6. Briefly describe the general options for creating a new sheet set.
7. What are subsets in relation to a sheet set?
8. How do you add a custom property to a sheet set?
9. Explain how to create a subset in an existing sheet set.
10. How do you modify a sheet name or number?

Drawing Problems

Start AutoCAD if it is not already started. Follow the specific instructions for each problem.

▼ Basic

1. Create a new sheet set using the **Create Sheet Set** wizard and the New Sheet Set example sheet set. Name the new sheet set My Sheet Set.

2. Create a new sheet set using the **Create Sheet Set** wizard and the Civil Imperial Sheet Set example sheet set. Name the new sheet set Civil Sheet Set.

▼ Intermediate

3. Use a word processor to write a report of approximately 250 words explaining the purpose of sheet sets. Cite at least three examples from actual industry applications of using sheet sets to help manage a set of drawings. Use at least two sketches to illustrate your report.

▼ Advanced

4. Design and draw an arbor press. Prepare layouts for the assembly and each component. Then create a new sheet set to organize the drawings and layouts.

Answer the following question. Write your answers on a separate sheet of paper.

1. Which of the following actions can you perform using the **Sheet Set Manager**? *Select all that apply.*
   A. add sheets to a sheet set
   B. change the name of a sheet set data file
   C. organize files related to a project
   D. open drawing files that contain layouts referenced in a sheet set
   E. purge unused blocks from a sheet set

Additional Sheet Set Tools

Learning Objectives

After completing this chapter, you will be able to:

✓ Create and use sheet views with callout and view label blocks.
✓ Add model views to a sheet set.
✓ Create sheet list tables.
✓ Manage sheet set fields.
✓ Publish a sheet set.
✓ Archive a sheet set.

After you prepare a sheet set with subsets and sheets, you are ready to continue sheet set development using sheet views and objects specifically related to sheet set applications. This chapter explains this process and also describes options for publishing and archiving a sheet set. In order to work effectively with sheet sets, you must understand the elements of a sheet set such as drawing templates, blocks, and fields, as explained throughout this textbook.

Sheet Views

Use the **Sheet Views** tab of the **Sheet Set Manager** to group views by category, open views for viewing and editing, and add *callout blocks* and *view label blocks*. *Sheet views* provide a way to access specific views and to link drawing views to sheets in the sheet set. You must create sheet views to add callout and view label blocks from the **Sheet Set Manager**. Callout and view label blocks use attributes containing fields that link data between the sheet set and drawing views. The fields update automatically to reflect changes in sheet numbering and organization. See **Figure 33-1**.

One option to create a sheet view is to use the **VIEW** command in paper space to prepare a named view of a layout. This technique essentially forms a copy of a sheet as a sheet view that you can reference to add callout and view label blocks. The second option to create a sheet view is to add a *model view* to a layout. You will learn about model views later in this chapter.

callout block: A block that uses attributes containing fields that link the view number and sheet title between the sheet set and drawing (sheet) views.

view label block: A block that uses attributes containing fields that link the view name, number, and scale to drawing (sheet) views.

sheet view: A layout or model view saved for use in a sheet set; allows you to add views to layouts and insert callout and view label blocks.

model view: A drawing file or named model space view added to a layout to create a sheet view.

Figure 33-1.
Sheet views allow you to access specific views and add callout and view label blocks that link drawing views to sheets throughout the sheet set.

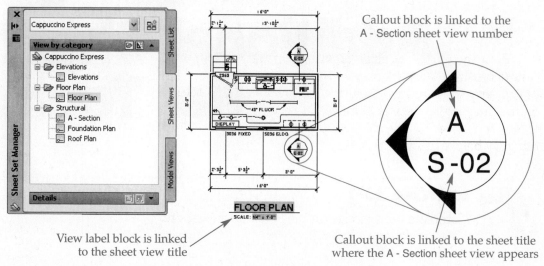

Callout block is linked to the A - Section sheet view number

Callout block is linked to the sheet title where the A - Section sheet view appears

View label block is linked to the sheet view title

Floor Plan Sheet View

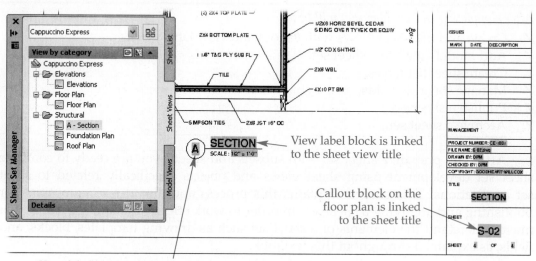

View label block is linked to the sheet view title

Callout block on the floor plan is linked to the sheet title

View label block is linked to the sheet view number, which is linked to the callout block on the floor plan

A - Section Sheet View

View Categories

View categories organize views in the **Sheet Views** tab, similar to subsets in the **Sheet List** tab. However, view categories group views, while subsets group sheets that might include several sheet views. For example, an Architectural subset includes an Elevations sheet and other architectural sheets. The Elevations sheet includes Front Elevation, Left Elevation, Rear Elevation, and Right Elevation views, added to the sheet using sheet views of the same name and grouped in an Elevations sheet view category.

One method to add a view category is to type a value in the **View category:** text box in the **New View/Shot Properties** dialog box of the **VIEW** command. To add a view category without using the **VIEW** command, pick the **Sheet Views** tab and select the **View by category** button. See **Figure 33-2**. Pick the **New View Category** button or right-click on the sheet set name and select **New View Category...** to open the **View Category** dialog box. See **Figure 33-3**. Type a name for the category in the **Category name** text box.

VIEW

Ribbon
View
> Views

Named Views

Type
VIEW
V

The **View Category** dialog box lists all available callout blocks for the current view category. Check the box next to the callout block to make the block available for all views added to the category. If a block is not in the list, pick the **Add Blocks...** button to access the **List of Blocks** dialog box, from which you can locate a drawing file with blocks to add. After you select the necessary callout blocks, pick the **OK** button to create the new category.

View categories are listed alphanumerically. To change the category name and add callout blocks, right-click on a view category in the **Sheet Views** tab and pick **Rename...** or **Properties...** to reopen the **View Category** dialog box. To delete a category, right-click on the category and select **Remove Category**. The **Remove Category** option is not available if the category includes sheet views. You must remove all sheet views from the category before deleting the category.

Exercise 33-1

Complete the exercise on the companion website.
www.g-wlearning.com/CAD

Creating Sheet Views from Layouts

Using the **VIEW** command in paper space creates a layout view that is added to the sheet set as a sheet view. You can then reference the sheet view to place callout and

Figure 33-2.
Create view categories in the **Sheet Views** tab of the **Sheet Set Manager**.

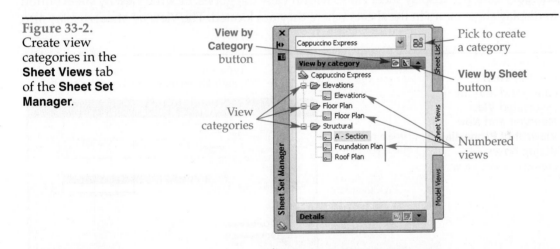

Figure 33-3.
The **View Category** dialog box allows you to name the category and select callout blocks for use with views.

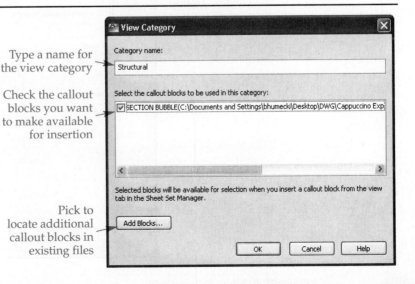

view label blocks. A sheet must be a part of the sheet set to include a sheet view. If the sheet set does not reference the layout you intend to capture as a named view, add the layout to the sheet set before continuing. Then open and activate the sheet set. Open the sheet that contains the layout you want to include as a sheet view. Ensure that the proper layout is active, and that no floating viewport is active.

Next, use display commands to orient the view of the layout as needed. Access the **VIEW** command to display the **View Manager**. Pick the **New...** button to open the **New View** dialog box. Select an existing category to associate with the view from the **View category** drop-down list, or type a name to create a new category. See Figure 33-4. Layout views are typically not as specific as model views. The main purpose is to assign the layout to the sheet set as a sheet view for callout and view label block requirements. Therefore, the **Current display boundary** option is often acceptable. Specify the remaining view settings and pick the **OK** button as needed to save the view and exit the **View Manager**.

The new layout view appears in the **Sheet Set Manager** in the specified view. To display the view from the **Sheet Set Manager**, double-click on the view or right-click on the view and select **Display**. If the drawing file is open, the view is set current. If the drawing file is not open, the file opens and displays the view.

Managing Sheet Views

Display sheet views by category or sheet using the appropriate button shown in Figure 33-2. Sheet views are listed alphanumerically in both formats. Pick the **View by category** button to display sheet views within view categories. Pick the **View by sheet** button to display sheet views in the sheets where they are located. The **View by sheet** display also

Figure 33-4.
Use the **VIEW** command and the associated **View Manager** and **New View/Shot Properties** dialog boxes to create a sheet view.

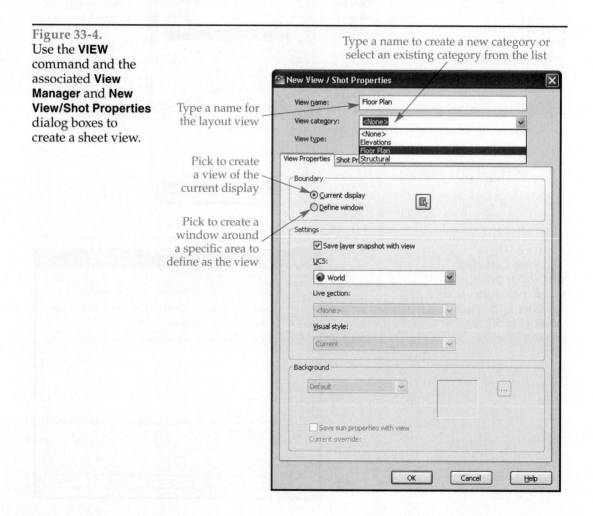

Type a name to create a new category or select an existing category from the list

Type a name for the layout view

Pick to create a view of the current display

Pick to create a window around a specific area to define as the view

provides an option to change the category assigned to a sheet view. Right-click on a sheet view and select a different category from the **Set category** cascading menu.

To change the name or number of a sheet view, right-click on the sheet view in the **Sheet Views** tab and select **Rename & Renumber...** to display the **Rename & Renumber View** dialog box. See **Figure 33-5**. Enter a number or letter for the view in the **Number** text box. A view number is typically not required for views that are not linked to other sheets, such as the Floor Plan view shown in **Figure 33-1**. However, the view number is critical for views referenced on other sheets, such as the Section view shown in **Figure 33-1**, which uses the letter A as the view number. The view number is displayed in front of the view name in the **Sheet Set Manager**. Modify the view name in the **View title** text box. Use the **Next** and **Previous** buttons to renumber or rename different views in the sheet or view category. Pick the **OK** button when you are finished.

Exercise 33-2

Complete the exercise on the companion website.
www.g-wlearning.com/CAD

Callout Blocks

A common example of a callout block is the section view information at the end of a cutting plane line on the floor plan shown in **Figure 33-6**. In this example, the upper portion of the callout block includes an attribute with a **SheetView** field that references

Figure 33-5.
Use the **Rename & Renumber View** dialog box to renumber or rename a sheet view.

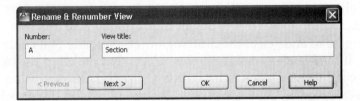

Figure 33-6.
Using a callout block assigned to a sheet view to link a cutting plane on a floor plan to a section view on a separate sheet.

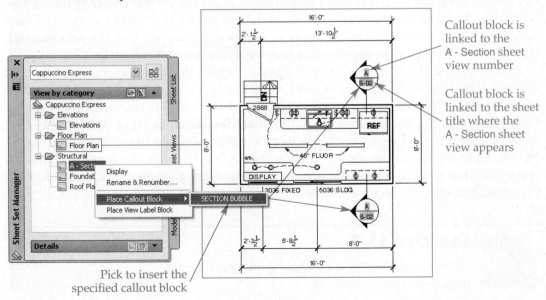

the **ViewNumber** property of the A - Section sheet view. The lower portion of the callout block includes an attribute with a **SheetSet** field that references the **SheetTitle** property of the 4 - S-02 sheet. The result is a link from the cutting-plane line on the floor plan to the sheet where the section is located and the view identification.

Callout blocks automate references between drawing content and views by updating to reflect changes in sheet numbering and organization. Other examples of callout blocks include viewing plane line symbols, such as exterior and interior elevation bubbles, and detail identification. You can also use callout blocks as balloons on an assembly drawing to associate components with a parts list, as found in a set of drawings. A callout block typically appears on a different sheet from the sheet view it references. Callout blocks can also include hyperlink fields, or *hyperlinks*, that provide convenient access to the reference sheet view.

PROFESSIONAL TIP

Use dynamic blocks with suitable parameters, such as rotational, flip, and visibility parameters, to reduce the number of blocks needed for a sheet set.

Before you can insert a callout block from the **Sheet Set Manager**, the block must be available to the sheet set and assigned to the view category. A sheet set or view category can reference multiple callout blocks. Callout blocks are often added to a sheet set during sheet set creation. To assign callout blocks to an existing sheet set, right-click on the sheet set in the **Sheet Set Manager** and pick **Properties...** to display the **Sheet Properties** dialog box. Specify the available blocks in the **Callout blocks** text box. Pick the ellipsis (...) button to access the **List of Blocks** dialog box. See **Figure 33-7**. Pick the **Add...** button to access the **Select Block** dialog box, and then pick the ellipsis button to navigate to and select the drawing or template file that contains callout blocks.

If the file consists of only the objects that make up the block, choose the **Select the drawing file as a block** radio button to use the file as a callout block. If the file includes blocks saved as blocks, pick the **Choose blocks in the drawing file:** radio button and select the blocks to use as callout blocks. Use the **Delete** button to remove a block from the **List of Blocks** dialog box.

After you add callout blocks to a sheet set, you must assign the appropriate callout blocks to each view category. To specify the callout blocks available for a view category, right-click on the category and select **Properties...** to open the **View Category** dialog box. See **Figure 33-3**. Use the check boxes to select the callout blocks to assign to the view category. Only check those callouts that apply to the specific category. Pick the **Add Blocks...** button to add new blocks to the view category using the **Select Block** dialog box.

To insert a callout block, open the sheet on which the reference is to appear and activate model space to add the symbol to model space. For example, open the floor plan shown in **Figure 33-6** to add the cutting-plane line callouts referencing the section view. Then pick the **Sheet Views** tab of the **Sheet Set Manager**, right-click on the sheet view to reference, and select the block from the **Place Callout Block** cascading menu. For example, right-click on the A - Section sheet view to create the cutting-plane line reference shown in **Figure 33-6**. Specify an insertion point for the block and follow the prompts to modify the block as needed.

Exercise 33-3

Complete the exercise on the companion website.
www.g-wlearning.com/CAD

Figure 33-7.
All of the callout blocks available to a sheet set appear in the **List of Blocks** dialog box.

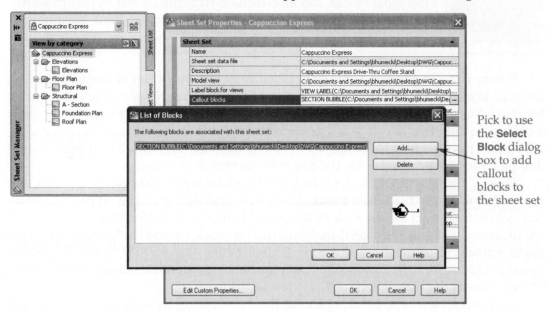

Pick to use the **Select Block** dialog box to add callout blocks to the sheet set

View Label Blocks

View label blocks automate reference between drawing views and view identification by updating to reflect changes in view number, title, and scale. A common example of a view label block is the view title and scale information displayed under the section view shown in **Figure 33-8**. In this example, the bubble portion of the view label block includes an attribute with a **SheetView** field that references the **ViewNumber** property of the A - Section sheet view. The title includes an attribute with a **SheetView** field that references the **ViewTitle** property of the A - Section sheet view. The scale includes an attribute with a **SheetView** field that references the **ViewportScale** property of the A - Section sheet view.

Figure 33-8.
Using a view label block assigned to a sheet view to title a section view on the sheet where the sheet view is located.

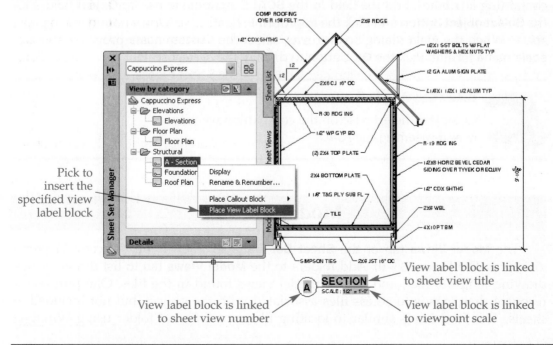

Pick to insert the specified view label block

View label block is linked to sheet view title

View label block is linked to sheet view number

View label block is linked to viewpoint scale

 View label blocks can also include hyperlinks, but hyperlinks are usually not useful in view label blocks because a view label block typically appears on the same sheet as the sheet view it references.

Before you can insert a view label block from the **Sheet Set Manager**, the block must be available to the sheet set. A sheet set can only reference a single view label block, and you cannot assign different view label blocks to each view category. View label blocks are often added to a sheet set during sheet set creation. To assign a view label block to an existing sheet set, right-click on the sheet set in the **Sheet Set Manager** and pick **Properties…** to display the **Sheet Properties** dialog box.

Specify the view label block in the **Label block for views** text box. Pick the ellipsis (**…**) button to access the **Select Block** dialog box, and then pick the ellipsis button to navigate to and select the drawing or template file that contains the view label block. If the file consists of only the objects that make up the block, choose the **Select the drawing file as a block** radio button to use the file as a view label block. If the file includes blocks saved as blocks, pick the **Choose blocks in the drawing file:** radio button and select the block to use as the view label block.

To insert a view label block, open the sheet containing the view to be labeled. For example, open the section view layout shown in **Figure 33-8** to add the view label referencing the section view. Items such as labels are often created in paper space, but you can insert a view label into model space if necessary. Pick the **Sheet Views** tab of the **Sheet Set Manager**, right-click on the sheet view to reference, and select **Place View Label Block**. For example, right-click on the A - Section sheet view to create the view label shown in **Figure 33-8**. Specify an insertion point for the block and follow the prompts to modify the block as needed.

If the view label block includes a **SheetView** field that references the **ViewportScale** property, AutoCAD only recognizes the viewport scale of sheet views created from model views. You will learn about model views later in this chapter. If you add a sheet view from a layout view, the scale appears as a series of pound (#) symbols to indicate an inability to reference the required information. To work around this problem, link the SCALE attribute value directly to the floating viewport object using the **EATTEDIT** command. See **Figure 33-9**.

A quick way to access the **EATTEDIT** command is to double-click on a block containing attributes. Edit the field in the **SCALE** attribute to use an **Object** field. Pick the **Select object** button and pick the appropriate floating viewport boundary in paper space. When the **Field** dialog box returns, select the **Custom scale** property and **Use scale name** format. Pick the **OK** button to display the correct scale.

 Exercise 33-4

Complete the exercise on the companion website.
www.g-wlearning.com/CAD

Model Views

The **Model Views** tab of the **Sheet Set Manager** allows you to access *resource drawings*. See **Figure 33-10**. Add folders to the **Model Views** tab to list drawing files, drawing template files, and named model views found in the files. One purpose of resource drawings is to access files associated with the project, but not included as sheets. This function is similar to locating files in a different folder using Windows

resource drawings: Drawing files that include named model space views referenced for use as sheet views.

Figure 33-9.

Using a field to display the correct viewport scale when referencing a layout view to place a view label.

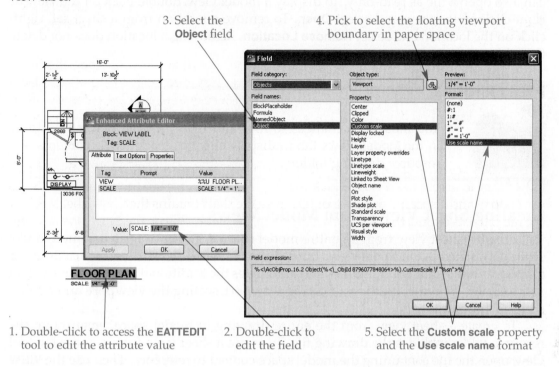

3. Select the **Object** field

4. Pick to select the floating viewport boundary in paper space

1. Double-click to access the **EATTEDIT** tool to edit the attribute value

2. Double-click to edit the field

5. Select the **Custom scale** property and the **Use scale name** format

Figure 33-10.
Use the **Model Views** tab to access resource drawings and create sheet views from model space.

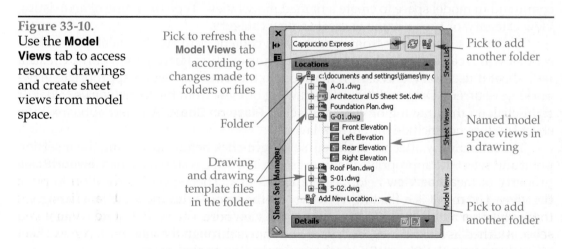

Pick to refresh the **Model Views** tab according to changes made to folders or files

Pick to add another folder

Folder

Named model space views in a drawing

Drawing and drawing template files in the folder

Pick to add another folder

Explorer. Another purpose of resource drawings is to create sheet views from model space content or named model views. You can use this function in addition or as an alternative to creating sheet views from layout views.

Adding Folders to Model Views

Folders are often added to the **Model Views** tab during sheet set creation. To add a folder to an existing sheet set, double-click on the **Add New Location** entry, pick the **Add New Location** button, or right-click on the **Add New Location** entry or a folder and select **Add New Location....** Then use the **Browse for Folder** dialog box to locate and select the folder containing the desired files. The folder and all of drawing and drawing template files found in the folder appear in the **Locations** list box. Named model space views are listed under the drawing files that contain them.

Right-click on a folder and pick **Collapse** to collapse the nodes of the folder. Right-click on a drawing and select **See Model Space Views** to expand the list of model space views in the drawing. This is the same as picking the + sign next to the drawing file. To open a file, double-click on the file or right-click on the file and select **Open**. You can also open a file as read-only. To display a model view, double-click on the view or right-click on the view and select **Open**. To remove a location from a sheet set, right-click on the location and select **Remove Location**. Removing a location does not delete the folder or the contents of the folder.

PROFESSIONAL TIP

You must save changes to files and then pick the **Refresh** button in the **Model Views** tab to display changes made to the contents of resource drawing folders.

Creating Sheet Views from Model Views

Create a sheet view from an entire model space drawing or from a named model space view listed in the **Model Views** tab. When you create a sheet view from a drawing or model view, AutoCAD combines and automates the traditional process of adding an xref to a new drawing, creating a floating viewport, scaling the viewport, and adding a view label block.

To create a sheet view from the entire drawing in model space, ensure that the resource folder includes the drawing file. To create a sheet view from a named model view, open the file containing the model space content to reference. Then use the **VIEW** command in model space to create a named model view. Type the name of an existing view category, or type a name to create a new category.

To create a sheet view from model space content, activate the sheet in the sheet set that will receive the sheet view, and erase any existing floating viewports. Create a new sheet if necessary, and assign the sheet to the sheet set. Set an appropriate viewport layer current. Drag and drop the drawing or view from the **Model Views** folder or right-click on the drawing or view and select **Place on Sheet**. An alert appears if the sheet set does not include the current layout.

The model view attaches to the cursor. Right-click before specifying the insertion point and select the appropriate viewport scale. The scale is stored as the **ViewportScale** property of the **SheetView** field. See **Figure 33-11A**. Then specify a location to place the view. The result is a floating viewport on the layout with the view label assigned to the sheet set at the lower-right corner of the viewport. The model space content you see is attached as an xref in model space and shows through the viewport. A new sheet view appears in the **Sheet Views** tab corresponding to the inserted model view. See **Figure 33-11B**.

Drag and drop a sheet view created from a model view into the appropriate view category. A view automatically appears in the correct category if you type the name of a view category when you create a model view using the **VIEW** command.

Exercise 33-5

Complete the exercise on the companion website.
www.g-wlearning.com/CAD

Figure 33-11.
Referencing a drawing as a model view to create a sheet view. The same basic process applies to forming sheet views from named model views. A—Drag and drop the drawing and select the viewport scale. B—Specify the location for the view.

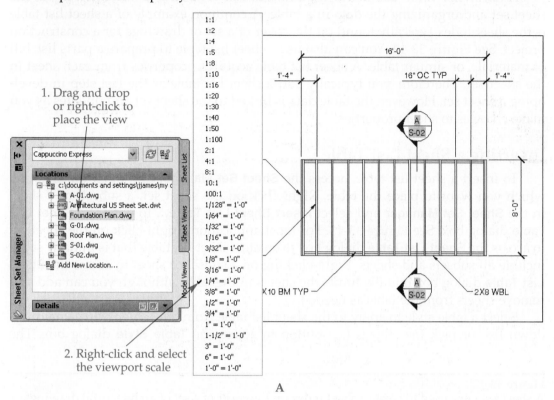

A

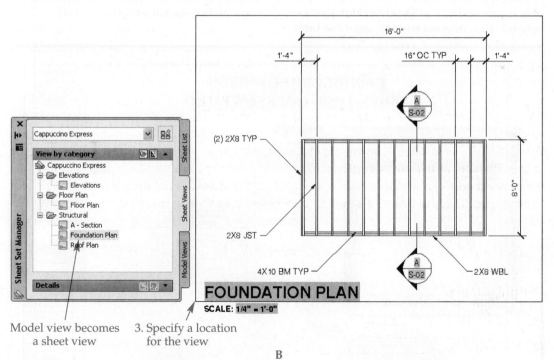

B

Sheet List Tables

sheet list table: An AutoCAD table that references a table style and selected items in a sheet set to create a list of sheets in the sheet set and related information.

A *sheet list table* automates the processes of collecting information about the sheet set and organizing the data in a table. A common example of a sheet list table is the sheet index typically found on the cover of a set of drawings for a construction project. See Figure 33-12. You can also use a sheet list table to prepare a parts list, bill of materials, or similar table. A sheet list table acquires properties from each sheet in the sheet set. Therefore, you typically add a sheet list table as the last step in developing a sheet set. However, the table data is linked to the sheet set and updates as you make changes to sheet properties.

Inserting a Sheet List Table

To insert a sheet list table, access the **Sheet Set Manager** and open the sheet on which you want to place the table. Right-click on the sheet set, a subset, or a sheet in the **Sheet Set Manager** and select **Insert Sheet List Table...** to open the **Sheet List Table** dialog box. See Figure 33-13. For most applications, right-click on the sheet set to access the **Sheet List Table** dialog box to create a sheet list table that is initially set to include all subsets and sheets. Right-clicking on a subset or sheet to access the **Sheet List Table** dialog box initially limits the sheets in the table, although you can add and remove sheets from the table as needed.

Select a table style to apply to the sheet list table from the **Table Style name** drop-down list, or pick the ellipsis (...) button to access the **Table Style** dialog box. The

Figure 33-12.

A sheet list table used to create a sheet index on the cover of a set of architectural drawings. This figure shows the **Sheet Set Manager** for reference. Notice the information linked between the sheet set and the sheet list table.

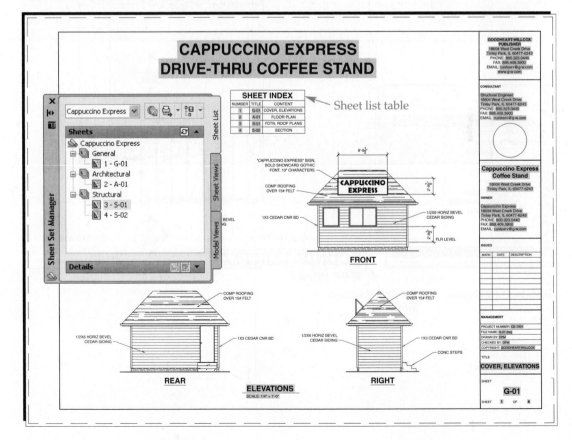

preview area displays a representation of the table. Select the **Show Subheader** check box to include subheader rows based on selected subsets. **Figure 33-12** shows a table without subheader rows. **Figure 33-14** shows the same table, but with subheader rows acquired from the General, Architectural, and Structural subsets.

Specifying the Title and Columns

Use the **Table Data** tab, shown in **Figure 33-13A**, to specify the title of the table, columns, and column organization. Type the title in the **Title Text** text box or accept the default Sheet List Table title. A sheet list table can include various types of information from the drawing file and the sheet set. The default sheet list table includes **Sheet Number** and **Sheet Title** columns corresponding to the **Sheet Number** and **Sheet Title** properties of each sheet.

Use the **Data type** column of the dialog box to specify the properties to reference as a column in the sheet list table. Pick the existing data type to select a sheet set or drawing property from the drop-down list. Add a custom property to the sheet set to add a different data type to the list. The **Heading** text column in the dialog box uses

Figure 33-13.
A—Use the **Sheet List Table** dialog box to set up a sheet list table. B—Specify the sheets to include in the table.

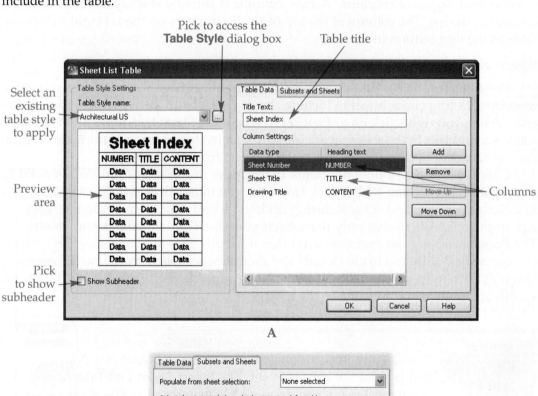

Pick to access the **Table Style** dialog box

Table title

Select an existing table style to apply

Preview area

Pick to show subheader

Columns

A

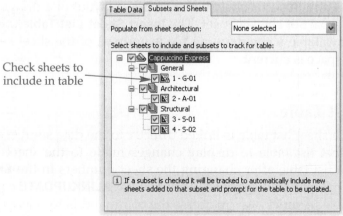

Check sheets to include in table

B

SHEET INDEX		
NUMBER	TITLE	CONTENT
General		
1	G-01	COVER, ELEVATIONS
Architectural		
2	A-01	FLOOR PLAN
Structural		
3	S-01	FDTN, ROOF PLANS
4	S-02	SECTION

the property name as the default heading. Select the name and type a new value if appropriate. The table shown in previous figures has headings that are different from the property names, which is common.

Use the **Add**, **Remove**, **Move Up**, and **Move Down** buttons as necessary to add, remove, and organize columns. A new column is initially displayed under the last column in the list. The column at the top of the list appears on the far right side of the table as the first column in the table.

Specifying Rows

The **Subsets and Sheets** tab, shown in **Figure 33-13B**, allows you to specify table rows by selecting check boxes to include sheets in the table. Each row is a sheet reference. When you right-click on a sheet set to access the **Sheet List Table** dialog box, all subsets and sheets in the sheet set are initially selected for addition to the table. Use the appropriate check boxes to add and remove sheets from the table.

Checked sheets are the only items that appear in the table. You do not need to check subsets to display subheaders. However, you must check sheet sets and subsets to include sheet sets and subsets during updates. If you make changes to the sheet set, a prompt appears to update only the subsets you check in the **Subsets and Sheets** tab. The **Populate from sheet selection** drop-down list provides access to *sheet selections*. Choose a sheet selection to check only the sheets associated with the saved selection. You will learn more about sheet selections later in this chapter. Pick the **OK** button and specify an insertion point to insert the table.

sheet selections: Groups of subsets and sheets that are often used to publish the same group of sheets.

> You can only insert a sheet list table into a layout of a drawing file that is a part of the sheet set. The **Insert Sheet List Table...** option is not available if the drawing file is not part of the sheet set or if model space is current.

Editing a Sheet List Table

Ribbon

Insert
> Linking &
Extraction

Download from
Source

Type

DATALINKUPDATE

The information in a sheet list table is linked directly to the data source, or property, field. Update a sheet list table to display changes made to the sheet set. For example, update a sheet list table after changing the sheet numbers in the **Sheet Set Manager** or after adding a sheet to the sheet set. Use the **DATALINKUPDATE** command to update a sheet list table. A quick way to access the command is to select the table to update, right-click, and select **Update Table Data Links**. You can also select inside a table header or data cell and then right-click and pick **Update Sheet List Table** from the **Sheet List Table** cascading menu.

To modify the properties for the table, select inside a table header or data cell and then right-click. Choose **Edit Sheet List Table Settings...** from the **Sheet List Table** cascading menu to reopen the **Sheet List Table** dialog box. Pick the **OK** button to update the sheet list table.

Use the same techniques to modify a sheet list table that you use to modify any other table object. For example, you can change text or add columns and rows. However, avoid unlocking cells and overriding values linked to the data source. When you use the **DATALINKUPDATE** command on the modified table, an alert indicates that manual modifications will be discarded.

Sheet List Table Hyperlinks

The **Sheet Number** and **Sheet Title** properties contain hyperlinks that you can use to access sheets. To open a sheet using a hyperlink, hover the crosshairs over a sheet number or sheet title until the hyperlink icon and tooltip appear. Hold down the [Ctrl] key and pick the hyperlink to open the selected sheet.

Exercise 33-6

Complete the exercise on the companion website.
www.g-wlearning.com/CAD

Sheet Set Fields

Throughout this chapter, you have experienced how text and blocks with attributes containing fields link text information to elements of the sheet set. Text and multiline text objects, title blocks, callout blocks, view label blocks, and sheet list tables use sheet set fields to automate documentation and the process of changing values during the course of the project. The following sections provide additional information on applying sheet set fields.

AutoCAD provides specific field types for use with sheet sets. In the **Field** dialog box, pick **SheetSet** from the **Field category:** drop-down list to filter fields specific to sheet sets in the **Field names:** list box. See **Figure 33-15**. Use fields in the **SheetSet** category to display values defined in the sheet set, subset, sheet, or sheet view. Some fields have several property and format options.

PROFESSIONAL TIP

When you prepare attributes with fields, avoid using the **Multiple lines** option, because multiple-line attributes with fields often display an inappropriate format. Use several single-line attributes and custom properties as needed. In addition, choose the **Preset** option in the **Attribute Definition** dialog box or **Properties** palette to have the attribute assume preset sheet set property values during block insertion.

Current Sheet Set Fields

Current sheet set fields, such as the **CurrentSheetSetProjectNumber** field shown in **Figure 33-15**, are linked to properties assigned in the **Sheet Set Properties** dialog box. The field value is specific to the current sheet set, which is where the field is located. For example, if you insert a title block onto a new sheet in a sheet set named Big House, the **SheetSetProjectName** field will display Big House. If you insert the same title block onto a new sheet in a sheet set named Little House, the **SheetSetProjectName** field displays Little House.

Current Sheet Fields

Current sheet fields, such as the **CurrentSheetTitle** field shown in **Figure 33-16**, is linked to properties assigned in the **Subset Properties** or **Sheet Set Properties** dialog boxes. The field value is specific to the current sheet, which is where the field is located. For example, a title block with an attribute containing the **CurrentSheetNumber** field displays 1 on sheet 1 and 2 on sheet 2 in a sheet set.

Sheet Set Fields

The various **CurrentSheetSet** and **CurrentSheet** fields are appropriate for most applications, but they link values only to the current project. Pick the **SheetSet** field to display options for linking values to any existing sheet set. See **Figure 33-17**. Use the **Sheet set:** drop-down list to display the contents of an open sheet set in the **Sheet**

Figure 33-15.
An example of a link between a current sheet set field and the corresponding sheet set property.

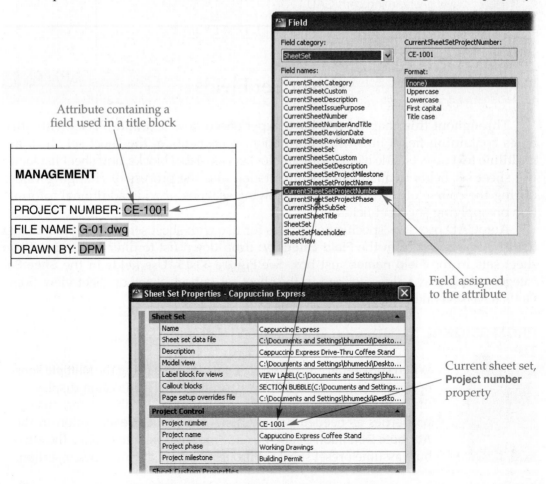

Figure 33-16.
An example of a link between a current sheet field and the corresponding sheet property.

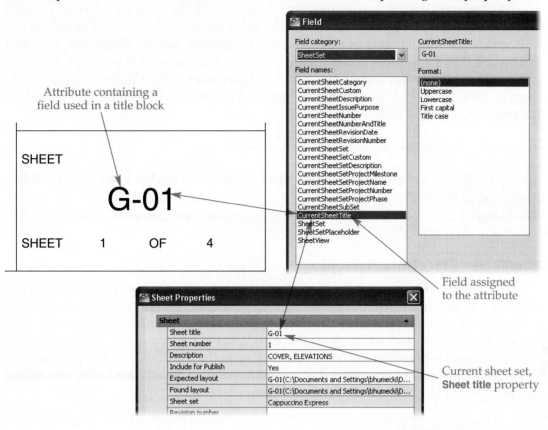

Attribute containing a
field used in a title block

Field assigned
to the attribute

Current sheet set,
Sheet title property

Figure 33-17.
The **SheetSet** field allows you to associate a value to a specific sheet set, typically related to a project different from the current sheet set.

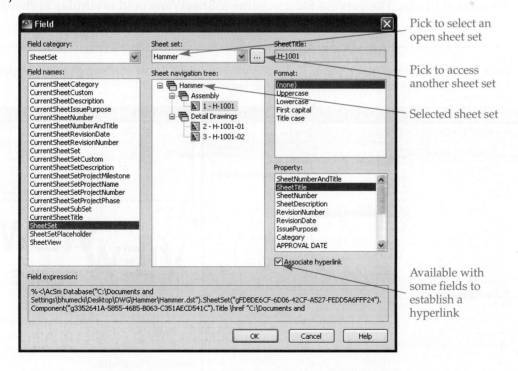

Pick to select an
open sheet set

Pick to access
another sheet set

Selected sheet set

Available with
some fields to
establish a
hyperlink

navigation tree, or pick the ellipsis (...) button to open another sheet set. Select the sheet set, a subset, or a sheet to specify the field value.

The **Property** list box, which appears when you pick the sheet set or a sheet, allows you to select a property to associate with the sheet set or sheet. Use the **Format** list box to override the format used in the **Sheet Set Properties**, **Subset Properties**, or **Sheet Properties** dialog box, depending on the selection. For example, if you type Assembly in the **Subset Properties** dialog box, pick the **Uppercase** format to force AutoCAD to display ASSEMBLY. Pick the **Associate hyperlink** check box to include a hyperlink to the property with the field.

 Custom sheet set and sheet fields are available from the **Property** list box.

Sheet Set Placeholder Fields

sheet set placeholder: A temporary value for a field that later references specific properties for values.

The **SheetSetPlaceholder** field inserts a *sheet set placeholder* that allows you to prepare text or a block without knowing the specific location of the property that the field will reference. Callout and view label blocks use **SheetSetPlaceholder** fields. See **Figure 33-18**. When you add a callout or view label block to a drawing in a sheet set, AutoCAD locates the information corresponding to the placeholders. The placeholders automatically become **CurrentSheet**, **CurrentSheetSet**, or **SheetView** fields, depending on the type of placeholder, and display the correct values.

Sheet View Fields

SheetView fields form when you use callout and view label blocks with attributes containing sheet set placeholder fields. Select the **SheetView** field to display options for linking values to any existing sheet view, or to place values without using callout or view label blocks. The process for assigning a **SheetView** field is similar to that for

---

Figure 33-18.
A view label block in the **Block Editor**. Callout and view label blocks use **SheetSetPlaceholder** fields to assign a field with the necessary characteristics.

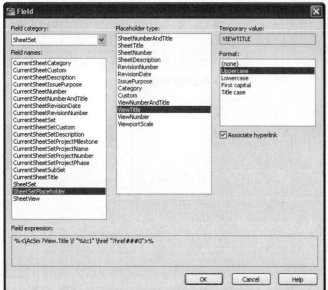
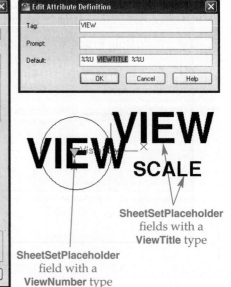

SheetSetPlaceholder fields with a ViewTitle type

SheetSetPlaceholder field with a ViewNumber type

---

assigning a **SheetSet** field, except that you select a sheet, view category, or sheet view from the Sheet navigation tree.

PROFESSIONAL TIP

If a field does not display the expected value, update the field manually using the **UPDATEFIELD** command, or use the **REGEN** command to apply an automatic update. In some cases, you may need to delete and then reinsert a block to display the correct field values.

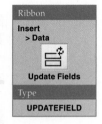

Ribbon
Insert
> Data

Update Fields

Type
UPDATEFIELD

Custom Properties

Custom properties are necessary to include additional information about a sheet set and sheets. For example, many of the properties referenced by fields in a title block are custom properties. To add or modify custom properties, access the **Sheet Set Properties** dialog box and pick the **Edit Custom Properties** button to display the **Custom Properties** dialog box. See **Figure 33-19A**.

Use the **Add** button to add custom properties using the **Add Custom Property** dialog box. See **Figure 33-19B**. Specify the name, default value, and owner of the property. The **Sheet Set** owner creates a custom sheet set property that applies to all sheets in the set. The **Sheet** owner creates a custom sheet property that might vary from sheet to sheet. Use the **Delete** button to delete custom properties. Custom sheet set and sheet properties appear in the **Sheet Set Properties** dialog box. Custom sheet properties also appear in the **Sheet Properties** dialog box.

Custom Property Fields

Choose the **CurrentSheetCustom** field to reference a custom sheet set property associated with the current sheet set. Use the **CurrentSheetSetCustom** field to reference a custom sheet property associated with the current sheet. When you are developing a block or template, and no sheet set is current, type the name of intended custom property in the **Custom property name** text box of the **Field** dialog box. AutoCAD correlates the name you enter with the property of the same name when you use the block or template in the current sheet set. Custom properties appear in the **Custom property name** drop-down list of the **Field** dialog box when a sheet set is current.

Figure 33-19.
A—Use the **Custom Properties** dialog box to add and delete custom properties. Study the examples of custom properties and default values added to this sheet set. B—Define a custom property using the **Add Custom Property** dialog box.

Existing custom properties

Pick to create a custom sheet set or sheet property

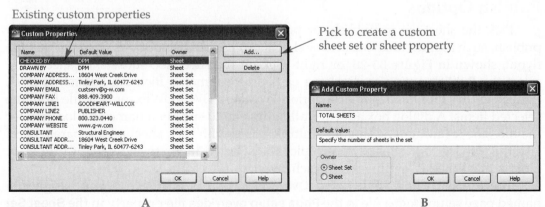

A   B

If you create a sheet set from an example sheet set, any custom properties in the example sheet set are added to the new sheet set.

AutoCAD does not calculate the total number of sheets in a sheet set. To display the total number of sheets, such as in a title block, create a custom **TOTAL SHEETS** property owned by the sheet set. At the end of the project, view the number assigned to last sheet and type the value in the **TOTAL SHEETS** property in the **Sheet Set Properties** dialog box.

Exercise 33-7

Complete the exercise on the companion website.
www.g-wlearning.com/CAD

Publishing a Sheet Set

A sheet set provides options to publish the sheet set, subsets, or individual sheets in the sheet set. Publishing a sheet set is a convenient way to create a hard copy or electronic version of a set of drawings, without plotting each layout separately. As when plotting a single layout, you have the option to publish hard copies to a printer or plotter or to create a portable document format (PDF) file or design web format (DWF or DWFx) file.

AutoCAD also provides tools for exporting drawings and publishing without referencing a sheet set using the **PUBLISH** command. *AutoCAD and Its Applications—Advanced* provides additional information on exporting and publishing drawings.

Publish Options

Pick the sheet set to publish, or press [Shift] or [Ctrl] to select specific items to publish, such as individual subsets or sheets. Access publish options from the **Publish** flyout, shown in **Figure 33-20**, or right-click on the sheet set, subsets, or sheets and select the **Publish** option to display a cascading submenu. Pick the **Publish to DWF**, **Publish to DWFx**, or **Publish to PDF** option to create a DWF, DWFx, or PDF file from the selected items. A dialog box appears, allowing you to specify a name and location for the file. The resulting file contains multiple pages with each sheet on a separate page. Use the **Publish to Plotter** option to plot the selected items to the default printer or plotter using the plot settings from each layout.

The **Publish using Page Setup Override** option is available if you assigned a named page setup from a file to the **Page setup overrides file** property in the **Sheet Set**

Figure 33-20.
The **Publish** flyout
provides options for
preparing a selected
sheet set, or subsets
or sheets within
the sheet set, for
publishing.

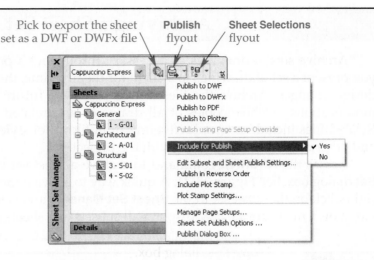

Pick to export the sheet
set as a DWF or DWFx file

Publish
flyout

Sheet Selections
flyout

Properties dialog box. Pick the option to publish all sheets using the settings specified in the saved page setup. Use the **Include for Publish** option to identify whether the selected items will be published. Pick the **Edit Subset and Sheet Publish Settings...** option to display the **Publish Sheets** dialog box. Check specific sheets to include for publishing.

The **Publish in Reverse Order** option publishes sheets in the opposite order from the order displayed in the **Sheet Set Manager**. With certain printers, publishing a sheet set in reverse order is helpful so that the last sheet is on the bottom of the stack, at the end of the set. Select the **Include Plot Stamp** option to plot the plot stamp assigned to the layout or page setup override with the selected sheets. Select the **Plot Stamp Settings...** option to open the **Plot Stamp** dialog box to specify the plot stamp settings. The **Manage Page Setups...** option opens the **Page Setup Manager**, allowing you to create or modify a page setup. The **Sheet Set Publish Options...** selection displays the **Sheet Set Publish Options** dialog box, which includes settings for creating a DWF, DWFx, or PDF file. The **Publish Dialog Box...** option opens the **Publish** dialog box, which lists the sheets in the current sheet set or the sheet selection.

Sheet Selections

Sheet selections provide a way to group specific subsets or sheets for publishing or for populating rows in a sheet list table. The most common application for a sheet selection is to create a selection set of specific subsets or sheets to publish. For example, create a sheet selection named Plans and Elevations from the Plans and Elevations subsets to publish only the plan and elevation sheets in the sheet set.

To save a sheet selection, pick the subsets or sheets to be included in the selection set. Then pick the **Sheet Selections** flyout on the **Sheet Set Manager** and select **Create...** to access the **New Sheet Selection** dialog box. Enter a name for the selection set and pick the **OK** button. The new selection set appears when you pick the **Sheet Selections** button. Pick a sheet selection from the **Sheet Selections** flyout to highlight the associated sheets in the **Sheet Set Manager**. To rename or delete a sheet selection, pick **Manage...** from the **Sheet Selections** flyout to display the **Sheet Selections** dialog box. Select the sheet selection and pick the **Rename** or **Delete** button.

Archiving a Sheet Set

Archive sets of drawings as necessary throughout a project. For example, when you present a set of drawings to a client for the first time, the client may want to make design changes. Archive the files at this phase for future reference before making modifications. Archive copies of all drawings and related files to a single location. Related files include external references, font files, plot style table files, template files, and other documents associated with the project.

Access the **ARCHIVE** command to archive a sheet set using the **Archive a Sheet Set** dialog box. See **Figure 33-21**. A quick way to activate the **ARCHIVE** command is to right-click on the sheet set in the **Sheet Set Manager** and select **Archive….** The **Sheets** tab, shown in **Figure 33-21A**, displays all subsets and sheets in the sheet set. Check the

Type

ARCHIVE

Figure 33-21.
The **Archive a Sheet Set** dialog box allows you to specify archive settings and files to archive. A—Use the **Sheets** tab to select sheets in the set to archive. B—Use the **Files Tree** tab to include other documents that relate to a project in the archive. C—The **Files Table** tab provides an alternative display for selecting documents to archive.

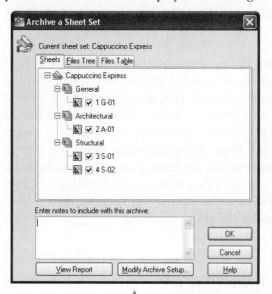

A

B

C

sheets you want to archive. The **Files Tree** tab shown in **Figure 33-21B** and the alternative **Files Table** tab shown in **Figure 33-21C** list drawing files and related files. Pick the **Add a File...** button in the **Files Tree** or **Files Table** tab to access the **Add File to Archive** dialog box. Use the dialog box to include files that are not part of the sheet set in the archive. You can include any type of file with the archive—the archive is not limited to AutoCAD files.

Use the **Enter notes to include with this archive** text box to type descriptive information about the archive, such as the design phase or items added to the archive. Pick the **View Report** button to list all files included in the archive in the **View Archive Report** dialog box. The **View Archive Report** dialog box includes a **Save As...** button that allows you to save the report to a text file.

Select the **Modify Archive Setup...** button in the **Archive a Sheet Set** dialog box to display the **Modify Archive Setup** dialog box for adjusting archive settings. See **Figure 33-22**. Use the **Archive package type** drop-down list to specify the archive format. Choose the **Folder (set of files)** option to copy all archived files into a single folder. Select the **Self-extracting executable (*.exe)** option to compress all files into a self-extracting *zip file*. Choose the **Zip (*.zip)** option to compress all files into a normal zip file. Use a program that works with zip files to extract the files.

> **zip file:** A file that contains one or more folders and/or files compressed using the Windows ZIP file format.

Select an earlier version of AutoCAD in the **File Format** drop-down list to convert the archived files to the selected version. The **Archive file folder** drop-down list defines where the archive is saved. Select a location from the list or pick the **Browse...** button to choose a different location.

The **Archive** file name drop-down list provides options for naming the archive. Choose the **Prompt for a filename** option to display the **Specify Zip File** dialog box so that you can specify a name for the archive package. Pick the **Overwrite if necessary** option to overwrite the file name if a file with the same name already exists. Select the **Increment file name if necessary** option to create a new file with an incremental number added to the file name if a file with the same name already exists. This option allows you to save multiple versions of the archive.

Figure 33-22.
Specify archive file settings using the **Modify Archive Setup** dialog box.

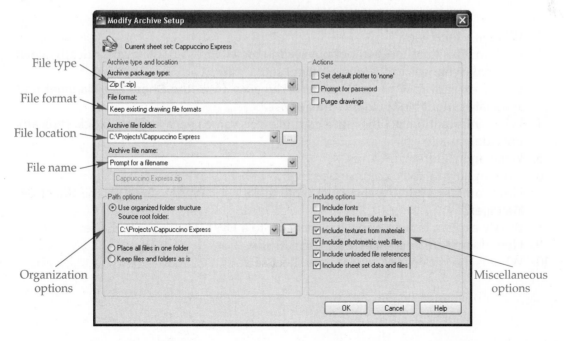

Select the **Use organized folder structure** radio button to allow the archive file to duplicate the folder structure for the files. The **Source root folder** setting determines the root folder for files that use relative paths, such as xrefs. Pick the **Place all files in one folder** radio button to archive all files into a single folder. Select the **Keep files and folders as is** radio button to use the same folder structure for all the files in the sheet set.

The **Set default plotter to 'none'** check box disassociates the plotter name from the drawing files, which is useful if you send files to someone using a different plotter. Select the **Prompt for password** check box to set a password for the archive. If you select this option, the password you choose is required to open the archive package. The **Purge drawings** action purges all drawings in the archive to eliminate unused content and reduce file size. Use the check boxes in the **Include options** area to specify items to include with the archive.

Right-click on a sheet set name and select **eTransmit** to display the **Create Transmittal** dialog box for use with the eTransmit feature. This option, which is similar to the **Archive** option, allows you to package drawing files and associated files for Internet exchange. The **Transmittal Setups** option displays the **Transmittal Setups** dialog box, which allows you to configure eTransmit settings. *AutoCAD and Its Applications—Advanced* provides additional information on transmitting drawings.

Chapter Review

Answer the following questions. Write your answers on a separate sheet of paper or complete the electronic chapter review on the companion website.
www.g-wlearning.com/CAD

1. What is a sheet view?
2. Explain why AutoCAD callout blocks and view labels update automatically when you make changes to the related sheet set.
3. Briefly explain how to create a view category from the **Sheet Set Manager** and associate callout blocks to the category.
4. What information do the upper and lower values in a callout block typically provide?
5. What are resource drawings?
6. Explain how to add a column heading to a sheet list table.
7. How can you update a sheet list table to reflect changes made in the **Sheet Set Manager**?
8. Briefly explain how to publish a sheet set to a DWF, DWFx, or PDF file.
9. How do you create a sheet selection set?
10. What is the purpose of archiving a sheet set?

Drawing Problems

Start AutoCAD if it is not already started. Follow the specific instructions for each problem.

Note: *Some of the problems in this chapter are built on problems from previous chapters. If you have not yet completed those problems, complete them now.*

▼ Basic

1. Open the arbor press sheet set you created in problem 32-4. Publish and archive the final sheet set.

▼ Intermediate

2. Complete the Hammer sheet set you started in Exercise 32-1. Save a copy of the drawing files H-1001.dwg, H-1001-01.dwg, and H-1001-02.dwg from the companion website to the Hammer folder you created during Exercise 32-1. Open the Hammer sheet set created in Exercise 32-1. Continue creating the sheet set as follows:

   A. Access the **Sheet Set Properties** dialog box and adjust the properties as shown in A, but use APPROVED BY, CHECKED BY, DRAWN BY, and COMPANY values appropriate to your drawings.

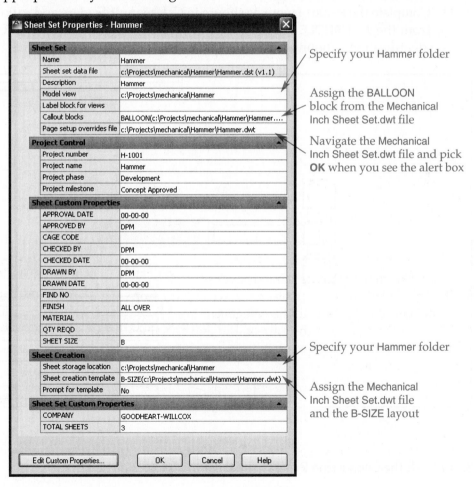

A

   B. Import the H-1001 layout into the Assembly subset and the H-1001-01 and H-1001-02 layouts into the Detail Drawings subset. Access the **Sheet Properties** dialog box for each sheet and adjust the properties as shown in B.

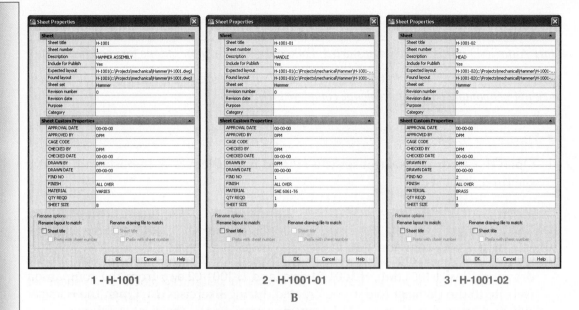

1 - H-1001   2 - H-1001-01   3 - H-1001-02

B

C. Create a sheet view of the 2 - H-1001-01 layout. Number the view 1 and name it HANDLE. Create a sheet view of the 3 - H-1001-02 layout. Number the view 2 and name it HEAD. Assign the BALLOON callout block to both views.

D. Complete the H-1001 layout as shown in C. Use multileaders and callout blocks from the 1 - HANDLE and 2 - HEAD sheet views in model space to create the identification balloons. Create the parts list using a sheet list table.

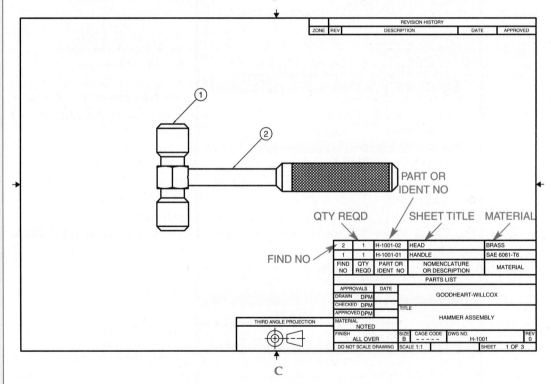

C

3. Publish the Cappuccino Express sheet set to a plotter or PDF file.

4. Publish the Hammer sheet set to a plotter or PDF file.

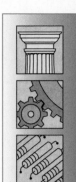

5. Archive the Cappuccino Express sheet set using the self-extracting zip executable (EXE) file format.

6. Archive the Hammer sheet set using the self-extracting zip executable (EXE) file format.

7. Create a new sheet set using the **Create Sheet Set** wizard and the **Existing drawings** option. Name the new sheet set Schematic Drawings. On the **Choose Layouts** page, pick the **Browse...** button and browse to the folder where you saved P29-14.dwg from Chapter 29. Import all of the layouts from the file into the new sheet set. Continue creating the sheet set as follows:
   A. In the **Sheet Set Properties** dialog box, assign the layout named ISO A1 Layout from the Tutorial-mMfg.dwt template file in the AutoCAD 2012 Template folder as the sheet creation template.
   B. Open a new drawing file using a template of your choice and create a block for a view label. Save the drawing file and then assign the block to the sheet set using the Label block for views setting in the **Sheet Set Properties** dialog box.
   C. Create a new view category and name it Schematics.
   D. Open the 3 Wire Control layout, create a new view, and add it to the Schematics view category. Double-click on the new view name in the **Sheet Views** tab and insert the view label block you previously created. Renumber the view and save the drawing.
   E. Add a custom property named **Checked by** to the sheet set and set the owner type to **Sheet**. Add another custom property named **Client** and set the owner type to **Sheet Set**.

8. Create a new sheet set using the **Create Sheet Set** wizard and the Architectural Imperial Sheet Set example sheet set. Name the new sheet set Floor Plan Drawings. Under the Architectural subset, create a new sheet named Floor Plan. Number the sheet A1. In the **Model Views** tab, add a new location by browsing to the folder where you saved the P17-16.dwg file from Chapter 17. Open the P17-16.dwg file and continue as follows:
   A. Create three model space views named Kitchen, Living Room, and Dining Room. Orient each display as needed to describe the area of the floor plan. Save and close the drawing.
   B. Open the A1-Floor Plan sheet. Create a new layer named Viewport and set it current.
   C. In the **Model Views** tab, expand the listing under the P17-16.dwg file. Right-click on each view name and select **Place on Sheet**. Insert each view into the layout. Delete the default view labels inserted with the views. Double-click inside each viewport and set the viewport scale as desired.
   D. In the **Sheet Views** tab, renumber the views. Insert a new view label block under each view.
   E. Save and close the drawing.

▼ Advanced

9. Plan a new shopping center for your area. Determine how many stores to include. If possible, obtain a copy of a survey for vacant land in your area suitable for building the shopping center. Determine the components of a complete set of plans for the shopping center, including a site plan, floor plans, foundation plans, roof plans, elevations, and any needed sections. Establish the components for a new sheet set to organize the drawings and layouts.

10. Plan a new residence with approximately 3500–4000 square feet, four bedrooms, three baths, a den/office, kitchen, dining room, nook, family room, and three-car garage. Create a complete set of plans for the residence, including a site plan, floor plans, foundation plans, roof plans, elevations, and any needed sections. Prepare layouts for each drawing. Then create a new sheet set to organize the drawings and layouts. Archive the final sheet set.

Answer the following questions. Write your answers on a separate sheet of paper.

1. Which of the following methods can you use to create a sheet view? *Select all that apply.*
   A. add a model view
   B. double-click an existing sheet
   C. right-click on an existing sheet
   D. use the **VIEW** command

2. To which of the following file types can you publish a sheet set? *Select all that apply.*
   A. DST
   B. DWF
   C. EXE
   D. PDF
   E. ZIP

Index